ALIEN SKIES
A Travelogue of the Universe

First Edition

AF681335

By Mario Mateo
University of Michigan - Ann Arbor

cognella® | ACADEMIC PUBLISHING

Bassim Hamadeh, CEO and Publisher
Kassie Graves, Director of Acquisitions
Jamie Giganti, Senior Managing Editor
Jess Estrella, Senior Graphic Designer
Sean Adams, Project Editor
Luiz Ferreira, Senior Licensing Specialist
Chelsey Schmid, Associate Editor

Copyright © 2017 by Cognella, Inc. All rights reserved. No part of this publication may be reprinted, reproduced, transmitted, or utilized in any form or by any electronic, mechanical, or other means, now known or hereafter invented, including photocopying, microfilming, and recording, or in any information retrieval system without the written permission of Cognella, Inc.

Trademark Notice: Product or corporate names may be trademarks or registered trademarks, and are used only for identification and explanation without intent to infringe.

Cover Image copyright © 2012 Depositphotos Inc./diversepixel.
© 2009 Depositphotos Inc./karelind.

Printed in the United States of America

ISBN: 978-1-5165-0632-3 (pbk) / 978-1-5165-0633-0 (br)

cognella® | ACADEMIC PUBLISHING

CONTENTS

DESTINATION 4: THE SKY FROM α CENTAURI

DESTINATION 5: THE 'EMPTY' SKY OF DEEP SPACE: AT 'ANTI-CENTAURI'

DESTINATION 6: THE ORION NEBULA

DESTINATION 7: EXTRASOLAR PLANETARY SYSTEMS: 51 PEGASI

DESTINATION 8: THE PLEIADES

DESTINATION 9: EXPLORING THE MILKY WAY

DESTINATION 10: INWARD BOUND: THE GALACTIC CENTER

DESTINATION 11: A BLAZING SKY OF STARS: THE GLOBULAR CLUSTER M 92

DESTINATION 12: THE MAGELLANIC CLOUDS: THE NEAREST GALAXIES

DESTINATION 13: THROUGH THE HEART OF DARKNESS: THE GALACTIC HALO

DESTINATION 14: THE NEARBY SPIRAL GALAXIES M 31 AND M 33

DESTINATION 15: THE GIANT ELLIPTICAL GALAXY MAFFEI 1

DESTINATION 16: THE VIRGO CLUSTER: A SKY OF GALAXIES

DESTINATION 17: THERE'S SOMETHING STRANGE HERE: THE SEYFERT GALAXY M 77

DESTINATION 18: THE BOÖTES VOID

DESTINATION 19: BACK IN TIME TO A DISTANT GALAXY CLUSTER

DESTINATION 20: HIGH REDSHIFT GALAXIES: THE YOUNG AND RESTLESS UNIVERSE

DESTINATION 21: QUASARS AT THE DAWN OF GALAXY FORMATION

DESTINATION 22: THE DARK AGES

DESTINATION 23: THE ERA OF RECOMBINATION

DESTINATION 24: THE FIRST ELEMENTS IN THE UNIVERSE

DESTINATION 25: TO A TIME BEFORE TIME

APPENDICES

ACKNOWLEDGMENTS

Well over 1000 undergraduate students at the University of Michigan have taken an introductory astronomy course based on the notes that became *Alien Skies*. Many of those students have directly provided me with invaluable and often quite detailed feedback regarding the destinations-based approach adopted by the book, as well as the readability and style of the manuscript, and the course itself. I am deeply grateful for these comments, suggestions and even (a few) complaints, all of which I hope have in some way or another helped to improve this book and to make it more effective and accessible to students.

Three of my colleagues—Edward Olszewski at the University of Arizona, Doug Richstone and Chris Miller at the University of Michigan—have used the notes for *Alien Skies* as the basis of their introductory astronomy courses over the past few years. I wish to thank them all for their comments/discussions about the approach and the text of the book, and also for the sharing with me the various, ingenious ways they tailored the course to their styles and their specific pedagogical goals. Their positive experiences helped convince me that *Alien Skies* was more than just some idiosyncratic reflection of my teaching methodology, but rather an approach that could be successfully adapted to a range of teaching styles by colleagues who value and enjoy undergraduate teaching. I want to also thank my many professional colleagues and friends within astronomy and other instructors who hail from a wide range of disciplines at the University of Michigan. These folks have over the years—and often without realizing it!—provided a wealth of information and an innumerable number of great ideas that have in some form or another made their way into *Alien Skies*. It is a pleasure and an inspiration to work with and learn from such knowledgeable people who, like me, aim to remain active students of their chosen disciplines—and of teaching—as long as possible.

I am happy to also acknowledge with gratitude the many astronomers—both professional and amateur—who have allowed me to use their exquisite photographs throughout these pages. Although my words attempt to convey what the alien skies we visit might look like, the pictures by these talented individuals and teams should remind us constantly of the exquisite beauty of the sky we really can see from Earth. To them all: Thank you and clear skies!

Thanks too to the folks at Cognella Academic Publishing—particularly Mark Combes, Jessica Knott, Sharon Hermann, Jess Estrella, Jamie Giganti, Sean Adams, and Chelsey Schmid—for taking a chance with the unusual approach embodied in *Alien Skies*. It has been a pleasure to work with people who clearly share my desire to convey the majesty of the Universe to a broader audience.

Although the writing of *Alien Skies* has been a somewhat protracted process, there have been lengthy episodes when it consumed my efforts like only a few tasks ever have. I thank my family, Nancy, Carmen and Emilio, for their support and especially for their patience with me at such times, even when it led to the occasional lost/postponed vacation. I also thank my parents, Luis and Rosa Mateo for providing me the opportunity and encouragement to pursue a not-so-practical career. I do not think they fully understood what attracted me to astronomy, but they certainly saw the passion I had for it and they supported my choice without question.

At the risk of being a bit late, I finally want to thank the staff at Mt. Rushmore National Monument. One beautiful summer evening many years ago, after the nightly show they promptly turned off the bright lights illuminating the presidents, allowing an impressionable 12-yr old from the big city to gaze in awe at the Milky Way in a truly dark sky. My first view of a glorious 'alien sky.'

FOREWORD

DON'T GET ME WRONG: The dark night sky from Earth is one of the most sublime sights humans can behold. But a lifetime of 'staring into space' to study the amazing things that go on in the sky has only whet my appetite for more. What would the numbingly distant objects we astronomers study from Earth look like if we could view them from close proximity? How would our perceptions of the Universe differ if our sky had been, for example, filled with so many stars that 'night' as we know it on Earth did not exist? What if the Sun were a double or even part of a triple system? In a world where mere planets were elevated to the stature of gods by our ancestors, how would humanity have responded to a sky like *that*? And what if we lived on a planet where the sky was essentially devoid of stars but in which a mere glance revealed the full extent and structure of an entire galaxy? Or if we lived at a location where the sky was filled with *dozens* of galaxies easily visible to the unaided eye? The socio-historic implications of alternative night skies—'alien skies'—are fascinating to contemplate in their own right. But my longing for 'something more' is based on a simple fact. There *are* alien skies out there. What would they look like? What could they tell us about how the Universe works?

In a world where books, television and movies—even games—make interstellar travel seem hardly different than taking longish airplane trips, the daunting challenges of a journey to even the closest stellar systems can easily be forgotten. Not because we are unable to understand these challenges. Rather, we instinctively do not want to abandon the hope that we—or our immediate descendants—may soon be able to carry out voyages to the stars. Humans have proven time and again to be driven by a desire to see new places, discover new worlds. We have demonstrated this innate desire even in the face of extreme discomfort, intense challenge and uncertainty, and the near-certain prospect of death. It is a trait that permeates virtually all of us, a restlessness that lies at the root of what has led humanity to look around and contemplate the Universe from which we arose.

And so my frustration about the night sky. Modern astronomy has developed to a level of sophistication that lets us describe in amazing detail the minute workings of objects so weird and so exotic that we cannot hope to ever reproduce their conditions here on Earth. Much of this advance reflects the use of new tools—immense ground-based telescopes, space-based astronomical satellites, advanced supercomputers to name a few—that allow us to explore the Cosmos in ways thought to be unimaginable not too long ago. But the same advances that have revealed myriad celestial wonders have come at the cost of a brutal realization: The Universe spans such a vast scale in both space and time that we humans have no hope of visiting in person any locations beyond the Solar System in the foreseeable future. Travel to even the closest places beyond the Solar System would require journeys spanning multiple human generations.[1] Most things we see in the Universe arise from locations so distant that, because of the immense, but finite speed of light, we view them as they were before humans existed, and even before the Earth, the Sun and our own Galaxy had been born. We have learned so much about so many places in the Universe. Yet they are virtually all out of reach.

But not *entirely* inaccessible. Modern astronomy is much more than an exercise of merely mapping or cataloging the details of what we see in the sky. Physical models, techniques and theories developed for applications on Earth have proven to be astoundingly successful at helping us to interpret and understand what we see going on in the sky, even out to the farthest reaches of the observable Universe. The success of these tools and of this approach lends confidence that we can make reliable predictions of what distant locations throughout our Universe must be like...even to the point of describing how the skies at such places would appear to any traveler lucky—or foolish?—enough to make the trip. And so the basis for this book: In these pages, we will embark on a 'voyage' to 'visit' a series of 'destinations' scattered throughout our remarkable Universe. This journey is not meant to be a flight of (pure) fancy, but it is without question a flight of the mind. As we shall discover, with a little thought and some imagination—and, not least, the right tools to help us interpret the views we encounter—we really can hope to 'see' the alien skies at these destinations. And often in surprising detail and at a high level of fidelity.

The organization of this book centers on the exploration of a wide range of core topics that comprise modern astronomical thought as revealed by what we encounter and see at our destinations. Each destination starts by describing the appearance of the alien sky at that location, taking special note of features that differ significantly from what we see in Earth's sky, as well as other features that—often rather surprisingly—don't differ nearly as much as we might have expected. At many of our destinations we will pause to 'look back' toward home even as it becomes harder and harder to identify Earth, our Sun, or even our Galaxy. And not simply because they are too far away, but sometimes because they have yet to come into existence! These views back toward where we started are meant to underscore the extraordinary scope of our travels in both space and time. Indeed, a central goal of the destination-based approach of this book is to try to convey an inkling of the titanic spatial and temporal scales of the Universe by revealing just how far removed they are from our everyday experiences. The hope is that this approach can make a strong impact on the reader by going to places that really do—or, in some cases, really did—exist in our Universe. Places that often appear as just faint points of light or dim smudges in Earth's sky, yet are as real as our home. And as fundamental a part of our amazing Universe.

Instructors using this book will notice a few novel features compared to traditional astronomy textbooks. First, the basic mathematical and physical tools that are used to understand destinations tend to be introduced only when needed and not all at once in a set of introductory chapters as is more typical. Consequently, the exposition of many basic concepts and physical tools occurs in a somewhat 'staged' manner. For example, we encounter Kepler's Third Law at many early destinations, each time introducing new details and expanding on its applicability as needed to understand what we are seeing in a given alien sky at the time.

1 Only a few manmade machines have left the Solar System. It has taken them decades to do so, and they still have tens of thousands of years to go before they will have traversed distances remotely comparable to the distances of the nearest stars.

Second, this book is *full* of footnotes.[2] These range from fairly light comments regarding some curious historical or practical aspect of the subject, all the way to discussions that reveal some of the deeper details associated with a given topic. The more technical footnotes aim to engage and challenge advanced readers. They may also be used to stimulate deeper discussions by the instructor on particular subjects; I have used the *Alien Skies* text to teach a course in precisely this manner. As a result this book can serve as an introductory survey of modern astronomical concepts for non-science majors or advanced high school level students—its primary intended audience—or as a text used, for example, to broadly describe basic astrophysical concepts in conjunction with the notes/lectures of a more advanced introductory course aimed at potential science majors.

Third, the destinations in *Alien Skies* are largely 'modular' in form. That is not to say they are fully self-contained since, as noted above, later destinations often build on concepts developed at earlier points in the text. But the broadly modular nature of the destinations means that instructors can—and indeed are encouraged to—introduce their own voyages to other 'destinations' they would like students to 'visit'. I would personally love to see an interacting binary star—say, a novae or some other cataclysmic variable—or perhaps spend some time at one of more notably unusual galaxies of the Local Universe—Centaurus A and the Antennae immediately come to mind. Those destinations, fascinating as they would be, simply did not fit within the structure and time constraints of the courses I have taught using this approach. Nonetheless, 'new' or 'alternative' destinations could certainly be exciting to students and they offer instructors an opportunity to address concepts that are not covered (or covered in little detail) in the current text. I have found that the effort to envision and accurately describe an alien sky is in itself a very instructive and illuminating exercise. And often a surprisingly challenging one.

Finally, this first edition of *Alien Skies* largely eschews the traditional emphasis of history and methodology that is common to most modern intro astronomy texts. The idea here is not to belittle the development of astronomical thought—indeed, some of that development *is* briefly covered in the text—but rather to place the spotlight squarely on the phenomena astronomers have uncovered and to convey a broad sense of how those phenomena can be explained. It is, after all, the wondrous things we see in the Universe that first excite people about astronomy and which inspire them to learn more about the subject and about science in general, all primary goals of this book. Students who choose to find out more will naturally become acquainted with the historical and technical developments of astronomical understanding as they delve deeper into the field. And the students who don't? Well, hopefully they will at least remember how exciting their journey of the Universe was.

To everyone set to embark on these journeys of discovery: *Bon voyage!*

2 Nearly 600 in all!

Destination 1

THE MOON: OUR FIRST STEP OFF EARTH

1.1 IN SEARCH OF ALIEN SKIES

Throughout much of human history, people could only dream about visiting exotic, far-off lands. The dangers, burdens, and uncertainties of travel meant that only the foolish—or extremely daring—would even attempt such adventures. Many who tried never returned. But some did, and their stories revealed an immense, diverse world full of surprises. A world of fantastical people, places, and things that must have seemed beyond belief or comprehension. Though few could ever imagine exploring that strange world in person, they could still catch a 'glimpse' of it through their imaginations, informed and guided by the vivid descriptions of those who had dared to take the journeys in their stead (Figure 1.1).

Fig 1.1

Not that long ago, explorers were the few who could travel the world. Their tales and experiences fueled the imaginations of others who could not directly follow in their footsteps. This image illustrates the progress of Marco Polo's caravan to the 'Indies.' When he returned, the memoirs of his trip—full of wondrous stories—became an instant sensation.

Image from the 14th century *Atlas Catalán* by A. Cresques.

Over the centuries, our insatiable desire to explore has steadily expanded humanity's horizons to the point where people have now set foot on virtually every inch of the Earth's surface. We can take guided tours to places—the North Pole, the top of Mount Everest, the heart of the Sahara Desert—that were utterly unreachable until very recently. Even the deepest trenches of the oceans can now be visited by adventurous and curious travelers who have achieved the challenging means to do so. To our ancestors—not even that long ago—journeys like these were fantasies. Today, virtually the entire world is just a plane ride away. The Earth has never been smaller.

Happily, despite these achievements, we have *not* reached the limits of exploration. Beyond our small planet lies a Universe that stretches out to the very limits of human comprehension. Only in the past 250 years have people begun to truly understand just how vast the ocean of space beyond our world really is. And only in the past 50 years have we taken the first baby steps into that deep, forbidding ocean.

In this book we will 'travel' through the Universe, visiting a series of carefully chosen 'destinations' that span the limits of space and time of our Cosmos. Some of these destinations are so distant that we could never hope to visit in our lifetimes. Some exist only in the past. Some of these places are so utterly hostile to life that humans could never visit even if we learned how to conquer the obstacles of time and space required to reach them. Indeed, it is safe to say that no humans will *ever* visit firsthand some of the destinations on our itinerary. But that will not stop us. We *can* venture to even the most exotic and most hostile destinations in the Universe if we allow our imaginations to take us there. Unlike our ancestors—who could construct their fuzzy conception of the far-off world only from the stories told by intrepid explorers—we have now reached a truly remarkable level of understanding of the physical laws that govern the structure and evolution of the Universe that we can make highly informed guesses about what we would see at each of our cosmic destinations. The sophisticated tools used in astrophysics, physics, chemistry, mathematics, statistics, and other sciences allow us to weave plausible descriptions of how myriad locations throughout the Universe might appear to us as 'cosmic tourists.' These tools also provide us with a means of understanding many of the processes that produce the diverse 'Alien Skies' we shall behold. As a result, our aim is not merely to *see* as a tourist might. Our goal is to try to *comprehend* what we encounter as any serious, curious explorer would hope to do.

The primary reason we can even hope to contemplate such 'voyages of imagination' is the fact that, to an astonishing extent, the physical phenomena we observe on Earth appear to be fundamentally 'universal' in nature. We have clear evidence—and strong, reasoned conviction—that the processes that shape even the most remote parts of the Universe arise from phenomena that are essentially the same as those we observe on or near Earth. This is not to say that the Alien Skies we shall encounter will all appear similar to what we see in Earth's sky. Far from it. Though the basic physical processes that shape the vistas we will encounter may be universal in nature, they manifest themselves in strange, often bizarre, ways in other environments and under different conditions. Of course, since—apart from the first one—no one has ever actually ventured to any of the destinations we shall visit, there will always be some uncertainty about what we might see. Some Alien Skies will be relatively easy to comprehend: Imagine a sky so filled with bright stars that it is never truly dark there, not even at 'night' (Destination 8). Others will present vistas that challenge our basic notions of how space and time function: Imagine a place where time flows at significantly different rates for people separated by distances smaller than our Solar System (Destination 17). Nonetheless, wherever we visit—whether the vista is comparatively familiar or absurdly exotic and disorienting—the same basic set of physical concepts that we have uncovered on Earth is generally sufficient to let us visualize and understand what we see. The universality of the physical concepts that underlie our voyages tie together the most fantastic Alien Skies we will visit to the familiar sky we see from Earth. This universality also represents one of the most fundamental, and, ultimately, most astonishingly beautiful mysteries of our Cosmos.

As explorers, the attraction of the unknown is in itself sufficient motivation for our travels. For many people, the more remote and unknown a destination, the more appealing and exciting it is to visit. We will visit places so far removed from home that the Earth, the Sun, even our own Galaxy—a structure of nearly incomprehensible size—will grow so minuscule that they will be utterly lost from view to our homesick eyes. Our travels will carry us far beyond our home on Earth, deep into the depths of space, past seemingly impossibly remote horizons of distance and time, all the way to the very birth of the Universe itself. We will visit places so remote from home that the Earth, the Sun, even our own Galaxy—a structure of nearly incomprehensible size—will grow so remote that they will be utterly lost to view to our homesick eyes. We face a long, exciting journey. Time to get started.

1.2 THE SKY FROM HOME

All voyages of exploration start with basic curiosity of what lies beyond the next bend of the road, the other side of the distant hills, or over the edge of the open sea. But to truly appreciate any journey, especially daring treks from which there may be no return, it is worth taking note first of how things appear from home. It is our familiar surroundings, after all, that define our horizons and which represent what we consider to be 'normal.' Our home will inevitably serve as a point of comparison for many of the destinations we shall visit. So, before we start, let's take a brief first look at the Universe from home. This means that we must begin by looking up at the sky. The sky from Earth.[1]

1 We'll come back to Earth briefly, in Destination 3, for a more detailed look at the sky before we start our explorations in earnest.

Fig 1.2

The end of day 1,750,000,000,000 (give or take a few billion) on Earth.

Photo by M. Mateo.

Most of us know something about the appearance of the sky; after all, we have lived under it our entire lives! A companion to all, the sky is one of the few things that all humans truly share. Perhaps the most familiar component of the sky is the Sun (Figure 1.2). Every day, we count on the Sun rising from below the horizon to supply the heat and light on which all life depends.[2] We also rely on its constancy. The Sun remains essentially the same brightness, the same size, the same color day after day, year after year.[3] Indeed, the Sun's properties and motions are so regular that they are easy to take for granted. The Sun rose yesterday. The day before. The day before that. None of us can remember a day when the Sun failed to rise. Nor have we ever met *anyone*—not even the oldest persons we know—who ever said that *they* endured a day when the Sun

2 Most of us make a distinction between solar power and, say, wind power, gasoline, or hydroelectric power. But, really, all of these are examples of 'solar power.' The wind is driven by solar heating of the Earth's surface. Gasoline comes from ancient plant life—which depended on sunlight to grow—that has been compressed over time by the slow accumulation of tons of sediments over millions of years. That accumulation and, more directly, hydroelectric power, relies on rivers fed by rains of water elevated into the air by evaporation by the heating of the Sun. About the only common sources of energy that do not come directly from the Sun are nuclear and some geothermal power. Both are based on the properties of elements that were formed *before* the Sun was born (though even these would not be on Earth had the Sun not formed nearby; we'll see why in Destinations 6 and 7). But much of what we use today to generate the energy on which our lives are sustained comes, ultimately, from the Sun.

3 Remarkably, some stars do change in these ways. Sometimes on timescales of just hours, or days. Sometimes predictably. Sometimes randomly. Sometimes violently. We will encounter many examples of these so-called variable stars at future destinations.

refused to rise.[4] Records from the oldest civilizations reveal the same pattern. The Egyptians could not have built pyramids in the dark and cold! Some of their contemporaries even left detailed records of the Sun's motions in the sky, revealing exactly the same behavior we see today. Many ancient civilizations revered the Sun's celestial motion and its amazing stability, fully cognizant that their survival relied on the constancy of its properties.

But the story of how the Sun behaves in the sky clearly goes far beyond the time people have existed to record it. The motion and properties of the Sun appear to have remained virtually unchanged for at least the past many millions of years. Long before humans inhabited Earth, the Sun must have shone much as it does today. The dinosaurs and all the various life forms that existed over the long history of the Earth needed sunlight not only to sense their environment, but also to stay warm and to produce the energy they needed to stay alive. Many species have existed virtually unchanged over many millions of years. They could not have survived for so long and with so little change had the Earth's climate—driven as it is by the Sun—been radically different at past times than it is today. What the fossil record tells us—even more dramatically than human historical records—is that the Sun is mind-bogglingly *stable*. It has poured out the copious quantities of light and heat that bathe the world today for at least many hundreds of millions of years. Indeed, as we'll discover later (Destination 8), the Sun has shone much as it does today for about the past five *billion* years. And it will continue to do so for about 5 billion years into the future.

As with so many astronomical phenomena, the age of the Sun is an example of a scale so far removed from human experience that it is hard to have a true sense of what it means. Yet we have to develop some understanding of these vast numbers to properly appreciate the scale of our travels and of the underlying nature of the things and phenomena that we shall encounter as we journey through the Universe. To get an inkling of the Sun's age—a *personal* sense of what it means—consider your age. *In seconds.* If you happen to be about 18 years old, you have lived about 0.6 billion seconds. You will not reach the ripe old age of 1 billion seconds until you are 32.3 years old. Now, assuming you were about 1 billion seconds old and you then replaced each second of your life with a full year—one year is about 31 million seconds long—you would still have lived only 20% of the current age of the Sun. A person would have to make it to an age of 160 years before claiming to have lived for as many *seconds* as the Sun has existed in *years*. Over this vast period of time, the Earth, the

4 Well, you can find some exceptions. Near the Earth's Poles—north of the Arctic Circle or south of the Antarctic Circle to be precise—the Sun might not set or rise over the course of a day, many days, or even months. But even at these locations, the Sun makes a complete circuit of the sky every day, sometimes above the horizon, sometimes, invisibly, below the horizon. And, if you wait over a full year, you would still find that the Sun is above the horizon 50% of the time, below it the other 50% of the time, same as at any location on Earth.

Sun, and the Solar System we see today were born and subsequently evolved to their present state. And yet, astonishingly, the age of the Sun does not define the limits of astronomical time. As we shall see (Destination 16), the Sun has existed for only a fraction of the age of the Universe itself.[5] We'll return to study some of the detailed properties of the Sun and other stars—properties that allow us to discern their ages and life cycles—as we travel around the Universe.

While located above the horizon, the Sun provides us with light. Lots of light. Even on days of heavy cloud cover, we can always see light from the Sun filtering through the clouds, reflected off things in our immediate environment. But even at night, sunlight percolates through our atmosphere from the opposite, sunlit side of the Earth. It may also be reflected off the surface of the Moon—assuming the Moon is visible at the time—proving that the Sun does not 'turn off' simply because it moves below our local horizon. On a clear day near the sea or in a snowy landscape, the intensity of the light from the Sun can literally blind us. Even our skin—the sensation of warmth on a sunny day—can reveal the presence of the Sun.[6] The sky's blue color results from sunlight scattering off molecules in our atmosphere. These simple observations are ones anybody can carry out simply by looking up at the sky, yet they reveal basic physical properties of the Sun, about the radiation we receive from it, and even about the existence and composition of our atmosphere.

At any given time when we see the Sun in the sky, it is easy to confirm that it is moving steadily to the west.[7] Every 24 hours—over the course of a full day—the Earth, along with us on its surface, completes one rotation about an axis that defines what we call the North and South Poles, the locations where this imaginary axis passes through the Earth's surface (more on this at Destination 3). This rotation carries the Sun across the sky until it eventually reaches the Western horizon. Daytime ends as the Sun slowly begins to set.

What happens next? First, the sky gets darker. This hardly seems profound. After all, the Sun is very bright and it is disappearing below the horizon. The sky *should* get darker. But we shall discover that the fact that the night sky is dark has long been a deep mystery (Destination 18), one we will not solve until we

5 That the Universe even has a well-defined age is yet another astonishing discovery that awaits us at Destination 16.

6 Light is one form of radiation—optical radiation—while the heat we feel on our skins from the Sun is another form of radiation—infrared radiation. In general, the term 'radiation' refers to any flow of energy from one place to another. In the case of the Sun, light is generated at its surface (we'll see how in Destination 6), travels through space, then reaches us on Earth. We will explore the fundamental nature of the most common form of radiation—Electromagnetic Radiation—in greater detail starting in Destination 5.

7 *NEVER LOOK AT THE SUN DIRECTLY* for any reason, including to confirm its motion across the sky. It is simply too bright and can rapidly burn your retinas. To add insult to (severe) injury, this isn't even the most effective way to track the Sun's motion. Instead, look at shadows, particularly long ones, and notice how *they* move with time. In a few minutes, you can readily confirm the Sun's continuous, smooth motion across the sky from east to west. Just remember that the shadow is moving in the opposite direction!

have traveled very far from home, nearly to the edge of the observable Universe and very far back in time (Destination 25). This is not the only time we will encounter that something 'simple' has, in fact, a complicated, often surprisingly profound, explanation.

But, back to the business at hand. As the Sun sets, the sky does indeed get darker as common sense dictates. As it does, we begin to appreciate that there is more in the sky than just the Sun. Celestial objects that we could not easily see during the day start to reveal themselves. On some nights, we can see the

Fig 1.3

Any time we see the Moon in the sky from Earth, we see the same physical features you see here, though the fraction of the moon illuminated by sunlight may be different. This is an example of the full moon. At other times the Moon may appear as a crescent or a 'gibbous' moon where more than half, but less than all, of the facing side of the Moon is illuminated. Any pictures of the Moon where the surface features appear significantly different from this view were taken from spacecraft. One of the most famous was obtained by the Apollo 11 astronauts on their way to lunar orbit. Only they and a handful of other astronauts have seen the far side of the Moon in person.

Photo by G. H. Revera.

Moon.[8] Sometimes it is a thin crescent. Sometimes about half of it is easily visible. At yet other times it appears fully lit as in Figure 1.3. There are nights when the Moon seems to be out every time we look, and others when we never see it at all. Like the Sun, the Moon also travels from east to west on a given night. But there is evidently more to its motion than just the simple movements we saw in the Sun. For example, whenever the Moon appears in the form of a crescent, we see it only soon after sunset or just before sunrise with the crescent side always facing where the Sun will set (or rise). And when the Moon is full, we

8 The Moon can often be seen during the day too. It is one of the most easily seen 'nighttime' objects visible when the Sun is up, but not the only one if you know where and when to look.

can see it all night since it rises as the Sun sets, and sets as the Sun rises. Even without going into details (yet), we can begin to appreciate from these simple observations that the Moon must be moving at a different rate across the sky than the Sun because sometimes it is near the Sun (and seen as a crescent) and sometimes it is on the opposite side of the sky, as far from the Sun that it can be (when it is seen to be full).

With a little patience and reasonably clear skies, it is easy to confirm that it takes about four weeks (actually, 29.5 days, give or take a few hours) for the Moon to go through all its phases. Throughout this time, the Moon moves steadily across the sky with respect to the Sun. But it is also apparent that something is *not* changing over this cycle. If we look closely at the face of the Moon over the lunar month, we will also notice that the pattern of the light and dark regions on the Moon's surface—actual regions of relatively lighter and darker rock—always appear essentially the same whenever the Moon is visible. That is, the Moon always presents the same face to us, implying that we can never see the 'back' side of the Moon[9] from the surface of the Earth. This behavior is called *Synchronous Rotation* and it tells us that the Moon is rotating about its axis at nearly the same rate that it is revolving around the Earth.[10]

As twilight is slowly swallowed up by night, individual stars become visible (Figure 1.4). From a dark, clear site (alas, there are fewer and fewer of these, particularly near cities; see Figure 9.2 of Destination 9), the number of stars in the sky can seem staggeringly large. In fact, a typical person at a dark location can see about 3000 stars at any given time, implying that there are about 6000 'naked-eye' stars visible over the entire sky (after all, we only see about half of the sky at a given time; the Earth itself blocks our view of the other half). But with even a few stars visible in the sky, it is easy to appreciate that they exhibit a wide range of brightness. And it is also immediately evident that all

9 Not the 'dark side,' mind you. The 'dark' side of the Moon is simply the part of the Moon that is not illuminated directly by the Sun. The 'back' (or 'far') side of the Moon is the part of the Moon that never faces Earth. The 'back' side of the Moon is not always dark. A question at the end of this destination addresses these distinctions.

10 It may seem like a fantastic coincidence that the Moon does this. But, in fact, synchronous rotation sets in comparatively rapidly in close-orbiting pairs of bodies—a moon and its planet, a planet and its star, or even two stars orbiting one another—due to strong gravitational *Tidal Forces* the two bodies exert on one another (we'll encounter tides in a different context at Destination 13). These effects cause the bodies to tend to a state where they present the same faces to one another. Since the Earth is much more massive than the Moon, the effect on the Moon is strongest, explaining why it has already become tidally synchronized in this way. But the Moon is also acting on the Earth, causing our planet to gradually slow its rotation (causing, incidentally, the Moon to move outward from its present position) at a rate of about 0.002 seconds per century. If uninterrupted, this process would cause the Earth and Moon to eventually present the same faces toward one another. It is kind of intriguing to imagine what things would look like if it happens that the current rate of rotational slowing continues indefinitely (a question at the end of this destination addresses this).

Fig 1.4

The night sky soon after sunset as seen from Earth. Ours is a sky of stars. To the naked eye, about 3000 can be seen at any given time from a dark site. And sometimes, as when this picture was taken, one might even spot a few things that appear distinctly un-starlike, including the Milky Way (see Destination 12), nearby galaxies (see Destination 13), and, if you are lucky, even a bright comet (in this case, Comet Lovejoy as seen in Dec 2011 from near the town of Roma in Queensland, Australia). Owing to the sensitivity of the camera, this image shows five to ten times more stars than one can see with the naked eye in this same part of the sky.

Photo by D. Liu.

stars are far, far fainter than the Sun or even the Moon at its brightest.[11] In dark sites where many stars are readily visible, it also becomes clear that there are many more faint stars than bright ones. This is why so few stars can be seen from cities where artificial lighting brightens the night sky appreciably. Only the brightest stars—which are comparatively few in number—are readily visible through the murk and glare around cities. The numbers and distribution of the brightnesses of stars are important clues to their underlying nature; we'll return to these characteristics later. But, for now, let's concentrate on the motions of stars in the sky. For indeed, with a bit of patience it is easy to see that the stars also do move across the sky (see Figure 1.5). Like the Sun and Moon, most (but not all!) stars typically rise in the east and set in the west. And, as with the Moon, the motions of the stars in the sky are not quite as simple as we might have first suspected. To appreciate this, we must study the stars over a longer period of time than just a few consecutive nights.

As autumn approaches in the Northern Hemisphere, one can go out on a clear night and see the characteristic form of the Summer Triangle to the west at sunset. The stars that define the Triangle are *Vega*, *Altair*, and *Deneb*, three of the brightest stars visible in the sky as seen from Earth. To the eye, all three appear mostly white in color. But, lest we conclude that all stars have similar, neutral colors, a bit further to the west, we see a very different star. This is *Arcturus*, a bright star with a distinctly red hue unlike the colors of any of the stars of the Summer Triangle. As it happens, three of these stars are relatively

11 Every now and then a star can become comparable in brightness to even the full Moon! These are known as supernovae, and we'll encounter them at a few future destinations. If a star were to get as bright as the Sun, on the other hand ... Well, that would be interesting, but very bad for life on Earth!

Fig 1.5

A long exposure image of a famous constellation (recognize it? See Figure 3.3 in Destination 3 to see it in more normal appearance). The long trails occur as the stars move relative to the camera in response to the Earth's rotation. The stars in this image were moving westward, toward the lower right of the figure, after having risen in the east (to the left in this image) several hours earlier. We can see plainly that stars vary in brightness and color as seen in Earth's sky. However, not everything easily visible in this picture is a star; we'll explore this non-stellar object in detail at Destination 6.

Photo by J. Orman.

nearby, while one, *Deneb,* is very much more distant.[12] As the seasons progress, another familiar constellation, Orion, comes into view, eventually dominating the winter sky. With its classic rectangular shape and distinctive 'belt' of three stars near the center of this rectangle, Orion is unmistakable as are its two brightest stars. In the upper left corner is the bright red star *Betelgeuse* (pronounced 'beetle-juice'; it is visible as the bright reddish trail in Figure 1.5), while the opposite corner in the lower right is marked by the bright, steely-blue star *Rigel* (also visible in Figure 1.5 near the bottom right).[13] Nearby shines *Sirius,* the brightest star in the sky after the Sun (Sirius is the bright star whose trail is partially hidden to the left in Figure 1.5). During the seasons when Orion dominates the sky, the Summer Triangle is no longer visible, having sunk further into evening twilight night after night as autumn flowed into winter. As spring approaches, the characteristic form of the constellation Leo dominates the nighttime sky, while Orion steadily moves into the invisibility of daytime. A few months later, the Summer Triangle returns to the night sky and the annual parade of stars repeats. Why do the stars change from season to season?

As we ponder this, it occurs to us that night reveals where the Sun is *not* located in the sky, so the seasonal changes in the visibility of stars and constellations must mean that the Sun is moving with respect to the stars.[14] As we study

12 As we shall discover when we begin to measure stellar distances (Destination 3), all stars are in fact tremendously far from the Sun. The descriptions here refer to *relative* distances.

13 These descriptions of the locations of Betelgeuse and Rigel—upper left, lower right—are somewhat imprecise and, technically, apply only to an observer in the Northern Hemisphere of the Earth. But the distinctive colors of the two stars would be apparent no matter where you are located.

14 If you think about it even more, we are also assuming that the pattern of the stars themselves is not appreciably changing. We'll soon see in detail that stars beyond the

this motion carefully over a few weeks, we conclude that the Sun is traveling from west to east with respect to the stars—exactly the opposite direction that we see the Sun travel during its daily circuit across the sky. With sufficient patience, we can confirm that the Sun requires about one year to travel from a given spot among the stars back to that same spot. Indeed, this is how we *define* the year. This also explains why, for example, we call it the *Summer* Triangle. The three stars that make up this pattern are prominent in the evening sky primarily during the summer months in the Northern Hemisphere.[15] What seemed pretty simple—the daily motions of the Sun, Moon, and stars—is getting a bit complicated!

1.3 OUR FIRST STEP

There are many more interesting things to see in the sky from Earth, even if we restrict ourselves to just what we can see with the naked eye. We will eventually understand the causes of these motions and learn about the nature of the stars and star patterns we see in the sky. But not quite yet. Let's instead take the time now to begin our travels. Cautiously. Our first destination—The Moon—is the only one that we shall visit that humans have actually traveled to. And for a simple reason: It is by far the closest celestial body to Earth (Figure 1.6). But make no mistake, the Moon is not 'close' by any normal standards. Nearly one quarter million miles away, the Moon's distance from Earth is equivalent to about 10 complete trips around our planet. We can certainly expect the view to be different after traveling so far. And, in many ways, it is.

For one thing, the lunar sky is black. Not just at night as on Earth, but even during the day! As we look above us, the inky blackness of the sky contrasts sharply with the blindingly bright Sun, itself essentially identical in appearance as seen from home. The very simple observation that the lunar sky is black at all times has observable implications. For one, this tells us there is no air here on the Moon. Recall that we noted earlier that the blue sky of Earth results from the scattering of sunlight by molecules in our atmosphere. The black sky

Sun are immensely further away than anything in our Solar System. So, even though stars are moving relative to one another, these motions are always tiny by virtue of the large distances to even the closest stars (Destination 3). As a result, our assumption that the patterns of the constellations do not appreciably change is a pretty safe one for our present discussion.

15 There is a bit of hemispheric chauvinism going on throughout this discussion. When it is summer in the Northern Hemisphere, it is winter in the Southern. But, on the other hand, the most northerly two stars of the 'Summer' Triangle—*Deneb* and *Vega*—cannot even be seen from much of the Southern Hemisphere!

Fig 1.6

A scaled image showing the relative separation and sizes of the Earth (*left*) and Moon (*right*). The width of North America is about equal to the size of the Moon, or about the size of the prominent V-shaped cloud pattern seen in the center of Earth's disk in this image. The brightness of the Moon is greatly enhanced in this image; if it were shown at a realistic brightness, it would hardly be visible against the black background of empty space.

Earth/Moon photos from NASA.

of the Moon implies there is no scattering in this sky, hence no air. That is, the Moon has effectively no atmosphere.[16] This simple conclusion drawn from the appearance of the lunar sky is also indirectly apparent from the spectacular landscape that surrounds us. We see rounded 'hills' pockmarked by impact craters. A long sinuous 'valley' runs nearby, its sides smooth. Yet these modest-looking lunar hills are really mountains, many rising tens of thousands of feet above us. The long valley is actually an ancient lava stream that extends over

Fig 1.7

A panoramic view at the Apollo 15 landing site. The long, sinuous Hadley Rille—an ancient lava river or collapsed lava tube—is visible from the left to the center of the image. The ramparts of the Apennine Mountains are visible to the right. This spectacular lunar valley is 1,300 feet deep and these mountains tower over 15000 feet above the surrounding plain. But they bear little resemblance to comparable geologic structures on Earth, appearing much smoother than we might expect for similarly large features on our planet. The black lunar sky tells us why. There is no air here, and hence no wind, ice, or water to erode and sculpt these structures into the kinds of valleys and mountain forms we see on our home planet.

Photo montage by E. van Meijgaarden using NASA images.

many miles and is over a thousand feet deep (Figure 1.7). The craters—we see them everywhere here on the Moon (see Figure 1.8)—result from impacts of small rocky bodies on the lunar surface. On Earth, our atmosphere causes most of such bodies to disintegrate long before they reach the surface. The few that do leave recognizable impact craters that are rapidly eroded by ice,

16 To be fair, had we not known this ahead of time, we would have discovered it quickly if we had tried to go out of our spacecraft without a pressurized suit. But even if we saw a picture from the lunar surface during daytime, we could infer the lack of an atmosphere from the appearance of the sky.

Fig 1.8

Impact craters dominate the Moon's surface (*Left*; the large crater in this image is called Clavius). By contrast, such features are rare on Earth (*Right*; these two impact craters are known as the Clearwater Lakes in Canada) despite the fact that the Earth has certainly been impacted far more than the Moon has. The lack of atmospheric weathering by water, ice, or winds on the Moon is apparent here, while weathering and even glacial erosion effects—the diagonal 'grain' in the image of the Clearwater Lakes region—are apparent in the terrestrial craters. About the only erosion that lunar craters suffer is from the formation of later craters.

Photos by H. Raab, Johannes-Kepler-Observatory (*left*) and Image Science and Analysis Laboratory, NASA/Johnson Space Center (*right*).

wind, and water (see Figure 1.8). This is why the lunar mountains and valleys appear so smooth. There is essentially no erosion on the Moon that can sculpt mountains or carve canyons. These forces are absent on the Moon due to the lack of an atmosphere, a feature we were able to deduce from the blackness of the lunar sky.

But what about other celestial objects in the lunar sky? How do these appear and how do they move as we view them from the surface of the Moon? The Sun's motion is no surprise. As from Earth, we can readily observe that the Sun rises in the east and sets in the west from just about any location on the Moon.[17] But rather than taking 12 hours on average to complete a trip from horizon to horizon, the Sun as seen here on the Moon typically takes a much more leisurely two weeks to travel from sunrise to sunset across the black lunar sky. Since we can see stars in the dark daytime sky from the Moon—though not as many as we might expect, due to the intense glare from the bright lunar surface itself—it is much easier to observe the Sun's motion with respect to the stars here, something not so easily done from Earth. Sure enough, over time we see that the Sun moves steadily from west to east, requiring one year to go fully around the sky. This is very nearly the same motion we inferred for the Sun as seen from Earth based on the seasonal changes of the nighttime constellations.

17 Why? Well, we humans are creatures of habit. Whenever we visit new worlds, we will certainly always define the direction where the local 'sun' (if such exists) rises as 'east', while the direction where it sets will be 'west'. There are likely to be some interesting exceptions, such as skies where the local sun never rises or sets (!), or places where there are more than one sun!! We'll encounter some of these cases later.

Of course, there is one obvious and awe-inspiring difference between the lunar sky and the sky from home. Assuming we traveled to, say, one of the six Apollo landing sites of the late 1960s and early 1970s (see Figures 1.6 and 1.8) our view of this sky would be graced by the beautiful hues of Earth all mingled, jewel-like, in the lunar sky. From the Moon, the Earth appears just under four times larger in diameter—about 13.5 times larger in area—than the Moon does from Earth. But since our planet reflects light about ten times more efficiently than the dark lunar surface—those oceans and clouds reflect light far better than the dark rocks that make up the lunar surface—the Earth appears far brighter to us in the lunar sky than the Moon appears from Earth. Around 41 times brighter.

The Earth also appears big enough in the sky that we can easily distinguish surface features such as oceans, weather systems, and even continents all at a single glance. The vibrant colors would contrast sharply with the monotonous shades of gray of the Moon itself. Even with binoculars or a very small telescope, the contrast is striking. From Earth, the gray Moon is pockmarked by craters and ancient volcanic flows, essentially a dead world. The vivid blues of the oceans, the browns, reds, and greens of the continents, the whites of the poles and clouds all reveal the Earth to be a glorious, living world. Nearly all the astronauts who traveled to the Moon discovered that getting to see the Earth from this vantage point was one of the highlights of their remarkable voyages (Figure 1.9).[18]

Fig 1.9

The Earth as seen from the Apollo 17 landing site. This photo was taken on December 12, 1972. The astronaut shown here is Gene Cernan, the human to stand most recently on the surface of the Moon. If we were to visit the Apollo 17 landing site, we'd find the Earth in very nearly the same spot in the sky. By comparison, the Moon as seen from Earth appears about one quarter the diameter of the Earth and is about 41 times fainter.

Photo by H. Schmidt/ NASA.

As we study it further, the behavior of the Earth that we see in the lunar sky bears some resemblance to what we see the Moon do as viewed from Earth.

18 One exception was Harrison Schmidt of Apollo 17. Schmidt, the only scientist to fly to the Moon was, as a geologist, in awe of being able to study the Moon and its rocks from up close. When his companion, Eugene Cernan, told him to look up and take in the beautiful view of the Earth, Schmidt said 'Ah, you seen one Earth, you've seen 'em all!' But we can forgive him; he was kind of busy at the time!

Like the Moon, the Earth undergoes phases over a period of about 29.5 (Earth) days. But there is a subtle difference. The phases of the Earth are not coincident with those of the Moon that we would see at the same time from Earth. For example, when the Earth appears full from the Moon, the Moon is new—and therefore essentially invisible—as seen from Earth. When the Moon is a thin crescent, the Earth is nearly full or 'gibbous.' When the Moon is half illuminated from Earth, the Earth also appears to be half illuminated from the Moon. And when the Moon is gibbous, the Earth exhibits a crescent shape. The reason for all this is that the source of illumination of both bodies is the Sun. When we on Earth see a full Moon, the Sun and Moon are on opposite sides of the sky. But from the Moon, the Earth would at that time be located in the same part of the sky as the Sun and so its illuminated side would be facing away from us. We would see a 'new' Earth. If we contemplate this reasoning and apply it to a full cycle of phases, we can appreciate that the phases of the Earth and Moon will always be perfectly out of sync over the course of the lunar month.[19]

But there is another important—and much more striking—difference in the Earth's appearance from the lunar surface. Recall that from any given location on Earth, the Moon moves around in the sky, rising in the east and setting in the west day after day. But the Earth behaves quite differently in the lunar sky. Rather than travel across the lunar sky over the course of the lunar day, the Earth hangs eerily in (nearly) the same location in the sky. All the time. From the lunar surface, there are no Earthrises. No Earthsets.[20] Even stranger, had we have traveled to the 'back' side of the Moon—the part we cannot see from Earth—the Earth would *never* be visible in the lunar sky, since it would always be located below the horizon, blocked from view by the Moon itself.[21] From a

19 There is actually a way to observe this for yourself, even without traveling to the Moon. When you see a thin crescent Moon, look carefully at the 'unlit' part of the Moon. You will immediately see that the part of the Moon not directly facing the Sun is also visible! Keep in mind that when we, on Earth, see a crescent, the side of the Moon facing Earth sees a nearly full—and *very* bright—Earth. The reflected light from our planet illuminates the 'dark' part of the Moon, which we then see in the light of what is known as 'Earthshine.' As the Moon becomes more illuminated as seen from Earth, the amount of Earthshine becomes smaller as the Earth now becomes a crescent as seen from the Moon. Thus, as the sunlit part of the Earth gets smaller as seen from the Moon over the course of a lunar month, the 'dark' part of the Moon becomes increasingly harder to see.

20 You may be familiar with the famous 'Earthrise' pictures and movies taken by the Apollo astronauts and more recently by some unmanned spacecraft orbiting the Moon. But these images were obtained from lunar orbit, not from the lunar surface. The changing location of the orbiting spacecraft results in the apparent motion of the Earth relative to the Moon's horizon. From the *surface* of the Moon, the Earth would appear essentially motionless (though see the next footnote).

21 There is a small effect known as 'libration' that arises from a lack of perfect synchronization between the Moon's rotational and orbital periods. This causes the Earth to oscillate slightly around a fixed position in the lunar sky. Though this means the Earth is not really stationary in the lunar sky, the comparatively subtle movement associated with libration is nothing like the steady east-to-west motion of the Moon across Earth's sky.

vantage point on the far side of the Moon we would experience only the grays of the moon, the monthly passage of the Sun, and the deep blackness of a sky in which our home planet never appears. The only colors in our lives would be the faint hues of the stars! Better to be on the near side and enjoy the glorious view and colors of Earth (Figure 1.9).

Before leaving the Moon, it is worth noting an important implication of the things that *don't* change at all in the lunar sky compared to the sky from Earth. Though they move across the sky at a somewhat different rate in the lunar sky, the Sun and the stars themselves appear very similar when viewed from either the Moon or Earth. From either location, the Sun has essentially the same size in the sky, the same intense brilliance. Likewise, the stars exhibit the same brightnesses, the same colors, the same relative locations regardless if they are viewed from the Earth or the Moon. There would be no need for new sky maps: Celestial charts designed for Earth's sky would work perfectly well to describe the sky as seen from the Moon. Evidently, traveling a quarter of a million miles—over 30 times the diameter of the Earth—makes very little difference in our perception of these objects. The only way to understand this is that the Sun and the stars must all be located much, *much*, further away from Earth than the Moon is. This is our first inkling of the vastness of the Universe we aim to explore. And it underscores just how tiny this first journey to the Moon is by comparison. It is a mildly unsettling thought. After all, the Moon is the farthest any human has ever traveled. Yet it is clear that we have barely moved at all.

DESTINATION 1 QUESTIONS

1.1. The Earth does appear to move a small bit in the sky from the near side of the moon, oscillating steadily around a single point in the sky. This motion is called 'libration' (it is described in one of the footnotes of this destination).

a) Given that the Earth moves this way in the Moon's sky, explain how libration alters our view of the Moon as seen from Earth.

b) Let's imagine that the Earth *does* rise and set from a location on the lunar surface. Explain how that would alter our view of the Moon from Earth.

1.2. Interview at least two people who were alive at the time, July 20, 1969, about where they were and what they were doing during the Apollo 11 lunar landing and Moon walk. Include a description of any feelings that they may recall having experienced at the time. Even many decades later, you will discover that just about everyone that age *will* remember. It was a unique event that tied all people of Earth together, if only for a moment.

1.3. We have described what the Moon does over the course of a *lunar month* as seen in Earth's sky. Summarize that information, including how the Moon's phases evolve and the lunar position changes over the course of the lunar month. Use that information to generate a 'model' of the Earth/Moon/Sun system. The model can be a precise verbal description or a set of figures that illustrate what the three bodies are doing to account for the observations. In this model, show clearly how the three bodies change their relative locations over the course of a lunar month in a manner consistent with the information you have summarized. Use this model to explain why there is a near/far side of the Moon relative to the Earth, but that there is all locations on the Moon will, at some time during the lunar day, receive direct sunlight. That is, there is no permanent 'dark' side of the Moon.

1.4. How old are you in (a) *years*, (b) *seconds*, and (c) *lunar months*? Determine these times to an accuracy of about one day.

1.5. The typical transit time from the Earth to the Moon during the Apollo missions was about 2.5 *days*.

a) Given that the Moon is located about 240000 *miles* from Earth, what is the average speed that the Apollo spacecraft had to travel to cover the distance in that amount of time? Report your answer in (a) miles *per hour*, (b) *miles per second*, and (c) *kilometers per second*.

b) What is the longest single trip you have taken? Estimate the distance you traveled. How long did that trip take? How long would it have taken had you traveled at the average speed of the Apollo spacecraft when they traveled to the Moon?

1.6. On Earth, the Sun takes approximately 12 *hours* to traverse 180 *degrees* across the sky as it moves across the sky from sunrise to sunset. On the Moon, this same motion takes about 14.8 *days*.

a) Calculate the angular rate of motion of the Sun as seen from Earth in *degrees/hour*.

b) Calculate the angular rate of the motion of the Sun as seen from the Moon in the same units.

c) How much faster does the Sun travel across the sky from Earth compared to its rate across the lunar sky?

1.7. We noted that the Earth appears about four times larger in the sky from the Moon than the Moon appears in the sky from Earth. Assuming both bodies are spherical in shape (not precisely true, but pretty close), what is the ratio of their ...

a) Surface areas?

b) Volumes?

1.8. Look up on the web information about the Far Side of the Moon. Describe how that hemisphere is similar and how it is different from the side that we see from Earth. Include in your description some of the history of humanity's exploration of that 'hidden' face of the Moon.

1.9. Tidal forces between the Earth and Moon are causing our planet to slow its rotation at a rate of about 0.002 *seconds per century*.

a) At this rate, how long will it take for the Earth's rotation period to double to 48 *hours*?

b) When the rotation period of the Earth reaches about 40 *days*, it will be synchronized with the Moon's orbital motion and the two bodies will permanently have the same faces oriented toward each other. Assuming the Earth slows down at the rate specified at the start of this problem, how long will it take for synchronization to occur?

c) When the Earth is synchronized to the Moon's orbital period, describe how the appearance of the Moon in the sky from Earth will differ from what we observe today. How will the appearance of the Earth from the Moon change compared to what we could see today from the lunar surface?

1.10. Some of the Apollo astronauts left a set of 'retro-reflectors' at their landing sites. This device contains a mirror consisting of reflective surfaces of the inside of a cube. Any ray of light illuminating such a device is reflected in the same direction from which it came. Consequently, by shining a laser from a telescope on Earth and then recording the time of return, it is possible to use these reflectors to very precisely determine the distance to the Moon at any given time.

a) Assuming a mean distance of 240000 *miles* (386000 *km*), how long would it take the light to travel from the telescope, to the Moon, and back to the telescope?

b) It is possible to measure the distance to the Moon with this method to a precision of about 1 *cm*. How precisely must the return signal be measured to make this possible? (**HINT:** How long would it take light to traverse a distance of 1 *cm*?)

c) It now takes on average about 1×10^{-8} (or, 10 *nanoseconds* where a *nsec*, the abbreviation for *nanosecond*, is one-billionth of a *second*) longer for the laser to reach us than it did when these experiments began in earnest in 1969. How much further away on average is the Moon today compared to 1969?

Destination 2

THE VIEW FROM MARS

2.1 TERROR AND FEAR IN THE SKY OF MARS

High above us, the Sun shines brightly. And yet, it is so cold here. Even at mid-day, the Sun provides virtually no warmth under this pale red sky. No human has ever set foot here, though a few man-made machines have visited this place. Some survived for hours or days. Some for a few years. But in the end, they all eventually succumbed to the cold, the dust, the wind.

Welcome to Mars. Though it's true we have longed for home from the moment we arrived, there is no denying the stark beauty of this place (see Figure 2.1). The largest volcano and the largest canyon known in the entire Solar

Fig 2.1

(*Left*) A panoramic image of Mars created from a mosaic of Viking orbiter images. The blotches near the left limb of the planet are two of Mars' majestic volcanoes. Each of these soars higher than Mauna Kea rises from the seabed in the Pacific Ocean. Near the center of the picture is the immense canyon, Valles Marineris, named after the first space probe to photograph the structure in detail. This gargantuan canyon—which is so long it could stretch across the entire United States—represents one of the many telltale signs that Mars had a watery past even though it is almost completely dry today. (*Top right*) The view from the Viking 2 landing site. A cover from one of the Viking instruments is visible in the photo, not some litter from an untidy alien civilization. (*Middle right*) A view of a Martian crater with what is likely water ice on its floor. This crater is located near the North Pole of Mars. (*Bottom*) A panorama of Mars taken by the *Spirit* Mars Exploration Rover from 'Lookout Point.' A small portion of a solar panel of the rover can be seen near the bottom of the picture.

Photos from JPL/NASA.

System are found here. But so are planet-wide dust storms that can last for months, and completely obscure the sky over the entire planet [1] Fortunately for

1 Martian dust storms typically kick off during summer in the planet's Southern Hemisphere. This is, coincidentally, also about the time when Mars is closest to the Sun. On both the Earth and Mars, the seasons are caused principally by the tilt of the planet's rotation axis relative to its orbit plane. When the tilt of either planet is oriented so that the southern pole faces the Sun, it is summer in the Southern Hemisphere. On Earth, we

us, it is perfectly clear today. And the outlook is good for the next few months. That should give us plenty of time to get used to the red daytime sky!

The duration of a day on Mars is about 24.6 hours, something that would be relatively easy to get used to since it's coincidentally similar to the 24-hour duration of a day on our planet. As on Earth, the Sun as seen from Mars rises in the east and sets in the west. But this is *not* a coincidence; it is a choice. No matter where we travel, whenever possible we will always define east as the direction where the local sun rises, and west where it sets. North would then be to the left as we face the rising sun, just like on Earth. But, in fact, the direction of the spin and the orientation of the spin axes of both Mars and Earth are quite similar when we measure these relative to a fixed reference point such as the center of the Sun or distant stars.[2] Consequently, the general path of the Sun in the sky and its motion relative to distant stars are remarkably similar to what we would see from Earth. But there are certainly differences. For example, it takes little time to appreciate that the Sun appears eerily smaller here. On average, the apparent diameter of the solar disk—the circular form the Sun takes on in the sky—is about 65% its size as seen from Earth. That's about the same as the relative sizes of the diameters of a quarter and a penny. Also, if we carefully track the Sun's motion relative to the stars, we find that it moves somewhat differently than from Earth. The Sun still travels steadily west to east with respect to the distant stars—just as we observe from Earth—it just does not move as quickly. Also, the Sun's path relative to the stars is very slightly different than its path as seen from Earth. These are not drastic differences, but they reveal fundamental differences between the skies of these two neighboring worlds.

When the Sun sets, darkness sweeps over us on this Martian landscape close on the heels of a remarkable pinkish-red twilight (Figure 2.2). As the night deepens, it soon becomes apparent to us that the same stars we were used to seeing from Earth are visible here. And because the tilt of Mars is similar to that of Earth, pretty much the same stars we see on Earth in the Northern sky would

---

are fortunate that we are closest to the Sun in January, when it is winter in the Northern, and summer in the Southern Hemisphere. Since there is so much more water south of the equator and water requires much more energy to heat up than most rocks do, the southern part of the Earth does not drastically overheat in response to the increased energy input even though the South Pole faces the Sun when the Earth is closest to it. This coincidence helps to moderate the temperature changes caused by Earth's tilt. If, on the other hand, we happened to be closest to the Sun during the Northern summer (that is, when the North Pole faces the Sun), the large amount of land in the Northern Hemisphere would heat much more rapidly than an equivalent-sized body of water. Our Northern winters and summers would probably be more severe than we are used to. It seems likely that some sort of instability on Mars—though not related to bodies of water, since the Martian surface is currently bone dry—plays a key role in driving the planet's periodic, massive dust storms.

2 This is somewhat of a coincidence. We'll see at Destinations 6 and 7 that the formation of process of the Sun and its planets likely caused the planets to all orbit in the same sense around the Sun. But the close similarity of the rotation tilt of Mars and the Earth is just a lucky chance.

Fig 2.2

Sunset on Mars. Imagine that for a moment. Our Sun. Setting on the horizon of another planet. This photograph was obtained by the *Spirit* Mars Exploration Rover from the ramparts of *Husband Hills*. No human has yet beheld this scene in person. The size of the Sun in this image is at the same scale as in Figure 1.2 to illustrate its smaller apparent size from Mars.

Photo from JPL/NASA.

be visible in what we would call the Northern sky of Mars. At first this all seems comforting. But then we think: We see *exactly* the same stars ... in *exactly* the same relative locations! As we ponder this, our initial sense of comfort begins to evaporate. Given the tremendous amount of effort required to get here, the vast distance we had to travel, the numbing duration of the trip, it dawns on us that the fact that the stars appear no different compared to Earth is actually quite unnerving.

We don't need to get into the details of how we reached Mars[3], but we do know one thing. Typically, when one takes a long trip, the view should change! The longer the trip, the more extreme these changes might be. If we were to travel for days in a car, for example, we'd certainly not be surprised to see snow-capped mountains even if we had started the journey in a prairie, a forest, or even a hot, dry desert. Even if we were to imagine having *walked* steadily for 10–20 *days* (and a trip to Mars using current technology would take much longer than this) we'd expect things to change around us. Certainly, the surface of Mars, like the surface of the Moon, is unlike any but the bleakest, most desolate regions of the Earth. The changes in our immediate surroundings—the place where we are located—is undeniable. But when we focus on the Martian sky, any sense that the view has changed compared to what we see from Earth is far less apparent. As we look up at the stars from Mars, we find *no* difference from the terrestrial view of these same stars. To the naked eye, the positions and

3 People have been dreaming for centuries of how to travel to Mars. There are some concrete plans being developed to do so, perhaps in the next 20–30 years. But there are also many severe practical obstacles that we do not know yet how to overcome. See www.space.com to see the latest news on how these plans are progressing.

brightnesses of the stars are indistinguishable from what we saw back home. What this is telling us in dramatic terms is that the Universe—defined by the stars we see in the sky—is *big*! Much, much bigger than the few tens of millions of *kilometers* that separate Earth and Mars (Figure 2.3). The scale of this enormity was something we could barely sense by traversing the comparatively short distance to the Moon. But that feeling of vastness becomes palpable now as we start to take in the eerie similarity of the skies of Mars and Earth.

Fig 2.3

A beautiful color portrait of Mars beside our Moon as seen from Earth. This image provides a sense of how much farther Mars is compared to the Moon. Keep in mind that Mars is about two times larger, yet only appears about 1/70 as large in this view. At the time this photo was obtained, Mars was about 200 times more distant from Earth than the Moon, or about 50 million miles away.

Photo by R. Wayman.

And yet, not everything in the Martian sky is the same as back home. As we study this new sky over time, we do discover some differences. Here comes one now: A small dot of light rising in the east. This is one of Mars' two moons, Deimos (see Figure 2.4). Its name comes from the Greek word for 'terror'—an apt companion to Mars, the god of war. Unlike our Moon—which, as seen from Earth, moves rather sedately relative to the stars—Deimos zips through star fields, traveling over 20 times faster in this sky than the Moon moves in Earth's sky.[4] So fast, in fact, that after just a few minutes, we can readily appreciate that Deimos has moved from west to east with respect to the stars. After an hour, that motion is equivalent to an angle approximately equal to the span of your open hand at arm's length, an angular span of about 15 *degrees* (recall, there are 360 *degrees* in one full circle). At this rate, Deimos needs only 30.3 *hr* to

4 'Speed' here refers to how rapidly an object such as Deimos or the Moon appears to move in an angular sense. That is, since the Moon requires about 29.5 days to traverse the sky defined by the stars and as seen from Earth, it travels 360/29.5 = 12.2 *deg/day*. Deimos appears to move 23.3 times faster across the Martian sky.

Fig 2.4

(*Left*) An image of Deimos (from the Greek word for 'terror'). (*Right*) An image of Phobos ('fear'). The large crater on the left edge of Phobos—named Stickney—is about 8 km across. The images of the moons are shown to the same scale. It is apparent that Deimos is about half the 'diameter' of Phobos, but that neither moon is spherical in shape. Both are puny compared to the Earth's Moon, which has a diameter over 200 times larger than Phobos. These images were obtained by the Mars Reconnaissance Orbiter as it passed relatively close to the moons at different times during its mission.

Photos from U. Arizona/ JPL/NASA.

complete an entire circuit of the stars in the sky, not the 29.5 *days* it takes our Moon to do the same thing as seen from Earth. But that should not obscure that in some fundamental ways the motions of Deimos and of our Moon are similar. Both rise in the east and eventually set in the west as seen from the surface of their respective planet. And both move the opposite direction—west to east—relative to the stars (see Destination 1).

Unusual as it is, the movement of Deimos in the Martian sky hardly prepares us for the behavior of Mars' other moon. Tiny Phobos—from the Greek word for 'fear'—zooms across the sky, requiring only 7.65 hours (!) to complete one lap relative to the stars (Figure 2.4). This period is actually shorter than the time it takes Mars to rotate, that is, the length of the Martian day. Consequently, Phobos does something we have never seen before. Unlike Deimos from Mars, unlike the Moon from Earth, unlike the Sun from both the Earth and Mars, we immediately see that from our vantage point here on Mars *Phobos rises in the west and sets in the east!* And yet, just like Deimos, the Moon and the Sun, we can also see that Phobos travels from west to east with respect to the stars. Quickly. In just 15 minutes, Phobos travels the same distance relative to the stars that Deimos does in one hour. The 30.3-hour period of Deimos and the 7.65-hour period of Phobos are known as the *sidereal* periods of the two moons since these are the times it takes each to complete a circuit of the sky, relative to the stars. The word 'sidereal' derives from the Latin word *siderus* meaning 'belonging to the stars,' an apt name for these periods measured relative to the distant stars.

But the periods of both moons are distinctly different if we measure them relative to our location here on the surface of Mars. In particular, if we determine the time, say, from when either moon rises, crosses the sky, sets, then rises again, we find that both take longer than their sidereal periods to complete one such circuit across our sky. For Deimos—which rises in the east—this motion takes nearly 5.5 (Earth) *days*. Speedy Phobos—which rises in the west—completes this full circuit of the sky in a bit over 11 *hr*, though, from our vantage point, in a seemingly opposite direction than Deimos. Although we did not stress this at the time, Earth's Moon behaves in an analogous way. The sidereal period of the Moon—the time it takes our Moon to make one complete circuit of the sky relative to the stars—is about 27.3 *days*, but it only takes 1.04 *days* (on average) for the Moon to travel from one moonrise to the next from a given location on Earth (see Table 2.1).[5] Notice how in all these cases, the time it takes the moons to travel around us on the surface of the planet can be very different from the time it takes to travel around the sky from the perspective of the stars.

TABLE 2.1 Some Properties of the Moons of Earth and Mars

MOON	PARENT PLANET	SEMI-MAJOR AXIS (*km*)	ORBITAL PERIOD (*days*)			MOON MASS (*g*)	MOON RADIUS (*km*)	ESCAPE VELOCITY (*mph*)
			SIDEREAL	SYNODIC (1)	HORIZONTAL (2)			
Moon	Earth	384400	27.32	29.53	1.04	7.4×10^{25}	1737	5350
Phobos	Mars	9380	0.3190	0.3191	−0.46	1.1×10^{19}	11.1	25.5
Deimos	Mars	23500	1.262	1.264	5.49	1.5×10^{18}	6.2	12.7

Notes to Table 2.1:

(1) The 'synodic' period is the period measured relative to the Sun. Thus, these periods would approximately equal the time from one new moon to the next, say. They are not precise at all times because the rates at which the Sun and the moons move vary with time. (2) The 'horizontal' period is the time required for the moon to travel from one rise to the next as seen from the surface of its parent planet. The negative value for Phobos signifies that it appears to move the 'wrong' way (west to east) as seen from the surface of Mars.

Why are these timescales so different? And why, in particular, do the two moons of Mars appear to move in opposite directions as we view them from the Martian surface? The answers involve understanding the important concept

5 Whoa! You may be wondering how the 29.5-*day* period of the Moon that we have been talking about so far fits into this. That is the approximate time—it varies a bit month to month—from one new moon to the next (or one full moon to the next full moon, etc.). Since the phases of the Moon are produced by the way the Sun illuminates it as seen from Earth, the 29.5-*day* period reflects the motion of the Moon relative to the Sun, also known as the *Synodic* period. The sidereal period is relative to the stars, while the 1.04-*day* period we mentioned here is relative to a position on the surface of the Earth, and is sometimes known as the *Horizontal* period.

Fig 2.5

(*Left*) This is a composite of five images that show both Deimos and Phobos in the Martian sky as observed by the Mars surface rover *Spirit* in 2005. In these images, the camera was not moved, so the images illustrate the apparent changes in the positions of the two moons over time as seen by a stationary observer on the Martian surface. The sequence of images was taken at different intervals; first three images close together in time, followed by two separated by longer intervals. We can see that, from the surface of Mars, Deimos is moving comparatively slowly since the first three images are blurred into a single elongated streak. In contrast, Phobos is moving rapidly enough that all five images are clearly visible. The order of the images in each sequence (and the arrows) indicates that Phobos appears to be moving in the *opposite* direction as Deimos from a viewer fixed on the Martian surface. (*Right*) This image shows a similar sequence to the one at left. The critical difference is that the camera was moving between exposures in such a way as to track the stars. Consequently, in this image the motions of Phobos (the brighter object) and Deimos (the fainter one) are relative to the stars, not the Martian surface. From this reference frame it is clear that both moons are moving in the *same* direction (as also indicated by the arrows)! And we

of 'relative motion.' In the case of the Moon, the Sun, or Deimos, one has to carefully distinguish motion relative to the observer and motion relative to the stars. In our case, we—the observers—are viewing the sky from the surface of a spinning planet, a motion so smooth and comparatively gentle that it is easy to forget it is actually happening. As we have already appreciated, the stars are, on the other hand, extremely more distant. Compared to a rotating planet or a moon moving in its orbit, the stars are essentially standing perfectly still. They act somewhat like a vast, fixed structure surrounding us. The stars—or the horizon around us—are said to define a *Reference Frame* that we can use to define the motions of objects. Thus, when we say that the Moon moves 'relative to the stars,' we are referring to the absolute motion of the Moon in comparison to the positions of the (seemingly) stationary reference frame of the stars. If we speak of the motion of the Moon relative to us—that is, to the horizon our location defines—then we act as if *we* define the reference frame. The different periods and strange motions all arise from the fact that we typically observe motions of celestial bodies from reference frames that are themselves moving relative to the stars and even the objects we want to study. For example, the spinning surface of the Earth. Or Mars. Because of this, the 'apparent' movements of bodies in the sky can appear quite confusing, and often fundamentally different from the way they are 'truly' moving relative to the distant stars.

Phobos and Deimos illustrate this nicely. As we have seen, both orbit Mars in the same direction—west to east in the reference frame defined by the stars—but they *appear* to move in opposite directions in the Martian sky. Deimos is moving in its orbit more slowly than the surface of Mars is rotating about its axis (which also happens to be in the same direction), so it takes a comparatively long time for the moon to overtake then pass us on the (moving) Martian surface. Consequently, although the sidereal orbital period of Deimos is about 30.3 *hr*, it takes it much longer—about 5.5 *days*—for it to traverse the sky from our reference frame on the moving surface of Mars. On the other hand, Phobos is moving so fast that it returns fairly quickly—in just over 11 *hr*—to a given location in the sky defined by the reference frame of our local horizon. In Phobos' case, this period is only a bit longer than its sidereal orbital period of 7.65 *hr* (see Table 2.1).[6]

Don't be daunted by all this. Ultimately, the gory details of the complex motions of the Martian moons are not the key points here. Rather, the lesson that Phobos and Deimos teach us is that it can be difficult to interpret the motions of objects when we, the observers, are also moving. From the surface of Mars we can easily be fooled into thinking its two moons are traveling in opposite directions and that they have orbital periods very different from their true sidereal periods. And all because we, viewing the sky from the surface of Mars, are moving too. We'll need to keep this in mind as we now consider another example of movement in the Martian sky. One so strange it seems impossible to understand in any simple way ... until we account for our own motion.

can also see clearly that Phobos—the closer of the two Martian moons to Mars—moves more rapidly than Deimos. Six stars—all in the constellation Sagittarius—are visible in the image. These stars show no motion in this image, confirming that the camera was moving to track the stars. Amazingly, the pattern of these stars in the Martian sky (shown by the lines) appears *exactly* the same in this image as it would be from Earth. Yet, the camera that took this picture was many tens of millions of miles from our home planet!

Photos from JPL/NASA.

2.2 THE RIDDLE OF PLANETARY MOTION

Eventually, we get used to the weird Martian moons. After all, we see them often and we humans are remarkably adaptable. But even as the motions of Phobos and Deimos become commonplace, we start to appreciate that there are other objects in the sky that we had not noted before. On this particular night, soon

6 This is definitely a confusing concept, so an alternative analogy might help, Consider watching two cars moving around a circular track but in the same direction but different speeds, say, 60 *mph* for the slow car, and 120 *mph* for the fast one. If we were watching them pass as we stood by the track, both would appear to move in the same direction, one very clearly much faster than the other. We would also see the faster car pass us twice as frequently as the slower car. But things would appear drastically different if we were to view these cars from a third car moving in the same direction at, say, 80 *mph*. The fast car would now appear to approach us from behind, but at the modest speed of 120 – 80 = 40 *mph*. Thus, it would pass us a bit less frequently than the slower car, moving at 60 *mph*, passed us while we stood beside the track. The slower car, however, would appear to approach us from ahead simply because we are overtaking it at a speed of 80 – 60 = 20 *mph*. It would clearly take a *long* time, comparatively, for us to pass that car again since it is moving, relative to us, so slowly. On Mars, Phobos is the speedy car, Deimos the slow one, and we, moving on the surface of Mars, are in the intermediate-speed car.

after sunset, we notice three of these. They were certainly not visible at this time of day when we first arrived at Mars, but now it is impossible to miss them. To the west, near the Sun is a brilliant white 'star,' one that's easily visible even though we view it during twilight. It shines more brightly than any of the stars we have seen in the Martian sky. Just above it—further from the horizon and also in the western sky—hovers a gorgeous 'star' unlike any other we've ever seen. It shines sapphire blue—unique in this sky—and with a brilliance even greater than its nearby companion, able to easily cast shadows on the Martian soil as the night darkens. Turning, we look to the east to spot another dazzling 'star.' This one emits a warm, creamy-brown color as it shines almost as brightly as the two others we see to the west. Now that we notice these prominent objects, we start to study them systematically night after night. After just a few days, it becomes apparent that they—like Phobos and Deimos—are moving with respect to the reference frame of the stars. The two in the west move, as everything has up to now, from west to east relative to the stars. But the third—the creamy-brown objects we see to the east—moves in the opposite direction! From east to west relative to the stars.

These certainly are not stars, but planets.[7] Even a small telescope confirms that it is Jupiter that dominates the sky to the east, the one currently moving the 'wrong' direction. It gets much closer to Mars than to Earth, and hence it appears considerably brighter here than it ever does at home. The brilliant white planet in the west is Venus, aptly named after the goddess of beauty. Which means the blue one must be ... Earth! If nothing about this journey had made a strong impression on us, this observation can't help making us ponder. There is our home. A single, beautiful, bright dot in the Martian sky. The realization makes the vast chasm that separates us from home somehow 'personal'. All that we know, all our history, all lives of all the people who have ever existed, are there, in this tiny point of light we see so prominently in the sky of a neighboring planet (Figure 2.6).

But why do the planets move as they do? Since the Sun appears smaller here at Mars, we can safely conclude we are further away from the Sun than when we were on Earth (we'll get into more detail on this point at Destination 3). It is also clear from observing Venus and Earth (and Mercury too, but that planet is much harder to see since it is always very close to the Sun), that at least those two orbit the Sun itself and not Mars. As we observe them over time, we see that Venus and Earth never venture far from the Sun. In particular, we never see them, say, on the opposite side of Mars from where the Sun is located. Since we see Earth

7 The word 'planet' derives from the Greek word *planetai*, meaning 'wandering,' which was always used with the word for 'star,' *asteres*. Thus, to the Greeks, planets were literally 'wandering stars.' And, of course, given that we had to have traveled between planets to get to Mars, we presumably already knew the nature of these objects in the sky! But the point is that we did not have to. Their motions reveal that planets are not stars and almost certainly not moons similar to Phobos and Deimos.

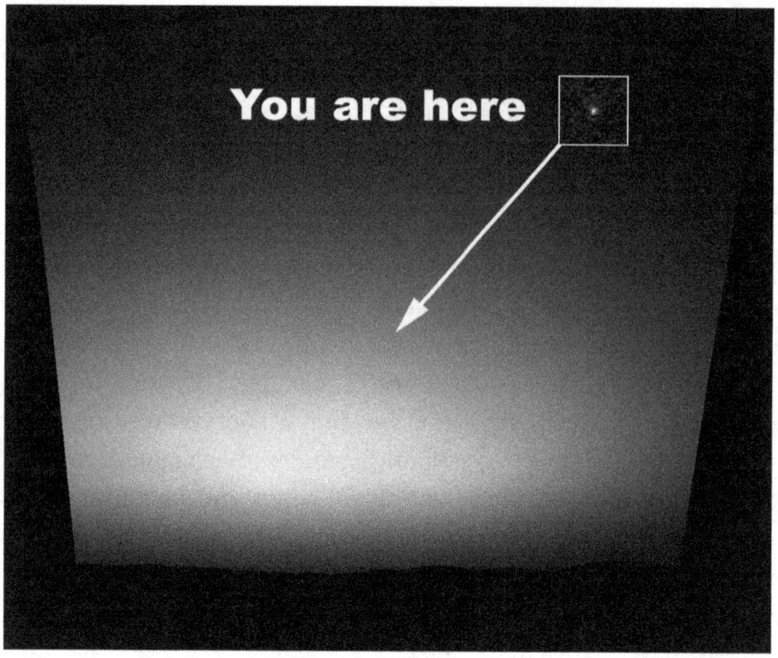

Fig 2.6

A portrait of Earth from the surface of Mars seen just after sunset by the *Spirit* Mars Exploration Rover. The totality of all of our life experiences have occurred on the tiny blue dot we see in the Martian sky.

Photo from JPL/NASA.

moving relative to the stars from Mars, it must be that Mars would appear to move in the sky relative to the stars as seen from Earth. And indeed it does. The only way this can be true is if both planets are in motion relative to the stars. And that implies that both must be orbiting the Sun. It seems reasonable from this argument to conclude that *all* the planets orbit the Sun.

But not everything about these motions appears capricious. As we study Venus and Earth as they travel through the Martian sky, we see clearly Venus generally moves much faster relative to the stars than Earth does. Venus is also closer to the Sun than the Earth. In the same way, we can easily determine from observations of Mars from Earth (or from observations from Mars' surface) that Phobos is closer to the planet than Deimos. And that Phobos moves faster in its orbit. If this is a genuine trend[8], we can begin to appreciate that the motions of the planets—and of the Martian moons—obey some sort of systematic rules. In this case, objects in smaller orbits seem to move faster. There seems to be some order here. A pattern.

But what about Jupiter? Unlike the Earth or Venus, we can see Jupiter nearly overhead at Martian midnight. This simple observation means that Jupiter must be further from the Sun than Mars (just as the full Moon as seen from Earth is, at that time in the lunar orbit, a bit farther from the Sun than the Earth). But we have seen too that Jupiter's motion appears to be fundamentally different from that of any object we have studied so far. Up to now, every moving body that we have studied has been observed to travel from west to east relative to the stars.[9]

8 Spoiler Alert! It is a genuine trend.

9 Notice that this is true even if we define east and west in a single reference frame—for example, the reference frame defined by Earth's rotation axis.

But Jupiter appears to be moving *backwards*, from east to west relative to the stars. What might solve this particular riddle?

Well, maybe Jupiter just moves in the opposite direction around the Sun. That would certainly be the simplest explanation for its peculiar motion. But, in fact, we'll soon see that all the planets of the Solar System—including Jupiter—orbit in the same direction (the reason that this is so will become apparent as we learn how stars and planets form in the first place at Destinations 6 and 7). To be more precise, if we could look down on the Solar System from a point far above the Earth's North Pole, all the planets move along their orbit in a counter-clockwise direction.[10] The simple explanation to account for Jupiter's odd motion is not the correct one.

In astronomy in particular—and science in general—one of the best ways to address a riddle is to just keep on watching what Nature does. So, in this spirit of patient observation, we continue our studies of the motions of the planets in the Martian sky. Eventually, we see something bizarre, unlike anything we've seen before. After a few (Earth) months of our initial studies of Jupiter's motions as seen from Mars, we notice that Jupiter begins to slow down relative to the stars. Then, to our amazement, *the planet stops moving, then commences to travel from west to east with respect to the stars!* That is, Jupiter stops going in the 'wrong' direction in the sky (east to west) and begins to move in the 'normal' direction (west to east)! This change in the direction of travel of a planet in the sky is known as *Retrograde Motion*. What is going on to account for this unusual behavior?

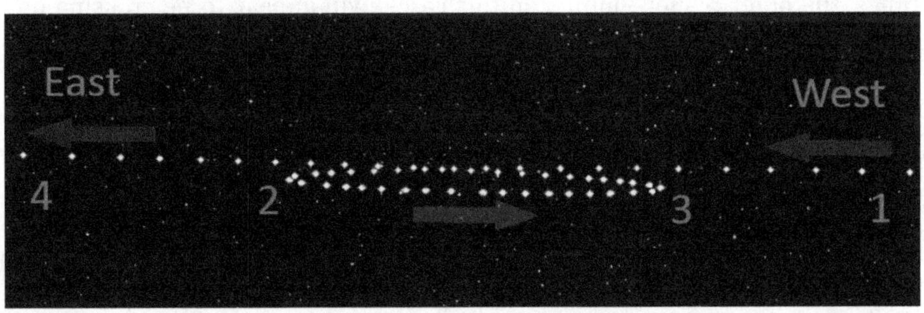

Fig 2.7

This 'photograph' shows the movement of Jupiter relative to the stars across the Martian sky. Each 'image' of Jupiter corresponds to its location at intervals of 7–10 Martian days; the entire sequence covers about 1.6 Earth years. Jupiter is first seen at right (position '1') moving in a prograde direction (red arrow) from west to east. It then eventually stops that motion (position '2') and begins to move in a retrograde direction from east to west relative to the stars (blue arrow). Many weeks later, it stops again (at position '3') and resumes its original travel from west to east (red arrow) until it leaves the field of view at left (position '4').

Image produced by M. Mateo using *Starry Night* software.

First, it is certainly not the case that Jupiter is actually changing its direction of travel. We have already emphasized that the planets orbit the Sun in the same direction. But even if we ignored that fact, to actually bring something the size of Jupiter to a stop in the time we observe—a few months—would require a phenomenal force that would radically affect Jupiter's structure. No, the answer is more subtle and involves the confusion that can arise from observations from moving reference frames. Recall from our earlier discussion that the planets not only all move in the same direction, but also that those closest to the Sun

10 Most of them, but not all, also rotate that direction too. In particular, Venus and Uranus rotate in the opposite sense, the only planets that do so.

seem to always move the fastest. This is not to say that inner planets merely go around the Sun in a shorter time. That would be expected even if the planets all moved at the same speed given that the orbit of a close-in planet is smaller than the orbit of one that lies further from the Sun. Rather, the inner planets really do move faster—in *miles per hour*, or, to use a more common unit used by astronomers, *kilometers per second* (usually abbreviated as *km/s*). This is true of Mars relative to Jupiter. Mars is closer to the Sun than Jupiter, so Mars travels at a higher speed around the Sun than Jupiter. And it is true in general.

Now, when we see Jupiter to the east soon after the Sun sets in the west, we—on Mars—must be more or less directly between the Sun and Jupiter. But, at all times, we—on Mars—are also moving in our orbit faster than Jupiter is moving in its orbit. At this particular time—when we on Mars lie between Jupiter and the Sun—both planets also happen to be moving in about the same direction on temporarily nearly parallel paths. A car we pass on the freeway can appear to be moving backwards from our perspective even though we know that both cars are moving in the same forward direction. In a similar way, Jupiter appears to be moving backward—that is, in the 'wrong' direction in the Martian sky—simply because we, on Mars, are overtaking it.

Does this explain why Jupiter appears to stop, then change direction later on? Partly. Both Mars and Jupiter orbit the Sun in what are approximately circular orbits, though Jupiter's orbit is much larger than that of Mars (see Figure 2.8). Consequently, during part of its orbit, Mars will be traveling more or less directly toward or away from Jupiter (the green and purple lines in Figure 2.8). At such times, the outer planet—Jupiter in this case—will appear to be crossing our line of sight much as cars in a distant cross-street appear to move across our path. But during other parts of Mars' orbit, the planet will begin to move along a direction that is more nearly parallel to Jupiter's motion, either traveling in the same direction when the planets are closest together, or in the opposite direction when furthest apart (see Figure 2.8). When Mars is overtaking Jupiter (the blue lines in Figure 2.8), the slower outer planet appears to move 'backwards.' It is because of the change in direction of a planet as it traverses its orbit *and* the fact that inner planets move faster that we see retrograde motion. Although we've focused here on Mars and Jupiter, we would see retrograde motion like this whenever we view any planet more distant from the Sun.[11] From Earth, for example, Mars exhibits retrograde motion very similar to what we observed of Jupiter from Mars (see Figure 2.8).

Whether it is the weird behavior of Phobos and Deimos, or the bizarre motions of planets such as Jupiter, it is important to remember the underlying point: When we observe moving objects from another moving object, the

11 In fact, even inner planets (those closer to the Sun than the one we are on) exhibit retrograde motion. However, it is a lot easier to map and understand the retrograde motion of outer planets (planets further from the Sun than the one you are on).

Fig 2.8

This figure shows the motion of an outer planet (*orange orbit*) as viewed in the sky from an inner planet (*red orbit*) following a smaller orbit about the Sun (the central yellow dot). Both planets are moving in a counter-clockwise direction as indicated by the arrows. Each dot along the orbits represents the positions of the planets over constant time intervals. Because the red planet is closer to the Sun, Kepler's Third Law tells us it must be moving faster in its orbit than the outer orange planet. The colored lines show the direction one must look to see the orange planet from the red one. The image at top illustrates the retrograde motion of Jupiter as seen in the Martian sky over the course of a few months (see Figure 2.7 for details). From Earth, both Mars and Jupiter would exhibit this sort of motion in the night sky since both travel along orbits that are larger than the Earth's.

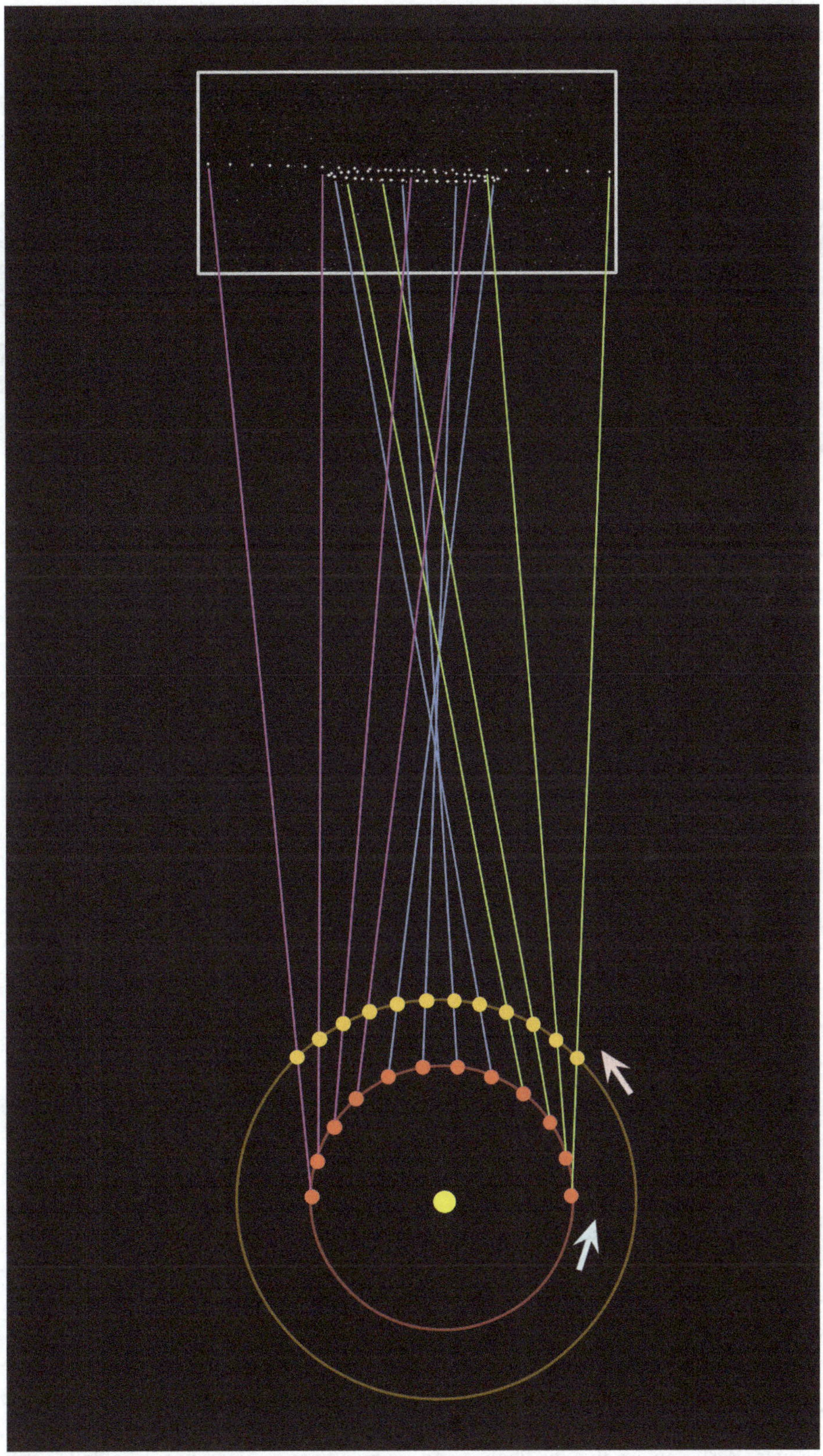

resulting motions can be complicated, even quite misleading. But we have also seen that the true motions of the planets—the motions we deduce after removing our own motion—seem to reveal some systematic behavior in how they travel around the Sun.

2.3 KEPLER'S THIRD LAW: THE TRIUMPH OF SIMPLICITY

The riddle of retrograde motion took humans *millennia* to figure out, so don't feel bad if you are confused! However, the solution, once it came, evolved rapidly and led to even more profound realizations. It all started when one person—Nicholas Copernicus—figured out that these bodies—including the Earth itself—orbit the Sun and not the Earth as had generally been believed. Fairly soon after, Johannes Kepler took Copernicus' idea and a series of planetary observations by Tycho Brahe and others to uncover an underlying pattern in planetary motions that helped to explain why they appear to move as they do in the sky. After thirty years of painstaking analysis—remember, he had no calculators to help him carry out the number crunching—Kepler found a connection between the sizes of planetary orbits and the speeds that the planets travel along their orbits.[12] We refer to this relationship as Kepler's Third Law of planetary motion. It is so important, and so fundamental to helping us interpret what we will see in many future destinations, that it is necessary to get a little technical about it here.

In mathematical terms, Kepler's Third Law states that

$$a^3 \propto P^2.$$

In this relation, the symbol a refers to the *Semi-Major Axis* of a body's orbit—that is, half the long axis of the orbit—of a planet or moon about its parent body. The symbol P is the *Orbital Period* of that body around its parent as measured with respect to the stars. For a circular orbit—and most planetary orbits are nearly

12 And not just planets around the Sun. Kepler's Third Law holds for moons around planets. We'll explore this for Phobos and Deimos here at Mars, but it holds for the moons of other planets too: Jupiter, Saturn, Uranus, and Neptune. Indeed, it holds for any system where small bodies orbit around a much bigger, more massive, central body.

circular—the semi-major axis is just the orbital radius. Also, in case it is new to you, the symbol '$\propto$' means 'proportional to.' This simply means that if the quantity a^3 varies by some factor, then the quantity P^2 changes by the same factor. That is, the quantities change in proportion to one other. Another way to say the same thing is to realize we can write Kepler's Third Law as an equation of the form

$$a^3 = CP^2.$$

Here, the symbol C represents a constant that makes the relation true for at least one planet, or moon, or satellite, depending on what system we apply the equation to. According to Kepler's Third Law, the same constant can then be applied to all other bodies orbiting the same central body (the Sun or some planet).

To use this relation properly, we must be careful to be consistent with our units. For example, let's say we measure the period, P, in (Earth) *years*, and the orbital semi-major axis (or, often, the radius), a, in units equal to the mean distance that the Earth is from the Sun. This distance is so important that it is known as the *Astronomical Unit*, or *AU*. For these units, Kepler's Third Law must be consistent with $1^3 = C1^2$ since for Earth, the period is 1 *yr*, and the orbital semi-major axis (radius) is 1 *AU*. In this case, it's easy to see that the

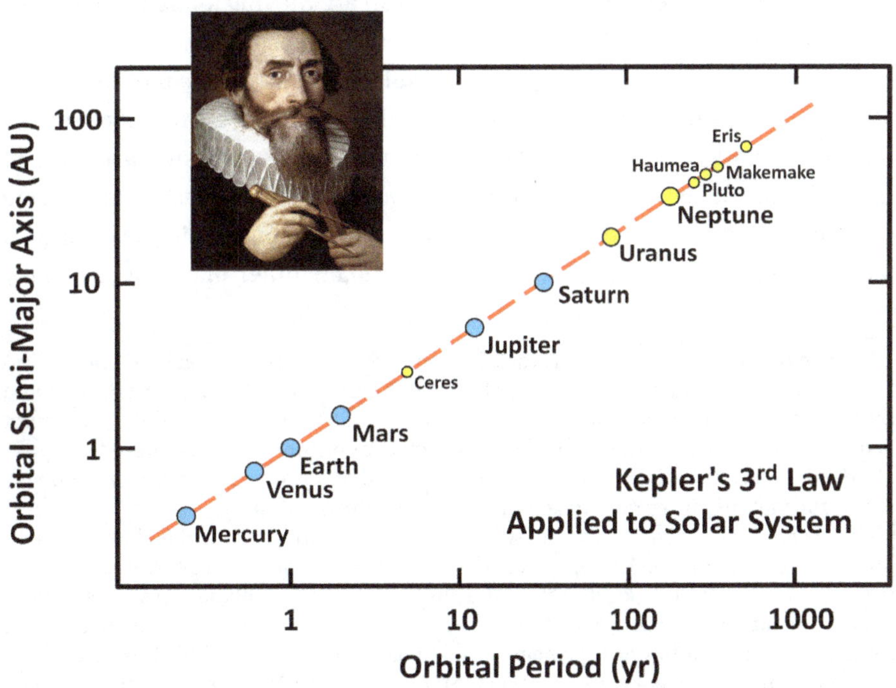

Fig 2.9

A plot of the semi-major axis of the orbits of the planets of the Solar System (y-axis) versus their orbital periods (x-axis). This is a graphical illustration of Kepler's Third Law where a is the semi-major axis of the orbit of a given object in Astronomical Units (*AU*) and P is the orbital period in *years*. For a circle, the semi-major axis is its radius. For an ellipse, the semi-major axis is half the long dimension of the ellipse. This graph is an example of a logarithmic plot that is used to make it easier to see the individual points over the large range of periods and semi-major axes. The planets denoted with yellow symbols were not known to exist when Kepler stated his laws of planetary motion. Yet they are in perfect agreement with the prediction of the Third Law shown as a dashed red line. (*Inset*) A contemporary portrait of Johannes Kepler.

constant C must be equal to 1 precisely. Since Mars orbits the Sun—same as the Earth—and it is located at an average distance of 1.52 *AU* from the Sun (that's its orbital radius), its orbital period—in Earth years—must also satisfy this equation with the same constant, as long as we employ the same units. Thus,

$(1.52\,AU)^3 = P^2$, or $P = 1.88\,yr$. The constant C is still equal to 1.0 because we are still considering objects orbiting the Sun and using the same units for a and P. But don't be fooled. The constant in Kepler's Third Law can take on other values if it is applied to other situations or if other units are used. For example, if we consider the orbital sizes and periods of the moons of Neptune, we'd discover that C has a different value if we impose the same units of time and distance we applied to the Earth and Mars. Though the constant changes, the key point is that the basic proportionality relationship between a^3 and P^2 embodied in Kepler's Third Law remains valid for Neptune's moons just as it does for the planets orbiting the Sun.

As we saw earlier, it is not merely enough that our speed around the Sun—here on Mars—is greater that Jupiter's. We also have to follow a curved path. Kepler also gets the credit for first understanding the true nature of those paths (Copernicus came close). In particular, Kepler's First Law of planetary motion states that planets should orbit the sun along a set of very well-defined types of paths, so-called *Conic Sections*. Two such paths are well known: the circle and the ellipse.[13] In fact, all planets do closely follow nearly circular, but strictly elliptical paths as they orbit the Sun.[14] And because they do, the angle with which we view Jupiter (and all the planets) changes continuously because the planets are moving at different speeds and because all are moving along these closed, curved paths (see Figure 2.8).[15]

In one critical way, Kepler's work to explain the enigmatic motions of the planets ushered in the modern technological world of today. It demonstrated that mathematical relations could be used to make sense of at least some of the phenomena seen in Nature. And it helped to inspire Isaac Newton some years later to go one step further and discover a way to explain the totality of planetary motions (all of Kepler's Laws) and many other phenomena—many

13 The other two conic sections that describe orbital paths allowed by Kepler's First Law are the parabola and the hyperbola. In both cases, bodies on such orbits would pass by the central object only once, never to return. Since planets and moons do not generally behave this way, we can for most practical purposes ignore these two types of orbital paths. Also, Kepler did not formally state that all conic sections were allowed. Credit for that discovery goes to Isaac Newton (see Destination 4).

14 See Figure 4.6 to get a sense of what noticeably elliptical orbits look like.

15 Just to be complete, Kepler's Second Law states that a planet's speed along its orbit is related to how far it is located from the Sun. In particular, a planet must move faster along its orbit when closer to the Sun, slower when further away. This behavior results from a fundamental physical principle, the 'Conservation of Angular Momentum.' We will return to this concept in Destination 6. Since the orbits of planets are never perfectly circular, the distance to the Sun is typically changing all the time. As noted earlier, the Earth, is closest to the Sun in January (seems ironic!) and it is most distant from the Sun in July. As a result, Earth moves more rapidly in January than in July. We can actually see this by noting that the Sun moves across the sky relative to the stars more rapidly in January than in July, an effect that causes the length of the solar day to vary by up to eight minutes over the course of the year. The Sun also appears larger in the sky in January than in July, confirming that we are also closest to it in January.

unknown to Newton—in terms of a single 'Universal Law of Gravity.' Most importantly, Kepler's results showed people that something as baffling and complex as the retrograde motions of the planets could be understood in terms of much simpler mathematical concepts. This realization that an underlying set of principles in Nature can connect diverse phenomena drives to this day much of scientific inquiry in general, and astronomy in particular. We'll see many further examples of the triumph of simplicity when we journey to more exotic locales far beyond the Moon and Mars.

Of course, since we are standing on Mars studying its sky, we must have actually known all this before we left home. We would have had to rely heavily on Kepler's Laws and Newton's Universal Law of Gravity (which we will explore in more detail later on at Destinations 9 and 10) to get here in the first place. But the beauty here is that we can *see* these laws in action as we watch the planets dance across the sky. To be fair, we did not have to go to Mars to appreciate this particular display. If we carefully track the positions of planets such as Mars, Jupiter, and Saturn in the night sky from Earth, we can see for ourselves—with a little patience and a decent star chart—the retrograde motions of the planets. However, by going to Mars we also got to visit the biggest volcano and the longest canyon known to exist in the Solar System (Figures 2.1 and 2.10).

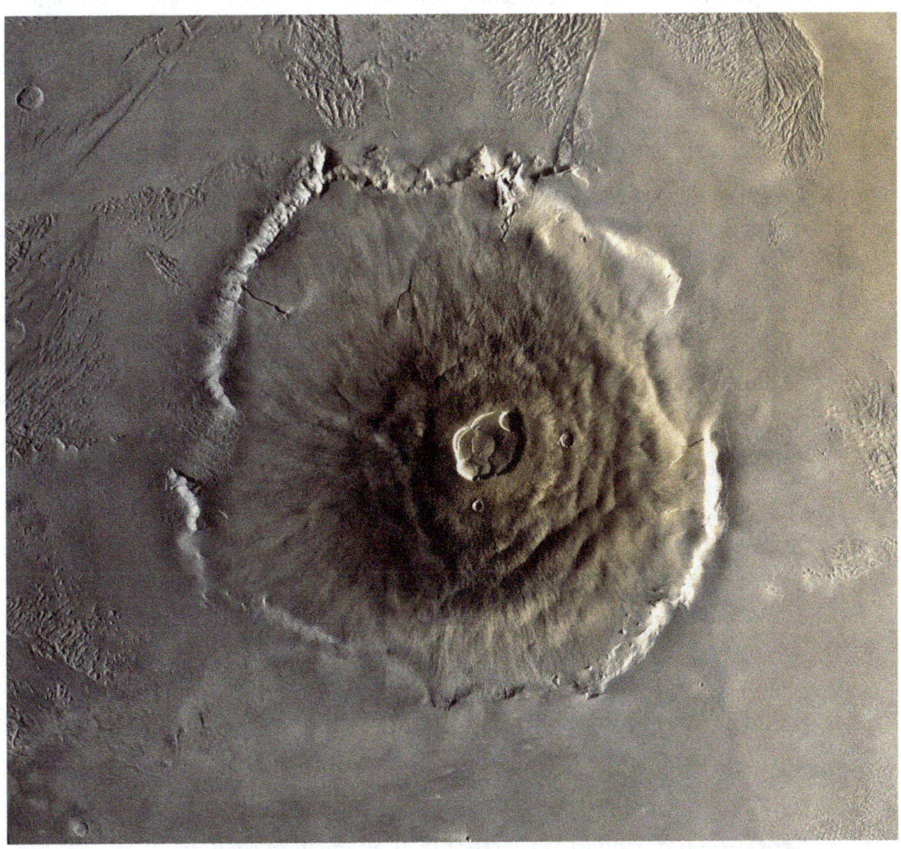

Fig 2.10

An overhead view of Olympus Mons, not only the largest volcano on Mars, but the largest volcano known to exist in the Solar System! The caldera at the top is about 60 miles in diameter. Some of the cliffs on the near side of this image are nearly 7 *kilometers* high!

Photo from NASA.

DESTINATION 2 QUESTIONS

2.1. Let's say we find three planets orbiting the star Zeta Capricornus (a real star); they are called Arnold, Betty, and Charlie (not real planets!). If Arnold moves fastest about Zeta Capricornus—its sun!—and Betty is slowest, and Charlie is in between, which planet is closest to the star? Which is farthest?

2.2. In a car analogy to planetary motions in footnote 6, how long does it take the slow and fast car to go around the track, assuming the track is one mile long? Now, how long does it take each car to overtake the observer's car? The second question is tricky, so consider it carefully!

2.3. You may have noticed that Kepler's Third Law involves the orbital periods and orbital radii of the objects orbiting a central body. But I kept talking about the orbital *speed*. Prove that Kepler's Third Law in fact does mean that objects with smaller orbits and, hence, shorter periods, must also be moving faster than objects with larger orbits/longer periods. You may assume circular orbits. **HINT:** Find a way to relate the distance a body travels over an orbital period, and then recall that speed = distance/time. The rest is algebra.

2.4. One clever way to reach Mars that uses a minimal amount of fuel is to place your spacecraft on an orbit where the point closest to the Sun (known as 'perihelion') is at the distance between the Earth and the Sun, while the point furthest from the Sun ('aphelion') is the distance between Mars and the Sun. If you can make sure that Mars is at the right place when the spacecraft reaches that distance, you can reach that planet. If you do this right, you need only to accelerate to get into this orbit at the Earth, then decelerate when you reach Mars to be able to land or get into orbit around that planet.

a) Draw a picture that illustrates how this orbital path might work to get you to Mars.

b) What is the 'semi-major' axis distance of the orbit of your spacecraft? This is half the sum of the minimum and maximum distances from the Sun. Show why this is the case in your figure of part (a).

c) Strictly, Kepler's Third Law applies to the semi-major axis of an orbit. Given that you calculated this in part (b), what is the period of the orbit for which you calculated the semi-major axis, a, in part (b)?

d) Referring to your figure from part (a) and your result from part (c), how long does it take to reach Mars along this orbit?

2.5. At what distance (semi-major axis) from the center of Mars would a satellite have to be located to have an sidereal orbital period exactly equal to the sidereal rotational period of the planet? All you need to know to answer this question is (a) the orbital period orbital radius for the two Martian moons (Phobos and Deimos; see the Destination 2 notes) and (b) the rotational period of Mars (that is, the length of its day).

Any satellite at this special distance around Earth (the distance is different from that for Mars, but the concept is the same) is referred to as a 'geosynchronous' satellite. They might be called 'Mars-synchronous' in the case of Mars. Why are these appropriate names for such satellites? How would a Mars-synchronous satellite appear to you from the surface of Mars assuming you have a good enough telescope to see it?

2.6. Use the data in the accompanying table to draw a graph to prove that the moons of Neptune conform to Kepler's Third Law. **HINT:** Remember that Kepler's Third Law states that, where $a \propto P^2$ is the mean orbital size of the orbiting body (a planet or moon), and P is the orbital period. One way to show if the Third Law is valid is to calculate a^3 and P^2 and plot these against each other. If they lie on a straight line, the Third Law is valid. You can use Figure 2.9 as an example of how to produce a logarithmic version of such a plot.

Data for Neptune's Moons

NAME	DISTANCE FROM NEPTUNE (10^3 *km*)	ORBITAL PERIOD (*days*)
Naiad (NIII)	48.227	0.294
Thalassa (NIV)	50.075	0.311
Despina (NV)	52.526	0.335
Galatea (NVI)	61.953	0.429
Larissa (NVII)	73.548	0.555
Proteus (NVIII)	117.647	1.122
Triton (NI)	354.76	5.877

2.7. From knowledge about their orbit sizes and periods, how much faster does the Earth travel on average in its orbit than Mars? Calculate the difference in speeds in units of *km/s*.

2.8. For the planets orbiting the Sun, calculate the value of the constant, *C*, in the more general form of Kepler's Third Law that we introduced at this destination for the following cases:

a) If *a* is in *AU*, and *P* is in *yrs*.

b) If *a* is in *km*, and *P* is in *sec*.

c) If *a* is in *furlongs*, and *P* is in *fortnights*.

2.9. Notice that Figure 2.9 shows how Kepler's Third Law applies to the planets and the *Dwarf Planets* of the Solar System. Look up on the Web information regarding the properties of any one of the five dwarf planets shown in that figure. Include their basic orbital information and at least one fact regarding the size, nature, number of moons, or other properties for the dwarf planet you choose to report on.

2.10. Explain with a figure why Jupiter is visible at all times of night from Mars but the Earth and Venus (and Mercury) are seen only near evening or morning twilight.

2.11. Look up and write a short report on some specific, interesting surface feature of Mars other than Valles Marineris (Figure 2.1) or Olympus Mons (Figure 2.10).

2.12. There have been numerous missions to Mars by various space agencies. Look up one of these missions and write a short report on the details of the mission, when it occurred, how it studied Mars—or how it failed to study Mars, in case the mission was not entirely successful—and any other information you find interesting about the mission.

2.13. From the information given in the text and Table 2.1, calculate the orbital speeds of Phobos and Deimos. You may choose whatever units are most convenient.

2.14. As noted in question 1.6 from Destination 1, the typical transit times of Apollo lunar missions was about 2.5 *days* to cover the approximately 240000 *miles* from the Earth to the Moon.

a) Assuming that the trajectory from the Earth to Mars is about 60 million miles in length, how long would it take to travel there if we had to go at the same speed as the Apollo spacecraft averaged on their way to the Moon?

b) For a crew of 10 people, how much total food, air, and water would be needed for this journey? Look up or estimate (with details) how much of these consumables a typical person uses each day. And don't forget to include enough supplies to get them back to Earth!

2.15. Retrograde motion would occur in different situations than the one we explored in detail at this destination.

a) Describe how differently retrograde motion of an outer planet would appear if the two planets moved in opposite directions in their orbits (that is, one moves clockwise and the other counter-clockwise from a given perspective above their orbits). You might want to use a figure to help illustrate your answer.

b) As pointed out in a footnote in this destination, inner planets also exhibit retrograde motion. Show why it is considerably more difficult to see this directly than it is for an outer planet when all the planets are revolving around the Sun in the same sense.

Destination 3

THE SKY FROM EARTH: A LAST LOOK FROM HOME

3.1 PREPARATIONS FOR A LONG VOYAGE

Exploration is tough; it's entirely understandable if anyone wishes to turn back at this point. Traveling to Mars was not for the fainthearted, but the journeys to our next destinations—far beyond the Solar System—will be incomparably more taxing. We will be going to places where the Sun, when we look back trying to find it, will appear as a faint point of light. If it is visible at all. There will be no returning home for a long time.

Obviously, we'll need supplies. The logistical requirements of interstellar travel are daunting.[1] But we need to also remember that ours is a voyage of discovery, so we must not forget to bring things that will help us to fully appreciate and understand the views that await us at our alien destinations. For example, we will want to bring along devices that can aid our vision such as binoculars and telescopes. We should also bring spectrographs—instruments that break up light into its constituent components—a powerful capability we shall exploit often at our destinations. And we need to bring tools that allow us to detect radiation that our bodies cannot perceive, much like a radio antenna can pick up radio signals we are otherwise oblivious to. This is quite important because some of our destinations will seem unremarkable to our eyes. Only by 'looking' with instruments designed to extend our senses will we be able 'see' phenomena that our eyes cannot detect. The details of how these instruments work are not important at this point—we will explore this a bit as we go along—but having them with us is essential for the core purposes of our voyages: To explore. To observe. To understand.

Before we go it would also be prudent to try to make some educated guesses as to where we should be going, how long it will take, and how we might know we have arrived. The Universe is so vast, yet so empty, that a random approach would not work—the odds of stumbling onto interesting locales in the vastness of the Cosmos are minuscule. Most of the great explorers of the Age of Discovery knew this (see Figure 3.1). Columbus could not have ventured from Europe to the New World had others—mostly Portuguese and Spanish explorers—not already learned enough about navigation to ensure that he could find his way with some level of confidence. He also understood—even if his estimate was not that

---

1 Try keeping an inventory of everything you consume during the course of one week. That includes food, water, supplies, air ... *every*thing. Now imagine what you—and your many shipmates—would require for an interstellar journey that could not be resupplied en route and that could last decades!

Fig 3.1

Having *some* idea of what lies ahead is better than being completely in the dark. This map of the New World from 1550 may not be very precise, but it was probably still of use to early explorers. For most mariners, having a chart even this crude was certainly better than just guessing and hoping for the best.

accurate due to his poor guess of the size of the Earth—the approximate duration of the trip. That simple information allowed him to suitably stock his fleet with the necessary provisions for his crew. He also knew that by heading west from Europe he would ultimately find land, not sail forever through an endless sea. Like most explorers, he knew that stepping blindly into the unknown is not courageous. It is foolhardy.

With this in mind, let's return to Earth and spend some time understanding what our next voyage—a trip to the nearest stars—will entail. In the most basic terms, this means we need to have some idea *where* to go. And we need some idea of *how far* we can expect to travel to get there. Of course, we could have figured all of this from Mars given enough time and suitable instrumentation. Apart from the Sun, all stars appear *exactly* the same from the two planets, so the distance we covered to get to Mars is clearly negligible compared to what we must traverse to reach even the closest stars. But Mars is cold and inhospitable. It is certainly much more comfortable to carry out these measurements back at home!

3.2 NAVIGATING IN THE SKY

We've already noted a few things about how stars appear in Earth's sky. First, they exhibit a large range in brightness. Urban dwellers who visit a dark site away from lights especially appreciate this as they marvel at how many stars—most of them fairly dim—they can see in a truly dark sky. For a person with reasonably typical eyesight, the faintest stars visible at a location far from city lights are about 500 times fainter than the brightest stars[2]. We've also noted that the stars can differ significantly in color; many appear noticeably redder or bluer than the Sun. Assuming that color has something to do with the stars themselves, this tells us that the stars must be *intrinsically* different from one another in some fundamental way to account for such color differences. Since we already saw that the Sun appears dimmer from Mars, this tells us that distance can make stars appear fainter (we know that from everyday experience, of course, but we'll explore this effect in a more precise manner in Destination 4). If some stars are at least somewhat similar to the Sun (and some are), the fact that the stars are at least billions of times fainter than the Sun offers a first hint of how far away they must be. This same line of argument suggests that the large range in stellar brightness implies there may be a large range in stellar distances. These simple observations of the night sky already reveal to us that stars probably vary quite a lot in their intrinsic properties, that they must all be very distant compared to the Sun, and that they *may* vary quite a bit in their distances from Earth. This also gives us a first hint of the magnitude of distances we must cover to reach even the closest destinations beyond the Solar System. Far enough that our Sun will be transformed into a mere nighttime star!

As it happens, determining how distant stars are requires precise determinations of their locations in the sky. So, let's turn to understanding better *where* we must go to reach our destinations. To do this, we need to devise a way to navigate in the sky. For various reasons, people have long felt a need for the means of specifying celestial positions. Early in the history of civilization, it became inadequate to merely say that there was a bright new star or a comet somewhere in the sky. These sorts of events, along with the enigmatic motions of the planets, were considered important omens, potential portents of future events. As sky watchers began to systematically track celestial events, they needed a more precise way of specifying the locations of celestial objects that everyone could understand. For the earliest astronomers, 'precise' may have meant that it was sufficient to specify localized regions of the sky. One way to do this is to subdivide the sky in some accepted manner, much as land is subdivided into specific

---

2 This excludes the Sun, of course. We will see as we visit other destinations that the Sun is a fairly normal star. But, because it is so close, and because the term 'brightness' involves the *perception* we have of a star's appearance, it is obvious the Sun is much, much brighter than any other star in the sky. In fact, about 10 billion times brighter than the next brightest star in the sky, *Sirius.* To put this in perspective, recall from Destination 1 just how big a billion is. This does not mean the Sun emits more energy than other stars. That quantity is known as 'luminosity' and we will define it precisely later. We shall encounter many stars that emit far more total light than our Sun—that is, stars with larger luminosities—and many with far lower luminosities than the Sun. Some of these high-luminosity stars would appear millions of times brighter in our sky if located at the distance of the Sun! In contrast, the lowest-luminosity stars would appear only a bit brighter than the full Moon at the Sun's distance.

regions. In the sky, these sorts of subdivisions are called *Constellations* and they are generally defined by specific patterns or groupings of stars. Every early civilization apparently had the need to subdivide the sky in this way because each invariably produced a set of constellations that could be used to help convey to other observers the locations of objects in the sky.

The Chinese liked numerous, small constellations. This allowed Chinese astronomers to more precisely describe the locations of objects in the sky but at the cost of requiring a fairly complex celestial subdivision that consisted of a large number of constellations. The Babylonians and, later on, the Greeks preferred to define larger regions of the sky as constellations. These celestial patterns generally represented single entities, often characters associated with important historical or mythical events. For example, in one part of the sky—best visible in Northern autumn—we find King *Cepheus* and Queen *Cassiopeia* trying to sacrifice their daughter *Andromeda* to avoid the wrath of the terrible sea monster *Cetus*—really, he is just a whale, though presumably a fairly psychotic one like Moby Dick. The daring Prince *Perseus* rides to Andromeda's rescue on the winged horse *Pegasus*, unchaining the princess from her sacrificial altar and then scampering off to safety and, one presumes, happiness. Some constellations represent common, exotic, or entirely imaginary creatures. The sky contains a number of Centaurs (*Centaurus, Sagittarius, Capricornus*), a Giraffe (*Camelopardalus*), a Pony (*Equuleus*), a Dragon (*Draco*), even a Fox (*Vulpecula*) to name a few. Some constellations are inanimate objects, such as *Corona Borealis* (the Northern Crown) and *Sagitta* (the Arrow). Other constellations commemorate professionals (at least ones from ancient times), such as *Boötes* (the Herdsman) and *Auriga* (the Charioteer). Many of the ones in the Southern Hemisphere—in parts of the sky not visible to the Greeks—were named as recently as the early 19th century. Some of these constellations reflect their origin at the start of the Industrial Revolution: *Microscopium*, *Telescopium*, and *Antlia* (the Air Pump) all grace the Southern sky.[3] Some 'constellations' really aren't. The Big Dipper, for example, is part of the constellation *Ursa Major* (the Large Bear), while the Little Dipper is part of *Ursa Minor* (the Small Bear). The handles of both dippers—distinctive patterns known as *asterisms*—represent the long tails of these two celestial bears. Obviously, the Greeks had a different perception of what bears look like than we do! All told, there are 88 modern constellations in the sky (see Figure 3.2).

Some constellations are very well known. *Orion* the Hunter prowls the (Northern) winter sky (see Figure 3.3). *Leo, Virgo, Libra, Scorpius, Sagittarius, Capricornus, Aquarius, Pisces, Aries, Taurus, Gemini*, and *Cancer* make up a special

3 As modern as the names of these constellations seem, it was actually weirder at one time. Sometime during the 19th century, some sky atlases included such technological marvels as a typewriter and a sewing machine in the southern sky! These are now officially gone.

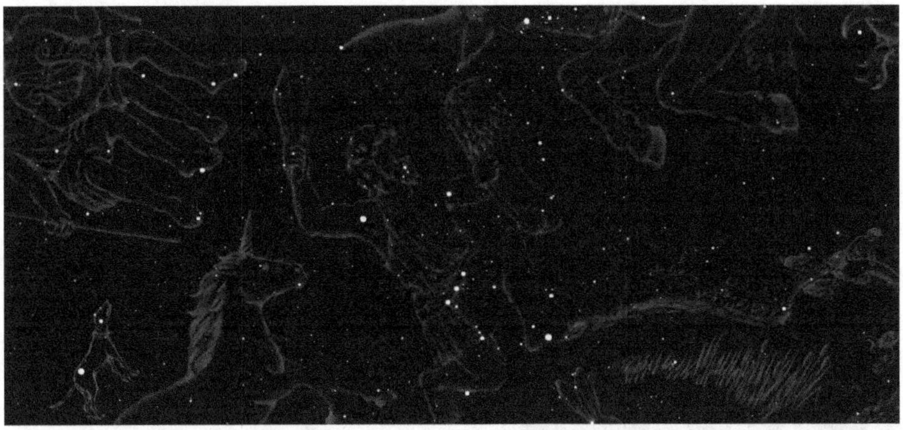

Fig 3.2

Orion and his companions in the night sky are depicted as drawings of the celestial creatures, people, and things the constellations are meant to represent.

Image generated by M. Mateo using *Starry Night* software.

set of well-known constellations known as the *Zodiac*. These constellations have long been considered particularly important because the Sun, Moon, and all the planets travel exclusively within their boundaries.[4] The Sun itself traces out a path, known as the *Ecliptic*, which passes through the constellations of the Zodiac. The location of the Sun on special dates, such as the first day of spring (the *Vernal Equinox* located in the constellation *Pisces*), the first day of autumn (the *Autumnal Equinox* in *Virgo*), and the starts of winter (the *Winter Solstice* in *Sagittarius*) and of summer (the *Summer Solstice* in *Gemini*) are all found on the Ecliptic, and all within Zodiac constellations.[5] We will visit later the *Andromeda Galaxy* (Destination 14), the *Virgo Cluster* (Destination 16), and the *Boötes Void* (Destination 18), all places that are denoted by the constellations in which they reside in Earth's sky. In this sense, the constellations do their job reasonably well. They can indeed be used to help localize features in the sky.

Ultimately, however, we require greater precision than what the constellations can provide. After all, we have to *travel* to the stars and a constellation is too big to be useful as a means of specifying a destination. That would be like trying to localize a hotel you'd like to stay at by just knowing the state it is located in. Worse, as we shall see, the stars within a given constellation are not generally physically related to one another. They merely produce random patterns in the Earth's sky.

The Greeks again came up with a system that we still use, one that provides more precision in some ways. Since individual constellations are defined by groups of stars, it was natural to name the stars within each grouping in a

4 Not quite. The Sun in fact is occasionally found in a thirteenth constellation, *Ophiuchus*, the snake bearer. That constellation is located near *Scorpius*, one of the official constellations of the Zodiac.

5 The term 'equinox' means 'equal day' because when the Sun is at those locations, the duration of night and day are nearly equal everywhere on Earth. The term 'solstice' literally means 'sun stop' and refers to the fact that on these dates the Sun halts its motion north or south, and begins to move in the opposite direction. The names of the equinoxes and solstices are biased to Northern Hemisphere inhabitants. Though the Winter Solstice may be the first day of winter in the north, it is actually the first day of summer south of the equator. Likewise, the 'Summer Solstice' is winter in the south, while the 'Vernal Equinox' and 'Autumnal Equinox' are the starts of fall and spring, respectively, in the south.

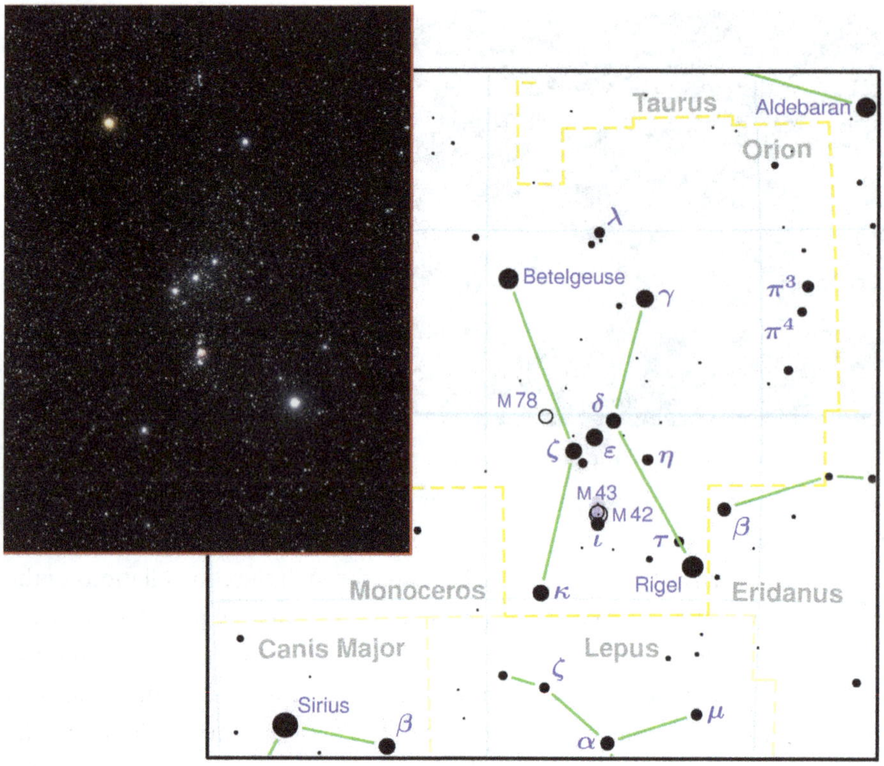

Fig 3.3

(*Left*) A photographic view of Orion in the night sky. Many more stars are visible here than can be seen by the naked eye due to the higher sensitivity of photographic film and the ability of a camera to collect more light over a longer time than our eyes can. (*Right*) A chart of Orion and its environment. The names and designations of many of its stars are shown in this chart.

Photo by A. Fujii.

systematic manner (Figure 3.3). For example, the star Spica—it has long had a proper name—was christened Alpha Virginis, or α Vir.[6] The first letter in the Greek alphabet, alpha (α), usually denotes the brightest star in a constellation. The letter beta (β) denotes the second brightest star, and so on. Many years later, astronomers replaced this scheme with a pure numbering system within each constellation. After all, there are 6000 visible stars, and 88 constellations, so there must be many constellations with more than 24 stars, the number of letters in the Greek alphabet. Thus, Spica, which is also α Vir, is also named 67 Virginis, or more simply, 67 Vir. Over the years, many subsequent catalogs of stars have been produced, and each one—for perfectly good reasons—tended to give yet another designation to previously named stars. Thus, Spica is *also* called HR505, HD116658, BD-10 3672, SAO157923, HIP65474, FK5-498 ... you get the idea. All these names refer to just one star!

But even this approach, though better than specifying just the constellation, has serious limitations. Knowing the name of a star is useful only if we have a chart and only if the star (or other object) happens to have already been named. We've improved our ability to specify locations to a level comparable to specifying the state and the (large) city where we might find our hotel. Better, but still

6 Each of the 88 modern constellations has a three-letter abbreviation. For example, *Orion* is Ori, *Camelopardalis* is Cam, *Ursa Minor* is UMi, and *Telescopium* is Tel. These shortened names are often used as parts of the names of stars or other celestial objects located within constellations. See Appendix F for a full list of the constellations visible in Earth's sky.

not great. Ideally we want a more general scheme that allows us to specify the position of anything, anywhere in the sky, whether it has been named or not, even if it is yet to be discovered! This is particularly important to us because few of our travels will take us to destinations associated with known stars.

Earthbound travelers sometimes have similar needs to ours. What if one has to travel to some random position far from any city, nation, or even land? Terrestrial cartographers addressed this problem by producing a coordinate system on the surface of the Earth. We refer to the two coordinates defined by this system as *Longitude* and *Latitude*, which measure positions in the east–west and north–south directions, respectively. Armed with the knowledge of its latitude and longitude, we can find—and travel to—any location on the Earth's surface. For example, there is a classroom at the University of Michigan located at 83.7348 west longitude and 42.2762 north latitude. This means that that room is 82.7348 degrees west of the prime meridian that passes through Greenwich, England, and 42.2762 degrees north of the Earth's equator.[7] Greenwich is the currently accepted location that is used to define longitude ZERO (see Figure 3.4)[8], while the equator has long been taken as the location of ZERO latitude.

Can we apply this idea to the sky? Yes, but doing so is slightly trickier. To measure north–south positions, we need an equator just as on the Earth's surface. One way to identify this is to go to the Earth's equator and chart the location of the point directly above us during the course of a day. But a better way—one that does not require that we travel to the equator—is to notice that from any point on Earth, one can see at least one spot on the sky that appears to stand still during the course of the night. In the north, we are lucky to have a star near that spot: *Polaris* (also known as α Ursa Minoris, 1 Ursa Majoris, HD8890—OK, we get it!).

7 The details of determining these angles are actually a bit tricky. Latitude is the shortest angle from a given location to the equator. So, for the example of a classroom in Michigan, this is the angle as seen from the center of the Earth (one has to imagine this, of course) from this classroom, due south to the equator. The poles are at 90 degrees north or south latitude. Longitude is measured as the shortest angle from a location east or west to the prime meridian (the one that passes through Greenwich; see Figure 3.4). The longitude of the Michigan classroom we are using as an example is the angle of the arc from that classroom, due east until we reach the prime meridian at the same latitude that we started (in this example, we'd be about 600 miles or 1000 km south of Greenwich). Because of this, a degree of longitude at the equator is about 70 miles, but as you approach the poles, a degree of longitude becomes steadily smaller until it becomes zero at the poles (see Figure 3.4). The longitude of the poles is not even defined since all lines of longitude meet at those two locations (you can see this directly in Figure 3.4).

8 Over the past few centuries, many places have tried to lay claim to the prime meridian, including Paris, Lisbon, and Stockholm to name a few. After a few wars and a few treaties, Greenwich was finally chosen as *the* Prime Meridian at an international conference in 1884. The true prime meridian is located about 100 meters to the east of the Greenwich Observatory where the meridian was defined to be located at that conference (see Figure 3.4). This is partly due to plate tectonics that have moved Greenwich, England, and indeed all of Europe relative to other parts of the world, but mostly due to various technical issues involving the precise definition of the latitude/longitude coordinate system on the (slightly) non-spherical surface of the Earth.

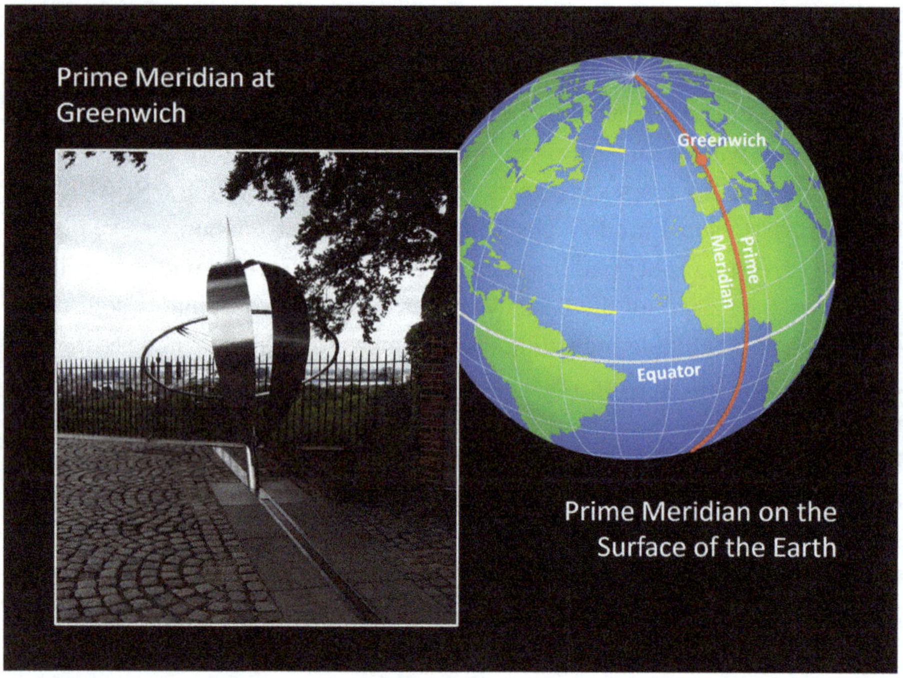

Fig 3.4

(*Right*) A map showing the path of the prime meridian (in red) passing through the Northern Hemisphere and part of the Southern Hemisphere of the Earth. Angular positions west (left in this figure) and east of this meridian define the longitude of a location on the face of the Earth. Other lines of longitude that travel north–south and pass through the North Pole are also shown at 15-degree intervals. Lines of latitude that travel east–west and are parallel to the Earth's equator (the thick white line) are also shown at 15-degree intervals. Notice that the linear distance between lines of longitude (denoted by the short yellow arcs in the Atlantic Ocean) become smaller as one moves to the pole.
(*Left*) A photo of the marker of the prime meridian—the strip along the ground in the photo—at Greenwich, England. Though handsome in appearance, this marker is just for show. The accepted prime meridian is actually located about 100 meters (or, more appropriately, metres) east of the location of the one shown in the photo. Map from the National Maritime Museum, London.

Photo by M. Mateo.

If we watch *Polaris* during any night, it seems to hover above the horizon as all the rest of the stars spin around it (Figure 3.5). Since this spot is so easy to identify, it is more convenient to find it—the North Celestial Pole—and then simply measure 90 degrees from it to determine where in the sky the Celestial Equator is located[9]. Once we know where the equator is located, we need to only measure the angle of a star north or south from this celestial line to figure out its *Declination*, the term astronomers use to refer to latitude in the sky[10] (see Figure 3.6).

The other coordinate, the one that measures east-west positions in the sky, is called *Right Ascension*[11] (see Figure 3.6). As on Earth, this coordinate is tricky in that it requires that we specify a unique spot in the sky that can serve the same purpose that Greenwich does on Earth. How do we accomplish this? We cannot simply pick the spot *above* Greenwich, since this continuously changes as the Earth rotates and carries the sky from east to west. We could pick a specific star as the reference point, but astronomers realized long ago that stars actually *move* with respect to one another (we'll see this at Destinations 4 and 9). Though this motion is slow—nothing like the comparatively rapid motions of the planets that one can observe in a day or two—the stars nevertheless do move. That disqualifies them as suitable reference markers for a celestial coordinate system.

9 There is no bright pole star in the Southern sky (see Figure 3.5), but it is still easy to apply the same procedure once you note the motions of the stars and find the Celestial South Pole.

10 Astronomers are pretty notorious about giving a new name to *everything*, whether necessary or not.

11 See the previous footnote!

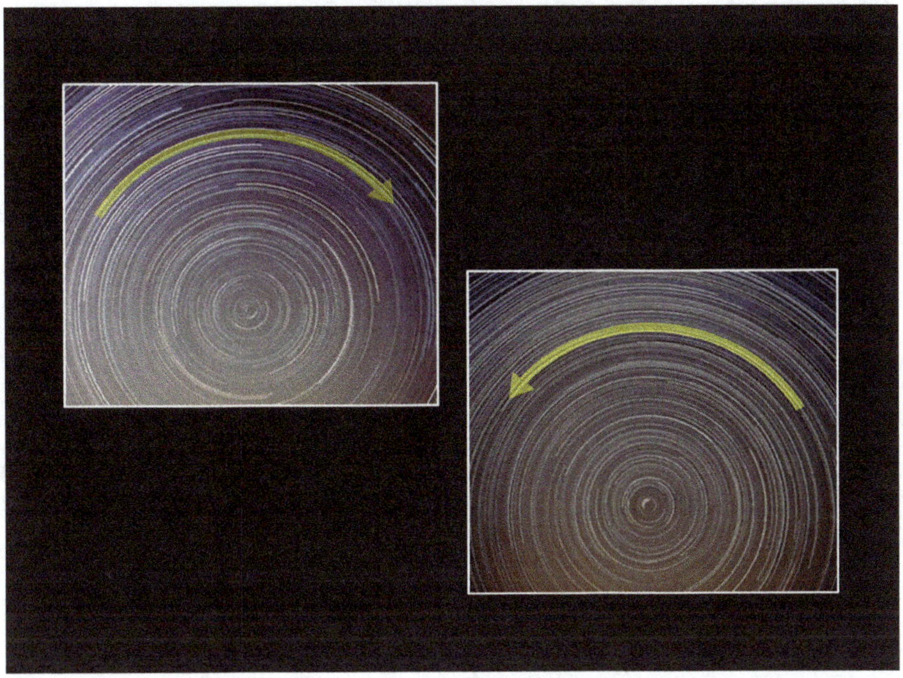

Fig 3.5

These are long exposures of the polar regions of the sky obtained by keeping a camera's shutter open for a few hours. As the Earth rotates, the stars appear to move around two specific points in the sky, the South Celestial Pole (*left*) and the North Celestial Pole (*right*; see Figure 3.6). Note that by taking an image like this, or carefully measuring the motions of the stars, one can determine the location of the poles quite precisely. With the poles identified in this way, it is easy to define the Celestial Equator as the imaginary path in the sky that is everywhere precisely midway between the poles. Though there a bright star that just happens to be near the North Celestial Pole—*Polaris*—the South Celestial Pole isn't so fortuitously marked. But it is still easy to tell where the pole is located in both pictures. The arrows indicate the direction of motions of the stars with time for an observer in the corresponding hemisphere.

Photos by D. Malin/AAO.

For better or worse, astronomers decided that the best way to define the zero of Right Ascension is to use one of the two locations in the sky where the Sun's path—the Ecliptic—crosses the Celestial Equator (see Figure 3.6). The dates when the Sun is at these locations are close to March 21 and September 21 each year. Because the Sun is at the equator, the duration of nighttime and daytime is effectively equal on those two dates and hence these are known as *Equinoxes*—a word that derives from Latin words that mean 'equal-night.' On the September date, the Sun is moving from the Northern to the Southern Hemisphere in the sky and so this date is often taken to be the first day of autumn in the Northern Hemisphere. For that reason, this location is known as the *Autumnal Equinox* (and as we noted earlier, it is in the constellation *Virgo*). In March, however, the Sun is moving from the Southern to the Northern Hemisphere, signifying the start of (Northern) spring. Hence, the location in the sky where the Sun is located on that date—in the constellation *Pisces*—is known as the *Vernal Equinox.*[12] Arbitrarily, astronomers have chosen to *define* the Right Ascension of the Vernal Equinox to be ZERO. That is, this nondescript location where the Celestial Equator and the Ecliptic cross in the constellation Pisces serves the same purpose in the sky as Greenwich does on Earth (see Figures 3.4 and 3.6).

Now that we have our reference locations—the Celestial Equator/Poles and the Vernal Equinox—the positions of stars in the sky can be measured as angles between these locations and the stars themselves. If this seems odd, imagine

12 As noted in footnote 5 of this destination, this is actually the start of autumn in the Southern Hemisphere.

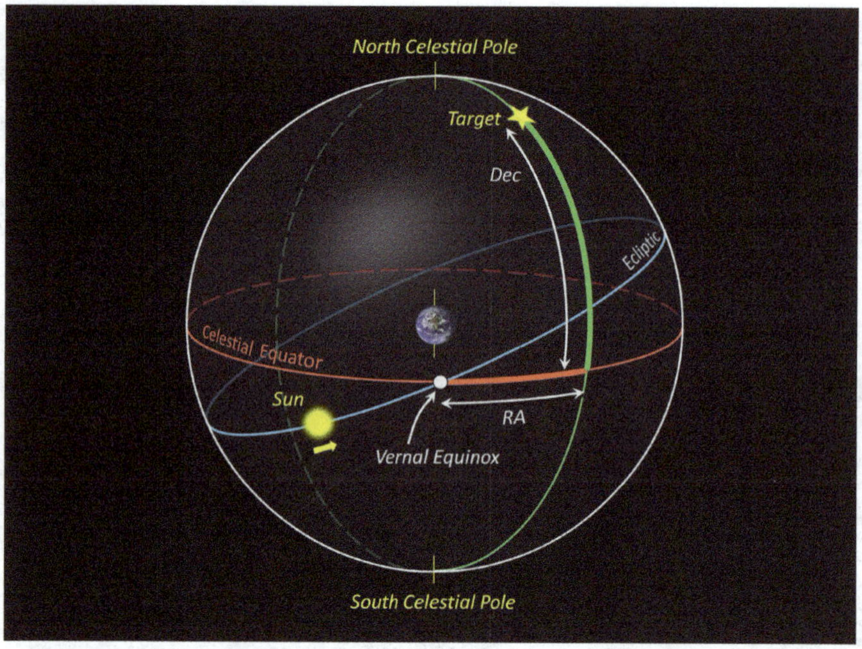

Fig 3.6

For many purposes, the sky can be thought of as an immense, transparent sphere centered on the Earth and on which all the celestial objects are located. This is not, of course, how celestial objects are actually distributed in space, but this concept is useful nonetheless to define coordinates of objects in the sky. The approach is based on determining the location of the Celestial Equator (in red), most easily done with respect to the Celestial Poles (in yellow; see Figure 3.5), and the path of the Sun along the Ecliptic (in blue). Every year, the Sun crosses the Celestial Equator twice, once heading from the Northern Hemisphere to the Southern Hemisphere. The Sun's position on the equator causes the length of daytime and nighttime to be approximately equal on that date. Since this specific event occurs in September, it is known as the Autumnal Equinox. Six months earlier/later, the opposite event occurs—the Sun travels from the Southern Hemisphere to the Northern Hemisphere. The location on the sky when this occurs is known as the Vernal Equinox (shown as a white dot at one of the crossings of the Ecliptic and Equator). The north–south angle from a target location in the sky (shown here as a yellow star) to the Celestial Equator (the thick green arc) is the target's Declination. The angle between where that N–S line from the target crosses the equator and the Vernal Equinox is known as the Right Ascension (shown as a thick red arc in the figure). With this system we can determine the coordinates of any object in the sky.

pointing with one arm at one point in the sky—say, the approximate location of the Vernal Equinox if it happens to be above the horizon—and the other pointing at the star or location you are interested in. The angle between your arms is closely related to the star's Right Ascension and Declination. Generally, astronomers measure such angles in terms of *degrees*, *arc minutes*, and *arc seconds*. There are 360 *degrees* in one full circle, 60 *arc minutes* in one *degree*, and 60 *arc seconds* in one *arc minute*. The standard abbreviations for these angular units are *deg*, *arcmin*, and *arcsec*, respectively. Just to confuse things, astronomers often measure Right Ascension in units of angular hours (abbreviated like the time unit as *hr*). Thus, one full circle equals 24 *hr* which means one hour of Right Ascension is equivalent to 15 *deg* (= 360/24). For our purposes, we'll mostly stick with just *deg*, *arcmin*, and *arcsec*. Using such *Celestial Coordinates* we can now specify the position of *any* object—not just stars—in the sky. For example, *Spica* is located at a Right Ascension of 13 *hours*, 25 *minutes*, 11.58 *seconds*, or, equivalently, 201.298 *deg* (by convention, this angle is always measured east of the Vernal Equinox), and a declination of –11 *deg* 9 *arcmin* 40.8 *arcsec*, or, equivalently, –11.161 *deg* (the minus sign means it is south of the Celestial Equator). These celestial coordinates are a key element of what we require before we can hope to undertake our journeys through the Universe. They tell us where to point our spacecraft.[13] But we need something more.

13 And not just abstractly. When interplanetary space probes are launched from Earth, they aim for a specific location in the sky defined by a particular Right Ascension and Declination. Once the rocket engines turn off, if the spacecraft is heading toward those coordinates at the right speed and if everyone has calculated things properly, it should coast to its final destination.

3.3 THE DISTANCES OF STARS: STELLAR PARALLAX

We still do not know how *distant* the stars are. Even after traveling to Mars, we saw no change in the properties or positions of any stars in the sky, suggesting that the problem of finding their distances will be daunting. This is true, but there is in fact a way. And, as a bonus, we can do it without having to leave the Earth.

We have seen from the phenomenon of retrograde motion that the planets all orbit the Sun in a similar sense and in accord with Kepler's Third Law (and the other two laws, but we need only consider the Third Law for now). We will discover that we can use Kepler's Third Law to tell us how large the Earth's orbit is if we know the orbital periods of Mars and Earth, and the size of Mars' orbit.[14] This distance—the mean radius of the Earth's orbit—is so important that it has a special name: the *Astronomical Unit*. To keep the notation manageable, this is typically abbreviated as *AU*. Since the Earth is in orbit about the Sun, this means that during the course of one year it—and we!—travel some 910 million *km* along the Earth's orbit, equal to the circumference of a circle of radius 1 *AU*. Clearly, after 6 *months* we will find ourselves at the opposite side of Earth's orbit. This turns out to be as far away from our starting point as we can ever be while remaining on the planet. In general, any place on the Earth's orbit where the planet is found is located about two *AU* (about 185 million *miles* or 300 million *km*) away from the Earth's position six months earlier. This is a pretty respectable distance. For example, depending on the specific trajectory we might take to reach Mars and how long we stayed there, two *AU* can exceed the distance we would have to traverse to get to Mars! Even though we are barely aware of it as we reside on Earth, we travel a gargantuan distance around the Sun every year. Despite this movement, the changes in the sky caused by the Earth's orbital travels remain far too small to produce changes in the stars that can be perceived by the naked eye. But, it turns out, not too small to detect with optical aid.

One way we can enhance our vision is by using telescopes that permit us to see subtle changes in stellar positions caused by Earth's motion around the Sun. Telescopes succeed where the unaided eye fails because larger optical elements can distinguish finer angular separations in the sky.[15] Astronomers say

14 This is actually what Kepler did.

15 This improvement is related to a phenomenon called diffraction, which waves exhibit. As we shall see at Destination 5, light and other forms of radiation have wave-like properties. When analyzed carefully, one discovers that the angular resolution of a telescope depends on the ratio D/λ, where D is the diameter of the principal optical element of a telescope and λ is the wavelength of the radiation.

that larger telescopes have higher *Angular Resolution*. The pupil of the eye can open up to at most a diameter of about 3 *mm*. For optical light, this allows us to see details as small as about 2–3 *arcmin*. For comparison, the Moon is about 30 *arcmin* in diameter in the sky as seen from Earth. Modern optical telescopes, some up to 8 to 10 *m* in diameter, can discern detail about 3000 times finer than this, equivalent to a few thousandths of an *arcsec*. But even a fairly small telescope—one only a few inches in diameter—can see details under an *arcsec* in size. This is a vast improvement over the eye, and quite remarkable by any standard. After all, one *arcsec* is the angular size of a dime located 1.6 *miles* away!

But how can we use the motion of Earth and the high angular resolution of telescopes to detect changes in the properties of stars that allow us to estimate stellar distances? One approach can be easily demonstrated with a simple experiment. Identify something stationary that is located fairly nearby and that can be viewed against a distant background. A mark on a window, or a nearby telephone pole or a tree are all perfect (see Figure 3.7). Open one eye and note

Fig 3.7

Parallax is apparent in everyday life. Notice how the palm tree changes its apparent position relative to the distant background structures due to the change of position of the camera (Look carefully: You can see many other parallax shifts in these photos). If the distance separating the camera when the two photos were taken is known, we can use the parallax equation to estimate the distance to the tree. Of course, in this case we could just use a measuring rule to get the distance to this nearby object. When it comes to stars, we do not have that option! These two photos happen to constitute a stereoscopic pair: If you relax your eyes so the images merge, it will pop out at you in apparent 3D. This is not a subtle effect; you'll definitely know when you see it!

Photos by M. Mateo.

where this nearby object is located relative to more distant objects in the background. Close that eye and open the other or, just shift your head a few inches to the side. The change in the apparent position of the nearby object relative to the distant background that you see is called *Parallax*. Notice that this shift in position does not come about due to any motions of either the nearby object or the objects in the background. The *only* thing that changed as you did this experiment was the angle of your view relative to the background. If you had chosen relatively remote objects to serve as your background, you will also appreciate their positions changed very little as your viewing angle changed. If you want to make these more remote objects appear to change position, you may have to move your head a considerable distance or, if that's not enough, move to a new location. This simple demonstration illustrates two important points about parallax. First, closer objects exhibit more parallax than distant ones. Second, we can make all objects exhibit a larger parallax by increasing the distance between our viewing positions.

As we have seen, the Earth conveniently carries us 2 *AU* every six months. Are the stars still too distant for us to detect parallax? For our eyes, certainly they are. Since we can discern angles only as small as about 2–3 *arcmin* without optical aid but we cannot detect stellar parallaxes directly, we can conclude that stars must all be located more than about 1500 *AU* from Earth.[16] Because of their greater angular resolution, telescopes allow us to do better. A lot better. The first telescopes from the early 17th century were not quite good enough to detect stellar parallax. But by the early decades of the 19th century, the first parallaxes of stars were finally measured, followed by many others in subsequent years.[17] One of the first results to come out of systematic measurements of stellar parallaxes is that most stars exhibit no parallax at all, a result that is still true using even the most modern measurement tools. But many stars do exhibit parallax shifts, and for these important cases we can determine their distances directly.

To understand how, let's consider the geometry of parallax in more detail by referring to a simple model of the Earth's orbital motion (Figure 3.8). Note that, due to the change in position of the Earth over its orbit—let's say over the six-month interval from January to July—there is a maximum angle between the two lines connecting the Earth to any given nearby star. This angle can be measured quite accurately relative to 'background stars' which, by comparison to the nearby star, will likely have a very small or essentially zero parallax shift. Astronomers define the *Parallax Angle* to be equal to half the full angle between the Earth, the nearby star, and the Earth six months later. This angle is denoted as p in Figure 3.8. If we draw a line between the Sun and the nearby star that bisects the original triangle into two identical right triangles (one of these is shown in red in Figure 3.8), we can prove that

$$D = d/\tan(p),$$

where D is the distance of the star from the Sun, d is the radius of the Earth's orbit (1 *AU*), and p is the parallax angle. The sides of the triangle on which this result is based are highlighted in red in Figure 3.8. Notice that all we need to get a star's distance in this way is a measurement of the angle p, and knowledge of the size of the Astronomical Unit, which is denoted in this equation as d.

16 This will be explained shortly and is the subject of a problem at the end of this destination.

17 This is common in science in general, and in astronomy in particular. When someone figures out how to do something hard—sometimes a task that had been considered 'impossible'—everyone else soon learns how to do it too.

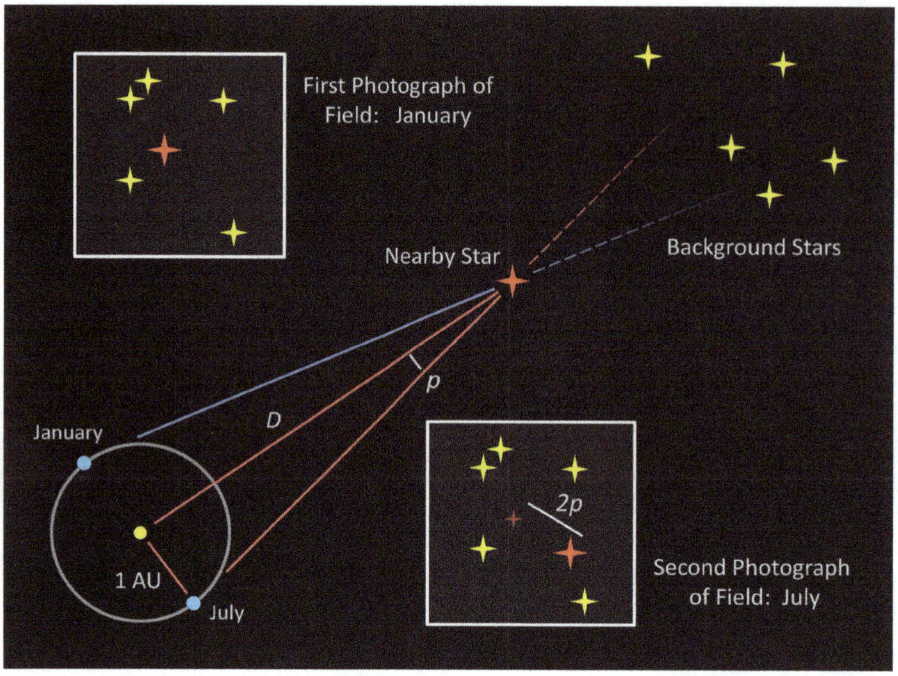

Fig 3.8

A schematic description of stellar parallax and how it is measured. The gray circle represents the orbit of the Earth (blue dot) about the Sun (yellow dot) and has a radius of 1 *AU* (*Astronomical Unit*). The location of the Earth is shown at two times six months apart (here chosen to be January and July). When we view a 'Nearby Star' (in red) at these two times, we see its position change with respect to 'Background Stars' (yellow). This can be measured as shown in the light-gray squares that represent photographs of the star field containing the Nearby Star and a number of Background Stars in January (*upper left*) and July (*lower right*). The July image—the second one obtained of the field—shows as a pale red star the location where the Nearby Star had been in the January image. Half the full angle by which the Nearby Star shifts its position (shown as the white bar in the July image and labeled '*2p*') due to our changing position along Earth's orbit is known as the *parallax angle*. This angle is denoted as *p* in the figure showing the relative locations of the Earth, Sun, and stars. Based on analysis of the triangle with the red sides, it is apparent that one can use simple trigonometry to determine the distance, D, of the Nearby Star, assuming the length of the *AU* is known. Keep in mind that this diagram

It is important to understand two key things about the geometry of parallax. First, the actual triangle involved in a stellar parallax measurement is vastly different from the one shown in Figure 3.8. If one were to print the parallax geometry to scale and set one *AU* equal to one inch, the *closest* star would then be about 4 *miles* away! Because of this, it is clear that the hypotenuse of the red triangle in Figure 3.8 is essentially equal in length to the other long side that makes up the triangle. What this means is that we can make no practical distinction between the distance of the star from the Sun and the distance of the star from Earth. As a result of this, astronomers normally refer to stellar distances as if measured from the Sun. Second, because the parallax angle is so small, we can simplify the equation for parallax as

$$D = d/p,$$

where p is still the parallax angle, but now measured in units of '*radians*.' By definition, there are 2π *radians* in a circle, and since a circle covers 360 *degrees*, one *radian* must equal $360/2\pi = 57.3$ *degrees*. The simplification in the equation comes about because $\tan(p) \sim p$ if p is a small angle (as all stellar parallax angles are) that is measured in *radians*. We can simplify things even further if we take the additional step of defining the distance between the Sun and Earth to be exactly 1 *AU*. This lets us write the *Parallax Equation* as

$$D = 1/p.$$

is grossly out of scale. If an *AU* were reduced to 1 *inch*—similar to what is shown here—the nearest star would be located about 4 *miles* away! The background stars would be, on average, *hundreds* of miles away. Mars' orbit would be 1.4 inches in radius.

As written, this equation implies that *D* is also to be measured in *AU* since *p* is in *radians*. But astronomers go even one step further, taking advantage of the fact that parallax angles are typically measured in *arcsec*. In that case, the distance, *D*, though still the distance to the nearby star, is defined to be measured in a new unit called a *parsec* and abbreviated as *pc*. This strange word is a modern invention that comes from the phrase '**par**allax of one arc**sec**.' Notice from this equation that if the parallax of a star is equal to one *arcsec*, its distance is by definition one *parsec*. Stars with parallax angles smaller than one *arcsec* must be more distant than one *parsec*. Not coincidentally, one *parsec* equals 206264.8 *AU*, precisely the number of *arcsec* in one *radian* (see the questions at the end of this destination to understand why this is so). In any case, we should appreciate that a *parsec* is a vast distance. Mars is only 0.4 *AU* away at its closest, a distance of about 2 *millionths* of a *parsec*, immensely larger than any distance humans have conquered so far. One *pc* equals about 3.3 *light-years*. This latter unit equals the distance light travels in one year while traveling at its considerable speed of 300000 *km/sec* (or 186000 *miles/sec*). The light from the Sun takes about 8 *minutes* on average to reach us on Earth, meaning that the Sun is 8 *light-minutes* from Earth.

As people began to measure stellar parallaxes for many stars, it soon became apparent that there are no stars that are even as close as one *parsec* from the Sun. That is, all stars exhibit parallax angles *smaller* than 1 *arcsec*! The closest star known beyond the Sun is called *Proxima Centauri*, and has a parallax of 0.76 *arcsec*. From the parallax equation, this result immediately tells us that *Proxima* is located 1.3 pc away (since 1/0.76 *arcsec* = 1.3 *pc*), or about 4.3 *ly*. Merely sending a radio signal to *Proxima* and waiting for the return answer—if there were anyone there who could reply—would take over 8.5 years! Don't be fooled. By no means is *Proxima* 'close.' But it is the *closest* star beyond the Sun, the obvious first destination for us to visit as we venture beyond the Solar System.

We mustn't for a moment underestimate the magnitude of this next journey. One could guess that humans *might* reach Mars in the first half of the 21st century. It will be a technically daunting, grueling, and extremely dangerous trip. Yet, getting to Mars is just the very first tiny step for humans on their way to the stars. If we equate Columbus' journey from Spain to the New World with the distance between the Earth and *Proxima Centauri*—that is, if we set 4.3 *light-years* equal to 3000 *miles*—the trip to Mars is the first 30 feet! The trip to the Moon—the only journey humans really *have* done from the face of the Earth—is 1.4 *inches*! In this respect, we haven't even left port. Really, we haven't even gotten out of bed! We had better get going soon or we are sure to lose our nerve.

DESTINATION 3 QUESTIONS

3.1. The most precise parallax measurements are accurate to about 0.003 *arcsec* (this is likely to improve greatly in the next 10 years). Given this precision, how far is the most distant star for which a distance can be determined using parallax?

3.2. Given that we do not see stellar parallax with the naked eye and that the eye can make out angles only as small as 2–3 *arcmin*, explain why this places a lower limit of about 1500 *AU* on the distances to the stars.

3.3. This is a web research question. List the names of ten constellations and what these constellations represent. You may *not* use any of the constellations I refer to in the text, including the familiar constellations of the Zodiac. That still leaves a lot of constellations, many you've likely never heard of before.

3.4. Prove from the various versions of the parallax equation given in this destination that one *pc* is equal to 206265 *AU*.

3.5. How would the parallax of a star measured from Earth differ from the parallax of the same star measured from Mars? How would the determination of the parallaxes differ on the two planets? If a star is located 10 *pc* away, what is its parallax angle as measured from Earth and Mars?

3.6. It has been proposed in the past to send a space probe into deep interplanetary space to measure parallaxes of stars.

a) Imagine we can build and launch such a satellite and send it outward from Earth at a constant speed of 100000 *mph*. How long would it take to reach a distance of 10 *AU*?

b) At that distance, how far could a stellar parallax be determined if we assume we can measure angles to a precision of about 0.003 *arcsec*?

c) What practical issues do you think such a mission as this would have to overcome to be successful?

3.7. Find the proper names of the brightest stars in the following constellations. You can assume that the brightest star is named α for each constellation.

a) Taurus
b) Scorpius
c) Eridanus
d) Canis Minor
e) Cygnus

3.8. Imagine that we devise a human spacecraft that can go 100 times faster than the Apollo spacecraft described in question 1.6. How long would such a spacecraft require to reach Proxima Cen from Earth?

3.9. Use a calculator to determine at about what angle the approximation $\tan(p) = p$ becomes valid. To answer this, assume that you want the two sides of the equation to be equal to better than one part in one thousand. Be sure to use radians as your angular unit when you do this!

3.10. As noted in the text, a given change in longitude can correspond to very different linear distances depending on how far one is from the equator.

a) Imagine you are 100 *km* from the North Pole. If you were to travel the full circle around the pole at this distance, how many *km* are there in each degree of longitude? You can assume this small region near the pole is essentially flat.

b) Calculate the circumference of the Earth at the Equator. Assume the Earth is a sphere of radius 6380 *km*. Along the equator, how many km are there in each degree of longitude?

3.11. *Nautical miles* are equal to the distance corresponding to one *arcmin* along Earth's equator. How long is a *nautical mile* in *km* and in *statute miles*? Use information given in the previous question. You may also want to use the following conversion factor: 1 *statute mile* = 1.61 *km*.

3.12. In Figure 3.7 assume that the base of the palm tree is 1 *deg* across. If the two images were obtained at locations that were 1 *m* apart, how far away is the tree from the photographer? **NOTE:** It might not be a fair approximation to use the simplified forms of the parallax equation to answer this question.

3.13. Estimate the approximate Right Ascension and Declination of the star shown in Figure 3.6. Explain your reasoning.

3.14. Let's say that you develop a means of accelerating an interstellar spaceship to about 0.5 times the speed of light.

a) If the time to accelerate is fairly short (months), how long would it take to get to Proxima Cen at this speed?

b) If the spacecraft has a crew of 25 adults, describe how many supplies—particularly consumables such as food and water—would be needed for the trip, assuming no recycling or regeneration. You will have to estimate how much one person consumes per day, including food, water, and even air.

c) What techniques might you suggest to make the answer to part (b) more manageable?

3.15. How long is a *parsec* in *km* and *miles*? What is the distance to the Sun in *light-years*?

3.16. What is the approximate longitude and latitude of your birth location? Be sure to specify this location in your answer. Project these coordinates to the sky. Read off these coordinates from an on-line or hard-copy sky chart to determine what constellation you were 'born' in.

DESTINATION 4

THE SKY FROM α CENTAURI

4.1 A FAMILIAR LIGHT IN AN ALIEN SKY

Under typical circumstances it may be true that 'Getting there is half the fun'. But whoever first said that never had to endure an interstellar voyage. Traveling through the Solar system is challenging enough, but in many respects, the next step—reaching the nearest stars—is the hardest of all as we set out to venture through the Universe. Even after we master the ability to move between the planets, the relative distances to the stars are so much greater—thousands of times greater—that we will still need to advance our technology to a significantly higher level before we can begin to seriously imagine exploring in person beyond the Solar System. On the other hand, when (if?) we ever learn how to span the vast, empty gulf that separates us from the closest stars, subsequent visits to more distant stars will be of comparable difficulty.

But those are details we'll let the future address. For now, we find ourselves at the end of an epic voyage through deep space that has brought us to the closest 'island' beyond the Solar System: the α *Centauri* system—or just 'α *Cen*' for short.[1] As the word 'system' implies, this is not a single star—though it looks like one to the naked eye from Earth—but rather a group of three stars that form a gravitationally bound family. Nearest the center of this triple system is α *Cen A* and its companion α *Cen B*, the two brightest stars of the system. With even a modest telescope, both are readily visible from Earth (or from Saturn; see Figure 4.1). It is from a vantage point near α Cen A that we've completed our immense trek. And it is where we begin our tour of this alien place. But, where are we precisely? By chance, we have come across a small planet here. One with a solid surface—unlike the 'gas giants' Jupiter, Saturn, Uranus, and Neptune of the outer Solar System—and with an atmosphere similar to Earth's. Even more remarkably, we find that this planet has a rotation period—and hence, a 'day'—of about 24 *hr*, and that it orbits at a distance of one *AU* from α Cen A.[2] In

1 Pronounced 'Alpha Centauri' or 'Alpha Cen' with a soft 'C.' As you may recall from Destination 3, the name 'α Centauri' implies this is the brightest star—as seen from Earth—located within the constellation Centaurus, the Centaur. 'Centauri' is the Latin possessive (genitive) form of the constellation name. In future destinations we will encounter other stars with similar kinds of names consisting of a Greek letter or a number, along with the abbreviated name of the constellation in which the stars reside. By the way, α Cen is also sometimes known as Rigel Kentaurus (with a hard 'K'), a name that is very rarely used by astronomers.

2 Hmmm. Perhaps you are thinking that this hypothetical planet of α Cen A is a bit too good to be true. It may well be. As of mid 2015, we do not know for certain if any planets

Fig 4.1

α Cen A and B as seen by the *Cassini* spacecraft in orbit around Saturn. The two stars show up as a pair of dots just above an arc of the ring system of Saturn. Getting to Saturn—a seven-year voyage for *Cassini* and a major technological feat—gets you 0.004% of the way to α Cen. That's equivalent to traveling the first quarter mile down the runway on a plane trip from Chicago to Beijing.

Photo from Space Science Institute/JPL/NASA/JPL.

other words, this is a perfect place to compare the properties of α Cen A with the only other star we know intimately and that we commonly view from a similar vantage point: The Sun.

As we begin to study α Cen A in detail, it becomes clear that our new 'sun' is remarkably similar in appearance to *the* Sun as seen from home. In brightness, apparent size in the sky, even color, α Cen A appears to be a near twin to our home star (see Table 4.1). But not an identical twin. We soon establish that α Cen A is about 1.5 times brighter than the Sun as seen from Earth, and about 1.2 times larger in diameter. Considering the enormous range of properties that we'll discover stars can exhibit (see Destinations 8 and 12), these differences between α Cen A and the Sun are relatively subtle. It requires fairly careful measurements to notice them at all.

TABLE 4.1 Properties of the Sun and the Stars of the α Cen System

STAR	MASS (*g*)	MASS ($M_\odot$)	LUMINOSITY (*ergs/s*)	LUMINOSITY ($L_\odot$)	RADIUS (*cm*)	RADIUS ($R_\odot$)	COLOR
Sun	1.99×10^{33}	1.00	3.84×10^{33}	1.00	6.96×10^{10}	1.00	'Yellow'
α Cen A	2.10×10^{33}	1.06	5.82×10^{33}	1.52	8.53×10^{10}	1.23	'Yellow'
α Cen B	1.77×10^{33}	0.89	1.92×10^{33}	0.50	6.01×10^{10}	0.86	'Orange'
Proxima Cen	2.45×10^{32}	0.12	6.52×10^{30}	0.0017	1.01×10^{10}	0.14	'Deep Red'

like this exist in the α Cen system though a planet has been claimed to orbit very close to α Cen B. Recent ideas regarding how planets form (Destination 6) suggest that it is entirely plausible that an Earth-like planet could exist in the α Cen system. Certainly, some stars other than the Sun definitely do have planets (Destination 7) and searches for Earth-like planets are in full swing. It is therefore not crazy to assume such a planet exists here to serve as a base for our observations. In any case, finding Earth-like planets has been a staple of science fiction from before the work of Jules Verne, continuing right up through the latest books and movies. We are justified in continuing that tradition. At least we do not have to assume the natives of this planet, if there are any, speak English!

As our new home planet rotates about its axis, α Cen A steadily sinks low toward the western horizon, producing a 'sun'set that we find reminiscent with a 'normal' sunset that we might have seen back on Earth.[3] As on Mars (Destination 2), we'd *define* the direction where α Cen A—our new 'sun'—sets to be 'west.' The opposite side of the sky would be 'east,' while 'north' and 'south' are the same relative directions as on Earth. Consequently, we would define north to be 90 degrees to our right as we face west toward the waning light of day.[4] As α Cen A first reaches then dips below the horizon, an entirely new Alien Sky blossoms from the twilight. The first anyone has ever seen from a location outside the Solar System!

Expecting to see something wholly new, we are slightly surprised—perhaps a bit let down?—as we begin to realize that many of the same stars and even most of the same constellations that we remember from Earth are visible here. For example, there's Orion, still instantly recognizable in this sky so far from home. But not *exactly* the same. One very bright star—which we come to realize is Sirius—is now located well inside the boundaries of Orion near the star Betelgeuse and far from its 'normal' location in Earth's sky within the constellation Canis Major, the large dog. The Little Dipper—part of the constellation Ursa Minor—is clearly visible too, though it also appears somewhat odd. One of the stars near the bowl of the dipper is clearly in the wrong place, but not so far off that we fail to recognize the distinctive, slightly distorted form of this familiar celestial asterism. Indeed, many other star patterns that we recall from Earth's sky remain clearly visible from this world. Though nearly all of the constellations in this sky are notably distorted—some more than others, some less—the overall impression is that the night sky here at α Cen A is remarkably similar to Earth's. Disappointing? Maybe. But also profound. Once again, we find that after completing a long journey—in this case, a numbingly long journey millions of times longer than any humans have made so far—we find that the view around us has hardly changed. As far as we have traveled to reach the α Cen system, it is clear that we have only barely moved compared to the evidently far, *far* larger expanse of the Universe.

There are, however, two features in this sky that have no counterparts to what we can see in Earth's nighttime sky. The first is visible near the constellation Cassiopeia. There it is now coming slowly into view in the east. This

3 Or were told about. It is unlikely any of the explorers of this planet in the α Cen system were alive when the expedition left Earth.

4 What's different here compared to Mars is that there is no reason to suppose that the spin axis of our planet is even roughly aligned with that of Earth. We are now in an entirely independent stellar system, one that may have formed far in time and space from the Sun and its planets. And, as we shall see at Destinations 6 and 7, while it is reasonable to expect planets orbiting a single star to be roughly aligned, there's no reason for planets of different stars to have similarly oriented rotation axes. But, regardless of this, we can still define the cardinal directions—north, east, south, and west—just as we do at home. All we do is note where our local 'sun' sets and rises.

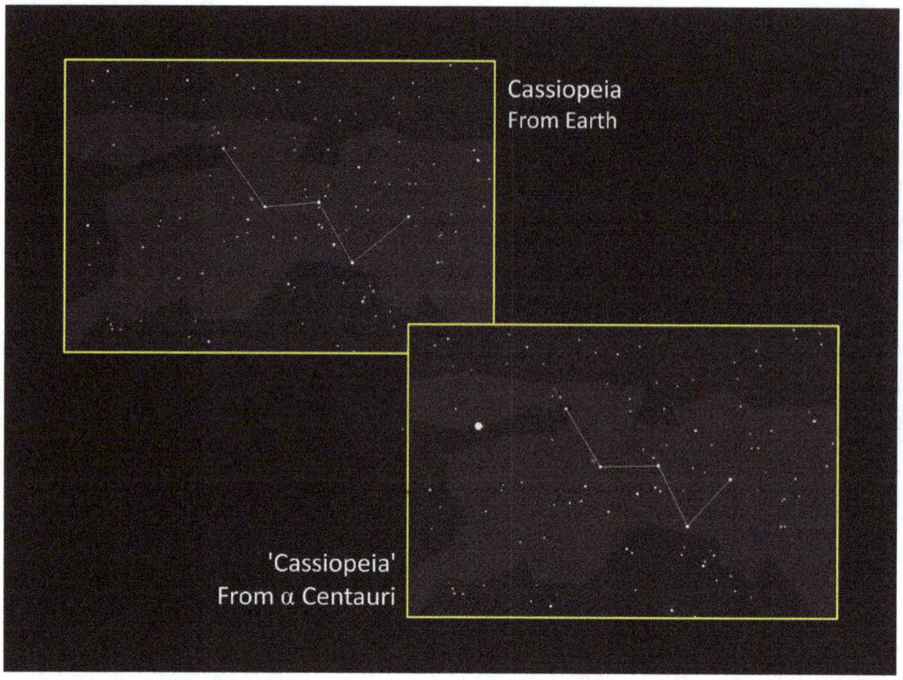

Fig 4.2

(Left) Cassiopeia as seen from Earth. The basic 'W' form of the constellation is evident. *(Right) Cassiopeia* as seen from near α Cen. The basic shape of the constellation is still evident, but there is also another star nearby not visible in the nighttime sky from Earth: The Sun. The lighter gray regions represent the approximate location of the Milky Way that we will explore in detail at Destination 9.

Images generated by M. Mateo with *Starry Night* software.

constellation forms a distinctive 'W' (or an 'M,' depending when and from where you see it) that is readily visible near the Celestial North Pole from Earth and it is easily spotted in our new sky (see Figure 4.2). But something about Cassiopeia is different here. Back on Earth, we are used to seeing five fairly bright stars in this constellation, each located at the vertices of the 'W' (or 'M') that comprise Cassiopeia's body. But in this sky we notice a sixth star (Figure 4.2). A star that is hard to miss since it is among the brightest stars in the sky, and certainly the brightest star in Cassiopeia. Yet, somehow, we never noticed this bright star from Earth. As we study it, we soon discover that it does not move relative to the other stars. This is not a planet of the α Cen system. And night after night, it shines steadily in the same way, the same brightness, the same color.

Then it hits us. Here in the alien sky of the α Cen system, within the boundaries of a familiar constellation, is an even more familiar star. The Sun! As the realization burrows into our consciousness, we begin to truly sense just how far we have traveled to reach this place. The vast distance from home has diminished our Sun—the blindingly bright source of all light and energy back on Earth—into a mere nighttime star. And what about the Earth? Our home planet has become far too faint and is now located far too close to the glare of the Sun to be visible. It is an odd feeling. And somehow very lonely.

There is an important point to appreciate here. It is safe to assume that the Sun itself did not change significantly while we traveled to the α Cen system. Yes, it was a long trip. But there is ample evidence (see Destination 1) that the Sun has changed very little over the past few *billion* years. It is entirely reasonable to suppose that its bulk properties have remained essentially constant over the comparatively short duration of our voyage (years? decades?). This means that the difference in the brightness we perceive between the Sun as seen from α Cen and the brightness of the Sun viewed from Earth must be related to the change in the distance at which we view it. This is not entirely surprising. We all appreciate from everyday experience that objects appear fainter (and smaller) as we (or they) move farther away. Our journey to α Cen provides us with an opportunity to more precisely determine just how brightness changes with distance.

To do this, we compare the brightness of the Sun as seen from Earth—when the Sun is 1 *AU* distant—with the brightness of the Sun as seen from here near α Cen A, 1.26 *pc* away. What we discover is that the Sun is now 68 billion times fainter than it appears from Earth. And it is 260000 times more distant. No doubt you immediately spotted that $260000^2 = 68$ billion (!), implying that the brightness of the Sun has diminished by a factor proportional to the inverse of the square of the distance. That is, since we are now 260000 times further from the Sun than when we were on Earth, the Sun appears $(1/260000)^2$ times as bright as it did from Earth.

This is no coincidence. Figure 4.3 shows that as light pours outward in all directions into empty space from some point-like source (for example, a light bulb, a torch, a star), it spreads out in such a way that the same total amount of light passes through ever-larger spherical surfaces centered on the source.[5]

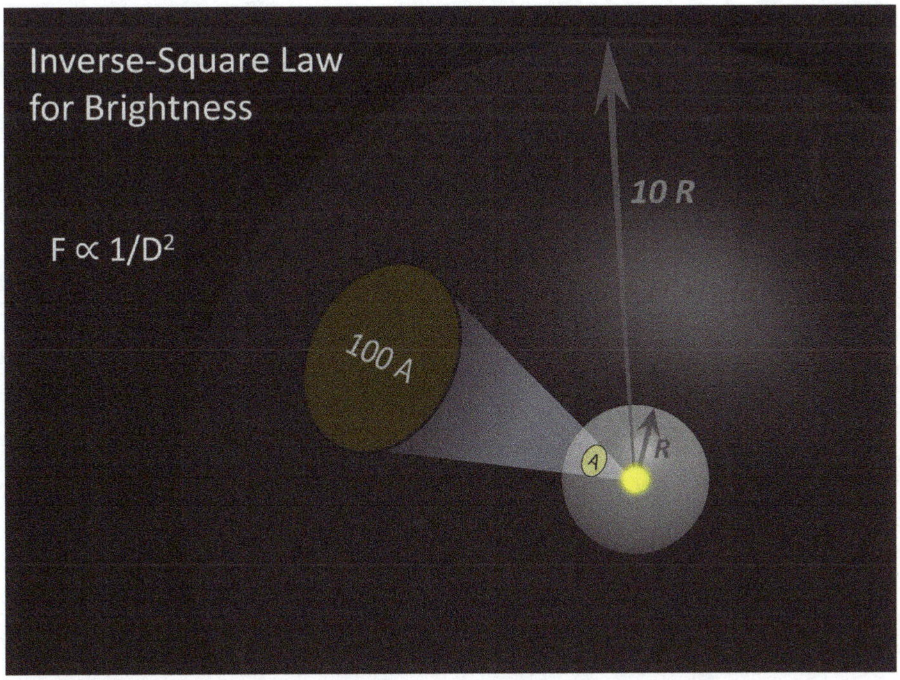

Fig 4.3

As radiation flows away from a source—shown here as a yellow star—it spreads out over an ever-larger area as represented by the two spheres, both of which are centered on the star. The small sphere has a radius *R*, while the larger sphere has a radius that is ten times larger. If we focus on a cone of light from the source, we can show from simple geometry that this cone intersects a region of area *A* on the inner sphere, but an area that is $10^2 = 100$ times larger on the outer sphere. Another way to think about this is that the flux, *F*, decreases at precisely the same rate that the area of a sphere centered on the star increases with distance, *D*. This implies that the flux times the area is constant (one decreases as D^2 while the other increases as D^2 so the product does not change). It turns out that flux times area is related to the total energy a star produces. The inverse-square law simply states that the energy emitted by a star is conserved as it flows from the star into space.

5 We are assuming here that the central source of light really does emit equally in all directions. Virtually all stars seem to do so but some types of stars and other astronomical sources definitely do not. The same basic principles apply in all cases, but the details of how to measure brightness can get a bit more complicated.

The specific example in Figure 4.3 illustrates that the amount of light passing through a sphere of radius, R, around a star must equal the total amount of light passing through a sphere that's ten times larger in radius. But since the surface area of a sphere is proportional to R^2, where R is the radius of the sphere, the light is being 'stretched out' over an ever-larger area. That's important. What we perceive as 'brightness' is actually related to a quantity called flux, the rate at which energy passes through a given area. Since—for a given amount of time—the total energy passing through each larger sphere, E, remains constant[6] (as shown in Figure 4.3) but the surface area grows as R^2, the flux—which is proportional to $E/Area \propto E/R^2$—must decrease as $1/R^2$.

For this reason, the fundamental relationship between distance and the brightness of a source is known as the *Inverse-Square Law for Brightness,* or just the *Inverse-Square Law.*[7] It can be written symbolically as

$$F \propto \frac{1}{D^2},$$

where F is the flux of a source, and D is the source's distance. As a specific example, the Inverse-Square Law tells us is that if one travels to a location twice as far from a star from the location one began, the star will appear four times fainter from the more remote location. On the other hand, if one moves to a location three times closer, the star will appear nine times brighter. You may recall that we saw this effect when we ventured to Mars. But in that case, the changes were not particularly dramatic. Because we started at a distance of 1 *AU* from the Sun and then traveled only a fraction of an *AU* to get to Mars, the Sun became only slightly fainter than it appears from Earth. As a result, even from Mars, the Sun remains dazzlingly bright, nothing at all like the comparatively much fainter Sun we see shining in the nighttime sky from the α Cen system.

There is a very important consequence of this discussion. Since we know how bright the Sun *really* is from our observations within the Solar System, we can use the Inverse-Square Law to figure out how far the Sun is from us *just by measuring its brightness*! This does not apply just to the Sun. If we know the 'intrinsic brightness' of any star—a quantity known as the star's *luminosity*—then we can figure out how far away it is just by measuring its brightness. Up to now, our only way to get distances to stars was to rely on tedious and difficult parallax measurements. We see now that we can use something that is much easier to measure—the brightnesses of stars—to infer stellar distances. This is not a *direct* distance estimate like parallax, but it can be very precise *if* we

---

6 This assumes that nothing absorbs the energy as it travels through space. As we'll see at future destinations, this is generally—though not always—true.

7 But be careful. Other things vary as the inverse-square of the distance from a source, so be sure to understand the context when the term 'inverse-square law' is used.

can make a good guess of a star's luminosity. That, of course, is the hard part: Figuring out a star's luminosity *before* knowing its distance. We will see at later destinations how this is possible.

4.2 AN ALIEN LIGHT IN AN ALIEN SKY

So far, we've studied two stars in some detail: The Sun and α Cen A. They have turned out to be very similar in most obvious respects, including their total luminosity, size, and color. Based on this (small!) sample, we might be tempted to conclude that all stars are Sun-like. But the second remarkable feature of the sky from α Cen A quickly rids us of this notion. Sometime after 'sun'set—or should we call it α-set?—a glow begins to spread from the eastern horizon. The growing pale twilight reminds us of what we see back on Earth when the full Moon is about to rise. But, as we shall soon discover, this glow does not come from a moon. As we ponder the view, a dazzlingly bright point of light *suddenly* appears at the horizon and slowly rises in the eastern sky.[8] We immediately notice that this object shines brilliantly with a dull reddish-orange color that reminds us of the Sun at sunrise or sunset back on Earth. But this object's color does not become whiter as it rises higher into the sky the way the Sun appears to from most places on Earth.[9] The object rising in our sky really is significantly redder than the Sun or α Cen A. This new object is α Cen B, the stellar companion to α Cen A that we already knew about from having observed it from Earth (Figure 4.1). Though approximately 230 times brighter than the full Moon as seen from Earth, from our current vantage point alpha Cen B shines only 1/1000 as

8 It really would be quite 'sudden.' Since we assume the planet we found orbiting α Cen A has a rotation period similar to that of Earth and given the size and distance of this new object (later in this destination), it would only take 4–5 seconds for it to fully appear over the horizon. From Earth, the Sun takes a comparatively leisurely 2 minutes to fully rise or set and so generates a more gradual change in brightness as it rises and sets.

9 The change in color of the Sun as it rises from or nears the horizon is due to the *scattering* of the light from the Sun in the Earth's atmosphere. This process is most efficient for blue light and least for red light. During the day, the scattered blue light colors the sky. But at sunrise and sunset, the process is more pronounced because the light from the Sun has to pass through more of the atmosphere to reach us. On typical days, only the reddest light makes it—the rest is scattered away—causing the Sun to appear red. The fact that the new, bright object we see in this Alien Sky remains red even as it moves away from the horizon means that it really is that color. Of course, if the planet we are on has no atmosphere, the colors would not change as they rise into the sky. We knew how that works from having visited the airless Moon (Destination 1).

brightly as the Sun when viewed from Earth. Or about 1/1600 fainter than α Cen A appears from here. As we look around us, we also appreciate that faint stars have disappeared, hidden in the glare generated over the entire sky from α Cen B. Only a handful of stars, including our Sun, remain readily visible in this 'pale daytime'—or is it 'bright nighttime'? The sky itself now appears a dull battleship gray. Even though it is technically 'nighttime'—α Cen A is below the horizon—there is light everywhere now that α Cen B has risen. Not quite enough to perceive colors vividly as we can during daytime, but certainly plenty of light for us to get around. It is a sky unlike any visible from Earth. The sky of a binary star!

As we grow accustomed to this strange spectacle—a second sun!—we start to study what's going on in this binary system in some detail. After just a few days it becomes evident that α Cen A and α Cen B both move relative to the 'fixed' stars in the sky. But that is only part of the story as we notice that A and B are also moving relative to one another. As we watch, α Cen A—the brighter 'sun' in this Alien Sky—travels through the fixed pattern of the stars in about one Earth year, equal to the orbital period of the planet we have encamped on. This motion is virtually identical to the way the Sun travels around the sky along the Ecliptic as viewed from Earth. But star B acts quite differently, moving across the sky more like a planet does as seen from Earth (or Mars). As we track this motion, we observe that α Cen B even exhibits retrograde motion (see Figure 2.7)! When this is happening, there is never really any nighttime on our planet because α Cen B is in the sky when α Cen A is below our horizon, and vice versa. About six months later, the two stars are close together in the daytime sky (as in Figure 4.4), producing a truly alien sight: A sun accompanied by a tiny point of light nearby, easily visible in the daylight glare. Only during these times does this planet have true nighttime when both stars spend a significant amount of time each day below the horizon. When α Cen B eventually returns to the nighttime sky about six months later, we notice that it has not returned quite to the same place as we saw it the year before. Relative to the stars, it has moved a bit from west to east. At the rate it's going, it requires nearly 80 years to make a full circuit around the entire sky. Though we did not stress this at the time, the planets as seen from Earth also do this, each one taking a different amount of time to make its slow circuit of the sky. These timescales correspond to the orbital periods of the planets in Earth's sky. Here in this Alien Sky, this motion tells us the orbital period of α Cen B. The orbital period of a binary star.

As we study the motions of the two stars carefully over a long time[10], we discover that α Cen B is typically located about 23.2 *AU* from α Cen A and that

10 A trip to the *a* Cen system is likely a one-way proposition, at least for the first expeditions. Not only is it a long trip, but having to carry all the consumables and fuel needed to return might be next to impossible. It really would be good to know if there is a habitable planet there *before* we depart!

Fig 4.4

The view of the sky in a binary star system. This image shows the sky from a vantage point near a hypothetical (human?) base on an Earth-like planet orbiting α Cen A, the brighter star closest to the horizon visible through a thin layer of high clouds. Its much-fainter companion, α Cen B, is also visible. Because the orbit of our planet is smaller than the separation of the two stars (Figure 4.5), the fact that we see both stars close together in the sky implies that α Cen B is on the far side of the brighter star in this image. When both stars set, the night sky will be as dark as it ever gets from this planet. However, in about six months when the planet is on the opposite side of its orbit, it will be located *between* the two stars (though still much closer to α Cen A). At that time, α Cen B will be rising about the time that α Cen A is setting. The nights on this planet will never get truly dark at such times, a feature of life on a planet in a binary star system. From a base such as this we'd be able to study the motions of both stars in detail.

Image from Firaxis Corp.

its orbital period—to be more precise than earlier—is 79.9 *yr* (see Figure 4.5).[11] But we also notice that α Cen B is not orbiting α Cen A the way that the planets of the Solar System orbit the Sun. Indeed, it becomes apparent that α Cen B is not orbiting any specific object at all. Instead it travels around a point that is located directly *between* the two stars, but where nothing obvious is located. As we study this carefully, we find that our new 'sun'—α Cen A—also orbits this same point in space.[12] The size of α Cen B's orbit around this point is 12.6 *AU*, while the size of α Cen A's orbit about this point is 10.6 *AU*.[13] But there is also something disturbing here. We notice that *a* Cen B has an orbital period similar to the 84-year period of Uranus—one of the outermost planets of the Solar System. Yet, the average separation of the two stars here (23.2 *AU*) is considerably larger than the size of the orbit of Uranus in the Solar System (19.2 *AU*). What does all this mean? At face value, this suggests that Kepler's Laws—which describe the motions of planets in the Solar System extremely well—seem to 'fail' here in the α Cen system.

As it turns out, the odd behavior of the motions of α Cen A and B does not mean that Kepler's Third Law is 'invalid' but rather that it is 'incomplete.' Kepler's accomplishment of coming up with his three laws of planetary motion

---

11 We are assuming here that the stars in the α Cen system follow circular orbits (see Figure 4.5). This is not actually the case (see Figure 4.6), but this assumption greatly simplifies our discussion. And, remarkably, it does not alter any of our basic results or our most important conclusions.

12 This would be trickier to determine—though certainly possible—than establishing the motion of star B from our vantage point in orbit near star A. The problem is similar to tracking the motion of the Sun relative to the stars when the stars are not readily visible whenever the Sun is up.

13 Notice that these distances are rather big compared to the distance of 1 *AU* our planet is from α Cen A. This is why our planet probably has a fairly stable orbit. If the two stars were closer together or our planet further from α Cen A, it is likely that the planet would be ejected from the system in just a few million years. As it is, our new home planet probably will be ejected from the α Cen system, but not for a few *billion* years! There's no need to rush. This situation is similar to the way that Jupiter has many moons that orbit it, even as Jupiter and its moons orbit the Sun.

was remarkable, but keep in mind that he was able to study in detail only the motions of the *Sun's* planets. Had he known about the orbits of binary and triple stars (he didn't even know such objects existed), he might have realized that his Third Law was missing something. A few years after Kepler, Isaac Newton solved the problem when he developed a more universal (literally and figuratively) form of Kepler's Third Law that can be written as

$$a^3 = (M_1 + M_2)P^2.$$

Notice that there is a new term here, $(M_1 + M_2)$, where M_1 and M_2 are the masses of the objects orbiting one another and measured in a new unit: *Solar Mass*. This special unit—which we will use quite a lot—is typically denoted $M_{\odot}$, which uses the classical symbol for the Sun, '⊙'. When adopting this specific form of Kepler's Third Law, the orbital separation, α, must be measured in Astronomical Units (*AU*), and the period, *P*, is measured in Earth years (*yr*). Remarkably, if we are careful about the units, this single equation works for *any* binary system—the Earth and Moon, the Sun and the Earth. Or α Cen A and B.

We have described the stars in the α Cen system as 'orbiting one another,' but in the Solar System we've always spoken of planets 'orbiting the Sun.' In real systems, objects never orbit other objects, but rather a special point in space known as the *center of mass*. Though we can—and do—say the Earth orbits the Sun, in fact, *both* orbit about their common center of mass. But since the location of the center of mass is always closest to the more massive object (we'll see why shortly) and because the Sun is about 340 thousand times more massive than the Earth, it turns out that the Earth–Sun center of mass is located deep inside the Sun. Not *quite* at the Sun's center, but pretty close. Moreover, since the Sun is much more massive than *all* of the planets combined, that factor, $(M_1 + M_2)$, in Newton's modified form of Kepler's Third Law is never larger than 1.001 $M_{\odot}$, even when we consider Jupiter, the most massive planet of the Solar System. We should not be too hard on Kepler for failing to notice this effect. And we can certainly be forgiven for thinking that the center of mass of the Solar system is essentially coincident with the center of the Sun. For most practical purposes—even for many precision astronomical applications—this is essentially correct.

But here in the α Cen system, this is no longer a valid assumption because α Cen A and α Cen B have comparable masses. As a result, the center of mass of the system—a point we can localize by measuring the motions of the two stars—is about 46% of the way between α Cen A and α Cen B, closer to star α Cen A. This immediately tells us that A is more massive than B. But we can say more than this. By using Newton's modified form of Kepler's Third Law and being more careful about what we mean by 'center of mass,' we can actually determine the masses of the two stars.

To illustrate how, imagine that we have a long, uniform rod. To balance it, we'd need to place a finger near the rod's center. If we were to add a weight to one end of the rod, we would discover that the balance point has moved closer to the heavier end of the rod. Every time we add or remove a weight from the rod, we must re-determine the location of the center of mass by searching for the balance point. This is true of the center of mass of the *a* Cen system. In order to remain balanced, the two stars must travel not about the center of either star, but about the center of mass, the balance point of the system. As with the rod with weighted ends, that center of mass point is always located along the line connecting the centers of the stars (see Figures 4.5 and 4.6).[14] It turns out that if the distance from the center of mass to each star is denoted a_A and a_B, then the center of mass is located at a point such that the relation $a_A M_A = a_B M_B$ is always true.

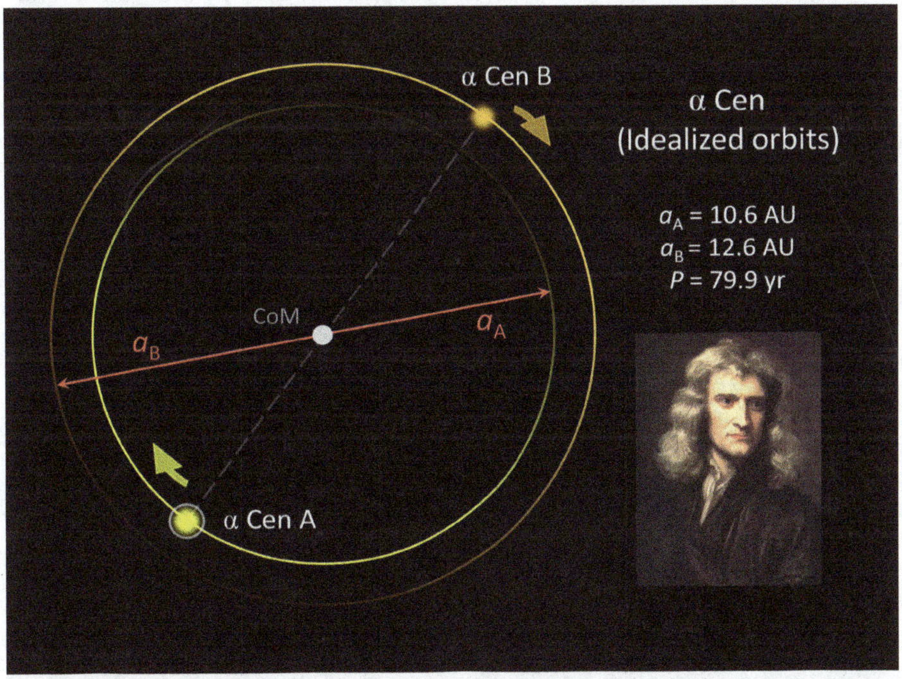

Fig 4.5

An idealized representation of the α Cen system in which the orbits of the two brightest stars are assumed to be circular and are shown centered the system's center of mass, denoted here as 'CoM' and shown as a white dot. The semi-major axis (radius) of the orbit of α Cen A is denoted by a red arrow labeled a_A, while the semi-major axis (radius) of the orbit of α Cen B is denoted by the red arrow labeled a_B. The values of these distances and the orbital period of the system are listed in the figure. The small white circle around α Cen A represents the orbit of a (hypothetical) planet at 1 *AU* surrounding the star. That planet, in that orbit, is our current destination. The inset image is a portrait of Isaac Newton, the person who extended Kepler's Third Law so that it could be applied to a binary system such as α Cen and not just the planets of the Solar System.

We can use this definition of the center of mass along with Kepler's 'new and improved' Third Law to *calculate* the masses of the two stars. We start by writing the Third Law as

14 The reasoning here is a bit more complicated and involves Newton's Laws of Motion. In particular, the First Law of Motion states that objects will not change velocity (neither their speed nor direction of motion) if there are no forces acting on them. Although we shall see that stars in binary systems such as the α Cen system are exerting strong gravitational forces on each other, there are no (significant) forces on the system as a whole. Thus, the system must remain moving at a constant velocity, something the individual stars are clearly not doing. Consequently, the center of mass must satisfy Newton's First Law of Motion and the only way to do this is if the stars are always balanced about that point.

$$(M_A + M_B) = \frac{a^3}{P^2},$$

where the two masses denote the masses of the stars in the α Cen system, and $a = a_A + a_B$. That is, the average separation of the two stars is equal to the sum of the average separation of the two stars from their common center of mass. For the orbital parameters of α Cen B, we can calculate that the sum of the masses of A and B is about 2 times the mass of the Sun ($1.95M_\odot$ to be a bit more precise). By introducing the information we discovered regarding the location of the center of mass of the system, we also know the mass ratio from the relation $a_A/a_B = M_B/M_A$. This information can be combined to determine the individual masses of the two stars. It works for the idealized representation of α Cen that we've been considering (Figure 4.5). And it works for the way the system really is, non-circular orbits and all (Figure 4.6).[15] By simply measuring the orbital properties of the stars in binary systems such as α Cen, we can derive the stars' masses! That's really remarkable. This is a tool we will use at future destinations to help us explore the fundamental nature of stars.

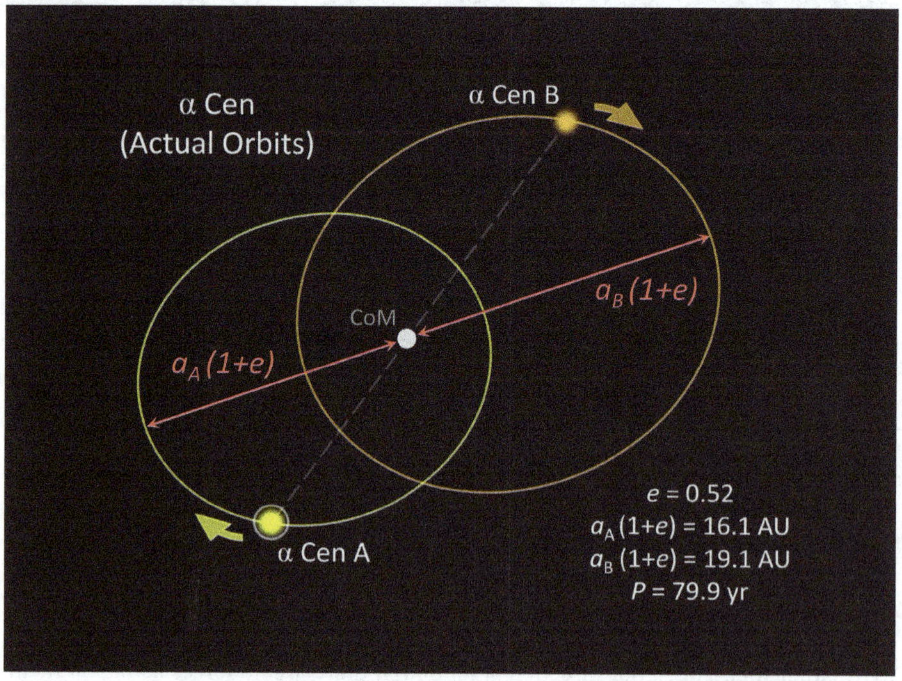

Fig 4.6

A plot showing the actual, elliptical orbits of the stars in α Cen about the system center of mass (the white dot, denoted 'CoM') as seen from directly above the orbital plane of the binary. Although the *eccentricity* of the orbits—a measure of how non-circular the orbits are and denoted by the symbol '*e*' in the Figure—complicates the analysis compared to our idealized binary with circular orbits (Figure 4.5), the basic results are the same. In particular, the ratio of the distance of α Cen A from the center of mass and the distance of α Cen B from the center of mass is always equal to M_B/M_A, just as in the case of circular orbits (Figure 4.5).

15 Since the actual orbits of the stars in α Cen are elliptical (Figure 4.6), the view is actually a bit more complicated than we've described. In particular, the two stars vary in distance over their 79.9-yr orbital period. At their closest, they are only 11.3 *AU* apart, while at their most separated, they are 35.4 *AU* apart. You can explore some of the consequences of this in a question at the end of this destination.

4.3 A FAINT LIGHT IN AN ALIEN SKY

And what about Proxima Cen (Figure 4.7)? Remember? The star closest to the Sun (see Destination 3)? Oddly enough, this third member of the system hasn't really made much of an impression on us in the sky of our new home. It takes some time before we even notice that a faint star—dimly visible to the naked eye—is moving steadily, slowly across this Alien Sky at a rate of about 1.3 *arcsec/yr*. We can use the technique of parallax (Figure 3.8; Destination 3) to determine that Proxima is about 12700 *AU* from α Cen A (or from α Cen B since 12700 *AU* is a lot bigger than 23.2 *AU*, the average separation of α Cen A and α Cen B). This distance is so large that it implies Proxima is only very weakly held by the gravitational force of α Cen A and B combined. So weakly, in fact, that if Proxima were gravitationally jostled—even slightly—by some passing star, it could easily wander off into interstellar space, never to return to the α Cen system (see Figure 4.7).

It turns out that Proxima is also distinctly different from α Cen A or α Cen B. Using the Inverse-Square Law, we realize that Proxima must have a very low luminosity to be so faint while being so (relatively) close. Since we know its distance, we discover that Proxima is visually about 590 times less luminous than the Sun, over 900 times less luminous than α Cen A.[16] It also has a deep red color

Fig 4.7

An image of Proxima Cen as seen from Earth. Proxima is the obviously red star near the center of the image.

Photo by D. Malin/AAO/UKST.

(Figure 4.7), unlike anything we've seen so far, and considerably redder than α Cen B. Normally, a third star would essentially invalidate Kepler's Third Law, even the new modified form, but only if the third object is both massive and nearby. Proxima is so distant and, as we eventually determine, so small in mass that it does not significantly alter our estimates of the masses of stars A and B. When we adapt Kepler's Third Law to estimate Proxima's mass[17], we find it is

16 These values refer to the total energy output of Proxima compared to the Sun and α Cen A. If we consider only visible light output, we find that Proxima is about 19000 times less luminous than the Sun, nearly 30000 times less luminous than α Cen A. Any way we look at it, Proxima is a faint star.

17 And you can do this yourself in one of the problems at the end of this destination.

Fig 4.8

This image overlays a series of photographs of Proxima Cen taken from 1985 to 2010. The motion of Proxima—which is identified by the labels of the first and last years in the sequence—is clearly evident. The length of the arrow is approximately 1 *arcmin*. It was the knowledge of this rapid *Proper Motion* that led people to believe that Proxima might be nearby even before its parallax could be measured.

Photo by S. Quirk/Frog Rock Observatory.

about 12% as massive as the Sun, only about 6% as massive as α Cen A and B combined.

Since it is now comparatively nearby, we can also tell how large Proxima is. We first measure its angular size—even from our location next to α Cen A, we'd need a very good telescope to do this—and its distance; we can calculate Proxima's true size with the relation

$$R = D\theta,$$

where R is the true radius of the star, D is the distance to the star, and θ is the angular radius of the star in *radians* (see Destination 3 to review the properties of *radians*). The units of R and D can be any units of length that we want—*miles*, *kilometers*, *furlongs*, etc.—as long as they are the same. Indeed, this equation works for measuring the sizes of α Cen A and α Cen B too, assuming we can measure their angular sizes as well.[18] Because this equation breaks down if the angle θ gets too large (say, more than about 0.1 *radians*, or about 5 *deg*), it is often referred to as the *Small-Angle Formula*. Notice how similar this equation is to the Parallax Equation we derived in Destination 3. Indeed, the Parallax

18 Which, from our new vantage point, we can. But for those of us forced to remain on Earth, there are very few stars, apart from the Sun, whose radius can be measured directly as described here. We will discover in Destination 8 that there is another, slightly more indirect but still accurate, method to determine stellar radii.

Equation is a specific application of the Small-Angle Formula, which is certainly reasonable given that the parallax angles for stars are always very small. When we apply this to the α Cen stars and the Sun, we find that α Cen A is somewhat

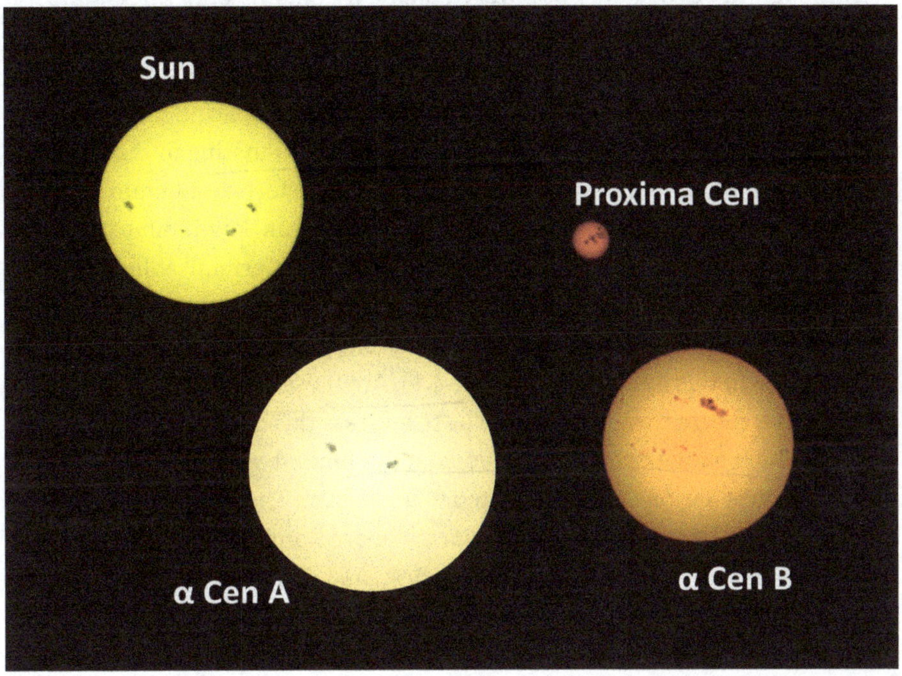

Fig 4.9

The true relative sizes of the Sun (*left*) and the three stars of the α Cen system (*right*). These are the four closest stars to us, yet they are all notably different from each other. The Earth is about the size of the smallest sunspot visible on the image of the Sun. The other stars also probably have sunspots, but their disks have never been directly imaged from Earth.

larger than the Sun, α Cen B somewhat smaller, and Proxima is the smallest of all (Figure 4.9).

As we study Proxima over subsequent months, we are startled to see it occasionally brighten by a factor of two or more in just a few seconds, then fade back to its original brightness minutes later. These events are called *Flares* and are triggered most likely by complicated interactions of the star's strong magnetic field with matter near its surface. The Sun—and very likely α Cen A and B—exhibit very similar flares, but these stars are so luminous to begin with that the additional light from the surface flare is easy to miss.[19] Not so for

19 Most solar flares are observable only with specialized instruments that zero in on specific colors (wavelengths of light; see Destination 5) in which the flares emit most of their energy. But occasionally a solar flare *is* visible directly in simple optical telescopes. This is not good. One such flare occurred on September 2, 1859 and is known as the 'Carrington Event' after the astronomer who visually spotted the flare. These eruptions on the solar surface can eject energetic particles at speeds of a few thousand *km/s*, and if oriented correctly, those particles will hit Earth a day or two later. The energies and quantities of the particles from the Carrington Event were so great that they drastically compressed the Earth's magnetic field—the structure that protects us from such particles normally—and the electronic world of that day, still in its infancy, was almost completely knocked out (telegraphs mostly). The skies were ablaze with Aurorae—the 'Northern Lights' and the 'Southern Lights'—that are generated as particles from the Sun are directed by the magnetic field and collide with the upper atmosphere. Normally confined to near Earth's poles, Aurorae associated with the Carrington Event were easily

Proxima. Its flares can release as much as or more light than the rest of the star for a few minutes at a time. Other *flare stars* share many of the properties we see in Proxima. They have very low luminosities and are very red in color.

Although the α Cen system is very small—it contains only three stars—it seems to be telling us some important things about stars in general. First, either stars like to group together or we've just been remarkably fortunate that the closest star to the Sun is a triple system. Second, by comparing the properties of the three stars that make up the α Cen system—and tossing in the Sun—we seem to have identified some trends in stellar characteristics. Luminous stars appear to be larger, more massive, and yellower than low-luminosity stars, which are smaller, lower in mass, and red (Figure 4.9). Certainly when we compare α Cen B and Proxima, we see these tendencies clearly, but even α Cen A and the Sun, two stars with only slightly different properties, appear to also be consistent with this behavior (see Table 4.1). Are these trends a general feature of stars? Our future travels will allow us to address this question directly.

It has been a long trip to α Cen—one of the longest single voyages any humans have ever seriously contemplated to undertake. The view back to the Sun is awesome in the true sense of the word. We are humbled by the vastness of the Universe that this immense, yet tiny, first step demonstrates. But we have really only started. As we travel even further from the Sun, we shall no longer be able to distinguish our home star from the myriad other stars of our Galaxy any more than we could distinguish an individual grain of sand on a beach from high above in an airplane. And the Earth? It is so tiny, so faint, so close to the Sun, that we can no longer see our home planet. And we never will see it again as we venture deeper into the Universe.

---

seen as far south as Mexico and Cuba. A similar event today would cause major, long-lasting power and communication disruptions.

DESTINATION 4 QUESTIONS

4.1. The angular diameter of the Moon is half a degree in the sky as seen from Earth. It is located an average distance of about 240000 *miles* from Earth. From this information and using the Small-Angle Formula, what is the true diameter of the Moon in *miles*?

4.2. Use the new and improved version of Kepler's Third Law to determine what the Earth's orbital period would be if it orbited a star of twice the mass of the Sun but still at a distance of 1 *AU*. What would the period of the Earth be if it orbited a star of half the mass of the Sun? What would the period of the Earth be around a star of one solar mass, but if it were 0.5 *AU* from that star? What if it were 2 *AU* from a one–solar mass star?

4.3. Assume that Jupiter has a mass of 0.001 $M_\odot$. Also assume that Jupiter's orbital semi-major axis is 5.2 *AU*. What is Jupiter's orbital period, in *yr*? How far from the center of the Sun is the center of mass of the Sun-Jupiter system? Is this inside or outside the Sun?

4.4. Imagine that the Earth is replaced by a star with a mass of 2.2$M_\odot$ The Sun is still present and still has a mass of 1 $M_\odot$ We'll ignore all the other objects in the Solar System. Let's call the new star B and the original Sun star A.

a) What is the value of a_A and a_B in this new binary star system, assuming that separation of the stars is 1 *AU*?

b) What is the orbital period of this binary?

c) What is the maximum angular separation these stars would appear to have from one another when viewed from Alpha Cen? (Which, you may recall, is located 1.26 *pc* away.)

4.5 The information about Proxima Cen given in Destination 4 allows us to determine some of its basic properties.

a) What is the orbital period, in *years*, of Proxima Centauri around the center of mass of the Alpha Cen system? **HINT:** There are two independent bits of information in Destination 4 that allow you to determine Proxima's period.

b) Given this period, calculate the mass of Proxima Cen using the modified version of Kepler's Third Law. You can check your result with the data in Table 4.1, but be sure to show how you obtained your answer.

4.6. From the information in Table 4.1 and the text in Destination 4, estimate the angular sizes of the Sun and of Proxima Cen from our vantage point near α Cen A. Good telescopes can make out details as small as 0.01 *arcsec*. Can either star's diameter be measured with this precision from near α Cen A?

4.7. Compared to our view here on Earth, by what factor is the Sun fainter or brighter from

a) The planet Mercury?

b) The planet Neptune?

c) The current approximate location of the deep space probe Pioneer 10 (about 95 *AU*)?

d) A star 2.5 *pc* away?

4.8. Why can solar panels be used to power spacecraft that explore the inner Solar System (out to Mars), but not for the outer Solar System (from Jupiter on out)?

4.9. In reality, α Cen A and α Cen B follow elliptical orbits as shown in Figure 4.6. At their closest—known as periastron—the two stars are 11.2 *AU* apart. At their most distant—known as apastron—they are 35.2 *AU* apart.

a) What is the angular size of α Cen B as seen from α Cen A at both periastron and apastron?

b) By what factor is α Cen B brighter as seen from α Cen A when the stars are at periastron compared to apastron?

c) How far are the individual stars from the system's center of mass at periastron and apastron?

4.10. Imagine that we observe the α Cen system from Earth. How far apart are the two stars in angular units when they are at periastron? How far apart are they when at apastron? See Question 4.9 for details on the periastron and apastron distances.

Destination 5

THE 'EMPTY' SKY OF DEEP SPACE: AT 'ANTI-CENTAURI'

5.1 NO PLACE IN PARTICULAR

Normally, one expects a destination to be, well, a destination. But our next stop is, literally, no place in particular. As a guide to our navigator, we've chosen to go to a point in space that is located 1.26 *pc* from the Sun—the same distance as α Cen—but in a direction opposite where that star is located. We'll call this place 'Anti-Centauri' or just 'Anti-Cen' for short. This seems like an odd place to visit, but in fact, due to the vast distances between our destinations, it is in seemingly empty regions like this where we would spend most of our time during interstellar journeys. As such, Anti-Cen represents a typical region of *Interstellar Space.*[1] What's here?

Apparently nothing. Even light seems to shun this place. We are used to images from TV shows and movies that depict brightly lit spacecraft hurtling through deep space (Figure 5.1). What we appreciate now is how *dark* it is here between the stars. A darkness unlike anything we've experienced before. Even on a moonless night on Earth, the air itself emits light.[2] Though any given patch of sky produces only a tiny bit of light—the equivalent of a single very faint

Fig 5.1

Hmmm. What's illuminating the *Enterprise* deep in interstellar space? Out here at Anti-Cen our spacecraft would appear virtually invisible, illuminated dimly by the feeble light of the distant stars. Its onboard lights and inky black silhouette would be the only things betraying its presence to our eyes.

Photo by M. Padilla and M. Mateo

1 Actually, it turns out that the space around the Sun, out to at least 100 *pc* or so, is in many ways atypical, comprising a region known as the 'Local Bubble.' For our discussion here at Anti-Cen, we shall assume that space at this location is similar to what's typical throughout all interstellar space in our Galaxy, the vast collection of stars in which the Sun resides (Destination 12).

2 Why? On Earth the ultimate source of energy for this light are particles from the Sun that interact with atoms in our atmosphere and, like a neon tube, respond by emitting light. The full details of how this process works await us at Destination 6—the Orion Nebula—where we will find that the very tenuous matter in seemingly empty space can emit light, basically for the same reason the nighttime sky on Earth does.

star—when we combine the contributions of all the patches visible above the horizon at a given time, it adds up. If you let your eyes become well-adapted to the night—the extensive chemical and physiological changes that occur in the eye and brain to fully adapt to darkness require about 30 minutes—it is surprising how much you can see on 'dark' nights. On Mars (Destination 2), the effect is less pronounced, but the sky there also emits light. On both planets, scattered sunlight makes the daytime sky so bright we see virtually nothing but the Sun itself. From the near side of the Moon (Destination 1), we saw that the nighttime sky is lit by the bright, nearly full Earth shining above us at night. During the lunar day, the sky may be dark but we still receive reflected sunlight off the Moon's surface. And from all these locations within the Solar System—even the airless Moon—sunlight is scattered by an extensive cloud of dust particles orbiting between the planets.[3]

But here at Anti-Cen there is no dominant light source to illuminate objects around us. Just the feeble light of the stars, none close enough to dominate the sky the way our Sun does on Earth or α Cen A and B did at Destination 4. When we venture outside our ship, all we see are the stars around us and an inky blot beside us. Only the weak reflections of stars off the ship's hull offer any direct visual clue that it is present at all. The view of this alien sky produces a distinctly disconcerting sensation. The relentless darkness of this place is unnerving.

The Sun is visible from here, of course. When we visited α Cen, we found our home star located in the direction from which we had come, in the constellation Cassiopeia. But from Anti-Cen, the Sun is now plainly visible in the constellation Centaurus. With α Cen and β Cen it produces a beautiful trio of bright stars in this sky (Figure 5.2).[4] Many of the other bright stars we know from Earth are still easily recognizable, most in about the same locations we would see them from home. Constellations such as Orion, composed of mostly distant stars, seem almost unchanged, while some constellations—ones that contain nearby stars such as Sirius in Canis Major—are much more noticeably altered in appearance. We've seen similar sorts of changes in the patterns of the stars before when we visited α Cen. After all, we are now at the same distance from

---

3 From Earth, sunlight scattered by such dust is visible from dark sites as the 'Zodiacal Light,' a glowing band that stretches along the ecliptic which is delineated by the constellations of the Zodiac. Sometimes, the point opposite the Sun can be seen to glow faintly as well from scattered sunlight as a ghostly feature known as the 'Gegenschein.' All these features that light up the nighttime skies of planets such as Earth and Mars and even the Moon would not be present at Anti-Cen.

4 Yes, we know that α Cen is actually a triple star. From our present location, two of its stars—α Cen A and α Cen B—are readily visible in even a small telescope. The third—Proxima—is visible only with a fairly large telescope. As it turns out, β Cen also consists of three stars, all much more massive than the Sun, and in a configuration somewhat similar to α Cen (two stars close together with the third farther out). However, from our present location at Anti-Cen, both α Cen and β Cen—and the Sun—appear to be single stars to our unaided eyes. It is indeed a beautiful trio of stars that we see as we look back in the direction from which we have come.

Fig 5.2

The view from Anti-Cen looking back from the direction from which we came. Visible together in the sky from this new destination deep in interstellar space are α Cen (*left*), β Cen (*right*) ... and the Sun (*lower*).

Sun added to image by M. Mateo.

the Sun, just a different direction. As far as the distant stars are concerned, only the details of the sky are different here. What really makes this place distinct from any we've visited so far is that there is no star that we could call our 'sun.' Here, deep in interstellar space, there seems to be nothing at all.

5.2 THE MATTER OF 'EMPTY SPACE'

But this impression of emptiness is an illusion. There *is* something here. In fact, we have relied on its presence to power our spacecraft.[5] As you may now appreciate, traveling to nearby stars in anything like a 'traditional' manner is as much a feat of endurance as it is of technology. But the technology does matter, and it boils down to accelerating our ship on the way to our destination, coasting, then decelerating when we arrive. For this we need some sort of propulsion system. Some sort of engine. And engines need fuel. Now, one can take all the fuel needed for both the acceleration and deceleration phases of the journey right from the start. But that's inefficient. Why use a lot of fuel early in the trip just to accelerate the fuel we need to slow down at our destination ... particularly if we can pick it up along the way!? This sort of thinking has led some people to suggest that the best way to get the fuel needed to slow down and arrive at an interstellar destination is to collect it en route. Now, there are no gas stations, of course, but—like a baleen whale trawling for food with its

5 In general, we will not speculate about the specific details regarding *how* we might carry out our fantastic journeys to remote destinations. But here at Anti-Cen, it is useful to do so a bit in order to describe what 'empty' space contains. And after all, in this case, 'empty space' *is* our destination.

Fig 5.3

A model for an interstellar spacecraft? Just as this baleen whale trawls for food as it swims through the water, we can imagine 'trawling' for fuel as we cruise through interstellar space. But for this to work, there had better be something to collect!

Image by Carl Buell and John Gatesy.

mouth agape (Figure 5.3)—it may be possible to do this by sweeping up fuel that exists in space itself as we cruise to our final destination. For this idea to work, there had better be something in interstellar space for us to collect as fuel. As passengers, this may seem risky—after all, we keep talking about *empty* space. But after a short time, we are relieved—and somewhat amazed—to discover that our fuel tanks are indeed steadily filling up with stuff. Familiar stuff. But not in familiar quantities.

On Earth, a typical sample of air (at sea level) is dominated by a two elements, Nitrogen (N, about 80% by number) and Oxygen (O; about 20% by number).[6] And what about all the water on Earth? Well, it contains a lot of H_2O (!) which is composed of Hydrogen (H) and Oxygen, the two elements that dominate the watery parts of our world. That includes us humans of course, since we are made mostly of water (see Table 5.1). But air and water (and humans!) comprise a tiny fraction of the mass of Earth. The vast majority of our home planet is in the form of rocks and minerals, including those found near the surface (the Earth's crust) and within its voluminous interior (the mantle and core of the Earth). A chunk of the Earth's crust—something we can easily study and measure the chemical composition of (see Table 5.1)—contains a lot of O, Silicon (Si), Aluminum (Al), Calcium (Ca), and Magnesium (Mg). The inner

6 Every element has a standard abbreviation consisting of one or two letters. Some are easy to figure out (H = Hydrogen, He = Helium, Ne = Neon), but some may seem strange because they derive from Latin or other languages (Fe = Iron, W = Tungsten, Pb = Lead). As we discover new elements during our journeys, we'll introduce the appropriate abbreviations and use them in subsequent descriptions. If you lose track of what's what, however, refer to Appendix G which lists the standard abbreviations of the naturally occurring elements.

TABLE 5.1 A summary of the relative abundances of the elements in the Earth's crust, the Interstellar Medium (ISM), and in the human body.

ELEMENT NAME	SYMBOL	ATOMIC NO.	NUMBER PER SAMPLE OF 1 BILLION ATOMS		
			EARTH (CRUST)	ISM	HUMAN BODY
Hydrogen	H	1	26 million	919 million	630 million
Helium	He	2	<1	79 million	—
Lithium	Li	3	69000	1.7	—
Beryllium	Be	4	5300	0.02	—
Carbon	C	6	2.6 million	320000	94 million
Nitrogen	N	7	34000	95000	13 million
Oxygen	O	8	550 million	740000	260 million
Neon	Ne	10	<1	80000	—
Sodium	Na	11	24 million	2100	270000
Magnesium	Mg	12	27 million	34000	130000
Aluminum	Al	13	70 million	2700	—
Silicon	Si	14	230 million	36000	—
Phosphorus	P	15	770000	270	2.4 million
Sulfur	S	16	220000	15000	390000
Chlorine	Cl	17	770000	290	350000
Argon	Ar	18	<1	2700	—
Potassium	K	19	10 million	120	320000
Calcium	Ca	20	30 million	2100	2.4 million
Iron	Fe	26	25 million	26000	34000
Cobalt	Co	27	11000	79	27000
Nickel	Ni	28	29000	1700	—
Copper	Cu	29	22000	14	20000
Zinc	Zn	30	30000	37	19000
Silver	Ag	47	18	0.009	—
Iodine	I	53	94	<0.0001	9900
Gold	Au	79	0.2	0.008	—
Lead	Pb	82	1100	0.06	—
Thorium	Th	90	590	0.002	—
Uranium	U	92	160	0.003	—

parts of the Earth—naturally a lot harder to study—are dominated by Iron (Fe) and Nickel (Ni). Together, these two elements comprise perhaps as much as 50% of all the atoms in our planet. This basic survey reveals that the Earth consists mostly of what astronomers refer to as *Heavy Elements*[7], especially Iron, Nickel, Oxygen, and Silicon.

Here in interstellar space, we soon find that the distribution of elements is drastically different. If we carefully study the composition of the stuff that we've collected during our journey (see Table 5.1), we find that by far most of the atoms in our tank are of Hydrogen, an element that is comparatively rare on Earth (Table 5.1). But not here. It turns out that almost 92% of all the atoms—by number—we have collected as we cruise through interstellar space are H atoms.

What's noteworthy about this is that Hydrogen[8] is the lightest, simplest element. Here in Anti-Cen, virtually all H atoms are composed of two electrically charged particles. The central nucleus of the atom consists of a single relatively massive particle called a *Proton*. Surrounding it is a much lighter particle called an *Electron* (we'll explore how the electron orbits the nucleus in detail at Destination 6). The electric charges of the two particles are precisely equal in strength, but they differ in sign. We say that the proton has a *Positive Electric Charge*, while the electron has a *Negative Electric Charge*. Since opposite electric charges attract, this means the proton and the electron in a H atom are bound together with a force that is far stronger than the minuscule gravitational force they exert on one another. This *Electrostatic Force* is also called the *Coulomb*

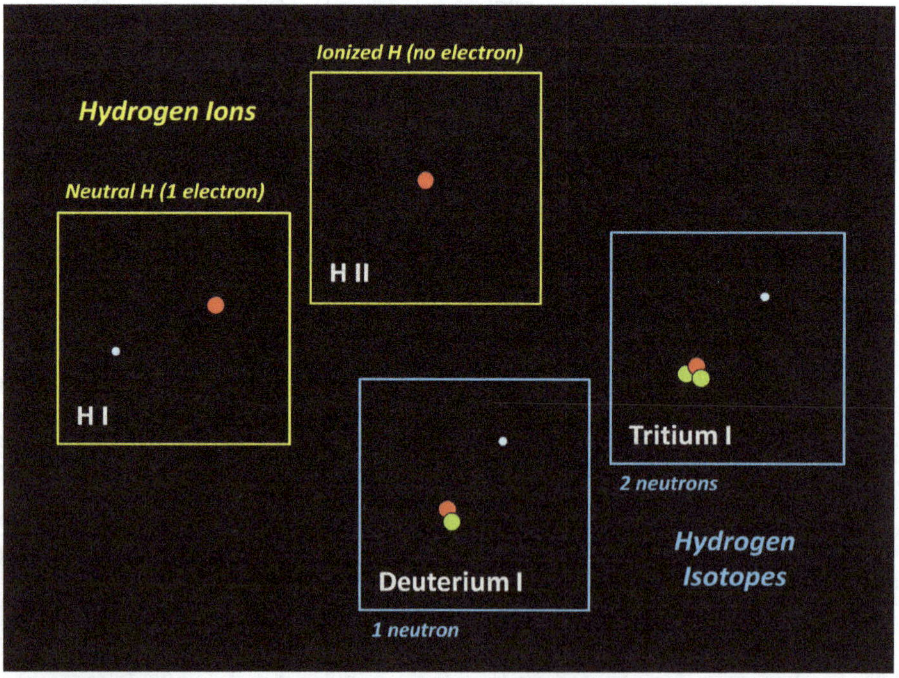

Fig 5.4

The yellow boxes show the two most common ions of the H atom. On the left is neutral H, consisting of one proton (*red dot*) comprising the atom's nucleus, plus one electron in orbit about the nucleus (small blue dot). If we were to remove the electron from this atom, that is, if we were to *ionize* a neutral H atom, the result would be a bare proton (*right yellow box*). Neutral H is denoted as H I (pronounced 'H-one') while ionized H is referred to as H II ('H-two'); the details of this strange nomenclature are described in Section 5.4 and Figure 5.7. The H I atom is electrically neutral because the positive charge of the proton precisely balances the negative charge of the electron. The atoms of H II are clearly positively charged since they contain only the single, positively charged proton. The blue boxes show two neutral isotopes of H. Deuterium contains a nucleus consisting of one proton (*red*) and one neutron (*green*). Tritium has a nucleus consisting of one proton and two neutrons. By definition, H, Deuterium, and Tritium are all 'Hydrogen' atoms because they contain a single proton in their nuclei. They represent examples of different *isotopes* of the same element. It's often important to distinguish between different ionization states of atoms (*yellow boxes*) and different isotopes (*blue boxes*).

7 That is, any element composed of atoms more massive than those of the two lightest elements, Hydrogen (H) and Helium (He).

8 Not surprisingly, *Hydro*gen got its name from its association with water.

Force in honor of the French physicist who helped characterize its properties. When there are equal numbers of electrons and protons in an atom, the atom is said to be *Neutral*. When there are unequal numbers, it is an *Ion* (technically, a neutral atom is an ion with no net charge). Most ions have fewer electrons than they have protons, so they have a net positive charge (see Figure 5.4). This is also why the process of removing electrons from a neutral atom or ion is called *Ionization*. As we take careful stock of the atoms we are slowly amassing in our tank, we discover that virtually all the H atoms have a neutral charge. That is, they consist of an electron *and* a proton.

But not everything in the tank is Hydrogen. The next most common element we find is another surprise: Helium (see Table 5.1). This is the second lightest element and is typically composed of a nucleus consisting of two protons and two *Neutrons*—particles with nearly the same mass as protons, but that have no electric charge—along with two electrons that orbit about the nucleus. We know this is Helium because elements are *defined* by the number of protons their nuclei contain. An atom with a nucleus with one proton is Hydrogen, regardless of how many (if any) neutrons are present (see Figure 5.4). Likewise, atoms with two protons are Helium, regardless of the number of neutrons they may contain.[9] As we count the atoms we collect in the tanks, we notice that for every billion atoms we have collected—that's less than a trillionth of a gram of material!—there are about 79 million He atoms. That comes out to about twelve H atoms for each He atom, or a bit under 8% by number. By comparison, Helium is extremely rare on Earth (see Table 5.1).[10]

But we seem to be missing something. Where are all those elements so common on Earth? If we look very carefully at the contents of our tank, we notice

9 The number of protons in a nucleus is so important that elements are often referred to by this number as the *Atomic Number*. The atomic number of H is 1, He is 2, O is 8, Fe is 26, and so on. Atoms of a given element but with different numbers of neutrons are known as *Isotopes* of that element. There are three common isotopes of H, for example (see Figure 5.4). One isotope, consisting of a nucleus with only one proton, is just standard Hydrogen. *Deuterium* is an isotope of H that consists of two particles in the nucleus, a proton (this is why it is H) and a single neutron. Yet another isotope of Hydrogen contains two neutrons in the nucleus and is known as *Tritium*. These isotopes of Hydrogen will turn out to be very important to understand how the elements formed in the first place, a process we shall explore at destinations 'located' near the dawn of the Universe. For most elements in most environments, there is one dominant isotope. For Hydrogen, the single-proton isotope is by far most common. Deuterium and Tritium comprise a tiny fraction of the H atoms we are collecting here in interstellar space.

10 Indeed the name of Helium hints to how rare it is on Earth. The root *helios* comes from the Greek work for 'Sun.' When people first started to study the detailed properties of the Sun in the 19th century, they noted the presence of an element that had never been detected on Earth using 'spectroscopic' techniques we shall start to investigate in some detail at Destinations 6 and 8. This new element was proposed to exist only on the Sun and so was named after our home star. It was only somewhat later that Helium was discovered in significant quantities on Earth. The relative rarity of Helium on our planet is the subject of one of the questions at the end of this destination.

that indeed there is something here other than just H and He. For every billion atoms in our collection tank, we count about 740000 Oxygen atoms. Carbon is next most common: about 320000 C atoms are present for each billion we collect. And Iron? We count a measly 26000 Fe atoms per billion atoms here in interstellar space. And so on (see Table 5.1). If we study the contents of our tanks thoroughly, we discover that most of the heavy elements we find on Earth are indeed present here in interstellar space, though generally in far lower quantities relative to the very large number of H and He atoms. As we shall see, most of the Universe resembles the gas we find here in interstellar space: H and He plus a bit of other stuff. It turns out that it is the composition of the Earth—heavy elements plus some H and almost no He—that is unusual. The matter found out here in 'empty space' is referred to as the *Interstellar Medium*, or, to use the acronym most astronomers adopt, the *ISM*.

The composition of the ISM is handy for us from a practical point of view: in many respects, H is an excellent fuel.[11] But the distribution of elements in the ISM—and, as we'll see, essentially, the composition of the Universe as a whole—raises a lot of questions. For example, why is the Earth so different from the ISM? Why are some specific elements so rare (Lithium, Beryllium) compared to others in the ISM? Why do the quantities of H and He dominate over all other elements? A full accounting of the formation and evolution of the elements turns out to be a long and complicated story. Knowing the distribution of elements in the ISM is just the first chapter. To piece the story together fully, we will have to visit numerous destinations, including some that take us to the dawn of the Universe.

The mixture of elements we have collected tells us something about the relative numbers of elements within the ISM. But we also need to know *how much* we can expect to collect, particularly if we are counting on using this stuff to decelerate our spacecraft at the end of the journey. We need to know: Will we collect a few pounds of matter? Or many tons? In practice, we can figure this out from measuring how much stuff we've collected in our tank, how big the opening of our tank is, and how long we've been traveling.[12] The result of this analysis tells us the *Density* of gas in the ISM, which we can measure as

---

11 You might imagine that we would 'burn' the H chemically, producing energy—and water—in the process. But this is unlikely what our spacecraft does. For one thing there is very little O in the ISM as we have seen, so if we used this chemical reaction to generate energy, we'd use very few of the vast number of H atoms we collect. Second, this reaction does produce energy, but not as much as we require (there is a problem at the end of this destination that lets you explore this). No, it is much more likely that we are collecting the H to carry out 'nuclear fusion' reactions similar to those that power stars. We will explore how this works in detail starting at Destination 8. Fusion processes produce *far* more energy than any chemical reactions. This source of energy could supply us with the power we require to slow down at our destination given the amount of H we can realistically collect en route.

12 You can find out how this is done in one of the problems at the end of this destination.

a mass per volume, or as the number of atoms per volume. Either way, the result is startling. At sea level on Earth, air contains about 5×10^{19} particles per cubic centimeter (a unit abbreviated as cm^{-3}), a volume about equal to the size of a sugar cube. Think about this. That's 50 billion billion particles! Yet out here in interstellar space, we find that there is, on average, approximately *one* particle—generally a H atom—per cubic centimeter! One!! It is not only lonely for us out here. Even the atoms are few and far between.

One way to appreciate just how rarified the conditions are here at Anti-Cen is to calculate how often atoms collide in interstellar space. To get a sense of how this is done, consider trying to move through a crowd of people—say, at a train station or in an airport terminal—to get to the other end of the room or hall. It would clearly be easiest—not to mention fastest—to pass through the room if it were empty. But we've already said that other people are present, so the next best thing would be to hope that everyone is moving slowly. Even better, standing still. That would allow us to find a direct or near-direct path to the other side that involves as little deviation from a straight line as possible. But what if the people in the room are unwilling to slow down? After all, they have places to go, too. We could still get across the room relatively easily if the people in the crowd were small. Kids. Or very thin people. That would also cut down on the chance of a collision, or of near-collisions that would require us to slow down or alter our path.

The same basic points apply to the particles in a gas. As atoms move around in the gas, they collide most often when (a) there are lots of atoms, (b) the atoms are moving rapidly, and (c) the atoms are comparatively large. Since any given atom has no control over its size and the mixture of atoms in the ISM is fairly uniform, the only two practical variables here are their *number density* of particles—that is, the number of particles in a given volume—and their speed. On Earth, we can use these principles to calculate that a typical Oxygen molecule[13] will collide with another particle after about 10^{-10} seconds. That is, a given molecule will collide with other molecules in the air about 10 *billion* times every second. But here in interstellar space, a H atom typically must wait about 1500 *yr* before colliding with another atom! That's a factor of over 10^{19} longer than the time a typical molecule in air has to wait before colliding.

In order to figure out these collision rates in air the ISM, we needed to know the density of atoms and have some idea of how fast the atoms are moving. Recall the analogy of the crowded terminal. Collisions are more likely when people in a crowded room are packed together densely and/or moving around

13 In the part of Earth's atmosphere where humans can live, Oxygen is typically found as a molecule composed of two Oxygen atoms (and denoted O_2). The same is true of Nitrogen (N_2)—the most common gas in Earth's atmosphere—and some of the other gases in our atmosphere. There are molecules in space, as well, and we will discuss these as we encounter them in future destinations. But out here at Anti-Cen, most of the gas is H, and most of the H is in atomic form. that is, single atoms.

rapidly. How do we apply this idea to a gas? We are not accustomed to measuring the speed of particles in a gas. But we are quite familiar with measuring something that is directly related to that speed: *Temperature*. This works because temperature turns out to be a measure of the *Kinetic Energy*—the energy of motion—of particles in a body or a gas. We can therefore use temperature to tell us how fast the particles are moving.

To see how we do this, we can write the formal expression for the average kinetic energy, *KE*, of particles of mass *m*, moving at an average velocity *V*, and contained in a gas of temperature *T* as

$$KE = \frac{1}{2}mV^2 = \frac{3}{2}kT.$$

The constant *k* is the *Boltzmann Constant*, which has value of $k = 1.38 \times 10^{-16}$ *erg/K*. Notice that for these units we are measuring the energy of the particles in units of *erg*, and the gas temperature in units of *degrees Kelvin* (typically abbreviated simply as *'K')*. A change of 1 *K* is equivalent to a change of 1 *C*—that is, one *degree Centigrade*, or 1.8 *degrees Fahrenheit, F*. What this equation tells us is that the kinetic energy of the average particle motion in matter is what we perceive as temperature.[14] When we feel that something is hot, what we are actually responding to is the average comparatively large speed of the atoms or molecules in the object. Consequently, the particles have high kinetic energy. At the other extreme, cold objects consist of relatively slow-moving particles. This suggests that if we cool matter down to the point that the motions of the particles have essentially ceased, the temperature can go no lower. This is indeed the case, and the temperature at which this occurs is known as *Absolute Zero*. It is denoted as 0 *K*, and is equivalent to –274.2 *C*, or about –458 *F*.

On Earth, when the temperature is a comfortable 291 *K* (about 17 *C*, or 63 *F*), the typical speed of individual molecules—mostly in the form of molecular Nitrogen, N_2—in the air is around 300 *m/s*, or about 750 *mph*.[15] Out here in interstellar space, we can use a number of techniques to measure that the gas temperature is about 200 *K*, or –74 *C* (brrr). Since H is about 30 times less massive than N_2 molecules, the typical speed of the H atoms in the ISM is actually higher, about 2000 *m/s*, or nearly 5000 *mph*. Though this exceeds the speed of the

14 You may fairly wonder where this equation came from. Back in the 19th century, physicists discovered that they could describe the properties of a gas extraordinarily accurately by assuming that it is composed of particles—atoms or molecules—that move around with a range of kinetic energies but a well-defined *average* energy. When they did this, they discovered that this average energy was directly related to the temperature of the gas as described by this result. Hence, temperature is a measure of the typical energy of the particles in matter.

15 This speed may be familiar to you as the *Speed of Sound* in air. What it means is that sound—which is a compression wave that propagates through air—travels at the speed typical particles are moving at a given temperature and pressure.

heavier, hotter particles in the air at sea level on Earth, the kinetic energy—which, remember, depends on the speed *and* mass of the particles—is on average lower for the atoms in the ISM, including the comparatively speedy H atoms. We shall visit destinations containing other components of the ISM—technically known as *Phases* of the ISM—later on. Some of these phases are composed of colder gases than the gas we find here. Some hotter. And some much, much hotter. Because of this, the ISM here at 'Anti-Cen'—with a density of about 1 cm^{-3} and a comparatively balmy temperature of around 200 *K*—is referred to as the *Warm ISM.*[16]

5.3 SEEING THE INVISIBLE: NEUTRAL HYDROGEN AND 21-cm RADIATION

The fact that we built a ship designed to collect ISM gas suggests that we actually knew about this stuff *before* we set out. How? It seems to be both invisible and so rarified that it should not be detectable. On Earth one way to 'see' air is to note how it dims the light from distant objects. Some days, such objects—for example far-off mountains or buildings in a city—appear distinct, seemingly close at hand. On murkier days, these same objects may appear much less distinct, perhaps invisible. The gas of the Warm ISM does *not* reveal itself easily in this way[17], primarily because there is just too little of it. If we were to collect all the gas of the Warm ISM inside a (*very* long) cylinder 4.2 *ly* long but only about 1 cm^2 in area—about the area of a penny—we'd have only 10^{18} atoms. That sounds like a lot—and it certainly is a big number—but this is equivalent to only about 10% of the total number of molecules in one cubic centimeter of air at sea level. Light can generally travel very easily through a lot more air than this, so it can usually travel quite efficiently over very long distances—hundreds,

16 In reality, the Warm ISM spans a very large range in properties. Some of the gas of this phase of the ISM can reach densities of 20–50 particles cm^{-3} and have temperatures as low as 80–100 *K*. Other parts of the Warm ISM can get to 6000 *K* or hotter and may have densities as low as 0.1 particles cm^{-3} or lower. As we shall see shortly, the defining feature of this phase is that it is dominated by neutral atoms, so the Warm ISM is often referred to as the *Neutral ISM*.

17 We shall see later, starting at Destination 6, that some phases of the ISM are quite effective at blocking the light from background objects, much as dense clouds of water vapor can render the normally transparent gas of our atmosphere completely opaque.

even thousands of *pc*—through the Warm ISM without much problem.[18] No, what reveals the presence of the gas of the Warm ISM results from the detailed structure of the principal constituent of this gas: Hydrogen.

As we've already seen, a neutral H atom consists of two particles, an electron and a proton. If we imagine that these particles are tiny balls, it is not hard to further imagine that they are spinning about their axes like planets do. But the subatomic world is very weird, and it turns out that particles cannot spin any

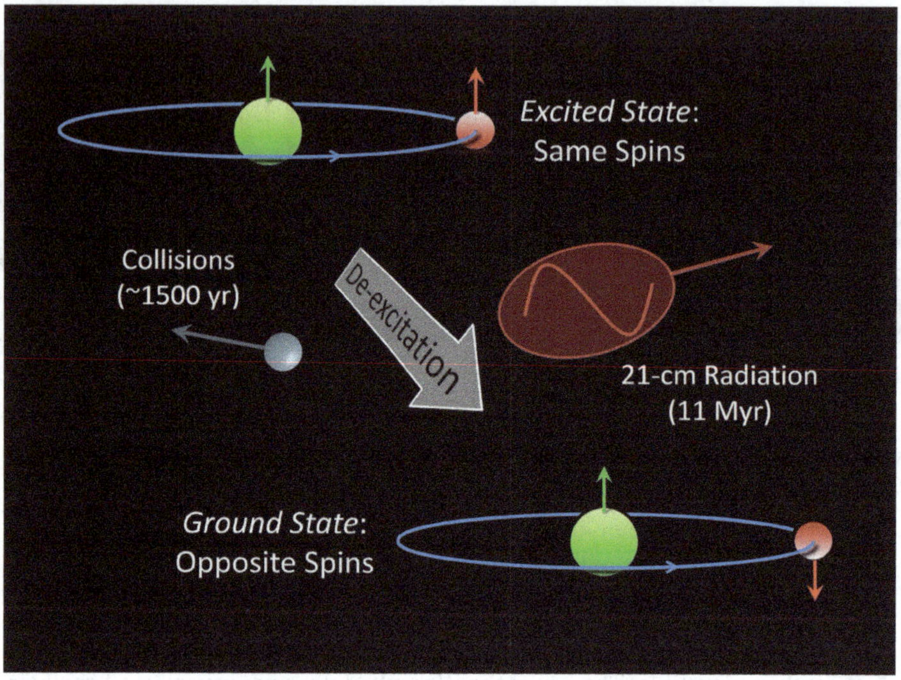

Fig 5.5

A schematic of the excited (*upper left*) and ground (*lower right*) states of the neutral H atom. In the excited state, the proton (*green*) and electron (*red*) in the atom spin in the same direction (here the spin is denoted by an arrow; the rotation is counter-clockwise when you look at the particle from its lighter-shaded side from where the spin arrow points). In the ground state, the two particles spin in opposite directions. Note that to get from the excited state to the ground state, the atom must lose energy through a process of 'de-excitation.' In the example shown, the atom can do this by colliding with another particle (*center left*) or by emitting energy in the form of a 21-*cm* radio photon (*center right*). Since the typical amount of time an atom has to wait to radiate its energy spontaneously is about 11 million *yr*, it is clear that a typical H atom has to be left alone a *long* time for it to emit 21-*cm* Radiation. By comparison, neutral H atoms collide with other particles (usually other H atoms) about every 1500 *yrs* in the Warm ISM.

random way that they might want. Instead, if the electron is part of a H atom, it can spin only with its rotation axis aligned to that of the proton. Despite this constraint, the electron still has two choices for the *direction* of its spin. It can either spin in the same sense as the proton or it can spin in the opposite direction. It turns out that when the two particles spin in the same direction—what is known as *Parallel Spin*—the atom has very slightly more energy than when the spins are in opposite directions, in what is known as *Anti-parallel Spin*.[19] This means that the parallel spin case represents what is known as an *Excited State* of the

18 This is not true of for all radiation. It turns out—for reasons we will begin to explore in detail at Destination 6—that some elements preferentially absorb certain colors of light. The pattern of this absorption reveals the presence of even minuscule quantities of these elements in the ISM.

19 This is due to the interactions of the very weak magnetic fields of the two particles. If you try to place two bar magnets side by side, one configuration—when the same poles are opposed—is obviously much easier (lower energy) to achieve; indeed, the magnets are attracted strongly to one another in this case. The other configuration—when the same poles are of the two magnets are brought near to one another—requires that you push hard to bring the magnets close together, a higher-energy state. If you let go of the

neutral H atom (Figure 5.5). As it turns out, the anti-parallel spin case represents the least energy possible for the electron in a neutral H atom. Consequently this configuration is known as the *Ground State* of the atom (Figure 5.5).

Now there are two important things to appreciate about this. First, physical systems generally strive toward the lowest possible energy state.[20] A pitcher of water falling from your hand onto the floor will dramatically remind you of this tendency. As the pitcher falls, it loses energy—in this case gravitational energy—compared to how much it had at the start of its fall. And it has even less energy when it is broken and scattered all over the floor. Given the chance—by letting go of it—the pitcher sought a lower energy state. The same is true of a H atom. If it is excited to a higher energy state, the atom generally responds by dropping to a lower energy state. The analogy of the pitcher illustrates why it is apt to call this lowest-energy state the ground state of the atom. The second key point to appreciate is that the rate at which the atom tries to lose energy to reach the ground state is generally faster if the energy difference between the states is large. An unsupported pitcher falling from your hand responds immediately. One sitting upright on the floor may take quite some time to tip over or may even require some provocation—a nudge—to reach a lower energy state. In the H atom (Figure 5.5), the energy difference between the parallel spin (high energy) and anti-parallel spin (low energy or ground) states is very small so the typical timescale for it to lose energy and reach the ground state is very long. How long? If we were to excite a H atom to the parallel spin state and then leave the atom alone, it would take *11 million years*—on average—for it to slip to the ground state where the spins are anti-parallel (Figure 5.5).

What happens when the atom switches states? One of the most fundamental laws of physics is that energy must be conserved in isolated systems. Though we will stretch this concept to its limit as we encounter extreme conditions at other destinations, we will find that this principle of *Conservation of Energy* always manages to be valid. For exploding stars. For planetary systems. And for individual atoms. Consequently, if a H atom does switch energy levels, the loss of energy within the atom must be accounted for by a gain of energy elsewhere. In the case of the pitcher we described earlier, the Conservation of Energy was evident in the speed at which the pitcher was falling—that's kinetic energy—as it hit the ground. This is a form of energy conservation you can really appreciate if the pitcher falls on your toe! The H atom has different options by which it can conserve energy as its electron switches between the parallel spin and anti-parallel spin states. One way it can do this provides the key to detecting

magnets in this case, they would move apart, or, if able, one would spin around to get the two magnets into the lower-energy configuration.

20 As fundamental as this feature of our Universe is, the underlying reason for this relentless drive of physical systems to move toward lower energy states is unknown.

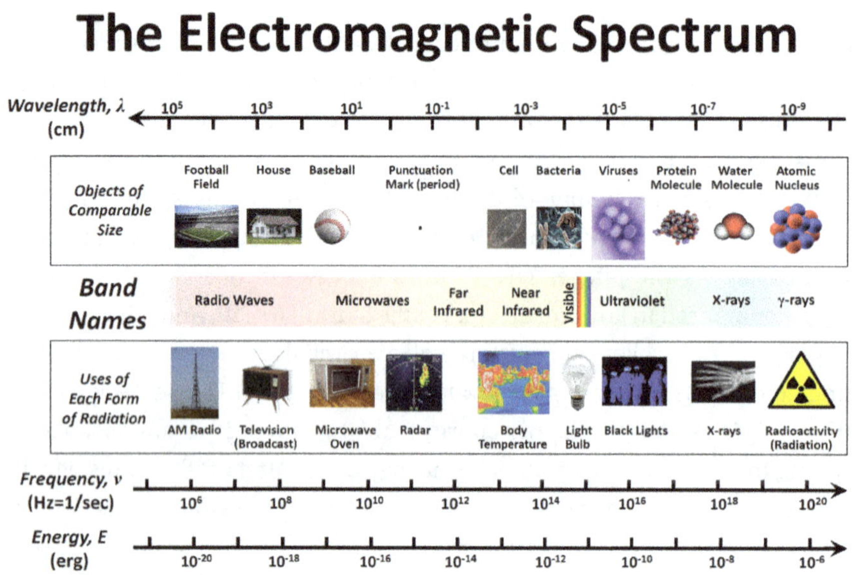

Fig 5.6

The Electromagnetic (EM) Spectrum. The names of the various bands of the spectrum are labeled along the middle with the corresponding wavelengths (in *cm*, *upper scale*), the corresponding frequencies (in *Hz*, *middle scale*), and the corresponding energies (in *erg*, *lower scale*). Notice that as the frequency and energy increase, the wavelength of the radiation decreases. If we know *any* of these quantities—frequency, wavelength, or photon energy for EM Radiation—we can determine the other two. The approximate boundaries between the different types of EM Radiation—Radio Waves, Microwave, Far Infrared, etc.—are entirely arbitrary and generally refer back to historical differences in how each type of radiation was originally detected. All forms of EM Radiation propagate through empty space at the speed of light, $c = 3.0 \times 10^{10}$ *cm/sec*. For reference, the sizes of things that are of comparable size to EM Radiation of different wavelengths are shown, as are some sources of different types of EM Radiation. Notice too that visible light—the rainbow region in the 'Band Names' part of the figure—represents a tiny fraction of the full range in wavelength/frequency/energy of EM Radiation.

the Warm ISM from Earth. And to understand this mechanism, we have to first discuss how an atom—and in particular the H atom—can emit energy.

In general, whenever electrically charged particles—such as electrons—are accelerated or change their energy state, *Electromagnetic Radiation* (which we will abbreviate as *EM Radiation*) is produced. Visible light is just one form of EM Radiation. X-rays, ultraviolet light, infrared radiation, and radio waves are also all examples of EM Radiation. Together, the full range of EM Radiation defines what is known as the *Electromagnetic Spectrum*, or *EM Spectrum* (see Figure 5.6 and Table 5.2).

All of the radiation of the EM Spectrum can be characterized by a few basic properties. Light, for example, can be shown to exhibit wavelike properties. The *Refraction*, or bending, of light in lenses, for example, is most easily understood by considering light to be a wave. For any wave there is a fundamental relationship between its *Wavelength*, *Frequency*, and *Speed* given by

$$\nu\lambda = c.$$

In this relation, ν (the Greek letter 'nu') denotes the frequency of the radiation, and λ (the Greek letter 'lambda') represents the wavelength. Frequency is defined as the rate at which the wave oscillates. A frequency of 1 *cycle per second*, or 1 *Hertz* (typically abbreviated as *Hz*) means that the wave goes through an entire cycle—from trough, to crest, and back to trough—once per second. The wavelength measures the extent of the wave. For simple waves this could be, for example, the distance from one peak to the next. The constant c in this equation represents the speed of the wave. When we apply this equation to EM Radiation, c represents the *Speed of Light*, equal to about 3.0×10^{10} *cm/*

TABLE 5.2 The Electromagnetic Spectrum

SPECTRAL REGION	WAVELENGTH (*cm*)		FREQUENCY (*GHz*)		ENERGY (*eV*)	
Radio	∞	10.0	0.0	3.0	0.0	1.2×10^{-5}
Microwave	10.0	0.01	3.0	3000	1.2×10^{-5}	1.2×10^{-2}
Infrared	0.01	7.5×10^{-5}	3000	400000	1.2×10^{-2}	1.7
Visible	7.5×10^{-5}	3.8×10^{-5}	400000	790000	1.7	3.3
Ultraviolet	3.8×10^{-5}	10^{-7}	790000	3.0×10^{8}	3.3	1200
X-Ray	10^{-7}	10^{-9}	3.0×10^{8}	3.0×10^{10}	1200	120000
Gamma Ray	10^{-9}	0.0	3.0×10^{10}	∞	120000	∞

Notes to Table 5.2:

The boundaries between spectral regions are arbitrary. One *giga-Hertz* (*GHz*) equals 1 billion *Hertz*. An electron volt (*eV*) is a unit typically used to describe energies on atomic scales. It is also a convenient unit to describe the energies of photons. One *eV* is a tiny fraction of an *erg*, which is itself a tiny unit of energy. Specifically, 1 $eV = 1.6 \times 10^{-12}$ *ergs*. Notice these are not all the same units as used in Figure 5.6.

sec. All EM Radiation—not just visible light—travels at the speed of light when propagating through a vacuum or near-vacuum like the one we find ourselves at Anti-Cen, but also in the atmosphere of the Earth.[21]

The key point about all this is that waves carry energy. This fact is well known to anyone who has tried to swim in heavy surf at a beach or just watched large waves crash against a shoreline. Since EM Radiation has clear wave-like properties, this implies that it must also be able to transmit energy. This is true, but the way in which EM Radiation transmits energy is fundamentally different from, say, water waves. At the beach, gentle waves of a given wavelength that are only one foot in height might produce a soothing rumble as you walk along the sand. On the other hand, waves with the same wavelength but that are 20 feet high would be, by comparison, highly destructive. This illustrates that the energy of a water surface wave depends mostly on its *Amplitude*, the difference in height from its trough to its peak. But EM Radiation does not behave this way. The energy of EM Radiation depends only on its wavelength or, equivalently, its frequency.

Odd as this is, EM Radiation exhibits other properties that are not characteristic of waves at all. X-rays, for example, can travel along a straight line directly from a source, through your body, and onto a specific spot on a detector that allows doctors to decide which bone you broke when you dropped that pitcher

21 This is not quite true. EM Radiation travels somewhat more slowly through matter—air, glass, plastic, wood (in the case of x-rays)—than it does in a vacuum. For our purposes, however, the differences can be ignored. For us, the speed of light is always about 3.0×10^{10} *cm/sec*.

on your toe. This sort of behavior (and others) is reminiscent of how particles move—for example, similar to billiard balls or bullets—and represents part of the evidence that suggests EM Radiation is composed of discrete particles. Particles known as *Photons.*

These seemingly contradictory characteristics of EM Radiation were perplexing to people as they first struggled to understand the nature of light. Is EM Radiation a wave? Is EM Radiation composed of particles? It took German physicist Max Planck—at about the turn of the 20th century—to realize that the answer to both questions is 'Yes'! Light—indeed all of the radiation of the EM Spectrum—has both wave-like *and* particle-like properties. Simultaneously. This *Wave-Particle Duality* reveals itself in a fundamental relation that Planck discovered that ties together the energy, frequency, and wavelength of photons of EM Radiation:

$$E = h\nu = \frac{hc}{\lambda}.$$

Here, E represents a photon's energy, and h is known as *Planck's Constant,* which has a value of $h = 6.63 \times 10^{-27}$ *erg* × *sec.* Notice how very odd this result is. The energy in EM Radiation seems to come only in discrete 'packages,' or *Quanta* (singular: *Quantum*). A photon is a 'particle' of EM Radiation that has no mass but that carries a well-defined quantum of energy. At the same time, however, it is characterized by a discrete wavelength and a frequency, properties we typically ascribe to waves. We shall discover that EM Radiation is not the only thing that behaves in this schizophrenic manner. Ordinary matter also exhibits particle-like and wave-like properties, but its duality is not so evident until one inspects matter at the atomic and sub-atomic scales.

The fact that EM Radiation can carry energy provides the critical clue we require to understand how we can detect the gas comprising the Warm ISM. When the electron flips from the excited state to the ground state in an isolated H atom, a photon must be created to conserve energy. Not just any photon, but one whose energy equals the energy difference between the two spin states of the electron in the atom. This turns out to be a very small fraction of an *erg,* itself a tiny unit of energy.[22] Since photons are also characterized by a frequency, and because frequency and wavelength are directly related for EM Radiation, by knowing the energy difference of the energy states of the atom, we can calculate the wavelength of the associated radiation. This turns out to be about 21 *cm.* That's about the *width* of a typical sheet of paper. For comparison, the wavelength of visible light is about 5.0×10^{-5} *cm,* nearly 1 million times shorter than this *21-cm Radiation,* and about 1/200 the *thickness* of that same piece of paper. The EM Radiation that the H atom emits in this case is not visible light, but rather radio radiation. It's this radiation we can detect from neutral H. It's

22 A flea expends about 1 *erg* of energy per jump.

this radiation that reveals the presence of the Warm ISM. And it's because of this radiation that we could be so certain *before* undertaking our journey that interstellar space was filled with matter that we could potentially use as fuel.

You may have noticed that there is something slightly fishy about all this. If the average time between collisions is around 1500 *yr*—as we determined earlier—but the typical time required for an H atom to flip its electron's spin is about 11 million *yr*, why do we see this radiation at all? Another way to extract energy from or inject energy into an atom is through collisions. Think of the collisions of billiard balls or between players during an ice hockey game: It is obvious collisions can transfer energy between 'particles.' Couldn't the energy of the spin state leak out of the atoms during these collisions and in the process cause the rebounding particle to speed up? Or, to state this in a slightly different way, to gain kinetic energy? Not only is this possible, but it is also likely. By far the most common way for neutral H atoms in the Warm ISM to lose (and gain) energy between their excited and ground states is through collisions with (usually) other H atoms. But even given the relatively high frequency of collisions, it is still possible that some H atoms in a given region of interstellar space might, by chance, emit 21-*cm* Radiation *before* they collide with other atoms. This is rare. Very rare. But H atoms in the Warm ISM are abundant. Very abundant. The result is that the Warm ISM as a whole emits a vast amount of energy in the form of 21-*cm* Radiation, all arising from intrinsically rare emission which becomes collectively visible due to the immense number of neutral H atoms present in the gas.[23]

This 21-*cm* Radiation was first detected by *Radio Telescopes* in the 1950s after predictions—carried out during World War II in the Netherlands—that such radio emission should be detectable. This radiation has proven to be one of the most important tools for understanding the structure of our Galaxy, as well as other galaxies beyond our own; we will be using it extensively on some of our future journeys (Destinations 9 and 14). Because 21-*cm* Radiation can be produced only by Hydrogen atoms that contain one electron—that is, by neutral hydrogen atoms—it is also sometimes referred to as *Neutral Hydrogen Radiation*, or just *H I* (pronounced 'H-one') *Radiation.*[24]

---

23 A question at the end of this destination allows you to explore this conclusion in a bit of detail.

24 The designation for ions is strange, even by astronomers' standards. All neutral atoms are denoted by the name of the element followed by the Roman numeral I. Thus you have H I, He I, O I, Fe I, for neutral Hydrogen, Helium, Oxygen, and Iron, respectively. If an atom loses one electron, it gets Roman numeral II: H II, He II, O II, Fe II, etc. Lose a second electron? Use the Roman numeral III. And so forth (see Figure 5.7). This is hardly an ideal form of notation. For example, it cannot handle the case of an atom gaining an *extra* electron, such as H^- (pronounced 'H-minus'). Even so, the use of Roman numerals to designate the ionization level of atoms remains quite common practice.

5.4 HOT MATTER IN COLD SPACE

There is something else here in this lonely space between the stars. Not all the particles have energies consistent with a temperature of only a few hundred or even a few thousand degrees *K*. Some particles appear to move *much* faster. So fast, in fact, that their energies correspond to temperatures of 100000 *K* to 1 million *K* or higher! But these particles are extremely rare, even by comparison with the low density of the gas in the Warm ISM. This hot gas has a typical density of around one particle every 1000–10000 cm^3, corresponding to a density of 0.001 to 0.0001 cm^{-3} or lower. The composition of this hot gas is similar to that of the Warm ISM; in particular, H is still the most abundant element. But unlike the Warm ISM, these fast-moving atoms are nearly all ionized, having lost one or more electrons from their normal complement. Consequently, virtually all these atoms are charged ions created from neutral atoms that have lost some or all of their electrons. Some of these ions are extreme. For example, we can detect the presence of Oxygen atoms that have been ionized five times (!) in this very hot gas. Such ions contain only three of their normal complement of eight electrons and are denoted as O VI (this is pronounced 'Oxygen-six' or 'Oh-six'; see footnote 24 and Figure 5.4 to review the weird nomenclature astronomers use to describe ions). Although the vast majority of the H atoms here are neutral atoms associated with the Warm ISM, the few H atoms associated with the very

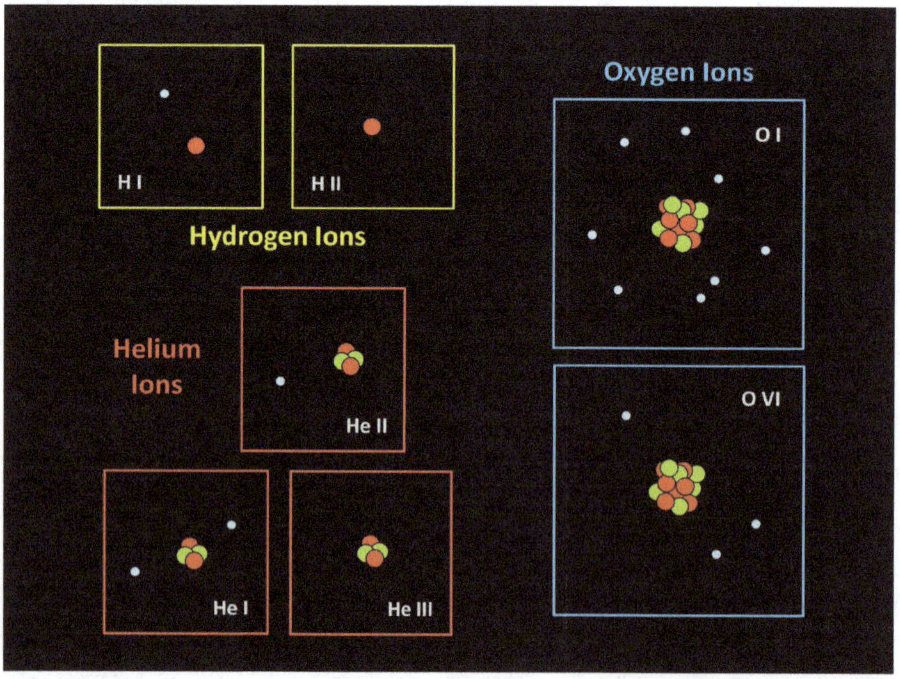

Fig 5.7

Examples of neutral atoms and charged ions. The yellow box on the left illustrates a neutral H atom—denoted H I—consisting of a nucleus with a single proton (large yellow dot) with a single electron (small blue dot) in orbit around it. The most common H ion, H II, is simply a bare proton as shown in the yellow box on the right. The red boxes illustrate He ions. In each case, the nucleus is taken to consist of two protons (large yellow dots) and two neutrons (large green dots). He I, or neutral Helium, has two electrons, while He II has only one electron, and He III is fully ionized and has no electrons. The blue boxes illustrate an example of a neutral Oxygen atom—denoted O I—which consists of a nucleus of 8 protons and 8 neutrons with 8 electrons in orbit around it. If five electrons are stripped from this atom, the result, shown in the lower blue box, has only three electrons in orbit and is denoted O VI. Keep in mind that these images are grossly out of scale. The electrons would be over a mile away from the nucleus on the scale of the drawings for the neutrons and protons.

hot gas have lost their single electrons. These are *H II* (pronounced 'H-two') ions, which cannot emit 21-*cm* Radiation.[25] Energy is required to remove electrons from atoms and create ions. Sometimes, quite a lot as, for example, in the case of atoms of heavy elements from which multiple electrons must be stripped. The existence of this hot but rarified gas—known as the *Very Hot ISM* or the *Coronal ISM*—suggests that something very violent has occurred to produce particles with such high energies, and therefore very high temperatures. We will track down some of the possible heating sources for this gas at future destinations.

For now we are resigned that interstellar *travel* means spending most of one's time in interstellar *space*. But we've used our time here well to discover that space is not empty after all. There *is* matter out here between the stars. Indeed, the ISM appears to be everywhere. Because of this and since the volume of space between stars is so large, the total amount of matter in the ISM turns out to be considerable; we'll obtain a detailed census of this material in our Galaxy and in other galaxies at Destinations 9 and 14. We've also learned that the ISM is composed of at least two phases—the Warm and Coronal/Very Hot gas—each characterized by distinct temperatures and densities. But this is just the start of that story. We will soon discover that there are other phases of the ISM, each with its own characteristic set of properties. And all co-existing in a more-or-less stable configuration throughout interstellar space. We've also seen that the abundances of the elements in the ISM are nothing like what we are used to on Earth. We may as well start to get used to this. Very little of the Universe that we will explore from this point on will remind us of home.

25 It should be clear to you why this is the case, but if it isn't, a question at the end of this destination explores why 21-*cm* Radiation can be produced only by neutral H atoms.

DESTINATION 5 QUESTIONS

5.1. We saw that He makes up 8.5% of the total number of atoms in a typical sample of interstellar gas. What fraction of the *mass* of the ISM is made up by He? Keep in mind that each He atom weighs on average four times as much as each H atom. You can ignore the contributions of all other elements apart from H when you do this problem.

5.2. What is the frequency of 21-*cm* Radiation? What is the energy of a 21-*cm* photon?

5.3. Explain why one cannot produce a H III ion.

5.4. Let's explore the time between collisions of atoms in the Warm ISM. For this problem, assume that the number density of H atoms is 1 cm^{-3} and the temperature of the gas is 200*K*.

a) The 'radius' of a neutral H atom is approximately 5.3×10^{-9} *cm*. What is the 'cross-section,' *A*, of such an atom? This can be calculated as the area of a circle with the radius of the atom.

b) What is the typical speed of a H atom in the gas?

c) What is the typical volume that each atom has to itself?

d) How long will it take a typical atom to trace out a cylinder of the size given in part (c)? **HINT:** Note that the volume of this cylinder is $Vt \times A$. Set this equal to the volume from part (c) and solve for the time, *t*.

e) Explain why the answer in part (d) is a reasonable approximation for the time between collisions in the gas.

5.5. Explain why 21-*cm* radiation is also sometimes called H I radiation.

5.6. Calculate the frequency, *v*, and the energy, *E*, for photons corresponding to the following types of EM Radiation:

a) 21-*cm* radiation.

b) Green light with a wavelength of $\lambda = 500$ *nm*.

c) X-ray radiation with a wavelength of $\lambda = 0.1$ *nm*.

5.7. Assume a gas has a temperature of 1000 *K*. What are the speeds of the following particles?

a) A neutral Hydrogen atom.

b) An Argon atom.

c) An atom of Iron.

5.8. List the ten most common elements—by number—in the Earth's crust, the ISM, and the human body (see Table 5.1). Does the human body most resemble the Earth's crust or the ISM?

5.9. Give the astronomical-style designations for the following ions (see footnote 24). Beware! There is one trick case.

a) H that has lose one electron.

b) Neutral Neon.

c) Iron that has lost eleven of its electrons.

d) Twice-ionized Helium.

e) Oxygen that has lost nine of its electrons.

5.10. What is the speed of a Hydrogen atom in the Coronal gas, assuming the gas has a temperature of 1 million *K*? Compare this to the escape velocity from the Sun, which is given as $V_{esc} = \sqrt{2GM_{\odot}/R_{\odot}}$, where *G* is a constant known as the *Gravitational Constant.* The mass and radius of the Sun are provided in Table 4.1. Based on this comparison, where do you think this hot gas goes?

5.11. Imagine a pitcher, a baseball, and a hitter. Describe how the energy of this system is distributed as the pitcher throws the ball and the hitter then knocks it out of the ballpark. In particular, explain how the total energy remains conserved throughout the process.

5.12. Let's say our interstellar spacecraft has an opening in front that is 1 *km* in diameter (that's pretty big!). If the mean density of matter in the ISM is 1 cm^{-3}, and, assuming that matter is all in the form of Hydrogen, how much matter would we collect after traveling one *parsec*? **HINT:** What is the volume of a cylinder that is 1 *km* in diameter and 1 *pc* long?

5.13. Let's explore how 21-*cm* Radiation can be abundant even when the fraction of neutral H atoms that can emit this radiation is minuscule in a typical region of the Warm ISM. For our purposes, let's (incorrectly) assume the probability of a collision or spontaneous emission for a neutral H atom in the ISM is uniform. That is, the ratio of the number of atoms that lose energy through collisions and through radiation will just be the ratio of the relevant timescales for each process.

a) For the Warm ISM, what fraction of H atoms in a typical region of interstellar space would lose energy via radiation? Use Figure 5.5 as a guide to the timescales for radiative and collisional energy losses in neutral H atoms of the Warm ISM.

b) A radio telescope searching for 21-*cm* Radiation can scan about 1 *square degree* on the sky. What is the volume of a cone of this angle that extends 100 *pc* from Earth? **HINT:** Use the small-angle formula to calculate the size of the base of this cone.

c) Assuming that the mean density of neutral H atoms is 1 cm^{-3}, how many H atoms are in this cone?

d) Given the answers in parts (a) and (c), explain why it is plausible that something as intrinsically rare as 21-*cm* Radiation emission can be detected from the Warm ISM.

Destination 6

THE ORION NEBULA

6.1 SECRETS OF A CELESTIAL HUNTER

Instantly recognizable, the winter[1] constellation Orion is one of the most familiar sights in the night sky (Figure 6.1). On cold, clear nights, its two brightest stars—steely-blue Rigel and red Betelgeuse—help to outline the celestial hunter's unmistakable rectangular form. At the center of the constellation, almost precisely between Rigel and Betelgeuse, are the three stars that make up the famous 'belt' of Orion. Since it is located on the celestial equator (Figure 3.3), these iconic features of Orion are fittingly visible from virtually all parts of the Earth.

No hunter stalks his prey without weapons, and Orion is no exception. If we look carefully, we can spot his sword hanging off his belt to the south (Figure 6.1), represented by three faint stars. On close inspection, a keen-eyed observer can see that that middle 'star' is different. Rather than appearing as a sharp point of light, it shines as a small smudge. The earliest telescopic observers of the sky discovered here what appears to be an extended, glowing patch of

Fig 6.1

The constellation Orion as seen from Earth. (*Left*) A chart of the region shows the celestial hunter's basic outline in the sky. The three stars of his belt are particularly obvious. Hanging 'below' the belt is his sword. (*Right*) A photograph of Orion. Red Betelgeuse is apparent in the upper left, while bright blue Rigel shines at the opposite corner of the rectangular pattern of Orion's body. The three stars of Orion's belt are apparent, as are the 'stars' of his sword. The middle object of the sword is no typical star, however. This is the Orion Nebula—known also as M 42—and our current destination. The faint red patches arise from hot gas in and around the Orion Nebula. The streak is a meteor caught by chance in this photo.

Photo by J. Hackmann; Image generated by M. Mateo using *Starry Night* software.

1 There's some hemispheric chauvinism again. Of course, for people who live in the Southern Hemisphere, Orion is a *summer* constellation.

Fig 6.2

The Orion Nebula. A broad view (*left*) shows the environs of the nebula, including the extent of its emission and the stars associated with it. This particular image is in the near-infrared portion of the EM spectrum. Our destination is near the brightest portions of the central (reddish-green) part of the nebula, shown in more detail in the image on the *right*.

Photo (*left*) by J. Emerson, VISTA/Cambridge Astronomical Survey Unit/ESO. Photo (*right*) from STScI/NASA.

light (see Figure 6.2). Using the Latin word for 'cloud,' people began to refer to all obviously non-stellar objects as *nebulae*, the plural of *nebula*. Some observers found nebulae to be annoying since they mimicked the appearance of more 'interesting' objects, particularly comets. An astronomer could become instantly famous for finding comets.[2] But nebulae seemed designed just to fool even veteran comet chasers. Charles Messier, a particularly avid comet hunter of the 18th century, got so exasperated with these imposters that he catalogued all the nebulae he could see from Paris[3], over 100 in all. His aim was to ensure that he and others would not be fooled by these 'false comets.' The nebula in the middle of Orion's sword ended up as number 42 in his catalog.[4] Today most sky watchers still refer to that fuzzy central 'star' of Orion's sword as *Messier 42*, or, more typically, just *M 42*. But for many people, this spectacular region is known more grandly as the *Orion Nebula*. It is our next destination.

---

2 To this day, comets are almost always named after their discoverer. The most famous exception is Halley's Comet, named after the person who first used Newton's Universal Law of Gravity (see Destinations 2 and 9) to predict that comet's return. The original 'discoverer' of that bright comet would have been some ancient—now unknown—sky watcher.

3 Not so crazy as it sounds. This was the 18th century, so Paris was not yet the bright 'City of Lights' that it is today. Virtually all of Messier's nebulae would be lost in the glare of the abundant artificial lighting of modern Paris, particularly for the telescopes Messier had access to in his heyday.

4 It is unclear whether Messier would have found it ironic or not to discover that his long-term fame comes not from the thirteen comets he discovered, but from his catalog of non-stellar objects. The Messier Catalog remains famous and contains some of the best examples of certain types of astronomical visible to small telescopes in Earth's sky. A number of our destinations—some quite spectacular—are objects listed in Messier's catalog.

Having reached the nebula—and after a journey of about 500 *pc* from home—it is hard to remember what the 'normal' nighttime sky of Earth looked like. We've chosen to stop near the center of the nebula where, by chance, we find a small, airless chunk of rock to use as a base for our observations of the nebula. This particular planetoid does not seem to be in orbit around any single star, but is simply floating independently in interstellar space.[5] From this convenient base we look up and behold a new Alien Sky unlike any we have encountered so far.

So many stars! They seem to fill our new sky completely. But on closer inspection, this impression turns out to be a bit mistaken. There indeed are many stars surrounding us, but we soon appreciate that one side of the sky in particular is dominated by a number of unusually bright stars[6], each about 100 times brighter on average than the brightest nighttime stars we can see from Earth. It is also obvious that most of these bright stars shine with an intense steely-blue color reminiscent of a welder's arc. It is a glorious sight: A constellation of 'frozen' strobe lights spread over half our sky. But there is more than starlight here. Between these spectacular stars, we can plainly see that the *sky itself* shines with a pale reddish-green/gray hue. It is as if the blackness of space is straining to match the color of the Earth's daytime sky. But remember, there is no air on the rock we've taken as our base. This light is not produced by atmospheric scattering as Earth's blue sky is. The glow we see in this sky is from the Orion Nebula itself. It surrounds us from our vantage point, though is becomes soon evident that the light from the sky is most prominent on the same side of the sky where we see the brightest stars. The weirdest features of this sky, however, are the holes. Peppered around us are a few dozen small, distinctive regions in the sky where some celestial carpenter seems to have drilled holes into the nebular light, allowing us to see a pitch black sky that lies beyond. We've seen nothing like these celestial ink blots before (Figure 6.3).

Each of these features so clearly visible in this unusual sky—the ghostly nebular light, the dark 'holes,' the gloriously bright stars—represent distinct, but interrelated components of this amazing place. We shall focus on each component individually, but the elegant interconnections between them will

---

5 Based on observations from Earth before we left, we do not actually know with certainty that there are planetoids like this within the volume of space containing the Orion Nebula. But, as we shall see, stars are actively forming in the Orion Nebula. And, almost certainly, intimately associated with these new stars are newly-formed planets. As the stars move around and interact gravitationally, some of the planets are bound to be stripped from their original parent star, either to float freely between the stars in interstellar space or perhaps to be swapped from one star to another. Though we know of no definitive example of free-floating planetoids anywhere, the outermost bodies of our Solar System—collectively known as the Kuiper Belt—are so weakly gravitationally bound to the Sun that they represent good examples of what free-floating planetoids might look like and where they might come from.

6 Recall from Destination 3 that brightness is what we perceive. It depends on how far we are from a light source—the Inverse-Square Law for Brightness (Destination 4)—and how much light the source truly emits—that is, the source's luminosity (Destination 7).

Fig 6.3

The view of the sky from within the Orion Nebula would be dominated by the pale, reddish-green background of the nebula itself, and the light of many bright stars. Though too small to see in the wider-field image, some of the dark 'holes' visible in the sky of the nebula are shown in detailed images. If you look carefully, you will see that the clouds are not perfectly dark; some faint, usually reddish objects are visible near their centers.

Wide-field photo by J. Christensen; smaller photos from STScI/NASA.

become plain as we learn more about what we are seeing. As we begin to comprehend the fundamental nature of the Orion Nebula.

6.2 THE BRIGHT SKY OF THE ORION NEBULA

In many ways, the key to understanding the Orion Nebula—and certainly the primary reason for its cloud-like appearance—resides in the pale, reddish-green glow we see around us in this sky. We've already noted that this has nothing to do with the small, airless planetary body on which we are based. This glow is a fundamental feature of the nebula itself. What is the nature of this background light and how does it relate to the other things we see in this unworldly sky? To start to answer this, we must analyze the light in more detail than possible using only our unaided eyes.

We saw earlier (Destination 5) that light—indeed all EM Radiation—can be characterized by wavelength, frequency, and energy. These properties are functionally equivalent because if you know one, you know all three

($c = \lambda\nu$ and $E = h\nu = hc/\lambda$; see Destination 5). But we can also measure the *amount* of light we receive at different wavelengths, frequencies, or energies. This sort of fundamental analysis is called *Spectroscopy* and is one of the most powerful tools available to us to probe the nature of astronomical objects. Any time we use spectroscopic techniques, we produce a *Spectrum*, which is a measure of how much EM Radiation is present over a range of wavelength, or frequency, or energy (or any combination of the three). Figure 6.4 presents some examples of spectra, emphasizing some of the ways that spectra can be observed and displayed. It is worth careful study.

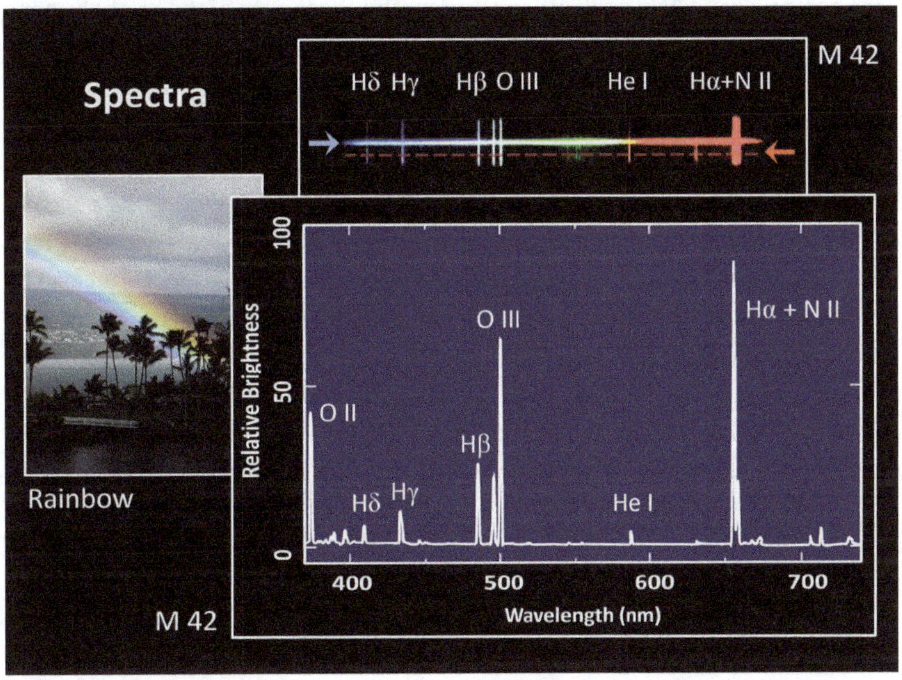

Fig 6.4

This figure shows some examples of *Spectra*, and in particular *Optical Spectra*, corresponding to the visible part of the EM Spectrum (see Figure 5.6 of Destination 5).

At the *top* is a *Slit Spectrum* of a star in M 42 and the surrounding nebula. A horizontal cut at the blue arrow (to the left) is an example of a *Continuous Spectrum* of a star, basically the same sort of spectrum we see when a rainbow is formed by raindrops splitting the light of the Sun into its constituent colors (*left*). Notice how all colors are present at all wavelengths in the continuous spectrum and in the rainbow. A horizontal cut through the top spectrum along the dashed red line (indicated by the red arrow on the right) reveals the *Emission Spectrum* of the Orion Nebula itself. Notice that light is present only at a few narrow wavelength intervals. The *lower* figure—which is lined up with the top image—represents another method of displaying the same emission spectrum, but this time as a plot. Notice how the brightness of the spectrum—in this case an emission spectrum—is nearly zero at all wavelengths except for the few regions where the bright emission lines are seen. These images illustrate just a few of the ways in which spectra can be presented. We will see many other examples of spectra here at M 42 and at many future destinations.

Photo of slit spectrum from Okayama Observatory/NAOJ.

Without having said so at the time (Destination 5), we've already used spectroscopy to detect and measure 21-*cm* radiation from neutral H, an example of a spectrum in the radio region of the EM Spectrum. We will use spectroscopy extensively to study the compositions, motions, and natures of a wide range of phenomena at virtually all the destinations we'll be exploring. Naturally, we knew this before we started our voyages, and so one of the key tools we brought with us is a *Spectrograph*, or, as it is sometimes called, a *Spectrometer*. This fundamental astronomical instrument is designed to spread the light—more generally, the EM Radiation—from an astronomical source into its constituent wavelengths, frequencies, or energies. By doing so, we produce a spectrum that we can then study in detail using a range of techniques (Figure 6.4).[7] In the

7 Exactly how spectrographs do this is not centrally important to us now. But, broadly speaking, spectrographs employ some sort of 'disperser' to split light into its constituent frequencies/wavelengths/energies. The details of how this is done can

visual part of the EM Spectrum—the part our eyes can detect—different wavelengths/frequencies/energies of light are interpreted as different colors by our brain. In all other parts of the EM Spectrum—radio, infrared, ultraviolet, x-rays, and gamma-rays (see Figure 5.6 of Destination 5)—we must rely on electronic, thermal or chemical detectors to record spectra.

We decide to attach our spectrograph to one of the optical telescopes we brought and use it to observe the pale reddish-green light that fills the sky here at the Orion Nebula. What we discover from the resulting spectrum is that the light from the nebula is concentrated at only a few, specific wavelengths (see Figure 6.4) with essentially no light at all other wavelengths. Since each of the bright regions in the spectrum appears as vivid, distinct color to our eyes, we now can really appreciate that *different colors correspond to different wavelengths of radiation in the optical portion of the EM Spectrum*. The spectrum of the Orion Nebula is an example of an *Emission Spectrum*, and the specific wavelengths where light is visible are known as *Emission Lines* (Figure 6.4). Because this spectrum covers the visible portion of the EM Spectrum, the colorful lines are known as *Optical Emission Lines*. Similarly, 21-*cm* Radiation (see Destination 5) can also be detected as an emission line, specifically a *Radio Emission Line*. But the distinction between Radio, Optical, or any other type of emission line is artificial. Emission lines can occur in any portion of the EM Spectrum, from the radio to gamma-rays (see Figure 5.6), a crucial fact that we shall confirm—and exploit—in our later studies of the Orion Nebula and at many other future destinations.

It is worth stopping to appreciate the beauty of the emission spectrum of Orion (Figure 6.4). At the long-wavelength end of the spectrum—the red end of the spectrum—we see a very bright, intensely ruby-red emission line whose wavelength we determine to be about 656 *nm* (that's the abbreviation for *nanometer*, a unit of length equal to 1 billionth of a *meter*, or 10^{-9} *m*). Two very strong emission lines shine with a piercing emerald green color at a wavelength of about 500 *nm*. These three bright lines dominate the spectrum and are clearly the principal reason our eyes perceive the nebula to have a reddish-green hue. At about 486 *nm* we see a weaker, but gorgeous turquoise line. At shorter wavelengths, we can make out a series of increasingly bluer, then purple emission lines. The lines of this characteristic sequence steadily crowd closer together as they terminate at an intensely purple limit with a wavelength of about 360 *nm*.

Recall that 21-*cm* Radiation (Destination 5) arose from a change in the spin state of an electron in a neutral H atom, a process that produces a distinct (radio) emission line. The formation of optical emission lines like those in the spectrum of the Orion Nebula also involves electrons in atoms, but there are clear differences. For one, the optical emission lines we see in the nebula are

---

differ greatly from the radio to the optical to the x-ray portions of the EM Spectrum, but the result is essentially the same: A spectrum is produced from the radiation for an astronomical source.

composed of photons with much higher energies, shorter wavelengths, and higher frequencies than 21-*cm* Radiation—in each case, differing by nearly a factor of one million. In addition, while 21-*cm* Radiation involves just one emission line, the optical spectrum of the Orion Nebula clearly consists of many emission lines (see Figure 6.4). Finally, as we shall soon see in detail, the optical emission lines we detect in the nebula do not all arise from H atoms. How then, do the optical emission lines we see here in the Orion Nebula come about?

Let's start addressing this question by again considering the properties of the H atom, the simplest of all atoms. Neutral H consists of only two electrically charged particles: A negatively charged electron and a positively charged proton. To explain 21-*cm* Radiation (Destination 5), we had to consider only how the spins of the electron and proton are oriented with respect to one another. But to explain the optical emission lines of H, we need to consider how the two particles *orbit* one another. As it happens, a proton has a mass that is over 1800 times higher than that of an electron. For our purposes, it is therefore entirely reasonable to imagine a neutral H atom as a tiny electron orbiting a (nearly) stationary proton, the nucleus of the atom.[8] This image for the H atom is misleading, however, since it reinforces the false notion that the electron and proton are like a minuscule planet—the electron—orbiting a larger, but still tiny sun—the proton. To get a better sense of what is actually happening within H atoms, we need to take into account the bizarre features of matter on scales comparable to the diameters of individual atoms. That's about 10^{-8} *cm*, or less than about one tenth of a *nm*.

We saw earlier (Destination 5) that EM Radiation exhibits a schizophrenic wave-particle duality that allows it to simultaneously possess wave-like and particle-like characteristics. One of the most revolutionary results of 20th century physics was the realization that *matter* also behaves this way: Objects that we normally consider to be particles can also exhibit wave-like properties. We normally do not notice this in everyday life because, as we will see, the wavelengths of large particles—for example, baseballs or raindrops—are minuscule. The wavelengths of subatomic particles are tiny too, but the particles themselves are also very small. So small, they are comparable in 'size' to their wavelengths. As a result, the wave-like properties of subatomic particles can dominate how they interact with other particles and with radiation.

It turns out that we can calculate the wavelengths of particles with a very simple equation first proposed by French physicist Louis de Broglie:

$$\lambda = h/p.$$

8 This is reminiscent of how stars in binary systems orbit their common center of mass (see Destination 4). In the case of a H atom, the center of mass is essentially at the location of the proton.

In this equation, h is Planck's constant (which we first encountered at Destination 5) and p is the particle's momentum. For a particle with mass m traveling at a velocity V, the momentum is just $p = mV$. De Broglie's equation also works for EM Radiation, which is composed of photons, particles that transmit momentum but that have no mass.[9] It is clear from de Broglie's equation that a particle will appear more wave-like when its momentum is small. Tiny particles with tiny masses—like an electron—most readily exhibit wave-like behavior because their masses—and hence their momenta—are generally tiny.

The wave-particle duality for subatomic matter is important because this allows particles to act in strikingly 'unparticle'-like ways. For example, we normally think of particles as objects that can bounce off one another, in the process careening to and fro. But they never disappear. That's just what particles *do*! Waves, on the other hand, can travel right through one another. For example, when surface waves produced from the wake of a ship 'collide' with swells in the ocean, the wake passes right through the swells—and the swells pass through the wake. Such behavior is distinctly 'unparticle-like.'

When waves do interact, the results, though familiar, are still rather odd. In particular, waves can interact in such a way that they *Interfere Constructively* and in the process produce a wave that is larger than the original waves. Or they can interact so that they *Interfere Destructively*, momentarily canceling the original waves out completely! In constructive interference, the peaks of two (or more) waves coincide to produce, momentarily, a much larger wave. Perhaps you have been surprised by 'rogue waves' at a beach, very likely produced through the constructive interference of two smaller waves that just happened to reach you simultaneously. Destructive interference occurs when the peak of one wave coincides with the trough of another. At the beach, this might cause a short lull in the waves even on a day of choppy seas. If particles can have wave-like properties, does this mean they can also behave in these strange ways?

In the early 20th century, Dutch physicist Niels Bohr produced a very successful description for the H atom—dubbed the *Bohr Model*—that relies on electrons having wave-like properties to define where they can exist within a stable atom. To understand the basics of the Bohr Model, let's return to our

---

9 What this means is that although photons are massless, they do have momentum. As noted in the text, a normal particle of mass m and moving at some speed V has a momentum equal to mV. For a photon, this definition of momentum fails since its mass is zero; instead, the de Broglie equation must be used. Broadly speaking, a photon's momentum can be calculated in terms of its wavelength, λ, its frequency, v, or its energy, E as $p = h/\lambda = hv/c = E/c$.
An intriguing way the momenta of photons can be harnessed is through 'solar sails.' The idea is to attach huge, highly reflective 'sheets' on a spacecraft traveling in the inner Solar System. As photons from the Sun bounce off these shiny 'sails', they impart momentum that can push the spacecraft in a direction 'upwind.' That is, away from the Sun. Though the idea sounds crazy, some real-life spacecraft have used this form of propulsion in limited ways.

picture of the electron orbiting the proton in the H atom (see Figure 6.5). According to our discussion regarding de Broglie's equation, the electron has a wavelength defined by its momentum as it orbits the proton. Because of the way waves constructively and destructively interfere, Bohr realized that the electron—exhibiting its wave-like characteristics—can exist only in specific orbits within the atom. In particular, he argued that an electron in a H atom *must* abide by the rule

$$\lambda_e n = 2\pi R_n,$$

where λ_e is the electron's wavelength (given de Broglie's equation), and n is any positive whole number (that is, n = 1, 2, 3...). The quantity R_n is the radius of the electron's orbit corresponding to that particular value of n. In other words, the circumference of the orbit of the electron in the atom, $2\pi R_n$, must equal a whole number of wavelengths of the electron, $\lambda_e n$. If this condition is met, the wave associated with the electron constructively interferes (with itself, essentially) and the electron is able to 'exist' in orbit about the atom's nucleus. Any other orbital radius will cause the electron's wave to interfere with itself destructively, meaning that the electron cannot exist in an orbit of that size!

Since the electric force that keeps an electron in an atom is similar in form to that of gravity (though the gravitational force is *much* weaker), an electron at a specific orbital radius, R_n, will have a specific velocity and a specific total energy, E_n (see Figure 6.5). When Bohr analyzed this in detail, he discovered that this energy could be calculated as

$$E_n = -C\left(\frac{1}{n^2}\right).$$

That is, the energy of the electron in its orbit depends *only* on the value of n and a bunch of known physical quantities that comprise the constant C. None of this discussion is meant to imply that an electron cannot change its orbit within the H atom. It can. For example, if an electron in a H atom were to move to a larger orbit—n gets bigger—it must *gain* energy (and consequently the energy tracked by this equation becomes more positive). If the electron shifts to a smaller orbit, it must lose energy (so E becomes more negative). Since total energy must be conserved—as we saw when we discussed the Inverse-Square Law at Destination 4 and 21-*cm* Radiation at Destination 5—one way for an electron in a H atom can alter energy is to *absorb* EM Radiation from outside the atom. The gain in energy from the radiation places the electron into a larger orbit. On the other hand, the electron can *emit* radiation if it shifts to a smaller orbit (Figure 6.5). What's most important is that the energy of the radiation must precisely equal the energy *difference* between the two orbits the electron traverses. That is,

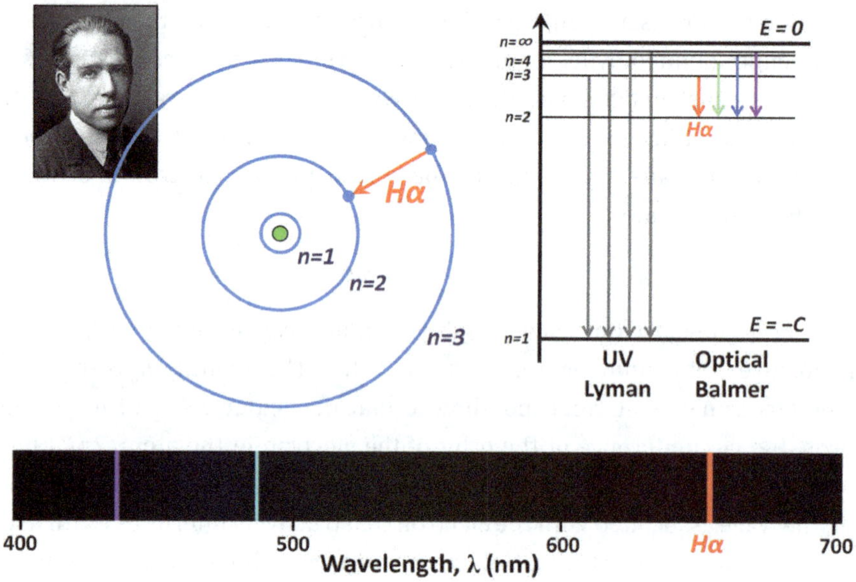

Fig 6.5

(*Top left*) A schematic view of the orbits in a Bohr Model of the H atom (that's Niels Bohr in the inset photo). Each blue circle represents orbits for different values of *n* for an electron in the atom. The arrow shows the transition from $n_1 = 3$ to $n_2 = 2$, corresponding to the production of the red Hα emission line of the Balmer series. The energy that is lost to the atom through emission of a photon is $\Delta E = h\nu$. The orbit sizes for $n = 1,2,3$ are approximately to scale, though the central green dot representing the proton that makes up the nucleus of the H atom is grossly oversized. The $n = \infty$ orbit is infinite in size and so cannot be plotted in a diagram such as this. (*Top right*) An 'energy-level' diagram for the H atom. Each horizontal line corresponds to the energy of an orbit of value *n* with each higher line corresponding to a higher energy, somewhat like a staircase of ever smaller steps. The arrows indicate the energy, ΔE, that would be radiated away in each transition. The transitions to the $n = 2$ level correspond to the Balmer line series visible in the optical part of the EM spectrum and are drawn with the approximate color of the emission lines they correspond to. The specific—red!—transition for the Hα emission line is noted. The left set of arrows corresponds to transitions to the $n = 1$ level, a sequence of UV emission lines known as

$$\Delta E \equiv E_{n_2} - E_{n_1} = -C\left(\frac{1}{n_2^2} - \frac{1}{n_1^2}\right).$$

If n_1 (the initial value of *n* for the electron) is smaller than n_2 (the final value of *n* for the electron), the electron has moved to a higher orbit and a positive amount of energy must be added to it by, say, absorbing EM Radiation with photon energy equal to ΔE. On the other hand, if n_1 is larger than n_2, the energy difference is negative, so the electron must lose energy, which it can do by emitting EM Radiation with photon energy equal to ΔE.

The fact that the energies are specified by integer values of *n* means only photons of specific wavelengths—remember, energy, wavelength, and frequency are equivalent for EM Radiation—will be absorbed or emitted by the atom (Figure 6.5). That beautiful red emission line we saw in the spectrum of the Orion Nebula? It turns out to have a wavelength *exactly* equal to Bohr's prediction for an electron transitioning from the $n_1 = 3$ orbit to the $n_2 = 2$ orbit. That turquoise line? It corresponds to an electron moving from the $n_1 = 4$ orbit to the $n_2 = 2$ orbit. Most of the lines we see in the blue and purple end of the spectrum of the Orion Nebula correspond to electrons shifting from orbits with large values of n_1 down to the $n_2 = 2$ orbit. This produces a sequence of emission lines, each closer to the previous one as the value of n_1 increases. This set of lines is known as the Balmer Series and was familiar to physicists from laboratory experiments long before Bohr proposed his model. But no one had understood why these lines—or any other emission lines for that matter—should form in the *Lyman Series*. On this diagram the $n = \infty$ level is easily plotted since it has a finite energy—equal to the *Ionization Potential*—relative to the ground level at $n = 1$. (*Lower*) A representation of the emission spectrum of the H atom. The lines in the spectrum correspond to some of the emission lines of the Balmer series shown in the energy-level diagram. The red Hα line is indicated.

Spectrum image adapted from artwork by J. Köppen.

the first place, nor why the lines have the specific wavelengths they do.[10] Bohr's model, quite understandably, was a sensation.

Notice that something unique happens when $n_1 = 1$ and $n_2 = \infty$. In this case, $\Delta E = C$. To raise an electron in a H atom from its smallest orbit—or *Ground State* corresponding to $n_1 = 1$—to an *infinitely* large orbit ($n_2 = \infty$) requires a *finite* amount of energy equal to *C*. The process where an electron is removed entirely from the atom is known as *Ionization*, and any atom that has lost an electron is said to be *Ionized* (a concept we have already encountered at Destination 5). The energy required to ionize an atom from its ground state is referred to as the *Ionization Potential.* Ionized atoms—or *Ions* for short—have a net electrical charge because the total number of electrons and protons are no longer equal (Figure 5.7). Although it is not yet obvious, the strong emission lines of the Orion Nebula rely on the fact that the H atoms are mostly ionized. This may seem puzzling. After all, in order to generate the lines of the Balmer Series, a H atom must be neutral. That is, the electrons must be bound to the atom in order to transition from one orbit to another. We'll return to this apparent contradiction later in this destination.

The properties of the glowing gas we see here in the Orion Nebula differ from the other phases of the ISM we have encountered so far. Not only are the majority of the atoms ionized (there's that paradox again), but the gas temperature—about 10000 *K*—is much higher than the gas temperatures we encountered the Warm ISM (about 100–6000 *K*). The gas density is also much *lower*—about 0.01–0.1 H atoms cm^{-3}—compared to a density of about 1 atom cm^{-3} characteristic of the Warm ISM. The gas in the Orion Nebula represents yet another distinct phase of the ISM, one we shall refer to as the *Hot ISM.* An isolated region of the Hot ISM—such as the Orion Nebula—is often referred to as an *Ionized Nebula*, or, more commonly, an *H II Region* (pronounced 'H-two' region). Both of these names[11] emphasize the key role that ionized H atoms have in producing the emission lines seen in the spectra of regions such as the Orion Nebula.

And what about the bright green lines that are so prominent in the spectrum of Orion? The Bohr Model of the H atom does *not* predict the presence of these strong emission lines. Nor of numerous other emission lines that we see in the spectrum of the nebula (see Figure 6.4). The reason is simple: These emission lines are not produced by Hydrogen. The bright green lines, for example, arise

10 The problems were even deeper than this. From classical theory, if an electron is placed in an orbit about an atomic nucleus, it should immediately lose its energy as radiation and spiral into the nucleus. There'd be no stable atoms. No matter. No us. Bohr's atom beautifully solved this problem by implying that there is a smallest possible stable electron orbit, the one corresponding to $n = 1$.

11 As a reminder, the Roman numeral II signifies that the H atom is once ionized. You might want to refer to footnote 24 of Destination 5 and Figure 5.7 for reminders of the strange system of naming ions that astronomers employ.

from Oxygen atoms that have lost two of their original eight electrons, ions that are denoted as O III (pronounced 'Oh-three'). Unlike H, the pattern of emission lines from atoms and ions such as O III can be quite complicated. The multiple electrons in most ions or atoms interact with one another, as well as with the nucleus, which itself is much larger and has more protons—and hence more electric charge—associated with it. The energy levels that in H atoms are so regular (see Figures 6.5 and 6.6) become a jumble in larger atoms, ions, and molecules. But the same principles apply. Electrons can exist only within an atom, ion, or molecule at *specific* orbits with specific energies. As the electrons change orbits, the wavelengths of EM Radiation that are emitted or absorbed correspond to the specific energy differences between these orbits. Calculating the energy levels of an ion such as O III is far more complex than for the comparatively simple H atom. Molecules, which we will encounter soon, are trickier still, in part because the individual atoms that comprise the molecules can themselves be vibrating or rotating relative to the other atoms.

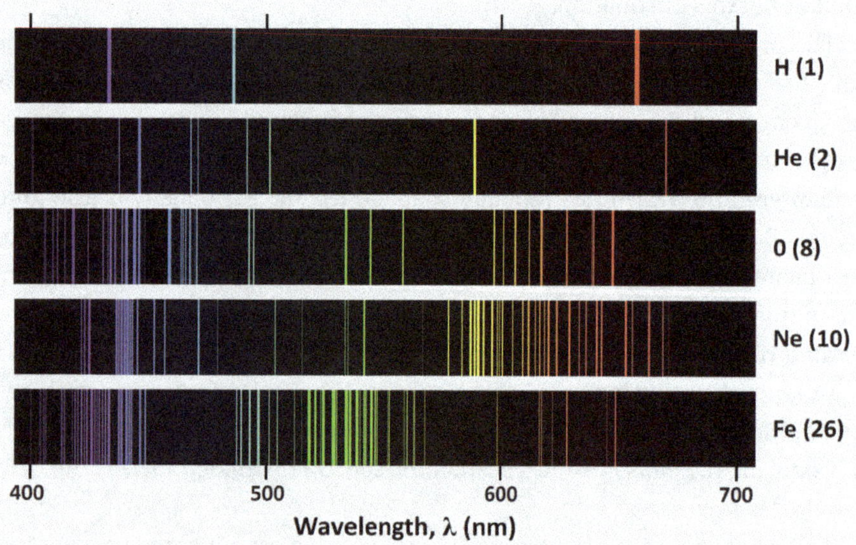

Fig 6.6

Emission spectra of five well-known elements. The bright vertical lines are always seen at the same wavelengths (colors) for a given element. From the top, the elements shown here are Hydrogen (repeated from Figure 6.5), Helium, Oxygen, Neon, and Iron, all in the form of neutral ions. The numbers in parentheses indicate the number of protons (and electrons in these neutral ions) for each element. The Bohr Model—due to its simplicity—accurately predicts only the locations of the lines produced by H atoms. But it is important to appreciate that emission lines can arise from changes in distinct energy states of the electrons in atoms, ions, and molecules.

Image adapted from artwork by J. Köppen.

But there is a beneficial consequence to this complexity. Every atom—indeed, every ion and molecule—has a unique pattern of emission or absorption lines (Figure 6.6 shows some examples). This serves as a sort of 'fingerprint' that allows us to identify individual elements, ions, and molecules just by analyzing their spectra. What this means is that we can tell what the ISM, stars, and galaxies are composed of without having to physically collect material from them! And they don't even have to be nearby!! We will take full advantage of this powerful capability here as we explore the Orion Nebula further, as well as at many future destinations.

6.3 THE DARK CLOUDS OF THE ORION NEBULA

Perhaps the most enigmatic features of the sky of the Orion Nebula are the dark 'holes' (Figure 6.3). As we stare at them intently, we realize we can say rather little about these regions from visual observations. Though these objects may look like holes in the sky, they must actually be some sort of comparatively 'compact' structures localized in space. If not, it would be a spectacular coincidence for the holes to all line up directly with us to produce such uniform features in our sky.[12] Indeed, these 'holes' reveal the locations of *Dark Clouds* spread throughout the Orion Nebula. Like water clouds occasionally do in Earth's atmosphere, the dark clouds in Orion block our view of the brighter background light of the nebula.

There are dozens of these dark clouds visible here. The bright core of the Orion Nebula—the region where the sky glows most intensely—is about 5–10 *pc* in diameter and spans about half the visible sky from our vantage point. Most of these compact clouds are concentrated in the same region of the sky, clearly associated with the brightest parts of the nebula. Assuming they are at the same distance, we can use the small-angle formula (Destination 4) to estimate that the sizes of the clouds range from 5–50 times the diameter of the Solar System, or, equivalently, a few hundred to a few thousand Astronomical Units (*AU*; see Destination 3).[13] As we saw earlier (Figure 6.3), most of the dark clouds have fairly smooth elliptical outlines—some even appear nearly perfectly round—but there are other shapes as well. A few of the clouds exhibit intriguing teardrop shapes as if they are reacting to a wind[14], while others appear extremely

12 Of course, having traveled here from our previous destinations, we would have been able to see directly that the dark 'holes' are actually localized clouds inside the nebula. But the point is that even from a fixed location—say from Earth—it is reasonable to conclude that these dark objects are not actual holes through the nebula, but rather distinct structures within the Orion Nebula.

13 The Solar System does not have a distinct 'edge,' of course, so it has no well-defined size. For purposes of comparison, we'll adopt a diameter of around 100 *AU* for the Solar System. It's a nice round number, easy to remember, and certainly it contains all known major bodies that orbit the Sun. For reference, the outermost planet of our Solar System—Neptune—is about 30 *AU* from the Sun. Some smaller bodies are now known to exist out in the Kuiper Belt (see footnote 5 of this destination) to nearly 50 *AU*. The key point is that even the smallest of the dark clouds in Orion would fully envelop the Sun, all the planets, the Kuiper Belt, and even a sizeable volume that extends far beyond these objects. But as large as that is, the clouds are still 'compact' relative to the much larger scales that typically separate stars.

14 As indeed they are. Some of these clouds are close enough to luminous stars—which we will start to discuss later at this destination—that *Stellar Winds* can interact with the clouds and sculpt them into unusual shapes. These winds are broadly similar to

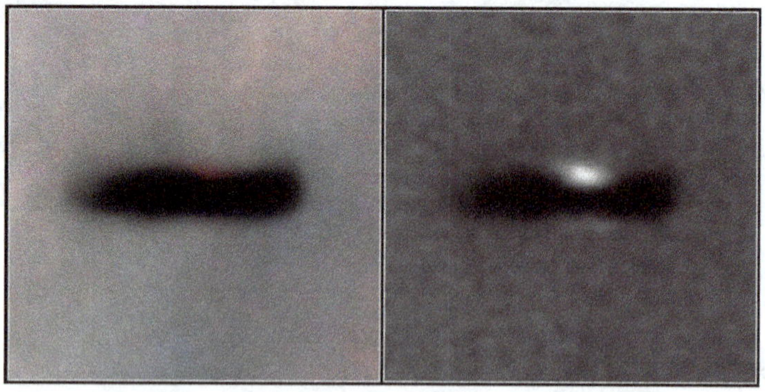

Fig 6.7

Two views of an Orion dark cloud. (*Left*) The optical appearance makes the cloud look like a slit cut into the bright sky of the nebula. (*Right*) The infrared appearance reveals some structure in the cloud in the form of a compact source near the center of the dark region. As we shall see, these characteristics are consistent with the interpretation of these clouds as extended disks surrounding stars in their initial stages of formation.

Photos by C. R. O'Dell/ STScI/NASA.

Fig 6.8

Some of the unusual compact clouds of the Orion Nebula. The two dark clouds (one inside a bright region) appear as 'holes' in the otherwise bright sky of the nebula, which is evident in many of the images.

Photo from STScI/NASA.

elongated, almost like a sausage or cigar in outline (Figures 6.7 and 6.8). Not all of the clouds are completely dark. A few harbor faint, extremely red star-like sources close to their centers. Some are surrounded by bright filaments that extend well beyond their dark cores (see Figure 6.8).

But as we study these clouds more systematically, it soon becomes apparent that we can't say much more than this by simply looking at the clouds in visible light. To delve into the true nature of these strange objects, it's clear that we'll need to look *inside* them. And to do this, visible light won't do the trick. We must inspect them using radiation from a different part of the EM Spectrum.

We've already had success at this. Radio observations revealed the presence of neutral Hydrogen in the Warm ISM through its emission of 21-*cm* Radiation (Destination 5). Since that approach worked so well, it makes sense to try it again here. And so, when we point our radio telescopes at the clouds we discover, oddly

---

the Sun's *Solar Wind* (see Destination 1), except much stronger in the kinds of luminous stars found here in Orion.

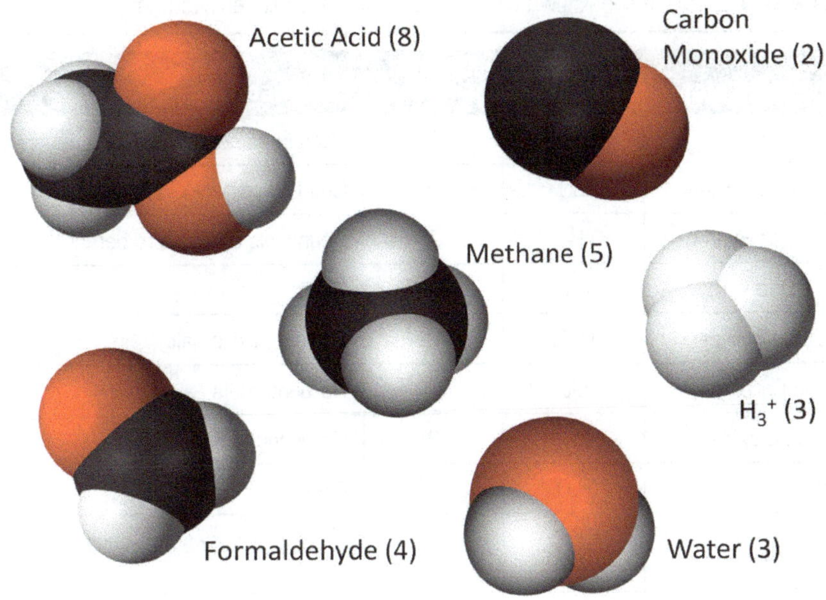

Fig 6.9

Schematic drawings of models for six molecules found in cold molecular clouds such as the dark clouds in the Orion Nebula. In these drawings, H atoms are shown in gray, Carbon (C) in black, and Oxygen (O) in red. Acetic acid (CH_3COOH), a molecule that gives vinegar its pungent taste, contains a total of 8 atoms and illustrates how complex the chemistry of the Interstellar Medium can be.

enough, that there is virtually no radiation from *atomic* Hydrogen associated with the dark clouds. But we are persistent and so we try other radio—and even microwave—frequencies.[15] And soon ... success! We *do* detect radiation from the clouds, though not from atoms or ions. Instead, we see emission lines that arise from *Interstellar Molecules.* One of these molecules—Carbon Monoxide (CO; Figure 6.9)—is particularly bright and easily identified from its characteristic pattern of microwave emission lines. As we explore a broader range of frequencies, we discover that other simple molecules are present too. Molecules such as H_2 (molecular Hydrogen), FeO (Iron Oxide, or rust!), NaCl (Sodium Chloride, better known as table salt), and even O_2 (molecular Oxygen). More complex molecules are also present, including H_2O (water; Figure 6.9), NH_3 (ammonia), N_2O (Nitrous Oxide, or 'laughing gas'), CH_4 (methane; Figure 6.9), H_2CO (formaldehyde; Figure 6.9), CH_3CH_2OH (ethanol, or 'drinking' alcohol), CH_3COOH (acetic acid; Figure 6.9), and NH_2CH_2CN (amino acetonitrile, a likely precursor of amino acids and, hence DNA). In all, over 120 different molecules have been detected in dark clouds.[16] This is also why these objects are more commonly referred to as *Molecular Clouds.* As you ponder the list of some of the known interstellar molecules (Table 6.1), keep in mind these molecules form

15 Recall that microwaves correspond to EM Radiation with wavelengths/frequencies/energies intermediate between radio and infrared radiation (see Figure 5.6). To study the interiors of the dark clouds in Orion, we require specialized detectors and spectrographs sensitive to microwave radiation. Fortunately, we are very well equipped on this journey and we have the tools needed to detect and analyze microwave radiation.

16 Molecules have been detected in various dark clouds visible in Earth's sky. Not all of the known interstellar molecules have been identified in Orion's clouds.

TABLE 6.1 Some of the Known Molecules in the Interstellar Medium

MOLECULE NAME	CHEMICAL MAKEUP	NUMBER OF ATOMS	COMMENTS
Cyanogen	CN	2	Seen in some stars as well as the ISM
Carbon Monoxide	CO	2	Good tracer of Cold ISM
Molecular Hydrogen	H_2	2	Abundant, but hard to detect
Iron Oxide	FeO	2	Rust
Sodium Chloride	NaCl	2	Ordinary table salt
Silicon Carbide	SiC	2	Hardens tools
Carbon Dioxide	CO_2	3	Greenhouse gas on Earth
Water	H_2O	3	Essential for life
Nitrous Oxide	N_2O	3	Commonly known as 'Laughing Gas'; can also be used as an anesthetic
Sulfur Dioxide	SO_2	3	Common chemical reagent; preserves biological tissue
Formaldehyde	H_2CO	4	Smells like rotten eggs
Ammonia	NH_3	4	Used as cleaner and fertilizer
Methane	CH_4	5	Common in atmospheres of outer planets of Solar System
Formic Acid	HCOOH	5	Common in insect stings and bites
Methanol	CH_3OH	6	Wood alcohol; useful as antifreeze
Ethanol	H_2CHCOH	7	Drinking alcohol
Acetic Acid	CH_3COOH	8	Gives vinegar its pungent taste
Amino Acetonitrile	NH_2CH_2CN	8	Basis for amino acids, DNA, life
Acetone	$(CH_3)_2CO$	10	Versatile solvent and cleaner
Benzene	C_6H_6	12	Used to make other chemicals; has distinctive ring molecular structure

and survive in interstellar space. Not on the surface, atmosphere, or ocean of some planet!

What's so striking about the ease with which we find molecules in these clouds is their utter lack in the other phases of the ISM such as the Warm and Coronal ISM we first encountered at Destination 5, as well as the Hot ISM that comprises most of the Orion Nebula. There is a good reason for this. Recall that the typical temperature of the gas in Warm phase of the ISM ranges from about 100–6000 *K*; the other phases are even hotter. This implies that the average particle speed in the Warm ISM—mostly in the form of neutral H atoms—is

TABLE 6.2 Basic Properties of the Interstellar Medium

PHASE	TEMPERATURE PARTICLE DENSITY	CHARACTERISTICS
COLD ISM (Molecular ISM)	5 to 30 K 10 to >10^6 cm^{-3}	• Often associated with star-forming regions. • Typically associated with dust, producing optically opaque regions. • Highly clumpy structure consisting of clouds ranging from $1 M_\odot$ to over 1 million $M_\odot$. • Detectable from radio, microwave, IR molecular emission lines and thermal dust emission. • May represent 10–20% of the mass of the ISM in our Galaxy.
WARM ISM (Neutral ISM)	100 to 6000 K 0.1 to 50 cm^{-3}	• Dominated by neutral atoms. • Most common phase of ISM, by volume, near Sun and in Galactic Disk (see Destination 9). • Fairly uniformly distributed but with frequent large 'bubbles' and extended filaments. • Detectable from radio 21-cm Radiation emission and absorption. • By mass, the most common phase of the ISM in our Galaxy.
HOT ISM (Ionized ISM, H II Regions)	10000 K 0.01 to 1 cm^{-3}	• Emits strong emission spectra from UV to IR dominated by ionized atoms or recombination lines such as those from H atoms. • Generally associated with star-forming regions containing hot stars that ionize the ISM. • Detectable from radio to x-rays through a variety of emission mechanisms. • Short lived; once sources of UV ionizing radiation die off, these regions fade away. • Highly clumpy distribution, often associated with molecular clouds of the Cold ISM. • Represents a tiny fraction (<1%) of the mass of the ISM in our Galaxy.
CORONAL ISM (Very Hot ISM)	500000 to 10 million K $\leqslant 0.001 cm^{-3}$	• Hottest component of the ISM. • Most extended and rarified phase of the ISM, filling the entire volume of our Galaxy. • Mass is highly uncertain, but probably only a small fraction of total mass of ISM. • Dominated by X-ray emission lines from extremely highly ionized atoms and absorption lines in the UV portion of the EM Spectrum. • Name reflects similarities of properties to the very hot and rarified corona surrounding the Sun.

around 2 *km/sec* (Destination 5). The kinetic energy of particles moving that fast (Destination 5) can, in a collision, easily break the fragile bonds that hold the atoms of a molecule together, just as the kinetic energy of a well-placed bowling ball can scatter the fragile configuration of pins. The only way we can imagine molecules surviving in the ISM is if they are found in regions where the particles move much more slowly. That is, in regions that are considerably colder than the Warm ISM. Regions as frigid as 10–20 *K* or even colder. It turns out that molecular clouds are fairly common, and although their properties vary considerably, all of them are indeed quite cold. These clouds, including the dark, compact clouds here in the Orion Nebula, constitute yet another distinct phase of the ISM—the *Cold ISM*—the last of the four principal phases of the ISM that we shall encounter during our journeys (see Table 6.2).

But why are clouds associated with the Cold ISM so dark in the visible region of the EM Spectrum? We surmised earlier that the clouds are blocking our view of background light. If that is true, a lot of material must be present within the clouds to produce such dark blots in the sky. Whereas the Warm ISM contains about 1 atom cm^{-3}, dark clouds have comparatively high densities that range from hundreds to millions of particles cm^{-3} and sometimes even considerably higher. But high density is not enough—after all, the number of particles in a cubic centimeter of air is vastly larger than this, yet light can easily pass through miles and miles of air.[17] The reason is that EM Radiation is most effectively scattered by particles similar in size to the radiation's wavelength. Since the dark clouds of Orion block visible radiation, they must contain many particles that are roughly 0.0001 cm (that's 10^{-4} cm) in size. That's much bigger than most molecules—which are typically about 1000 times smaller—but comparable in size to the very fine particles in smoke. Astronomers refer to these tiny particles—somewhat misleadingly—as 'dust.'

The presence of dust in these clouds is fascinating but also frustrating since we want to peer inside of them. Though microwave radiation can detect molecules within clouds, it tells us only about a tiny fraction of the matter that comprises these clouds. Blueward of microwave radiation in the EM Spectrum[18] lies infrared (IR) radiation (see Destination 5), characterized by wavelengths ranging from 2 times to 1000 times longer than the wavelengths of visible light. Because these wavelengths are generally much longer than the sizes of dust particles, IR (and microwave) radiation can pass more easily through optically

17 Well, *some* air. Extreme fog or smog can lower the visibility in air dramatically, down to a few feet sometimes. But that just reinforces the point. Air alone, even though its density is high compared to the ISM, can be quite transparent. But adding other stuff—water droplets, hydrocarbons, etc.—can make that same air very murky.

18 Strictly speaking, we can sense 'color' only in optical wavelengths. However, it is very common for astronomers to talk about sources being 'bluer' or 'redder' even when referring to portions other than the optical band of EM Radiation. Thus, a bluer IR source is one that is brighter at shorter IR wavelengths than a redder source. In each case, the objects may only be visible in the IR, but we can still in this manner refer to one as being bluer or redder than the other. In this sense, color is just a relative wavelength of the radiation being detected. 'Blue' corresponds to shorter wavelength, higher frequency, higher energy. 'Red' refers to longer wavelength, lower frequency, lower energy.

Fig 6.10

A schematic illustration of how the Doppler shift operates. Imagine the *top* car in this figure emitting sound or light (the argument works for either type of wave) approaching a stationary observer (at *right*) at a speed V that is slower than the speed of the wave, c. From the car's point of view, the waves it is emitting—denoted by the dashed red line—have wavelengths of λ_0. But the observer sees something different. If we follow a wave from its start at a time $t = 0$ (the red dot in the upper figure) it will travel to the location shown in the lower figure, at which time a new wave is being emitted. If we choose that time to be equal to the time it takes one wave to be produced, that is $t_0 = 1/\nu_0 = \lambda_0/c$, then the next wave is produced at a distance $d = Vt_0 = V/\nu_0 = V\lambda_0/c$ closer to the observer as denoted by the green dot from the *bottom* car (this distance is denoted by the black arrow between the locations of the car at times $t = 0$ and $t = t_0$). The observed wave will have a wavelength equal to the distance between the red and green dots of the lower wave shown in the figure. The observer will therefore detect a wavelength equal to $\lambda = \lambda_0 + Vt_0 = \lambda_0(1 + V/C)$. *Motions toward the observer are defined to have negative velocities*. In the example shown in the figure, the velocity, V, is negative and so the observed

'dark' clouds.[19] This is important for us because the conditions within these cold, dark clouds allow matter inside of them to *produce* IR and microwave radiation that we can then detect from outside. We've already seen that molecules do this, mostly in the microwave portion of the EM Spectrum.

Fortunately, we have also brought IR telescopes and spectrographs with us, instruments that allow us to peer inside Orion's dark clouds in a different manner than microwave instruments can. What do we see? Strikingly, virtually every small cloud contains a compact, infrared-bright source. Some of these cores are extremely red and faint. Some are comparatively much bluer and brighter. But they are nearly always present, and significantly more compact than the clouds in which they are embedded. These observations are starting to suggest what might be going on here in the Orion Nebula. But to be sure of our suspicions, we need to explore one further important feature of the nebula's dark clouds.

As we have discussed, EM Radiation—radio, microwave, infrared, optical, or any other (see Table 5.2)—can be considered to be a wave with a well-defined wavelength. Up to now, we have simply denoted this wavelength as λ. But, strictly speaking, we should denote it as λ_0. Notice the subscript. When the wavelength is represented in this way it signifies that we are referring to the wavelength of the radiation as emitted by the source, a parameter often referred to as the *Rest Wavelength*. As its name implies, the rest wavelength for 21-*cm* Radiation, λ_0, equals, well, about 21 *cm*! But there are factors that

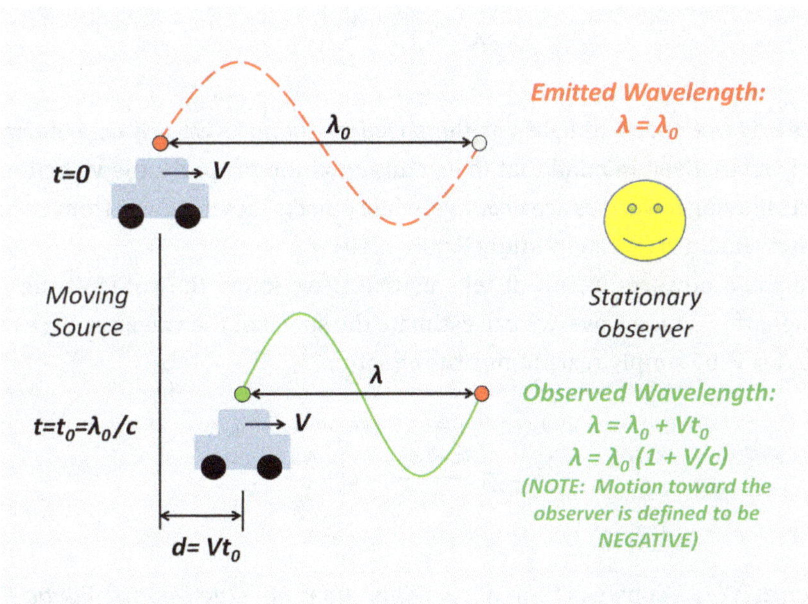

19 This is the same basic reason that the Sun can appear red when close to the horizon. At such times, light from the Sun must pass through a lot of air—and particles in the form of dust and smog—in our atmosphere. Redder light is more effective at passing through particle-filled air, causing the Sun to take on a distinct red color. All other visible colors, which have shorter wavelengths, are scattered away from our line of sight. IR and microwave radiation are even better able to pass through gas filled with tiny particles, even when the particles make the gas completely opaque to visible light.

can change the wavelength of radiation.[20] In particular, an observer in motion relative to a source emitting radiation at its rest wavelength λ_0 can measure a wavelength that differs considerably from the rest wavelength value.

To understand why, consider the situation depicted in Figure 6.10. In this example, a wave of EM Radiation—though it could be sound since it too is a wave phenomenon—is produced by a source that is toward us at some speed *V*. We can imagine that at some specific time, the source produces a wave with a well-defined rest wavelength λ_0. Since the product of wavelength times frequency equals the speed of a wave, the wave also has a well-defined frequency (the *rest frequency*) equal to $\nu_0 = c/\lambda_0$. (Be careful: The symbol for frequency, the lower-case Greek letter ν (nu), looks a lot like the symbol *V* (vee) used for velocity.) From the perspective of the source of the wave, the next wave will be produced after the passage of a time interval equal to $t_0 = 1/\nu$. However, during that interval, the source will have *moved* a distance equal to $Vt_0 = V/\nu_0$, or $\lambda_0 V/c$ (recall from Destination 5 that $\lambda\nu = \lambda_0\nu_0 = c$, so $1/\nu_0 = \lambda_0/c$). Notice that for a source moving away from us, we would measure a wavelength for the radiation that is *longer* (redder) than the rest wavelength, λ_0, while for a source moving toward us (as in Figure 6.10), we would detect a wave with a wavelength shorter (bluer) than λ_0. By combining these results, we can derive that the observed wavelength, λ, is related to the rest wavelength, λ_0, through the equation

$$\lambda = \lambda_0\left(1 + \frac{V}{c}\right),$$

where *c* is the speed of light (or the speed of sound if we are considering a sound wave). Keep in mind that to use this equation properly, the velocities of objects moving *toward* us are *negative*, while objects moving *away* from us have *positive* velocities (carefully study Figure 6.10).

This is a powerful result. It tells us that if we know the value of the rest wavelength, λ_0, for a wave, we can estimate the speed of the wave's source relative to us, *V*, by simply rearranging the equation:

$$V = c\left(\frac{\lambda - \lambda_0}{\lambda_0}\right) = c\left(\frac{\Delta\lambda}{\lambda_0}\right).$$

The term $\Delta\lambda$ is a shorthand for $\lambda - \lambda_0$. This equation—the *Doppler Equation*—describes what is known as the *Doppler effect* or *Doppler shift*. The velocity we measure from this shift is called the *Radial Velocity* because it refers only to the

wavelength is shorter than the emitted wavelength, an example of a *blueshift*. *Motions away from the observer are defined to have positive velocities*. Had the source been moving away from the observer, the velocity, *V*, would be a positive number and the observed wavelength would be longer than the emitted wavelength, a change known as a *redshift*.

20 Some of these are considerably more subtle than the effect we are about to explore. The key point is that the wavelength of radiation depends on many factors, some of which involve properties of an observer that the source otherwise doesn't interact with at all!

motion of a source toward or away from us (that is, along a 'radial' direction relative to us). When a source is moving away from us, the observed wavelength, λ, *increases* relative to the rest wavelength, and so the velocity is defined to be positive. Such motion away from the observer is known as a *Redshift* because the radiation is shifted to longer wavelengths and because red light has a longer wavelength than blue light. Motion toward the observer results in negative velocities, producing a wavelength that is shorter—bluer—than the wavelength at the source. This results in a *Blueshift*.

Although our discussion about the Doppler effect seems a bit esoteric, we are actually quite familiar with this phenomenon. As a source of sound approaches us at a speed c from a source with a speed $-V$ (the negative sign means the source is coming toward us), the wavelength of the sound wave is shorter than the emitted wave. Our ears interpret this sort of blueshift as a rise in the pitch of, say, a siren on an ambulance. As the source of the sound passes us and the speed of the source becomes $+V$, the wavelength of the sound wave gets longer, resulting in what we perceive as a lower pitch. In this respect the Doppler Equation is really quite general. It applies to any wave with a well-defined speed, c, and rest wavelength, λ_0, moving at a speed V relative to an observer. We shall discover at future destinations that for EM Radiation the Doppler Equation changes form rather drastically as the source speed, V, approaches the speed of light, c (sound waves behave differently as the source speed approaches the speed of sound). But, for relatively slow-moving objects, the form of the Doppler Equation given here is sufficiently accurate for our purposes.

In order to measure radial velocities using the Doppler Equation, we clearly need to know the rest wavelength, λ_0, of radiation emitted by the source we are observing. For 21-*cm* Radiation, we obviously know the value of λ_0 (after all, it's in the name!). But, in fact, the value of that wavelength was initially calculated crudely, then refined by laboratory measurements. Our earlier discussion about the Bohr atom suggests that we can calculate the wavelengths of the emission lines that arise in H atoms from changes in the orbits of electrons in the atoms. The same is true of more complex atoms, ions, and molecules, though these calculations are far more complicated than for the simple H atom. But even if we cannot calculate values of λ_0 for emission lines of a given atom, ion, or molecule, we can usually measure their properties in a laboratory to get experimental estimates of the relevant values of λ_0.

As we've seen, molecules in the dark clouds in Orion emit a wealth of emission lines that are visible from IR to radio wavelengths. When we train our telescopes and spectrographs at these clouds and measure the wavelengths of their emission lines, we frequently find that the patterns of lines mimic the patterns seen for known atoms, ions, or molecules, but the observed values of the wavelengths differ significantly from the rest wavelengths. In many of the dark clouds in Orion, we notice that the behavior of these shifts is not random. Often, one side of the cloud appears to be approaching us—the emission lines are shifted to shorter wavelengths, that is, blueshifted—while the other side of the cloud is moving away—the lines are redshifted to longer wavelengths. If these shifts reflected internal motions happening by chance—say, turbulence—the clouds would soon disperse, implying that our chances of seeing them in this state would be very low in the first place. But there are many dark clouds here in Orion, and many of them behave this way. The motion we see from the Doppler shifts must represent real and persistent motions within the clouds. Motions that are fairly long-lived. If we ponder this for a while[21], we realize that the only way this motion can be understood as a stable configuration is if the clouds are *rotating*. This in itself is a fundamental result. But to understand what this rotation implies about the clouds, we need to explore some basic properties of spinning objects.

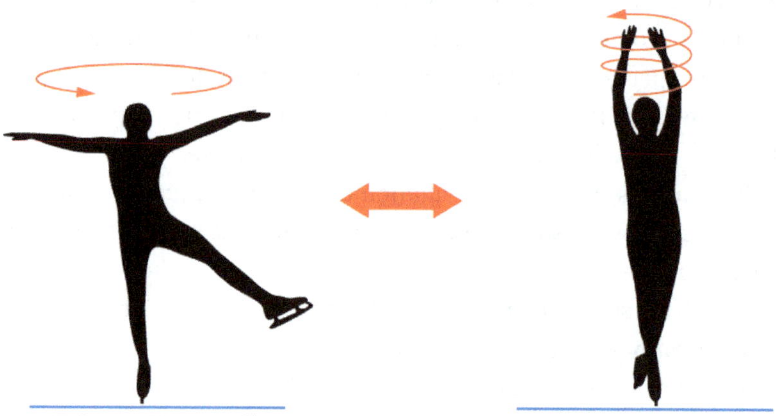

R large, V_{rot} small   R small, V_{rot} large

Angular Momentum, $J = mV_{rot} R$, is the *same*.

Fig 6.11

A figure skater starting his spin with his arms and legs spread out can rotate faster by simply pulling them closer to his axis of rotation. This process exhibits the conservation of angular momentum and applies equally to a contracting cloud of gas thousands of Astronomical Units (*AU*) in size!

One good example is a figure skater. Talented skaters not only know how to spin about a fixed point on the ice, but they also know how to control the rate at which they spin. By extending their arms or legs, they can slow down their rotation rate, while they can speed up their spin by pulling their arms or legs closer to the axis of their rotation (Figure 6.11). This behavior occurs because a fundamental property of spinning objects—known as *Angular Momentum*—is conserved in a rotating system. Like the Conservation of Energy we used in Destination 5 to predict the emission of a 21-*cm* Radiation from neutral H, the *Conservation of Angular Momentum* tells us that the total angular momentum in an isolated system remains unchanged even as the system reconfigures the distribution of its mass. To apply this concept, we must first know how to determine the angular momentum of a system—after all, that's what we are conserving. Fortunately, this is not hard to do.

21 You can do so in a question at the end of this destination.

In general, the angular momentum (denoted J) of a spinning body can be calculated as

$$J = m\, V_{rot}\, R,$$

where m represents the object's mass, V_{rot} is the rotation speed of the object about its spin axis, and R is some measure of how far the mass of the object is from the rotation axis.[22] For our skater (Figure 6.11), this implies that when he is spinning on his skates at some speed V_{rot} with his arms extended (as in the left side of Figure 6.11), R is relatively large. But as he pulls his arms inward (as in the right side of Figure 6.11), R decreases. Since his mass, m, remains constant during this process, the conservation of angular momentum requires that his rotation speed, V_{rot}, must increase. That is, he spins at a faster rate.[23]

How do we apply all this to the dark clouds of the Orion Nebula? Since we have already seen that much of the ISM—particularly the Warm ISM—has a mean density much lower than these clouds, it is plausible that compact clouds collapsed from an initially much larger region of the ISM. Since 'mass density' is defined as mass divided by volume, the density for a region of constant mass will clearly increase as the volume decreases. If the original, large cloud was rotating, even very slightly, this contraction would also cause it to spin faster as it conserves angular momentum. Just like the skater as he draws in his arms and legs to spin faster.

What could cause this collapse? Gravity is the obvious culprit, but gravity alone does not dictate what happens. Recall that the particles in the warm ISM move at relatively high speeds. This motion produces an outward force, known as *Pressure*, which can potentially equal or even overcome the inward pull of gravity in a given region of the ISM. For the cloud to collapse due to gravity, it must either be very dense—and hence have a lot of mass—or must be very cold so that its internal pressure is low. Or *both* dense and cold. Only under these circumstances can gravity overcome the internal pressure of the cloud and cause it to collapse. The specific conditions for this to occur were first figured out early in the 20th century by Sir James Jeans. Here in the Orion Nebula, much of the ISM is composed of gas that is distributed in relatively clumpy, low-temperature clouds that are dense

---

22 This is a generic form of the equation that gives the angular momentum of any source. In practice, the detailed shape of the body that is rotating is important. A uniformly solid sphere with a radius R, mass M, and surface rotation speed of V_{rot}, for example, has a total angular momentum of $\frac{2}{5} M V_{rot} R$. These distinctions are not important for our purposes, so we will stick to the simpler version of the equation.

23 There are many ways to experience this even if you do not know how to skate. Perhaps the easiest is to get on a playground merry-go-round while it is spinning at a reasonable speed. If you move inward on the merry-go-round, it (and you!) will start to spin faster. If you move back to the outer edge, you slow down. This is a weird sensation because although you may have had to push the merry-go-round to get its spin started, these changes in the spin happen without any effort on your part. The merry-go-round and you are simply conserving angular momentum.

enough and cool enough to collapse under the influence of their own gravity as Jeans described. But the clouds must still conserve angular momentum. Like our skater, they will spin faster and faster as they become smaller and smaller.

In reality, a cloud of gas collapsing in this manner does not do so uniformly. The central, dense regions of the cloud collapse faster than the outer parts, a behavior that helps explain the existence of the compact cores we see in so many of these clouds. As these cores become denser, they heat up as the energy from gravity is first converted to inward motion of the gas particles, then to random motions as these particles increasingly collide with one another. These faster, random motions imply higher gas temperature as described by the relationship between particle energy and temperature that we discovered at Anti-Cen (Destination 5).[24] Given the sizes of the dark clouds and their approximate masses—which, by adding up the total matter we see in these clouds is typically around 10–100 $M_{\odot}$—it's possible to use Jeans' results to show that the typical dark clouds here in the Orion Nebula will require about 0.1–1 million years to contract down to the size of a normal star. This turns out to be a timescale that is quite short compared to the typical stellar ages as we'll discover at Destination 8. At this stage of their collapse, the cloud cores are nearly as hot and dense as the cores of normal stars, but because they are still contracting and adding mass from the infalling cloud, they are not true stars yet (which we will define at Destination 8). To make the distinction, these objects are referred to as *Protostars*. The fact that there are so many protostars here in Orion suggests that we have discovered a site where stars are born. The Orion Nebula is a gargantuan stellar nursery!

But there are still some important mysteries. For one, where does the angular momentum of the collapsing cloud go? Although stars do rotate, it is easy to show that they should rotate *much* faster than they do if all of the original angular momentum in a collapsing cloud ends up in a central star (see the questions at the end of this destination). Also, though we can see why gravity can induce a cloud to collapse, what *stops* the collapse? Since stars like the Sun appear to be very stable (see Destination 1), the obvious answer is that the collapse ends with the formation of a star. But what is it specifically about stars that halts the contraction of the cloud from which they formed? And what becomes of all the gas surrounding protostars? The masses of the dark clouds in the Orion Nebula are on average far more massive than the stars that have formed in the region. The clouds must have some means of shedding mass—sometimes quite a bit of mass—before it can collapse onto the central protostar. We will need to visit other destinations beyond Orion before we can fully address these questions.

---

24 This behavior of the gas is consistent with the so-called *Perfect Gas Law*, which relates the pressure, density, and temperature of a gas, and the fact that the energy that heats the cloud comes from gravity. In a collapsing interstellar cloud, this means that the cloud temperature increases as it shrinks in size. We'll encounter this behavior again in various contexts at future destinations.

6.4 THE STARS OF THE ORION NEBULA

Shining so brightly, so steadily in our sky, the stars here in the Orion Nebula seem like eternal fixtures in this spectacular sky. It is hard to appreciate that they are all—by stellar standards—very young (Figure 6.12). Yet they must be. We have come to the realization that the dark clouds of the Orion Nebula are very likely sites where stars are being born. The large concentration of so many 'normal' stars in this compact region nearby is too coincidental to be chance. The stars here—and not just the protostars that are still in the process of forming—must *all* be young. But they are hardly identical. As we look around us we see stars within the Orion Nebula that exhibit a tremendous variety of properties. The brightest stars—those that dominated our attention even before we first arrived—shine with the steely blue-white light of a welder's torch. Some of the fainter stars are more nearly white in appearance, some yellowish in color (like the Sun!), while many of the faintest stars in this sky shine with a dull orange, even deep red light. There are certainly some exceptions, but the trend is clear. The bluer a star is, the brighter it appears. And it is not mere coincidence. As we factor in the distances of the stars around us, we confirm this fundamental feature (we'll see how to do this at Destination 7). The blue stars really are the most luminous of the stars in the Orion Nebula, while the redder stars are the least luminous. This is similar to the mini-trend we noted when comparing the stars of the α Cen system and the Sun (Destination 4). But here in Orion we now see the same trends being obeyed by hundreds of stars that exhibit a far larger range in color and luminosity than those four stars of the solar neighborhood.

Before we can fully understand what these trends mean, we have to figure out just what the colors of the stars are telling us. A first step is to observe their optical spectra. It is immediately apparent that these stellar spectra could not be

Fig 6.12

The stars of the central part of the Orion Nebula. The four brightest stars—which together are known as *The Trapezium*—are also among the most luminous and hottest stars of the region; their obvious blue colors are clear indications of their high surface temperatures. Various compact clouds, some dark, some bright, many oddly shaped, can be seen near the brightest star. These clouds may represent some of the remnant gas from the original molecular cloud from which these stars recently formed. The orientations and strange shapes of these clouds strongly suggest that they are still interacting with these hot, luminous stars.

Photo by J. Bally/STScI/NASA.

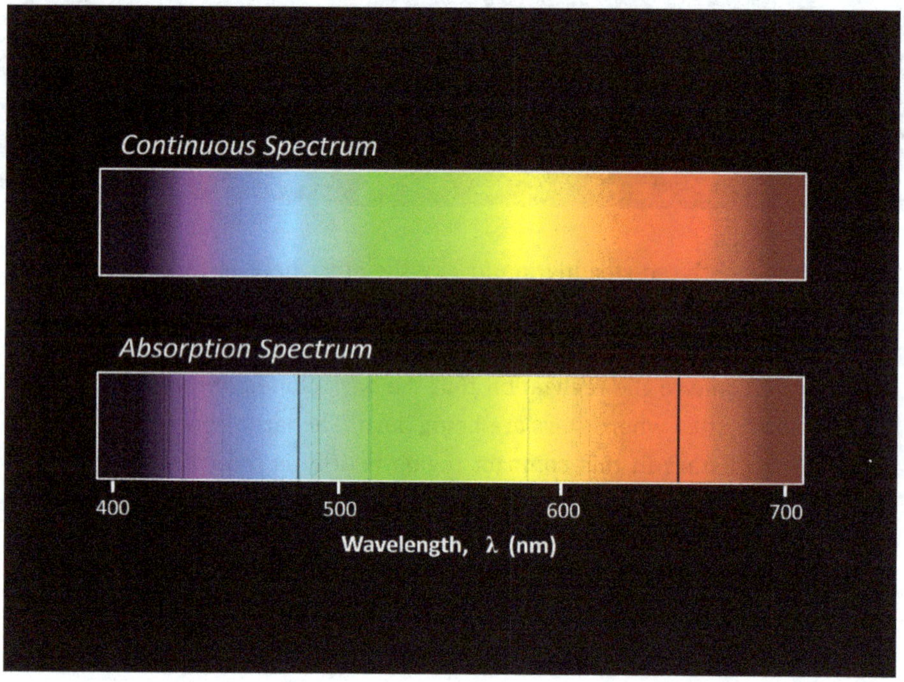

Fig 6.13

A *Continuous Spectrum* (*top*) reveals light at all colors (wavelengths). The spectrum of the Sun (*bottom*) is a good approximation to a continuous spectrum apart from a few specific wavelengths where absorption lines are visible. This is an example of an *Absorption Spectrum*. Note that the solar spectrum appears brightest in the green but trails off to the red and the blue. This reflects the underlying thermal nature of the continuous radiation from the Sun.

more different from the spectrum we observed from the nebula itself. Whereas the nebular spectrum was dominated by emission lines with virtually no light between the lines, the stars seem to emit light at *all* wavelengths (Figure 6.4). They represent examples of what are known as *Continuous Spectra* (Figure 6.13). Almost. If we look at them carefully, we see that at some isolated wavelengths the stellar spectra appear dark (Figure 6.13), precisely the opposite of the appearance of an emission-line spectrum (Figure 6.4). What we discover is that most stars exhibit what are known as *Absorption Spectra* (Figure 6.13). We will look more carefully at what the dark absorption features in stellar spectra tell us (starting at Destination 8). For now, let's focus on the overall *shape* of the stellar spectra, ignoring the absorption lines. In effect, we will treat the spectra as if they were continuous, which they very nearly are (see Figure 6.13).

The bluest stars are a good place to start. Apart from a few weak absorption features, these stars really do exhibit nearly continuous spectra. As we measure the intensity of the light at different wavelengths, we find that these spectra peak in the ultraviolet region of the EM Spectrum, radiation with wavelengths shorter than the bluest—actually, 'purplest'—optical light we can detect with our eyes (see Figure 5.6). The redder stars in the sky here are generally much fainter, but also quite numerous. Though these tend to have lots of dark absorption features, we can still plainly tell that their spectra peak in the red or infrared portion of the EM spectrum. Blue and the red stars emit light at all wavelengths, but these observations tell us that we *perceive* stellar color because some stars emit relatively more blue light (the blue stars), while others emit relatively more red light (the red stars).

It dawns on us that we have encountered other things that behave in this manner. Back on board our ship, we head to the kitchen and ask the cook to fire up the stove. As the electric burner warms up, we can feel heat on our skin without even touching the burner. We are sensing infrared radiation. As the burner heats up further, it emits a faint, deep red glow. As it gets hotter still, the burner begins to shine a bright red-orange color, then a very bright, pure orange color. If we turn the stove up to its highest setting, the burner glows with a dazzlingly bright, yellowish light. If it could survive even higher temperatures, we'd see the burner glow blue-white, then steel blue in color, steadily becoming brighter until we can no longer comfortably look at it directly. Notice the trend. The burner changes color and brightness as its temperature rises. It is reddest, and faintest, when coolest. Bluest and brightest when hottest. Do the stars here in Orion behave the same way? Certainly we see that the bluest stars are most luminous, the reddest least luminous. But does this mean that the blue stars are hotter than the red stars?

It turns out that all matter emits radiation in a manner that is very similar to what we saw in the burner, including the stars here in the Orion Nebula. Comparatively cool stars emit mostly red light. For very cold objects, most of the radiation is emitted as infrared, microwave, or even radio radiation. The human body is fairly cold compared to the temperature of the burner when we first saw it emit visible light. Like the burner, we also glow, but most of the radiation we emit is far in the infrared, which our eyes cannot detect. If we could see that part of the EM spectrum, we would all appear as glowing beings to one another (Figure 6.14). What we see when we observe people (and most

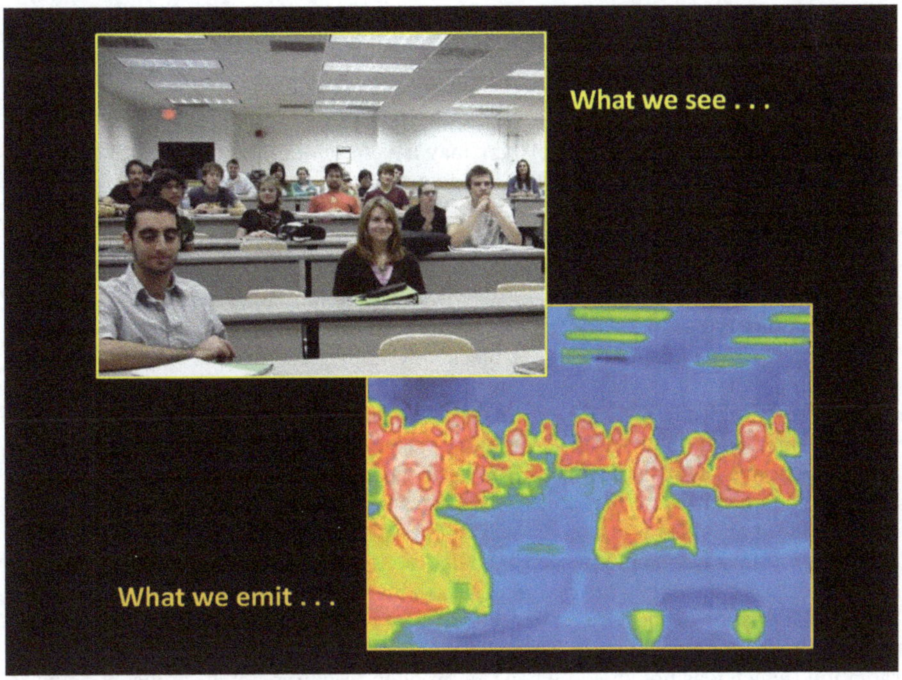

Fig 6.14

We see the world in reflected light (*upper left*). But we emit radiation in the infrared (*lower right*), where we are all glowing beings. Precisely what Wien's Law predicts. The *amount* of radiation we emit can be determined by knowing our body temperature and using Planck's Law.

Photos by M. Mateo.

objects) is not the radiation they emit, but the visible light that reflects off their bodies from sources such as light bulbs, the Sun, or, in our present case, from the light of the stars and from the sky in the Orion Nebula.

Hotter bodies emit comparatively more EM radiation further to the blue—that is, at higher frequency, higher energy, shorter wavelength. At temperatures of about 500 *K*, we can just start to see some of this radiation in the form of a faint deep red glow. By the time we reach 5000 *K*, a body emits bright yellow light. At 10000 *K*, we perceive white or blue-white colors. Objects much hotter than this have a distinctly blue tint to our eyes. This holds for the stars here in Orion, implying that the colors can indeed be related to the stars' surface temperatures. The brightest blue stars are hottest, well above 10000 *K*. The reddest stars are comparatively cool[25] with temperatures of 'only' 1000–3000 *K* or less. Stars of intermediate color have intermediate temperatures. Our yellow Sun—a faint star we can just see with binoculars from the Orion Nebula—has a surface temperature of about 5780 *K*. The continuous EM Radiation that hot bodies emit—known as *Thermal Radiation* or, for rather technical reasons, *Blackbody Radiation*[26]—and is emitted by *all* bodies simply because they are hotter than absolute zero.[27] To a very good approximation, the continuous spectrum of thermal radiation does not depend on what the heated bodies are made of. The burner in the ship's kitchen (composed mostly of Carbon) and the stars of the nebula (mostly Hydrogen) can exhibit similar colors despite having vastly different chemical compositions.

People have known for centuries that hot bodies emit light. But at the start of the 20th century it was still a deep mystery why the spectrum of such bodies had their characteristic form that peaked at some intermediate wavelength and that depended so systematically on temperature.[28] The German physicist Max Planck solved this problem by postulating a trick similar to the one Bohr

---

25 Keep in mind that even a 'cool' star may have a temperature of well over 500 *K*, and usually much more than 1000 *K*. These are not 'cool' temperatures by everyday standards, but compared to other stars, these are indeed relatively low temperatures. Still, it would be best to keep our distance!

26 I will use the term 'thermal' to describe how hot bodies emit radiation throughout our voyages because it so aptly describes why matter emits this way. Astronomers and physicists more typically use the term 'blackbody' to describe this radiation for a rather obscure reason. As it happens, any true 'blackbody' is, by mathematical definition, perfect at absorbing radiation as well as emitting it. Thus, a blackbody can appear perfectly black to our eyes (specifically, if it is very cold). Of course, if it has a temperature of a few thousand *K*, it can appear blindingly bright, too! It is this somewhat contradictory behavior that makes the term 'blackbody' confusing. The term 'thermal' underscores that the properties of the radiation emitted by hot bodies depends principally on their temperatures.

27 A body at absolute zero would, in fact, emit no thermal radiation.

28 This mystery was so profound it even had a name: The Ultraviolet Catastrophe. According to classical theory, a hot body should exhibit a spectrum that gets more and more intense the further to the blue (in the ultraviolet) one observes. Consequently, a body hotter than absolute zero should immediately emit all its energy as ultraviolet and X-ray radiation (which was not known when the problem was first stated) and cool down to

would use later to understand the H atom. Planck considered the possibility that light, which he knew has obvious wave-like properties, could also behave like particles—photons—with well-defined energies. By making this radical (at the time) assumption, Planck could accurately predict the shape of the thermal spectrum for the first time. He also found that the basic properties of a thermal spectrum depend on only one variable, the temperature of the emitting body. Specifically, the peak of a thermal spectrum obeys the simple relation

$$\lambda_{max} T = 0.290\, cm K.$$

This is known as Wien's Law, and, although it was known before Planck came along, he was the first person to offer a plausible explanation for it. In Wien's Law, λ_{max} is the wavelength at which a thermal source of temperature T emits the most radiation. Note that as T increases, the peak wavelength gets smaller (see Figure 6.15), consistent with our observations that the spectra of the hottest stars in Orion peak in the ultraviolet, while the coolest peak in the infrared. Planck also discovered that a given area on the surface of a hot body emits energy at a rate given by

$$F = \sigma T^4.$$

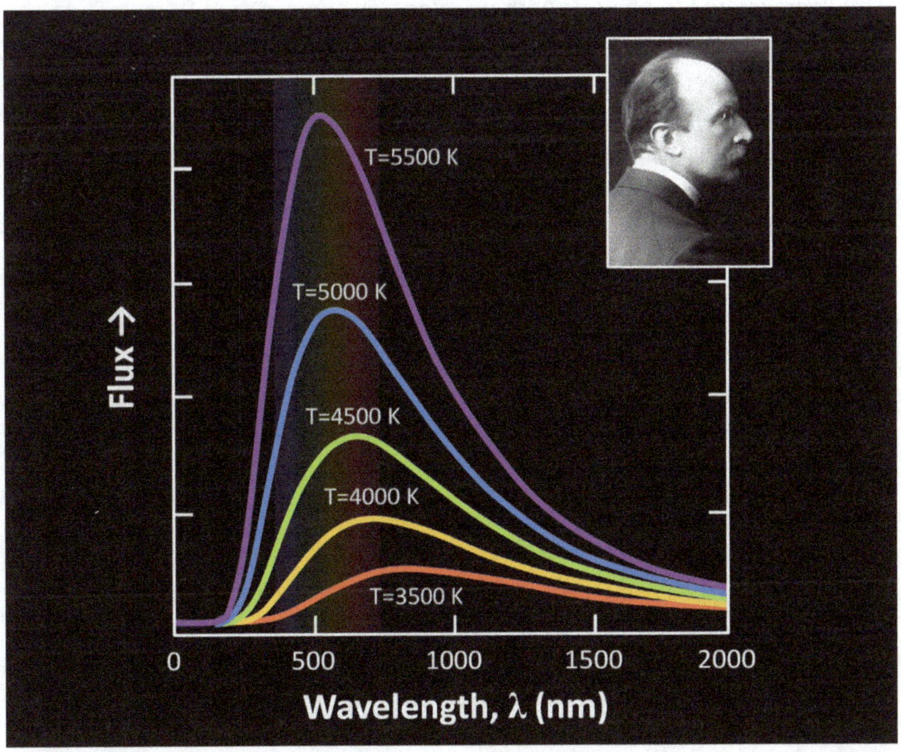

Fig 6.15

The thermal spectra for five bodies emitting radiation at the temperatures shown. Notice that the peak of each curve moves to shorter (bluer) wavelengths as the temperature increases (Wien's Law), and that the total area under each curve also grows as the temperature increases (this is Planck's Law, $F = \sigma T^4$, in action). Stars emit radiation approximately in this way. You can imagine that as the temperature of the source increases, much of the radiation will be emitted in the ultraviolet ($\lambda < 300$ *nm*). Indeed, for $T > 20000$ *K*, quite a lot of UV radiation is produced, enough to ionize the H atoms in the ISM of the Orion Nebula. The range of visible light is shown by the pale rainbow colors. Max Planck, who first determined how to explain the form of a thermal spectrum, is shown in the inset photo.

absolute zero rapidly. Not only would this be bad for our ability to exist in the Universe, but more to the point, that is also not how hot bodies actually radiate their energy.

In this relation—which we'll refer to as *Planck's Law*—the constant σ is the Stefan-Boltzmann constant, which has a value of $\sigma = 5.7 \times 10^{-5} erg/cm^2/K^4/sec$. The quantity F measures how much energy passes through a fixed area on the surface of a body per time, and is known as the *Flux* (we introduced the concept of flux at Destination 4 and will discuss it again in detail at Destination 7). Planck's Law is remarkable: If you double the temperature of a body, it emits *sixteen* times as much total energy (see Figure 6.15)! This is why the burner not only changed color as we heated it, but also became considerably brighter.

Our observations of the stars in the Orion Nebula now make a bit more sense. The blue stars are blue because they are hot, and because they are hot, they emit a large total amount of radiation due to Planck's Law. This is the principal reason that such stars are very luminous. At the other extreme, the red stars are cooler and dimmer, over 60000 times less luminous than the most luminous stars in the nebula. And the key reason for this is that they have lower surface temperatures.[29]

The thermal nature of stellar spectra also means that the brightest, hottest stars emit considerable ultraviolet radiation with wavelengths shorter than 91 *nm*. Why is this important? Recall that for the H atom a finite amount of energy can ionize the atom, an energy we defined as the ionization potential. The energy required to ionize a H atom that has its electron in the smallest ($n = 1$) orbit corresponds to photons with wavelengths shorter than 91 *nm*. Thus, the bluest—and only the bluest—stars in Orion produce large numbers of photons energetic enough to ionize H and other elements within the Orion Nebula. The result is an electrically charged gas—known as a *Plasma*—where electrons and ions continuously interact. Sometimes a free proton and a free electron will collide in such a way that the two *Recombine* to form a neutral H atom. Since the most stable configuration of the electron in the atom is the ground state—the $n = 1$ orbit—the electron rapidly 'cascades' down to that level much like a jostled ball rolls to the lowest level of a set of stairs. Each transition the electron makes within the atom, say, from $n_1 = 3$ to $n_2 = 2$ generates a photon. This particular transition produces a photon with a wavelength of 656 *nm*, which corresponds to the bright, ruby-red emission line we observed earlier in the nebula's spectrum (Figures 6.4 and 6.5). This line is known as Hα (pronounced 'H-alpha'). The next bluer one is Hβ ('H-beta'), and corresponds to the transition from $n_1 = 4$ to $n_2 = 2$. Further lines in this sequence—the Balmer series of lines in the H spectrum—are named Hγ, Hδ, and so on[30], correspond to larger values of n_1, but always $n_2 = 2$ (Figure 6.5). Fortunately, this process does not take long in a typical H atom since the newly created neutral atom is almost sure to soon be

29 But not the *only* reason, as we will see at Destination 8.

30 In fact, there are an infinite number of emission lines possible since n_2 can continue to get larger and larger without bound. As a result, the last 'Greek' line in the Balmer series is Hε ('H-epsilon'). Thereafter, the lines are denoted by the larger value of n in the transition. As a result, the line after Hε is known as H8.

Fig 6.16

The Orion Nebula as seen in near-Infrared radiation where the effects of obscuration by dark clouds is much less than in optical images of the region (see Figures 6.2 and 6.3). The stars here—all comparatively young—formed from the dark clouds of the cold ISM, and ionize the gas of the hot ISM that pervades the nebula.

Photo by M. McCaughrean/ESO.

ionized again when it absorbs an ultraviolet photon from a nearby hot star. It is only during these very short time intervals when a H atom is neutral that the prominent H emission lines in the spectrum of the Orion Nebula are formed.

The bright sky we gaze at here at the Orion Nebula comes about because the hottest stars produce enough ultraviolet radiation to ionize the gas, which then produces a rich emission spectrum as atoms briefly recombine before being ionized again (Figure 6.16). The hot stars are present because star formation has been going on recently in the nebula. It is a delicate balance. Once the hottest stars disappear—and, as we shall see at Destination 8, they will be the first to die—the Orion Nebula will fade from view very soon afterwards. From that time onward, when H atoms in the nebula recombine they will not be reionized because the sources of the energetic, ionizing photons—the hottest stars—will be gone. The gas of the ISM, the star-forming regions, the young stars, all work together in a finely tuned way to maintain the beautiful, but undeniably alien sky of the Orion Nebula.

DESTINATION 6 QUESTIONS

6.1. Prove or convincingly explain why the series of emission lines of Hydrogen that involves transitions to the $n = 1$ orbit—known as the Lyman series—all have shorter wavelengths than all of the Balmer emission lines. The latter are the ones that terminate at the $n = 2$ energy level.

6.2. Imagine that you observed a round molecular cloud and you used the Doppler shift to measure the radial velocities of different parts of the cloud. Describe how the pattern of inferred radial velocities over the face of the cloud would appear in the case of (a) a rotating cloud and (b) and expanding cloud.

6.3. Use the results from Destination 5 to estimate the typical speed of a CO molecule in a molecular cloud with a mean temperature of 10 *K*.

6.4. You tell the police officer that you ran the red light because you were going so fast that it appeared green! Use the Doppler Equation to determine how fast you would have to be moving for this to occur. Since this was a 10000 *km/s* zone, are you guilty of speeding?

6.5. Let's explore this idea that stars form from the collapse of gas from the ISM.

a) How big would the Sun's radius be if its matter were spread out into a volume so that its mean density becomes that of the ISM, 1 particle per cubic cm? Feel free to assume the Sun is composed *only* of H atoms.

b) An expression for how long it takes for things to fall due to their own gravity is $\tau \approx 1/\sqrt{G\rho}$, where $G = 6.67 \times 10^{-8}\ \mathrm{s^2\ cm^3/g}$ is the Gravitational Constant. In this equation, ρ is the mass density in $\mathrm{g/cm^3}$. If the gas is indeed made entirely of H atoms, what are its mass density, ρ of the gas, and the approximate time it would take to collapse? The answer you get will be in seconds if you use the units given in this problem. You might want to convert that result to years to get a better sense of the timescale involved.

c) If the original cloud of gas were rotating at a speed of 1 cm/s, use the expression for angular momentum in Destination 6 to calculate how fast it should be rotating when it gets to be the size of the Sun, or about 700000 *km* in radius.

d) How does the answer to part (c) compare to how fast the Sun actually rotates? You will have to look up the rotation speed of the Sun or figure it out from basic solar information you find on-line and/or in this problem. What do you think this comparison implies probably must happen during this phase of the star-formation process?

6.6. Calculate the ratio of the orbital angular momentum of Jupiter with the rotational angular momentum of the Sun. For this problem assume that you can calculate angular momentum as $J = mV_{rot}R$. Determine the relevant values of R and rotational velocity, V_{rot}, for each case. You can assume that the Sun is 1000 times more massive than Jupiter. What does this tell you about how a protostar may 'solve' its angular momentum problem during the star-formation process?

6.7. Knowing that the wavelength of the Hα emission line of the Balmer series of Hydrogen is 656 *nm* and that this line represents a transition from $n_1 = 3$ and $n_2 = 2$, what is the wavelength of the Ly-a line of the Lyman series for which $n_1 = 2$ and $n_2 = 1$?

6.8. Photons with wavelengths shorter than 91 *nm* can ionize H atoms. What is the ionization potential—an energy—of the H atom?

6.9. Only hot stars produce enough ultraviolet photons to ionize a significant volume gas in the ISM. This question explores how this process operates.

a) What is the energy of a photon of 91 *nm* that can ionize a H atom?

b) Assume that a hot star has a luminosity 600 times that of the Sun, all emitted in the form of 91-*nm* photons. How many photons are produced per second from this star? Recall from Destination 4 that luminosity is the total radiated energy output of a star.

c) It takes H atoms about 100 *yr* to recombine in a region like the Orion Nebula. How many of these ionizing photons are produced in 100 *yr*?

d) If the density of H atoms is 1 cm^{-3} and each photon can ionize just one of these atoms until they recombine, how far out from the star will the gas be ionized, assuming the density is constant to at least this radius? Notice how this result implies that a star can ionize out to a fairly large, but not unlimited, radius.

6.10. Use Wien's Law to figure out the peak wavelengths of the thermal emission from

a) A person;

b) A cold molecular cloud with $T = 20K$;

c) The Sun with $T = 5860K$;

d) A star with a surface temperature of 20000 *K*.

6.11. Compare the luminosities and wavelengths of the peak of the thermal emission of a ball that is heated up to temperatures of 200 *K*, 400 *K*, 1000 *K*, 2000 *K*, and 10000 *K*. For this purpose, assume that the luminosity of the 200 *K* ball is 1.0 in some arbitrary units. Plot the luminosity you derive against the peak wavelength.

6.12. On the web, look up information on some H II region other than the Orion Nebula. Provide information on properties such as the age, size, distance, stellar content, chemical composition, or anything else you might find interesting about the nebula.

Destination 7

EXTRASOLAR PLANETARY SYSTEMS: 51 PEGASI

7.1 A STAR IN PEGASUS

Viewed from Earth, our destinations so far have included some of the more prominent targets visible in the night sky: The Moon (Destination 1), second in brightness only to the Sun in our sky. Mars (Destination 2), which can appear as bright as—and much redder than—any star visible from Earth. And α Cen (Destination 3), one if the brightest stars visible from Earth. The Orion Nebula (Destination 6) is one of the most easily visible non-stellar objects in Earth's sky. Even when we visited 'Anti-Cen' (Destination 5), we used α Cen as our reference point. Like tourists visiting a large city for the first time, we have hit many of the big attractions located in the Sun's immediate neighborhood.

But today we find ourselves approaching a star that, from Earth, appears quite ordinary. *51 Pegasi—51 Peg* for short—is one of the fainter 'naked eye' stars in the constellation Pegasus, the Flying Horse (Figure 7.1). The star's name—or, really, its designation—comes from the fact that it was the 51st star listed in a sky catalog produced by the Astronomer Royal John Flamsteed in the early 18th century.[1] To the naked eye, 51 Peg is barely visible to people with

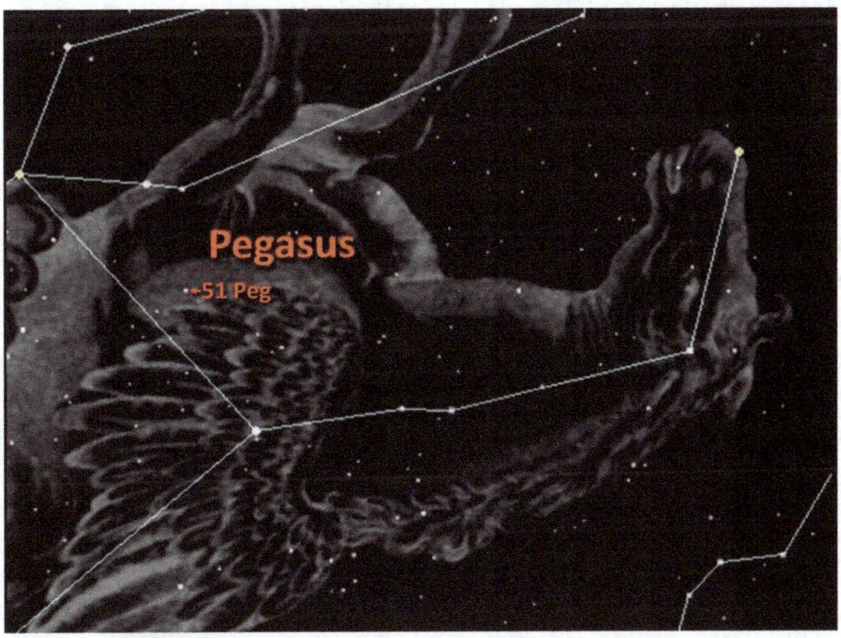

Fig 7.1

A map of a portion of the region of Earth's sky within the constellation Pegasus, the Winged Horse. The labeled star near the base of Pegasus' wing is 51 Pegasi. It is not a particularly prominent star in the Earth's sky, or even of its constellation. From 51 Peg itself, the Sun would appear similarly faint, an inconspicuous star in the inconspicuous constellation Crater, the Chalice.

Image generated M. Mateo using *Starry Night* software.

1 As we noted at Destination 3, astronomers can't seem to help but name, rename, and rename again, individual stars. So, to be (somewhat) complete, other designations for 51 Peg include HR 8729, HD 217014, HIP 113357, SAO 90896, BD + 19°5036, and GC 32003. These designations—and a few others not listed here—all refer to 51 Peg. Alas, 51 Peg is so faint and so inconspicuous, it was never graced with a more poetic name by ancient astronomers.

good vision from clear, dark sites on Earth. From most cities, the star's feeble light cannot compete with the glare of nearby artificial lights and typically murky skies. Lowly 51 Peg doesn't even represent a key part of the constellation's (Figure 7.1). Why, then, are we visiting this seemingly unremarkable star?

Perhaps we are becoming a bit homesick. We can apply some reasoning—based on information gleaned from our earlier journeys—to surmise that 51 Peg must be very similar to the Sun. For example, the colors of the two stars are nearly identical. In Orion (Destination 6) we discovered that a star's color reveals its surface temperature through Wien's Law. Stars with similar colors must therefore have similar surface temperatures. In this regard, the Sun and 51 Peg are near twins: The surface temperature of 51 Peg is about 5650 *K*, only a couple percent cooler than the Sun (about 5780 *K*).

But the stars are similar to one another in more ways than just temperature. To reach 51 Peg we have had to travel 14.7 *pc* from the Sun, a distance we knew from parallax measurements before we set out on our journey. A long trip, certainly, but we are still very much in the immediate neighborhood of the Sun.[2] So it is not surprising that, if we look *very* carefully, we can just see the Sun from the vicinity of 51 Peg. There it is, right there. In the unremarkable—but still recognizable—constellation Crater, the Celestial Chalice. The Sun's not too bright in this sky, barely visible to the naked eye ... in fact, just what we said about 51 Peg's appearance from Earth. Clearly, the light output of the two stars must be similar if, at the same distance, they appear to be nearly the same brightness.

There is something important in this simple concept. To appreciate this, let's consider some typical star. It is reasonable to think of stars as immense spheres that emit light uniformly over their surfaces.[3] This is not always true. Rotation or tidal forces from companions can force stars to be non-spherical in shape. Most stars also likely have *starspots*—or *sunspots* in the case of our Sun—that cause parts of their surfaces to emit a bit less light than average. But these are typically minor effects and our approximation that stars are uniformly lit spheres is not a bad one. This assumption also makes things a lot simpler since the geometry of spheres is well understood. For example, the surface area, A, of a sphere is given by the equation $A = 4\pi R^2$, where R is the sphere's radius. We can then use this result to define a quantity known as *surface flux*, denoted as F_s, which measures the rate at which energy passes through a given surface area of the star. If we apply this to a spherical star of radius R, we can write

2 We'll really start to appreciate the truth of this point as we explore our home Galaxy in Destination 9.

3 There are two reasons to accept this assumption. First, the Sun is a star and it certainly appears to be closely approximated as a sphere emitting light uniformly over its entire surface. Humans have actually sent probes over the Solar Poles to observe this directly. Second, models of normal stars—we'll explore these later at Destination 8—effectively demand that they be closely spherical in shape in the absence of any other external forces. If they weren't, gravity would quickly readjust their structure until they become spherical in shape and uniformly bright over their surfaces.

$$F = \frac{L}{A} = \frac{L}{4\pi R^2}.$$

The term L is the *Luminosity* of the star, a measure of the total rate at which the star emits energy from its surface.[4] In more familiar terms, luminosity is equivalent to *power*, the rate at which something produces energy. A typical star produces a lot of energy. The Sun, for example, has a luminosity of 3.89×10^{33} *erg/sec*, a value known as a *Solar Luminosity* and denoted as $L_\odot$ (recall the symbol for the Sun is '$\odot$').[5] To put this into perspective, *all* the energy produced by *all* humanity over *all* history amounts to about 10^{-11}—or, if you prefer, 0.000000001%—of the energy the Sun emits in *one second*!

Now let's consider an imaginary sphere centered on a star of luminosity L, but now with a radius we'll denote as D (similar to the spheres we considered at Destination 4; see Figure 4.3). As long as nothing absorbs the radiation on its way to this imaginary sphere, the *same* total energy produced at the surface of the star must eventually pass through this larger sphere. This represents the same principle of *Conservation of Energy* that we introduced at Destinations 5 and 6, and it lets us generalize the previous equation in terms of D instead of R as

$$F = \frac{L}{A} = \frac{L}{4\pi D^2}.$$

If we take D to be the distance to the star, we can see that the flux diminishes as the inverse-square of the distance to the star. Since 'flux' is equivalent to 'brightness,' this result confirms the Inverse-Square Law for Brightness—namely, $F \propto 1/D^2$—that we encountered at Destination 4. But there is a new factor now, the luminosity. This makes sense. A star can appear faint to us for two reasons. It may simply be far away. Or it may truly have a low luminosity. Or some combination of both. The fact that the brightness—that is, the flux—of the Sun as seen from 51 Peg is similar to the brightness of 51 Peg as seen from the Sun means that the two stars must have similar luminosities. After all, we are comparing how the two stars appear when viewed from the same distance (51 Peg as seen from the Sun, the Sun as seen from 51 Peg). With this powerful new tool that relates distance, brightness, *and* luminosity, we can now be more precise about what 'similar luminosity' means. By carefully noting their relative fluxes (brightnesses), we can infer the luminosity of 51 Peg to be about 20% higher than that of the Sun[6], or approximately $1.2L_\odot$. Compared to α Cen A, it turns out that 51 Peg is even more nearly a twin of the Sun.

---

4 We introduced the general concept of luminosity at Destinations 4 and 6, but did not define it as precisely as we have here.

5 That luminosity is equivalent to 3.89×10^{26} *W*, where '*W*' stands for the fairly commonly known unit *Watts*, which is equivalent to *joules/sec*. That's a lot of light bulbs!

6 This may not seem all that similar to the Sun. After all, we can readily notice a difference between someone who is 5 *ft* tall and one who is 6 *ft* tall, a 20% variation in height. But as we shall discover starting in Destination 8, stars can be millions of times more

The relation between distance, brightness, and luminosity lets us do something more. Something very useful. If we have information that allows us to infer the luminosity of a star without measuring its distance—say from its color or spectrum, using methods we shall start to explore at Destination 8—then by simply measuring the star's brightness, *we can derive its distance*. This relationship represents a critical link between how stars *appear* to us and their *intrinsic* properties. It also greatly expands the number of stars for which we can estimate distances compared to the much smaller number for which parallaxes can be measured (see Destination 3).

And so here we are. At 51 Peg. A star that appears to be a near-twin of the Sun. Seems a long way to have come to visit something so familiar. But we did not journey here to study this star. We came to 51 Peg to see its planet. The first planet ever discovered orbiting a star beyond the Solar System.[7]

7.2 WHAT WE REALLY CAME TO SEE

Now that we're here, it's obvious. This planet is huge! Its mass is at least half that of Jupiter, the Sun's largest planetary companion, at least 150 times more massive than Earth.[8] But any similarity with the Planets in the Solar System ends there. Jupiter is over 5 *AU* from the Sun and has an orbital period of nearly 12 *yr*. This planet is entirely different. It is nearly on top of its sun[9]

---

luminous than the Sun, and tens of thousands less luminous. The comparatively mild 20% difference in luminosity between 51 Peg and the Sun really is not very big at all.

7 This is not quite true. A group of three planets was found a few years before the planet orbiting 51 Peg was discovered. Those planets orbit a highly unusual star known as a *Pulsar*. We'll discuss pulsars in detail at Destination 12, but suffice it to say that they are remnants of stellar explosions! The planets of this system are unquestionably unusual (see PSR B1257 + 12 in Table 7.1). How they survived—or formed in—the explosion of their parent star is unknown; they remain the only examples of planets in such a bizarre habitat. It is therefore fair to say that the planet of 51 Peg was the first known example of a planet outside the Solar System orbiting a *normal* star.

8 In everyday usage, the word 'massive' has come to mean 'large in size.' But for astronomers, 'massive' very precisely refers to how much mass an object has. We shall discover some objects that are small in size but that are nonetheless massive. When referring to the actual size of an object, as measured using a ruler of some sort, we'll use the terms 'large' and 'small' as appropriate.

9 We need to be a bit careful now that we are talking about planetary systems beyond the Sun. For our discussion, Capital Letters will be used to refer to our Sun and to the Planets of our Solar System, while lowercase letters will refer to suns and planets of alien extrasolar systems.

(see Figure 7.2), orbiting a mere 0.05 *AU* from the center of 51 Peg, about 100 times closer than Jupiter is from the Sun. And with a period of only 4.2 *days*. By comparison, Mercury—the closest Planet to the Sun—is about 0.4 *AU* from the Sun and requires 88 *days* to complete its orbit. In just about every respect, this *Extrasolar Planet*[10]—a body known as *51 Peg b*—is like nothing we've seen or imagined before. But the presence of 51 Peg b—and the fact we are visiting it—begs an obvious question: How did we know 51 Peg b existed *before* we set out on our journey to see it?

Fig 7.2

A view of 51 Peg b as it orbits its sun, 51 Peg. The relative scale of the two objects is approximately correct in this rendering. At maximum elongation—when it appears further from its sun in our new alien sky—51 Peg *b* would be seen at a distance equal to about ten times the radius of the star. From our hypothetical vantage point 1 *AU* from 51 Peg, at such times the planet would appear as a semi-circle, similar to the way a first- or third-quarter Moon looks from Earth. The detailed appearance of 51 Peg b would be strongly affected by the intense heating on its sun-facing side and its view of the deep cold of empty space on the opposite—and perpetually dark—side of the planet.

Certainly, now that we reached a convenient vantage point near 51 Peg[11], we can make out the planet directly (Figure 7.2). But not easily. Even from close range—let's assume we are located about 1 *AU* away from 51 Peg[12]—we have to take care to view 51 Peg b visually due to its proximity to the *star* 51 Peg in our sky. We've already seen that 51 Peg is about the same luminosity as the Sun, so it is no surprise that from this distance the star appears very bright. Indeed, to our eyes, blindingly bright. Just as we cannot stare at the Sun from Earth, we cannot look at 51 Peg from this distance or it could permanently damage our eyes. The brilliant glare often hides 51 Peg b from view. There are times, particularly when 51 Peg b appears most separated from 51 Peg from our vantage point—a configuration known as *Maximum Elongation*—that we can see 51 Peg comparatively easily. At such times, it appears to us as a 'half-planet,' similar in

10 Planets outside the Solar System are also sometimes generically referred to as *Exoplanets*. We will exclusively use the somewhat longer term *Extrasolar Planets* to emphasize that these planets are not associated with our Sun.

11 Technically, the star 51 Peg should be called '51 Peg a' once its planet, '51 Peg b', was discovered. This can get somewhat confusing, as it suggests the star has somehow changed merely because we humans found a planet (or planets) associated with it. For our discussion, we'll refer to the stars by their original names, but keep in mind that, strictly speaking, we should add an 'a' to the end of the names of the stars too.

12 This is our standard assumption. However, it is worth stressing that no planet is known (yet) to exist at a distance of 1 *AU* around 51 Peg. But 51 Peg b certainly does exist! Unfortunately, it is too hot (only 0.05 *AU* from 51 Peg), has too poisonous and violent an atmosphere, and is far too massive for us to visit safely. It's much more comfortable at 1 *AU* where we can still get a good view of 51 Peg b!

shape to the first- and third-quarter Moon seen from Earth[13], but only about 1/5 as large in the sky. If we really want to study this strange planet in detail, we need specialized instruments designed to reduce the glare from 51 Peg. It's not surprising that we didn't actually see 51 Peg b easily until we arrived here.

But wait. If it's tricky to see 51 Peg b *when we are near the planet,* how could we have possibly seen it from Earth, over *3 million times further away*? In fact, we didn't. No one has ever seen 51 Peg b from Earth.[14] It is too deeply hidden in the glare of 51 Peg. And too faint. So we wonder again: How *did* we know 51 Peg b was here?

When we visited α Cen (Destination 4), we introduced a version of Kepler's Third Law—first stated by Newton—that included the masses of the individual objects orbiting around one another in a binary system. As we noted at the time (Destination 4), Kepler did not discover this version of his Third Law because he never had to. The masses of the Sun and *any*—or even *all*—of the Planets of the Solar System is very nearly equal to $1M_\odot$. Consequently, Newton's modification of Kepler's Third Law effectively reverts to Kepler's original version of the Third Law when we consider the Planets of the Solar System (Destination 2). But it is important to appreciate that while Kepler's version may be adequate in some instances, Newton's version is more precise in all circumstances.[15] And it is this improved precision of Newton's version of Kepler's Third Law that explains how we knew 51 Peg b existed despite never having seen it directly from Earth.

Just like stars in a binary system—recall α Cen A and α Cen B from Destination 4—planets and their suns orbit their common centers of mass (Figure 7.3). In the case of a planet–sun pair—such as Jupiter and the Sun or 51 Peg b and 51 Peg—the center of mass is usually very close to the center of the much more massive object. As a result, the orbital path of the sun is *much* smaller than the

---

13 From Earth, the planets Mercury and Venus, both closer to the Sun, also appear as little 'half moons' when they are located at maximum elongation as seen from Earth. Like 51 Peg b from our hypothetical base 1 *AU* away, both Mercury and Venus have orbits that remain entirely inside the Earth's orbit. Because of this, they are known as *Inferior Planets* (nothing personal!).

14 This certainly does not stop people from trying. Some cases where we have directly observed extrasolar planets are noted in Table 7.1 and described elsewhere in this destination.

15 There is a subtle but important distinction between 'accuracy' and 'precision.' An accurate result is one that is close to the 'truth.' A 'precise' result is one that has a small uncertainty. However, a precise result—say a number determined with a calculator to 11 or 12 decimal places—may be highly inaccurate if it was determined using false or improper assumptions. The original form of Kepler's Third Law could be used to get highly precise results for a binary system, but it might be inaccurate because that relation does not account for the masses of both objects in a binary system. On the other hand, Newton's version of Kepler's Third Law is more accurate for such cases since it does account for the masses of both objects. Whether it is more precise depends on whether other effects are in play, such as the presence of magnetic fields that could affect the motions of the two objects, or if the objects are moving at speeds close to that of light (when really bizarre things happen; see Destinations 12 and 17).

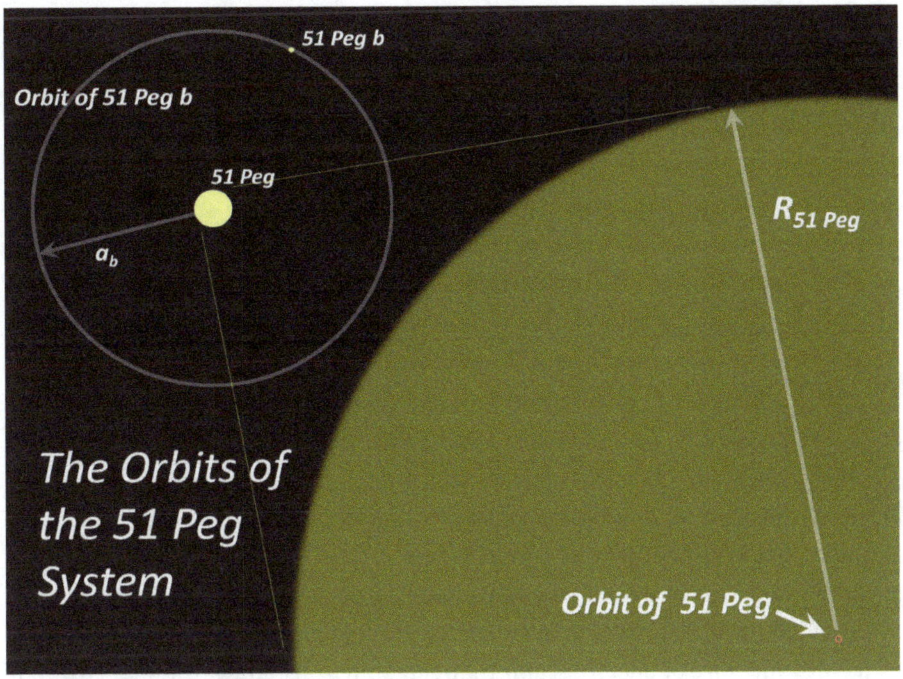

Fig 7.3

An image showing the orbital path of 51 Peg b and its sun, 51 Peg, as they both revolve around their common center of mass. At left, the planets orbit is shown as a gray circle with a radius a_b. The relative sizes of 51 Peg, 51 Peg b and the orbit are all rendered approximately to scale. The orbit of the star about the center of mass of the system is much smaller, shown here as a tiny red orbit inside the zoomed-in view of 51 Peg.

orbital path of the planet (Figure 7.3). Yet both sun and planet require precisely the same amount of time—the orbital period *P*—to complete their respective orbits. This must be true since the center of mass is, by definition, always located between the two bodies. Having further to go but the same amount of time to do so, the planet must move relatively rapidly about the center of mass. On the other hand, the star—which has just as much time to traverse its far shorter orbit—moves much more slowly in response to the planet's movement.

To illustrate this, let's use Jupiter and the Sun as an example. The mass of Jupiter is about one thousandth that of the Sun. Thus, Kepler's Third Law—in its expanded form—can be written as $a^3 = (1+0.001)P^2 = 1.001P^2$, where a is in *AU*, and *P* is in (Earth) *years*. We can then use the center-of-mass equation, $a_\odot M_\odot = a_J M_J$, and the fact that $a = a_\odot + a_J$ (again, '⊙' refers to the Sun, while '*J*' refers to Jupiter) to show that the center of mass of Jupiter and the Sun is located about one thousandth of the way between the two bodies. The Sun is obviously the more massive of the two, so the center of mass is located closer to it and always directly on the line between it and Jupiter. Since Jupiter's orbital radius is about 780 million *km*, this tells us that the Jupiter–Sun center of mass is about one one-thousandth of this distance—about 780 thousand *km*—from the center of the Sun. The Sun's radius is about 700 thousand *km*, so this places the center of mass of the Jupiter–Sun system just outside the Sun, about 80 thousand *km* above its surface. This is the location—the center of mass of the Sun–Jupiter system—that both bodies orbit.

We already understand what this means for Jupiter. Since its distance from this center of mass location is essentially equal to its distance from the Sun's center, this has little effect on its inferred orbital properties. Jupiter still requires about 12 *years* to orbit the center of mass of the Jupiter–Sun system. Given the circumference of its orbit ($2\pi a$, where a is the semi-major axis of the orbit), and recalling that *velocity* × *time* = *distance,* this implies Jupiter must travel at an average orbital speed of about 13 *km/s*. But what does this imply for the Sun? When we—and Kepler—considered the Sun to be stationary relative to the Planets, it was, well, stationary. But now that we see the Sun is following an orbit—admittedly a tiny orbit—then it too must be in motion! We've seen that the Sun's orbit due to Jupiter's motion is 780000 *km* in radius, so the Sun must travel the circumference of its orbit—about 4.9 million *km*—in the same 12-*year* period that Jupiter requires to complete its orbit. For the Sun, that's an average speed of 13 *m/s*, about 1000 times slower than Jupiter travels. To put these speeds in perspective, when you are traveling in a car at 70 *mph*, you are zipping along at about 31 *m/s*, about 2.5 times the speed of the Sun as it saunters around its center of mass with Jupiter!

This is a tiny speed for a star. Yet, amazingly, it is possible to measure such minuscule motions. To do this from Earth requires three things. First, you have to have some sort of spectrograph that can split starlight into its constituent colors, each color defined by a unique wavelength/frequency/energy of EM Radiation (see Destination 5). Second, the spectrum of the star must have features—either absorption or emission lines—that can be positively identified as arising from electron energy jumps of specific atoms (see Destination 6; we shall explore stellar spectra in more detail in Destination 8).[16] And third, we need to measure the orbital speed of the star through the Doppler effect (Destination 6). Light from the star will periodically appear to have shifted to the blue then to the red relative to the rest wavelengths as it is approaching us or receding from us, respectively, during its orbital motion. The magnitude of the shift can be converted to radial velocity variations using the Doppler Equation (Destination 6). Careful measurements of tiny velocity changes over time can be used to reveal the presence of a planet and even provide information that can be used to measure its mass.[17]

---

16 This second step sounds harder than it is. The patterns of absorption and emission lines from a very wide range of neutral and ionized atoms have been thoroughly mapped in the lab and from observations of celestial objects. Moreover, it is usually not even necessary to identify the various absorption or emission lines. One has to assume only that the relative positions of the lines remain reasonably constant with time.

17 There are many practical complications with this process. But the most important has to do with the orientation of the orbit of the planet. Imagine we are looking straight down on a planet's orbit so that its motion is entirely perpendicular to our line of sight. Then, no matter how massive the planet, we'd never see a radial velocity change. On the other hand, if we are in the same plane as the orbit of the planet, we get to see the full orbital velocity of the star (and we should see *Transits* as the planet periodically passes in front of its sun; see Figure 7.6). Unless a transit occurs, we therefore have a fundamental uncertainty regarding its mass. For this reason, many of the masses listed in Table 7.1, including for 51 Peg b, are 'lower limits.' The true masses may be larger, perhaps even significantly larger. If too large, however, we'd see radiation from the planets directly.

Even though it is easy enough to list the steps involved, these measurements are not easy to carry out. The Doppler shifts induced by a planet as it orbits its sun correspond to changes in the wavelengths of the stellar absorption lines of about one part per 10 million. In the case of 51 Peg, the star's velocity varies by about 55 *m/s* in response to the orbiting planet (see Figure 7.4). Since this

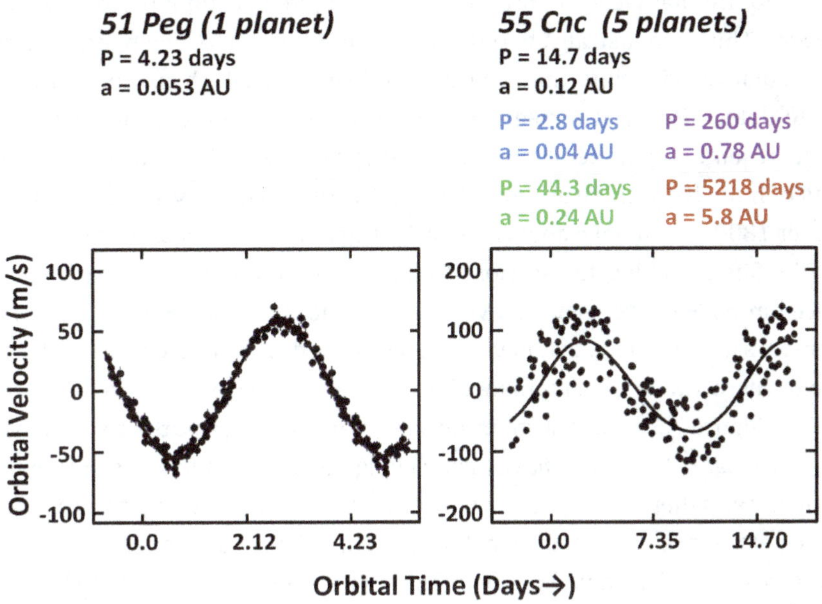

Fig 7.4

(*Left*) A plot of the velocities measured for 51 Peg as a function of time (denoted here as 'Orbital Phase'). The variations in the star's radial velocity reveals the presence of the planet 51 Peg b as it orbits the star with an orbital period 4.23 *days*. The predicted velocities due to the planet are shown as a black line, though it is very hard to see the line under the individual data points. Each data point in this plot refers to an individual measurement of the star's velocity as measured using the Doppler effect (Destination 6). One can, in principle, drive a car as fast as 51 Peg is moving due to the motion of the planet 51 Peg b (55 *m/s*, or about 115 *mph*), though legally perhaps only on a race track or on the Autobahn in Germany! (*Right*) The radial velocity variations over time for the star 55 Cnc. As for 51 Peg (*left*), the plot shows the data fit to a model involving a single planet orbiting the star (this fit is denoted by the solid black line). However, it is clear that in this case there is a lot more scatter in the velocities than expected given the precision of the data (which is essentially as good as the data for 51 Peg). The simplest interpretation is that there are multiple planets in this system. A careful analysis of these data reveals a total of *five* planets in the 55 Cnc planetary system (see Table 7.1).

shift does not require that we actually *see* the planet, it is possible to use this approach to detect planets in any star, in principle at any distance.[18] In recent years it has become possible to measure stellar radial velocities to a precision of 1 *m/s* or better, comparable to the typical walking speed of a person! This makes it feasible to find lower-mass planets, or planets in larger orbits, both of which induce smaller stellar velocity variations than massive, close-in planets.

To illustrate these points, let's consider Jupiter again. This time, though, let's imagine we are trying to detect Jupiter from our current vantage point at 51 Peg. How would we go about this now that both Jupiter and the Sun are 14.7 *pc* away? Direct detection is nearly impossible. Jupiter is located only about 0.3 *arcsec* from the Sun as seen from 51 Peg, and is about 10 billion times fainter than the Sun. It would be much harder to distinguish Jupiter under these

When we don't—as in the case of 51 Peg—that implies its mass is probably pretty close to the lower limit we derive from the Doppler shift.

18 This is strictly true, but in practice, achieving the sort of high precision needed to measure planets requires pretty bright stars that are also fairly small in size (that is, with radii similar to the Sun). All things being equal, the brightest, solar-like stars tend to be fairly close by, so it is no surprise that most extrasolar planets known are comparatively close to Earth.

conditions than it would be to see a firefly's flash in front of a bright spotlight ten miles away. On the other hand, we can certainly measure the Doppler Shift of the Sun as it orbits its center of mass with Jupiter. With a measurement precision of around 1 *m/s*, it should be relatively easy to see the Sun's Doppler Shift of 13 *m/s*. But there is catch. To see the full range in the Sun's velocity—a minimum requirement to ensure we are really seeing evidence of a planet—would require that we wait for at least half of Jupiter's orbital period, or about six *years*! This demonstrates two things. Planets such as 51 Peg b induce large, rapid radial velocity shifts in the spectra of their suns, which are comparatively easy to detect. But it also means that comprehensive searches for extrasolar planets resembling Jupiter require patience, patience, and, yes, more patience.

As of mid-2014, there are over 1500 extrasolar planetary systems containing over 1800 individual planets known. 51 Peg b was the first to be discovered[19] in 1995. Since it is located so close to its sun, the radial velocity variations it induces in 51 Peg are comparatively easy to measure (see Figure 7.4). Many other extrasolar systems that have been discovered closely resemble the 51 Peg system where a fairly massive planet is located very close to an otherwise normal star (see Figure 7.5). Some systems possess planets that orbit their suns in very nearly circular orbits; some have highly elongated orbits. A few have more than one planet and there is one system now known—55 Cancri A—that has a total of five planets, a true extrasolar planetary *system*. A list of some other noteworthy cases can be found in Table 7.1. Complete and more up-to-date lists can be easily

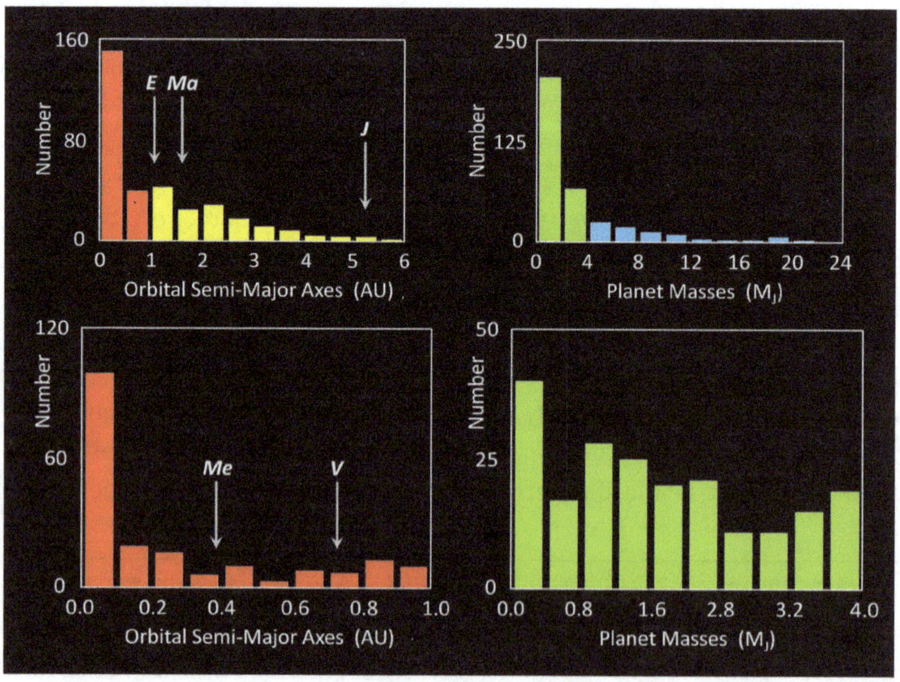

Fig 7.5

(*Left*) The upper-left histogram (red and yellow) shows the distribution of orbital semi-major axes—the radii for circular orbits—for the sample of all extrasolar planets as of mid-2014. At that time, there were about 10 known or suspected systems that have semi-major axes greater than 6 *AU*. The values of the semi-major axes of the orbits of the Earth (*E*), Mars (*Ma*), and Jupiter (*J*) are noted. The lower-left plot (red) shows the distribution between 0 and 1 *AU*. Notice how most of the planets are located very close to their parent suns. The semi-major axes of the orbits Mercury (*Me*) and Venus (*V*) are indicated. (*Right*) The upper-right histogram (green and blue) shows the distribution of masses, in units of the mass of Jupiter, M_J, of known planetary candidates. The lower-right histogram (green) shows the distribution of masses between 0 and 2 M_J. For comparison, the Earth has a mass of about 0.001 the mass of Jupiter.

19 Not quite; see the disclaimer in footnote 7.

TABLE 7.1 Properties of Some Noteworthy Extrasolar Planetary Systems

STAR NAME	DISTANCE (*pc*)	PLANETS	METHOD OF DISCOVERY (1)	PLANET MASSES (M_J) (2)	PERIODS (*days*)	ORBITAL SIZES (*AU*)	COMMENTS
51 Pegasi	15.5	b	RV	>0.46	4.23	0.0521	First extrasolar planet discovered associated with a normal star.
Upsilon (υ) Andromedae	13.5	b c d	RV, Transit RV RV	0.669 >1.92 >4.12	4.62 241.3 1278	0.059 0.830 2.53	One of the first multiple-planet systems discovered. Planet 'b' transits the star so its mass is reasonably well determined (Figure 7.6).
PSR B1257 + 12	300	A B C	Timing Timing Timing	0.00006 0.014 0.012	25.26 66.54 98.21	0.169 0.321 0.417	Planets orbit a pulsar, a remnant of an exploded massive star (Destination 12). These were the first extrasolar planets discovered.
Gliese 876	4.7	d c b e	RV RV RV RV	>0.0184 >0.612 >1.95 >0.039	1.94 30.1 61.1 124.3	0.021 0.130 0.208 0.333	Gliese 876 has a mass much lower than the Sun.
HR 8799	36.8	d c b	Direct Direct Direct	~10 ~10 ~7	~36500 ~69000 ~170000	24 38 68	One of the first examples of planets detected via direct images. Periods all over 100 years.
55 Cancri A	12.5	e b c f d	RV RV RV RV RV	>0.026 >0.801 >0.165 >0.173 >3.54	0.737 14.7 44.4 260.7 4909	0.0154 0.113 0.237 0.781 5.48	Five planets! Figure 7.11 compares this system to the Solar System and Figure 7.4 shows the complexity of the radial velocity variations all these planets induce in the star.
47 Ursae Majoris	14.1	b c	RV RV	>2.55 >0.546	1078 2391	2.10 3.57	A solar-like star with planets that have orbital periods of 3 and 6 years, similar to range in Solar System.

(*CONTINUED*)

TABLE 7.1 *(CONTINUED)*

Epsilon (€) Eridani	3.2	b	Astrometry	1.05	2500	3.38	Closest known extrasolar planet to Earth. Found from 'wobble' of star in sky.
OGLE-2005-BGL-071L	2900	b	ML	3.5	3600	3.6	Discovered using microlensing (Figure 7.7).
Gliese 581	6.2	e b c	RV RV RV	>0.006 >0.050 >0.0168	3.15 5.37 12.9	0.028 0.041 0.073	The innermost planet, Gliese 581 e, may be only 2–3 times more massive than the Earth. But see note (2).
Pollux	10.3	B	RV	>2.76	589.6	1.76	One of brightest stars in sky. Also known as a β Geminorum.

Notes to Table 7.1:

1. Discovery methods: 'RV' = radial velocity measurements; 'Transit' = transit of planet in front of star (see Figure 7.6); 'Timing' = measurements of changes of pulsar arrival times (Destination 12); 'Direct' = direct detection; 'Astrometry' = measurements of 'wobble' of star as it moves across the sky; 'ML' = microlensing (see Figure 7.7 and footnote 20 of this destination).
2. Lower limits are possible only for planets detected through radial velocities and for which the orbital inclination is not determined. See footnote 17 of this destination.

found on-line. Many extrasolar planets have been detected using the Doppler techniques described here. But there are other methods. One of these—known as *Planetary Transits*—involves the detection of a tiny drop in the brightness of a star as its planet periodically passes in front of it (Figure 7.6). This has become one of the most common ways of finding extrasolar planets. Another more esoteric technique is known as *Gravitational Lensing*. This approach takes advantage of the fact that the mass of a planet or star bends the light of a background source: As a star (and possibly one of its planets) passes between us and a remote star (see Figure 7.7), it can act as a *Gravitational Lens* that causes the

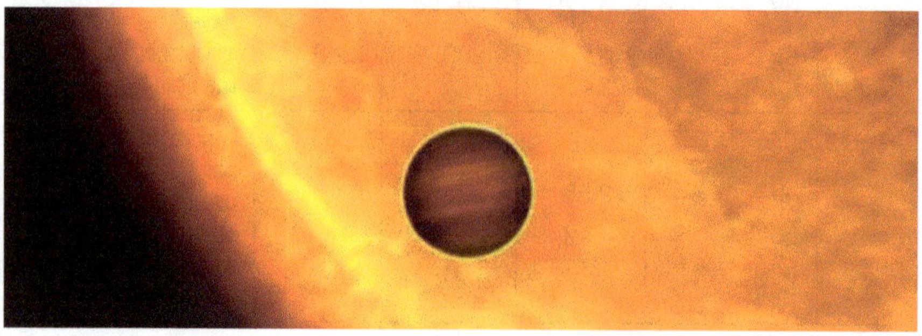

Fig 7.6

Planets can also be detected as they pass between us and their sun during events known as *Transits*. Though we do not see the disk of the star or of the planet—they are far too distant—we do see the light output of the star dip by a tiny amount when the planet blocks part of the stellar disk. If the process repeats at a precise interval, we can reasonably infer something dark is orbiting the star to periodically block its light. Detailed studies of transits can provide information on the sizes and masses of planets, and sometimes, on the structures and compositions of extrasolar planet atmospheres.

Image by G. Bacon/ESA/NASA.

more distant object to rapidly brighten then fade in a highly distinctive manner.[20] The details of the brightness changes not only can reveal the presence of an extrasolar planet associated with the intervening star, but can also provide information regarding the planet's mass (Figure 7.7). A few extrasolar planets have even been found by *Direct Detection*; that is, from specialized imaging that shows the reflected light of the planet directly (see Table 7.1).

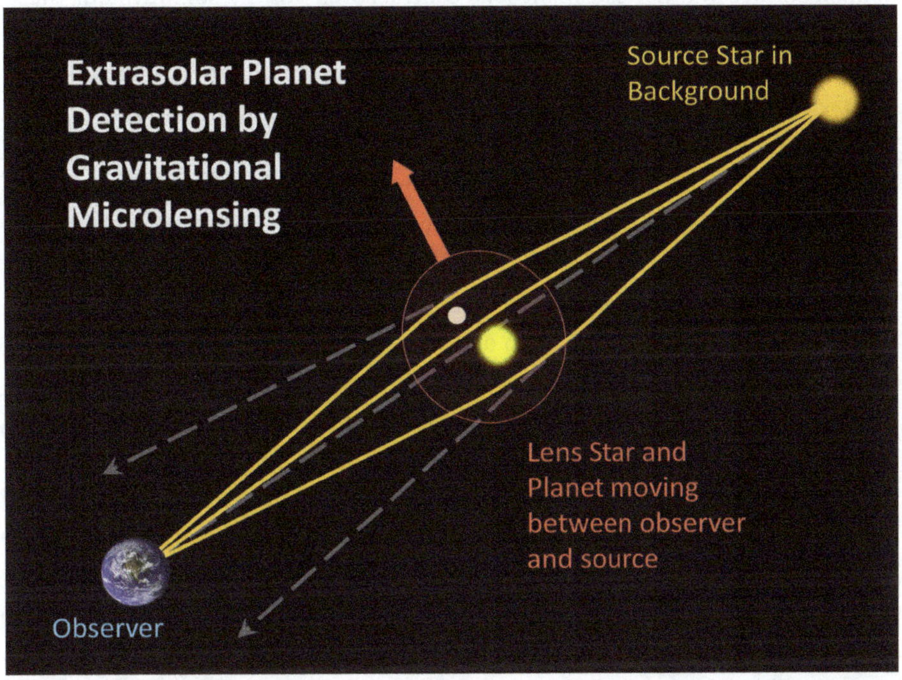

Fig 7.7

This figure illustrates schematically how planets can be detected through a process known as *Gravitational Microlensing*. Imagine first that we have an observer on Earth looking at a star labeled here as the 'source star' (shown in orange at upper right). If nothing lies between the observer and the source, three rays of light from the source would travel as shown by the gray dashed lines. Note that only one of these would reach Earth and we'd see only one image of the source star. If a star and planet happen to move between us and the source—denoted here within a red ellipse and moving along a direction denoted by the red arrow—then they act as a 'gravitational lens' in which their gravity bends the light from the distant source star. When this happens, multiple rays from the source reach the Earth (as denoted by the various orange lines). This causes the light of the source to brighten in a characteristic manner, then fade as the intermediate lens objects pass by. Since the angles involved here are very, very small (micro-*arcsec*, hence the name of the effect), the brightening of the source

7.3 THE PROPERTIES AND FORMATION OF PLANETARY SYSTEMS

In the Solar System all the *Gas Giant Planets* are located fairly far from the Sun in a region known as the *Outer Solar System*. Only *Rocky*, or *Terrestrial*,

20 Gravitational lensing is a concept from Einstein's theory of *General Relativity* (or simply *GR*), a model for how light and matter behave in the presence of gravitational effects. One astounding prediction of GR is that light can be deflected by the gravitating mass, a result that was observed directly during a solar eclipse in 1919, and which serves as the basis for planet detection via gravitational lensing.

Planets—Mercury, Venus, Earth, and Mars—are found in the *Inner Solar System*. Before extrasolar planets were discovered, this all seemed to make sense. The Sun's heat would boil off the gases surrounding any massive planet located too close to the Sun, eventually leaving behind only the rocky core. Thus, the inner planets have thin atmospheres and low masses compared to the gas giants Jupiter, Saturn, Uranus, and Neptune, which are *mostly* atmosphere and have no well-defined surface and, at best, small solid central cores. For the Solar System, these ideas offered a seemingly obvious—and presumably universal—framework to understand what we might see in planetary systems surrounding other stars.

will change on short timescales—days, even hours—as the lens star and planet pass between the observer and the source. This is a difficult method to implement for many reasons, but some extrasolar planets have actually been discovered in this manner (see Table 7.1).

Here at 51 Peg we find something completely different and totally at odds with the distribution of planets in our Solar System. The planet we see before us in this Alien Sky is certainly a gas giant—perhaps twice the diameter of Jupiter—but one that is continuously bathed by the intense radiation and of its nearby sun. Tidal effects have almost certainly synchronized the rotation of 51 Peg b with its orbital period so that—just like the Moon (Destination 1)—it spins on its axis in the same time it takes to orbit 51 Peg, about 4.2 days. Consequently, one side of the planet always faces the relentless heat of its sun and the other forever faces cold, empty space. The planet's thick, deep atmosphere responds to this extremely uneven heating by generating ferocious, never-ending winds that draw hot material to the backside of the planet, where it cools and returns back to the sunlit side via deep currents in the atmosphere (Figure 7.8). It is a wild, truly alien world. But it is not unique. Many other systems resemble 51 Peg (Figure 7.4 and Table 7.1), repeatedly contradicting[21] expectations based on our knowledge of the Solar System.

As we study 51 Peg and its bizarre planet, we can't help wondering, Why is this system—and indeed many other extrasolar systems—so strikingly different from our Solar System? The place to start to try to understand this is one we've already visited: The Orion Nebula (Destination 6). There, we discovered that star formation occurs in close association with compact molecular clouds. But we also discovered that these clouds have a problem. They must somehow shed angular momentum or else they would fail to produce stable stars

Fig 7.8

The sunlit crescent of 51 Peg b is intensely bright from the light of its nearby sun. Due to the planet's synchronous rotation, that side is perpetually heated by 51 Peg, while the other side remains forever in darkness. The imbalanced heating would induce powerful flows of hot material from the hot side of the planet to the cooler side. That motion is apparent here in the form of strong flows to the

21 One might say 'mocking.'

planet's dark side. The gas is so hot it actually glows from its thermal radiation emission (Destination 6), so one can directly see the temperature patterns in the atmosphere. These motions would be accompanied by powerful winds and strong down currents that allow material to flow back to the front side of the planet (as it must to keep the atmosphere stable). The rings visible in this image come from reflections inside our 'telescope' that we use to view 51 Peg b in detail from our safe base 1 *AU* from 51 Peg.

Artwork by G. Laughlin/ UCSC.

that spin as slowly as we observe.[22] Some of the gas clouds in Orion had the distinctive shapes and internal motions (measured from their Doppler Shifts) characteristic of disks surrounding a dense protostar. It does not take too much effort to imagine the Planets of the Solar System formed in a similar sort of disk long ago when the Sun was in the process of forming. The Planets (and most of their moons) lie in very nearly the same plane and virtually all orbit in the same direction (most also rotate in the same direction, though there are some exceptions). Most importantly, the Planets contain a very large fraction of the angular momentum of the Solar System. Even more than the Sun itself. In this respect, the Sun—where most of the mass of the Solar System is located—could not have formed without the Planets. The fact that we see remnants of disks surrounding other, typically young stars provides strong circumstantial evidence that disks must play a crucial role in forming planetary systems (see Figure 7.9).

But disks are not planetary systems. While the space between the planets in the Solar System is effectively empty, a disk around a protostar is presumably filled with stuff, initially in the form of gas, dust, and molecules from the natal cloud in which it formed (see Figure 7.10). Such disks will likely be clumpy, and—in a manner similar to the initial collapse of the cloud—the densest, coolest parts of the disk may start to collapse gravitationally. Or perhaps dust particles in the disk begin to easily stick to one another at this stage[23], in the process forming progressively larger chunks of material known as *Planetesimals*. Either way, significantly massive objects begin to form that, due to their gravity,

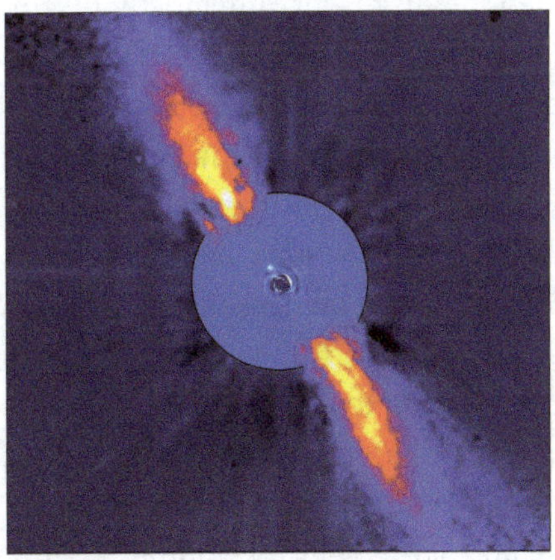

Fig 7.9

The disk of material surrounding the star β Pictoris. The yellow-red material is what is seen when the bright light of the star is subtracted from an image taken in the infrared. From our vantage point on Earth, we are seeing the disk almost from the side. The extent of the disk around β Pictoris is *much* larger than the Solar System, but the star is also much younger than the Sun and may still be in the process of forming its planetary system. As confirmation, the central region of the disk—shown here in the light-blue circular region closest to the star—shows a tiny point of light, one of the planets in the β Pictoris system!

Image by J. L. Beuzit et al./Grenoble Obs/ESO.

attract even more material, which accelerates the process further. Eventually, an object deserving the name 'planet' may form in the disk.

If a large amount of material is still filling the disk, a newly formed planet may 'migrate' to smaller orbits as it trades orbital energy with that of the gas,

22 More to the point, if stars did not shed angular momentum, they would not be able to form at all since they would rotate at a rate so fast that they would fly apart long before reaching their final size.

23 A key problem in our understanding of this process is figuring out just how particles in a protostar's disk can stick together in the first place to start the process.

Fig 7.10

A disk surrounding a newborn star. Much of the original angular momentum of the gas cloud from which the star formed is now in this rotating, flat disk. As molecules and particles bump around in the disk, they start to clump together to form bigger and bigger objects until large chunks of matter—known as planetesimals—begin to form. Some of these are depicted in this image. As even larger objects such as full-fledged planets start to form around the star, the disk can become evacuated of gas and larger particles as is apparent in the region closest to the star. This phase of evolution—from the birth of the star to the formation of planets and the ultimate dissipation of the disk—occurs in just a few tens of millions of years.

Image by L. Calçada/ESO.

molecules, and planetesimals still present in the disk (this is essentially due to drag forces between the planet and the stuff in the disk). This process heats up and helps to disperse disk particles as the planet's orbit shrinks. In some cases—perhaps most—this is a fatal trip. The planet's orbit shrinks so rapidly and completely that it is swallowed up by the central star. But under a surprisingly wide range of conditions, it appears that the planet's migration can be halted. Sometimes just at the last minute. Just above the star's surface. The result of this process might look a lot like the 51 Peg system: A massive planet orbiting very close to its sun. In other cases—perhaps most of the time—the planet may form when the disk is already mostly dissipated, perhaps through collisions of a few large planetesimals than through the aggregation of a huge number of tiny particles. There will then be very little drag to cause the planet's orbit to shrink. Such planets might not migrate far from where they were formed. This could explain why Jupiter resides over 100 times further away from the Sun than does 51 Peg b from its nearly identical sun.

Our voyage to 51 Peg has spawned another, somewhat strange thought. What if it's the *Solar System* that is unusual? Are most planetary systems like the one we see here in 51 Peg? To answer this we have to recall what it took to find planets in the first place. One commonly detected signature of a planet is the radial velocity changes it induces in the spectrum of the star it is orbiting. What favors the detection of this velocity wobble? Well, it is obvious that a large velocity change is easier to detect than a small one. At some point—due to the limited precision of our instruments—we become insensitive to planets below a certain mass simply because they don't induce sufficiently large motions in their parent stars. It's also obvious that it is easier to find planets with short orbital periods compared to planets with longer periods. The planet orbiting 51 Peg takes a very reasonable 4.2 days to complete one orbit. That can be measured with ease in under a week. But, as already noted to do the same with Jupiter would take years. And what if we happen to be looking straight 'down' onto the orbits of the planets surrounding a star? In that case, the motion that planets induce in their parent star would be neither toward nor away from us. We could *never* be able to measure a Doppler Shift to detect planets in this case no matter how precise our measurements.

The net result is that we do not necessarily get a representative picture of the properties of extrasolar systems on the basis of our first sets of observations. Instead, our techniques probably favor the detection of high-mass planets in small orbits that happen to be oriented in just the right way that we can see Doppler Shifts. Astronomers and other scientists refer to such favoritism as *Selection Effects.*[24] In this particular case, the selection effects involved in detecting extrasolar planets can grossly distort what we think of as a 'typical' solar system.

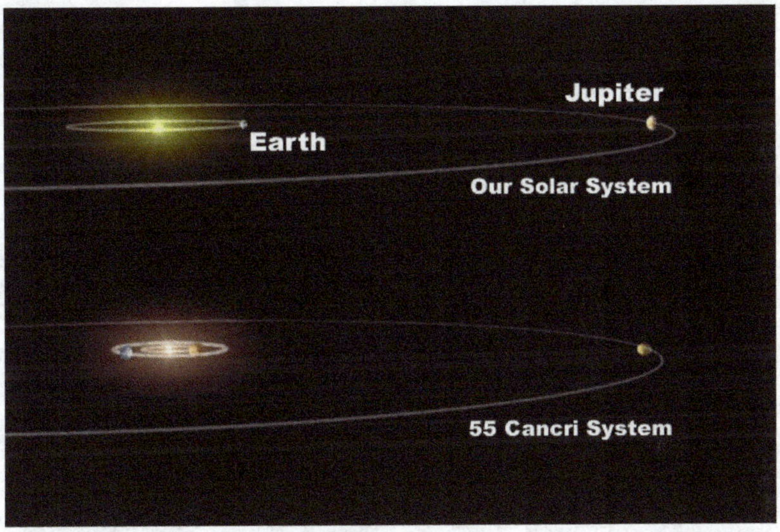

Fig 7.11

A comparison of our Solar System with the extrasolar system of the star 55 Cancri. Maybe our Solar System is not so unusual after all. A detailed summary of the 55 Cancri system is provided in Table 7.1. In this figure, the sizes of the planets relative to their suns and their orbits are grossly out of scale! Figure 7.4 shows how the radial velocity variations observed in 55 Cancri become quite complex when there are so many planets, each with a different mass and orbital size.

Image from NASA.

Over time, extrasolar systems have been found that more closely resemble our Solar System. Systems with multiple planets. Systems with planets that have orbital periods of years, not just days (see Figure 7.11 and Table 7.1). Though many systems are like 51 Peg, there is now strong reason to suppose that they do not represent the most common type of extrasolar planetary system. It is also important to appreciate that only a few percent of all stars being searched for planets contain close-in gas giants like 51 Peg b. Most stars reveal no Doppler shifts at all. Perhaps they contain no planets. Or perhaps they do harbor planets, but planets too low in mass, too distant from their sun, or with orbits too nearly perpendicular to our view for them to induce significant Doppler shifts. But one thing seems certain: More systems will be found. And as they are, we steadily

24 Selection effects routinely—and sometimes severely—afflict all manner of sampling experiments, ranging from political polls, weather forecasts, automobile crash statistics ... and planet detection!

improve our chances of finding extrasolar systems that look more like our own Solar System. Systems that contain planets like the Earth.

7.4 DO PLANETS IMPLY LIFE?

Fig 7.12

Just because planets exist does not immediately imply they support conditions suitable for life. We have already seen how hostile conditions are on 51 Peg b. But conditions can be extreme in other ways. This figure shows what it would be like to be on a planet with a highly non-circular orbit. When the planet is close to its sun (*third image from left*), it is likely to be a very hot world, but when more distant from it sun, the planet is likely to turn into a frigid world (*first image on left*). Can life really develop and thrive on such a world?

Image by C. Thrower.

Most of us signed up on these voyages because we felt that the fantastic sights we were bound to see in Alien Skies would alone be worth the long and arduous journeys. But now that we have discovered extrasolar planets, we are encouraged to broaden our expectations. Will we—can we—find *life*? Over the history of humankind, no evidence of direct contacts with extraterrestrial beings has ever been recorded.[25] But here *we* are, exploring the Universe and finding that

25 This statement repudiates a huge body of literature and the testimonials of a large number of individuals who feel that we *have* been visited by extraterrestrial beings. But, have we? Let's be blunt. Where is the proof? Real, physical proof. On the other hand, we have ample proof that people hallucinate, lie, carry out cons, or simply like to stage elaborately planned practical jokes just for fun. Only people who have never met a scientist would think researchers would cover up such proof. The fame and glory (not to mention research grants!) that would accompany the detection of evidence of an extraterrestrial civilization would be a dream to most scientists. Their methodology would also ensure that anyone could replicate their observations to confirm their claims (or not). But even scientists who are not into the quest for fame or notoriety (and there are many) would still care strongly about scientific integrity. They'd publicize their findings

planets appear to be extremely common. Certainly 51 Peg b itself is not a very hospitable place. It is too hot and too massive for humans—or likely any form of life we'd easily recognize—to live on. Other known extrasolar planets have their own weird obstacles that make them unlikely to support life (Figure 7.12). But the realization that planets exist at all beyond the Sun offers hope of finding life somewhere.

People have been thinking of this for centuries, even millennia. As we've begun to understand more about the Universe, it is possible to at least try to make the discussion a bit more precise. In this spirit, American astronomer Frank Drake developed an equation that he felt could help illustrate the factors that are involved in trying to estimate how many extraterrestrial civilizations might exist in our Galaxy.[26] The so-called *Drake Equation* can be written as

$$N = R_* f_p \, n_e \, f_l \, f_i \, f_c \, L.$$

The symbol N denotes the number of advanced civilizations in our Galaxy that might realistically be able and willing to try to communicate with other civilizations. The factor R_* is the rate at which solar-like stars form in our Galaxy. This may seem like an unnecessary bias—why do civilizations require solar-like stars? We'll soon see, starting in Destination 8, that there are good reasons for this assumption. The term f_p is the fraction of stars that have planets, while n_e is the number of planets per solar system that are Earth-like. Again, this represents another bias towards 'life as we know it,' but one has to wonder seriously if life on a burnt cinder like Mercury or on a frozen gasbag like Jupiter is possible. If so, such life would certainly be utterly unlike anything we're able to imagine. The other factors include f_l, the fraction of planets on which life develops, f_i the fraction that develops intelligent life, and f_c the fraction that develops the technology and desire—as we have—to send messages and travel into space. Finally, the term L is a measure of the lifetime of such a civilization, probably the most important factor in the equation.

The beauty and the danger of the Drake Equation is that we have limited knowledge of the values of most of these factors. We do know that $N \geq 1$ since

for this reason if for no other. The military? As one Defense Department official once said, 'The Army would give its left arm to be able to maintain a secret for half as long as the UFO community believes it has covered up evidence of past alien encounters.' Finally, it seems odd that a civilization able to unlock the secrets of interstellar travel would use that advanced knowledge to cover vast distances to a brilliant blue planet orbiting a non-descript star and then ... deface corn fields! The detection of extraterrestrial life would be an extraordinary discovery requiring extraordinary evidence. And it would likely bring us face to face with technologies vastly beyond anything we've dreamt up so far.

26 We've not yet discussed the Galaxy in any detail (see Destination 9). For now, all we need to know is that the Sun is a member of an immense stellar system—the Galaxy—that contains around 60 billion stars (at least) and stretches some 60000 *pc* across. We've barely ventured out into this system, but we'll learn a lot more about it soon.

the Earth itself has developed a more-or-less technologically advanced civilization. But what of the other factors? Both R_* and f_p can be determined directly, in principle. For example, if we carefully count the number of stars forming in the Orion Nebula and in other similar star-forming regions, we can compare that total to the number of stars in the entire Galaxy. If we also have some information on the lifetimes of stars—and we do, as we shall see later at Destination 8—we can start to get some idea of the rate at which stars form. In our Galaxy, this rate is about $R_* = 1$ star per year.[27] Searches for planets are starting to suggest that many—perhaps all—stars have planetary systems of some sort, so it is probably fairly safe to take the factor f_p to be approximately equal to one.

Now for the wild(er) guesses. The value of n_e is hardly constrained by observations.[28] We do not yet have the precision to detect Earth-like planets around large numbers of other stars (though we are getting close). For now, let's guess that only one in ten planetary systems have planets that resemble the Earth. That is, $n_e = 0.1$. An upper limit to N would come from assuming that $f_l f_i f_c = 1$. And what about L? This factor is particularly hard to estimate. After all, *we* represent the only example of a technological civilization that we know about, and our duration as one is tricky to estimate. Humans have had the capability of space travel for only approximately 100 *yr* (being very generous), ominously similar to the timescale that we've had the power to obliterate ourselves. Consequently, exactly how long a period of time L can be remains unclear. Perhaps one could imagine humans existing as a technological civilization for 10000 *yr*, remaining more or less, as we are now. How about 100 thousand years? That's harder to believe, but may still be possible. One million *yr*? That seems very unlikely. So, as a start, let's imagine that L = 10000 *yr*. If we plug these numbers into the Drake Equation, we find that $N_{Civilization} = 1000$. This also suggests that the factor $f_l f_i f_c$ can't be much smaller than about 0.001. Otherwise, our own existence would not be predicted by the Drake Equation.

If we accept this result—and we are free to question its validity or the values of most of the factors that went into it—what does it mean? Will we likely run into other civilizations on our journey? The volume of the Galaxy is immense: about 10^{12} pc^3 (that's one trillion cubic *parsecs*, an essentially incomprehensible volume!). That would mean that the typical separation between civilizations is about 2000 *pc*. That's a large distance compared to the typical separation between stars near the Sun, about 1 *pc*. Merely sending a message back and forth to a civilization at that distance using a radio signal would take over 13000 years, perhaps longer than humanity will be around to receive the reply! It seems fair to conclude that the mingling of space-faring, intelligent species is

---

27 Sound like a lot? Just remember, this is the total number of stars formed over the *entire* Galaxy, and the Galaxy is really, *really* big! Sound like a little? Well, remember how big a star like the Sun is, and imagine forming one *every* year!

28 Though maybe not for long. People are actively looking for Earth-like planets in many stellar systems using a variety of techniques.

extremely rare in a Galaxy such as ours. But it's even worse than it seems. The number of civilizations we infer from the Drake Equation is the total number to have existed over the history of the Galaxy. For us to converse with alien 'neighbors,' they must exist at the same time we do, or at least at the same time they receive our signals. If the Galaxy is about 10 billion *yr* old (a reasonable guess as we shall see in Destination 20), then a typical technological civilization needs to exist for about 10 million *yr* to ensure that two would be present simultaneously in our Galaxy. And they would *still* have to find one another!

This discussion has focused on intelligent life. But our initial question was if we'd find any life at all. We can still use the Drake Equation, only now we calculate the number of planets that contain any form—primitive or otherwise—of life. This requires that we remove the factors f_i and f_c and that we replace L with the time that life is present on a planet. Instead of a mere 10000 *yr* that a civilized society may exist, we can now consider a time interval of more like 1 billion *yr*, comparable to the amount of time life has existed on Earth.[29] We further set $f_l = 1$, in essence assuming that all Earth-like planets will develop life. With these new factors we get N_{Life} = 100 million, implying that the average separation between life-bearing planets in our Galaxy might be as small as about 5 *pc*!! If this is correct, many stars closer to the Sun than 51 Peg might possess planets that harbor life.

The likelihood that extraterrestrial civilizations are rare is probably just as well. After all, they may not be friendly. Worse, they might even find us to be

Fig 7.13

Toss in some fries and a shake with my human! And supersize it!!

Image from by M. Mateo and J. Bailey

culinary delicacies (Figure 7.13)! On the other hand, finding alien life of any form—not just intelligent life—would be a defining moment in human history. But even this could still be very dangerous, as alien microbes might not be too good for our health. We'll certainly need to be careful when we start visiting

29 Life has existed on Earth for about 3.8 billion *yr*, but only in the past 1 billion *yr* have multi-cellular lifeforms existed. We assume this last value in our estimates of the number of life-bearing planets in our Galaxy.

alien worlds. Still, that would not deter us. As we gaze at this bizarre extrasolar planet in the remarkable Alien Sky of 51 Peg, it dawns on us that we now *know*: Planets do exist beyond the Solar System. And among them, there are likely some planets on which life might form. The fantastic quest to find life beyond our Solar System—long just a dream—has begun.

DESTINATION 7 QUESTIONS

7.1. From 51 Peg how would the flux from the Sun change if the Sun's *luminosity* increased by a factor of two while its distance is unchanged? How about if the Sun's *distance* increased by a factor of two while the luminosity was unchanged?

7.2. Imagine that the flux (brightness) of the Sun is measured from a distance of 10 *pc*. For the purposes of this problem, we will define that flux to equal 1.0 precisely.

a) What is the flux if the Sun is replaced by a star with a luminosity of $0.5\ L_{\odot}$?

b) What is the flux if the Sun were moved to 20 *pc* distance?

c) What is the observed flux if the Sun were replaced by a star of $10^6\ L_{\odot}$ and located at a distance of 1000 *pc*?

7.3. In analogy to what we found for Jupiter, how fast does the Sun move in response to the motion of Saturn? Be sure to determine the Sun's velocity and period induced by Saturn's motion. You will have to hunt for some information regarding Saturn to answer this question.

7.4. Write a short description of the properties of an extrasolar planetary system *other than any of the ones listed in Table 7.1*. Be sure to include basic information about the planetary system (use Table 7.1 as a guide), how and when the system was discovered, and any other interesting tidbit you discover about the system. Also, provide enough background information so that your sources and basic facts can be checked.

7.5. Calculate the periods and orbital speeds of two planets. The first is located 0.1 *AU* from a star of mass $0.8M_{\odot}$. The second is 6 *AU* from a star with a mass of $1.4M_{\odot}$. Comment on the relative detectability of each planet and outline how one might go about discovering each.

7.6. Let's assume that the Earth is 1/340000 the mass of the Sun and that the Earth is the only planet in the Solar System.

a) Recalling that $M_1a_1 = M_2a_2$, where M denotes the mass of a body and a is the distance of the center of a body to the center of mass of the system, how far is the center of mass of the Earth–Sun system from the center of the Sun? Express your answer in kilometers. Is this inside or outside the Sun?

b) How far does the Sun have to travel—in kilometers—as it traverses its tiny orbit about the center of mass during a year?

c) What is the average speed of the Sun during this journey? How fast do *you* travel when walking at a typical speed? Since modern instruments can detect motions as small as about 1 *m/s*, could we detect an Earth-like planet 1 *AU* from a Sun-like star using existing technology?

7.7. Planetary transits (Figure 7.6) are becoming a common way to find extrasolar planets.

a) When Venus passes in front of the Sun—as it did most recently in 2004 and 2012—what fraction of the solar disk does it cover? **HINT:** You will need to find out information about the size and distance of Venus when it is between us and the Sun and use the Small-Angle Formula from Destination 4.

b) If Jupiter were located 0.1 *AU* from the Sun and passed in front of the Sun as seen from Earth, what fraction of the solar disk would it cover?

c) Let's say you observed a star transited by a planet similar to Venus and Jupiter. If you can detect changes of 1% in the star's brightness, which of these planets (either, neither or both) would be detectable from the dip in the star's brightness during a transit?

7.8. Refer to Figure 7.5. What fraction of all extrasolar planets are located within 1 *AU* of their parent sun? What fraction of these planets have masses less than the mass of Jupiter?

7.9. Discuss how you might alter the reasoning behind the parameters used in the Drake Equation or even the reasoning that led to the equation itself. If you feel that some factors could legitimately be different from the ones we adopted, specify what different factors you would adopt and how those alter the results we'd derive from the equation.

7.10. Let's assume that life-bearing planets are found around stars that are, on average, 6 *pc* apart. Let's further assume that stars, in general, are 1 *pc* apart. From this information, how many stars do we have to explore before we can reasonably expect to find one that contains a life-bearing planet?

7.11. You may have noticed that most of the planetary masses listed in Table 7.1 are lower limits. That is, the reported masses represent the smallest possible value and the true masses may be larger. The reason for this is that, apart from a few cases, we rarely know the angle that a planet's orbital plane makes relative to our viewing direction (see footnote 17 of this destination). This problem explores the magnitude of this effect.

a) Calculate the orbital speed of a star of mass $1.5M_\odot$ about its center of mass with a planet of mass $0.002M_\odot$ located 0.1 *AU* from the star.

b) If we, on Earth, are located in the plane of that planet's orbit, what is the full velocity range that we would observe?

c) What is the full velocity range that we measure if the orbit is inclined by 30 degrees to our line of sight?

d) Draw a plot of how the observed velocity range varies with the angle between the orbital plane and our line of sight. Assume that that angle varies from 0 (we are in the orbital plane) to 85 degrees (we are almost perpendicular to the orbital plane).

7.12. Consider the data in Table 7.1 for the extrasolar planetary system 55 Cnc A.

a) Calculate the mass of the central star in the 55 Cnc A system using the data for each of its five planets.

b) Do these value all agree to better than 1–2 percent? If not, speculate why they differ.

DESTINATION 8

THE PLEIADES

8.1 A SKY WITHOUT NIGHT

Even at 'night' it is never really dark here. Our new 'sun' set long ago in the western horizon, but the sensation now is like being in a well-lit stadium. Our shadows radiate from our feet like pale flower petals. But unlike the artificial lights of a stadium, the brilliant sources of light that surround us all have ancient names. High above us stands *Alcyone*, shining nearly as brightly as the full Moon as seen from Earth. To the east we see the glorious grouping of *Electra*, *Maia*, *Celeano*, *Taygeta*, and *Asterope*, the latter actually two stars (Asterope 1 and Asterope 2) that we can individually distinguish close together in the sky. Though fainter than Alcyone, all of these stars shine with a steely blue-white brilliance surpassing the brightness of Venus as seen from Earth. Setting to the west are *Atlas* and *Pleione*. From our current vantage point, these two shine as brightly as Jupiter back home. Welcome to the *Pleiades* (Figure 8.1).

The name of this group of stars derives from stories involving Seven Sisters—the Pleiads (Figure 8.1). The celestial Pleiades—a well-known feature

Fig 8.1

The Pleiades as seen in Earth's sky (*large image*). From any location within the cluster, we are surrounded by hundreds of bright stars in our 'night' sky, some of which are labeled by name in the picture. Although the mythology of the Pleiades (*lower left*) is a story involving seven sisters (the original Pleiads), most people can see only six of the stars in the cluster (Atlas, Alcyone, Merope, Maia, Electra, and Taygeta) with the naked eye. Eagle-eyed observers can just make out Calaeno and Pleione, as well, and some can even see Asterope (which actually consists of two stars). Interestingly, few people see precisely seven stars in a cluster named after seven sisters.

Pleiades photo from NOAO/NSF; *Pleiades* painting by Elihu Vedder/MOMA.

of Earth's winter sky[1]—are located northwest of Orion, initially appearing to most people as a rather large fuzzy patch. On closer inspection, anyone with reasonable eyesight can see that this blob of light is actually a compact grouping of stars. Few people can make out more than six bright stars, though persons with really good eyesight tend to see eight or nine stars here. Why the number 'seven' is always associated with the celestial Pleiades remains a bit of a mystery. But here, from our vantage point inside the Pleiades, the question is not about whether we see seven or six or even nine stars. Along with the handful of extremely bright stars we've already introduced that dominate our view, there are *hundreds* of stars in this sky that rival the brightest stars visible from Earth. The reason is that we are now very close to these stars. *All* of them. The Pleiades is not some lucky projection of unrelated stars in Earth's sky.[2] This is a genuine physical grouping of hundreds of stars, all confined to a small volume of space. The Pleiades is an example of a *Star Cluster*—a group of stars with a common origin and history. At a distance of 140 *pc* (about 426 *ly*), the Pleiades is one of the closest star clusters to Earth.[3] We can use the small-angle formula (Destination 4) to show that over 90% of the stars of the Pleiades lie within about 1 *pc* of the cluster center (Figure 8.2 provides a sense of scale for the Pleiades). When we recall that the nearest star to the Sun is over 1 *pc* away, it's not surprising why the sky here appears so different: This is a comparatively crowded region of space.[4] The result is a sky that is stunningly beautiful—even hypnotic. We could stare at the spectacle for a long time.

---

1 The Pleiades is also frequently referred to as M 45. That's the same 'M' that named the Orion Nebula (M 42; Destination 6): Charles Messier. We'll learn more about Messier later (Destination 11), but recall that he was mostly interested in finding comets. His catalog was a way to warn other comet hunters of celestial objects that might reasonably mimic the appearance of a comet. The inclusion of the Pleiades in Messier's catalog is somewhat strange since it is so large and so obviously composed of individual stars that no one would have mistaken it for a comet. The object just before the Pleiades in Messier's catalog—M 44, also known as the Praesepe—is another obvious grouping of stars that would never have been mistaken as a comet by anyone even casually familiar with the sky. It seems likely that Messier added these two objects to reach what he considered to be a nice round number, 45, in the first published edition of his catalog.

2 As we've noted before, such lucky groupings of stars are called *Asterisms*. The individual stars of most asterisms are far apart in space, completely unrelated to one another. They just happen to appear close together on the sky just as other groups of unrelated stars form patterns we call constellations. In contrast, the stars that comprise the Pleiades are physically associated.

3 The closest star cluster to Earth is the Hyades, which is located quite near the Pleiades in Earth's sky. Both clusters are in the constellation Taurus, though the Hyades are about 30 *pc* away, or about five times closer than the Pleiades. Not surprisingly, most cultures had ancient myths about the Pleiades that intersected with stories involving the nearby Hyades. In Greek mythology, the two groups had the usual jolly interactions that included political intrigue, murder, illicit affairs, and so forth. A veritable celestial soap opera.

4 But don't be fooled. Even here in the Pleiades, the stars are still quite far apart. On average, a given star in the cluster is about 0.05 to 0.1 *pc* from its closest neighbor (not counting companions in binary systems). That's still 10000 to 20000 times the distance from the Sun to the Earth, over 4 million times the distance from the Earth to the Moon!

Fig 8.2

As seen from Earth, the Pleiades are located near the Ecliptic, the path of the Sun across the sky (Destination 1). The Moon is often located near the Ecliptic, so it occasionally passes close to, or even in front of the Pleiades. This image of the Moon—with its face mostly lit by 'Earthshine'—passing close the Pleiades offers a good sense of the cluster's apparent size in our sky. The angular size of the Moon corresponds to about 1.2 *pc* at the distance of the Pleiades. Since virtually all of the bright stars of the cluster can be covered by a circle about two lunar diameters in size, we can gauge the cluster to be a couple of *pc* across. In the solar neighborhood, stars are typically about 1–2 *pc* apart; it is clear that the Pleiades is a comparatively crowded place.

Photo by Sylvain Weiller.

As beautiful as it is, we have to admit that this really isn't such a great place to view a dark night sky. The stars themselves produce as much glare as we'd expect from a moderate-sized city. But we also notice a pale haze over the entire sky filling in the 'empty' sky between the stars. This does not arise from the atmosphere of the planet we've chosen to use as a base to study the Pleiades.[5] The diffuse glow in this sky comes from the light of the stars of the Pleiades that has been reflected off dust particles in the vicinity of the cluster, producing what is known as a *Reflection Nebula* (Figure 8.3). This name helps distinguish this sort of nebula from emission nebulae like Orion, which produces light directly from the atoms making up the ISM of the nebula (see Destination 6).[6] The scattered light from the reflection nebula surrounding the Pleiades makes the sky here even brighter than the stars alone. Which is why there's no true nighttime anywhere within the Pleiades.

Now, generally speaking, a bright sky is not so good for astronomical studies since most problems require study of very faint objects. But for us—for now at least—the bright sky is ok. We did not come here to look at faint, distant objects. We came to the Pleiades specifically to study *its* stars. Here, at the center of this spectacular cluster, we have a front-row seat from which to do so. Star

In addition, the reason the sky here in the Pleiades is so spectacular is not merely that the stars are comparatively nearby. As we shall soon see, many of these stars are also very much more luminous than the Sun.

5 Come to think of it, we do not even know if this planet has an atmosphere! In fact, we don't even know if this planet exists!! But because we do need a base from which to study the Pleiades, we have for our purposes 'invented' this hypothetical planet orbiting a Sun-like star within the Pleiades. We may not be certain such a planet exists, but based on our explorations of 51 Peg (Destination 7), we'd be entirely reasonable to suppose that it may. Certainly, the Pleiades do contain a number of Sun-like stars, so at least that part of our assumption is justified.

6 Interestingly, the dust near the Pleiades that produces this reflection nebula does *not* appear to be related to the cluster. It is likely some diffuse part of the Cold ISM that just happens to be passing by the cluster at the present time.

Fig 8.3

Part of the reflection nebula surrounding the Pleiades. This particular patch of the nebula is known as IC 349 ('IC' stands for 'Index Catalog') and is located near the star Merope (you can see the glare from the star in the upper right). The nebula is visible due to reflected light from Merope and the other stars in the Pleiades, not from emission processes similar to those that we explored in the Orion Nebula (Destination 6). The complex structure of the nebula probably results from gravitational interactions and stellar winds—streams of particles emitted by the stars of the cluster (we'll encounter stellar winds again at Destinations 11 and 12). The nebula is not part of the cluster, but is composed of clouds of gas and dust that happen to be passing near the Pleiades at the present time.

Photo from STScI/NASA.

clusters like the Pleiades (Figure 8.4) turn out to be ideal places to determine basic stellar properties. And to address the most elementary question of all: What makes stars shine?

As we look at the spectacle around us and ponder the nature of the Pleiades as a cluster, we can start to appreciate that its existence is perhaps not so surprising. In Orion (Destination 6) we found a strong concentration of dark, star-forming clouds—many containing protostars—confined to a fairly small

Fig 8.4

Some star clusters, such as M 11 (*upper left*) and NGC 2158 (on right of *upper right* image) are fairly compact, while others, such as M 35 (on left of *upper right* image) are loose in structure, similar to the Pleiades. Some clusters such as NGC 3603 (*lower*) remain associated with the gas from which they formed. M 35 and NGC 2158 only appear to be close to one another. In fact, NGC 2158 is about 4–5 times more distant. As in the Orion

Nebula (M 42; see Destination 6) the 'M' in M 11 and M 35 refer to Charles Messier, while 'NGC' stands for 'New General Catalog,' a compilation of astronomical objects from the late 19th century (well, it was new at the time!). These clusters range in distance from about 800 *pc* (M 35) to over 6000 *pc* (NGC 3603). The Pleiades—a 'mere' 140 *pc* from Earth—is a much shorter trip!

Photos from J.-C. Cuillandre/CFHT (M 11), N. Sharp/NOAO/NSF (M 35) and STScI/NASA/ESO (NGC 3603).

volume of space. Could the nebula be in the process of forming a star cluster, perhaps one similar to the Pleiades?[7] The very first star we visited beyond the Sun—α Cen (Destination 4)—turned out to be a member of a triple system. Even the Sun and 51 Peg have moderately massive companions in the form of planets (Jupiter and 51 Peg b, respectively). For whatever reason, stars seem to congregate into groups. Sometimes with other stars. Sometimes with massive planets. And sometimes, both.

Which reminds us: What about home? From the Pleiades, the Sun—140 *pc* distant—appears too faint to be seen with the naked eye, even for the sharpest-eyed observers. We would need a good pair of binoculars to see our home star shining weakly through the glare of the nearby stars of the cluster and the surrounding reflection nebula. From here at the center of the Pleiades, it becomes starkly apparent that the Sun is just one star among the myriad stars we see surrounding us. And not a particularly prominent star, at that. Our home has never looked smaller. And we have only begun our travels.

8.2 THE STARS OF THE PLEIADES CLUSTER

As we start to study the stars of the Pleiades in detail, we soon appreciate that they possess some surprising systematic characteristics, some of which we had begun to glean at earlier destinations. The brightest Pleiads—including many of the named stars described earlier—are all steely blue in color. Since we carefully surveyed the cluster upon our arrival, we know the individual distances to each star. Combining that information with the observed brightness of each star (or 'flux' to use the more technical term), we can infer their luminosities as described in Destination 7. When we do this, we discover that the blue stars are not only generally quite bright in our sky, but they are also the most luminous stars of the cluster.[8] At the other extreme we find that most of the fainter stars

7 Whether or not the Orion Nebula cluster will really form a long-lived stellar cluster like the Pleiades is uncertain. As a star-forming region like the Orion Nebula evolves, stellar winds and stellar explosions (see Destination 12) can blow out much of the gas surrounding the stars. What appears to be a gravitationally bound cluster today may disperse after its mass decreases when the gas of its ISM is dispersed. Since most stars today are not in clusters—but most probably formed in clusters like the one in the Orion Nebula—it is clear that most star clusters do disrupt. And probably often pretty soon after they were born.

8 From Earth, one can easily measure the brightnesses of stars in a cluster like the Pleiades. Since the cluster is very distant from Earth compared to its size (140 *pc*

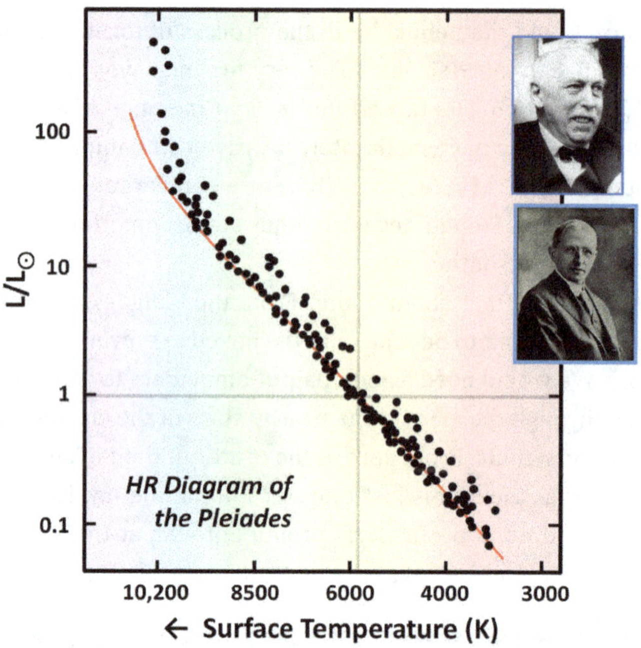

Fig 8.5

The H-R Diagram of the Pleiades, a plot of the luminosities (*y*-axis with luminosity increasing upward) and temperatures (*x* axis, but with temperature increasing to the left) of the stars in the Pleiades. There is a clear relationship between luminosity (increasing upward) and color (redder to the right) defining a main sequence known as the, uh, *Main Sequence* (red line)! Most stars spend most of their lives with properties that place them on the Main Sequence. In practice, star luminosities are determined from the observed brightness (flux) and known distances to a star, while the temperatures are inferred from a measurement of a star's color. Notice that the Pleiades stars shown in this plot exhibit a range of more than a factor of ten thousand in luminosity and a bit more than a factor of three in temperature. As we encounter at future destinations more extreme stars than the ones in the Pleiades, we shall discover that the range in stellar luminosity extends over a factor of about 10 million, and by more than a factor of 20 in temperature. The insets show the portraits of the two astronomers for whom the H-R Diagram is named: Enjar Hertzsprung (*top*) and Henry Norris Russell (*bottom*).

in the sky are considerably redder in appearance. Again, this is no illusion. When we determine their luminosities we discover that these stars generally have very low luminosities. The trend is clear. Luminous stars in the Pleiades are blue, while stars with low luminosities are red. We can even measure the colors of the stars quantitatively by taking the ratio of the brightness of the red and blue light that they emit.[9] In this way we can more precisely define and compare the colors of different stars.

If we plot the luminosities and colors of the Pleiades stars, we discover that the trend we noted earlier is more pronounced than we may have imagined (Figure 8.5). Virtually all of the stars in the Pleiades lie within a narrow band that spans a luminosity range of over a factor of 10 thousand, and a broad range of color. We've already seen that color is an indicator of temperature (see Wien's Law, Destination 6), so the trend in Figure 8.5 confirms our initial impression that the most luminous stars are the *hottest* (bluest) stars in the cluster, while the lower-luminosity stars are the *coolest* (reddest). By measuring the wavelength at

distance compared to 1–2 *pc* size in the case of the Pleiades), the relative brightnesses of the stars are directly related to their relative luminosities. That is, since all the stars are effectively at the same distance from Earth, the brighter stars as seen from Earth are also the more luminous stars. Now that we are in the cluster, the relative distances matter and so we have to know the brightness and distance of each star to infer their luminosities from the distance-brightness-luminosity relation from Destination 7.

9 One way to do this is to take multiple images of the stars, one through a red filter, say, and one through a blue filter. Red stars will appear comparatively brighter when viewed through the red filter while the opposite will be true of blue stars. If we measure the fluxes of the stars in the various filters, we can determine a color for each star. Generally, astronomers define *Color Indices* so that blue stars have more negative values, and red stars more positive values of the indices.

which the stars emit the most radiation, we can apply Wien's Law to determine that the temperatures of the stars in the Pleiades range from about 12000 *K* at the hot end, to well under 2000 *K* for the coolest known stars of the cluster (the coolest, lowest luminosity stars in the Pleiades are not plotted in Figure 8.5).

This simple plot of the luminosities and colors of stars turns out to be so important that it has a name—the *H-R Diagram*—in honor of the two astronomers who first used it extensively: Enjar Hertzsprung and Henry Norris Russell (see Figure 8.5). The diagonal sequence of stars in the H-R Diagram is such a fundamental property of stellar ensembles that it is known simply as the *Main Sequence*. Perhaps not the most imaginative name, but one that, in this case, is aptly descriptive. We shall discover on our journeys that about 90% of *all* stars are found on the Main Sequence at any given time. Certainly this is true here in the Pleiades where virtually all the stars are located on the the Main Sequence in the H-R Diagram. And it was true at home: The Sun's luminosity and color place it solidly within the Main Sequence (see Figure 8.5). As we shall see, it has been on the Main Sequence for a very long time. It will remain on the Main Sequence for a long time to come.[10]

The H-R Diagram turns out to be one of the most useful tools in understanding stars; we'll return to it many times at future destinations. We can start to appreciate its utility by noting that we can use the H-R Diagram to figure out how *big* stars are. To understand this rather odd claim, recall that the total luminosity of a star is related to its size—more specifically, its surface area—and to its surface flux through the relation

$$L = A \times F_{surface},$$

where A is the surface area of the star and $F_{surface}$ is the flux passing through the star's surface.[11] For spherical stars (and we are very well justified in assuming that most stars are mostly spherical in shape; see Destination 7) the surface area equals $4\pi R^2$. That's the easy part. But what about the surface flux? For actual stars, this term is hard to calculate precisely with any techniques we've discussed so far. Since it is impossible to stand on the surface of any star[12], we can't measure the surface flux directly even if we could travel to the stars to try to do so.

But we *can* make a crucial—and reasonable—assumption that helps us estimate the surface flux fairly precisely. We've already used Wien's Law

10 We shall soon get more quantitative about stellar lifetimes. But if you are simply dying to know, the Sun has been on the Main Sequence for close to 5 billion years. It should remain on the Main Sequence for about another 5 billion years. We have plenty of time to pack up if we feel we need to move.

11 Keep in mind that stars do not have solid surfaces like planets such as Earth. However, the final layer of a star—its so-called *Stellar Atmosphere*—is generally very thin compared to the full stellar radius. For most practical purposes, we can treat the star as if it did have a luminous outer surface.

12 See the previous footnote!

(Destination 6) to estimate the temperatures of stars based on the wavelength where the peak radiation is emitted in their spectrum. Stars with spectra that peak in the blue are hotter than stars whose spectra peak at redder wavelengths. This is a bit of a cheat, since, strictly speaking, Wien's Law applies only to thermal emission sources and we saw in Destination 6 that stars do not emit radiation precisely as thermal sources. But they are close. It turns out that the continuous spectra of most—not all!—stars actually closely resemble the spectrum of a thermal source of appropriate temperature. This suggests that we can treat stars *as if* they are thermal sources and not be too far wrong.

This lets us take advantage of a fundamental property of thermal sources—Planck's Law—that we discovered in Destination 6. That is, the surface flux of a thermal source is given by the remarkably simple relation

$$F_{surface} = \sigma T^4.$$

If this can truly be applied to stars—and, really, it's not a bad approximation in most cases—then we can combine these various relations to calculate a star's luminosity as

$$L = 4\pi R^2 \sigma T^4.$$

What's great about this is that when we plot a star in the H-R Diagram, we already know its luminosity and temperature, the latter through the stars' colors. If we hadn't had this information, we could not have plotted the star in the H-R Diagram in the first place! But if we know L and T, this equation tells us that we also know the radius of the star. We can see this most clearly by rearranging the terms in the last equation to solve for the star's radius:

$$R = \left(\frac{L}{4\pi\sigma T^4} \right)^{1/2}.$$

Oddly enough, when we calculate stellar sizes in the H-R Diagram (Figure 8.6) we find that the most luminous blue Main Sequence stars are only about 50–100 times larger than the smallest red Main Sequence stars, despite the fact that the bluest, hottest Main Sequence stars can be up to about 10 million times more luminous than the reddest, coolest stars on the Main Sequence.

Now that we are here in the Pleiades, we notice something else that's rather evident about many of the stars in the cluster. In particular, stars we thought were single as seen from Earth turn out to be systems containing two or more stars. The H-R Diagram of the cluster reveals this, though more subtly. In Figure 8.5, notice how the lower edge of the Main Sequence seems to be well defined, while there appears to be a more indistinct scattering of stars above it. The lower branch of the Main Sequence—the branch that contains most of the stars plotted in this H-R Diagram (Figure 8.5)—is what really defines the true Main Sequence for *single* normal stars. The less distinct scattering of points above it reveals the existence of multiple—usually binary—star systems. Their

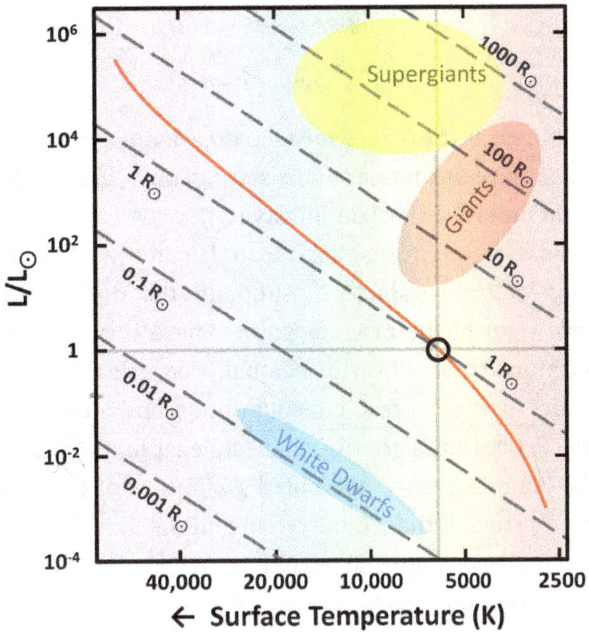

Fig 8.6

A schematic H-R Diagram showing the Main Sequence as a curved solid line (in red) and the axes labeled in terms of stellar luminosity (y-axis, in solar units), *vs* surface temperature (x-axis, in K). The dashed lines show the locations of stars of constant radii, assuming they shine as thermal sources and are spherical in shape; the radii are labeled in units of solar radius. It is clear that the stars in the *lower left* of the H-R Diagram *must* be small, while stars in the *upper right* of the H-R Diagram *must* be large. Once luminosity and temperature are specified, a star's size is non-negotiable! The luminosity and surface temperature of the Sun are noted with the horizontal and vertical gray lines. Note how the Sun is located on the Main Sequence and, naturally, on the $R = 1R_\odot$ line.

anomalous locations in the H-R Diagram reflect the fact that their brightnesses and colors are the sum of the light of two (sometimes more) stars. We can see from the H-R Diagram that these so-called *Photometric Binaries* appear to be relatively common, perhaps amounting to 10–20% of the stars in the cluster.

It turns out that binary stars are important because they can be used to determine the *masses* of stars using Kepler's Third Law. And there are, as we've seen, lots of binaries in the Pleiades. We already measured the masses of the stars in α Cen with one variant of this technique that took advantage of the fact that we could actually plot the orbits of the two stars over time (Destination 4; see Figures 4.5 and 4.6). We have also used Kepler's Third Law to interpret the velocities we measured using the Doppler effect (Destination 6) and derive the masses of extrasolar planets (Destination 7). In the cases of eclipsing binaries—where the orientation and the size of the orbits can be determined precisely—we can use Kepler's Third Law to derive rather precise masses for stars once we measure their orbital periods and orbital velocities from the Doppler effect.

As we carry out these measurements for binary systems in the Pleiades, we uncover another remarkable trend. The bluest, hottest, and most luminous stars are also the most massive, while the reddest, coolest, and least luminous stars turn out to be the least massive. The relationship between the luminosity and the mass of the stars is so pronounced, we can write a fairly precise equation relating the two:

$$L = \begin{cases} M^{3.6}, \; for \; M > 1 \, M_{\odot} \\ M^{2.5}, \; for \; M \leq 1 \, M_{\odot} \end{cases} .$$

This equation—known as the *Main-Sequence Mass-Luminosity Relation*—works as written if we measure the masses of the stars in units of Solar Mass ($M_{\odot}$) and the luminosities in units of Solar Luminosity ($L_{\odot}$).

This discussion has gotten somewhat technical and we have touched upon many interesting stellar parameters. Although the details are important, we must not lose sight of the broader point. There is a surprising—indeed, astounding—harmony between various global properties of stars on the Main Sequence. Hot stars are the largest, most luminous, and most massive stars on the Main Sequence. Cool stars are the smallest, least luminous, and least massive stars on the Main Sequence. We noted a glimmer of this trend when we compared the three stars of the α Cen system and the Sun (Destination 4). But, apart from that hint, there was little reason to suspect that stars would exhibit such broad correlations among so many properties and over such a large range in those properties. What this tells us plainly is that the stellar characteristics we've discussed so far—luminosity, radius, surface temperature, and mass—must all be closely coupled to one another. That is, though stars may differ in detail, it is generally the case that hot, massive, large, and luminous stars located at one end of the Main Sequence are—in some fundamental way—the same sorts of structures as the cool, low-mass, small, and low-luminosity stars we see at the other extreme of the Main Sequence. What we do not yet know is which one of these parameters—temperature, mass, size, or luminosity—is the most influential one that dictates the overall properties of an individual star. We'll return to this fundamental issue a bit later on. But there is another important observable feature of stars to explore first.

8.3 THE SPECTRA OF STARS

We have taken advantage of spectroscopic observations of stars to detect and measure the properties of extrasolar planets at Destination 7 (51 Peg), and, here in the Pleiades, to derive the masses of stars in binary systems. The secret in both cases was to measure the changes in the wavelengths of features in a star's spectrum so we could then use the Doppler effect to infer the presence and measure the masses of a companion object, whether a planet or star, or, as we'll see at Destination 12, things that are far more exotic. This is how we knew that 51 Peg had a planet *before* we had arrived to the system. It is also how we determined stellar masses in binary stars to establish the Main-Sequence

Mass-Luminosity Relation. But what kinds of features are visible in stellar spectra that allow us to measure Doppler Shifts? And why do such features exist in stellar spectra in the first place?

To address this, we must return to the Bohr Model of the H atom that we introduced at Destination 6. There, we saw that electron transitions between orbits in an atom can produce emission lines in the hot, low-density gas of the Orion Nebula when electrons 'drop' from larger, high-energy orbits, to smaller, low-energy orbits. But what if the opposite occurs? To raise an electron to a higher-energy orbit requires energy input. If that energy comes from radiation, the photons 'consumed' by the atom disappear as they pump energy to the orbiting electron. This represents an example of what is technically referred to as *Absorption* by

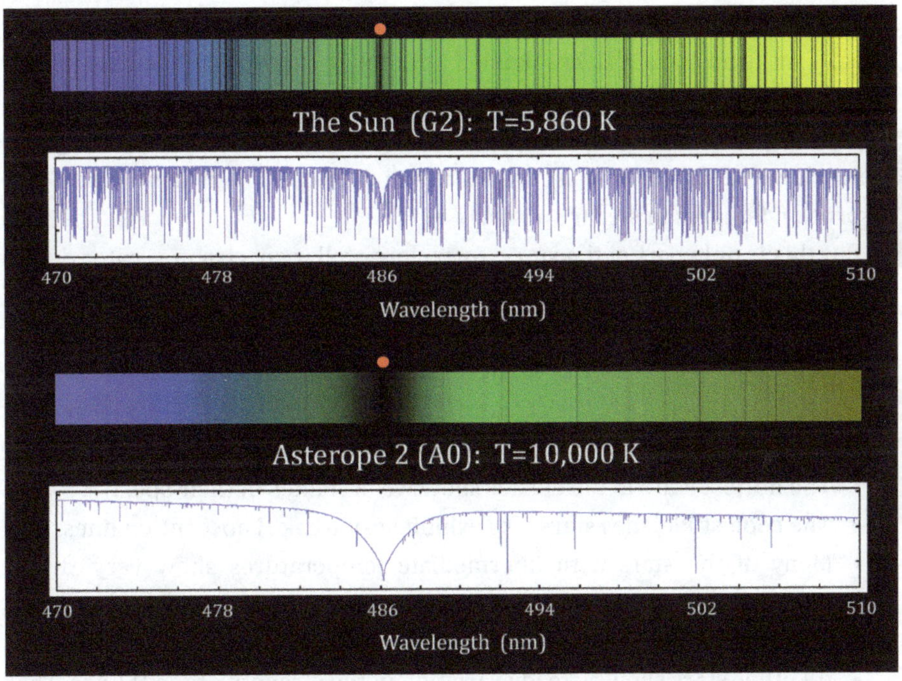

Fig 8.7

A section of the optical absorption spectra of two stars, the Sun, and a bright star (Vega = α Lyrae; see Destination 3), which is very similar to Asterope 2 in the Pleiades. For both stars, the spectrum near the Hβ line (corresponding to the transition of the electron from the $n_1 = 2$ to $n_2 = 4$ orbital levels in the H atom; see Destination 6) is depicted in two ways. The colorful bands show what the spectra would look like to our eyes when viewed through a suitable spectrograph. The lower graphs depict the spectra as plots of brightness (along the *y*-axis) as a function of wavelength (along the *x*-axis). For both the Sun (a G-type star) and Asterope 2 (an A-type star), the Hβ line is clearly visible in either depiction of the spectrum (identified with a red dot for both stars) but obviously much stronger in Asterope 2. Other, weaker absorption lines are also visible in both spectra, arising from atoms of elements such as Iron (Fe), Chromium (Cr), Titanium (Ti), Calcium (Ca), Barium (Ba), Magnesium (Mg), Manganese (Mn), and Nickel (Ni), to name just some! Notice how the spectrum of the cooler star—the Sun—shows a much richer array of weak lines than the hotter Asterope 2. The designations of the spectral types for the two stars

the atom. The features we most commonly see in stellar spectra are examples of such absorption processes, and they are appropriately known as *Absorption Lines* (Figure 8.7) in analogy to the emission lines emitted by the ISM in Orion.

Can the Bohr Model for Hydrogen (H) tell us something useful about the nature of absorption lines seen in the spectra of *stars*? Keep in mind that H is a particularly useful choice for this purpose not only for its simplicity, but also because we have already determined that it is the most common element in the interstellar medium (ISM). Since we know that stars form out of the ISM, it stands to follow that stars probably also contain a lot of H. And since we also know from the Bohr Model what the wavelengths of H emission (and absorption) lines are, we can easily identify them in spectra (Destination 6). As it happens, we do see absorption features in many stellar spectra at precisely the same wavelengths predicted for H from the Bohr model. This tells us that Hydrogen atoms in the

shown here—G2 for the Sun and A0 for Asterope 2—employ the subclasses described in footnote 20 of this destination.

Images and plots by M. Lemke, U. Sternwarte Bamberg, and S. Jeffery, Armagh Obs.

outer parts of stars must be absorbing some of the thermal radiation from their hot interiors to produce the distinct absorption lines we already know arise from H atoms (Figure 8.7). In many stars, the portions of their spectra between the absorption lines resemble the continuous spectrum typical of a thermal source (see the lower panel, Figure 8.7). This is the thermal radiation from the hot stellar interior that escapes into space without being absorbed.

A key result that emerges from these observations is that stars are essentially thermal bodies. This helps to explain why Wien's Law and Planck's Law work so well to account for bulk stellar properties (for example, sizes, surface temperatures, and luminosities). But, it is also true that the presence of absorption features means that stars are not *perfect* thermal emitters like the ones Planck hypothesized. Instead, we see that just before the thermal radiation from the inner, hotter parts of a star can leave a star's surface, it passes through a final, semi-transparent layer—known as a *Stellar Atmosphere*[13]—where some of the radiation is absorbed by atoms present in this layer. It is because of this process—thermal radiation from the hot interior of a star passing through its cooler, lower-density outer atmosphere—that we see both continuous and absorption features in stellar spectra.

But this is only part of the story. To explore stellar spectra in greater detail, we must study how absorption lines behave in the spectra of different types of stars in the Pleiades. We can summarize the behavior that we observe by noting how H absorption lines appear in the spectra of a variety of Pleiades stars that span the Main Sequence of the cluster (Figure 8.5). What we find is rather odd:

- The hottest stars in the cluster show weak H lines in their spectra.
- The coolest Pleiades stars also exhibit very weak H absorption lines.
- Many of the stars with intermediate temperatures show very strong H absorption. Moreover, *all* of the stars that do show strong Hydrogen absorption have very nearly the same colors.
- All other stars show a steady variation in the strength of the H lines. That is, stars just a bit hotter (bluer) than the stars with the strongest H lines exhibit weaker H absorption lines. Spectra of stars that are a bit cooler (redder) than the H-strong stars also exhibit weakened, though still clearly visible, H absorption.

13 Detailed mathematical models of stars show that most stars are gaseous throughout their entire volume. Thus, a stellar 'atmosphere' is not the uniquely gaseous layer we associate with Earth-like planets, but rather just the outermost structure of the star. It is only once photons reach the atmosphere of a star that they have a good chance of escaping the star altogether. Deeper in a star, photons have very little chance of escaping the star and instead are much more likely to be absorbed and re-emitted or scattered. For an astronomer, stellar atmospheres are, by definition, where the absorption features we see in stellar spectra are formed.

The peculiar thing about this behavior is that, up to now, all the stellar properties we have studied have varied systematically along the Main Sequence. Temperature, mass, radius, and luminosity either rise or fall from one end of the sequence to the other. This sort of behavior is what mathematicians refer to as *Monotonic*. But the behavior of the H absorption lines in stellar spectra is decidedly *not* monotonic. The strength of these lines[14] peaks in stars with intermediate temperatures, radii, luminosities, and masses[15], while stars located at the extreme ends of the Main Sequence—the hottest, largest, most massive, and highest-luminosity stars at one end, *and* the coolest, smallest, least massive, and lowest-luminosity stars at the other end—have spectra in which the H lines are virtually invisible.

We can see specific examples of this among the Pleiades stars visible to us in this sky within the cluster. To the east, rising up from the horizon, is Asterope 2. Its spectrum is dominated by H absorption lines (see Figure 8.7). But this star is not the most luminous or hottest star in the cluster. That distinction belongs to Alcyone. Its spectrum shows clear H absorption lines, but ones that are significantly weaker than the Hydrogen lines seen in Asterope 2. One of the fainter red stars of the cluster—its name is HZ 1306—shines like a ruby in our sky. It is not bright like the blue Pleiades members, but it is clearly visible to our unaided eyes from our location inside the cluster. Its spectrum contains a great many absorption lines. So many that it becomes harder to see the continuous spectrum. In contrast, the absorption lines from Hydrogen, though still visible, are now noticeably weak. How can we explain the unusual way that H absorption lines behave in these stars?

The most obvious solution—that these stars have different quantities of H—is not the correct one. We know from studying star formation in the Orion Nebula (Destination 6) that stars form from the ISM, gas that is dominated by H (both in terms of numbers of atoms or total mass). It therefore seems entirely reasonable that young stars will be composed of a similarly high fraction of H atoms. And yet, the stars in the Orion Nebula show the same behavior we see here in the Pleiades: Hot and cool stars in the Orion Nebula exhibit weak H absorption, while stars with intermediate temperatures have the strongest Hydrogen lines.

The secret to understanding stellar absorption spectra is more subtle and depends sensitively on the structure of atoms that create the absorption lines. Recall that spectral lines are formed when an electron moves from one energy level to another (Destination 6). These levels correspond to very specific electron orbits or spin states in the Hydrogen atom. This is true of all atoms, in fact,

14 The 'strength' of an absorption line in a spectrum refers to how much of the light is removed from the continuous portion of the spectrum. In Figure 8.7, the Hβ line in the lower spectrum is said to be 'stronger' than in the same line in the upper spectrum because it has clearly removed a larger amount of the light from the continuous spectrum.
15 Remember, if you specify any of these four parameters, you can figure out the other three for *Main Sequence stars.*

but Hydrogen—as usual—is simplest to understand (see Destinations 5 and 6). If the electron moves from a higher to a lower energy level within the atom, the excess energy is carried away as a photon, which we see as an emission line. But if the electron in a H atom is elevated to a larger, more energetic orbit, it must *absorb* energy. This energy may come from a collision with other atoms, or it may come from a photon of light that has a wavelength that precisely corresponds to the energy difference between the two orbits.[16] If the radiation is the source of the energy, then a photon is lost to the radiation field and an absorption line is produced. It's apparent that this process operates in stars since their spectra are filled with absorption lines (Figure 8.7).

But what if there are very few or no photons around that can elevate the electron in a Hydrogen atom from the smallest orbit to even the next largest orbit? In this case all of the Hydrogen atoms remain in the ground state and *no absorption lines are created!* You can throw as many photons as you want to at the atom, but if none are energetic enough, there will be no excitation of the atom. Hence, no absorption.[17] What about the other extreme? Imagine now that there are many photons interacting with an atom and that the photons are so energetic that they not only can excite the electron within a Hydrogen atom, but they also supply sufficient energy to *ionize* the electron, stripping it completely from the atom. If the electrons have little or no chance of recombining with the bare H nucleus—perhaps because the density of the gas is low, or the number of energetic photons is very large—we'd again see *no absorption lines because the H atoms have no bound electrons.*

Between these extremes, conditions exist where there are a healthy number of photons present to excite the Hydrogen atoms but not so energetic to ionize most or all of the atoms. Under these circumstances, intermediate-energy photons will be efficiently absorbed by the H atoms. Wien's Law tells us that hot thermal sources produce lots of energetic photons, photons that mostly ionize H. At the other extreme, cool thermal sources produce very few photons that can excite H atoms. But intermediate-temperature thermal sources are 'just right,' able to produce many photons that excite but do not ionize H. Photons that can be *absorbed* by the Hydrogen atoms. It is therefore *temperature* that dictates how strong the Hydrogen lines will appear in a star's spectrum.

But not just Hydrogen. All elements—and even molecules—behave this way. For example, Helium—by virtue of the two protons in its nucleus—is much harder to excite or ionize than H. This is especially true for once-ionized He[18]: The second

---

16 Recall that the energy of the photon is given by $E = h\nu = hc/\lambda$, where ν and λ are the frequency and the wavelength of the radiation, respectively.

17 Since we know a lot about the Hydrogen atom now, this may seem pretty obvious. But in fact it was this realization that it was the energy of the photons in radiation that dictated whether or not spectral lines could form—not the intensity of light—that led to the one Nobel Prize that Einstein was awarded.

18 Recall that He can be ionized twice since it contains two electrons in its neutral state.

electron is very hard to strip off given the strong positive charge of the nucleus. As a result, only very hot stars show neutral He lines, and only the absolutely hottest stars—ones with surface temperatures in excess of 30000 *K*—exhibit He II lines (recall that this is the designation for ionized Helium; see Destinations 5 and 6). Calcium has a lower ionization energy than H, and it is relatively easy to excite its outer electrons (the strong positive nuclear charge from its 20 protons is effectively screened by its many negatively charged electrons). Consequently, hot and even intermediate-temperature stars show little Ca absorption since these atoms are highly ionized. But cooler stars—stars like the Sun—have surface temperatures that allow Ca atoms to be excited. The spectra of such stars show Ca absorption. In very cool stars, the surface temperatures not only excite a wide range of atoms, but they even allow molecules (a common one is TiO, titanium oxide) to survive. In such stars the complex absorption spectra of these and other molecules are visible. In hotter stars, molecules are disassociated—the molecular equivalent of being ionized—so such stars do not exhibit molecular absorption features. Again, it is primarily *temperature* that dictates how the spectra appear. Not only for H, but also for all atoms and molecules.

Now, if you think about this a bit, this should strike you as rather weird. Our natural intuition is that if a star has strong Calcium absorption lines in its spectrum, it must contain relatively large quantities of Calcium. It seems 'obvious.' But our discussion of how absorption line strengths vary as the surface temperature of a star changes indicates that our intuition is seriously mistaken in this case. *The appearance of stellar spectra depends primarily on a star's temperature* because the star's temperature dictates the typical energy of the photons that are produced to excite, ionize, or do nothing to the atoms within the star. Surprisingly, the chemical composition of a star has far less effect on the appearance of its spectrum! Of course, if a star hasn't got a single Hydrogen atom in it, it won't show any Hydrogen absorption no matter what its temperature might be. Likewise if it has no Calcium. But the quantities of H or Ca would generally have to drop to extremely low levels before we can expect to detect a strong drop in the strength of their absorption lines. The same is true of most other elements. The strength of the absorption lines of a given element (either in its neutral or ionized form) depends primarily on the star's surface temperature and only secondarily on the quantity of that element in the star. As we will see later on, we *can* use stellar spectra to determine the abundances of elements in stars by measuring the strength of absorption lines. But to avoid being badly fooled, we must very precisely determine the star's temperature first.

If this behavior confuses you, imagine the mystery for astronomers when they first started to record stellar spectra photographically in the late 19th century. At that time there was no Bohr Model to help explain the intimate relation between temperature and stellar spectra. Yet there were literally thousands of stellar spectra that had been obtained as part of extensive photographic surveys.

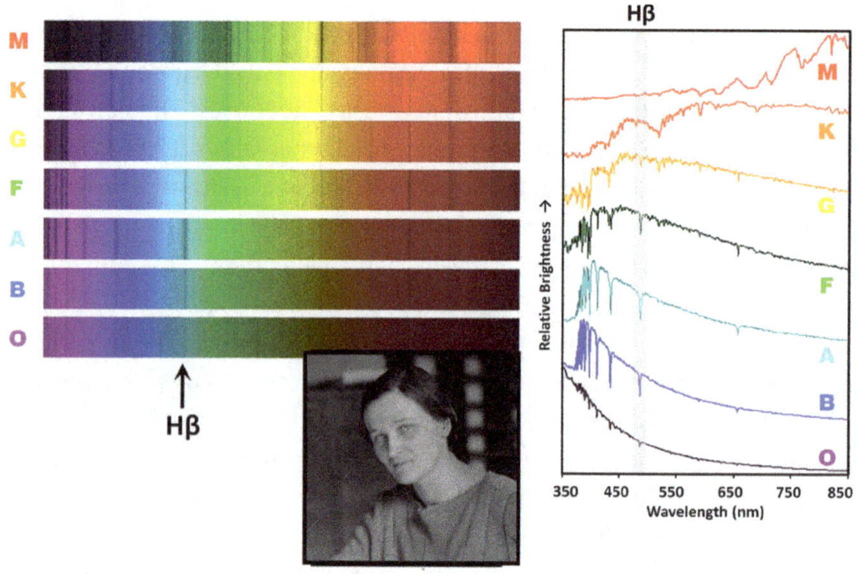

Fig 8.8

Examples of stellar spectra. (*Left*) The principal spectral classes are depicted as they would appear to the eye through a spectrograph in order from coolest (top) to hottest (bottom). The arrow shows the location of the Hβ absorption line (see Figure 8.7). Note how its strength (darkness) varies with spectral class and is strongest for spectral class A. (*Right*) The principal spectral classes are depicted as plots showing the relative brightness (*y*-axis) as a function of wavelength (*x*-axis). The systematic increase in surface temperature from M to O stars is clear in these plots by the way an increasing fraction of the light is emitted at shorter wavelengths along the spectra sequence. The pale vertical line highlights the position of the Hβ absorption line. Plotted spectra (as on the *right*) are generally easier to measure and analyze, though perhaps not as beautiful to look at to the untrained eye! The portrait is of Cecilia Payne-Gaposchkin, the first person to demonstrate that the key variable that dictates spectral types is surface temperature.

Images of spectra from CSIRO and T. O'Brien/U. Manchester.

Spectra that had to be organized in some way. Since the Hydrogen lines are easy to spot in most stellar spectra (see Figures 8.7 and 8.8), it seemed natural to order stellar spectra on the basis of H line strength. Thus, stars of spectra class 'A' have the strongest H absorption lines, 'B' the next strongest, and so on until the letter 'O', which was used to designate the stars that have the weakest H absorption lines in their spectra. When it was later understood that the physical basis for the appearance of these spectra was the temperatures of stars[19], the classification scheme was modified to reflect this new knowledge. As it happens, 'O' stars are the bluest—and therefore the hottest—of all stars, so they became the first *Spectral Type* in the new sequence. 'B' stars, slightly cooler, became second in the sequence. This continued right up to the coolest, reddest stars denoted as class 'M.' The resulting classification scheme—still in use today (Figure 8.8 and Table 8.1)—reflects this ordering by temperature, and, mercifully, the omission of some of the redundant spectral types: O, B, A, F, G, K, and M. Two new classes—denoted 'L' and 'T'—have been added to the sequence in recent years to include extremely cool objects, some of which may be stars, some not (more on this later). Since OBAFGKM (LT) is kind of an odd ordering of letters (you think?), some people find it easier to use a mnemonic than to memorize the sequence itself. One famous—but somewhat silly—example is 'Oh, Be A Fine Girl/Guy, Kiss Me (Lotta Times)'. But, regardless of how one chooses to remember the progression of stellar spectral types[20], the key is to appreciate the sequence is ordered by *temperature*.

19 Primarily due to the work by Cecilia Payne (later, Payne-Gaposchkin; see Figure 8.8).

20 The spectral sequence as described here turns out to be a bit crude. Astronomers can easily see differences in the spectra of stars classically classified as A-type stars, for example. To account for this, most spectral classes are further subdivided into so-called *subclasses* denoted by numbers usually ranging from 0 to 9. For each spectral

TABLE 8.1 Summary of Stellar Properties Ordered by Spectral Types

SPECTRAL CLASS	SURFACE TEMP RANGE (*K*)	COLOR	H LINES	OTHER SPECTRAL FEATURES/ *REPRESENTATIVE STARS*	LUMINOSITY RANGE ($L_\odot$)	RADIUS RANGE ($R_\odot$)	MASS RANGE ($M_\odot$)	MAIN SEQUENCE LIFETIME
O	28000 to > 50000	Blue	Weak	He II and He I lines; Strong UV radiation. *ζ Orionis, δ Orionis.*	90000 to > 1 million	9 to 20	20 to 200	< 10 *Myr*
B	10000 to 28000	Blue-White	Medium	Neutral He lines. *Alcyone; Deneb*	90 to 90000	3 to 9	3.0 to 20	10 to 400 *Myr*
A	7500 to 10000	White	Stron-gest	No He lines; Ionized metal lines. *Asterope 2, Vega*	10 to 90	1.7 to 3	2.0 to 3.0	0.4 to 3 *Gyr*
F	6000 to 7500	Yellow-White	Medium-Strong	Weak ionized metal lines; some neutral metal lines. *Procyon*	1.8 to 10	1.2 to 1.7	1.1 to 2.0	3 to 7 *Gyr*
G	4900 to 6000	Yellow	Medium	Strong ionized Ca, neutral and ionized Fe and other metals. *Sun, 51Peg, α Cen*	0.7 to 1.8	0.85 to 1.1	0.85 to 1.1	7 to 15 *Gyr*
K	3500 to 4900	Orange	Weak	Weaker ionized metals, stronger neutral metal lines, molecular features. *Arcturus*	0.1 to 0.7	0.65 to 0.85	0.65 to 0.85	15 to 30 *Gyr*
M	2000 to 3500	Red	Very Weak	Strong molecular features, strong neutral metal lines. *Proxima Cen, HZ 1306*	0.001 to 0.1	0.17 to 0.65	0.07 to 0.65	> 30 *Gyr*
L,T	< 2000	Deep Red Infrared	Ex-tremely Weak	Strong molecular features, neutral metals, 'weather' (e.g., clouds!) *Gliese 229B*	<< 0.001	0.05 to 0.17	< 0.07 (H burning limit)	No fusion, so not on Main Seq.

Notes to Table 8.1

1. For Main Sequence lifetimes, *Myr* = Megayears = 1 million years; *Gyr* = Gigayears = 1 billion years.
2. L,T 'stars' include Brown Dwarfs and the intermediate objects between them and planets.

As we look around us at the brilliant steely-blue stars of the Pleiades, we now appreciate that the most luminous of these—Alcyone, Electra, and Maia (see Figure 8.1)—are all B-type stars. Asterope 2 is an example of a slightly cooler A-type star. And the Sun? That faint, utterly inconspicuous star in our sky is a G-type star (Figure 8.9) almost completely lost in the continual glare of the hot, luminous Pleiades members.

8.4 WHY DO STARS SHINE?

Our efforts to understand the nature of the stars of the Pleiades has gotten us quite far. But we have not yet explored the most fundamental puzzles that they pose: Why do stars shine? What is their energy source? Why do nearly all stars lie close to the Main Sequence?

In our daily experience, getting energy out of something—making it shine—usually involves burning some sort of fuel. The electricity in our homes back on Earth is generated largely by burning fuels such as oil or coal at a power plant. If the Sun were doing the same thing, it would require about 10^{27} *tons* of fuel and a comparable amount of oxidizer for combustion every second to generate the observed luminosity of the Sun—which you'll recall is $L_\odot = 3.9 \times 10^{33}$ *erg/sec* (see Table 4.1). At that rate, the Sun would consume all its fuel in about 10000–100000 *yr*, depending on the fuel efficiency of the process. Unfortunately for this idea, the Sun is not composed of Carbon (or some other combustible material) and does not have sufficient oxygen to serve as chemical fuel. Moreover, back on Earth, fossils and even ancient human artifacts are known that are much older than this. Since it seems reasonable that the Sun must be at least as old as the Earth (consider our discussion of planet formation at Destination 7), we can immediately rule out chemical burning as the source of the Sun's luminosity.

What about gravity? We do not ordinarily think of gravity as a source of energy, yet we are all well aware that it is. How else can we understand the inevitable lack of volunteers to stand under a brick dropped from a building? Or to jump off a building? The energy inherent in gravity can maim and kill. But gravity can also be used constructively. For example, one can generate electricity by funneling water from the top surfaces of reservoirs into deep chutes that allow the water to fall under the influence of gravity into turbines located in

---

class, smaller subclass numbers correspond to hotter stars. Thus, the Sun is a G2 star, and Asterope 2 is an A0 (pronounced 'A-zero') star. The Sun is therefore a bit hotter than a G4 star and Asterope 2 is a bit cooler than a B8 star. The most luminous stars of the Pleiades—the ones we can easily see with our naked eyes from Earth—range from spectral types B6 to A0.

Fig 8.9

The complexity of stellar spectra is apparent in this stunningly beautiful image showing the solar spectrum from the extreme blue end of the optical region (*bottom*) to the near infrared (*top*). The image shows the colors our eyes would detect if we viewed the Sun's light through an appropriate spectrograph. Each row is a small section of the spectrum stacked up so we can see the full range of colors produced by the Sun in this comparatively compact image (plotted end to end on standard letter-sized paper, this spectrum would be over 25 *ft* long!). Notice the myriad absorption lines present; over 50 different elements are readily identifiable in this image. Some of the strongest (darkest) lines correspond to Calcium (Ca; its lines are visible in the far blue part of the spectrum), Hydrogen (H; many of its lines are seen here, but Hα in the red (Destination 6), is particularly prominent), Magnesium (Mg; a set of three lines from this element are visible in the green-blue region of the spectrum), Sodium (Na; the two strong lines in the yellow part of the spectrum arise from this element). Notice too that despite the immense number of absorption lines, there is light at all intermediate wavelengths. We can see that the Sun's spectrum is essentially the continuous spectrum of a thermal source (Destination 6) punctuated by a large number of fairly narrow absorption features.

Image from NOAO/NSF.

hydroelectric powerhouses at a lower elevation. To get the most energy in this manner, we must allow the water to fall from as large a height as practical. We'd get even more energy if we could increase the gravitational field by increasing the Earth's mass (not a practical option).

The Sun *is* both extremely large and extremely massive, and so its *Gravitational Potential Energy* is enormous. Can we figure out a way to determine if this source can provide its luminosity? Back in the 19th century, the prominent physicist Lord Kelvin[21] actually did this for the first time by calculating how much energy could be extracted simply by using gravity to power the Sun. He first determined the total gravitational energy of the present-day Sun, E_{grav}, from Newton's Universal Law of Gravity (Destinations 2 and 9)

$$E_{grav} = GM_{\odot}^2 / R_{\odot} .$$

He then divided this total energy content by the Sun's luminosity—the rate at which the Sun emits energy, $L = L_{\odot}$—to derive the time over which the Sun would have to shrink to very small size while shining at its observed luminosity

$$\tau_{\odot, gravity} = \frac{GM_{\odot}^2}{R_{\odot} L_{\odot}} .$$

In Lord Kelvin's highly influential opinion, the resulting timescale—about 30 million years—seemed more than adequate to account for the Sun's evolution and the formation of the Earth.

But some people dissented. Darwin and a few famous geologists of the era considered 30 million years to be far too little time. They countered that life could not have developed on such a short timescale on Earth, while some rocks

21 Yes, the same person for whom the Kelvin temperature scale is named.

appeared to show signs of far greater ages. Although Kelvin was able to dodge most criticisms of his concept in his heyday, over time, these and other objections became fatal to Kelvin's idea. The Sun must be vastly older than tens of millions of years since the Earth, and life on Earth, have been around for *billions* of years. The Sun—presumably most stars—must use some form of energy other than gravitational energy as a power source.[22] The fifty years following the demise of Kelvin's explanation for the Sun's energy source must have been an uncomfortable time for astronomers. They could not answer the most basic of all questions: Why do stars shine?

After considerable thought and a few failed explanations, the first part of the solution came in the early 20th century from a crucial contribution by Albert Einstein. His famous equation[23]

$$E = mc^2$$

Fig 8.10

Einstein was correct. This is especially evident when mass is converted spectacularly into energy as through the process of nuclear fusion (*left*). But *Einstein's Equation* operates all the time, even when minuscule amounts of energy are generated by an ant (lower right).

Photo of leaf-cutter ant by Scott Bauer/USDA.

22 Probably the simplest and most convincing argument against Kelvin's idea would have been if one could actually rule out the contraction of the Sun by direct measurement. Unfortunately, the rate of contraction would be about 0.0002 *arcsec*/year. If Kelvin were correct, the Sun would now be about 0.04 *arcsec* smaller than it was 200 years ago, about the time when people first started carefully measuring its apparent size. Unfortunately, the precision of measurement of the Sun's diameter have typically been about 0.5 *arcsec* over that time. So, to be fair, we cannot yet directly rule out that the Sun is contracting at the rate that Kelvin suggested. The age arguments have to suffice for now.

23 Which is surprisingly easy to prove. Just imagine, as Einstein did, what happens if a photon is produced at one end of an isolated box, travels to the other end of the box, then is reabsorbed, keeping in mind that both the box and the photon carry momentum (see Destination 6). The analysis of this situation proves that $E = mc^2$.

demonstrated that a tiny amount of matter of mass, m, is equivalent to an immense quantity of energy (since the speed of light, c, is already a large number, the factor c^2 is huge). The subsequent development of atomic and nuclear weapons brought home to everyone the fundamental truth of this equation (Figure 8.10). What's less clear is that Einstein's Equation also applies to more mundane situations. Indeed, *any time* energy is generated—be it through chemical, mechanical, gravitational, or nuclear processes—mass is converted into energy in accordance with Einstein's Equation. For most normal cases—burning fuel, explosions, crashing objects together, breaking objects by dropping them from great heights—the amount of mass transformed into energy is so small that we cannot measure it directly. Consequently, we don't normally appreciate the continual conversion of mass into energy that occurs all around us all the time (Figure 8.10).

The early 20th century was also a time when the nature of atoms was first being understood in detail. We've already explored Bohr's contribution to the understanding of emission and absorption lines in atoms (Destination 6). His work was based on findings that atoms are composed of ultra-compact nuclei containing all the positive charge (in the form of protons) and nearly all the mass (in the form of protons and neutrons) of atoms. As people started to measure the properties of atomic nuclei, they noted some curious features. For example, although Helium (He) atoms contain four nuclear particles—two neutrons and two protons—the mass of these four particles is less than the mass of four Hydrogen atoms—that is, four protons.[24] For our purposes, we'll denote this mass difference as Δm. Around 1930, people began to realize that *if* it were possible to merge four Hydrogen atoms into a single Helium atom—and there were some ideas how this might happen—a certain quantity of energy, $E = (\Delta m)c^2$, would be released. The reason has to do with the familiar requirement that reactions must conserve energy. Since Einstein's Equation tells us that mass and energy are equivalent, if there is an apparent loss of mass, an equivalent amount of energy *must* be generated. The process of merging atoms to make new atoms and energy is called *Nuclear Fusion* or, often, just *Fusion*.

Keep in mind that the mass difference Δm is tiny in this case. Only 0.8% of the mass of the original four Hydrogen atoms has to be converted to energy to produce one Helium atom and ensure that the overall energy is conserved. This sounds like very little, and it is if you consider only a handful of nuclei. But if you scale this process to an object as big as the Sun, it can become a gargantuan quantity. To see this, let's assume that the Sun started out with only Hydrogen atoms. Our knowledge of the composition of the ISM (Destination 5) suggests that this is not a bad first guess. If these H atoms figure out a way to transform themselves

24 You might notice that we are ignoring the electrons in this discussion. These particles have masses that are about 1/1800 as massive as the individual protons and neutrons that comprise the nuclei. Consequently, we can always safely ignore their contribution to the atomic mass in calculations such as these.

into He nuclei, then up to $0.008 \times M_{\odot} \approx 10^{31}\ g$ could be converted to energy. Using Einstein's Equation, this much mass is equivalent to $E_{\odot,\, nuclear} = 10^{52}\ erg$. If we then divide this total energy content by the energy at which the Sun produces energy—its luminosity—to get the maximum solar lifetime of

$$\tau_{\odot,\, nuclear} = \frac{E_{\odot,\, nuclear}}{L_{\odot}} \approx 10^{11}\ yrs.$$

That's *one hundred billion years.* Now this is an interesting number. As noted earlier, the Earth is at least 3–4 billion years old. The maximum age of the Sun based on using the fusion of Hydrogen into Helium as its energy source is 20–30 times longer. In this regard, nuclear fusion is clearly a viable energy source for the Sun. We'll see that the actual lifetime of the Sun is not quite this long, instead *only* about 10 billion years. Still plenty of time for the Earth—as well as life on Earth—to reach its present state, while short enough for H fusion to supply the total energy required to keep the Sun shining as we see it today.

But wait! There is a 'little' problem with all this. Just *how* does one force four Hydrogen atoms to combine to form one Helium atom in the first place? The H nuclei—protons—are positively charged and particles with the same electrical charge repel. How, then, can protons combine into a single atom? This question raises an even more fundamental one. How do the atoms of *any* element avoid disintegration given the tight packing of protons in their nuclei? They should repel each other with tremendous force since they all have positive charges and are nearly in contact with one another in atomic nuclei. The solution to this riddle comes from the presence of an additional force—the so-called *Strong Nuclear Force*—that can attract protons and neutrons toward one another *if* the particles are in extremely close proximity to one another. In the case of nuclear fusion, that means separations that are not much larger than the sizes of the neutrons and protons themselves.[25] If two protons—equivalently, two H atoms—are located significantly further apart than this, their positive electrical charges dominate and the two are strongly repelled from one another just as we would have guessed. But if the particles can somehow be brought together close enough so the strong nuclear force takes over, even protons can be made to 'stick.' How can this happen?

One approach is to somehow cause H atoms to collide at sufficiently high speed so that they get close enough for the attractive strong nuclear force to overcome the repulsive electric force of the protons. And one way to achieve high speeds for the particles in a gas is to heat the gas. How high a temperature is required? In the case of H fusion, one can calculate the kinetic energy—given

25 Keep in mind that protons and neutrons are not solid little balls with well-defined radii. Still, their effective sizes are minuscule, around $10^{-13}\ cm$. By comparison, the radius of the $n = 1$ orbit of the Bohr atom (Destination 6) is around $5 \times 10^{-9}\ cm$, some 50000 times larger!

by $\frac{1}{2}mV^2$ for a particle of mass m and velocity V—that is required for a H nucleus to overcome the repulsive electric force of another H nucleus. From our discussion of the temperature of particles in the ISM (Destination 5), we know that the average kinetic energy (KE) of particles in a gas of temperature T is given by $KE = \frac{3}{2}kT$, where k is the Boltzmann constant. The surface temperature of the Sun—about 5780 K—does not even come close to heating H nuclei up to the kinetic energy required for fusion. One million degrees K? Still too cool! Temperatures of 15 *million* degrees K are needed to ensure that some H nuclei have enough kinetic energy to be able to fuse into He nuclei at a rate that can account for the Sun's luminosity. This is an inconceivably high temperature. But we need to get used to such hot conditions. We shall encounter even more extreme temperatures in the cores of other stars and also in the core of the Sun itself during future stages in its life cycle. The process of fusing nuclei in a hot gas is known as *Thermonuclear Fusion*, a nicely descriptive name that underscores the key role that high temperatures play for this sort of fusion to occur.

How do stars attain such high central temperatures in the first place? Remember that in the Orion Nebula (Destination 6) we inferred that stars form through a process of collapse in molecular clouds. One immediate and obvious result of this process is that the material in the cloud becomes compressed—due to the gravitation attraction of the mass in the cloud—into a higher-density gas[26]. The particles in the gas accelerate as they fall toward the cloud center, but as the density increases, they also have a much greater chance of colliding with other particles. As these collisions become more frequent, the inward motion due to gravity is converted to more random motions, which we measure as increased gas temperature. Thus, the collapse of the cloud provides energy to *heat* the cloud, especially at its center. This process would continue indefinitely with more and more gravitational energy converted into faster and faster random motions of the gas particles—higher temperatures—in the cloud core. But as the temperature reaches the threshold where Hydrogen fusion can begin, the energy generated from the fusion reactions can now maintain the high gas temperature and high pressure needed to resist the inward pull of gravity. At this stage, gravity is no longer needed to provide the star's energy, and so the onset of fusion reactions can actually *halt* the collapse while providing the star's new source of energy. At least for now.[27]

There are limits to this process. In particular, the 'ignition' of nuclear reactions in a star's core requires a certain minimum temperature. To attain that

26 Density is defined as mass divided by volume. If the cloud does not initially lose much mass, then as it contracts, the volume gets smaller. Consequently the density increases. Even if the cloud loses mass through stellar winds, disk formation, or some other process, the cloud cores can still increase in density significantly since the centers of the clouds collapse more rapidly than the outer regions do.

27 If the Sun were gravitationally unstable, its structure would change radically on timescales of *minutes*! We know the Sun, and other stars, are more stable than this.

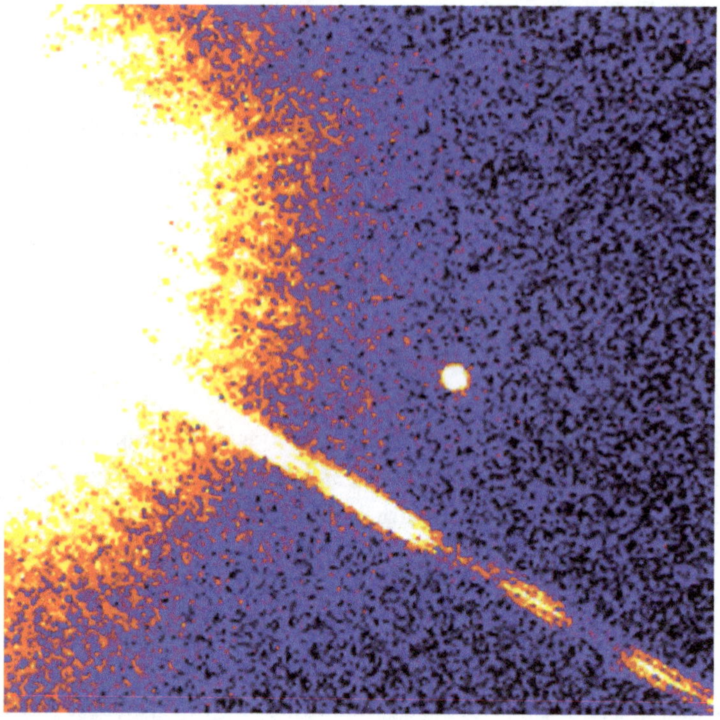

Fig 8.11

One of the first brown dwarfs discovered was this one, a companion of the low-luminosity M-type Main Sequence star Gliese 229 (the brown dwarf is known as Gliese 229 B and is the small dot near the center of the picture). Keep in mind that Gliese 229—the bright star in this image—is itself about 60 times less luminous than the Sun, but the brown dwarf is over 1000 times less luminous than its companion. The long streak in the image is an artifact of the telescope that took this picture. Gleise 229 B is about 30 times more massive than Jupiter, well under the limit of $0.07M_\odot$ (about 70 times the mass of Jupiter) below which objects can initiate and sustain core H thermonuclear fusion reactions. Because they are in some sense 'failed stars,' brown dwarfs such as Gliese 229 B shine from thermal radiation while they slowly cool off from when they formed through gravitational collapse. Gliese 229 B has a surface temperature of about 1000 *K* and exhibits a T-type spectrum (see Table 8.1).

Image from STScI/NASA.

temperature, the star must release sufficient gravitational energy to heat the core to this threshold temperature. Since the gravitational energy depends on the mass, *M*, of the star and how far the cloud collapses, *R*, through the equation $E_{grav} = GM^2/R$, and since most molecular cloud stars are roughly the same size, the main variable here is the star's mass. As a result, there is a minimum mass—corresponding to about $0.07M_\odot$—about 70 times the mass of Jupiter—for stars to initiate thermonuclear reactions. Of course, a collapsing cloud does not know this (!), so smaller objects certainly can form. Such 'failed' stars are known as *Brown Dwarfs* (see Figure 8.11). These strange objects slowly fade and cool as they age while the last bit of gravitational energy radiates from their surfaces. But they do stop contracting. Not because fusion reactions begin, but because they become so dense that they can resist further contraction through gas pressure or other means, much as the mechanical strength of the inside of the Earth keeps it from shrinking. The distinction between Brown Dwarfs and planets—particularly gas giants—can be murky. Even Jupiter emits more radiation than it receives from the Sun, probably because it is still leaking away some of the gravitational energy from its formation many billions of years ago. At the other extreme, a star may become so luminous that its own radiation causes it to blow apart (we'll visit stars near this limit at Destination 10). A careful analysis of this effect—known as *Radiation Pressure*—suggests that no stars should exist with masses higher than about 200 $M_\odot$.

It's pretty clear that gravity is calling all the shots when it comes to dictating what a star will be like (or even if an object is to become a star!). For example, it is the *mass* of a star that determines how rapidly the various steps occur during the formation process. The most massive stars collapse most quickly. They also require the highest central temperatures to generate sufficiently high random motions of the atomic nuclei within the star's core to produce enough energy from fusion to halt the collapse. Since the rate of energy production from H fusion reactions rises very rapidly as temperature increases, these stars must also produce much more energy—that is, they have much higher luminosities—than lower-mass stars. In the same way, low-mass stars take longer to collapse and their luminosities and central temperatures are lower. Too low a mass and the 'star' never forms, becoming a Brown Dwarf or planet instead. This interplay between the mass, the internal temperature, and the luminosity of stars is the fundamental reason that we see a well-defined mass-luminosity relation for Main Sequence stars. Because the star must end up as a stable object, the mass of a star also controls how large the star must be. Since a star's size and luminosity fully dictate its surface temperature (see our earlier discussion at this destination and Figure 8.6), the thermonuclear fusion reactions in the cores of stars are the ultimate reason for the existence of a Main Sequence in the H-R Diagram. A simpler way to say this is that *the mass of a star is the principal parameter that determines virtually all of its global properties.* Consequently, if we specify the mass of a star and we verify that the star is on the Main Sequence, we practically know everything about it. We can use the star's mass to derive its luminosity, its size, its surface temperature, its core temperature. Even its spectral type!

There is a very important implication of this discussion. If the total energy of a star is related to its mass—through Einstein's Equation—and the total luminosity is given by the Main-Sequence Mass-Luminosity relation, then we can calculate the total *lifetime*, τ_*, of a star of mass M as

$$\tau_* \propto \frac{M}{L} = \begin{cases} \dfrac{M}{M^{3.6}} = M^{-2.6},\ for\ M > 1\ M_\odot \\ \dfrac{M}{M^{2.5}} = M^{-1.5},\ for\ M \leq 1\ M_\odot \end{cases}.$$

This relation states that a star with a total energy reservoir that is proportional to its mass can emit energy at a rate given by its luminosity for a finite time, τ_*. One interesting consequence of this relation is that massive stars have significantly shorter lifetimes than less-massive stars. It's true that these massive stars have more H fuel to start with. But they must burn that fuel at a much higher rate to achieve the extremely high central temperatures they need to halt their collapse. Cars, among other things, operate roughly the same way. Big ones generally get poorer gas mileage and have less range than small cars, despite having larger gas tanks.

Sophisticated physical models can be calculated for the Sun that reveal its total lifetime to be about 10 *Gyr*, with the current Solar age to be about half that, around 5 *Gyr* (the new unit, *Gyr*, is an abbreviation for *Gigayears*, or 1 billion *years*). Independent age indicators from the Earth, the Moon, and from other bodies of the Solar System confirm that the Sun has lived about 5 *Gyr*. The Sun's expected life expectancy of 10 *Gyr* is significantly less than the 100 *Gyr* we estimated earlier, and for a good reason. Only the gas in the Sun's core ever gets hot enough to fuse H into He through thermonuclear reactions. In practice, this means that only about 10% of the Sun's mass is ultimately involved in generating energy through fusion processes. If we take this detail into account, the 100 *Gyr* lifetime we estimated earlier is reduced by a factor of 10, or about 10 *Gyr*, much closer to the lifetime estimated from Solar models.

What does this all imply about the Pleiades? Since the most massive stars live the shortest time, we can estimate how long the cluster has been around by determining the ages of these stars. The hottest, most luminous, most massive star in the Pleiades is Alcyone. It is definitely on the Main Sequence (well, nearly so; see Footnote 28 below), and its luminosity is about 910 times greater than the Sun's. The Main-Sequence Mass-Luminosity relation tells us that its mass is $910^{1/3.6} = 6.6\ M_{\odot}$. Its lifetime is therefore $6.6^{-2.6} = 0.007$ that of Sun, or about 73 *Myr* (an abbreviation for *Megayears*, or 1 million *yr*). This rough estimate is close to what more precise calculations give for the age of the Pleiades, around 100 *Myr*. Stars more massive than Alcyone almost certainly did form in the Pleiades when the cluster was formed. They are now gone. They ran out of fuel—the H in their cores—at which point they did whatever it is that stars do when this happens.[28] We'll discuss this process—known as *Stellar Evolution*—in more detail at Destinations 11 and 12. But we mustn't miss the key point. By understanding how stars function—how they generate their energy—we can estimate their *ages*. This is an extraordinarily powerful tool.

As we'll see later, the age we derive for the Pleiades—about 100 *Myr*—is a relatively short time for most stars. Of course that's an inconceivably long time span for humans, and even compared to the time over which life has developed on Earth. But it is still wondrous to think that the Pleiades first graced our

Fig 8.12

Ceratosaurus: Master of the Jurassic Earth over 140 million years ago. Alas, he (or is this a she?) never knew the Pleiades. The cluster had yet to be born!

planet's sky late in the Age of the Dinosaurs[29], well before the first mammals began their journey through the hierarchy of life. Most of the dinosaurs never

28 This is why our age estimate for the Pleiades was a bit off. Alcyone (and a few of the other most luminous Pleiads) has already begun to evolve away from the Main sequence (see Figure 8.5) and so the Main-Sequence Mass-Luminosity and Mass-Age relations do not strictly apply to them.

29 Technically, at about the start of the Late Cretaceous Period, the last of the major epochs during which dinosaurs inhabited the Earth.

had a chance to look up in their sky—the Alien Sky of ancient Earth—and gaze at the Pleiades (Figure 8.12).[30] The cluster had not yet been born.

DESTINATION 8 QUESTIONS

8.1. An ionized Helium atom is nearly identical to a neutral Hydrogen atom except that the nuclear force is greater since there are two protons instead of one. Explain why absorption lines of ionized Helium are seen in the hottest Main Sequence stars only.

8.2. How long will the faint red Pleiades member HZ 1306 live? Assume that this star is on the Main Sequence and has a luminosity of 0.05 that of the Sun. Be sure to use the proper branch of the Mass-Luminosity relation to answer this question. You can also assume that the lifetime of the Sun is 10 billion years.

8.3. Look up the 'biography' of one of the Pleiads. Though they may have been mythical persons, they each had distinct 'lives' attributed to them. Note which Pleiad you have chosen, and summarize the key features of her story.

8.4. Suppose you own a 150 *hp* (= *horsepower*) car that has a tank that holds 20 gallons of gas. If one *hp* is about equal to 750 *Watts*, how much mass of gasoline do you burn by traveling at full power until you run out of fuel? Assume the tank was full at the start of your trip, you made no stops (or, that you maintained 70 *mph* when moving), that you are going 70 *mph*, and that your car's gas mileage is 35 *miles per gallon*. **HINT:** First, figure out from the information how long (in time) you can travel. Then calculate the energy you actually consume to propel the car. Then use Einstein's Equation to determine how much mass that energy corresponds to.

8.5. Use the Main-Sequence Mass-Luminosity relation to determine the masses—in Solar units—of the following stars:

a) α Cen A (basic data for this star can be found in Destination 4).

b) A star that, as seen from Earth, is 10^{10} times fainter than the Sun, but is located at a distance of 1 million *AU*.

c) A star with a radius of 0.5 $R_\odot$ and a surface temperature of 3700 *K* (the Sun's surface temperature is about 5780 *K*).

8.6. If stars produced no energy in their cores, the only source of energy to power their light output would be contraction due to gravity. The 'gravi-

30 This is certainly true, but another reason the Pleiades would not have been easily visible is that it probably was not that close to the Earth when the cluster was born. Due to the motions in our Galaxy that we shall explore at Destination 12, it is likely the Pleiades would have been a few thousand *parsecs* from the Sun 100 million years ago. It would have been far too faint to see even with sensitive reptilian eyes. That would have made it a double whammy for the dinosaurs!

tational potential energy' is given by the relation $E_G = GM^2/R$, where G is the gravitational constant ($6.67 \times 10^{-8}\ cm^3/gs^2$), M is the mass of the star, and R is the radius of the star. Since the luminosity of the star, L, is the rate at which it emits energy, gravity can supply that energy for a time approximately equal to $t_{KH} = (GM^2)/(RL)$. This is known as the 'Kelvin-Helmholtz' timescale.

a) What is the KH timescale for the Sun? Be very careful with your units! They have to be consistent!!
b) Given the current size of the Sun and the timescale you just calculated, at what rate should the Sun be shrinking if gravitational collapse is the source of its energy? **HINT:** This question is essentially asking you to calculate a speed. From part (a) you now know a timescale, and from knowledge of the size of the Sun, you know a distance. From these calculate a speed.
c) Do you feel this is a viable source of energy for the Sun? Why or why not?

8.7. The Pleiades, M 45, is just one of a number of so-called open star clusters that are found in Messier's catalog of comet-like astronomical objects. Some others include M 6, M 7, M 21, M 23, M 25, M 26, M 29, M 34, M 35, M 36, M 37, M 38, M 41, M 46, M 47, M 48, M 50, M 52, M 67, and M 103. At present, there are about 6000 cataloged open star clusters, so Messier's clusters are a tiny fraction of the total known! Pick one of these Messier clusters and look it up on the web. Make a copy (electronic or a sketch) of the HR diagram or some variant (such as the color-magnitude diagram) for the cluster you chose. Describe how you would go about estimating the cluster's age from these results. Is the cluster likely younger or older than the Pleiades? Is it younger or older than the Sun? Be sure to report the value of the cluster's age you found from your on-line research about it.

8.8. Calculate the surface temperature of the Sun given its radius and luminosity (see Table 4.1).

8.9. This problem outlines how we can derive the masses of stars by analyzing the properties of binary systems.

a) Let's say that we observe a star that exhibits two sets of absorption lines apparently arising from two distinct stars. After a while, we measure an orbital period for the stars of 1.6 *yr*. For one set of lines (we'll call this 'Star A') we measure, using the Doppler shift, an orbital speed of 35 *km/s*, while for the other ('Star B'), we measure 45 *km/s*. What is the mass ratio of the two stars? **HINT:** Relate the orbital speeds, semi-major axes, and period of the two stars in the binary system. Then refer to the Center of Mass equation provided in Destination 4.

b) Use Kepler's Third Law to determine the sum of the masses of the two stars. You may assume that we are close to the orbital plane of the binary system (why does that matter?), and that the orbits are essentially circular.

c) What are the individual masses of the two stars?

d) Assuming both stars are on the Main Sequence, how luminous is the binary system as a whole in Solar Units?

8.10. Compare the mass density of the Pleiades with that of the Solar Neighborhood. For the Pleiades, assume that the cluster has a radius of 1.5 *pc*, and contains 1000 stars with an average mass of $0.6M_{\odot}$. For reference, the mass density near the Sun is approximately $0.07\ M_{\odot}/pc^3$.

8.11. Although the Sun's luminosity is enormous, the Earth captures only a tiny fraction of it. We will see in this problem that that's still a lot of energy.

a) What is the flux of the Sun at the distance of the Earth? Give your answer in units of $erg/sec/cm^2$.

b) What is the total energy from the Sun intercepted by the Earth? **HINT:** From the Sun, the Earth looks like a circle (not a sphere) with a radius equal to one Earth radius.

c) What fraction of the solar luminosity is this?

d) For how long would we need to capture this energy to account for our current energy needs for a single year, equal to 5×10^{27} *erg*?

Destination 9

EXPLORING THE MILKY WAY

9.1 OUR HOME'S HOME

Roaming through the Solar neighborhood has revealed to us a variety of Alien Skies, some surprisingly similar to Earth's, some very different. But no matter where we have traveled, there has been a consistent feature visible in these skies: A gossamer band of silvery light meandering across the entire firmament (Figure 9.1). Most ancient cultures associated this celestial band with some sort of long, winding structure. Some revered it as a divine bridge or a river used by the dead and by the gods as a connection between our world and the heavens. For the ancient Greeks—who liked geometry—it was Galaxies Kuklos, the Milky Circle. To the Romans—who liked roads—it was Via Lactea, the Milky Road. For us, it is *The Milky Way*.

Fig 9.1

The Milky Way as seen from Death Valley, California. Because of distortions introduced by the camera in this all-sky image, the Milky Way appears as a U-shaped band in this picture. In reality, the Milky Way almost perfectly bisects the sky. Notice too how the luminous band seems itself split into two, the result of absorption of starlight by dark clouds that cause, in part, the band of light that defines the Milky Way to vary considerably in brightness along its length. From a dark location such as this, the Milky Way is a prominent, sublimely beautiful component of the night sky.

Image by D. Duriscoel, US NPS.

Few modern humans have the same personal acquaintance with the Milky Way enjoyed by our ancestors. Over the past two centuries, the steady increase in the amount of artificial lighting (see Figure 9.2) has caused the nighttime sky—and with it, the Milky Way—to gradually fade from most people's lives.[1] This is a shame. Those who make the increasingly difficult effort to escape the glare of the modern world will discover the Milky Way to be one of the most glorious sights of Nature, one sharable by all humans on Earth. The goal of the individual journeys that constitute the present destination is to learn what the Milky Way really is. In the process, we shall discover a structure of colossal scale and awesome order known simply as *The Galaxy*.[2]

1 It does not have to be this way, nor do we have to live in complete darkness at night. If you live in a city, look around you at night. How much of the light from the various light fixtures are aimed *upward*? And what do they accomplish other than lighting the bellies of birds and suppressing the night sky? Wisely designed lighting not only can return the night sky to city dwellers, but it can save energy too.

2 This is a subtle but important distinction. 'The Milky Way' is the band of light we see in our sky. 'The Galaxy' is the structure that leads to the appearance of the Milky Way. As we journey through the Galaxy, the detailed appearance of the Milky Way will change depending on our locations. Those changes will help to reveal the Galaxy's underlying

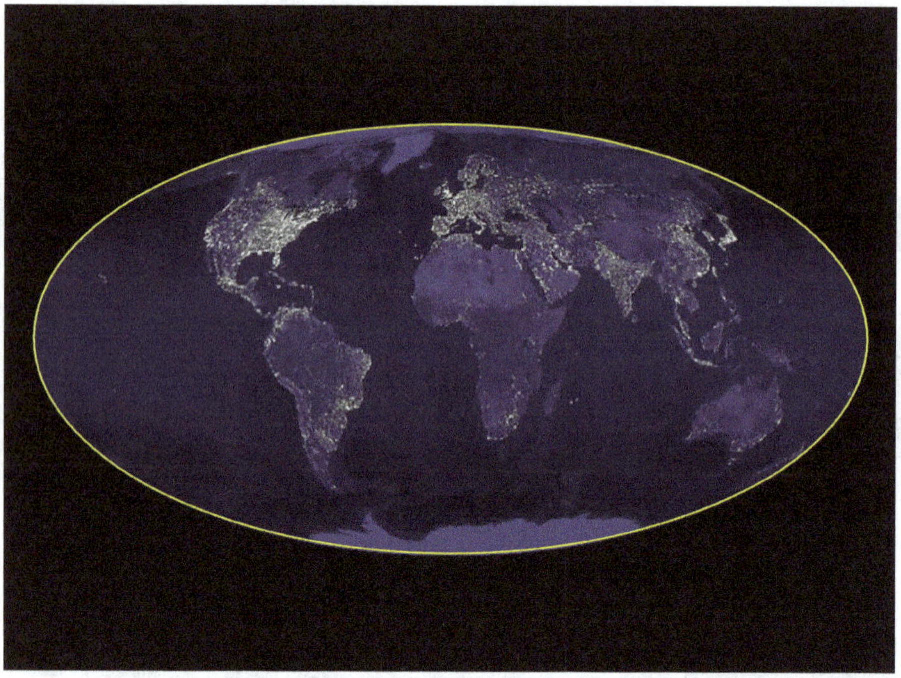

Fig 9.2

This is a composite image of the Earth at night showing the sources of artificial lights over the entire surface our planet. Even on this scale, the outlines of major population centers are obvious (for example, U.S. East Coast, Europe, Japan), including individual cities (for example, Denver, Sydney, Johannesburg, Lima, Paris). Some of the light sources are from petroleum flares (the Arabian Peninsula) or brushfires (Amazon Basin in South America, central Africa). Light from celestial sources has to compete with this human-made lighting. This figure also illustrates how we will represent the sky in numerous maps of this and future destinations. In general, any depiction of a spherical surface—such as the surface of the Earth or the Celestial Sphere (Destination 3)—onto a flat map is known as a *projection*. The projection illustrated here for the Earth, one that we'll use often, is known by cartographers as a Mollweide projection.

Image by C. Mayhew, R. Simmon, GSFC/NASA.

The form of the Milky Way already provides an important hint regarding the underlying nature of the Galaxy. Though we have now traveled hundreds of *parsecs* from the Sun to reach our various past destinations, the Milky Way itself has always been visible to us. More significantly, its appearance has remained essentially unaltered during these voyages. Following our earlier reasoning that led us to conclude the stars are very distant compared to the size of the Solar System (Destinations 1 and 2), this observation suggests that the Milky Way must be far away. And it is, in one sense. But, paradoxically, it is also very close. We shall discover that the Milky Way is the result of viewing the Galaxy from *inside* its structure. The constancy of the appearance of the Milky Way from the destinations we have visited so far makes us conscious that the scale of the Galaxy must be considerably larger than the distances we've traversed so far. Much larger than hundreds of *pc*. It is clear that to truly appreciate the scale and organization of the Galaxy, we need to travel much farther from home than we've gone so far.

But we won't embark in complete ignorance. There are bits of information we can glean from historical observations and others from simple observations that we can carry out of Earth's sky that can help us design a set of journeys that best unveils the nature of the Galaxy. For example, in the early 17th century the Italian astronomer/physicist Galileo Galilei[3] first described the view of the Milky Way

nature. Though many astronomers use the two terms interchangeably, we—as travelers through the Galaxy—have to be more careful.

3 Perhaps because of his 'rock star' fame, Galilei is typically referred to by his first name only. Or perhaps we do this simply because his names are so nearly identical, differing by only one letter. In any case, he is one of the few scientists commonly referred to with his given, and not his family, name.

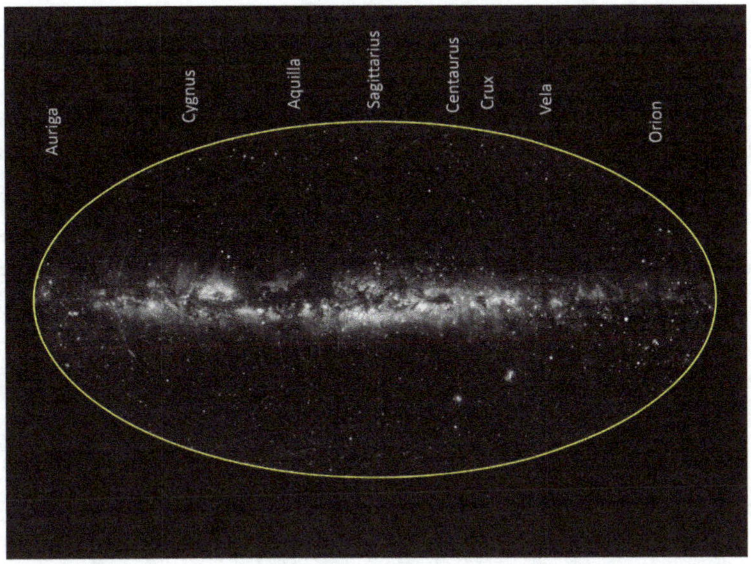

Fig 9.3

An image of the entire sky reproduced so that the Milky Way defines the equator along the horizontal axis of this picture (the cartographic projection employed here is similar to the one used to produce the map of the Earth's surface in Figure 9.2). This map illustrates very clearly that the band of the Milky Way effectively bisects the sky. Some constellations that lie on the Milky Way and that we'll use as references during our journeys at the present destination are labeled along the top. The Great Rift is apparent just left of the center of the image extending from Cygnus to Aquila. Numerous other dark clouds are also visible including the famous *Coal Sack* in Crux. This stunning painting (!), made in the 1950s, is located at the Lund Observatory in Sweden, where it still occupies a prominent location[6].

Image use by permission of Lund Observatory, Sweden.

through an astronomical telescope. He found that the Milky Way, though largely smooth in appearance to the naked eye, is actually composed of 'countless' faint stars.[4] Neither Galileo nor the Church authorities of his day could conceive what the creator might want with so many 'extra' stars. After all, it was clear that the Earth (and its Sun) was where all the important action was happening. But only the most extreme egotists who were willing to deny the evidence from their own eyes could avoid the obvious conclusion: The ethereal glow of the Milky Way is in fact the sum of the light of an immense number of very faint—and likely very distant—stars.[5]

We can learn more about the Milky Way from more detailed observations of its overall appearance (see Figure 9.3).[6] For example, in the constellation *Aquila*—the Eagle—the Milky Way splits into two parallel bands of light.

4 Saying the number of stars is 'countless' has almost become a cliché in popular astronomy articles and books. But, in fact, we *can* try to count the stars of the Milky Way, at least approximately. The number—around 10^{12}, or one trillion—is certainly large, but it is not infinite. If you tried to personally count the stars at a rate of one per second, it would take about 30000 years to complete the census. So, in this sense, maybe this number is 'countless' after all.

5 And you can easily see this for yourself. Even a pair of binoculars will reveal many faint stars in regions of the sky that the Milky Way traverses compared to regions away from the Milky Way. It is a beautiful sight and provides a personal appreciation of the scale and structure of the Galaxy.

6 The coordinate system used in Figure 9.3 is tailored to make navigation in the sky relative to the Milky Way easier than the usual celestial coordinates described in Destination 3. This system relies on knowledge of the path of the Milky Way in the sky and the location of the Galactic Center (more on this later in this destination and at Destination 10). While the path of the Milky Way was mapped reliably a long time ago, our understanding of the location of the Galactic Center has changed since the mid-1950s when this painting was made. As a result, there is a subtle shift between this map and all subsequent maps of the Milky Way in this and later destinations. See if you can discern this shift when you compare, say, Figures 9.3 and 9.5.

As we look northward along the Milky Way, these bands join into a single stream within the constellation *Cygnus*, the Swan. To the south, one of the bands fades to invisibility, while the other defines the main band of the Milky Way. This split in the Milky Way is known as *The Great Rift*. We also note that in most parts of the sky, the Milky Way is around 10–15 degrees wide, but in the constellation *Sagittarius*, the faint outer fringes of the Milky Way expand to about twice that width. Even farther to the south—in the constellation *Crux*, the Southern Cross—one section of the Milky Way appears to have been blotted out altogether. This inky hole is well known to sky-watchers in the Southern Hemisphere and is called the *Coal Sack*. Many bright stars—especially numerous blue ones—appear to congregate near this portion of the Milky Way in the Southern Hemisphere, most notably between the constellations *Centaurus* and *Vela*.[7]

As we think back to our earlier voyages, we can discern a connection here. The dark regions where the Milky Way splits or is blotted out *could* be regions where there are no stars. But it seems more likely that they are regions where molecular clouds—which we know also contain dust (Destination 6)—block our view of background stars. If this is the correct interpretation, we might expect to see evidence that young stars—which we know (Destination 6) are born in such clouds—are found preferentially near the Milky Way. And, indeed we do. Most star-forming regions such as the Orion Nebula and young star clusters, many considerably younger than the Pleiades, are preferentially found within or near the band of the Milky Way. And, as we have already noted, many blue stars—which we now know are likely young, massive stars (Destination 8)—are also commonly found along the band of the Milky Way.

The grand order of the Milky Way is not restricted to its optical appearance. When we map the structure of the Milky Way at wavelengths other than visible light, there certainly are differences in detail mostly because different wavelengths map different types of objects and different components of the ISM. But even so, no matter what wavelength we use to observe the sky, the Milky Way can be seen as a distinct band stretching across and bisecting the sky (Figure 9.4). There is clearly an awesome order here. An order that reveals the presence of a structure—the Galaxy—that is the 'habitat' for a wide variety of stars, star-forming regions, and all of the distinct phases of the ISM (see Figure 9.4). And, not incidentally, the home of our Sun.

7 This concentration of bright, blue stars is known as Gould's Belt and is quite noticeable to the eye when pointed out.

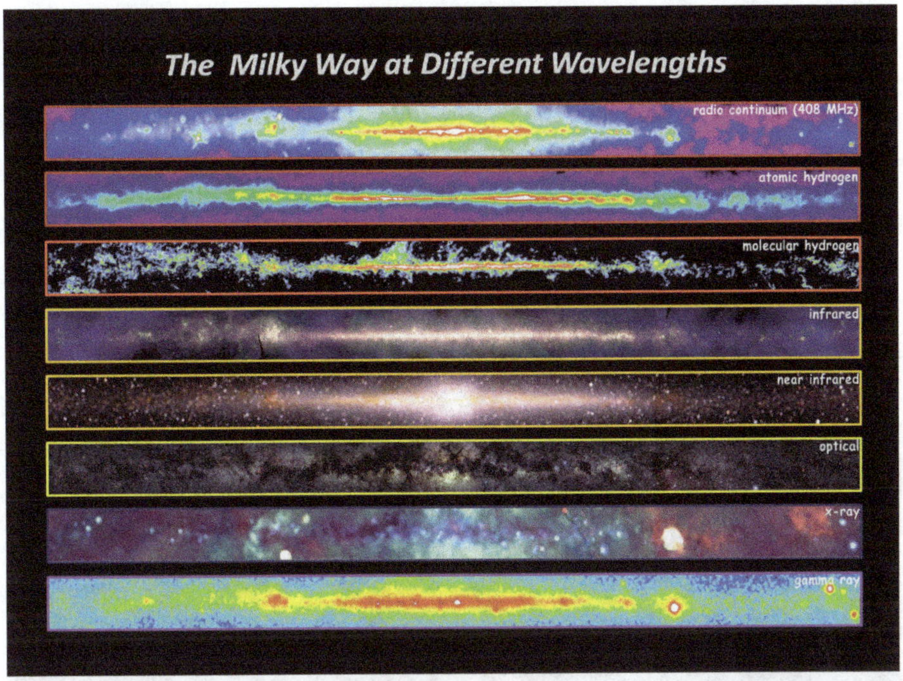

Fig 9.4

The appearance of the Milky Way is shown at increasingly shorter wavelengths of EM Radiation, ranging from low-frequency radio (at *top*), 21-cm Radiation (labeled 'atomic Hydrogen'), microwaves (labeled 'molecular Hydrogen'), infrared, near-infrared, optical, x-rays, and gamma-rays (at *bottom*). The orientation and approximate scale of these images is similar to what's shown in Figure 9.3, though in this image only a narrow strip around the Galactic Equator is shown for each wavelength. The various wavebands reveal different details about the distributions of stars and interstellar medium along the Milky Way. Together, they provide a remarkable and comprehensive map of the contents and structure of the Galaxy. But the key point is that the basic appearance of the Milky Way as a band that stretches completely across the sky is visible in every case. This structure is a fundamental characteristic of the Galaxy.

Composite image by GSFC/NASA.

9.2 RISING UP FROM THE MILKY WAY

Our goal is to explore the Milky Way to discern the underlying nature of the Galaxy. How do we do that? In the past, we've been able to identify specific locations as destinations. But now that we want to 'explore' the Galaxy as a whole, it is much less obvious where to go. The difficulty is akin to exploring a vast forest from a point deep inside it. From our vantage point in our Galaxy, as in a forest, all directions seem similar. There seems to be no 'best' direction in which to set out.

Or maybe there is. We have already noted that the Milky Way appears to trace a path that neatly bisects the sky into two equal hemispheres (see Figure 9.3). Such a path on a sphere[8] is an example of what is known as a *Great Circle*. We've encountered other great circles before (Destination 3). The Celestial Equator is one example. It bisects the sky into the Northern and Southern Celestial Hemispheres. The path of the Sun in the Earth's sky—the Ecliptic—also bisects the sky as it

8 Recall from Destination 3 that the sky can be thought of as a vast sphere with us at its center. The distant stars—and the Milky Way itself—can be thought of as 'attached' to this fictitious surface, the *Celestial Sphere*.

passes through the constellations of the Zodiac and is another example of a great circle in Earth's sky. The fact that the Milky Way follows a great circle is quite a bit more profound, however. The Celestial Equator is just the extension of the Earth's equator onto the sky, a great circle on the surface of the Earth that we use to define the Northern and Southern Hemispheres. The Ecliptic is simply the projection of the orbital plane of the Earth onto the sky. These two great circles can always be defined for any rotating planet orbiting any star. In contrast, there's no evident reason that the Milky Way had to follow a great circle in our sky. The fact that it does must be an important clue about the Galaxy's structure. More to the point, the great circle of the Milky Way defines two special points on the sky that can serve to guide us: The *Galactic Poles* (see Figure 9.5).

Fig 9.5

This image is a composite of 51 separate photographs that were stitched together to provide a true-color image of Earth's sky (the projection is the same as Figures 9.2 and 9.3). The North Galactic Pole (denoted NGP)—in the constellation Coma Berenices—is denoted by a blue star, while the South Galactic Pole (denoted SGP)—in the constellation Sculptor—is denoted by a yellow star. Our initial voyages through our Galaxy will take us in the direction of the SGP.

Photo mosaic by A. Mellinger.

If we consider the forest analogy again, it dawns on us that a reasonable plan to understand the layout of a forest would be to rise above it! By climbing the trees—particularly the tallest ones—we are more likely to gain a broader and less-obstructed view of the entire forest and the land beyond. A view from such a perspective might also tell us where we might want to go next. Perhaps we'd see a clear center or edge of the forest, or some other feature beyond it that would be worth visiting. Our realization that the Milky Way follows a great circle means that we have two options—the Galactic Poles—that define ideal directions that would allow us to 'rise above the forest.' One of these poles is in the constellation *Coma Berenices*, 'Bernice's Hair.' Because it is located north of the Celestial Equator, this pole is referred to as the *North Galactic Pole*, or *NGP*. The other pole is in the constellation *Sculptor* (you guessed it, 'The Sculptor') and is located south of the Celestial Equator. This pole is known as the *South Galactic Pole*, or *SGP* (Figure 9.5). Since it sounds a bit more exotic[9], we decide to head toward

9 Over the vast majority of the history of humanity, most people have lived in the Northern Hemisphere and crossing the equator used to be a noteworthy event, one that

the SGP to scout out the structure of the Milky Way. But, really, the choice is not critical. We would see the same basic features if had chosen instead to travel toward the NGP. As we shall see, the Galaxy is essentially symmetric in structure with respect to the great circle that the Milky Way defines in the sky.

Taking baby steps at first, we begin by traveling 10 *parsecs* from Earth in the direction of the SGP. Like the Anti-Cen location we visited at Destination 4, there's not actually anything here. Just a seemingly empty point in space defined relative to the position of the Sun. As we look around us, we notice that not much has changed in our sky. As we saw on earlier journeys of comparable distance, the constellations generally don't appear very different from how they look from home. We can still see the Sun dimly visible 'behind' us. It is now projected onto the still-recognizable constellation of Coma Berenices. This makes sense. Since we are traveling toward the SGP, the Sun will now always appear to be located in the direction of the NGP, and the NGP is in Coma Berenices as seen from Earth. The Milky Way itself—what we are aiming to study in this journey—remains virtually indistinguishable in appearance compared to what we see from Earth. Its delicate light still bisects the sky. It still exhibits regions where clouds of gas and dust block our view of the background light. The Milky Way still traces out a band of light about 15 degrees wide as it meanders around us along a great circle in the sky. In short, not much change.

Emboldened (or frustrated?), we set out to a location 400 *pc* from the Sun, making sure that we continuously travel in the direction of the SGP. This is now comparable to the distance we traveled to get to the Orion Nebula, but we are heading in a completely different direction. Orion is located close to the Milky Way; we are moving in a direction as far from where the Milky Way is located in the sky as possible. From this new vantage point, the constellations in the sky are now totally unrecognizable from those we see from Earth. And so are the stars. Few of the ones we know by name from home are visible here. The Milky Way is certainly still visible from this location, seemingly pretty much unchanged from what we saw before. Or is it? If we carefully measure the position of the band of the Milky Way in this sky, we notice, somewhat to our surprise, that it no longer follows a great circle! The change is pretty subtle, but the Milky Way is now definitely slightly offset toward the direction from which we've come. When we measure this offset we find that the Milky Way is now on average about 93 degrees from the SGP. It has shifted to the north, closer to the NGP!

There are other subtle differences in the sky here. Not only are the brightest stars different, but the total *number* of stars in this sky appears diminished compared to the number we see from Earth. If we make a careful census, we

---

only a few sailors could typically have claimed to have done. As a result, the Southern Hemisphere has always seemed a bit more remote and mysterious. It's only in the past few hundred years that it has become comparatively routine for people from northern latitudes to cross the equator and visit the Southern Hemisphere.

find that, instead of the 6000 or so naked-eye stars we can see in our sky from home, there are about 2000 stars visible in this sky. But it is not just the number of stars that is different. The characteristics of the brightest stars are not the same. From Earth, we can see many bright stars such as Vega, Spica, Sirius, and Altair in the night sky. These are examples of A or F type stars with strong H lines in their spectra and blue-white colors (see Destination 8 to review stellar colors and spectral types). Such stars are noticeably missing here. Instead, the brightest stars in this sky tend to be redder, similar in color to the Sun (a yellow G-type star) or even redder, more like K-type stars such as Arcturus and Antares.

Encouraged by these changes, we press on, moving even further in the direction of the SGP. At 2000 *pc*—or, to use a new unit more suitable for Galactic distances, 2 *kiloparsecs* or 2 *kpc*—the trends we saw at 400 *pc* become more pronounced. The Milky Way is now clearly skewed to the north, located on average about 105 *deg* from the SGP. Even to our eyes it is apparent that the band of the Milky Way no longer traces a great circle in this sky. As a result, there is much more 'open' dark sky south of the band of the Milky Way—the direction we are traveling—than to the north. The Sun itself is now so faint that we need a fairly large telescope merely to see it. But the most striking change is the eerie emptiness of this Alien Sky. As we look around us, it's hard to shake off a strong sense of loneliness here. There are not thousands, nor even hundreds, of stars visible in this sky. We count only 10–20 easily visible stars in the *entire* sky! There are no bright blue stars *anywhere* in the sky apart from maybe a few very faint ones just visible within the band of the Milky Way. The one or two brightest stars we do see in this sky are distinctly red. But these stars also differ in another important respect. Observing their spectra, we find that they exhibit absorption features—particularly of 'heavy elements'—that are clearly weaker than in the spectra of stars of similar spectral types located near the Sun. Though we've seen that spectral line strengths depend principally on temperature (see Destination 8), the fact that stars of similar temperature have weaker lines means that they must possess fewer atoms of the elements that produce those lines. That is, the chemical makeup of stars around us at this location 2 *kpc* from the Sun in the direction of the SGP is different! The stars here have, on average, lower chemical abundances than stars near the Sun.[10]

The appearance of the Milky Way is now drastically different than before. The main band of the Milky Way is still peppered with individual, but now very faint, blue stars. It is still interrupted by dark molecular clouds that

---

10 Two oddities of how astronomers talk are apparent here. First, to an astronomer, the term 'heavy element' refers to any element other than H or He. Second, the term 'abundance' has, to an astronomer, approximately the same meaning as the word 'quantity' to everyone else. That is, 'abundance' does not mean, as it does to us in everyday usage, 'a lot of,' but simply, 'the quantity of' something. Thus, terms such as 'low abundance' and 'high abundance' make sense to an astronomer. We'll use this terminology periodically at this and future destinations.

Fig 9.6

If you can't actually move out of the Galactic Plane, the next best way to see the Galaxy—and in particular the Galactic Center—is to take an infrared image of the entire sky. Because of its longer wavelength than optical light, IR radiation can mostly pass through dark clouds, allowing us to see what's behind them. Here we can clearly see how the density of stars in the Galactic Disk varies from the Galactic Plane, which goes horizontally across the middle of the image. It is also clear that there is an additional component of the Galaxy (at the center of the image) that is aptly named the Galactic Bulge. We'll explore that component of the Galaxy in detail as we travel to the Galactic Center at Destination 10.

Image from 2MASS/J. Carpenter, T. H. Jarrett and R. Hurt

produce rifts and blotches in the silvery band. But we now see something new. Toward the direction where the constellation Sagittarius is visible from Earth[11], the Milky Way now appears *much* broader. The light from this portion of the Milky Way appears quite smooth, apparently lacking molecular clouds that can block our view of what lies beyond. This region is also roundish in appearance with a clear central concentration of light, not at all like the long, comparatively uniform band of the Milky Way. It is as if we are looking at a side view of an immense football-shaped structure embedded within the Milky Way. This region is called the *Galactic Bulge* because it seems to protrude from the main band of the Milky Way (Figure 9.6). And it is not nearby. We can use a variety of methods[12] to determine its distance from the Sun to be about 8500 *pc*, or 8.5 *kpc*. Far more distant than the Sun's closest stellar neighbors—about 1 *pc*—and even considerably farther than the 2 *kpc* we have traveled to reach our current vantage point. As we shall see plainly on later journeys, the core of the Bulge is coincident with a unique location in our Galaxy known simply—and descriptively—as the *Galactic Center*.

The clear view we now have of the Galactic Bulge reveals another subtle change in our surroundings. Near the Sun, we found the ISM to be a ubiquitous feature of all our destinations. For example, the Warm ISM—consisting mostly of neutral H atoms—fills most of the volume of interstellar space around the Sun as revealed from its 21-cm Radiation emission (Destination 4; see also Figure 9.4, second image from the top). In other regions, we find dark molecular clouds of the Cold ISM or ionized gas such the Hot ISM near

11 That constellation is long gone from our sky, no longer visible at this location. But the direction that Sagittarius was located can still be defined here. Of course, with only a handful of stars visible, even the concept of 'constellations' is no longer very useful in this sky. That makes it even more important to keep track of directions as we travel through the Milky Way. In practice, astronomers do this by inventing coordinate systems to facilitate navigation in the Galaxy. We'll need to define one of these systems in a bit of detail later on in this destination.

12 We will explore one method of determining the distance to the Galactic Center at Destination 10. However, if we really *could* journey to a location where we see the Galactic Center, it should be clear we can use simple geometry to get its distance. You can explore one such method in a question at the end of this destination.

luminous young stars, such as those in the Orion Nebula (Destination 6). The patchy structure of the Milky Way reveals that molecular clouds—some enormous in size—must also be common throughout the main structure of the Milky Way (Figure 9.4; fourth image from the top). But here, 2 *kpc* from the Sun in the direction of the SGP, we seem to have left much of the ISM behind us. That's partly why we can see the Galactic Bulge so clearly here, far from the Sun. The lack of dust, in particular, means that space here is clearer, containing less material to block our view of distant objects. Returning to our analogy of the forest, we can now see what lies beyond the trees because we have risen above them.

But space here is not entirely devoid of gas. There is interstellar matter here almost entirely in the form of very fast-moving particles. We've encountered this stuff before as the 'Very Hot' or Coronal ISM. Its properties here are remarkably similar to what we saw near the Sun (Destination 4). The particle speeds correspond to temperatures of 1–2 *million K*, and the gas density is very low, under $0.0001 cm^{-3}$. Clearly, this material is not confined to the same regions of the Galaxy as the gas associated with the Cold (Destination 6), Warm (Destination 4), and Hot (Destination 6) phases of the ISM. Somehow, we've left those other components of the ISM 'behind' us. But not the Coronal ISM. That component of the ISM is still present here, some 2 *kpc* from the Sun.

9.3 THE GALACTIC DISK

What can we discern about the Galaxy from our reconnaissance toward the SGP? The most obvious conclusion is that by moving toward a Galactic Pole[13], we appear to have passed through some sort of transition. Not a transition with a sharp, abrupt edge, but a smooth change from one region to another. Within a distance of a few *kpc*, we traveled from a region in the Galaxy that contains many stars of all types (near the Sun), to one with far fewer stars of more limited types (2 *kpc* from the Sun toward the Galactic pole). The last point is particularly important: The transition differs for different types of stars and for various phases of the ISM. This is why the mixture of stars in our sky changed as we traveled further from the Sun toward the pole. The easiest way to interpret our observations is to conclude that by traveling toward the Galactic poles[14] we moved away from the mid-plane of some sort of immense, flattened structure. We have seen that the

13 Remember, it does not really matter which pole we had traveled toward. The Milky Way is very nearly symmetric in the north and south directions. From now on, we'll generally stop making any distinctions between the two poles as far as the overall structure of our Galaxy is concerned.

14 Again, north or south. It does not matter which.

virtually all of the components of the ISM, and most stars—including essentially all young stars—lie close to that mid-plane, better known as the *Galactic Plane*. Together, these stellar and gaseous components of our Galaxy comprise what is known as the *Galactic Disk*, the vast flattened structure that defines the Galactic Plane. From our initial journeys, it is already clear that the Galactic Disk is a highly organized structure. And an immense one.

We can be a bit more precise about all this. For example, because the Milky Way defines a great circle as seen from Earth tells us that Sun must be located quite close to the Galactic Plane. By moving toward either Galactic Pole—along what is typically referred to as the z-axis of the Galaxy—we traveled away from the Plane. Our motion caused the path of the Milky Way to appear to move in the opposite direction, in the process causing it to no longer define a great circle in our sky. But it is also important to appreciate that despite its prominence, not *everything* is associated with the Galactic Disk. The Galactic Bulge, for example, appears to represent a distinct component of the Galaxy—and indeed it is, as we'll see at Destination 10. Only by moving away from the Galactic Plane and the obscuring gas and dust it contains did our view of the Bulge improve sufficiently to let us surmise it is coincident with the Galactic Center (as we'll prove at Destination 10).

Our journey has revealed one other important result. The Sun's location in the Galaxy—though very special to us—appears to be a rather generic spot within the Galactic Disk. Though certainly located near the Galactic Plane, the Sun is not at the central point of the Disk. That special point appears to be coincident with the Galactic Bulge. And since we see that the Milky Way surrounds us as seen from Earth, we also know that the Sun is not at the outer 'edge' of the Galactic Disk either.

It's worth emphasizing that while a solid disk has a well-defined edge, the Galactic Disk does not. We know this because the number of stars did not suddenly go to zero as we rose from the Galactic Plane, but instead steadily declined. In this sense, though we were clearly moving out of the Disk, we never really left it. This odd characteristic of the Galactic Disk—its smooth transition from high space density near the Sun and lower space density far from the Galactic Plane—was first known from star counts that astronomers carried out to discern the true space density of stars around the Sun.[15] In particular, the number of stars does not fall off in some haphazard manner with distance from the Galactic Plane. The space density of stars changes in very close accordance to a specific functional form known as an *Exponential Density Law*. When applied to the Galactic Disk, this density law can be written as

15 We'll use the term 'space density' to refer to how many stars are located in a given (large) volume of space. This is distinct from the density of a star, which is the total mass of the star divided by its volume. Unfortunately, we are using the same symbol for both, so be careful about this distinction.

$$\rho_i(z) = \rho_{i,0} e^{(-|z|/h_i)}.$$

As we've seen, z is a measure of the distance from the Galactic Plane and $|z|$ refers to the absolute value of z—a quantity that is always positive—underscoring the fact that the behavior of the stellar space density is the same north (positive z) or south (negative z) relative to the Plane. The Sun is effectively located at $z = 0$, right in the Galactic Plane.[16] The term $\rho_i(z)$ refers to the space density of objects of type i as a function of the distance from the Plane, z. The term $\rho_{i,0}$ refers to the space density of objects of type i at $z = 0$, and is known as the *Local Space Density* for objects of type i. The other parameter in the exponent—denoted h_i—is called the *Scale Height*. It determines how rapidly the space density of objects of type i changes with increasing z.

What does this really mean? First, the concept of an Exponential Density Law is not as esoteric as it may seem. Here on Earth we live *within* an Exponential Density Law (see Figure 9.7). It has long been known that the density of Earth's atmosphere steadily drops as one moves to higher altitudes, obeying an Exponential Density Law similar to that of the Galactic Disk. Most of the gases in our atmosphere have scale heights, h_i, equal to about 7 *km*. The Exponential Density Law tells us that there is only about 56% as much Oxygen at the tops of 4000 *m* (about 13000 *ft*) mountains as at sea level, a fact that anyone who

Fig 9.7

The Earth's atmosphere as seen from low orbit. Notice how the density of the atmosphere slowly fades from tops of the clouds near the Earth's surface toward empty space. This drop in gas density obeys an Exponential Density Law; there is no sharp 'edge' to the atmosphere. The space density of stars in the Galactic Disk also obeys an exponential fall-off with increasing distance from the Galactic Plane. Like Earth's atmosphere, there is no abrupt 'edge' to the Galactic Disk, just a gradual drop in density from the Galactic Plane to extra-Galactic space beyond.

Photo by NASA.

16 In fact, very careful measurements suggest that we are located about 20 *pc* north of and moving away from the Galactic Plane (at about 10 *km/s*). But don't worry. The gravitational forces of the Galactic Disk will cause the Sun to turn around and go back toward the disk, only to overshoot and do the same on the north. And it will take some time to do this, about 50–100 *Myr*. This 'porpoising' motion is a characteristic feature of all stars in the Galactic Disk. For our purposes (not porpoises), we can very safely assume that the Sun is located in the Galactic Plane, precisely at $z = 0$.

has hiked in tall mountains is well aware of. At even more extreme altitudes, there is so little air that wings can no longer provide lift to aircraft. That altitude—about 20 *miles*—is often taken as the beginning of 'outer space.'[17] From the Exponential Density Law, we can calculate that the air density at that height is only about 2% the density of air at sea level.[18]

And so it is for the Galactic Disk.[19] As we traveled away from the Galactic Plane, the space densities of stars decreased as expected for an Exponential Density Law, though in a somewhat complicated manner. In particular, the space densities of some types of stars fell off more rapidly than for others. For example, O stars appear to be closely confined to the Galactic Plane, while K and M stars continue to be seen considerably farther from the Galactic Plane (see Figure 9.8). Even the ISM that's associated with the Galactic Disk is distributed differently than most of the stars (Figure 9.8). From the perspective of the Exponential Density Law, this tells us that different types of stars and the ISM have different values of local space densities, $\rho_{i,0}$, and scale heights, h_i. Table 9.1 lists the values of $\rho_{i,0}$ and h_i that we can use in the Exponential Density Law to describe the space density distributions for various types of objects associated with the Galactic Disk. The distributions for various types of stars and the ISM associated with the Disk are plotted in Figure 9.8.

The fact that the stellar and gas components of the Galactic Disk closely obey an Exponential Density Law helps explain in detail our observations during our journey toward the SGP (see Figure 9.8). As we moved from the Sun, we first 'left behind' the blue stars of the Galactic Plane. Notice that O- and A-type stars—relatively hot, young stars—have small scale heights of around 100 *pc* or less (Table 9.1 and Figure 9.8). Once we traveled significantly beyond that distance, the local space density of these objects fell rapidly. Most of the gas of the ISM is also concentrated toward the Galactic Plane at $z = 0$. Virtually all of the gas—but, as we'll soon see, not all—rapidly became more rarified as we

17 That's one definition of where space starts, but there are others. Spacecraft such as the Space Shuttle or the International Space Station fly (or 'flew' in the case of the Shuttle) much higher than this, typically at altitudes over 300 *km*. According to the Exponential Density Law, the air density at 300 *km* altitudes is about 1 particle/cm^3. At sea level, the air density is about 10^{19} particles/cm^3.

18 It is a bit more complicated than this. Recall (Destination 5) that temperature is a measure of the average kinetic energies of particles in a gas. At a given temperature, lower-mass particles will move faster on average than higher-mass particles. For our atmosphere, this means that lower-mass particles—molecular Hydrogen, H_2, for example—will rise to greater altitudes than higher-mass particles—such as molecular Oxygen, O_2. Thus, the composition of the atmosphere—just like the Galactic Disk—also depends on height, z.

19 The fact that Earth's atmosphere and the Galactic Disk obey Exponential Density Laws comes about because in both cases the 'gases' are confined by a nearly constant gravitational force. In the case of the Earth, this reflects the thinness of the atmosphere. For the Galactic Disk—where the 'gas' is composed of stars instead of molecules—the nearly constant gravitational force is from the Disk itself.

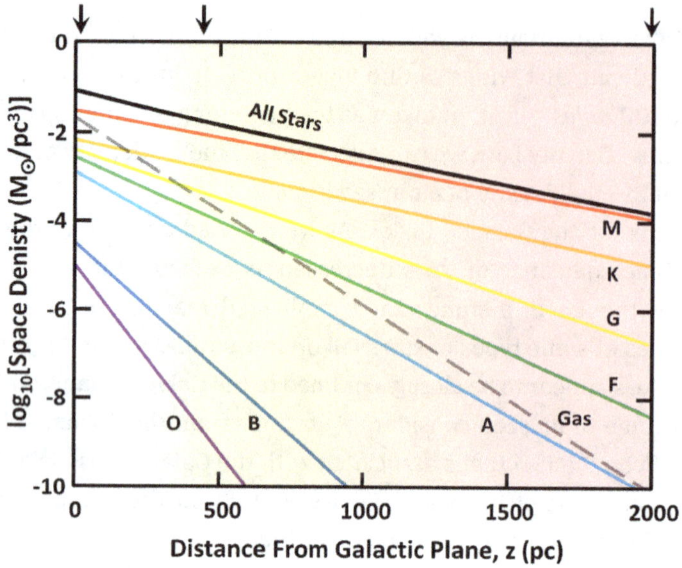

Fig 9.8

A plot showing the space densities for different types of stars and for the ISM—excluding the Very Hot/Coronal component of the ISM (see Table 6.2)—as a function of distance from the Galactic Plane, *z*. Keep in mind that the logarithm of the space density is plotted; each tick mark on the *y*-axis corresponds to a change of a factor of 100 in space density. In a plot like this, the steeper the line for a given type of star, the smaller the scale height; you can compare this plot with Table 9.1 to prove this is the case. The lines are labeled for the various types of stars in Table 9.1, and for the ISM (dashed line). The solid black line shows the sum of the space density of all components and shows that at $z = 2$ *kpc*, the space density of stars and the ISM is nearly 1000 times smaller than at the location of the Sun near $z = 0$. The

moved further from the Plane. Like the stars, much of the ISM also obeys an Exponential Density Law, though with a fairly short scale height (Table 9.1). By the time we reached $z = 2$ *kpc*, there was so little gas and dust around us that we had an essentially unobstructed view of the entire Galaxy, including the Galactic Bulge. Because redder stars—those with spectral types G to M—have larger scale heights, they appeared relatively more common compared to bluer stars as we traveled to larger distances, *z*, from the Galactic Plane (see Figure 9.8).

But this simple picture doesn't explain *everything* we saw as we journeyed out of the Disk. What about those rare, but bright red stars that 'dominated' the sky at $z = 2$ *kpc*? And what about the gas of the Coronal ISM (see Table 6.2) we detected far from the Galactic Plane? It turns out that these do *not* obey an Exponential Density Law. Ultimately, that's because they do not belong to the Galactic Disk but to a fundamentally different component of the Galaxy. One that we'll explore in detail at Destinations 11 and 13.

TABLE 9.1 Distribution of Stars and Gas in the Galactic Disk

OBJECT TYPE	h_i(*PC*)	$\rho_{i,o}$ ($M_\odot$/*PC*^{3})	COMMENTS
O Stars	50	1.0×10^{-5}	Rare but very luminous; Concentrated near $z = 0$.
B Stars	75	3.0×10^{-5}	Closely associated with O stars, but generally older and less luminous.
Gas and Dust	100	0.02	Mostly found in the form of the Cold and Warm ISM with the Galactic Disk.
A Stars	120	1.1×10^{-3}	Vega is a local example of such a star.
F Stars	150	2.6×10^{-3}	The oldest of these are 5–7 billion years old.
G Stars	200	4.2×10^{-3}	The Sun is a G star currently located about 70 *pc* south of the Galactic Plane.
K Stars	300	8.0×10^{-3}	Many of the Sun's closest neighbors are K stars.
M Stars	350	0.03	Very common, but such low luminosity that none are visible to the naked eye in Earth's sky.

9.4 THROUGH THE GALACTIC PLANE

arrows along the top of the figure show where we stopped during our journey toward the SGP. Notice that near $z = 0$, all types of stars are reasonably common. By $z = 400$ *pc*, O and B stars have become relatively much less common compared to other types of stars. At $z = 2$ *kpc*, even A and F stars are now relatively much rarer compared to G, K, and M stars and the ISM is now only a trace component. The color of the background of the plot emphasizes how the distribution changes from well-mixed near the Galactic Plane (denoted as white), to increasingly dominated by redder G-, K-, and M-type stars far from the Plane.

We've investigated the structure of the Galactic Disk perpendicular to the Galactic Plane, but what about the structure *within* the Disk? To get some idea of this of this, let's embark on a second voyage designed to explore this question directly. As tempting as it is to head toward the Galactic Center—which we have surmised, correctly, is located within the Galactic Bulge in the direction of the constellation Sagittarius (Figure 9.9)—let's defer that voyage for now (see Destination 10). Instead, let's travel in the *opposite* direction toward what astronomers call the *Galactic Anti-Center*. Starting from Earth, this takes in the direction of the constellation *Auriga*, the Charioteer (Figure 9.3). Our plan for this voyage is similar as before. While moving toward the Galactic Anti-Center, we will make frequent stops to look around us to determine what, if anything, has changed along the way.

At 10 *pc*, the sky—including the Milky Way itself—looks very much the same as from Earth. The Sun, now only faintly visible to the naked eye, is now in the constellation Sagittarius, still easily recognizable at this location. At 400 *pc* the individual constellations are completely different. But oddly, not much else has changed. When we traveled toward the SGP, at this distance we had begun to see quite significant changes in the appearance of sky, even in appearance of the Milky Way. But not here. The band of the Milky Way still bisects the sky, still follows a great circle. The number of stars in the sky is not noticeably different than what we see from Earth. Nor do we see any change in the mixture of stars. There are just about as many blue and red stars in this sky as we see from Earth. Really, nothing much has changed. Undeterred, we press on to 2 *kpc* before we stop again. But here, again, we see little change. A careful count of the stars reveals that perhaps there are a bit fewer than we can see from Earth—perhaps by 30–40%—though not so many fewer to significantly alter our impression of the sky's appearance. We still see the Milky Way; it looks amazingly similar as before.

Fig 9.9

The Galactic Center is noticeably the brightest part of the Milky Way. Our initial journey in the Plane of the Galaxy is *away* from this part of the Galaxy. We'll head to the Galactic Center at Destination 10.

Photo by Serge Brunier.

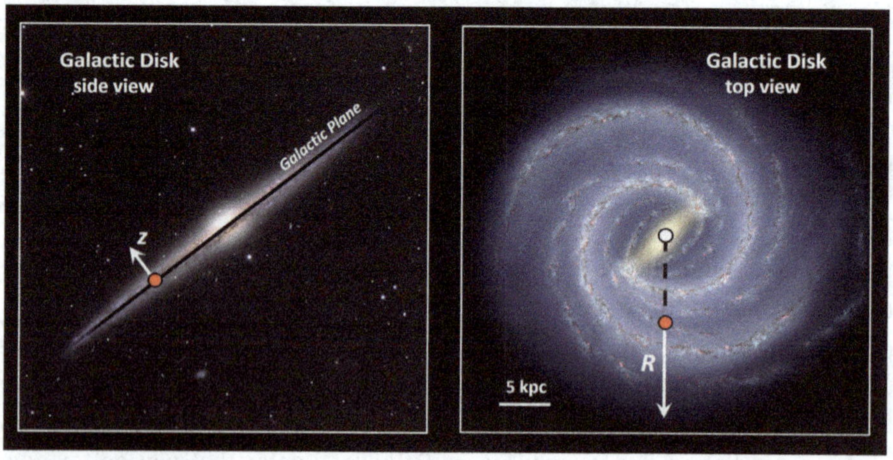

Fig 9.10

An external side view of the Galaxy (*left*; this is actually a photo of another galaxy, NGC 4565, believed to be very similar to our Galaxy in structure) and from above the Galactic Plane (*right*). The red dots show the location of the Sun in each view. Notice how the disk fades from the Galactic Plane in the side view (solid black line), and outward from the Galactic Center in the top view (white dot). The arrow in the side view (*on left*) is about 2 *kpc* long in the *z* direction, equal to the distance we will travel perpendicular to the Galactic Disk. The white arrow in the top view (*on right*) is about 8.5 *kpc* long from the Sun, about 17 *kpc* from the Galactic Center (dashed line plus the white arrow). This is the distance we will travel within the Galactic Disk outward from the Galactic Center in what is known as the *R* direction. Dust clouds and spiral arms within the disk are apparent in these images. So is the central Galactic Bulge, a distinct component of the Galaxy at the center of the Galactic Disk that we will explore in more detail at Destination 10.

Photo by K. Crawford (*left*); artwork of Galaxy from R. Hurt/NASA (*right*).

After the striking and comparatively rapid changes we saw on our trip toward the SGP, the very small changes we are encountering on this journey seem strange.

We clearly need to push onward, and as we do, we eventually arrive at a location in the Galactic Disk that's 5000 *pc* (5 *kpc*) from the Sun. Here, finally, we start to notice some significant changes in the sky. The Milky Way still traces a great circle in our sky. It is still interrupted in places by dark patches, revealing the presence of molecular clouds. But most of these features now seem to lie *behind* us, back in the direction from which we've been traveling. The direction where the Sun is located.[20] In front of us, the Milky Way appears dimmer, its light 'smoother' in appearance. Based on how we interpret the appearance of the Milky Way from home, this implies that there seem to be fewer stars in our direction of travel as well as fewer obscuring molecular clouds blocking our view. The total number of stars in the sky here is also now noticeably smaller than in Earth's sky. Only about 1500 are visible here, about 25% as many as we can see from home. But both blue and red stars are present here.[21] There is no evidence for the strong segregation of star types that we observed as we traveled toward the SGP.

We press on, stopping next at a distance of 10 *kpc* from the Sun. The effects we saw at our last stop are now magnified. The Milky Way still follows a great circle in the sky, but its appearance ahead of us—in the direction of the Anti-Center—has become very indistinct, its light is now considerably more concentrated behind us. All of the molecular clouds that darken the Milky Way are now confined to a fairly small section of the Milky Way located on the side of the sky where the Galactic Center—and the Sun—is located. Locally, the density

20 The Sun is now just about completely invisible to any telescope we could have brought along, but not merely due to its distance (Destination 4). Much of the gas and dust of the Galactic Disk is near the Milky Way where the Sun is located. This material blocks visible light within the Galactic Disk, limiting our view to perhaps 1–2 *kpc* within the Galactic Disk around our location. The Sun, about 5 *kpc* away in the Galactic Disk, is now completely hidden from view to optical EM Radiation.

21 The most luminous, bluest O and B stars are comparatively rarer here compared to Earth, but they are not totally absent from this sky.

of gas in the ISM is only about 0.1%—a decline by a factor of 1000—of what is seen near the Sun. The number of stars has also dropped, but not by so large a factor. About 300 shine in the sky here, around 5% as many as we see by eye from Earth. Apart from extremely blue stars—these really do seem to be comparatively rare out here—the mixture of star types here is not greatly different from what we see in the Solar neighborhood. And yet, the stars are not exactly the same either. Most appear to have weaker element absorption lines in their spectra compared to stars of similar temperatures near the Sun. Like the stars far outside the Galactic Plane, stars here in the outer Galactic Disk appear to have lower heavy-element abundances than the Sun. The difference is that we have had to travel *much* further to see this effect when traveling in the Galactic Disk compared to our earlier journey perpendicular to the Disk.

If we consider carefully what we've seen on our voyage through the Galactic Disk in the direction of the Galactic Anti-Center, we find that the general behavior of the star density is similar to what we saw as we moved toward the SGP. To be more precise about this, we need to define a new coordinate. Recall that as we journeyed toward the SGP, we traveled along what is known as the *z*-axis, defined to pass through the Sun and perpendicular to the Galactic Disk. In our more recent journey, we traveled in the Galactic Plane along what is referred to as the *R Direction*, defined as the distance in the Plane from the Galactic Center (which we assume for now to be at the center of the Galactic Bulge). With this definition, what we have discovered is that the space densities of most types of stars *within* the Plane appear to fall off exponentially as we move away from the Galactic Center according to the relation

$$\rho_i(R) = \rho_{i,0} e^{-(R-R_0)/H} .$$

Notice that this is an Exponential Density Law (!), similar to the one we saw earlier to describe the distribution of stars perpendicular to the Galactic Disk. But there are some important differences. For one, the distance is now denoted by *R*, the coordinate we just defined that corresponds to the distance from the Galactic Center in the Galactic Plane. The related symbol, R_0, represents the distance of the Sun from the Galactic Center, 8.5 *kpc* (we'll discuss how this is determined in Destination 10). The terms ρ_i and $\rho_{i,0}$ have exactly the same meaning as before, but there is a new term, *H*, known as the *Scale Length* (notice the capitalization of the symbol). Unlike the scale height h_i (lowercase) that we defined earlier, the scale length *H* does *not* have the index *i*. That's because we didn't notice such strong evidence that the space densities of different types of stars declined at different rates as we move outward from the Galactic Center (well, almost; we'll return to this point momentarily). Since the space density of (most) types of stars and (much of) the ISM in the Galactic Disk varies exponentially in both the *z* direction—perpendicular to the Disk—and in the *R* direction—outward from the Galactic Center—the Galactic Disk represents an example of what is known as an

Exponential Disk. Most types of stars in the Galactic Disk exhibit a scale length, H, of about 3.5 *kpc*. Recall that the typical scale heights of the ISM and different types of stars range from about 50 *pc* to a few hundred *pc*. This underscores a point we inferred earlier. The Galactic Disk is remarkably thin with an 'aspect ratio' (the ratio of its scale length to scale height) ranging from over 100:1 for A Main Sequence stars, to about 10:1 for lower-mass K and M Main Sequence stars (see Table 9.1).

But not everything in the Galactic Disk obeys an Exponential Density Law. We already saw that the density of the ISM, for example, seems to fall off more rapidly than that of most types of stars. Outside about 15 *kpc* from the Galactic Center, or around twice the Sun's distance from the Galactic Center, the density of the ISM takes a nosedive compared to an exponential law. The densities of O and B stars, the youngest, most massive Main Sequence stars, behave in an even more complicated manner. Their numbers fluctuate wildly as we travel outward within the Galactic Disk. If we carefully map the distribution of these stars throughout the Galactic disk—or if we map regions where the Cold and Hot ISM associated with star-forming regions such as the Orion Nebula—we find that these all trace out sweeping spiral patterns that appear to emanate from near the Galactic Center. We shall encounter such *Spiral Arms* again when we visit other galaxies similar to ours. For now, we simply note that these arms seem to be easiest to trace when we focus on the youngest stars or on star-forming regions within the Galactic Disk. Older stars show much less enhancement near spiral arms, consistent with our observation that they seem to obey a smooth Exponential Density Law.

Since the Exponential Density Laws for the Galactic Disk tell us how many stars are present in a given volume of the Disk, we can use them to estimate the total number of stars in that component of the Galaxy. Basically, we figure out the volume the Disk occupies then multiply that by the density of stars or ISM we derive from the exponential laws.[22] When we do this, we discover that the Galactic Disk contains about $3 \times 10^{10} M_{\odot}$ in visible matter, including at least 100 billion stars. Enough to give over a dozen stars to every person on Earth!

---

22 You may appreciate that since the density of stars in the Galactic Disk varies with position above or below the Galactic Plane (z), and with distance from the center of the disk (R), the calculation is not quite as simple as implied here. In fact, to do this correctly we need to use calculus to appropriately integrate the exponential functions over the distance variables, z and R, and over all the various types of stars and the ISM that you want to consider. Sounds complicated, but in fact it's a straightforward calculation and the result is relatively simple. For a given type of object denoted by the index i, the total number of that type of object is simply $N_i = 4\pi h_i H^2 \rho_{i,c} / m_{i,avg}$, where h_i and H are the scale height for objects of type i and their scale length, respectively, while $\rho_{i,C}$ is the density of these objects at the Galactic Center. Note that this is different from the parameter listed in Table 9.1, which gives the densities of objects near the Sun. Given the values in Table 9.1, however, the central density can be easily calculated from knowing the density near the Sun as long as you know how far the Sun is from the Center—that would be 8.5 *kpc*—and the scale length—typically about 3.5 *kpc*. The final parameter, $m_{i,avg}$ is the average masses of the stars of type i. A question at the end of this destination provides an opportunity to explore these results in a bit more detail.

9.5 ROTATION OF THE GALAXY

We have come to appreciate that the properties of the Galactic Disk lead to the distinctive appearance of the Milky Way in our skies. But the presence of such a large, organized structure begs the question: How does the Galactic Disk maintain itself? This is not the first disk we have encountered on our journeys. The Solar System almost certainly formed from a disk, and its current structure is reminiscent of a disk as well (see Destinations 2 and 7). We saw that many of the compact dark clouds in Orion are disk-like structures, very likely extrasolar planetary systems in the process of forming (see Destination 6). All of these disks or disk-like structures rotate. Does the Galactic Disk also rotate?

Answering this question is tricky for reasons that are similar to trying to understand the rotational properties of the Planets of the Solar System while observing from a moving Planet (Destination 2). But it is even worse than this. As we'll soon see, the timescales associated with rotation in the Galaxy are far, far longer than those of the Solar System.[23] And yet, remarkably, this doesn't mean we can't detect the effect. Rotation implies motion, and one way to measure the motions of stars is by using the Doppler effect (Destination 6) to determine their radial velocities (that is, motion toward or away from us). We can also—if sufficiently patient—measure the motions of stars *across* the sky[24] associated with Galactic Rotation. Since even the closest stars are quite far away, only precise measurements of the positions of stars over a long period of time can detect these very slow movements which are known as *Proper Motions.* Since such proper motions are a measure of angular speed (we are in effect measuring the change in the coordinates of stars in the sky over time), we need to know the distances of the stars for which these motions are measured[25] to convert them to physical units such as *mph* or *km/s.*

---

23 It is common in science fiction movies to show the Galaxy visibly rotating from some distant vantage point. The speeds implied by these sci-fi motions are over 10 *trillion* times faster than the actual motions in Galaxy like our own, and over 1 billion times faster than the speed of light!!

24 As we did earlier for the stars of the α Cen system as seen from Earth.

25 The distance matter because if a star is moving comparatively slowly across our line of sight but is relatively nearby, it can still have a high proper motiion. But something far away, really far away, can be moving at a much faster speed yet have no noticeable proper motion. We've already seen that the relation between parallax angle and distance is = $1/p$, where p, the parallax angle, is measured in *arcsec* and the distance, D, is in *pc.* There's a similar relation between distance, proper motion, and velocity given by the equation $V_t = 4.74D\mu$, where V_t is the actual velocity, in *km/s*, of an object across the sky, while D is again the distance, in *pc.* The term μ is the proper motion, measured in *arcsec/yr.* The subscript t on V_t refers to the fact that this motion is known as the *transverse*

These techniques let us measure the movements of stars and of gas in the ISM relative to us from any location in the Galaxy. What do we see? Around the Sun, our first impression is 'not much.' Some of the nearby gas and stars are moving toward us, some moving away from us. But, *on average*, just about everything nearby appears to be essentially stationary with respect to the Sun. This seems surprising at first. After all, the structure of the Galactic Disk suggests rotation, so why this apparent contradiction? Could the Sun and all local stars really be standing still with respect to one another? The problem with this interpretation is that—even if initially stationary relative to one another—stars still would attract one another gravitationally and begin to move toward one another. Indeed, any finite configuration of stuff—stars, gas, or, as we'll explore at Destination 16, galaxies—would move together in much the same way that a cold molecular cloud contracts under the influence of its own gravity (Destination 6).

If the stars around the Sun appear to be standing still but are actually moving, then it must mean that they are all moving *together*. Proving this

Fig 9.11

The Geminid meteor shower as seen from Las Campanas Observatory in Chile illustrates how our motion through a swarm of particles in the solar system causes them to move outward from a distinct divergent point, which happens to be near the bright image of the planet Jupiter in this picture. Stars in the Galaxy behave in a similar fashion—though at a much slower rate!—appearing to move outward from the divergent point of the Sun's motion located toward the constellation Cygnus. For similar reasons, stars appear to move toward a convergent point in the opposite part of the sky at the location of the constellation Vela.

Image by Yuri Beletsky, Las Campanas Observatory.

---

velocity—that is, the motion across our line of sight. This distinguishes it from the radial velocity measured from the Doppler effect. Watch your units carefully if you ever use this proper motion equation!

Fig 9.12

From a car, we can detect our movement in many ways. Nearby stuff (the roadbed in the green ellipse), is passing rapidly to our rear. To use an astronomer's lingo, it has a large rearward proper motion. More distant stuff (the hills in the red ellipse) is also moving to our rear but at a slower angular rate (see footnote 25); it has a smaller proper motion though is moving at the same relative speed as nearby stuff. Directly in front of us (in the blue circle), stuff is very slowly diverging away from the point we are approaching. If we could measure the Doppler Shift of what's in the blue region, we'd see it is rapidly approaching. On the other hand, the Doppler Shift of stuff to our sides (the green region) would be small since it is neither approaching nor moving away from us, but we would detect strong proper motions as the material streams toward our rear. Finally, at intermediate directions (the red region in the photo), we would see stuff that exhibits some blueshifts and is streaming backwards from our perspective. Just by analyzing these motions, we could determine how fast our car is moving.

is tricky, and to truly test if this really is the case, we need to look more carefully at the motions—radial velocities and proper motions—of stars and gas around the Sun. To get a sense of how we can do this, consider a situation where we're moving in a car along a freeway at a speed similar to other cars around us. If we focused only on those other cars, we would conclude we are moving very slowly, perhaps even standing still. But if we instead focus on the land around us—which is what we normally do—we'd see the nearby roadbed moving rapidly to our rear. Distant landmarks would also appear to move in that direction, though more slowly because they are farther away. As we look toward the front—in the direction of our motion—far-off features down the road appear to 'widen' as we approach them (see Figure 9.11 for a celestial example of these effects, and Figure 9.12 for a more familiar example).

In the case of the stars around the Sun, similar effects can occur. If we are moving within the Galaxy—and assuming there is some difference in the relative motions of objects in the Galactic Disk—then distant objects in the Disk should appear to move. And indeed they do. Using proper motions and radial velocities we discover that nearby stars appear as if they are *diverging* from a point in the constellation Cygnus, the Swan, while at the opposite side of the sky, in the constellation Vela, the Sail of Argo Navis, stars appear to be *converging* (see Figures 9.11 and 9.13). In the directions of the constellations Sagittarius and Auriga—that is toward and away from the Galactic Center, respectively—many stars appear to be moving across our sky, from the direction of Cygnus and toward Vela with very little net motion toward or away from us.

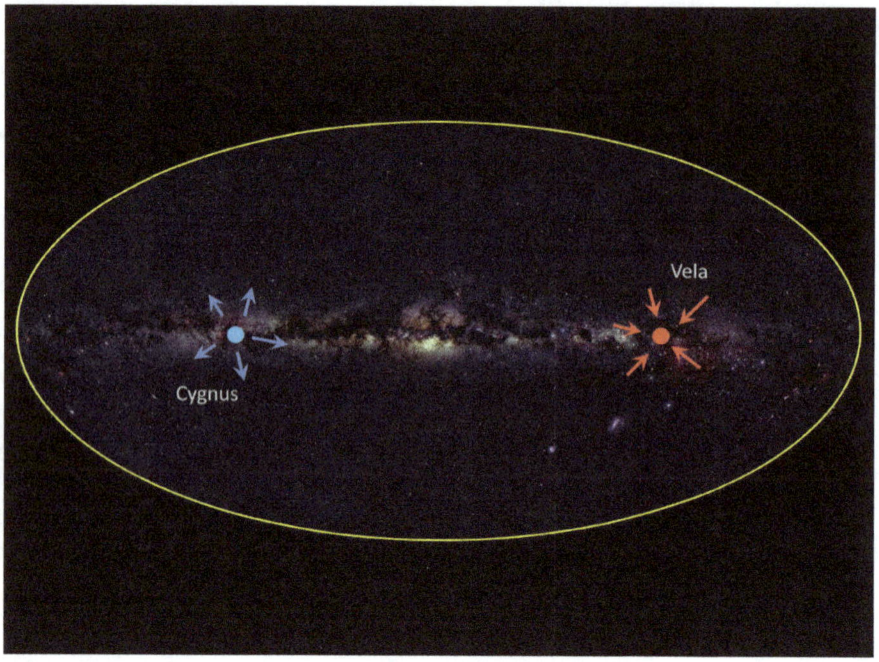

Fig 9.13

The effect of the Sun's orbital motion in the Milky Way. Keep in mind that this is an image of the *entire* sky using the same projections as in Figures 9.2 and 9.3. Toward Cygnus—the blue symbols on the left—we see blueshifts and proper motions that appear as if the stars are diverging from the point we are heading toward. On the opposite side of the sky, in Vela, the red symbols show that we would observe redshifts, with stars in that region appearing to converge to the point we are moving away from.

Photo mosaic by A. Mellinger.

Cygnus and Vela are located 90 *deg* from Sagittarius, and both lie within the band of the Milky Way (see Figure 9.3).

The most consistent interpretation of these motions is that the Sun, and virtually all the stars around it, is *moving about the Galactic Center along nearly circular orbits that lie close to the Galactic Plane.*[26] Moreover, the Galaxy's motion must be *Differential Rotation* in the sense that the Disk is not rotating as a solid body like a CD or DVD. It if were, we'd see no motions at all, just as two points on a solid DVD disk remain the same distance apart even as the disk rotates. A careful analysis of the proper motions and radial velocities of stars and, particularly, gas in the Galactic Disk, implies that the Sun—and most of the matter near it—travels about 220 *km/s* around the Galactic Center.

This kind of motion is not that different from motions in the Solar System where the Planets revolve about the Sun in nearly circular orbits (Destination 2). The stars in the α Cen system also move around their common center of mass in nearly circular orbits (Destination 4). The gas in the disks within the compact molecular clouds in the Orion Nebula also generally appears to travel in nearly circular orbits (Destination 6). And now we find that the Sun's motion within the Galactic Disk is similar. Keep in mind, however, that the Galactic orbit of the Sun is much bigger, with an orbital radius of about 8.5 *kpc*. Yet the Sun's orbital velocity is also larger than anything we've encountered

26 The first convincing measurement of Galactic rotation using the method outlined here was carried out by Dutch astronomer Jan Oort in the 1920s. At the time, Oort's analysis was an absolute *tour de force*, one of the most ingenious astronomical ideas of the generation.

so far. On reflection, this is rather startling. The Earth travels at about 30 *km/s* at a distance of 1 *AU* from the Sun. But the far more massive Sun travels about seven times faster around something that's 2 *billion* times further away!

We've noted many times that gravity is an attractive force—a relentless, merciless, attractive force. So it may be a bit perplexing why objects don't just fall into the things they orbit. There are two reasons. We discovered the first at the Orion Nebula (Destination 6): Angular Momentum. If the Earth's angular momentum were zero, it *would* fall straight into the Sun. But we've already seen that the angular momentum, J, is related to the orbital size, R, orbital velocity, V, and the mass M of the object through the relation $J = MVR$. For the Earth we know that the orbital velocity is fairly large (30 km/s), and so its angular momentum, J, is certainly not zero. The second reason has to do with the nature of gravity. Newton first related the force of gravity, F_{grav}, between two spherical objects in terms of the masses of the objects and the distance by which they are separated,

$$F_{grav} = \frac{GMM_0}{R^2},$$

where G is the gravitational constant, M is the mass of the central body, M_0 is the mass of the orbiting body, and R is the distance between the centers of mass of the two bodies. This relation is so important that it is often referred to as *Newton's Universal Law of Gravitation* (we introduced the term, but not the law itself, earlier at Destinations 4 and 6). How do these two concepts—angular momentum and gravity—conspire to produce stable orbits?

Newton answered this from combining his Law of Gravity with his knowledge regarding how things move along circular paths. In particular, a body of mass M_0 moving at a speed V_{circ} in a circular path of radius R must have a so-called *Centripetal Force* $F_{circ} = M_0 V_{circ}^2 / R$ applied to it to keep it on that trajectory. For a ball spun around a person on a string, the force that keeps the ball on its circular path is the tension in the string. For a person on a merry-go-round, the force may be friction or the effort the person exerts to hold onto the structure. For an orbiting body, the force must come from gravity. That is $F_{grav} = F_{circ}$. Since we have expressions for each of these forces, we can combine them and then solve for M to get

$$M = V_{circ}^2 R / G$$

(M_0 cancels out). This equation tells us that by knowing the velocity of an object in a circular orbit and the size of the orbit, we can derive the mass of the central body. Plugging in 30 *km/s* and 1 *AU* into this equation—and making sure the units are consistent!—we get 1 $M_\odot$. Precisely what you'd expect for the mass of the Sun given the orbital properties of the Earth. Though this latest equation may appear new, it is really just another form of Kepler's Third Law.[27]

The motion of the Sun about the Galactic Center suggests that we can use this relation to measure the mass of the Galaxy. We've established that the Sun appears to orbit the Galactic Center at a distance of 8.5 *kpc* and an orbital speed of 220 *km/s.* We also know from the properties of an Exponential Disk that most of the visible matter of the Galaxy must be located closer to the Galactic Center than to the Sun. This is not precisely the same way that mass is distributed within the Solar System—where virtually all the mass is the Sun—but it is not *that* different either since most of the matter in the Galaxy appears to be fairly strongly concentrated toward the center. If we plug in the values for the Sun's orbital parameters about the Galactic Center into the previous equation we find that

$$M_{Galaxy} = 10^{11} M_{\odot} .$$

That's 100 *billion* times the mass of the Sun.

What does this all imply for how the rotation velocity of the Galactic Disk should vary as we move further or closer to the Galactic Center? We've just seen that the mass of the Galaxy depends on the rotation velocity and the distance from the Galactic Center. But what if we now assume that all—or at least nearly all—of the mass of the Milky Way is located closer to the Galactic Center than to the Sun? Again, this is an approximation, but not a poor one. We can then solve for the orbital velocity as a function of radius to find that

$$V_{circ} = \sqrt{GM/R} \cdot$$

This equation is almost precisely correct in the Solar System because virtually all of the mass is in a central object: The Sun. Thus, if the Earth moves at 30 *km/s* at 1 *AU*, a hypothetical Solar System planet at 100 *AU* would travel 10 times slower, or 3 *km/s*. The point is that if the mass of a disk—be it the Solar System or the Galaxy—is concentrated toward the center, then the orbital velocities should *decrease* as you move outward from the Center.[28]

---

27 We could have derived this result from Newton's version of Kepler's Third Law, $a^3 = P^2(M + M_0)$, where P is the orbital period. If we assume that the mass of the orbiting object, M_0 is much smaller than the mass of the central object, M, then we can just ignore M_0. If we recall that $P = 2\pi a/V_{circ}$ for a circular orbit (this is using the basic relation *velocity × time = distance*), you can combine these relations to prove that $M \propto aV^2_{circ}$ (recall that a and R are just different symbols for the radius of a circular orbit). The problem is that it's not so obvious how to derive the constant with this approach. In this respect, using Newton's law is easier.

28 The behavior of the velocity as you move inward is more complicated. As long as most of the mass is closer to the Galactic Center than where you measure or calculate the velocity, then the orbital velocity will increase as you move inward. But once you get very close

Given the remarkable success we've had with Kepler's Third Law—and this decrease in the orbital velocity with increasing orbital radius is just another way of stating the law—this hardly seems worth confirming. It is *obvious* that the rotation velocity of the Galaxy will decrease as we move outward from the Center. But a strong sense of duty nags at us. As scientific explorers, we can't *really* say that we know something until we've measured it. And even 'obvious' results need to be checked. So we embark on one last outbound journey within the Galactic Disk. On this trip we will measure the rotation velocity at different locations of the Galactic Disk as we move outward from the Galactic Center.

At a location 10 *pc* from the Sun, we find that the rotational velocity is still around 220 *km/s*. This is no surprise since we would not have expected much change over such a comparatively small distance within the Disk. So we push on to 400 *pc* where we predict that the rotation velocity in the Disk, V_{circ}, should drop by about 2–3% compared to the velocity we measured at the Sun. When we determine V_{circ} here, we find it is still about 220 *km/s*. But that's ok. The uncertainty in our estimate of V_{circ} is about 10 *km/s*, so a 2–3% change would be hard to measure. So far, so good.

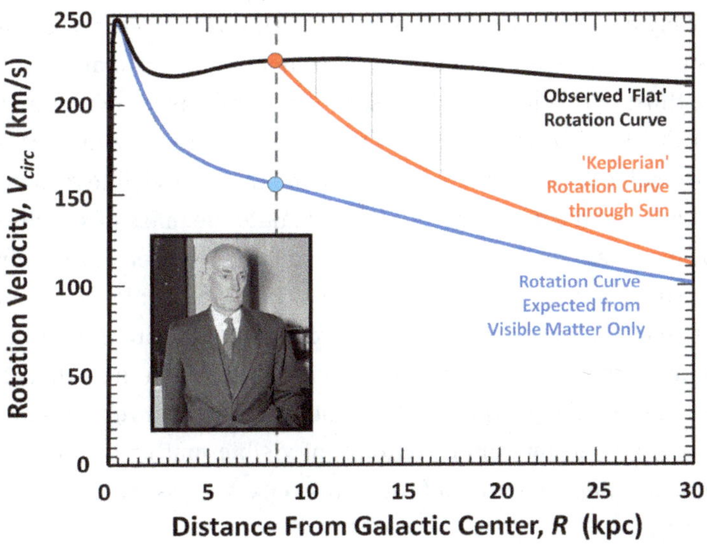

Fig 9.14

A plot of the observed rotation curve of our Galaxy (upper black curve). Because the rotation velocity, V_{circ}, varies relatively little beyond the location of the Sun (R = 8.5 *kpc*), this represents an example of a Flat Rotation Curve. The middle red curve shows what the Keplerian Rotation Curve of the Galaxy would look like if it were forced to pass through the Sun's rotation velocity at the Sun's location. The gray lines between the observed and predicted Keplerian curves are the velocity discrepancies we observe at the locations we are visiting as we travel outward in the Galactic Disk and measure the Galaxy's rotation curve. The lower blue curve shows the expected rotation curve for the Galaxy if the only mass in the Galaxy were in the form of stars and gas that we can detect. There are two distinct discrepancies here. First, the shape of the observed Flat Rotation Curve does not match the Keplerian Curve (the black and red curves). Second, the rotation velocities in the Galactic Disk are all too high (black curve) compared to the expected rotation curve derived from only the visible matter (blue curve). These anomalies are believed to indicate the presences of Dark Matter in the Galaxy that has a different distribution than visible matter (that's why the black and red curves disagree), and which constitutes a significant

As we push on, things start to get more interesting, but perplexing[29]. At a distance of 2 kpc from the Sun, we find that the rotation velocity has not changed. It is *still* around 220 *km/s*. From our earlier reasoning we would have expected a drop of about 10% in V_{circ} from its value near the Sun (Figure 9.14). That's a difference that we should have been able to measure fairly reliably here. But there's no apparent change. With some trepidation, we push on to

to the center, the orbital velocity will start to decrease because there is less and less mass inside the orbit. In other words, it's the mass *inside* an orbit that matters most in determining the orbital velocity. This very important result was proven early on by Newton himself.

a distance 5 *kpc* from the Sun, 13.5 *kpc* from the Galactic Center. When we measure V_{circ} here we again find it still hasn't significantly changed! From our simple expectations, the rotation velocity of the Disk should have dropped by about 30–50 *km/s* at this distance (Figure 9.14), a change we easily would have been able to measure. Astonished, we continue out to 8.5 *kpc* from the Sun—17 *kpc* from the Galactic Center. Here, at a distance twice that of the Sun from the Galactic Center, the rotation velocity *should* be about 30–40% lower (Figure 9.14). But, once again, we measure V_{circ} = 220 *km/s. No change!*

fraction of the mass of the Galaxy at least out to R = 30 *kpc* where the black and blue curves differ. The vertical dashed line shows the location of the Sun at R = 8.5 *kpc*. The observed rotation velocity of the Sun, V_{circ} 220 *km/s*, is denoted by the red dot. The blue dot illustrates the rotation velocity the Sun would have if there were no effects associated with the presence of Dark Matter. The inset image shows Dutch astronomer Jan Oort, one of the pioneers of the study of Galactic Rotation in the early 20th century.

The manner in which the rotation velocity of the Galaxy varies as a function of distance from the Galactic Center is known as the Galaxy's *Rotation Curve* (see Figure 9.14). On this journey, we have in effect determined the Rotation Curve of the Galaxy outside the Sun's orbit. From Kepler's Third Law, we had expected to measure a rotation curve that decreases as orbital radius increases. But instead of a *Keplerian Rotation Curve*—denoted as a red line in Figure 9.14—we discovered the Galaxy exhibits a so-called *Flat Rotation Curve*—the black curve in Figure 9.14. Contrary to our expectations, the orbital velocity of the Galactic Disk does not appreciably change as we move outward from the Galactic Center.

This enigma is even more vexing than it appears. The mass we derived for the Galaxy based on the motions in the Disk—10^{11} $M_\odot$—is interesting not only because of its sheer enormity, but also because we can account for only about one third of this mass in the visible stars and gas known to exist in the Galactic Disk (see Section 9.4). Part of the problem is that we did not consider the Galactic Bulge in our earlier estimate of the Galaxy's visible mass, and part has to do with technical issues of how to calculate the mass of self-gravitating disks. But even taking these points into account, we seem to require more mass to account for the motions within the Galactic Disk than we see in the form of visible stars and gas. In other words, beyond about 3–4 *kpc* from the Galactic Center, the Galactic Disk rotates *too rapidly* given the amount of matter we see in the Galaxy (compare the blue and black curves in Figure 9.14). Neither the distribution nor the amount of visible matter in the Galaxy can account for the shape and magnitude of the Galaxy's observed rotation curve (Figure 9.14).

Why do our predictions about how the Disk should rotate fail so spectacularly? Perhaps Kepler's Third Law is wrong? Or worse, have we perhaps discovered that Newton's Universal Law of Gravity is seriously flawed? These are disturbing possibilities since both laws appear to work very well to explain motions in our Solar System. We have used them to successfully send probes to virtually all the Planets. But the Galaxy is over 1 billion times larger than the Solar System. Perhaps gravity behaves differently than we expect when applied on such vast scales?

An alternate—and currently far more popular—possibility is that there is more mass in the Galaxy than we think. When one looks into the details of Newton's Law of Gravity, it becomes clear that we could produce a flat rotation curve if we assume the Galaxy's mass is distributed in an extended manner. An exponential disk won't do. It predicts too rapid of a falloff of mass outward from the Galactic Center. Instead, we'd need a structure that is more extended than the visible Disk, one that is more nearly spherical in shape. This component would also have to have about as much mass in it as the visible Galactic Disk and Bulge components do out to the location of the Sun at R = 8.5 *kpc*. The problem is that we don't *see* any such component in the Galaxy. Not from any form of EM Radiation it emits, nor as particles that it might be composed of. If this component really exists, it must be made up of some sort of stuff—dubbed *Dark Matter*—that emits no radiation and interacts with other matter only through its gravitational force. We'll encounter other instances that seem to require Dark Matter at future destinations, ranging from individual galaxies to the Universe as a whole! We'll discover too that the Dark Matter problem in our own Galaxy is considerably more extreme than we've described so far. Alas, none of our voyages will solve the riddle of what Dark Matter really is.

9.6 A SUMMARY OF A FANTASTIC JOURNEY

As befits the vast, complex structure of the Galaxy, our journeys have been grueling and, at times, rather confusing. It's worth reviewing some of the remarkable things we've discovered:

- The appearance of the Milky Way results from the detailed structure of the Galactic Disk. This component of the Galaxy obeys an Exponential Density Law both perpendicular to the Galactic Plane and outward from the Galactic Center within the Galactic Disk. The Disk contains at least 100 billion stars and most of the gas of the ISM associated with the Galaxy. The youngest, bluest, most luminous stars and active star-forming regions lie predominantly within the Galactic Disk.
- The center of the Disk—the Galactic Center—is located about 8.5 *kpc* from the Sun in the direction of the constellation Sagittarius (more on this at Destination 10).

- The Galactic Bulge—another significant component of the Galaxy—is distinct from the Disk but is centered on the same Galactic Center as the Disk.
- The thickness of the Galactic Disk is small compared to its width. The ratio of the Disk width and its thickness (as measured by the ratio of the scale length and scale height of different components of the Galactic Disk) range from about 100:1 to 10:1.
- The Galactic Disk rotates. The Sun partakes of this rotation, traveling at about 220 km/s along a nearly circular orbit about the Galactic Center. The Galaxy's mass that we derive from this rotation using Newton's Universal Law of Gravity is $10^{11}\ M_{\odot}$.
- Contrary to simple predictions based on Kepler's Third Law, the Galactic Disk rotation velocity does not decrease with increasing distance from the Galactic Center. Instead, the Galaxy exhibits a Flat Rotation Curve, meaning that the orbital speed does not change much with distance from the Galactic Center. Either Newton's Universal Law of Gravity is flawed—which most astronomers consider unlikely—or there is considerably more mass within the Galaxy than we can see in gas and stars. The extra mass is generally believed to exist in the form of Dark Matter, though the true nature of Dark Matter remains a deep and enduring mystery.

Back home, somewhere near the Sun, we look up again at the Milky Way. Before our voyages through the Galaxy, we had been content to enjoy the

Fig 9.15

The Northern Milky Way. One of the most sublime views in the sky, and one that everyone on Earth, and even far from Earth, can share.

Photo by M. Padilla.

sublime view at a mostly emotional level (Figure 9.15). And the stunning beauty of the Milky Way remains a good reason to search out—and protect—dark locations. But our new knowledge makes the view far more profound as we ponder the view. The immensity and exquisite organization of the Galaxy is there for us to see as we gaze on the Milky Way on clear, dark nights. It seems almost *alive*.

DESTINATION 9 QUESTIONS

9.1. What is the proper motion of the Galactic Center due the motion of the Sun in its orbit? You will need to read through the footnotes of this destination to find out how to calculate a proper motion given a known distance and relative velocity to a source unless you want to derive these relations on your own.

9.2. How massive would the Sun have to be for the Earth to have an orbital velocity of 220 *km/s* at 1 *AU*? How fast would the Earth move if it was in orbit around the Sun at a distance of 8.5 *kpc*? Of course, ignore the effects of all other stars in this latter calculation.

9.3. It turns out that the Sun is currently located quite close to the mid-plane of the Galactic Disk. Try to come up with and describe two observations you could make from Earth that would help to prove that this is true.

9.4. Look up and print an example of a projection of the surface of the Earth other than the Mollweide projection shown in Figure 9.2. Describe briefly the assumptions behind the projection method and outline how it distorts the actual distribution of features on the Earth's surface (all spherical projections onto a plane introduce distortions).

9.5. Let's say that A-type Main Sequence stars and G-type Main Sequence stars occur in equal densities in the plane of the Milky Way. If the scale height of the A-type stars is 200 *pc* and that of the G-type stars is 600 *pc* (*not* the actual values for the Milky Way!), what is the ratio of the densities of the two types of stars at a distance 1000 *pc* above the Galactic Plane?

9.6. These questions address issues related to the time it takes the Sun to orbit the Galactic Center.

a) The *Galactic Year* is defined as the time it takes the Sun to orbit the Galaxy once. If the Sun's orbital radius is 8.5 *kpc* and if its orbital speed is 220 *km/s*, how long is the Galactic Year?

b) If the Sun is 5 billion *years* old, how old is it in Galactic Years?

c) Use the age-mass relation (Destination 8) to determine the mass of a star that lives precisely one Galactic Year on the Main Sequence.

9.7. We discovered that the rotation curve of the Galaxy appears to be flat. This has interesting implications that we will deal with later. For now, let's explore the nature of rotation curves more generally.

a) Make a plot of what a flat rotation curve looks like. Be sure to label the axes!

b) Draw a plot showing what the rotation curve would look like if the Galaxy rotated like a solid body such as the Earth.

c) Draw the rotation curve if there were **no** matter beyond the location of the Sun (at about 8.5 *kpc* from the Galactic Center). **HINT:** This is the same as asking what the rotation curve of the Solar System is. In that case, all the mass is inside the Sun itself. Here we are assuming that all the mass of the Galaxy is inside the *orbit* of the Sun around the Galactic Center.

9.8. Let's explore some basic properties of an Exponential Density Law.

a) The total mass of a component in an Exponential Density Law is given by the relation $M_i = 4\pi\rho_{i,0}h_iH^2 \times e^{R_0/H}$. (Don't panic! This relation is very easy to evaluate.) Use the data in the Table 9.1 to determine the total mass in the Galactic Disk of the O stars, the ISM, A stars, G stars, and M stars. You can assume that the scale length is equal to 3.5 *kpc* for all of these types of objects but you will need to read off the values of $\rho_{i,0}$ (the space density at the location of the Sun) and h_i (the scale height) from the table. The term $e^{R_0/H}$ converts the value of the density to the value it would have at the center of the Galactic Disk at $R = 0$, and R_0 is the distance of the Sun from the Galactic Center. **SUGGESTION:** Notice that the only thing that changes as you apply this equation to the different types of stars and to the ISM is the value of the product $\rho_{i,0}h_i$. Calculate the term $4\pi H^2 \times e^{R_0/H}$ just once, then multiply this value by the term, $\rho_{i,0}h_i$, that changes. Also, be careful with the units! Remember, you want a mass in the end, and *kpc* is *not* equal to *pc*.

b) Assuming that the mass of the Disk is 10^{11} $M_\odot$, what fraction of that mass is in the form of each type of star and in the form of the ISM?

9.9. Oops. We've traveled so far through the Galaxy that we've gotten disoriented and have lost the Sun! Use your results from the previous question to determine how many solar-like (G-type) stars exist in our galaxy. If we could travel between these stars at the speed of light, how long would it take to survey them all as we search for the Sun? Provide some reasoning behind your choice for the typical distance between the stars that you must assume to answer this question. Also, comment on the implications of this result for the likelihood that we would be randomly visited by space travelers.

9.10. We found a relation between the rotation speed in the Galaxy, V_{circ}, at a distance R from the Galactic Center and the Galaxy's mass, M. Since the rotation curve of the Galaxy appears to be flat out to at least 30 *kpc* (see Figure 9.14), how does the total derived mass of the galaxy vary with increasing distance from the Galactic Center?

9.11. If the Galaxy were half as massive as we found in this destination, what would the Sun's orbital velocity be? If the Sun were located at a distance of 15 *kpc* from the Galactic Center but the Galaxy had the same mass as we derived in this destination, what would the Sun's orbital velocity be?

9.12. Describe how the motions of the stars would appear if it were only the Sun that was orbiting the Galactic Center and all other stars were standing still. Contrast that to the actual motions we see in our Galaxy as outlined in section 9.5 of this destination.

9.13. The orbital speed of the Sun about the Galactic Center, 220 *km/s*, seems like a high speed, and in some respects it is, while in others not so much.

a) How long would it take to go to the Moon at this speed?

b) How long would it take to get to Mars at this speed? (Adopt a reasonable distance from Earth to Mars.)

c) How long would it take to get to α Cen from the Sun at this speed?

d) How long would it take to get to the Galactic Center at this speed?

e) How much of a shift would we see in the Hα line (which has a rest wavelength of $\lambda_0 = 653\ nm$) for a body moving at this speed?

9.14. On the web, find information regarding the significance of the Milky Way to a specific culture, either one that exists today, or one from the past.

9.15. As a function of distance from the Sun, compare the orbital periods of Venus, the Earth, Mars, and Jupiter ...

a) As they really are.

b) As they would be if the rotation curve of the Solar System were flat with a mean speed equal to that of the Earth's orbital velocity.

9.16. Imagine that the Galactic Disk is a stubby cylinder with a radius equal to twice the distance of the Sun from the Galactic Center and a thickness equal to the scale height of G-type stars (see Table 9.1).

a) What is the volume of the disk in pc^3?

b) What is the average separation between O-type stars if there are 10000 in the Galaxy?

c) What is the typical separation between G-type stars if there are 1.5 billion of those stars in the Galaxy?

d) If a typical O-type star is visible with the naked eye out to a distance of 1 *kpc*, how many of these stars can be seen in this volume? (Take into consideration this is not a sphere.)

e) How many typical G-type stars can be seen, assuming they are visible to the naked eye out to a distance of 20 *pc*?

9.17. What is the space density of G stars at $R = 8.5\ kpc$ but at a position of $z = 250\ pc$ out of the Galactic Plane? What is the space density of M stars at a distance of $R = 15.5\ kpc$ and $z = -100\ pc$? Use the data in Table 9.1 to answer these questions.

Destination 10

INWARD BOUND: THE GALACTIC CENTER

10.1 TO THE GALACTIC CENTER: THROUGH THE INNER GALACTIC DISK

So far, our travels have taken us through many parts of the Galaxy. And, while we have seen much, one obvious region still beckons. It is also the evident nucleus of two other distinct components of the Galaxy: The Galactic Bulge (which we first encountered at Destination 9) and the Galactic Halo (which we will explore at Destinations 11 and 13). The entire Galaxy seems to be telling us that the *Galactic Center*—our next destination—is a special place. And indeed it is. Nothing we have seen so far has prepared us for what we shall find there.

Fig 10.1

Our path to the Galactic Center is clear, even if the view isn't due to interstellar dust. We start by heading toward Sagittarius (*lower right*), heading into the region highlighted in the *upper* image (the location of the Galactic Center is toward the top of this image). We will obviously encounter *a lot* of stars, gas, and dust as we progress on this journey (*left*), but what we cannot yet glean from these images are the bizarre things we will find at the Galactic Center itself.

Photos by S. Brunier (*lower right*), D. Malin/AAO (*upper left*), and S. Laycock (*left*).

Our survey of the Galaxy (Destination 9) revealed clearly to us what direction we need to travel to reach the Galactic Center: Toward the constellation Sagittarius (see Figure 10.1). This region lies in the South Celestial Hemisphere (see Figure 10.2), and is prominently visible from southern latitudes during wintertime there. From our explorations of the Galactic Disk, it seems likely that our route to the Galactic Center will be interesting too, so we shall stop to take in the sights en route. We'll mark our progress by noting how far we traveled from the Sun and how far we are from the Galactic Center. To be a

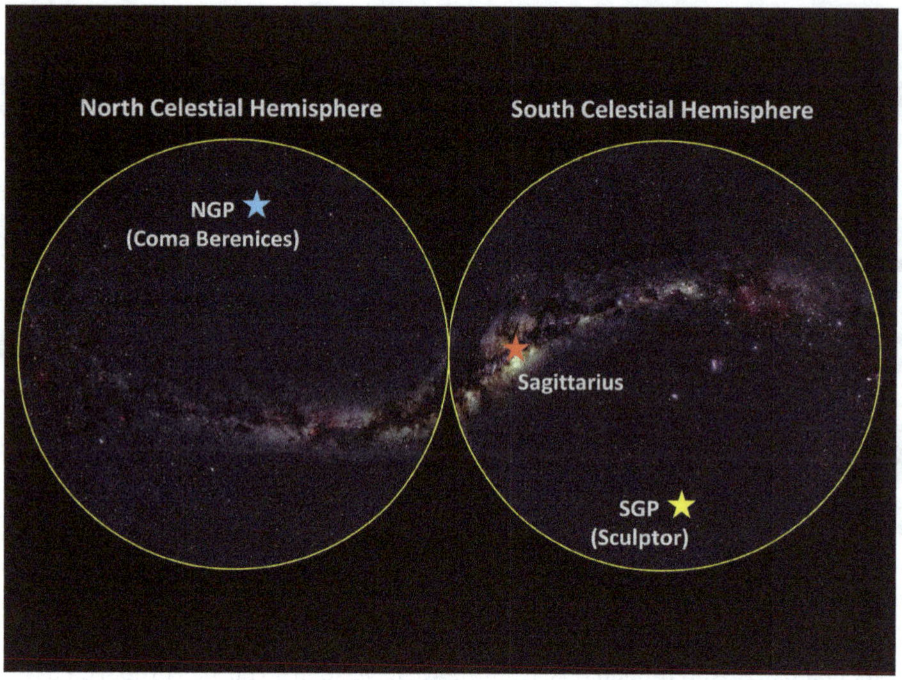

Fig 10.2

The full path of the Milky Way across Earth's sky is shown in this map (based on the images used for Figures 9.5 and 9.12) projected onto the North (*left*) and South (*right*) Celestial Hemispheres (if you look carefully, you can make out Polaris near the center of the left image). The location of the Galactic Center, in the constellation Sagittarius, is denoted with a red star. Sagittarius is a constellation of the Southern Hemisphere and so it is generally visible near the southern horizon during summertime from locations on Earth north of the equator. From the most populated latitudes of the Southern Hemisphere, Sagittarius—and the Galactic Center—appears glorious as it passes nearly overhead during the winter. The North Galactic Pole (blue star denoted NGP in the constellation Coma Berenices) and the South Galactic Pole (yellow star denoted SGP in the constellation Sculptor) are also noted. Compare their locations with Figure 9.5 where the Milky Way is projected horizontally across the long axis of the map. The Celestial Equator runs along the yellow circles around each hemisphere's map.

Photo mosaic by A. Mellinger.

bit precise about this, we shall follow the example from Destination 9 where we denoted the distance to the Galactic Center as *R*. Throughout the journey, we'll remain close to the Galactic Plane, or at $z \sim 0$ to use the notation for the distance above/below the Plane that we introduced at Destination 9. Starting at the Sun, we begin our travels at $R = 8.5$ *kpc*.

Our first stop is at $R = 7.5$ *kpc*, or about 1 *kpc* inbound toward the Galactic Center from the Sun (see Figure 10.3 to log our progress).[1] Veteran Galactic travelers that we are, we're not surprised to note that after traveling this distance, the familiar constellations of Earth's sky are no longer visible here. Nor are we surprised that the sky here is only subtly different than from home. We count about 8000 stars visible to the naked eye in this sky, a total close to the 6000 or so that can be seen from Earth. This small change—an enhancement of about 30% in the number of visible stars—does not make this sky seem appreciably more crowded than Earth's sky. Nor was this change entirely unexpected. We found from our travels in Destination 9 that the Galactic Disk obeys an Exponential Density Law both in the *z* direction (perpendicular to the Galactic Plane) and in the *R* direction (toward or away from the Galactic Center). Since we are staying close to the Galactic Plane as we move to the Center, the changes in stellar density along the *z* direction do not come into play. But radial density changes do, and the radial Exponential Density Law predicts that there

1 We've learned from our earlier journey through the Galactic Plane (Destination 9) that movements of less than a few hundred *pc* really make little difference in our view. We're taking much bigger steps right off the bat on this trip!

Fig 10.3

A map of our journey to the Galactic Center. We begin at the Sun, denoted as a yellow star, 8.5 *kpc* from the Galactic Center, which is denoted with a red dot. Our intermediate stops at $R = 7.5$, 4.5, and 0.5 *kpc* are marked. Note how—starting at the Sun—our first two stops (at $R = 7.5$ and 4.5 *kpc*), are dominated by stars and gas located within the Galactic Disk, while our stop at $R = 0.5$ *kpc* is deep inside the Galactic Bulge.

Image by R. Hurt/NASA.

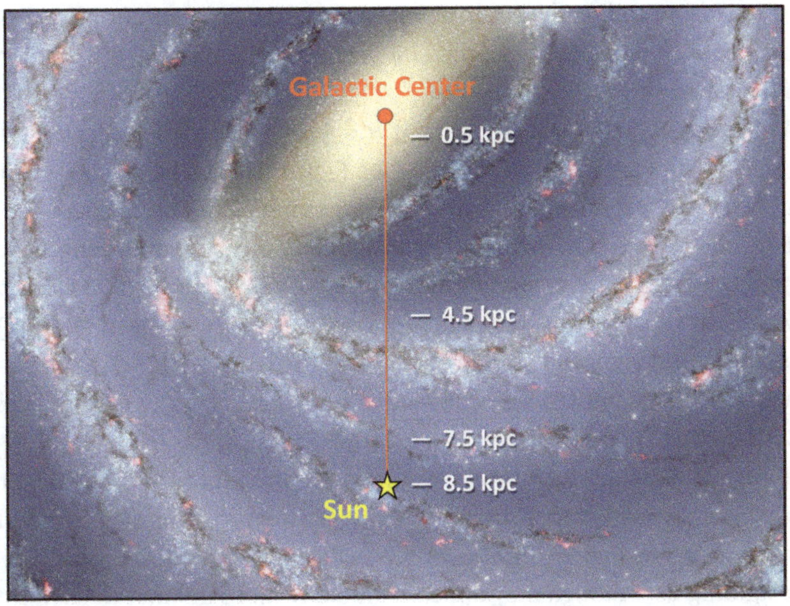

should be about $e^{1/3.5} = 1.3$ times more stars here in a given volume compared to near the Sun.[2] Precisely what we observe. What this tells us is that the stuff we see around us here at $R = 7.5$ *kpc*—that is, the stars and ISM—is still predominantly associated with the Galactic Disk. It's noteworthy too that the mixture of stars we see here appears similar to we observe around the Sun. That is, there seems to be about 30% more stars of all spectral types here. Or nearly. Perhaps the average number of bluest stars—stars of spectral types O and B—is enhanced a bit here compared to numbers of stars of spectral types A, F, or G (you can review spectral types at Destination 8). But the effect is subtle.

The gossamer band of the Milky Way is still clearly visible here. Some details differ of course—the specific stars, constellations, patterns of dark clouds—but many of its broad features remain evident. The Milky Way still traces a great circle in the sky, and it still appears patchy due to the presence of dark molecular clouds that obscure the view beyond. Is it only our imagination, then, that suggests the dark patches and long rifts in the Milky Way seem more distinct—maybe even more numerous—in the sky here? When we measure the properties of the ISM, we do find more gas here compared to near the Sun. There are more clouds of neutral Hydrogen associated with the Warm ISM here, more molecular clouds of the Cold ISM, and more ionized H II regions of the Hot ISM reminiscent of the Orion Nebula. These trends no doubt contribute to the distinct sense that the Milky Way is patchier here. But this too was to be expected. Apart from the outermost parts of the Galactic Disk, we found in our earlier travels (Destination 9) that the ISM also obeys approximately the same Exponential Density Law as the stars. The stellar density increases by

2 The '3.5' in this expression refers to the scale length of the Galactic Disk, which we know from Destination 9 equals 3.5 *kpc*.

30% compared to home, so it is not surprising the gas content here increases by a similar amount. As far as the ISM is concerned, it really *is* 'cloudier' here than at home.[3]

We press on, this time stopping at $R = 4.5$ *kpc* from the Galactic Center, a distance of about 4 *kpc* from the Sun (see Figure 10.3). The sky here is unmistakably different than what we are used to seeing from Earth. There are many blue O- and B-type stars peppered over much of the sky, some congregating in compact clusters, others in larger, more diffuse groupings called *OB Associations*. As we might expect—remember, these stars are young (see Destination 8)—we also see signs of vigorous star formation here. Huge ionized H II regions—some clearly associated with the clusters and associations of O and B stars—are easily visible to the naked eye over much of the sky.

Dark clouds, which we know are components of the Cold ISM (see Destination 6) are also obviously more common here than near the Sun. These clouds still block much of the light of the Milky Way—which we can still see as a great circle surrounding us—though they appear somehow more distinct to us in this sky, seemingly darker than we've seen before.[4] Observations of radio and microwave radiation confirm that these are indeed molecular clouds, their internal densities and temperatures similar to the dark clouds associated with the Orion Nebula (Destination 6). But they are different. For one, a number of these clouds are immense, weighing in at millions of solar masses and extending over hundreds of *pc* in size. Such *Giant Molecular Clouds* are very different in scale from the typically compact and comparatively low-mass clouds that we saw in the Orion Nebula (Destination 6). Like the smaller clouds within star-forming regions, these giant clouds generally appear to be dense enough and cold enough to be unstable to gravitational collapse (see Destination 6 for a review of the conditions that promote such collapse). Like moist air on hot days that's primed to form thunderstorms, these clouds appear ready to commence star formation at a more vigorous rate than we observe in the compact dark clouds of the Orion Nebula (Destination 6). And on a vaster scale. Giant molecular

---

3 There's a slightly confusing detail here related to one we raised at the last destination. The density in a cloud—measured in cm^{-3}—is generally about the same regardless if the cloud is near the Sun or in the outer or inner Galaxy. But, the ISM is patchy, usually clumped into distinct regions or clouds. Thus, we can talk about the larger-scale density of the clouds themselves. We do this on Earth all the time when we refer to cloud cover as 'partly cloudy' or 'mostly cloudy.' In both cases, the clouds themselves are essentially identical puffs of condensed water vapor; it is just that there are more of them in the sky on a 'mostly cloudy' day than on a 'clear' day.

4 Why? On average, these clouds are now closer to us because they are more common here than they were near the Sun. Being closer, they have fewer stars in front of them. Also, since the sky is brighter here due to the large number of stars, regions blocked by clouds seem by comparison even darker.

clouds appear to be where extensive complexes of star-forming regions—each equivalent to the entire Orion Nebula or larger—can be produced.

But there's a problem. There are too many of these 'on-the-verge' clouds here. From simple considerations regarding the rates at which these clouds should collapse into active star-forming regions[5]—a few million years—we'd expect to see a lot more young stars and ionized gas, and less material in the form of the Cold ISM in dense, dark clouds. It is as if some internal force—Magnetic fields? Stellar winds? Turbulence in the gas that makes up the clouds?—is inhibiting the collapse of these massive clouds beyond the role we expect from internal gas pressure. But don't be fooled. The rate of star formation here may be less than it *could* be given the number of molecular clouds present. But stars *are* forming vigiorously here in the Galactic Disk, some 4.5 *kpc* from the Galactic Center. And at a considerably higher rate than the star formation that is going on in the vicinity of the Sun.[6]

But what really gives us pause is how many stars of all types we see in this sky. Over 20000 stars are visible here. The sight produces an odd impression of being enveloped in a dark satin sheet of material that is embedded with bright glitter. Still, the increase in the number of visible stars here is consistent—for

---

5 Just about any process involves some sort of balance between an 'input' subprocess and a corresponding 'output' subprocess. A factory is designed to balance the input of raw materials with the output of finished products. If the rate at which materials enter a factory increases, then the output had better also increase. If it doesn't, then either more of the input raw material is being stored in the factory at an ever-increasing rate (and, typically, at increased cost), or some of that material is being secretly removed or diverted into making something other than the factory's output product. The same is true here with respect to these massive molecular clouds. If they were to collapse at the rate we'd expect on the basis of their observed densities and temperatures, we'd expect to see more star-forming regions here. The fact that we don't implies that something must be inhibiting the collapse of the clouds, something not directly related to the temperature and density of the clouds.

6 From our travels to the Orion Nebula (Destination 6), we know that the ISM plays a vital role in the process of star formation. On the scale of individual clouds, the only thing that seems to matter is if a cloud is sufficiently cool, or dense for its mass to start to collapse under the influence of its own gravity (see Destinations 6 and 7 to track what happens in detail during this process). But it turns out that the overall rate at which stars form in a region containing many dark clouds of the Cold ISM depends on how 'cloudy' the region is (see footnote 3 of this destination). Here, closer to the Galactic Center in a region dominated by stars and gas of the Galactic Disk, the 'cloudier' conditions seem to promote more vigorous star formation. This general behavior is seen throughout the Galactic Disk and, as we shall at future destinations, in other galaxies as well. But, as we've seen here, the increase in the rate of star formation is not directly proportional to how much cloudier a region is. Whatever inhibits the rate at which molecular clouds spawn star-forming regions seems to be related to some Universal property of the clouds themselves.

the most part—with what we would expect from the Exponential Density Law.[7] We conclude that here, almost halfway to the Galactic Center from the Sun, our view is *still* dominated by stars that belong to the Galactic Disk. The number of O and B stars here is again somewhat enhanced compared to our expectation from the Disk's Exponential Density Law. But that's consistent with the increased number of molecular clouds—regions that spawn star formation—at this location.

As we've traveled through the Galactic Disk en route to the Galactic Center, we have been dutifully measuring the rotation of the Disk as we did in Destination 9. Just as we saw as we moved out from the Sun toward the Galactic Anti-Center (Section 9.5 of Destination 9), the rotation speed we measure at each location as we move inward in the Galaxy is remarkably stable. Here at $R = 4.5$ *kpc*, the Galaxy's rotation speed has risen a bit to about 240 *km/s* (Figure 9.13), still consistent with a mostly flat rotation curve. If the inner rotation curve of the Disk was Keplerian[8], the Disk should be rotating even faster, around 290 *km/s*. In contrast to the outer Galactic Disk—which we discovered rotates too rapidly given the Sun's rotation speed of 220 *km/s* at $R = 8.5$ *kpc*—the inner Disk seems to rotate too slowly. Why the discrepancy?

A key feature we have discovered about the Galaxy is that its stars, its gas, and, most importantly, its mass, are not concentrated in one location, but spread out over an immense volume. For this sort of mass distribution, we actually do not expect a Keplerian rotation curve, particularly as we move inward. Newton's Universal Law of Gravitation implies that at any given point in the Galaxy, the mass interior (that is, closer to the Galactic Center) to one's location produces most of the gravitational force at that location.[9] Naturally, as we approach the Galactic Center there's going to be less and less mass between

---

7 Since we are now 4 *kpc* from the Sun, the stellar density here should be $e^{4/3.5} = 3.1$ times higher than at the Sun according to the Disk's Exponential Density Law. That's the primary reason we see so many stars here. However, there are mitigating effects. Dark clouds block the light of many more of the stars here than around the Sun. But the stars are on average closer together here, and so, because of the Inverse-Square Law for Brightness (Destination 4), they also typically appear brighter. These two effects roughly cancel on average, and the total number of stars we can see in the sky is about 3–3.5 times larger than the 6000 we see from Earth. In reality, we'd probably find the number of stars fluctuates significantly as we travel in the inner parts of the Galaxy. As we noted when we explored the outer Galaxy (Section 9.4 of Destination 9), the stellar density in the Galactic Disk, particularly for the younger, more massive O and B stars, varies due to the existence of large-scale spiral features in the Disk (see Figure 10.3). Nonetheless, the view we see here is likely pretty typical of what we'd see at a distance of about 3 to 5 *kpc* from the Galactic Center.

8 Recall (Destination 9) that a 'Keplerian' rotation curve is one that you get if all the objects rotate in accord with Kepler's Third Law. The Planets of the Solar System obey a Keplerian rotation curve, meaning that their orbital velocities drop in proportion to $1/\sqrt{a}$. In the Galaxy, this would imply that the rotation velocities vary as $1/\sqrt{R}$.

9 This statement refers again to 'Newton's Theorem,' a result we described in the last footnote of Destination 9.

us and the center. And as a result, the effective gravitational force we feel from the Galaxy itself will decrease. Since the rotation speed of the Galactic Disk has to be high enough only to balance gravity (see Destination 9), the rotation curve *should* deviate from a purely Keplerian curve as we move inward (see Figure 9.14). In that sense, the rotation speed in the inner Galaxy is approximately 'normal.' Some Dark Matter is required—there still is not quite enough visible matter to fully account for the Disk's rotation speed until we reach about $R \approx 2–3$ *kpc* (see Figure 9.14)—but far less than we need to explain the Galaxy's outer rotation curve (Section 9.5 of Destination 9; Figure 9.14).

10.2 TO THE GALACTIC CENTER: FROM THE DISK TO THE BULGE

Our journey has revealed reasonably good agreement with our expectations for an Exponential Disk, pretty closely continuing the trends that we observed in the outer Galactic Disk (Destination 9). We find that there are more stars in general—relatively more young stars in particular—and more gas and dust in the Galactic Disk the closer we approach the Galactic Center. But not all the way to the Center. Our last stop before reaching the Galactic Center itself is at $R = 500$ *pc*, or 8.0 *kpc* from the Sun. The Exponential Density Law for the Galactic Disk predicts that we should see a breathtaking 55000 stars in the sky here, about ten times the number we see in Earth's sky! But as we look in awe around us, we realize there are even more stars than that visible in this sky. Nearly 115000 stars shine down on us here like a gaudy, frozen, fireworks display.

And then, startled, we notice it. For the first time, something quite fundamental has changed in the sky. The Milky Way is different here. Indistinct. Nearly invisible. It is a jarring change. Everywhere we have visited so far, we have grown accustomed to being able to see the Milky Way shimmering in our sky. It has been a persistent feature of all the Alien Skies we have viewed from all the locations we have visited within our immense Galaxy. But no more. The vast array of stars in this sky appears nearly uniformly distributed, not like the band of the Milky Way meandering across the sky. Yet we do see something similar here. The 'empty' space between the stars here seems to be aglow. Not from the light of emission lines from the Hot ISM as we saw at the Orion Nebula (Destination 6), but from the light of countless stars too faint to see individually.

Fig 10.4

(*Top*) An optical view toward the Galactic Center (marked by the blue cross) from the Sun. (*Bottom*) The same view at infrared wavelengths penetrates much of the obscuring dust because longer-wavelength radiation is less easily absorbed or scattered by dust in the ISM. We really don't see much of the Galactic Bulge—which is mostly confined within the red dashed circle—in the optical view. Instead, the dominant contributor to the visible light is the inner Galactic Disk. We must get very close to the Galactic Center before the Bulge dominates our sky.

Images by A. Mellinger (*top*) and from 2MASS/NASA/NSF (*bottom*).

This reminds us of the gossamer glow of the Milky Way, which we discovered comes mostly from the combined light of stars that are, individually, far too faint to see with the naked eye. And so it is in this sky. The faintest stars here are evidently distributed more uniformly around us, not concentrated within a narrow band across the sky. It's rather ironic. Just as we are approaching the central, densest part of the Galactic Disk, the signature feature of that Disk—the band of the Milky Way—is starting to fade from view.

But that's not quite right. It's not that the Disk has faded away. Instead, we seem to now be inside a more symmetric structure, one that extends around us more or less equally in all directions. And one that contains *a lot* of stars. So many, that their combined light here overwhelms the light from the Disk. We have reached a structure that, evidently, is confined to the very central regions of the Galaxy. And it hits us! We have reached the *Galactic Bulge*. When we first viewed the Bulge with some effort from the vicinity of the Sun (see Section 9.2 of Destination 9; Figure 10.4) it seemed a remote, detached component of the Galaxy. But now, from our location deeply inside the Bulge, we gaze upon a sky that is filled mostly with stars of the Bulge itself. Stars belonging to the Galactic Disk—though still abundant here—no longer dominate the view.

The differences between the Bulge and Disk go far beyond the relative numbers of stars visible from each component. The uniformity of the distribution of stars here tells us that the density of dark molecular clouds must be low. Radio and microwave observations around the Galactic Center confirm this. The ISM, though not entirely absent, is far less prominent. This is particularly evident in the general lack of gas associated with the Cold and Warm phases of the ISM. Indeed, as far as the ISM is concerned, this central region is more reminiscent of the outer Disk, far beyond the orbit of the Sun (Destination 9). We seem to have passed into a 'hole' in the distribution of gas and dust surrounding the Galactic Center. Not surprisingly—given the lack of gas from which they could form—there are very few young O- and B-type stars visible here. But even A- and F-type stars are comparatively rare here, and these are stars that can live on the Main Sequence for many billion years (see Destination 8 on stellar ages). Most of the stars here have spectral types G, K, and M, the least luminous, reddest stars of the Main Sequence.[10] They are also the stars that live longest. When we estimate ages for these stars using the methods we developed at the Pleiades (Destination 8), we find that the majority have ages of about 8–12 *Gyr* (recall that *Gyr* stands for 'Gigayear,' which equals 1 billion, or 10^9, years). Compared to the Galactic Disk—

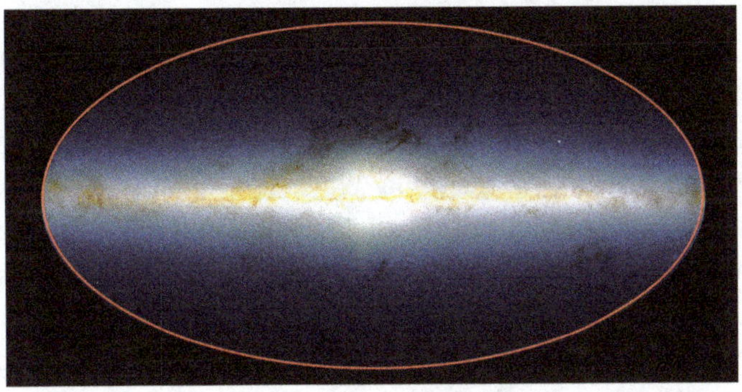

Fig 10.5

The difference between the Exponential Density Law of the Galactic Disk and the Power-Law Density Distribution of the Bulge is evident here. The Bulge is much more centrally concentrated and its density falls away from the Galactic Center faster than the Disk. It is also evidently much rounder than the flat Disk. But both the Bulge and the Disk co-exist all the way to $R = 0$.

Image from 2MASS/J. Carpenter, T. H. Jarrett and R. Hurt

where stars are still forming today—the Galactic Bulge is ancient and dormant.

The form of the Bulge also clearly differs from the Disk. Rather than a thin pancake-like structure like the Disk (Destination 9), the Bulge is in the form of a more nearly spherical distribution of stars. That's partly why the band of the Milky Way has been replaced by a more uniform distribution of stars over the sky (see Figure 10.5). We can be more precise about the distribution of stars in the Bulge by measuring their space densities as we near the Galactic Center. The first thing we notice is that—like the Disk—the space density of Bulge stars increases as we move toward the Center. But not as an Exponential Density Law (Destination 9). Instead, the distribution of Bulge stars appears to obey what is known as a *Power-Law Density Distribution* of the form

$$\rho_{Bulge}(R) = \rho_0 \left(1 + \left[\frac{R}{R_{Core}}\right]\right)^{-4}.$$

As in the Exponential Density Law, there are two parameters here: the central density of Bulge stars, ρ_0 and the *Core Radius* of the Bulge, R_{Core}. These parameters are analogous to the central density and scale height/length in the

10 Actually, many of the K and M stars in the sky here are examples of 'Red Giants.' These are what stars on the Main Sequence become after they run out of H to power nuclear fusion in their cores. We shall explore the nature of Red Giants in detail in Destination 11. But the main conclusion we draw from this view from inside the Galactic Bulge remains the same: The stars here are mostly very old.

Exponential Density Law (Destination 9). This equation tells us that the density of the Bulge varies quite rapidly as $1/R^4$ wherever $R > R_{Core}$. However, it also tells us that this behavior differs near the Galactic Center (for $R < R_{Core}$) where the Bulge density becomes nearly constant with a value near approximately equal to ρ_0. The core radius of the Bulge is about 1.5 *kpc*, while the central stellar density is $\rho_0 \approx 1.5 M_{\odot}/pc^3$, about twice the central density of the Disk.

This all sounds complex, but it should not hide an important point regarding the relative distributions of stars in the Bulge and the Disk. For $R > 1.5$ *kpc* from the Galactic Center—that is, beyond the core radius of the Bulge—the density of Bulge stars drops rapidly as we move away from the Galactic Center. At the location of the Sun, the Bulge contributes at most 2–3% of the stars we see in Earth's sky. But inside 1.5 *kpc*, Bulge stars become much more numerous compared to the number of stars in the Disk. At our current location 0.5 *kpc* from the Galactic Center—a location that is well inside the core radius of the Bulge—the Bulge contributes most of the stars visible in this Alien Sky. That's why the Disk and our view of the Milky Way seem to have 'disappeared.' Of course, the Disk has not disappeared at all. The Bulge and the Disk coexist here. It's simply much harder to discern the Disk among the more numerous stars of the Bulge near the Galactic Center.

The Power-Law Density Distribution for the Bulge is actually a simplification of how stars are distributed in the Bulge.[11] In particular, the relation implies the Bulge is spherical (notice that the only position variable is R while for the Disk we need two coordinates, R and z, to calculate the density). Had we traveled to the Galactic Center via a different path in the Disk—say, starting from a point on the Sun's orbit but many *kpc* away from where the Sun is currently located—we would have derived a similar density law for the Bulge, but likely with a different core radius. In other words, the detailed structure of the Bulge we measure depends on the angle at we enter the Bulge. This means that the actual shape of the Bulge is not spherical but more like that of a jelly bean (see Figure 10.6). In its most general form, this shape is known as a *Tri-axial Spheroid* because it can be defined by three distinct perpendicular axes, each with a different core radius.[12]

11 By contrast, the Exponential Density Law for the Disk is a remarkably good model for how stars are actually distributed in the Galaxy. There are some deviations, of course, but a pure exponential does an eerily good job when compared to the actual distribution of stars in the Milky Way.

12 A rugby ball, strictly speaking, is an example of a 'prolate spheroid,' an object in which one axis is longer than the other two, which are equal. A curler's rock (a web search can let you in on what that is if the term is new to you!) is an example of an oblate spheroid where one axis is shorter than the two other equal axes. The tri-axial case that best describes the Bulge is an ellipsoid where all three axes differ. Of course, if you've ever watched a rugby game, you know the ball often becomes tri-axial when heavy people fall on it!

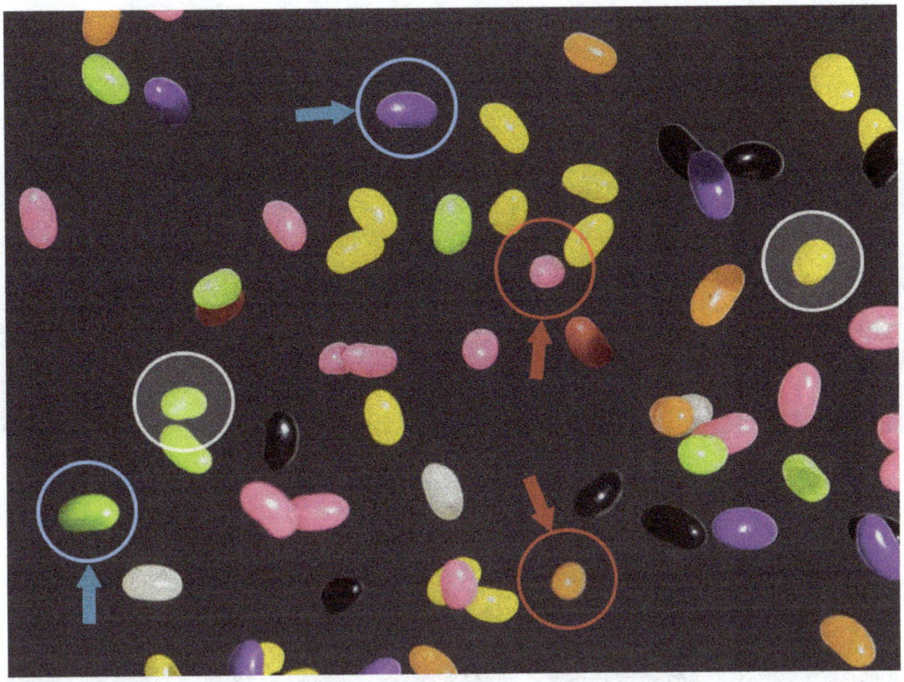

Fig 10.6

Jelly beans offer a reasonable model of the shape of the Galactic Bulge. Notice how the candies can appear quite elliptical in shape (two examples are noted by blue arrows and circles), but can also appear nearly circular if they happen to be lined up with the observer (red arrows and circles). Our view of the Galactic Bulge is about midway between a side view and a perfect end-on view; two examples of this shape are noted by the jelly beans in the white-rimmed 'spotlights'.

Image by M. Padilla and M. Mateo.

Like the Disk, we find that stars in the Galactic Bulge rotate about the Galactic Center. But whereas stars in the Disk all travel in about the same orbital plane, those of the Bulge don't. Instead, the individual stellar orbits appear to be tilted relative to one another, some moving up, some down, some left, some right. Also, the orbits of Bulge stars are generally less circular than those of stars in the Disk. The impression is that the stellar motions of the Bulge are somewhat more 'chaotic' than the fairly ordered circular motions in the Disk. We shall discover that most *Spheroid* stellar systems behave this way to some degree. The spectra we used to detect the motions of the Bulge stars reveal that, as a group, the stars in the Bulge also vary in chemical composition. As in the Disk, the outermost stars of the Bulge appear to be comparatively deficient in heavy elements. But, as we move closer to the Center, the abundance of heavy elements in Bulge stars increases to the point that many exceed the heavy-metal content of the Sun. For example, while only 2% of the mass of the Sun is in the form of 'heavy elements' (that is, elements other than H or He), some Bulge stars contain up to 20% of their mass in heavy elements. These are among the most 'metal-rich' stars known.[13] Comparisons of the motions and chemical compositions of stars in the Disk and Bulge are telling us something important about how different components of the Galaxy formed. We will have to wait until we travel back in time (!) to the era when galaxies like ours were

13 We've noted that astronomers refer to elements other than H and He as 'heavy elements.' They also refer to these elements as 'metals,' even elements that have no common metallic form! Thus, a 'metal-rich' star contains a larger fraction of its mass in the form of elements other than H and He.

forming—quite early in the history of the Universe (Destination 21)—before we can make some sense of these clues.

10.3 A STRANGE NEW WORLD: THE GALACTIC CENTER

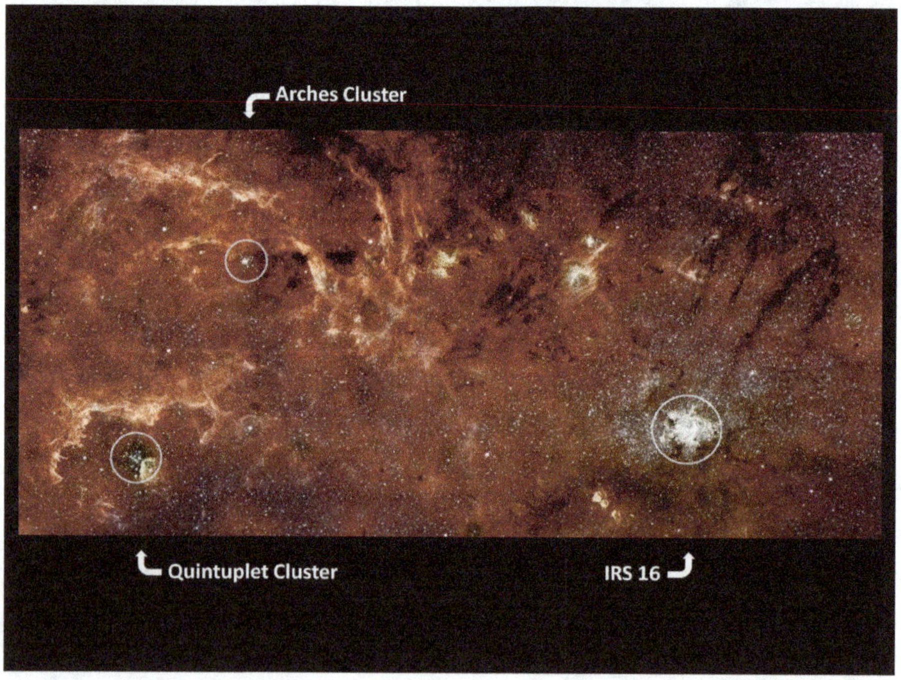

Fig 10.7

A panoramic view of the Galactic Center, a region that spans a bit more than the width of the full Moon in Earth's sky. For us, here at a location near the Galactic Center, this breathtaking view covers half our sky, filling a volume about 500 *pc* across! At the left side of the image, we see two star clusters of the Galactic Center, the 'Arches' Cluster (upper left circle), and the 'Quintuplet' Cluster (lower left circle; Figure 10.8 has more detailed views of these clusters). To the right is a large concentration of stars and gas. This is IRS 16 (circle at right), the home of Sgr A*, the central Supermassive Black Hole of our Galaxy (see Figure 10.10 for detailed view of IRS 16). This dense cluster is the hub about which the Galactic Disk rotates and the center of the Galactic Bulge. Even at a glance, we can plainly see that some of the gas and

Although our journey though the inner Galactic Disk and the Galactic Bulge has been fascinating in its own right, we finally reach the real goal of this voyage: The Galactic Center. We are now 8.5 *kpc* from the Sun, at the hub of the rotation of the Galactic Disk, the center of the Galactic Bulge. After finding a suitable place to 'park'—a random planet around a random star about 50 *pc* from the Galactic Center itself—we begin our survey of the fantastic scene before us (Figure 10.7).

The first thing we notice—we can't help but notice it!—is that the sky here seems incandescent with starlight. Over 240000 stars are visible to the naked

dust is distributed here into bizarre structures of enormous scale in this unique place in our Galaxy.

Image by L. Stolovy/JPL/NASA.

eye.[14] They are everywhere, spread fairly evenly over the entire sky. This makes sense for the same reason we described at our last stop: Nearly 75% of these stars belong to the Galactic Bulge. Since the Bulge is ellipsoidal—not a disk—we would expect its stars to be distributed around us fairly uniformly. The remaining 25% of the stars we see in the sky here are associated with the innermost part of the Galactic Disk. To our eyes—used to the comparatively anemic sky of Earth—it really does seem that there is a 'countless' number of stars around us.[15] And for good reason. From Earth, the average separation of stars in the sky is around 2.6 degrees, while bright stars—a small fraction of the stars visible—are typically tens of degrees apart in the sky. The stars here typically appear about 25 *arcmin* from their nearest neighbor, closer together than the angular diameter of the Moon as seen from Earth (about 30 *arcmin*). And there are many—in fact, hundreds—of stars visible here that are as bright as or brighter than any stars in Earth's sky. As in the Pleiades (Destination 8), there is no true nighttime here. The reason for this effect, however, is a bit different. In the Pleiades and similar clusters, individual stars dominate the brightness of the sky. But here, the combined output from all these stars generates so much light over so much of the sky that it is as if there is a continuous twilight. But, remarkable as this view is, it was not unexpected. The Exponential Density Law for the Galactic Disk (Destination 9) and the Power-Law Density Distribution for the Galactic Bulge reliably predicted the vast number of stars we see here at $R \approx 0$.

But any comfort we might gain from the 'predictability' of the scene soon evaporates. As our planet rotates about its axis, a new part of the sky of the Galactic Center slowly comes into view. We still see a uniform distribution of a vast number of stars, but there are other features visible here. Features we did *not* anticipate. The first surprise is the star clusters (Figure 10.7), some of the most remarkable in the entire Galaxy. In some respects, they are gigantic. By counting their stars, measuring their sizes, and determining their masses[16], we

14 Well, we see about half this number since we see only half the sky; our new home planet blocks the other half of the sky. But the point is, there are *a lot* of stars visible in this sky!

15 You may recall that at one time we felt that the numbers of stars visible from Earth seemed countless! Indeed, to many people who are unfamiliar with the dark sky, a typical first reaction to seeing it is that the dark night sky contains millions of stars, not merely the 3000 (half of the total of 6000 are visible at any one time, on average) we can actually see. The difference between the view of stars from Earth and at the Galactic Center—where we can see 50 times more stars in the sky!—is akin to the number of snowflakes in a gentle snowfall and the number during a raging blizzard.

16 How? We will explore this in more detail later (Destinations 11 and 16), but you might guess that it has something to do with measuring velocities and sizes. In every case that we have determined the masses of objects—for example, within the Solar System, Alpha Cen, protostars, the Galaxy itself—we have measured a velocity of some sort and a size of some sort. Kepler's Third Law, or more generally, Newton's Law of Gravity, then gives us the means of converting these results to a mass. Of course, in some cases, we used periods and orbital sizes, but you can appreciate that this is equivalent to using velocities and orbital sizes since we know that *velocity* $\propto$ *size/period*.

discover that these clusters have total masses of approaching $10^5 M_\odot$, far more massive than the Pleiades, which weighs in at about $1000M_\odot$. In size, however, these clusters are not all that different than the Pleiades—just a few *pc* in diameter (Destination 8)—and so they are much more densely packed with stars. There seems to be at least a half a dozen of these clusters here, all within 400 *pc* of the Galactic Center.[17] Amazingly, all of these clusters are very young, crammed with massive, young O- and B-type stars. From our nearby vantage point, they shine with a combined light as bright as the full Moon as seen from Earth. They are easily visible in this sky, even when competing against the light from the vast number of other stars spread around us.

But, hold on! As we traveled through the inner Disk and the Bulge, we noted how the density of gas and dust drops as we approach the Center, implying the presence of a gargantuan hole in the ISM around the Galactic Center. How could such massive and young clusters form here? A second surprise in this sky answers (in part) this puzzle. Right here, very close to the Galactic Center, we see a return—with a vengeance—of the ISM. Huge H II regions—gas ionized by the hot young stars of these clusters as in the Orion Nebula (Destination 6)—are spread throughout this region (Figure 10.7) along with other, unusual gaseous structures that seem to imply a more tortuous history (we'll return to these

Fig 10.8

Two rich, young clusters near the Galactic Center: The 'Arches' Cluster (*upper right*), and the 'Quintuplet' Cluster (*lower left*). Both contain many young, blue O and B stars, implying that the clusters are at most 2–5 million years old. The Pleiades is similar in size to these two clusters, but far less massive and luminous. Figure 10.7 shows where each cluster is located within the broader environment of the Galactic Center.

Images from STScI/NASA.

17 In fact, including the cluster right at the Galactic Center (IRS 16), there are only three young, massive clusters actually known to exist in this region of the Galaxy. Some astronomers who study the Galactic Center suggest that many other similar clusters must be present, but these are so heavily obscured by intervening dust and gas that they remain invisible from Earth. If we could really travel to the Galactic Center, we would know!

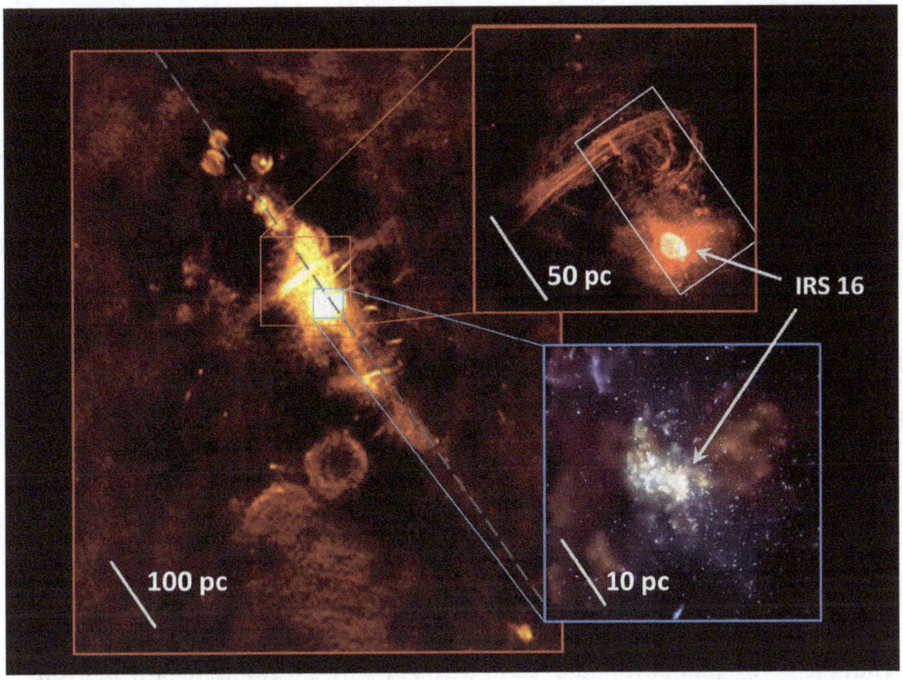

Fig 10.9

The Interstellar Medium (ISM) near the Galactic Center. The large-scale image (*left*) shows the inner 750 *pc* around the Center as seen at radio wavelengths that trace the Warm and Hot phases of the ISM. Many weird structures are visible here, including loops, rings, filaments, and rays as well as diffuse emission. These structures suggest that energetic events such as stellar winds and explosions (see Destination 12) and magnetic fields have sculpted the ISM in this region. The dashed line shows the location of the Galactic Plane. An enlargement of the brightest region (*upper right*) shows the bizarre distribution of the gas nearest the Center. The gray rectangle outlines the region shown in Figure 10.7, covering the spectacular view we have of this region as we gaze at it from our nearby vantage point. This richly structured radio radiation mapped here would be entirely invisible to our eyes. The image in the *lower right* shows where the clumpy distribution of the very hot/coronal ISM as mapped by its emission of very energetic x-ray radiation. Like the radio waves produced here, this x-ray radiation—both diffuse and clumped into bright arcs surrounding IRS 16—would also be invisible to our eyes. Notice that IRS 16, the star cluster surrounding the Supermassive Black Hole

sorts of structures at Destination 12). Radio observations tell us that there is also considerable molecular gas of the Cold ISM and neutral gas of the Warm ISM here as well. Even gas typical of the Very Hot (or Coronal) ISM is present here (see Figure 10.9 for a multi-wavelength portrait of the ISM of the Galactic Center).

As extraordinary as these clusters and the strange ISM near the Galactic Center appear, one specific region stands out (Figure 10.10). Using infrared observations, astronomers in the 1970's discovered this region, naming it IRS 16. We can now see that IRS 16 is yet another massive star cluster composed of young, blue, very luminous stars. It is the most luminous cluster we see here, and we soon appreciate that it is also the most unusual. Long, thin filaments of gas extend many *pc* around it. Some of these filaments glow with an eerie grayish-blue light unlike any gas we've encountered so far. The blue *continuous* spectrum of this light arises from radiation emitted by electrons spinning in magnetic fields—a form of light known as *Synchrotron Radiation* (the long streamers of gas visible in Figure 10.9 also suggest the presence of large-scale magnetic fields here). The region around IRS 16 also emits gamma-rays, the most energetic form of electromagnetic radiation (see Figure 5.6 of Destination 5). Some of the gamma-rays are emitted as an emission line that corresponds to what is ominously known as *Annihilation Radiation.*[18] This arises when electrons and their anti-matter partners—positrons—meet and annihilate one another in a process that converts their combined mass entirely into

18 It may sound ominous, but you have to admit, that's a pretty cool name!

energy.[19] The other gamma rays from this region exhibit a thermal spectrum, implying the presence of something absurdly hot, about 1 billion *K*. IRS 16 is surrounded by unusual blobs of neutral, and ionized gas, all distributed with little apparent correlation with the orientation of the Galactic Disk. Some of these clouds of gas have temperatures ranging up to many millions of degrees *K*, causing them to emit copious x-ray radiation (Figure 10.9). This is all pretty exotic stuff. Something very strange is happening here.

at the Galactic Center, lies in the brightest regions of the ISM visible in all of these images spanning most of the EM Spectrum (see Figure 5.6). Together, the visible, radio, and x-ray views of this region reveal intense activity that is producing young stars and creating great disturbances in the ISM.

Images from Galactic Center Radio Group/NRL (left); F. Zadeh et al./VLA/NRAO (upper right); K. Baganoff et al./Chandra/NASA (lower right).

To get a better sense of what's going on, we need to look at IRS 16 in closer detail. One particularly notable feature is an unusual ring—really, more of an annular disk—composed of dense molecular gas and clearly surrounding the cluster.[20] This disk has a well-defined inner boundary about 2 *pc* from the cluster center, but extends about 10 *pc* from the cluster core. Just as we've seen in every other disk we have encountered, this disk rotates. What's surprising in this case is how fast it is rotating. Using the Doppler effect (Destination 6), we can determine that the inner edge of this molecular ring—2 *pc* from its rotation axis—is moving at about 110 *km/s*, apparently centered on IRS 16. To put that result into perspective, Kepler's Third Law tells us that at a distance of 2 *pc* from the Sun, a planet would need to travel at only 0.05 *km/s* to stay in orbit.

Much of IRS 16 is located within this ring of cold molecular gas, so perhaps the source of the high rotation in the disk is the cluster itself. To check, we can measure the rotation speeds of stars inside the cluster, using both radial velocities and proper motions to determine their orbital paths (Figure 10.11). What we find is surprising. Stars located about 0.3 *pc* from the cluster center rotate at a speed of about 265 *km/s*. At 0.1 *pc*, the orbital speed about the cluster center has risen to about 460 *km/s*. At 0.03 *pc*, it is 840 *km/s*. As we probe even closer, at a distance 0.003 *pc* from the cluster center and a distance not much larger than the solar system (!), the stars are moving at an astonishing 2700 *km/s*—almost 1% the speed of light! It is clear that the velocities appear to keep increasing the closer we get to the center of IRS 16. The apparent center

19 All subatomic particles have corresponding *anti-particles* (although in a few cases, a particle's anti-particle is the particle itself). The anti-particle of the electron is the positron that has a positive charge but the same mass as the electron. For our purposes, the important point about *Anti-Matter*—as these particles are collectively known—is that whenever a particle meets its anti-particle, the two 'annihilate.' In this process, the combined mass of the particles is converted to energy via Einstein's Equation $E = mc^2$. In the Galactic Center, positrons are believed to be formed near the supermassive black hole in IRS 16 (more on this black hole a bit later in this destination). These particles, once formed, cannot go far before they encounter their anti-particle—a normal electron—causing the two to annihilate. From Einstein's Equation, the combined mass of the two particles corresponds to about 74000 times the energy required to ionize a H atom (see Destination 6). The result of this process is a well-defined emission line in the gamma-ray part of the EM Spectrum (see Destination 5).

20 How has a molecular ring survived in this violent region, a place dominated by hot stars and hot gas? Good question! But there it is. Just because we are traveling to all these destinations does not mean Nature will reveal her secrets willingly.

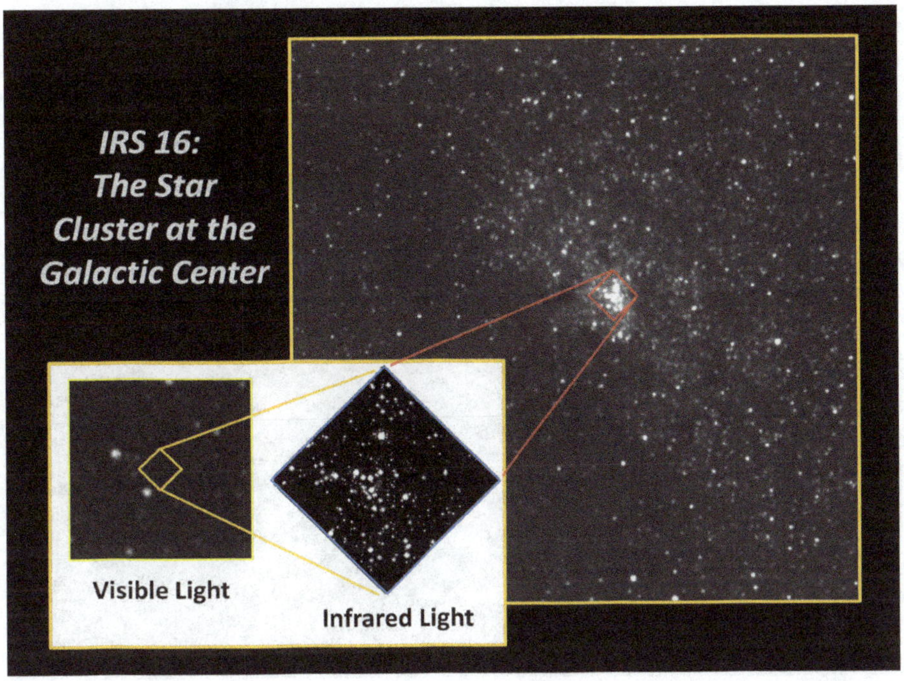

Fig 10.10

(*Right*) The immediate region around IRS 16, the cluster visible in the center of the larger image. The inset (*lower left*) shows the appearance of IRS 16 in the optical (*left* in the inset), where it is utterly hidden due to intervening dark clouds, and in the infrared (*right* in the inset) where it is clearly visible. This region is completely burned out in Figure 10.7. The Galactic Center resides in the heart of this cluster.

Images from G. and M. Rieke, University of Arizona.

of all this motion coincides with a strong source of radio emission known as Sgr A* (pronounced 'Sagittarius A-star' or 'Sag A-star') located very close to the center of IRS 16 itself.

Although the magnitude of the velocities is much larger, the relative behavior of the velocities of the stars at the core of IRS 16 is similar to the way that the velocities of the Planets in the Solar System vary. In both cases, the velocities increase as we move closer to the center. And they do so in accordance with Kepler's Third Law. The problem is that this is not what we would have expected inside a cluster. As in our earlier discussion of how the inner Disk rotation curve should eventually start to drop closer to the Galactic Center, we would expect that the velocities of stars located in the inner regions of an extended cluster like IRS 16 should start to appear to move *slower* as you get closer to the cluster center. Most of the cluster mass, after all, is *outside* the orbits of stars closest to the cluster center, and it is the interior mass that mostly dictates the orbital speeds. The fact that the stars inside IRS 16 show a Keplerian rotation curve (Destination 9) implies the presence of a lot of mass—comparable to the mass of the entire visible cluster yet utterly invisible—concentrated into a tiny volume about which the stars of the cluster orbit. If we turn to Newton's Universal Law of Gravitation and apply the same method we used to determine the Galaxy's mass (Destination 9), we discover from the innermost parts of IRS 17 where we can measure orbital velocities that

$$M_{Center} = \frac{V_{rot}^2 R}{G} = 5 \times 10^6 \ M_{\odot}.$$

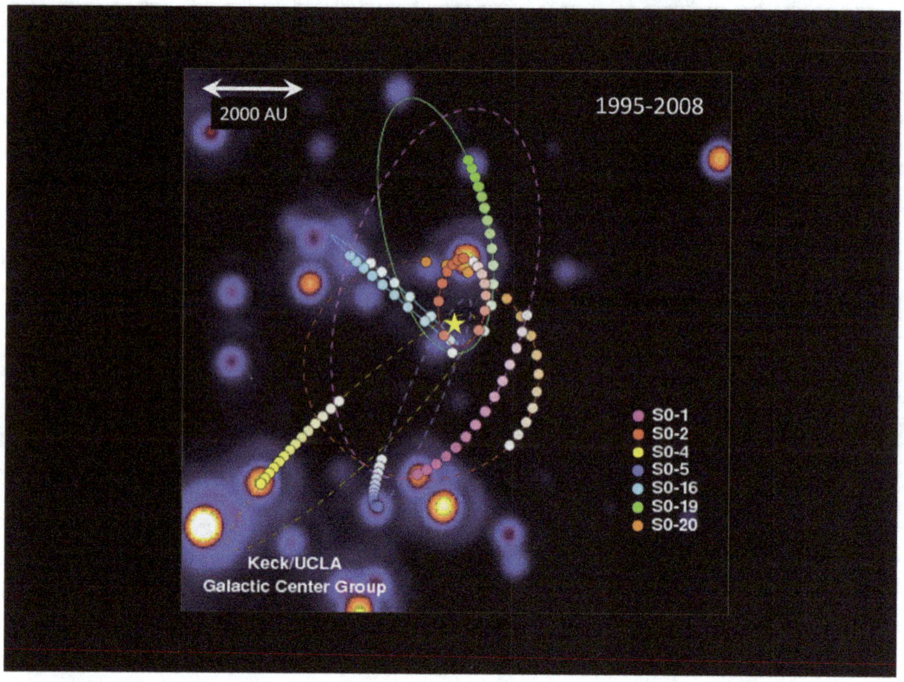

Fig 10.11

A plot showing the proper motions of stars in IRS 16 at the Galactic Center between 1995 and 2008. The individual stars are color-coded and identified at right and their positions are plotted as colored dots along their apparent orbital paths (dashed lines). The underlying image shows the positions of the stars at the last time they were observed to make this image. All of the stars here are clearly orbiting around the single point—denoted as a five-pointed yellow star—at speeds consistent with the presence of a compact body of mass $10^6 M_\odot$, the Supermassive Black Hole at the Galactic Center. The scale in the upper left shows that the orbits of these stars range up to 100 times the diameter of the Solar System as defined by its outermost planets.

Image from A. Gehz et al./ Keck Obs/UCLA.

This isn't the most massive thing we've encountered on our journeys—the Galaxy itself has a mass of at least $10^{11} M_\odot$. What's different here is that all this mass is crammed into a volume comparable in size to the Solar System. Perhaps even smaller! The only type of object we can guess that can pack so much mass in so small a volume is a *Black Hole.*[21] To contrast the gigantic black hole we find here at IRS 16 with much smaller black holes that have stellar-like masses[22], we refer to the one in IRS 16 as an example of a *Supermassive Black Hole.*

We will explore the nature of these strange objects in more detail later on (Destination 17). For now, we need to appreciate only that conditions can become extreme near black holes (supermassive or otherwise). Sufficiently extreme, in fact, to explain some of the more enigmatic features we observe from IRS 16. For example, very hot gas can be confined by the strong gravitational forces near a black hole in sufficient quantity to explain the thermal gamma-ray

21 The reasoning behind this conclusion is rather involved, but we can summarize it simply here. A *very* compact cluster of stars would be easily visible at the center of IRS 16 if it were present. But, as we can see for ourselves in Figures 10.10 and 10.11, there is nothing visible at the location around which the stars of IRS 16 appear to be orbiting. Likewise, if we imagine a rich cluster of massive objects such as stellar-sized black holes (Destination 12), white dwarfs (Destination 11), or neutron stars (Destination 12), it is possible to prove these would rapidly coalesce into a single, massive black hole due to the release of energy through gravitational waves. In short, there is so much mass in so small a volume at the core of IRS 16 that if it is not now a black hole, it soon will be. Just wait a few millions years.

22 As we'll see in Destination 12, these stellar-scale black holes usually weigh in at under $10 M_\odot$.

spectrum we see from the Galactic Center. Even more bizarre, black holes can generate positrons—the anti-matter particles of electrons—in their immediate vicinity.[23] These will annihilate when they encounter any of the numerous free electrons from all the ionized gas near the Galactic Center, producing the gamma-ray line emission we observe. There is so much circumstantial evidence—along with the concrete evidence of the compact mass distribution at the core of IRS 16—that it is hard to imagine anything other than a supermassive black hole explaining the full range of weird things we see here. Of course, being so close, we *could* go and look ... but, not yet! We'll have a chance to carry out a journey to the vicinity of a black hole at Destination 17.

But even from our moderate—and much safer!—distance from the supermassive black hole in IRS 16, we can still say some things about its detailed characteristics. To do this, we need to use Einstein's Theory of General Relativity, the standard description of gravity used to predict the properties of black holes. One important and remarkably simple result of Einstein's theory was discovered by Karl Schwarzschild, a brilliant contemporary of Einstein.[24] He noted that there should be a strict relationship between the 'size' of a black hole and its mass. The term 'size' does not refer to a physical surface around the black hole. Instead, 'size' refers to the radius at which any matter or radiation would have to travel at the speed of light to escape the gravitational pull of a black hole. This 'pseudo-surface' surrounding all black holes is called the *Event Horizon* for a simple reason: Under common circumstances, any event that occurs inside the horizon is effectively severed from the rest of the Universe. Nothing—not even light—can travel fast enough to communicate what's occurred from anywhere inside the event horizon to any place outside of it. Schwarzschild demonstrated that for 'simple' black holes[25] the radius of the event horizon depends only on the mass of the black hole. We now refer to this as the *Schwarzschild Radius*—typically denoted as R_S—and it can be written in a remarkably simple form:

$$R_S = \frac{2GM_{BH}}{c^2}.$$

23 Black holes are believed to be sites where anti-particles can be created through exotic processes driven by their associated intense gravitational and/or magnetic fields. Exactly how these processes work is neither certain nor is it obvious that anti-matter can be created only in the vicinities of black holes. But there is no question that anti-matter is distributed in a region centered on the supermassive black hole of IRC 16. There is little doubt that processes near the black hole must somehow produce at least some of the positrons found in this region.

24 Schwarzschild was one of only a handful of people at that time—the 1910's—who really understood and embraced from the start the concepts and mathematical underpinning General Relativity. There were no textbooks on the subject at the time!

25 That is, for black holes that do not rotate and that have no net electric charge.

If the mass we've found for IRS 16 is correct, the event horizon of the black hole here at the Galactic Center is only about 20 times the radius of the Sun! Yet crammed into this comparatively 'tiny' volume—about 10000 times the volume of the Sun itself—is five million times the Sun's mass!! This seems weird. Absurd. Ridiculous. Yet this result is consistent with our observations—there is certainly a compact, massive object at the center of IRS 16—and with what we know about the physics of gravity through Einstein's Theory of General Relativity. There is no better explanation for the full range of weird phenomena we see here at the heart of the Galactic Center.

This is not the last time we'll encounter black holes, not even supermassive black holes (Destinations 12, 17, and 21). But the lesson we have learned here—at the heart of the very first galaxy that we have explored in detail—is clear. The central regions of galaxies can be very strange places, characterized by extraordinarily large stellar densities, massive young star clusters, bizarre ISM properties, incredibly hot matter, even exotic stuff like anti-matter. We have discovered that if supermassive black holes can exist anywhere, they can exist at a place like this. At the center of a galaxy. At the center of *our* Galaxy.

DESTINATION 10 QUESTIONS

10.1. Uh-oh. The Sun just became a black hole. Of course, it gets dark without the sunlight. How does the orbit of the Earth after the Sun's collapse compare to the orbit before the Sun turned into a black hole? Assume that the Sun simply collapsed; that is, no mass was lost and nothing else fell into it to initiate the collapse.

10.2. For many centuries, astronomers understood that we lived in a disk-like Galaxy. But they thought we lived at the *center* of that Galaxy. What do you think is the principal reason for this fundamental error? **HINT:** To see the asymmetry of the appearance of the Galaxy from the Earth requires that we look within the band of the Milky Way. Think about what's present in the disk of the Galaxy that might make it hard to see the overall structure of the Galaxy from a single vantage point.

10.3. Calculate the Schwarzschild radii for the following objects, assuming they have been converted into black holes:

a) You.
b) The Earth.
c) The Sun.
d) The Galaxy.

10.4. We discussed the rotation curve of the galaxy back at Destination 9. However, it is clear that that discussion was incomplete since it did not include the effects of the central Black Hole. Draw a plot that shows our Galaxy's rotation curve in detail within the central 1 *kpc*. Assume that the center of the Galaxy is located at the position of the central supermassive black hole.

10.6. From the information given in this destination, what is the approximate mass density ratio of the clusters near the Galactic Center and the Pleiades? Precision is not the key here, but be sure to explain your reasoning and try to obtain an answer that gives a reasonable estimate of the mass density ratio of the clusters.

10.5. Consider the following figure from the European Southern Observatory.

a) From the information in the figure for star S02, estimate the mass of the supermassive black hole in the Galactic Center. **HINT:** Estimate the orbital semi-major axis from the image, figure out the period from the duration of the observations, and then use Kepler's Third Law.
b) Approximately when did Star 2 complete its first fully observed orbit? Can you find on-line information that confirms your calculation?

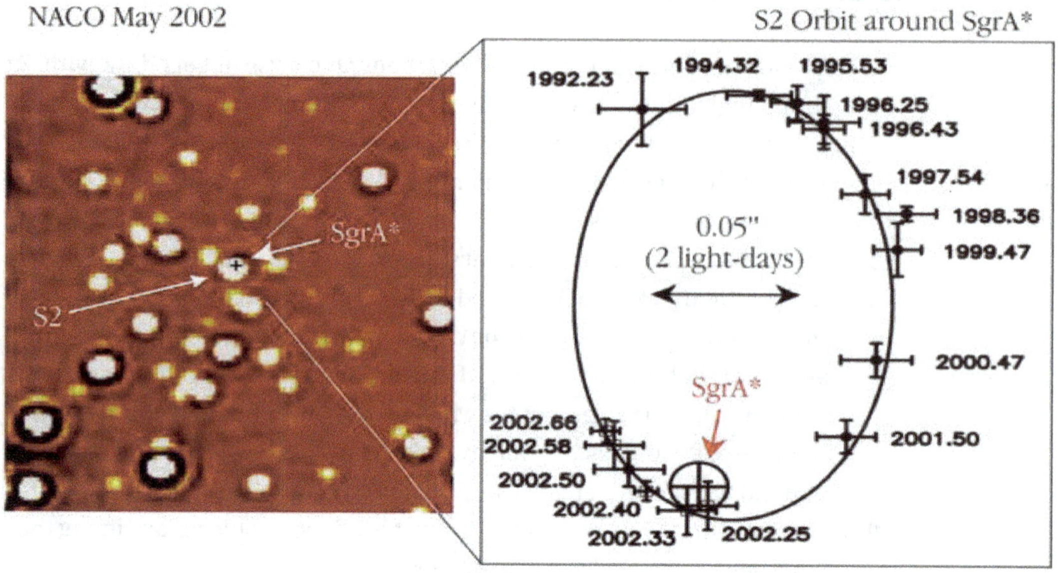

10.7. Make a single plot that shows the two following density profiles (in both cases, ρ_0 refers to the density at $R = 0$):

a) A power-law density distribution similar to that of the Bulge for which $\rho_0 = 100M_\odot/pc^3$ and $R_c = 1\ kpc$ and ...

b) ... an exponential density law for which $\rho_0 = 10M_\odot/pc^3$ and $R_c = 3\ kpc$.

c) From your plot, at what radius do the two profiles predict equal stellar space densities?

d) What is the ratio of the power-law to exponential space densities at $R = 8.5\ kpc$?

10.8. Imagine that the center of IRS 16 is composed of 100000 stars, each of $10M_\odot$, and all packed into a spherical volume of radius $R = 0.0005\ pc$.

a) How much volume does each star have 'to itself'? **HINT:** Calculate the total volume in this region and divide by the number of stars.

b) The volume one of these stars traces out as it moves is the size of an imaginary cylinder with a base of radius equal to the star's radius and a length of Vt, where V is the speed the star is moving in the volume and t is some amount of time. Assume that these stars all have radii of $3\ R_\odot$ and are moving at a typical speed of $3500\ km/s$. How long will it take for the volume of this cylinder to equal the volume you calculated in part (a)?

c) The time from part (b) is related to the typical 'collision time' of stars in a region of space. Given that the Galactic Bulge is likely over $10^{10}\ yr$ old, comment on the implications of your result from part (b) regarding the conclusion that the massive object at the center of IRS 16 is likely a supermassive black hole.

10.9. Look up on the web information on the Arches Cluster, the Quintuplet Cluster, or IRS 16 (see Figures 10.8 and 10.10). Report on some specific features of the cluster, either regarding something not mentioned in the text, or expanding on some feature that was noted in the text.

10.10. To get a sense of how the orbital velocity within IRS 16 should behave if there were no supermassive black hole at its center, let's consider how bodies would orbit within a sphere of uniform mass density.

a) What is the mass density of the Sun?

b) Assuming it was possible to place a body in such an orbit, what is the orbital speed of a body orbiting at 1 $R_\odot$ from the center of the Sun? **HINT:** Use the equation in this destination that we used to estimate the mass of the supermassive black hole at the Galactic Center.

c) Now, imagine that this body can orbit the center of the Sun at a distance of 0.5 $R_\odot$. How much mass is there in the Sun interior to this, assuming a uniform mass density from part (a)? Newton's Universal Law of Gravity implies that the mass exterior to this orbit has no effect on the body, so the orbital velocity depends on only the mass closer to the center of the Sun than where the body orbits. Given this, what is the orbital speed of the body? Repeat this calculation for a body orbiting at 0.1 $R_\odot$.

d) Is the behavior of the velocity Keplerian? If not, explain how it differs.

Destination 11

A BLAZING SKY OF STARS: THE GLOBULAR CLUSTER M 92

11.1 SUMMER IN PARIS

Every year it was the same, and Charles Messier must have hated it: Summertime in Paris. Messier's interest, pure and simple, was to discover comets. He considered this calling to be his ticket to fame and riches during his lifetime and—because comets are traditionally named after their discoverers—a way to achieve a touch of immortality. But summer nights in Northern France are very short, giving Messier little time to find comets during that otherwise pleasant season. Even worse, the sky itself seemed intent on fooling him. Over and over, he would find a fuzzy object with his telescope. One he had never seen before. Was it a comet? Far more often than not—and particularly during (Northern) summer—he'd be disappointed. Invariably, the new object would stubbornly refuse to move relative to the stars (Figure 11.1). Not after a few hours. Not after many nights. What he had found wasn't a comet after all, but something that lay far beyond the Solar System among the distant stars. Messier didn't really care much about the true nature of these objects; he was just annoyed that they were *not* comets. He certainly did not want to waste his time on them again. So, being a methodical person, he dutifully cataloged about 100 of these comet-like objects into a list that he published in installments over many years.[1] He—and other comet hunters—would know to ignore these masqueraders in the future.

As telescopes improved, people soon noticed that many of the objects in Messier's catalog—particularly ones most easily visible in the summer sky—were actually vast swarms of faint stars. Decades later, photographs revealed these objects contain thousands, sometimes millions of stars, all concentrated into a compact region of the sky (Figure 11.2). The generally round appearance of these clusters suggested that they must be approximately spherical in shape, like a globe. Because of these characteristics, these objects became known as *Globular Clusters*.

Around 180 globular clusters associated with our Galaxy can be readily seen from Earth[2], and twenty-nine of these are in Messier's catalog. The rest were

1 In case you are wondering, Messier did also find a lot of comets. Thirteen to be exact. But, ironically, it is his work on the 'annoying' non-comets that really gained for him the immortality he was seeking. As it happens, only two of his thirteen comets were discovered between mid-April and late September. This was likely a reflection of how the short summer nights and the large number of faux comets visible in the sky during that season complicated his ability to find real comets.

2 As we shall see soon, globular clusters are most commonly associated with individual galaxies. The 180 globular clusters referred to here are those that are known to

Fig 11.1

The real legacy of Charles Messier (whose portrait is shown in the *inset*) is illustrated in this beautiful image. The greenish fuzz ball with a faint tail (extending to the upper right) is Comet Tuttle, possibly similar in appearance to some of the thirteen comets Messier discovered. As it happens, none of Messier's comets have ever returned to the vicinity of the Earth, so they are largely forgotten. The other fuzzy object in this image, M 33, is an example of what Messier is truly famous for: The Messier Catalog, a list of about 100 of the most spectacular distant celestial objects seen from Earth. Many of our destinations are to objects in Messier's catalog; we've already traveled to M 42, the Orion Nebula, and to M 45, the Pleiades. The stars are smeared in this image because the camera followed the comet, which was moving relative to the stars during the exposure. M 33 is also smeared, implying that it doesn't move relative to the stars. Unsmeared images of M 33 can be found in Destination 14.

The Comet Tuttle/M 33 photo is by E. Kempen.

discovered subsequent to Messier's era, some as recently as the 2010's. But a key feature already apparent from Messier's small sample became even more pronounced as additional examples were discovered. Globular clusters are not spread uniformly over the sky (Figure 11.3), but instead are mostly concentrated into one region: The (Northern) summer sky. The sky in the direction of the constellation Sagittarius where the Galactic Center is located (Destination 10). Barring some extraordinary coincidence, this simple observation implies that globular clusters physically congregate around the Galactic Center. The *same* center defined by the Galactic Disk and the Galactic Bulge (Destination 9). It was this realization that the globular clusters are distributed about a remote location of the Galaxy—an idea energetically endorsed by American astronomer Harlow Shapley[3]—that led to the discovery of the Galactic Center in the early 20th century, about one hundred years ago. This helped to define for the first time the true structure of the Galaxy, and it ruled out any notions humans might have had that the Sun defined—or was located near—the Galaxy's center.

From Earth, or most locations within the Galactic Disk, the Milky Way does not show an overwhelmingly strong concentration toward Sagittarius when

be associated with our Galaxy. This number is a bit uncertain because the true nature of some objects known as globular clusters remains unclear. Some might be more similar to *Open Clusters* such as the Pleiades, and some may not be truly associated with our Galaxy.

3 As we shall soon see, Shapley did not base his arguments merely on the distinctive distribution of globular clusters in the sky; that had been noted for many decades. By introducing information on cluster distances, he concluded, correctly, that the clusters were defining the true, three-dimensional location of the Galactic Center.

viewed optically.[4] Due to the great distance of the Galactic Center and, more significantly, the large amount of dust and gas that blocks our view within the Galactic Disk, we simply cannot see the Center directly from optical radiation (see Destination 10). If the globular clusters were part of the Disk, but for some reason highly concentrated within hundreds of *pc* around the Galactic Center, they too would be difficult to see. But it is clear that the globular clusters are not part of the Galactic Disk. They are distributed in a more spherical distribution, one reminiscent of the shape of the Galactic Bulge (see Destination 10). But not quite. Recall that the Bulge is hard to see from Earth precisely because it is so concentrated around the Galactic Center; we cannot easily see it extending beyond the glow of the Milky Way. But the globular clusters trace out a more extended spherical structure, one that extends much further out from the Galactic Center than the Bulge. Consequently, many globular clusters are easier to see from Earth as they 'poke' into view above, below, and even in front of the band of the Milky Way along lines of sight that avoid much of the obscuring dust and gas residing in the Galactic Disk (see Figure 11.3).

Fig 11.2

Portraits of a globular cluster. A modern photograph of Messier 3 (*left*) reveals a cluster teeming with stars, which is clearly round in shape and exhibits a stellar density that falls steadily from the cluster center. This cluster contains over 1 million stars, though only a small fraction are visible here since most of the cluster's stars are too faint to show up in this photograph. Our current destination—M 92—is very similar in appearance to M 3. From Messier's perspective, the cluster would have appeared more like the view in the inset. He had no idea that objects such as M 3 were rich star clusters. But they didn't move in the sky relative to the stars, so he knew they were *not* comets, and that was enough for him!

Photo by B. Birket/SkyPi Observatory.

Since Shapley knew how to get reliable distances to individual globular clusters[5], he realized he could use the clusters to tell him the distance—not merely

4 This is evident from Figure 9.3, which, apart from showing more contrast than our eyes would detect, provides a reasonably accurate depiction of the appearance of the Milky Way to the naked eye.

5 One method that Shapley used is similar to one we will explore in some detail at Destination 14. The basic idea is to find classes of periodic *variable stars* (stars that change brightness on regular intervals), which obey what are known as *Period-Luminosity*

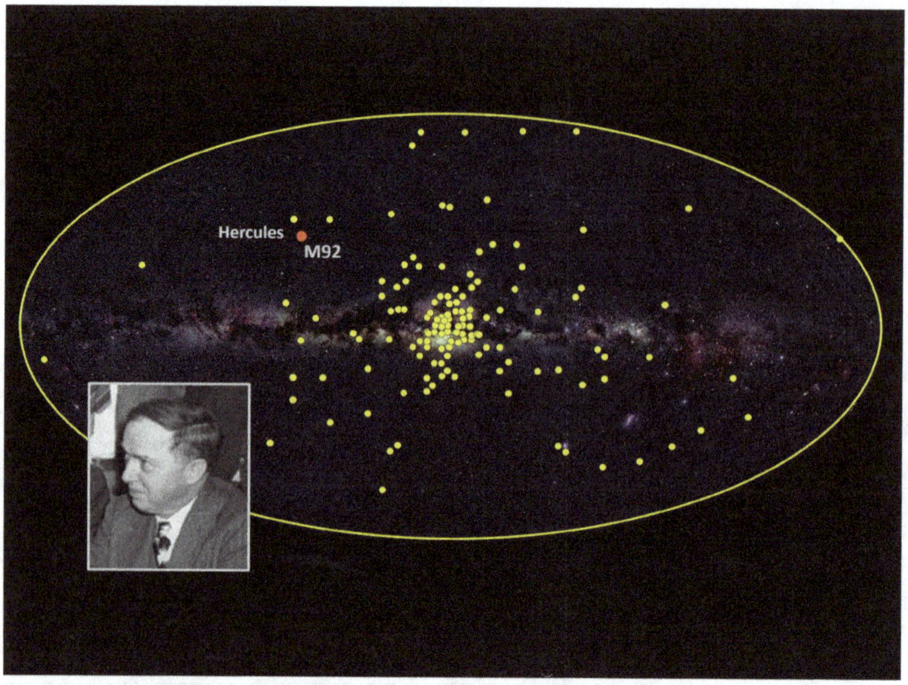

Fig 11.3

The distribution of globular clusters in the sky. The Galactic Center (toward Sagittarius) is at the center of this all-sky plot. Each point represents the location of an individual globular cluster. The clear concentration around the Center is obvious. The Milky Way extends horizontally across the middle of this map. Our current destination—the globular cluster M 92 in the constellation Hercules—is noted as a red dot. Harlow Shapley (*inset*) first recognized that the three-dimensional distribution of globular clusters in space could be used to localize the true center of our Galaxy.

The all-sky photo mosaic is by A. Mellinger.

the direction—of the Galactic Center from Earth. His rationale was simple. The average distance of the entire set of globular clusters should equal the distance to the center of their distribution, a location he argued must be coincident with the true Galactic Center. Shapley's method is still used—along with other approaches that have been developed since—to determine that the distance of the Sun to the Galactic Center is about 8.5 *kpc*, a result we used earlier to guide us during our journeys through the Galactic Disk (Destination 9) and to the Galactic Center itself (Destination 10).

11.2 THE SKY ABLAZE: THE VIEW FROM M 92

Our next destination is one of the globular clusters that Messier catalogued and, no doubt, cursed when he realized it was not a comet. Known simply as *M 92*—the

---

Relations. For such stars, one simply measures the periods of variability—which is relatively easy to determine and does not depend on distance—from which the stars' luminosities can be inferred. Combined with the brightness of the stars, one can determine their distances using the Brightness-Distance-Luminosity relation (Destination 7). Through no fault of his own, Shapley's original calibration of the Period-Luminosity relation was wrong. Fortuitously, this error was compensated by other effects he did not know about either and, in the end, he obtained a distance to the Galactic Center surprisingly close to what we adopt today.

92nd entry in Messier's list—the cluster can be found in Earth's sky near 'The Keystone,' a trapezoidal pattern of four stars that traces out the central region of the constellation *Hercules* (Figure 11.4). Though M 92 is not quite bright enough to be visible to the naked eye, it is easy to see from Earth even with rudimentary binoculars. A decent small telescope reveals immediately that it really is a rich star cluster composed of many individual stars (see Figure 11.4).

As it happens, M 92 is a particularly good choice for our travels because it is located about 8.5 *kpc* from the Sun *and* from the Galactic Center. It's also well outside the Galactic Disk—which we have explored thoroughly already (Destination 9)—about 4 *kpc* north of the Galactic Plane (see Figure 11.3). One consequence of the cluster's location is that our journey, though about as long as our earlier voyage to the Galactic Center (Destination 10), is extraordinarily lonely. Once we leave the region of space dominated by the Disk—that is, once we travel just a few hundred *pc* from the Galactic Plane (Destination 9)—we spend most of our time in a region where we can see only about 10–20 stars in the entire sky, and most of these are faint, near the limit of what our unaided eyes can detect. The inky emptiness that surrounds us during most of the journey underscores that M 92 is not part of the Galactic Disk where, even in its outermost reaches, we saw more stars than this in the sky (Destination 9). Our journey to the cluster also suggests that M 92 doesn't belong to the Galactic Bulge. After all, most Bulge stars reside considerably closer to the Galactic Center than along our path to M 92 (Destination 10).

But, despite appearances, we actually remain within the Galaxy during the entire voyage to M 92: We are traversing a third major component of the Galaxy, the so-called *Galactic Halo.*[6] We'll explore the properties of the Halo—the largest, most rarified, dimmest, yet most massive component of our Galaxy—as we explore it in detail at Destination 13. Like travelers trying to get across a vast desert or a great ocean, our goal for now is to simply endure the lonely passage through the Halo and get to our real destination, M 92.[7]

But we persevere through the emptiness of the inner Halo until, finally, we arrive near M 92. After some effort, we eventually find a suitable planet floating

6 The name of the Galactic Halo is somewhat unfortunate. For most people, the term evokes images of ring-like structures floating over the heads of saints or angels! But that's *not* at all like the true shape of the Galactic Halo. The Halo is actually structurally similar to the Bulge: Centrally concentrated about the Galactic Center, and roughly spherical in shape. The major difference with the Bulge is that the Galactic Halo is *much* larger in extent (Destination 13).

7 Let's not get blasé about the distances to our destinations simply because we have become seasoned interstellar travelers. At a distance of 8.5 *kpc*, it takes light about 30000 *yr* to cover the gulf between the cluster and the Sun. There are fundamental problems associated with traversing such distances—issues regarding the flow of time, survivability, and obtaining power for the trip. But the views of the Alien Sky that we describe at this destination—including the views of M 92 itself—*do* exist. These are real places, even if they prove to be unattainable for humans.

freely in the outskirts of the cluster[8] located some 10 *pc* from the cluster's center. This will be our base from which we will observe and study M 92 in detail. Though this planet is not orbiting any specific star—it has no 'sun'—the steady motions of stars from east to west[9] confirm that it is definitely rotating. Not that there are a lot of stars visible at first. Around 100 can be seen easily in the half of the sky above us, about five times more than we had become used to seeing during most of our voyage to this place.[10] And there is another difference. Curiously, many of the stars we see in this sky are rather bright and

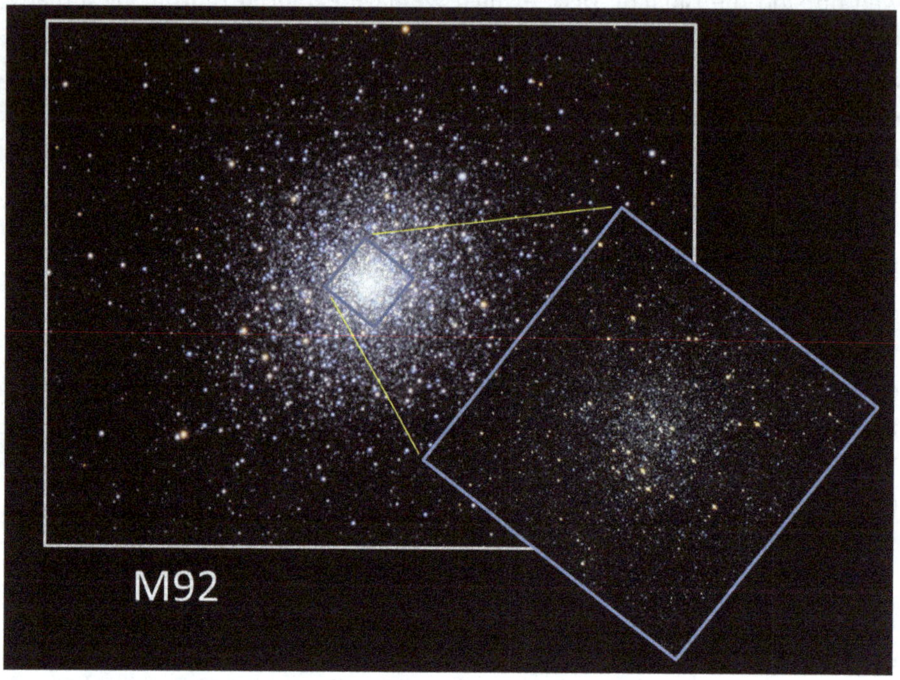

Fig 11.4

Two views of M 92 showing most of the cluster (*left*), and a blow-up of the central region (*right*). From a vantage point near the cluster, our view would resemble the one on the right. But, rather than seeing a few thousand stars as evident in this picture, over one hundred thousand stars would be visible in the sky! Notice how most of the brightest stars in both images exhibit a distinctly reddish hue.

Large photo by A. Block/ Mt. Lemmon SkyCenter/ UA. Inset photo from STScI/NASA.

8 For our purposes, we assume that the planet we've found here in M 92 is not associated with any star; it is just freely floating in the cluster (we made a similar assumption when we visited the Orion Nebula in Destination 6). Whether such orphan planets really exist in globular clusters is not (yet) known, though it is certainly possible.

9 Remember, there's no mystery or coincidence here. Given our familiarity with how directions are specified on Earth, we'd *define* east and west, and hence north and south, on the basis of the way the planet rotates. Stars rise in the east (which defines that direction), set in the west. And north is to the left as you face east. That's how it works on Earth. So that's how it works here.

10 Why do we see more stars here than during the journey? If you recall a subtle point from Destination 9, by the time we had reached a distance of 2 *kpc* from the Galactic Disk, there were only twenty or so stars visible in the sky. We are now twice as far from the disk, many scale heights further away. According to the Exponential Density Law for the Disk, we should expect to see no stars here at all. But there are two reasons for the stars we do see in this sky. First, the Galactic Halo is still the dominant component of the Galaxy here and it contributes a handful of stars to this sky. Second, and more importantly, even when we are currently looking away from the core of M 92 at our vantage point some 10 *pc* from the cluster center—as we currently are—the outer regions of the cluster itself contribute stars to the sky. Indeed, most of the stars we see in this part of the sky are cluster members that happen to be located further from the center of M 92 than we are.

obviously red in color. We certainly had not seen many stars like these as we journeyed here.

But the meager distribution of stars we see in this sky is not what commands our attention right now. As it happens, we've commenced our observations just as the bulk of the Galaxy looms above us. Since we are now located far outside the Galactic Disk, the band of the Milky Way is not visible here. At least not the Milky Way we are used to seeing from Earth. From our present location, the Galaxy—almost all of its visible light—spans much of the sky. As we look 'down' onto the Galactic Disk from this vantage point, its structure—so hard to discern from the inside—is laid out plainly before us. There's the Galactic Bulge. At a glance, we appreciate how it steadily dims outward from its unmistakable, bright core—the location where we know the Galactic Center resides (Destination 10). The Galactic Disk and its dark molecular clouds and bright H II regions (Destination 6) cover most of the sky. We can just make out a few individual luminous, blue stars, and even some compact star clusters. Together with the gas and dust—visible as fuzzy ionized H II regions and dark molecular clouds that blot out parts of the light from the Disk—the stars and clusters trace a vast spiral pattern that is simultaneously obvious, yet oddly indistinct, to our eyes.[11] Between these spiral arms, much of the sky glows with a smooth, grayish light that fades gradually with increasing distance from the Galactic Center. This ghostly light spans over 120 degrees in our sky, about two thirds of the angular extent of the sky from horizon to horizon. The vast number of stars that comprise the glow from the Galaxy—including the Sun itself—are now too distant to be seen individually without a large telescope. Their combined light is reminiscent of the gossamer glow of the Milky Way, though very different in form.

And there it is before our eyes: Our Galaxy. Nearly its full extent apparent in a single view. It is a magnificent, humbling sight, all the more awesome now that we have begun to learn just how vast and complex it really is.

But the view of the Galaxy from this place, beautiful as it is, is not the real spectacle of this remarkable Alien Sky.[12] As the Milky Way sets slowly, majestically in the west, we turn to the east where a few individual stars are rising from the horizon. Dozens at first. Then hundreds. And then, after an hour or two[13], thousands become plainly visible. But the procession does not end. Soon we see ... tens of thousands? Hundreds of thousands? More? It is a sight unlike any we've imagined. As the stars of M 92 rise into our sky, they blaze down on

11 We will return to this somewhat cryptic description when we start to learn how galaxies are classified at Destination 14.

12 Which is why, if I had to choose just one destination to visit, it would probably be this one.

13 Having assumed our planet base exists, we'll also make our usual assumption it rotates with a period similar to Earth's 24-hour day. If we are going to make up planets, we might as well make them convenient!

us like a frozen swarm of luminescent insects covering more than half of the visible sky. We see hundreds of stars—mostly distinctly red in color—that are all brighter than any seen from Earth. Thousands more stars, most only slightly fainter, are easily visible and clearly spanning a wider range of color. Many of these are also red, but some are distinctly more orange in appearance, some distinctly yellowish to our eyes. There are even a few that appear obviously blue in color. But the vast majority of the stars we can see from this vantage point—hundreds of thousands—are quite faint, only barely visible to our unaided eyes. We can tell there must be many more stars. Stars too faint to be seen individually but which impart a weird, almost creepy, mottled appearance to the sky.[14]

All the stars we see in this sky—the bright red ones and the fainter ones alike—are unmistakably concentrated about a well-defined center: The core of M 92 (Figure 11.4). It is also apparent that the distribution of stars is pretty uniform relative to this central region. There is no evidence of dust or molecular clouds blocking our view, just a steady drop in the number of stars in all directions outward from the cluster center (Figure 11.4). As we explore this sky more carefully using telescopes sensitive to radio and microwave radiation, we confirm there is virtually no ISM here apart from the hot, low-density gas of the Coronal ISM, which seems to pervade the entire Galaxy (Destinations 5, 6, and 13). The sky here really is dominated by what we see in visible radiation with our eyes: A riot of stars, concentrated toward the core of M 92 more densely than the overwhelming distribution of stars that filled the sky of the Galactic Center (Destination 10)!

This is also an altogether different sky from the one we saw at the other star cluster we have visited so far, the Pleiades (Destination 8). Not only are the Pleiades and M 92 obviously different in structure—mildly vs. strongly concentrated—and in total numbers of stars—about one thousand in the Pleiades vs. over one million here in M92—the two clusters clearly have a different mixture of stars. This is most apparent from the relative lack of bright blue stars[15] and

---

14 This is not merely some subjective impression, but a real effect. When large numbers of faint stars are crowded together, they appear as a faint glow rather than individually visible stars. We've seen this effect in the Milky Way (Destination 9) and just recently as we looked down on to the Galaxy from this planet. But because the light from such sources is made up of finite numbers of stars, there are regions that—just due to statistical fluctuations—are slightly brighter than others. The technical name for this effect is *Surface Brightness Fluctuations* and it is visible in many sorts of astronomical objects composed of faint stars. Because M 92 has a large number of stars just fainter than our vision can detect, the overall glow they produce in the sky has a distinctly mottled appearance due to such surface brightness fluctuations, particularly near the cluster center.

15 There actually are some fairly luminous blue stars in M 92 (evident in Figure 11.5), but these somewhat unusual stars—known as *Blue Horizontal Branch Stars*—are closely associated with the bright red stars we see here in such large numbers. The key point is that these Blue Horizontal Branch stars in M 92 are *not* Main Sequence stars. They are

the prevalence of bright red stars in M 92. This is precisely the opposite of what we saw in the Pleiades where the sky was dominated by the light from bright blue stars and no bright red stars at all. We already know how to interpret the lack of bright blue stars here in M 92. Recall from Destination 8 that stellar lifetimes are shortest for the most massive, most luminous, and bluest Main Sequence stars. The apparent absence of these stars here in M 92 suggests that they died off long ago.[16] M 92 is therefore not only larger than the Pleiades, it must also be older. But what about the bright red stars that are so common in M 92 but are completely missing from the Pleiades? These are a mystery. They tell us that we are missing something fundamental about how stars age and how—and why—stars die.

To fill in this gap in our understanding, we need to consider the properties of the stars in M 92 in more detail. Our tool—one we first used at the Pleiades (Destination 8)—is the H-R Diagram, which we can produce for M 92 by measuring the luminosities and colors (which are equivalent to temperatures) of the stars in the cluster[17], then plotting these appropriately as in Figure 11.5. We see immediately from M 92's H-R Diagram that our visual impression was correct. There really are no luminous blue Main Sequence stars in the cluster, in sharp contrast to what we saw in the Pleiades. But there *are* Main Sequence stars in M 92. They're just not very luminous. In fact—as shown in Figure 11.5—they are all less luminous than the Sun. From the Main-Sequence Mass-Luminosity

---

fusing He, not H, into heavier nuclei in their cores. Consequently, they do not obey the Mass-Luminosity or Mass-Lifetime relations for Main Sequence stars that we uncovered at Destination 8.

16 Or, perhaps such stars never existed in M 92? In fact, more massive—and therefore, more luminous and bluer—stars *did* exist in M 92. We know because, as we'll soon discover, we see their corpses!

17 To repeat a point first made in a footnote of Destination 8, it is actually easier to measure the data needed to plot the H-R Diagram of a cluster such as M 92 from Earth than it is from within the cluster itself! Since all the stars of the cluster are essentially the same distance from the Sun—in the same way that all the buildings in Manhattan in New York City are at essentially the same distance from Pike's Place Market in Seattle—all we have to do from Earth is measure the relative brightnesses of stars in the cluster to know their relative luminosities (see the Distance-Brightness-Luminosity relation of Destination 7). Since the stellar colors do not depend on distance at all—as long as there is no obscuration by dust that can make stars appear redder than they actually are—a simple plot of the brightnesses and colors produces a precise version of the H-R Diagram. If we really must know the true luminosities of the stars, we can easily calculate this for all of them by simply measuring the (one) distance to the cluster from the Sun. From *within* M 92, the distances to individual stars do affect their brightnesses significantly, just as the distances to individual buildings in Manhattan matter to anyone who also is in Manhattan. Some of the stars are close, some are significantly farther way. Consequently, we need to measure the brightnesses *and* distances of stars in the cluster to be able to determine their luminosities using the Distance-Brightness-Luminosity relation (Destination 7). Only then can we combine these results with the colors of the stars to plot the cluster's H-R Diagram. But the key point is the same. Regardless of how we do it, we can produce an H-R Diagram for the stars of M 92 with more or less effort using measurements we know how to carry out.

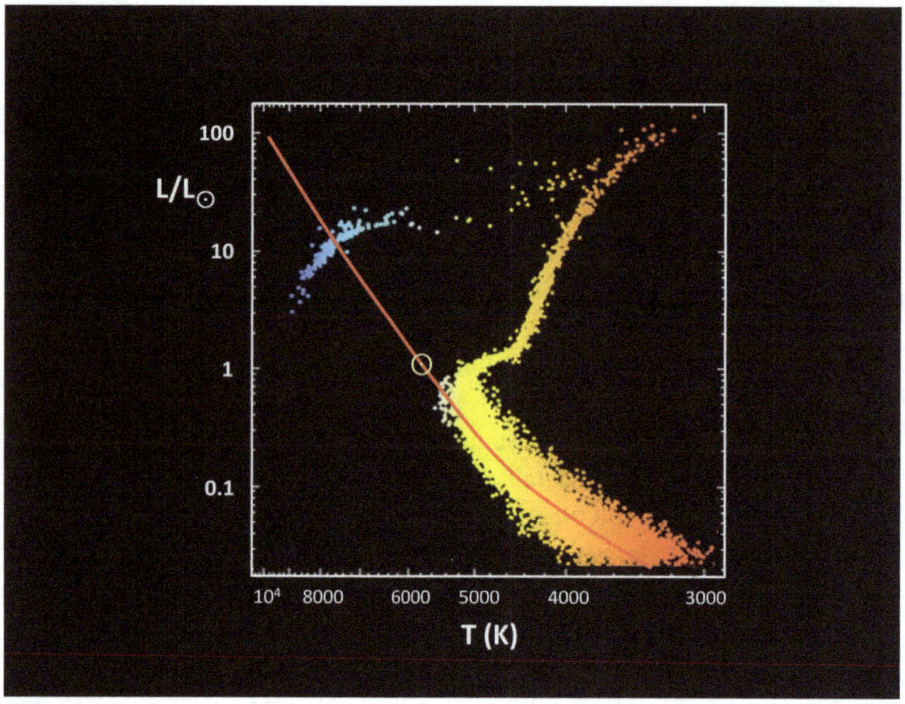

Fig 11.5

The H-R Diagram of a globular cluster similar to M 92 (the particular cluster shown here is M 55, but it is a very close match to M 92). The red line shows the location and extent of the Main Sequence in the Pleiades. The diagonal sequence of mostly red and yellow stars that extends from the lower right to the middle of the diagram is all that is left of the Main Sequence in this cluster. The most luminous stars still on the Main Sequence in the cluster are less luminous than the Sun (which is denoted at a yellow '⊙' symbol on the Main Sequence), meaning that M 92 is older than the expected total lifetime of the Sun. That is, the cluster is older than 10 *Gyr*. The blue stars that cross the Main Sequence at higher luminosities are not Main Sequence stars, but rather *Blue Horizontal Branch* stars. Unlike Main Sequence stars, they are fusing He in their cores, not H. Red Giants dominate the upper-right part of this diagram, as they extend along the Red Giant Branch.

Image from B. Mochejska and J. Kaluzny/CAMK.

relation (Destination 8), this implies that all the Main Sequence stars in M 92 are also less massive than the Sun. Since the mass of a star dictates how long it can live on the Main Sequence (Destination 8), this tells us that M 92 is not merely older than the Pleiades. It must be *older than the full life expectancy of the Sun*. The stars in M 92 are all older than 10 *Gyr*!

But what about the bright red stars in M 92? The ones that pepper this spectacular sky? From the H-R Diagram of the cluster (Figure 11.5) we can see that these stars define a distinct sequence that extends to the upper right of the plot. Even without understanding the nature of these stars, there is one thing we can be sure of. The stars in the upper right of the H-R Diagram of M 92 are considerably larger than stars on the Main Sequence (refer to Figure 8.6). Since the luminosity of a star depends on its temperature and radius through the relation $L = 4\pi R^2 \sigma T^4$ (Destinations 6 and 8), any star that has a large luminosity but a comparatively low surface temperature—such as the red, luminous stars of M 92—*must* have a large radius. For this reason, stars located in the upper right of the H-R Diagram are aptly referred to as *Red Giants* and the sequence they define in the H-R Diagram is known as the *Red Giant Branch* (Figure 11.5). As we see plainly from the H-R Diagram of the cluster—and also from simply looking up at all the bright red stars in the sky around us—M 92 contains thousands of red giants, nearly all of which lie along the cluster's well-defined Red Giant Branch (Figure 11.5). What are these stars? And what relationship do they have with Main Sequence stars?

11.3 STELLAR DEATH

We've discovered that stars located on the Main Sequence are converting H into He in their cores through stable thermonuclear fusion reactions (Destination 8). This simple description is essentially the *definition* of a Main Sequence star. And every bit of it matters: 'H,' 'He,' 'core,' 'stable,' 'thermonuclear,' 'fusion.' If *any* of these components are missing—say, stable H thermonuclear fusion into He but not in the star's core—then a star is *not* a Main Sequence star.

Throughout the lifetime of a Main Sequence star, fusion reactions produce the energy (by converting mass as expressed by Einstein's Equation; Destination 8) that is ultimately radiated away from its surface. This basic concept led us to derive a relationship between Main Sequence lifetime and stellar mass (Destination 8), revealing that stars have long—but certainly finite—life expectancies. In this respect, stars are like any sort of engine that produces power. Unless replenished, the fuel producing a star's power—its luminosity—will eventually run out. What happens then? Even without knowing the detailed answer yet, we can make some guesses based on the H-R Diagram. In the Pleiades (Destination 8) we found a Main Sequence that extended to luminosities far greater than the Sun's, but no red giants. Here, in M 92, the Main Sequence terminates at a luminosity considerably lower than that of the Sun and the cluster exhibits a prominent Giant Branch that merges smoothly into the Main Sequence in the cluster's H-R Diagram (see Figure 11.5). Compared to the Pleiades, it's as if M 92—the older cluster—has somehow gained red giants at the expense of high-luminosity Main Sequence stars. This strongly suggests that Main Sequence stars somehow become red giants as the stars get older.

But this can't be the whole story. When stars form in a region such as the Orion Nebula (Destination 6), their masses are not randomly distributed. For every massive star produced, a large number of lower-mass stars are invariably born at about the same time. This distribution of stellar masses at birth is known as the *Stellar Initial Mass Function*, often simply referred to as the *IMF*. We can use the IMF to estimate the total number of stars born in M 92 by simply counting the number of faint Main Sequence stars we see in the cluster today.[18] When we do this, we discover there are too few red giants in M 92 today to account for the number of Main Sequence stars we infer must have

18 Mathematically, this is not hard to do. But there is a trick. In particular, we are assuming that the IMF in a region such as the Orion Nebula—where the IMF can be measured reasonably directly as stars are forming today—is the same as the IMF when M 92 formed many billions of years ago. That's a big assumption, but it may actually be correct. All clusters—indeed, all star-forming regions—appear to obey the same IMF

existed in the past within the cluster. That is, while it seems clear that stars become red giants when they leave the Main Sequence, they do not remain red giants forever. They must change into something else. Something that is not yet evident to us from what we have seen in the cluster's H-R Diagram. The changes that occur in individual stars as they age are part of a process known as *Stellar Evolution*[19], a term we first encountered at the Pleiades (Destination 8). We've encountered two crucial phases of stellar evolution so far: Stellar birth at the Orion Nebula (Destination 6) and 'middle age' at the Pleiades (Destination 8). As we'll soon see, M 92 reveals important information on the final stages of stellar evolution that allows us to explore how stars die.

We know that stars convert H into He through stable thermonuclear reactions in their cores while they reside on the Main Sequence. As the H runs out—as it must—there's no fuel left to maintain the internal pressure that the star has required to resist the inward pull of gravity. At this point, the star takes on a somewhat schizophrenic nature. The inner region of the star where fusion reactions can occur—the *Stellar Core*—begins to act quite differently than the outer part—the *Stellar Envelope.* In particular, the drop in the energy production from fusion causes the core to collapse. Gradually at first, since H is not exhausted everywhere in the core simultaneously. By collapsing, the core taps a new energy source—gravitational energy—that causes it to heat up (see Destinations 6 and 7). In response to this comparatively rapid injection of energy at the core, the outer part of the star—the envelope—starts to expand. From our perspective outside the star, it appears to grow larger in size. In the H-R Diagram, this means the star must evolve to the upper right. That is, it begins to evolve along the Giant Branch.

Since the star's core is now mostly composed of Helium—the 'ash' of H fusion—it seems reasonable to imagine that this process of core collapse and envelope expansion will continue until *Helium* fusion can commence in the star's core. This is reminiscent of stellar birth where the collapse of a molecular cloud ends when a star begins to generate enough energy through H fusion to balance the inward pull of gravity. In the case of a red giant, what's different here is that *Helium* fusion—which transforms He nucliei into Carbon (C) and Oxygen (O) nuclei—is what provides the energy required to halt the core's collapse. But does this actually happen?

Well, yes, but the process is not quite so simple—or sedate—as this description implies. The reason is that it is much harder to fuse He into heavier elements than it is to fuse H into He. For a star like the Sun, H fusion requires

---

everywhere we look. It is not unreasonable to use the same IMF in M 92 to estimate the number of Main Sequence stars that have died over the cluster's lifetime.

19 It is important to emphasize that the term 'evolution' we are using here refers to the changes that occur to single stars. In this respect, 'evolution' is used differently by biologists for whom the term refers to changes of species over time, not individual organisms.

a comparatively 'mild' core temperature of about 15 million *K*. But initiating He fusion in a star of similar mass requires a much higher core temperature of about 100 million *K* (!), over six times hotter than the core temperature of the Sun. At this stage in the star's evolution, the only source of energy that could heat the core to such high temperatures is gravitational collapse. But before this happens, something very strange occurs.

To understand this, let's consider (again!) a neutral H atom consisting of a single negatively charged electron in orbit about the single positively charged proton in the atom's nucleus. We've seen earlier—from the Bohr model—that the electron can exist only at specific orbits within the atom (Destination 6). These orbits are denoted by a positive integer *n* known as the *Principal Quantum Number* (the same '*n*' that we introduced in the Bohr atom at Destination 6) and are characterized by a discrete energy, E_n. One of these orbits, $n = 1$, is known as the ground level because it represents the lowest-energy orbit possible in the atom. But what we refer to as the ground state of H is not really a single energy level. At Anti-Cen (Destination 5) we discovered that there are two very closely separated energy states associated with the ground level ($n = 1$). The 'true' ground occurs when the electron and proton spin in opposite directions. The other 'ground' state has very slightly higher energy (corresponding to a wavelength of 21-*cm* EM radiation), and occurs when the electron and proton spin in the same direction. What this tells us is that the energy level of an electron in a H atom—actually, an electron in any atom—is characterized by its orbit *and* spin, properties that can be characterized by a set of discrete *Quantum Numbers.*[20] One can think of these numbers as an electron's 'address' within an atom.

Now comes the weird part. If we consider an atom in which multiple electrons are allowed—say, any atom other than H—there is a fundamental rule that limits how electrons fill these levels. Known as the *Pāuli Exclusion Principle*, the rule states that no two electrons in a bound system can have the same set of quantum numbers.[21] That is, the Exclusion Principle limits how many electrons can exist at a given quantum address to one. It is this strange requirement that now causes so much mischief for a star as it strives to initiate core He fusion.

---

20 In addition to the principal and spin quantum numbers, there are two others that characterize the electrons in atoms. One describes the detailed shapes of electron's orbits in atoms, the other the orientation of the orbits. These details are not important for our discussion here. The key point is that the state of an electron is defined uniquely by a set of discrete quantum numbers.

21 The validity of the Pauli Exclusion Principle is well established. Its origin makes perfect sense within the mathematical framework of *Quantum Mechanics*, a theoretical model that quite successfully addresses the behaviors of subatomic particles. But the underlying basis for the Exclusion Principle is not understood. What is understood is that if this principle did not hold, the Universe—including us—would not exist in anything resembling the form we know!

As the core shrinks, its density increases so much that it begins to act somewhat like a single, gargantuan atom as far as the electrons are concerned. Due to the Exclusion Principle, this means that no two electrons in the *entire core* can share the same set of quantum numbers. As low-energy levels are filled, only higher and higher energy states are available. This overcrowding is known as *Electron Degeneracy*[22] and it produces a pressure—known reasonably enough as *Degeneracy Pressure*—that resists gravity, slowing the core's contraction.[23] But only in the densest parts of the core. Outside the central degenerate zone, material continues to fall in, piling onto the core and heating it up even further.

Degenerate matter has some very peculiar properties that now begin to dominate the core's properties. First, the density is phenomenally high, around 1.4×10^6 g/cm^3. For comparison, the density of water is 1 g/cm^3, while Lead is about 11 times denser. The core's density has become 1.4 *million* times higher than that of water, or equivalent to about one ton of matter crammed within a volume of a sugar cube! Second, the pressure of a degenerate 'gas' does not depend on temperature. In a normal star—which is typically fully composed of gases that obey the *Perfect Gas Law*—if the core is compressed the central temperature increases from the release of gravitational energy. This causes the pressure to increase, forcing the core to expand, cool down, and reduce the pressure. This is an inherently stable situation, one that can maintain a normal star's structure. But a degenerate core—which is composed of material that does not obey the Perfect Gas Law—reacts quite differently. When heated during collapse, the pressure doesn't change. Even as matter falls onto the degenerate region from outside and the core heats even further from the release of gravitational energy, the pressure barely changes. Third, degenerate matter conducts heat very effectively. This is the same characteristic that causes a metal rod with one end in a fire to heat up along its full length. For a degenerate stellar core, this means that the entire core heats up uniformly to a single temperature.[24]

---

22 The term 'degeneracy' here does not refer to any sort of moral shortcomings in the electrons. To a physicist, 'degeneracy' simply means that there is more than one thing sharing a given property. In the case of electron degeneracy, the term implies that the electrons in the core of a star, or any very dense material, start to try to share quantum numbers properties even though they cannot due to the Pauli Exclusion Principle. In this sense, the term 'degeneracy' is a bit of a misnomer. What's really happening in the core of the star is that the electrons must populate higher and higher energy levels to *avoid* becoming degenerate with—that is, having the same set of quantum numbers as—other electrons in the core.

23 It is tempting to think that the negative electric charges of the electrons are causing them to repel each other and producing this pressure. But this is not what's happening. By the time degeneracy sets in, gravity has compressed the core so intensely that the repulsive electric forces had long ago become ineffective at slowing the collapse.

24 For this reason, the degenerate core is often said to be isothermal. In contrast, the perfect gas in the core of a normal star is a lousy heat conductor. The core temperature in such stars is not uniform, but falls steadily outward. Definitely not isothermal.

The problem the core now faces is that it becomes degenerate *before* it is hot enough to support He fusion. But the temperature continues to rise until the core gets sufficiently hot to initiate He fusion. Since the degenerate core is a good conductor of heat, this happens simultaneously throughout the *entire* core. Now the real fireworks begin. Hold on to your hat!

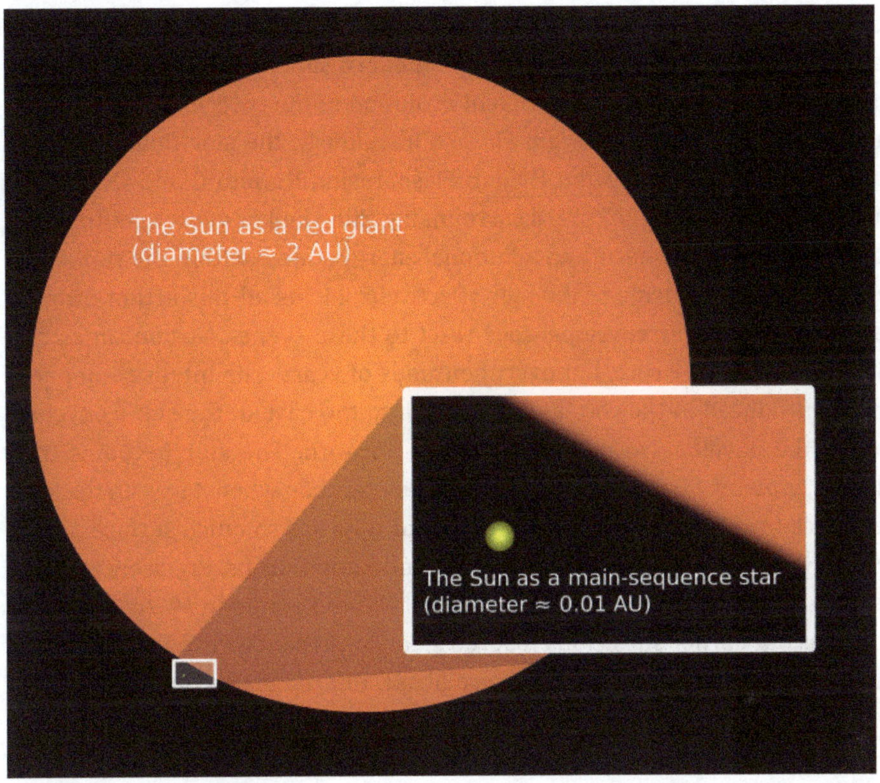

Fig 11.6

The Sun will one day swell to a size about as large as the orbit of the Earth when it becomes a red giant (*large red circle*), over 100 times larger than its present size (*inset*). If the Earth survives another 5 *Gyr*, this will truly mark the end for our planet. Some known red giants—technically known as *Red Supergiants*—are even larger than this. As a red giant, the Sun will be cooler (redder) and far more luminous than it is now due to its gargantuan surface area. When that happens, it will reside in the upper-right of the H-R Diagram where all red giants are found (see Figure 11.5).

With the sudden rise in temperature, He fusion commences throughout the core almost instantaneously. The rapid addition of energy causes the core to become even hotter. If the core were not in its bizarre degenerate state, it would expand, cool off, and reach a stable configuration. But it can't. Instead, it simply gets hotter, causing the rate of energy production from He fusion to skyrocket, causing the core to become hotter still. It is a violent runaway situation and, soon, so much energy is produced that the core cannot get rid of it without resorting to a process known as *Convection*, which causes mass movement of hot material deep in the core to cooler regions above. This is the same process that stirs the water in a boiling pot and percolates the Earth's atmosphere on hot, sunny days. For the star, the onset of convection allows the core to expand, in the process lowering its density. As this happens, new energy levels suddenly become available to the electrons within the core, lifting the degeneracy conditions. Although this sounds like a good thing, there's a problem. Since the core pressure did not change while the gas was degenerate, the suddenly non-degenerate gas finds itself far too hot for its (relatively) low pressure. And so the pressure increases. Suddenly. Violently. Explosively. If the rest of the star's mass surrounding the core—the star's envelope—was not present to act as a buffer, this rapid expansion of the core would blow it to bits.

What's astonishing about this is how rapidly these processes occur: The time from the initial ignition of He fusion in the degenerate core to the explosive expansion described in the previous paragraph takes about 5-10 seconds in a star like the Sun. That's probably much less time that it took you to read that paragraph!! *Much* less time than it takes modern supercomputers to calculate the details of the process. This violent evolution of the core as He fusion commences is known as the *Helium Flash*. Miraculously, the star finds itself in a stable configuration after the Helium Flash, fusing He into C and O stably in the core. Just outside the core, a surrounding thin shell of matter continues to support H fusion, producing an additional energy source within the star.

Although the outside of the star effectively shields all this action from direct view, the stellar envelope does react to these events. Not on timescales of seconds as in the core, but over thousands of years. The intense amount of energy produced in the core and dumped into the base of the star's envelope causes the envelope to puff even further outward. The star becomes truly bloated, growing 50–100 times larger in diameter than when it was on the Main Sequence (see Figure 11.6). Because the star now has so much surface area to radiate its higher luminosity, its surface temperature drops. Precisely what we need to account for the basic properties of the red giants in M 92. All of this confirms our earlier suspicions that red giants represent an evolutionary stage that closely follows the Main Sequence phase. But red giants do not represent the final stages in the lives of stars.

11.4 WHITE DWARFS: THE CORPSES OF LOW-MASS STARS

Relatively speaking, the red giant phase does not last very long. The reason has to do with fuel efficiency. Recall from Destination 8 that Main Sequence stars are powered by a reaction that involves the conversion of four H nuclei into one He nucleus. In a red giant—which is powered mostly by core He fusion—the primary reaction involves the conversion of three He nuclei into one C nucleus (sometimes, some O is also produced, but C nuclei must be created

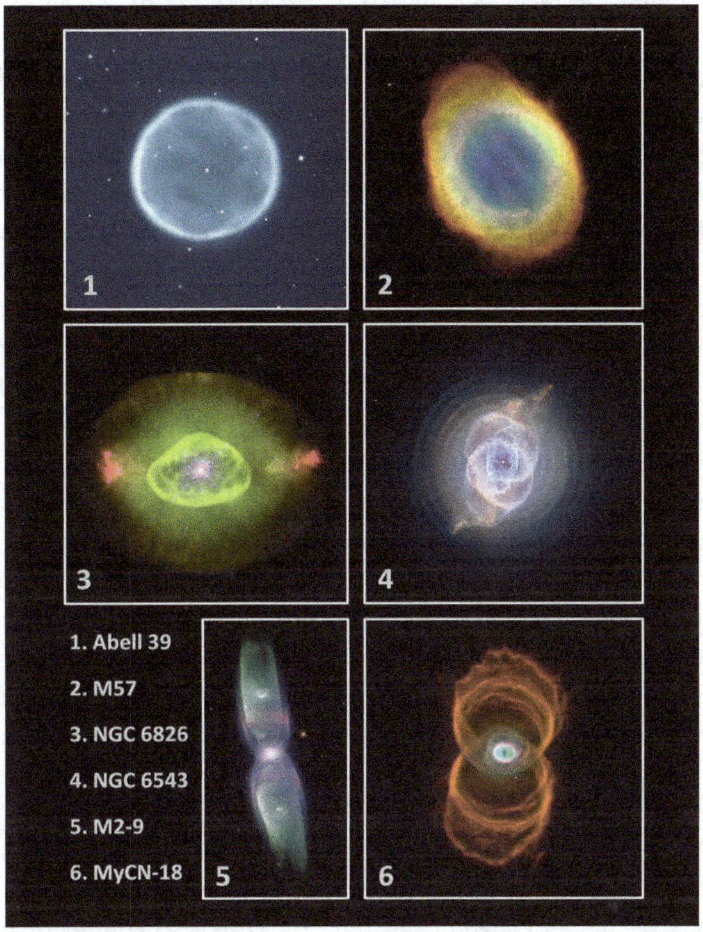

Fig 11.7

A gallery of Planetary Nebulae. The upper two nebulae, Abell 39 and M 57 (the Ring Nebula), are examples of relatively simple mass loss from the central star. The middle two nebulae, NGC 6826 (the Blinking Eye Nebula) and NGC 6543 (the Cat's Eye Nebula), are examples of cases where multiple shells have been ejected from the central star. The bottom two nebulae, M2-9 and MyCN-18 (no catchy names for these!) illustrate even more bizarre forms, possibly caused by rotation of the core as it ejected the gas that forms the nebula, or due to the presence of a companion star to the one with the dying core. Spectroscopy of the nebulae confirms they are *expanding* outward from the location of the central star. The stunning beauty of this phase of evolution is undeniable, but it is short-lived. All of these nebulae will disappear from view in a few tens of thousands of years. The central cores are well on the way toward becoming White Dwarfs.

Image from STScI/NASA.

first). Recall too that the reason we get any energy out of these reactions at all is that they are what are referred to as exothermic reactions. That is, the masses of the input particles exceed the mass of the output particle. For H fusion, the excess is 0.8% of the mass of the initial four H nuclei (Destination 8). Not much, but—as we found out at the Pleiades—more than enough to provide the energy needed to power a star over its lifetime. In the case of He fusion, the masses of the three input He nuclei exceeds that of the output C nucleus by a much smaller amount, only 0.06%. Since red giants get most of their energy through He fusion[25], they simply can't produce as much total energy as a Main Sequence star of similar mass. Worse, red giants are much more luminous than the Main Sequence stars from which they evolved, so they are expending the limited amount of energy they can produce at a furious rate. It's a combination of features that is not conducive to longevity. Once all the details are taken into account, it turns out that a given star will typically spend about 5–10% of the duration of its Main Sequence lifetime as a red giant. The Sun—whose Main

25 Though, not quite all. Remember there is a thin shell where H fusion can still occur surrounding the core. That shell contributes some of the energy output of a red giant.

Sequence life expectancy is about 10 *Gyr*—can expect to live for about 0.5–1 *Gyr* as a red giant. This helps explain why there are relatively few red giants in M 92—a few thousand—compared to the many hundreds of thousands of Main Sequence stars in the cluster. The red giants have shorter lifetimes.

But we seem only to have delayed the inevitable. What happens when *this* phase of a star's evolution ends? It's tempting to guess that the same basic process that occurs after the end of the Main Sequence phase repeats itself: Helium is consumed in the red giant's core, leaving behind Carbon and perhaps some Oxygen. The core collapses, heats up, becomes degenerate, suffers a 'Carbon Flash,' then moves on to the next stable configuration powered by C or O fusion. And so on. But this is *not* what occurs. Though the core does indeed collapse and becomes degenerate as He becomes depleted, the cycle is broken. The core never gets hot enough for Carbon or Oxygen fusion to commence. There simply isn't enough gravitational energy because the star is not massive enough. But it does continue to evolve. As the core collapses to a final degenerate state, the outer envelope of the star expands even further than before in the stages leading up to and after the Helium Flash. The result is a 'gentle' stellar wind as the star sheds its outer envelope into space, typically in a series of comparatively mild 'puffs.' For a star like the Sun, the mass loss during this stage can strip the entire envelope in 'only' a few hundred thousand years, a blink of the eye for a star that spent 10 *Gyr* on the Main Sequence and about one *Gyr* as a red giant.

The evolutionary stage during which stars lose their envelopes produces some of the visually most beautiful phases in a star's life. As more and more of the outer layers blow away, the core becomes more exposed. Though fusion reactions have largely ended by this time, the core is still quite hot, up to hundreds of thousands of degrees *K* at its newly-exposed surface. From Wien's Law (Destination 6), this implies the core will emit copious quantities of energetic blue and ultraviolet photons. These ionize the gas in the expanding shells—material that used to be the star's envelope. As in the Orion Nebula (Destination 6), when the ionized atoms recombine with free electrons, they emit a riot of emission lines that create emission nebulae surrounding the cores known as *Planetary Nebulae.*[26] Sadly, there are no planetary nebulae here in M 92 at the

26 Of all the bad terms used in astronomy, 'planetary nebula' might be one of the worst. The name—always a source of confusion—came about for historical reasons. In the late 18th century, William Herschel, a self-made English astronomer and, by day, musician, discovered the first new planet since antiquity: Uranus. Through a small telescope, Uranus appears like a small round blob, but with an unmistakable, intense, and quite beautiful turquoise color. Needless to say, the discovery of the first planet found with a telescope made Herschel quite famous and it spurred him to look for more. Soon afterwards, he thought he had found yet another Uranus-like object only to discover after a few days that it did not move with respect to the stars (Charles Messier would have sympathized). This was not a planet, but something far more distant. With time, he found many others, all of which resembled Uranus both in size and shape and, often, color. This uncanny resemblance gave these objects their name: Planetary Nebulae. But other than their appearance, they really have nothing at all to do with planets.

present time. It's disappointing, but not surprising. Even though all of the red giants that we see today in the cluster will 'soon' go through this evolutionary phase, planetary nebulae are so short-lived that the odds are small of seeing even one in a cluster of over one hundred thousand stars. Indeed, only four globular clusters of the Galaxy—out of a population of nearly 180 clusters, altogether containing over one hundred million stars—are known to have planetary nebulae (one each). On the other hand, the Galactic Disk contains many more total stars, and over 1400 planetary nebulae have been discovered in that part of the Galaxy. Detailed images of these nebulae reveal some of the most hauntingly beautiful objects in the sky (see Figure 11.7). And, if we look carefully at these images or any other images of planetary nebulae, we will always see near the very center the hot, blue object that used to be the star's core (Figure 11.7).

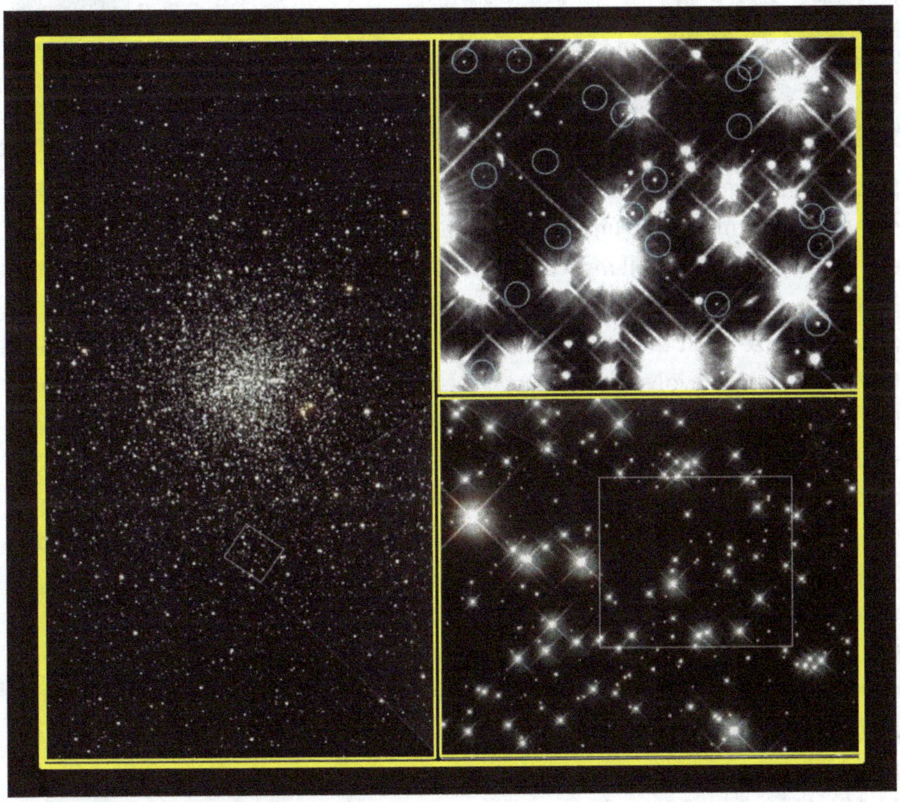

Fig 11.8

Where are all the white dwarfs? This composite image shows a wide-field picture of the nearby globular cluster M4 (*left*). The two images on the *right* show enlargements of a region near the cluster center. Individual white dwarfs are noted in the *upper right* image with gray circles. There are about 20 white dwarfs in this tiny area in the cluster, implying that there are many tens of thousands, perhaps more than one hundred thousand of these stars in the cluster. These are the stellar corpses of most of the Main Sequence stars that have lived—and died—in the cluster since its birth some 14 *Gyr* ago.

Images from NOAO and STScI/NASA.

Naked, supported only by electron degeneracy pressure, the exposed core of a star at this terminal phase of its evolution has no remaining energy source to tap. Not fusion. The core is not hot enough to support C or O fusion. Not gravity. The core is now static, no longer contracting, as its gravity is held in check by electron-degeneracy pressure. Initially, the core appears hot simply because the gravitational and fusion energy produced earlier in the core takes time to flow to the surface where it streams into space as thermal radiation. The star—really, the former core of the star—has no choice now but to grow cooler.

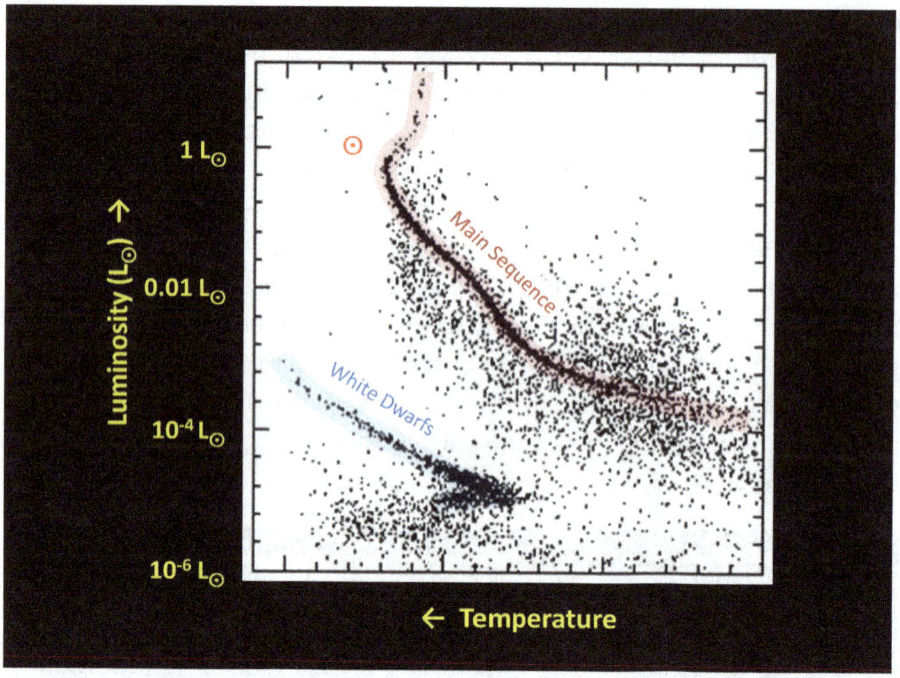

Fig 11.9

An H-R Diagram of a globular cluster (in this case, M 4, but M 92 would look similar) showing the Main Sequence and lower giant branch (the pale red band) and the sequence of white dwarfs stretching diagonally in the lower left of the diagram (the pale blue band). Each point represents a different star; there really are a lot of white dwarfs in globular clusters! But they are very, very faint. As shown in the plot, the most luminous white dwarfs are about 0.001 $L_{\odot}$, and most are considerably less luminous than this, making them hard to see. The red 'Sun' symbol denotes the location of the Sun in the H-R Diagram (compare with Figure 11.5).

Slowly, but steadily. And because the defunct core is structurally stable[27], unless something dramatic and unusual happens, such as a collision with another star, electron degeneracy will forever support it, regardless of temperature. For the first time, the star has finally—and eternally—won a truce with gravity.

As we gaze up at the spectacle of M 92, we have to wonder: Where are they? If so many stars evolved to this stage—all the stars that had Main Sequence lifetimes shorter than the age of the cluster—shouldn't M 92 be full of these stellar corpses? From our vantage point, we see little obvious evidence that objects like these exist in the cluster. But if we look more carefully, using larger telescopes, we find them. *Thousands* of faint stars (see Figure 11.8). Stars that, in the H-R Diagram, form a sequence in the lower left of the plot (Figure 11.9). When we derive their radii from their observed luminosities and temperatures (as in Section 11.2), we find these stars are typically 5000–10000 *km* in radius! *Earth-sized* (Figure 11.10)!! Stars with these properties were discovered early in the 20th century, a few close enough to the Sun that their distances could be determined using parallax measurements (Destination 3; see Figure 11.10). Many of these early cases turned out to have moderately high surface temperatures, so they were dubbed *White Dwarfs.*[28] Here in M 92, the name is

27 The structure of these degenerate cores is unusual. A normal star is typically gaseous throughout its volume. But the degenerate matter in these objects is more like a metallic crystal.

28 The more common—and usually more correct—plural form for 'dwarf' is 'dwarves.' However, not only do most recent dictionaries allow 'dwarfs,' but astronomers have been using the term 'White Dwarfs' for a long time. It would simply look wrong to refer to these stars as 'White Dwarves.' Ugh!!

Fig 11.10

White dwarfs are the outcome of the evolution of most stars in the Galactic Disk, not just stars in globular clusters. The faint object noted in the *left* image is Sirius B, the companion to Sirius A, the brightest star in the nighttime sky as seen from Earth. The illustration on the *right* shows that Sirius B is about the size of the Earth, despite having a mass comparable to the Sun. This drawing does show the correct relative scale, but it is incorrect in two ways. First the edge of Sirius B would, in fact, be very sharply defined. Second, a White Dwarf as hot as Sirius B would quickly incinerate and tidally tear apart the Earth if it was as close as shown! In the case of the Sun, the resulting white dwarf will be 1 *AU* away. But that won't happen for 5-6 *Gyr*, so it's probably nothing to worry about quite yet!

Images from H. Bond, STScI and M. Barstow/U. Leiscester/NASA/ESA (left), and S. Charbinet/ ESA and STScI/NASA.

a bit of a misnomer. Though all of these stars are small, some are so cool that they now appear rather red. Since it takes time for a White Dwarf to cool, these reddest objects must be the stars that have been White Dwarfs the longest time. They are the dead remnants of the most massive stars that produced White Dwarfs in M 92. Stars that died long ago in the cluster. Younger white dwarfs are both more luminous and hotter, and hence considerably bluer in appearance. They must have formed comparatively recently as the less massive—but long-lived—stars eventually died out. These hotter White Dwarfs simply have not had as much time to cool down.[29]

29 Star clusters are handy precisely because the stars in a given cluster all have the same age. But in, say, the Galactic Disk, stars can have a range of ages. After all, the Sun is 5 *Gyr* old, but nearby, in the Orion Nebula, stars are being born today. So, although it is easy to estimate the age of a cluster, it is trickier to determine the age of something that has a mixture of ages such as the Galactic Disk. One way to do this is to take advantage of the cooling properties of White Dwarfs. Since these stars generally become steadily redder with time, the reddest ones in the Galactic Disk provide a way to determine how long ago the first stars formed in the Disk. Interestingly, when people determine *White Dwarf Cooling Ages* in this manner, they find that the Disk is about 10–11 *Gyr* old. That's old, but, as we'll soon see, *not* as old as globular clusters such as M 92! We'll return to this puzzle at Destinations 20 and 21 where we go back in time to watch galaxies such as ours in the process of formation.

11.5 A WINDOW TO AN ANCIENT PAST

The most massive Main Sequence stars in M 92 today are about three times less luminous than the Sun (see Figures 11.5 and 11.10). Using the Main-Sequence Mass-Luminosity relation, we can infer that the masses of these brightest Main Sequence stars must be around 0.7 $M_{\odot}$ (see Destination 8). But we know that mass is also related to a star's lifetime. The Main-Sequence Mass-Age relation we derived at the Pleiades implies that the stars that are just about to die in M 92—the most luminous Main Sequence stars in the cluster—have lived almost twice as long as the Sun! But this is not quite correct. There is an important detail that complicates our estimate of the ages of the stars in M 92.

When we study the spectra of stars in the cluster, we discover that they exhibit very weak absorption lines compared to stars near the Sun—including the Sun itself—that have similar spectral types. From careful analyses of the spectra of nearby stars, we can infer that about 2% of the mass of typical stars in the Sun's neighborhood in the Galactic Disk is composed of elements other than H or He, elements astronomers refer to collectively as *Heavy Elements.*[30] But the stars in M 92 are different. A similar spectroscopic analysis reveals that only about 0.02% of the mass of a typical star in M 92 is composed of heavy elements. That's one hundred times lower than the Sun! The Mass-Age relation we developed at Destination 8 and that we used earlier to estimate the age of M 92 is strictly applicable only to stars with chemical compositions similar to the Sun. This is appropriate for the Pleiades (Destination 8), but for M 92 we have to account for the large difference in heavy-element content. For fairly involved reasons, the lower heavy-element abundance in M 92 causes its stars to appear more luminous and hotter (bluer) than they would if their chemical composition were more like the Sun's.[31] If we take these effects into account,

30 As we have noted, astronomers refer to elements other than H and He as 'metals.' They even have made up a word, 'metallicity,' that refers to the 'metal content' of a star. We'll try to stay away from this terminology, however, since many of these elements—for example, Oxygen, Neon, Iodine, etc.—are not really examples of what most people think of as metals.

31 There are two reasons that make stars with low heavy-element abundances more luminous and bluer, one related to processes in the stars' cores, one resulting from effects in their atmospheres. First, stars with relatively low heavy-element abundances need to be a bit denser in their cores compared to stars with more heavy elements. This causes their cores to run hotter, and, because thermonuclear reaction rates are incredibly sensitive to temperature, such cores produce more energy, making such stars more luminous. Second, many of the elements that produce absorption lines in the stellar atmospheres have mostly blue or ultraviolet lines. When heavy elements are relatively

the best current calculations suggest that M 92 is around 14 *Gyr* old! A similar analysis reveals that just about all of the other globular clusters of the Galaxy have comparably old ages.

We certainly have encountered nothing within the Galaxy this old. Nor will we. Globular clusters such as M 92 are among the oldest things known. As such, they reveal that the Galaxy was far different in the past. For one, the distribution of the globular clusters is nothing like the Galactic Disk that we see so stunningly in this sky—the Disk that gives rise to the glorious Milky Way back on Earth. Perhaps there was no Disk when the Galaxy—and its globular clusters—first formed. Another indication of great change since the time when M 92 was born is apparent from the chemistry of its stars. The lack of heavy elements in M 92 and nearly all globular clusters suggests that the quantity of these elements has steadily increased with time. But how? And where were these elements made? We have already seen that stars produce more He, C, and O than they started out with. And we've seen that some of these new elements are ejected back into the ISM through stellar winds and planetary nebulae. Any stars that might form later from this 'enriched' gas will have larger quantities of at least some of heavy elements compared to their ancestors. It even implies that the Sun—by virtue of its comparatively high heavy-element abundance—must have formed at least in part from gas that had already been processed in one or more previous generations of stars. And that means us. At least some of the C and O in our bodies likely was produced in some ancient red giant. One that lived and died over 5 *Gyr* ago within our Galaxy.

But we are still missing something. The heaviest elements—ones like Silicon (Si), Chlorine (Cl), Calcium (Ca), Iron (Fe), Nickel (Ni), Copper (Cu), Gold (Au), Silver (Ag), Iodine (I), Uranium (U) to name a few—could not have formed from any of the processes that we have described so far. Stars like the Sun—which astronomers refer to as *Low-Mass Stars*—can produce only some of the lighter elements. The stars we see today in globular clusters like M 92 clearly provide important clues to how the Galaxy formed, and how it evolved to become the astonishing structure we see today. But they obviously represent only part of the story (Destination 12).

---

scarce, more of the luminosity of a star can escape in the blue/UV regions of the EM Spectrum, causing the stars to appear bluer.

11.6 A DIFFERENT WAY TO STAY TOGETHER

Before we move on, it is worth pointing out one last thing. Something kind of odd but that has been in plain view since we arrived here. M 92 has no disk structure, neither in appearance nor in the form of any discernable rotation. It really is a 'glob,' nearly spherical in shape and with a strong concentration of stars toward its center (see Figure 11.11). But the fact that it does not rotate doesn't mean that it is static. The stars in the cluster *are* moving, though apparently at random. Surely the cluster is not just some accidental collection of stars. If that were the case, the H-R Diagram would reveal a random mix of unrelated stars, not the well-defined sequences we see (see Figure 11.5). Our analysis of these sequences reveals that the stars of M 92 have been together for billions of years. As we diligently study the cluster from our base, we notice that every star is in fact orbiting the center of mass of the cluster itself. It is just that—unlike the motions of the Planets in the Solar System—the orbit of every star in the cluster is oriented in a random way with respect to all the others. This was certainly not the case in the Galactic Disk. The stars near the Sun have orbits that are almost identical to the Sun's. This is why we get the impression that stars in the Disk appear to be flying in formation as they orbit the Galactic Center (Destination 9). The motions of the stars here in M 92 are more reminiscent of a swarm of insects that stay together for protection.

This is a profoundly important point. Now that M 92 has set and the Galaxy is again rising up into this Alien Sky, we appreciate that there are multiple ways a star system can maintain its stability. One method—the one that the Solar System and the Galactic Disk employ—is rotation in a plane. But objects like M 92 provide examples of another way. The apparently random motions of the myriad stars moving around their common center of mass can *also* produce stable stellar systems. Even systems as massive as M 92. As we gaze at the majestic view of the flat Galactic Disk surrounding the more nearly spherical Galactic Bulge, we realize—in some amazement—that different components of the Galaxy can *simultaneously* exhibit different types of internal motions. The Disk rotates, yet at its center the Galactic Bulge maintains itself, at least in part, through random stellar motions akin to what we've seen here in M 92. Alice's description of Wonderland seems appropriate: 'Curiouser and curiouser.'

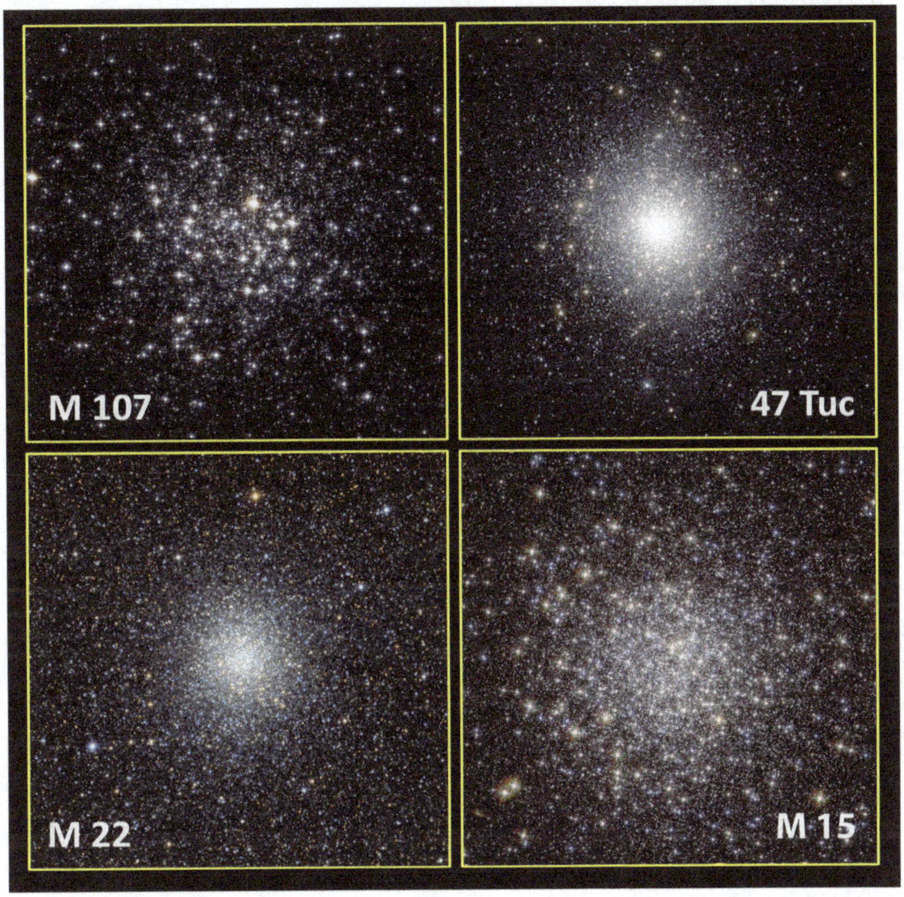

Fig 11.11

Snapshots in the lives of four globular clusters, M 107 (*upper left*), 47 Tucanae (*upper right*), M 22 (*lower left*), and M 15 (*lower right*), each with a distinctive size and central concentration of stars. At any given time, the stars of the clusters are all in motion around their respective cluster's center of mass. Some are moving toward us, some away, some toward the center of the cluster, some outward, some counterclockwise, some clockwise. At Destination 9 we saw that flat, disk-like systems like the Milky Way use coherent rotation in the disk to support themselves. But here in M 92 and in other globular clusters we have discovered an alternative way that stellar systems can support themselves: Via the random motions of their stars under their mutual gravitational attraction. Ten million years from now, each of these clusters will appear much the same as they do now, though every star will have looped a few times around its respective cluster's center. Like a swarm of insects, none of the stars—not one!—will be where we see them today. Yet the clusters will survive, essentially unchanged.

Photos from ESA/STScI/NASA (M 107 and M 15), Vista/ESO (47 Tuc), and B. Birket/SkyPi Observatory (M 22).

DESTINATION 11 QUESTIONS

11.1. We noted that planetary nebulae are shells of gas expanding outward from a dying red giant star. How can we use the Doppler effect to prove that what we see in a planetary nebula is expansion and *not* rotation as we saw in the dense molecular clouds in Orion?

11.2. Stars in globular clusters typically travel at about 2.5 km/s relative to the cluster's center of mass. How long would it take for such stars to cross a cluster once if we assume the cluster's radius is 10 pc?

11.3. From the globular cluster H-R Diagram in Figure 11.5, estimate the radius of the most luminous red giant in a typical globular cluster. You must read off the luminosity and temperature of the star, note the values, then estimate the radius using the relation between temperature, luminosity, and radius from Destination 8. Compare the result to the size of the Earth's orbit.

11.4. Use the web to find the following information about any Messier globular cluster other than M 92: Total luminosity, approximate total number of stars, age, chemical abundance, distance from Sun, distance from Galactic Center. Be sure you identify a *globular cluster* and not an open cluster like the Pleiades. You may have to estimate some of the cluster properties from other information you find on-line.

11.5. How large would the following things be if they had the same mass density as a typical white dwarf? In each case, assume that the final form is a sphere of uniform density.

a) You.

b) The Earth.

c) The entire cluster of M 92.

11.6. Let's imagine that M 92 has a total mass of about 300000 $M_{\odot}$, all contained within a radius of 20 pc from the center of the cluster.

a) How fast would a star located at 20 pc from the cluster center orbit around the cluster, assuming a circular orbit?

b) What is the orbital period of the star?

c) Show that this star's orbital properties are consistent with Newton's extension of Kepler's Third Law (Destination 4).

11.7. Because of the evolution of stars off the Main Sequence, when stars run out of H to 'burn' in their cores they 'turn off' the Main Sequence and head up the Giant Branch in the H-R Diagram. For this reason, the location of the most massive stars still on the Main Sequence in a cluster is known as the cluster *Turn-Off Point* or, simply, the *Turnoff.*

a) Draw a diagram of the H-R Diagram of a globular cluster such as M 92 and note the location of the turnoff.

b) Use the Main-Sequence Mass-Age relation (from Destination 8) to estimate the masses of stars located at the turnoff of M 92, assuming the cluster is 14 *Gyr* old.

c) Use the Main-Sequence Mass-Luminosity relation (also from Destination 8) to estimate the luminosity of turnoff stars in M 92. Compare this luminosity with the location of turnoff stars in Figure 11.5.

11.8. Why might it be tough to find planets in M 92? There are at least two fairly evident reasons that can be gleaned from the text.

11.9. Let's assume that M 92 has a mass of about 300000 $M_{\odot}$, and that half of that is located in the core of the cluster. We'll also assume that the core has a radius of 2.6 *pc*.

a) What is the volume of the cluster's core in pc^3?

b) If the typical star in M 92 has a mass of 0.7 $M_{\odot}$, how many stars are located with the cluster core?

c) On average, how much volume does a star have 'to itself' within the cluster core? **HINT:** That would just be your result from (a) divided by the result from (b).

d) If a typical star in the core of M 92 is half the radius of the Sun and is traveling at 23 *km/s* within the cluster (we'll discover in Destination 15 why this is the case), how much volume does the star traverse over the lifetime of the cluster, 14 *Gyr*? **HINT:** This volume is the 'cross-section' of the star (that is, $\tau V \pi R_s^2$, where τ is the time the star travels within the core, V is its velocity within the core, and R_s is the star's radius).

e) Compare results (c) and (d) and comment if you think that direct stellar collisions are rare, possible, or common in the core of M 92.

11.10. This question addresses why planetary nebulae appear to be somewhat rare in globular clusters despite being a standard phase in the final evolution of individual stars.

a) How long do typical red giants live in the cluster (information required to estimate this is given in the text).

b) What is the typical lifetime of a planetary nebula? Assume that such nebulae expand at 30 *km/s* and that they become invisible once they reach a radius of 1 *pc*.

c) The ratio of the results from parts (a) and (b) give you an estimate of how many red giants must be present to have a reasonable chance of seeing one planetary nebula. Given that M 92 contains about 2000–3000 red giants, is it unusual or reasonable that we see no nebulae in the cluster?

11.11. Find an on-line chart of the region of the sky that shows the location of M 92 within the constellation Hercules. What is the name of the (one) other globular cluster that is located within the confines of Hercules? Find a similar chart that denotes the globular clusters in the constellation Sagittarius, a region of the sky of similar size as Hercules. How many globular clusters are located in that constellation? Comment on the reason for the relative numbers of globular clusters in these two constellations.

11.12. Look up information on the masses of the ^{4}He and ^{12}C nuclei. From this information, determine the efficiency of C fusion, where 'efficiency' of a reaction is defined by the ratio of the amount of mass lost (converted to energy) divided by the input mass of the reaction (see Destination 8 for the formal definition of this). Comment on why this result implies the ages of red giants must be shorter than the ages of the stars when on the Main Sequence. **WARNING:** The atomic masses of He and C are *not* what you require for this problem. You need to find the actual masses of the specific nuclei (isotopes) used for this problem in consistent mass units. Atomic masses typically refer to the average nuclear masses in a standard Earth-based sample of a given element.

Destination 12

THE MAGELLANIC CLOUDS: THE NEAREST GALAXIES

12.1 THE MAGELLANIC CLOUDS: BEYOND THE GALAXY

Living in the Southern Hemisphere during winter—that is, *Northern* winter from December to February—has many perks. First, if you happen to come from north of the equator, you might miss some nasty weather and long, cold nights; it is summertime during these months in the South. But there is a compelling astronomical bonus too, something especially beautiful in the Southern sky that's visible at this time of year. Two silvery 'clouds' grace the Southern sky during this season, their delicate appearance giving the impression that some celestial giant tore out chunks of the Milky Way and absentmindedly tossed them into the sky near the South Celestial Pole (see Figure 12.1).

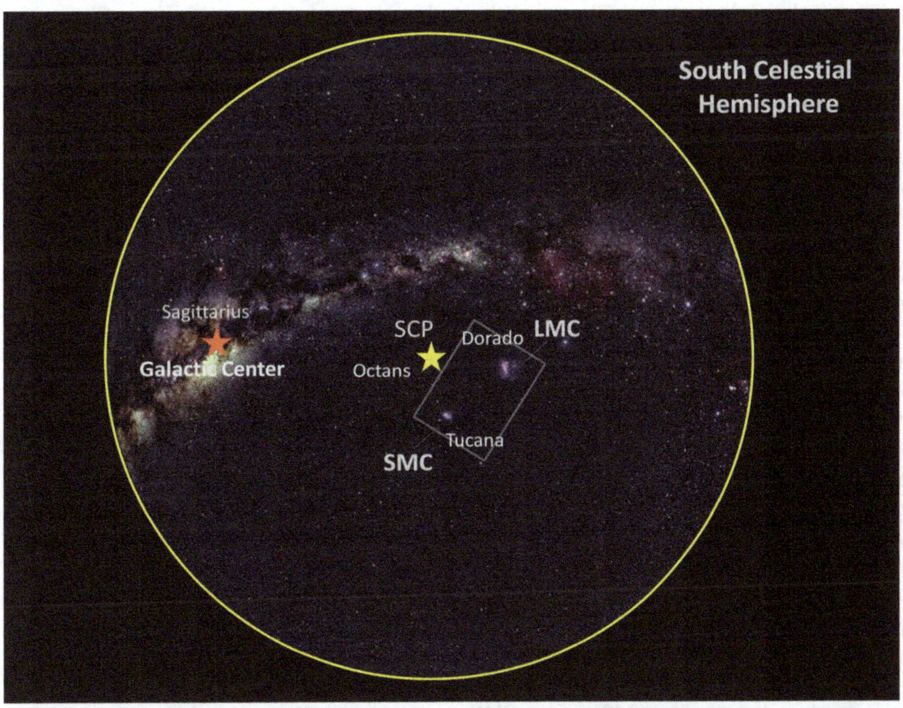

Fig 12.1

A view of the South Celestial Hemisphere as seen from Earth (see Figure 10.2 for a view of both hemispheres). The location of the Galactic Center in the constellation Sagittarius is noted as a red star, and the South Celestial Pole ('SCP'; located in the constellation Octans, the Octant, a tool used long ago for navigation by sailors) is denoted as a yellow star. The two Magellanic Clouds—the Large Magellanic Cloud (denoted 'LMC' in the image and located mostly within the constellation Dorado, the Golden Fish) and the Small Magellanic Cloud (denoted 'SMC,' located mostly within the constellation Tucana, the Toucan)—are clearly visible. The rectangle shows the extent of the field covered in Figure 12.2.

Photo mosaic by Axel Mellinger.

These delicate wisps of light in the Southern sky are known as the *Magellanic Clouds*. Since one of the two is quite clearly bigger and brighter that the other, it's known as the *Large Magellanic Cloud*—or simply, the *LMC*—while the smaller, fainter one is called the *Small Magellanic Cloud*—the *SMC* (Figure 12.2). These names honor the Portuguese navigator Ferdinand Magellan, who, while sailing under the Spanish flag in 1519, led the first expedition that circumnavigated the Earth (Magellan's posthumous portrait is shown in Figure 12.2). Of course,

Magellan and his crew were certainly not the first people to see these clouds. The native peoples of Australia, Southern Africa, and South America had woven the Magellanic Clouds into their folklore long before Magellan arrived. Magellan and his crew were not even the first Europeans to see these clouds. Other Portuguese navigators had been frequent travelers to the southernmost tip of Africa, but too secretive of what they saw. They knew of the shining clouds of the Southern sky, but failed to get credit for the discovery.[1] But Magellan did, largely because his logs were widely publicized when his expedition returned.[2] These included vivid descriptions of two celestial clouds, shining in the Southern sky with a light of 'reasonable bygnesse,' plainly visible as the expedition passed south of the Equator. Magellan's logs even noted that the Clouds had proven useful for navigation because they were located reasonably close to the South Celestial Pole (see Figure 12.1).[3] The LMC and SMC are sometimes—though rarely—referred to by their Latin names: *Nebecula Major* and *Nebecula Minor*; this is what Magellan himself called them. But for modern astronomers—and voyagers like us—Magellan's name is permanently attached to these beautiful celestial features.

The LMC is our next destination. Located 50 *kpc* from the Sun[4], our voyage to the LMC is by far the longest we have attempted so far. Even from a remote distance, it seems clear that the LMC is not a component of our Galaxy. Certainly not part of the Galactic Disk. That's evident already from its location (see Figure 12.1) and the fact that so much of our voyage to the LMC is through vast, empty

---

1 A classic case of 'publish or perish'! It's never about who's first. It's about who makes the first splash. Columbus 'discovering' America; the Wright Brothers 'first' in heavier-than-air flight; Edison 'inventing' the electric bulb; Lindbergh flying 'first' across the Atlantic solo. None of these people were the first to do these things. But they—or their supporters—made a lot of noise when they did. The lesson is clear. Secrets might be necessary sometimes, but explorers—and scientists—are best served, and best remembered, when they are open about their discoveries.

2 Alas, without Ferdinand himself. Magellan died at the hands of angry natives while the expedition was passing through the Philippines. The descriptions that secured for Magellan's expedition a place in history were written by Antonio Pigafetta, a Venetian scholar who idolized Magellan. He was one of just eighteen European men who, out of a crew of 240 who embarked with Magellan, actually completed the circumnavigation of the Earth. His logs were unusually detailed and engaging for his era, downright bawdy in places. He commented on the delicate smudges in the Southern sky, and it was largely due to Pigafetta's writings that these came to be known as the Magellanic Clouds.

3 Unlike the Northern Hemisphere where Polaris shines reasonably brightly near the North Celestial Pole, there is no pole star in the south. The Magellanic Clouds are useful guides in that Hemisphere.

4 We will discuss how we estimate the distances to something as remote as the LMC a bit later (Destination 14). For now, the main point is that we cannot use stellar parallax measurements directly. The LMC and the SMC are simply too distant for this approach to work given the precision with which we can currently measure stellar positions. Instead, we must rely on indirect approaches that are ultimately tied to parallax measurements. As travelers through the Universe, it is best to take any distances not directly tied to parallax measurements with a grain of salt. And it would be prudent to take along some extra fuel and supplies in case we need to go further than we expected!

Fig 12.2

(*Inset*) Ferdinand Magellan in a painting made after his voyage. Whether he really looked like this, however, is unclear since he died in the Philippines during the expedition. He remains immortalized not only in history, but also in the sky as the 'discoverer' of the 'Magellanic Clouds.' (*Main Image*) The Large (*left*) and Small (*right*) Magellanic Clouds in a wide-angle image of the Southern sky. The bright 'star' just to the right of the Small Magellanic Cloud is 47 Tucanae (see Figure 11.11), a globular cluster in our Galaxy (see Destination 11). The horizontal span of this image is about 30 *deg*, the angular distance is equivalent to how far the Sun appears to move in the sky in two hours. The full extent of this field is also shown in Figure 12.1 and illustrates the scale relative to the entire Southern sky.

Magellanic Clouds photo by M. Bessell, Mt. Stromlo Obs.

stretches of space, lonelier even than the regions we had to traverse to reach M 92 (Destination 11). For most of our trip, at most ten or so stars are ever visible, many at the very faint limit of our sight.

By the time we near the LMC, our view back to the Galaxy is startling. Our home! So faint. So small. From Earth, it stretched across the entire sky. A majestic, glowing band full of stars. Even from M 92 it dominated one half of our sky. But from here? All we see is a pale patch of light, about 20 degrees in diameter and elliptical in shape.[5] The center contains the obvious Galactic Bulge that extends about only 2 degrees outward from the Galactic Center from our new vantage point. Everything and everyone we've known, every destination we have visited, is there in the patch of light that is our Galaxy. The enormity of the rest of the sky—much of it seemingly empty—presses down on us, underscoring just how small our home Galaxy really is.

The view also confirms what we guessed from Earth and what we experienced during the journey. By traveling to the LMC, we seem to have left the Galaxy altogether. As we shall discover at our next destination, this impression is deceiving. The full extent of the tenuous outer regions of the Galaxy—the Galactic Halo—extends even beyond the distance of the LMC (Destination 13).

5 We are looking back at the circular Galactic Disk from a steep angle, making it appear elliptical in form. As for its size, 20 degrees is a bit further than the angular distance the Sun travels in Earth's sky in one hour. It is also about equal to the separation from the tip of the thumb to the tip of the pinkie when one's arm is fully stretched out. More than anything else, that conveys a sense of just how far we've come: Our entire Galaxy spanned by a single outstretched hand!

But now that we are near it, the LMC dominates our sky, appearing from here to be much brighter, much bigger, and much more complex than our Galaxy, the small, pale object we have left far behind. And we can see plainly: The LMC is a separate galaxy, distinct and independent from our Galaxy. And very different in form (see Figure 12.3).

Fig 12.3

A portrait of the inner portions of the Large Magellanic Cloud (LMC). This image shows a region about 3.5 *kpc* across in the LMC. Many star-forming regions are visible as pinkish blobs. These are H II regions like Orion (Destination 6), but most of the ones visible in this image are vastly larger. Dark clouds trace regions of future star formation. The central bar-like structure contains over 1 billion stars, while many stars from the disk of the LMC are readily visible over the entire field shown here. Many of the 'stars' peppered across this image are actually large star clusters. There are also extended groupings of blue stars known as OB Associations. One of these—dubbed 'Constellation III'—is visible left of top-center in the image. Our specific destinations within the LMC are labeled.

Photo by W.-H. Wang.

The most obvious difference between the two galaxies is size. The easily visible portion of the Galactic Disk extends at least 30 *kpc* from end to end. The main body of the LMC is considerably smaller, around 5–6 *kpc* across. If we carefully measure the distribution of stars within the LMC, we detect what appears to be an exponential disk, similar to the structure of the Galactic Disk (Destination 9). Though the scale heights of the various types of stars are not too different than in the Galactic Disk, the scale length of the LMC disk is much shorter, 1.5–2 *kpc* instead of 3.5 *kpc* as in the Galaxy. More striking is the fact that the LMC has no obvious bulge. Instead, there is a more diffuse concentration of stars visible that appears somewhat like a fat sausage or candy bar, forming an example of a structure that astronomers refer to as a *Galactic Bar*. Many galaxies exhibit bars[6]; indeed, as we'll see at Destination 15, bars are fairly common.

6 Indeed, the triaxial Galactic Bulge we discovered at Destination 10 is often cited as evidence that our Galaxy has a central bar. The distinction is not always clear and is particularly hard to discern in our Galaxy because the Bulge is hidden behind so much obscuring dust and gas in the Galactic Disk. By comparison, the bar of the LMC—which is slightly offset from the rotation center of the LMC—is obvious (see Figures 12.2–12.4). It is *not* a bulge. In many important ways—stellar content, stellar motion, age, gas content,

Because of their diminutive size relative to our Galaxy, the LMC and its even smaller sibling, the SMC, are considered to be examples of *Dwarf Galaxies.*[7]

Now that we are close to the LMC and can study it in detail, two other features stand out. First, it is *full* of star clusters. Many of these clusters rival M 92 (Destination 11) in size and in the number of stars they contain. In our Galaxy there are at most a few hundred clusters this big. But here in the LMC, there are about 500 bright clusters and well over 6000 in total, if we consider systems as small and faint as the Pleiades.[8] Despite its comparatively tiny size, it appears that the LMC has somehow been able to form and maintain a much larger fraction of its stars within star clusters than the Galaxy does.[9]

The other striking feature of the LMC is that it is very clearly full of gas. Some of this matter is in the warm (neutral) phase of the ISM detectable from its 21-*cm* emission (Destination 5). But quite a bit of the gas is in the hot phase of the ISM (see Figure 12.4), the same phase that dominated the glow of the sky in the Orion Nebula (Destination 6). Since the light from the hot ISM comes from emission lines of excited atoms and ions, we infer there must be many hot, young stars in the LMC, a point we confirm quickly as we look through the galaxy. O-type and B-type Main Sequence stars are indeed quite common, spread throughout the LMC's disk. Many of these stars are found in large groupings known as *OB Associations* (see Figure 12.3). Only stars as hot as these can emit the copious quantities of energetic, ultraviolet photons needed to keep this gas ionized (see Destination 5). Although the type and quantity of gas here

location relative to the disk—the central bar of the LMC is fundamentally different from the Galactic Bulge at the center of our Galaxy.

7 The distinction between dwarf and non-dwarf galaxies is arbitrary, and there are certainly many galaxies that range in size between our Galaxy and the LMC. We'll even visit one, M 33, at Destination 14. In practice, the Magellanic Clouds are often taken as examples of large-ish dwarf galaxies. To provide a sense of what this means, compared to our Galaxy, the LMC contains about 10% as much total mass in stars and gas. The SMC is about six times smaller than the LMC. We'll encounter much tinier galaxies at later destinations and discover that such dwarfs are in fact very common.

8 We certainly did not think the Pleiades was small or faint when we visited (Destination 8)! We've seen a lot since that journey!!

9 This is an intriguing effect. One possibility is that massive molecular clouds (Destinations 6 and 10) are less apt to break up into smaller clouds in the LMC than in our Galaxy. The reason has to do with how the luminous parts of each galaxy rotate. In our Galaxy, much of the star-forming regions are located where the rotation curve is flat (see Destinations 9 and 10). This causes strong 'shear' because the angular rotation speed changes significantly with radius: Since angular rotation is defined as V_{rot}/R; if the rotation speed, V_{rot} is constant as in a flat rotation curve, the angular speed decreases as the distance from the center, R, increases. In contrast, much of the star formation in the LMC occurs where the rotation obeys a so-called *Solid Body* rotation curve (there is a question at the end of Destination 9 that explores some properties of these different types of rotation curves). In this case, $V_{rot} \propto R$ so that the angular speed is constant as R increases. Consequently, large clouds experience much less shear and can form much larger, denser, and more massive clusters. These clusters not only form more frequently, but they also may survive longer because shear and other tidal forces can destroy clusters over time in our Galaxy more effectively than in a galaxy like the LMC.

Fig 12.4

(Left) An image of the Large Magellanic Cloud (LMC) that highlights the distribution of ion-ized gas in the Hot ISM. Some of these, such as the large region shown in the image at *right* (and highlighted in the rectangle in the larger image), are loca-tions where young stars are form-ing, the most massive, hottest ones ion-izing the local gas. Other regions—particularly the circular regions and arcs—represent places where stars are spewing gas out into the ISM of the LMC from stellar winds and stellar explosions.

Large image from C. Smith and S. Point and the MCELS project/ NOAO. Inset image by Don Goldman/Astrodon Imaging.

in the LMC is interesting in its own right, it's the *shapes* of the clouds that are particularly captivating (Figure 12.4). Though they can take on just about any conceivable form—somewhat like Earth clouds—it is clear that many of the emission regions form vast arcs or quasi-circular rings. Many appear to have been ejected from a central source in the past, reminiscent of planetary nebulae (Destination 11) but 10–100 times larger. Some of these arcs and rings contain obvious star clusters near their centers. There is clearly something systematic at work here. The gas and the stars are intimately related in more complicated ways than we've appreciated before.

12.2 THE CASE OF THE MISSING WHITE DWARFS

There are any number of interesting voyages we could take within the LMC itself. It is, as we've seen, a separate galaxy in its own right. But we need to limit ourselves or we will never have time to explore more of the Universe. So we dive into the LMC with the specific aim of learning more about how stars—particularly, young stars—evolve. The LMC is ideal for this because of its large population of massive star clusters, including many that are much younger than 100 *Myr*, the approximate age of the Pleiades (see Destination 8). As we saw both at the Pleiades and M 92 (Destinations 8 and 11), star clusters provide unique snapshots of stellar evolution. They allow us to identify and isolate groups of stars that were born at the same time, from the same gas, and under the same conditions.

Our first stop is the rich LMC star cluster known as NGC 1818.[10] Located close to the central bar of the LMC (see Figure 12.3), this cluster is a stunningly beautiful jewel box containing numerous bright red and blue stars (Figure 12.5). Its H-R Diagram bears a close resemblance to that of the Pleiades with two important differences. First, the brightest Main-Sequence stars within NGC 1818 are more luminous than those in the Pleiades. This immediately tells us that NGC 1818 is younger than the Pleiades. More precisely, the Main-Sequence Mass-Luminosity relation (Destination 8) can be used to reveal that the most massive Main-Sequence stars in NGC 1818—the stars that are next to evolve off the Main Sequence—have masses of about 8 $M_{\odot}$. Based on the Main-Sequence Mass-Age relation (also from Destination 8), we can infer that the cluster is about 50–55 *Myr* old, about half the age of the Pleiades. The second difference is that NGC 1818 contains about 30 red stars similar in color—and hence, temperature—to the red giants of M 92. The red stars we see in NGC 1818 are considerably more luminous, and consequently even larger in size than the M 92 red giants. For this reason, red luminous stars likes the ones in NGC 1818 are often referred to as *Red Supergiants*. By comparison, the Pleiades contains no red supergiants, but, on reflection, this is not as surprising as it might at first seem. The Pleiades contains about 1% as many stars as NGC 1818. Since NGC 1818 has only about 30 red supergiants, the odds of seeing even one red supergiant in the Pleiades are low if the stars form at roughly the same rate in both clusters.[11] In other words, we don't see these stars in the Pleiades because ... we were unlucky![12]

There's another important feature about NGC 1818 worth mentioning. If we look very closely, we discover that the cluster contains white dwarfs. Just a

10 The letters 'NGC' stand for 'New General Catalog.' This is a compilation of several thousand objects based on (mostly) visual telescopic observations from the 18th to late 19th centuries by many observers. The idea of the catalog was to have a listing of non-stellar objects similar to, but more complete than Messier's catalog. Together, the two catalogs contain only a minuscule fraction of the billions of non-stellar objects known in the sky. Nonetheless, many of the most spectacular and brightest objects seen from Earth are listed in the NGC and Messier catalogs, a big reason astronomers continue to use these compilations to identify and name many famous non-stellar celestial objects.

11 Which they do. Or would, if the Pleiades currently contained any red supergiants.

12 This kind of reasoning is common in astronomy and illustrates how sample size can strongly influence the conclusions we reach. After all, we cannot add stars to the Pleiades to allow us to compare it with a much larger cluster like NGC 1818. It also means that in around 1–2 *Myr*, the most luminous stars on the Main Sequence of the Pleiades—Alcyone—will likely evolve into a red supergiant.

few to be sure. But they are clearly present in the cluster's H-R Diagram. By the standards of White Dwarfs—which are intrinsically low-luminosity objects—the ones in NGC 1818 are comparatively luminous. But that's to be expected. Recently formed white dwarfs are more luminous simply because they are still fairly hot (see Figure 11.10). In the case of NGC 1818, the cluster itself is quite young, so naturally, the white dwarfs have had comparatively little time to cool off. All quite reasonable.

But, is it? On reflection, we soon realize there's actually a big puzzle here. If we calculate the degeneracy pressure in a white dwarf (not easy!), we discover that as the mass of the star increases, its radius gets smaller. This is precisely the opposite of what Main Sequence stars do where the more massive stars (the more luminous bluer ones) are larger than the less massive stars (the least luminous, redder ones). This is a curious result, but maybe nothing more. After all, the degenerate matter of a white dwarf reacts very differently to changes of pressure and temperature than the perfect gas of a normal Main Sequence star, so it is perhaps not so surprising that it behaves differently in this regard too.

Fig 12.5

The rich LMC star cluster NGC 1818. The blue and red luminous stars of the cluster are evident. Though similar in age to the Pleiades, NGC 1818 contains about 100 times more stars. The white dwarfs in the cluster are all considerably fainter than any of the stars visible in this image.

Photo from STScI/NASA.

However, there is an important consequence of this effect. If more massive white dwarfs are smaller, they are also denser. But higher density implies a stronger inward gravitational force at the star's surface. It turns out that at a specific white dwarf mass—about 1.4 $M_{\odot}$—gravity can actually overcome the internal electron degeneracy pressure that, at lower masses, can support a white dwarf. If a star's core mass exceeds this value—known as the

Chandrasekhar Limit—electron degeneracy *cannot* stop it from contracting further (see Destination 11). This is not an issue in M 92 where the most massive Main Sequence stars—which have masses of about 0.7 $M_\odot$—are already below the Chandrasekhar Limit. But that's not the case in NGC 1818 where Main Sequence stars as massive as about 8 $M_\odot$ are still present. Which poses an interesting question: How do the most massive Main Sequence stars in NGC 1818 become white dwarfs, stars that *must* have masses under 1.4 $M_\odot$?

The only conceivable answer is that these stars must lose mass before they become white dwarfs. Lots of mass. Based on what we've seen, this is not so far-fetched. Even comparatively low-mass stars in M 92 lose mass via stellar winds and planetary nebulae as they evolve after leaving the Main Sequence (Destination 11). What's different—and surprising—about NGC 1818 is the magnitude of the mass loss required. For a star that started out with 8 $M_\odot$ to get below 1.4 $M_\odot$ in a few million years requires much more violent mass loss. Not necessarily explosive mass loss, but at the very least as a relatively rapid and likely dense stellar wind or some sort of 'super' planetary nebula phase. It seems likely that some of the bubbles of hot gas that we noted as we approached the LMC represent matter that has been shed in some way from young, comparatively massive stars.

To see if we can gain any insight on this process, we travel to another LMC cluster, NGC 1850. This object is also located near the central bar of the LMC, not too distant from NGC 1818 (see Figure 12.3). It turns out that NGC 1850 is even larger and more luminous than NGC 1818, more similar to M 92 in structure and size (Figure 12.6). A concentrated ball of stars—hundreds of thousands of stars—NGC 1850 is packed into a volume extending only about 20 *pc* across. As we can plainly see from our vantage point, the cluster is studded with bright blue and red stars (Figure 12.6), all found in about three times greater numbers than in NGC 1818. The colors and dazzling brightness of the stars; the size and richness of the cluster. These features make NGC 1850 glorious to behold in this Alien Sky.

Fig 12.6

NGC 1850, another rich star cluster of the Large Magellanic Cloud. The blue wisps of gas are probably material left over from the era when the cluster formed some 10–15 *Myr* ago. This cluster is about three times more populous than NGC 1818.

Photo from STScI/NASA.

At first sight, the H-R Diagrams of NGC 1818 and NGC 1850 appear very similar. The most obvious difference is that the luminosity of the brightest Main Sequence stars in NGC 1850 is higher than in NGC 1818. That is, NGC 1850 is even younger. We can use the same techniques as before to estimate the masses of the most luminous Main Sequence stars in NGC 1850, as well as the cluster's age. What we discover is that NGC 1850 is about 10 *Myr* old, and its most luminous Main Sequence stars have masses of around 15 $M_\odot$. The H-R Diagram also reveals over 100 red supergiants, many of them more luminous—and hence even larger in size(!)—than the ones we found in NGC 1818. Overall, it appears the two clusters are similar in many respects, just different in size, age, and in the detailed properties of their stellar content.

But when we peer a bit harder at NGC 1850, we discover a difference that has more profound implications. As we search for white dwarfs in NGC 1850, it soon becomes apparent that the cluster contains none. We look again and ... nope. There are simply no white dwarfs in NGC 1850. The evolutionary stages that precede the formation of white dwarfs—the Main Sequence and red (super)giants—are certainly present in NGC 1850, just as they are in NGC 1818 and M 92 (Destination 11). But there are no white dwarfs in NGC 1850. This cannot be due to the kind of bad luck that explains why there are no red supergiants in the Pleiades. There are so many stars in NGC 1850 that we would certainly expect to find a few dozen white dwarfs here.[13] But there's not even one. What's going on?

12.3 THE EVOLUTION OF HIGH-MASS STARS

To understand, we need to explore what happens during the evolution of a 20 $M_\odot$ star. This is a somewhat higher mass than the most massive stars currently present in NGC 1850. But that's ok. We want to understand precisely why such massive stars—which must have existed when NGC 1850 was born—did not evolve into white dwarfs.

The story begins in a familiar way. Any star on the Main Sequence—whether it's a 0.1$M_\odot$ M-type star or a 20 $M_\odot$ O-type star similar to the one we are

13 This goes back to the concept of the Initial Mass Function discussed in Destination 11. To summarize, the existence of at least 100 stars in the mass range 10–15 $M_\odot$ in NGC 1850 implies that there were a few dozen in the range 15–20 $M_\odot$. The question we are raising here regards what's happened to these more massive stars' corpses. The cluster tells us they have not become white dwarfs.

considering here—is stably converting H into He via thermonuclear reactions in its core. After about 9 Myr the H is used up and the star's core—lacking a source of energy to maintain its internal pressure—starts to contract and heat up as gravitational energy is converted to thermal energy. In a 20 $M_\odot$ star, the core starts this process at a far higher temperature than the cores of less massive Main Sequence stars like the ones we encountered in the Pleiades (Destination 8), M 92 (Destination 11), or even NGC 1818. This core also has more gravitational energy available because it is more massive. It can use this energy to raise the core temperatures to even higher values as it contracts. As a result, the core of a 20 $M_\odot$ can reach a temperature of 100 million *K before* the matter in the core becomes degenerate. Consequently, Helium (He) can commence fusing into Carbon (C) *without* having to go through the Helium flash. Compared to lower-mass stars (Destinations 8 and 11) in which He fusion starts in degenerate conditions (Destination 11), a 20 $M_\odot$ star evolves from the Main Sequence to the red supergiant phase in a comparatively 'sedate' manner.[14]

After about 1 *Myr*—equal to approximately 10% of its Main Sequence lifetime—the star exhausts the He in its core, leaving behind a core composed mostly of C and Oxygen (O), but still too cool[15] for either element to fuse into heavier nuclei. So the core contracts again until the temperature reaches around 600 million *K*, at which point C fusion can begin. As before, this happens before the core becomes degenerate; there is no 'Carbon Flash'. And so the process continues. After only 1000 *yr*, the core runs out of C, having converted most into Neon (Ne). Another collapse, and another fusion reaction—this time involving Ne—begins as the core reaches a balmy 1.2 billion *K*. After just *one year*, Ne is exhausted. The core collapses again, heating up to over 1.5 billion *K* when O fusion commences.[16] This goes on for only six months(!), at which point Silicon (Si) fusion follows in a core that now has a temperature of 2.7 billion *K*. At this stage, the star consumes about 1$M_\odot$ of Si in *one day!!* Since new fusion reactions start preferentially where the core is hottest—that would be at the center of the star—the star's internal structure at this stage begins to resemble a gargantuan onion, each layer composed of different elements undergoing fusion reactions (see Figure 12.7).

---

14 It's not really that sedate. Even on the Main Sequence, massive stars are sometimes marginally unstable, losing mass through vigorous stellar winds, some of which border on miniature explosions in the star's envelope. This behavior typically gets more severe as massive stars evolve through subsequent phases of evolution until their grand finale (which we'll discuss soon). In many cases, this vigorous mass loss in the star's envelope can significantly alter the star's structure right down to its core.

15 That's a strange description for something with a temperature in excess of a few hundred million K!

16 For reasons related to its detailed nuclear structure, Oxygen (O) is harder to fuse into heavier elements than Neon (Ne) is, despite Oxygen's lower atomic weight. That's why O fusion occurs somewhat later in the sequence of reactions we are describing here.

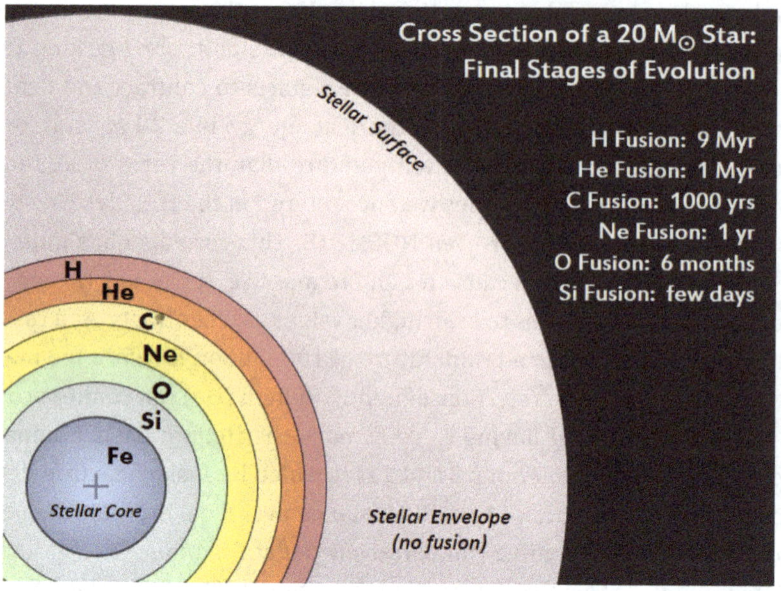

Fig 12.7

A schematic cross-section of the structure of a 20 $M_\odot$ star just as Si burning is ending. As the star evolves, distinct layers—similar to an onion—associated with different fusion reactions form around a core of Iron 'ash' from the Si fusion occurring nearest the stellar center. The typical timescales at the core of the star for each fusion reaction are listed. In reality, the various layers of the 'onion' shown here would be squeezed into a tiny central volume of a red or blue supergiant star, comprising about 5 $M_\odot$, or approximately 25% of the star's total mass. The rest of the mass is in the outer envelope of the star, which is still composed of mostly H and He and in which fusion is *not* occurring.

At this point it's clear that even if this basic cycle continues—contract, heat, fuse, deplete, contract, heat, fuse, deplete...—it can't go on forever. For two reasons. First, from simple mathematics we know that some infinite series of numbers can be added together to yield a finite sum.[17] The ever-shorter timescale of these fusion reactions seems to behave in this manner. Once Si fusion has commenced, the star has at most a day or two left even if its ultimate goal is to turn the entire core into one gigantic atomic nucleus! But something more fundamental happens first. The principal product of Si fusion is the element Iron (Fe). This nucleus typically contains 26 protons and 30 neutrons, much larger than any nucleus we've encountered so far. What's holding this nucleus together is the Strong Nuclear Force (Destination 8), a force that acts only over very short distances. As atomic nuclei get larger and larger, the strong repulsive electric forces between the many positively charged protons begin to counteract the tendency for the nuclei to stick together. Really large nuclei—elements such as Uranium (U) and Thorium (Th)—are only marginally stable and can spontaneously split into smaller, more stable nuclei, a well-known process referred to as *Radioactivity*. Iron represents a fundamental transition in nuclear stability. That's not to say that all smaller nuclei are stable—they aren't—or that all more massive nuclei are unstable—they aren't either. But Fe is at the threshold where it becomes relatively easier to break nuclei apart.

---

17 For example, it is evident that the infinite series $1+\frac{1}{2}+\frac{1}{4}+\frac{1}{8}+\ldots$ adds up to a finite sum, 2.0. This particular series is related to the famous *Zeno's Paradox*, which states that a runner should never be able to catch up with another runner who has even a tiny head start, regardless of how fast the two are running. The paradox is false because the gap between the runners—though expressible as an infinite series of ever-smaller gaps—still adds up to a finite distance that can be covered in a finite time.

A consequence of this is that when you fuse nuclei of atoms together to produce elements less massive than Iron, energy is *produced* by the reaction. At Destination 11, we noted that such reactions are referred to as *Exothermic Reactions*. Because of Einstein's Equation, $E = mc^2$, another way of saying this is that the total mass of all nuclei that go into an exothermic fusion reaction must exceed the total mass of the nucleus (or nuclei) present after the reaction, with the excess mass being converted into energy. But to produce nuclei more massive than Fe, we must *add* energy in what are known as *Endothermic Reactions*. In this case, the total mass of all nuclei going into the reaction is smaller than the mass of the final nucleus (or nuclei) produced by the reaction. This doesn't mean elements more massive than Fe can't exist—we know they do. But it does mean that energy must be provided to form such heavier elements through fusion reactions.

There is, of course, a very patient source of energy always present in the core: Gravity. Once the core becomes dominated by Fe nuclei and starts to collapse, energy is supplied in the form of gravitational energy. Unlike previous cycles where the energy heated up the core so that a new fusion reaction could kick in, the energy release now leads to a highly unstable situation, as the collapse feeds a plethora of new fusion reactions that rapidly 'cook' elements heavier than Fe within the core. Since this energy no longer initiates exothermic reactions that can release energy, there's no way for the core's gas pressure to increase and halt the collapse. And so the contraction accelerates. Within one quarter of a second, the entire core—which now comprises around $5M_{\odot}$—collapses to extraordinarily high densities. Some of the gravitational energy goes into producing heavy elements, but much of it heats the core to absurdly high tempertures of up to tens of billions *K*.

It certainly seems that as far as the star is concerned, the end is near. That's true, but the star *still* has some options. One it does *not* have is for the core to become a white dwarf. Within milliseconds of the start of this final collapse, the core density far exceeds what electron degeneracy can support. The options that are available depend on the core mass and the rate at which its gravitational energy is released.[18] In some cases, the rate can exceed what the star is capable of transporting to its surface. First the core, but then the entire star can be blown to bits in a spectacular explosion that leaves nothing behind. The gas that composed the star is expelled outward at thousands of *km/s*, all of its material flowing back into interstellar space to become part of the ISM. Of course, the gas is now thoroughly 'polluted' with the elements the star produced during its lifetime, including the heavy elements produced in those final eventful seconds before—and during—the explosion.

---

18 And many, many other variables, including details of the star's rotation, its detailed chemical composition, how much of the envelope has been shed through stellar winds, etc. A full analysis of the final few seconds of a dying, high-mass star is frightfully complicated and still only incompletely understood.

In other cases, the core can collapse to a density where the electrons and protons *merge* to form neutrons.[19] If the final mass of the collapsing core is less than about $3M_{\odot}$, these neutrons exert their own form of degeneracy pressure that can—quite suddenly—halt the collapse. The result is the formation of an ultra-dense object known as a *Neutron Star*. Whereas white dwarfs are Earth-sized, neutron stars are only 5–10 *km* or so in radius, or about the size of Manhattan! The suddenly neutron degenerate core acts like a very hard wall against which the rest of the collapsing core impacts causing the material to effectively bounce off the core at explosive speeds. As before, an expanding shell—made up of material that used to be the star's envelope—will speed outward at thousands of *km/s* to be ejected into the ISM, leaving behind only the material in the central neutron star. As in the earlier case, the matter expelled from the star is rich in heavy elements formed over the star's lifetime and during the explosion.

These scenarios probably do not play out if the core's mass exceeds $3M_{\odot}$. In that case, gravity is the ultimate victor, since neither electron degeneracy nor neutron degeneracy can halt the core's collapse. There's nothing known that can. And so the contracting core never stops: A black hole is produced, the ultimate collapsed object. Depending on details regarding how the energy is released around the black hole, this process can ignite a violent explosion of the star's outer layers, in some cases producing the most luminous sources—for a few seconds—in the entire Universe.[20] Or it can end as a gigantic 'fizzle' as most of the star's mass falls into the black hole. These various events—or at least the explosive ones—that occur at the end of the lifetimes of massive stars are known as *Supernovae* (the plural of *Supernova.*)

None of this evolution will happen to the stars currently evolving off the Main Sequence in NGC 1818. That cluster contains white dwarfs and so the basic evolutionary steps we outlined at Destination 11—though, to be sure, with different details—describe the future evolution of these stars. From careful studies of other clusters, it appears that white dwarfs occur only when the most massive Main Sequence stars are less massive than about $8M_{\odot}$.

---

19 Sounds strange, but this is a well-understood process. A free neutron will decay into a proton and an electron (this is an important issue that we will return to at Destination 24). At densities comparable to atomic nuclei, the opposite reaction—an electron and a proton combining to form a neutron—can also occur. In effect, the core is starting to become about as dense as the nucleus of an atom.

20 Such extreme explosions have been observed as γ-ray bursts, so named because of the wavelength range in the EM Spectrum where they are observed. These events occur with little or no warning and can evolve in just a few seconds to attain luminosities comparable to the entire energy output of all the stars in the Universe (!) for a brief time. Although there may be a number of different origins for these spectacular and rather mysterious events, one likely source is the formation of a black hole during the final collapse of a massive star, the scenario we are discussing here. In this respect it's a good thing there are no really massive stars near the Sun. A well-aimed γ-ray burst—even if many *pc* away—could completely strip off the Earth's atmosphere!

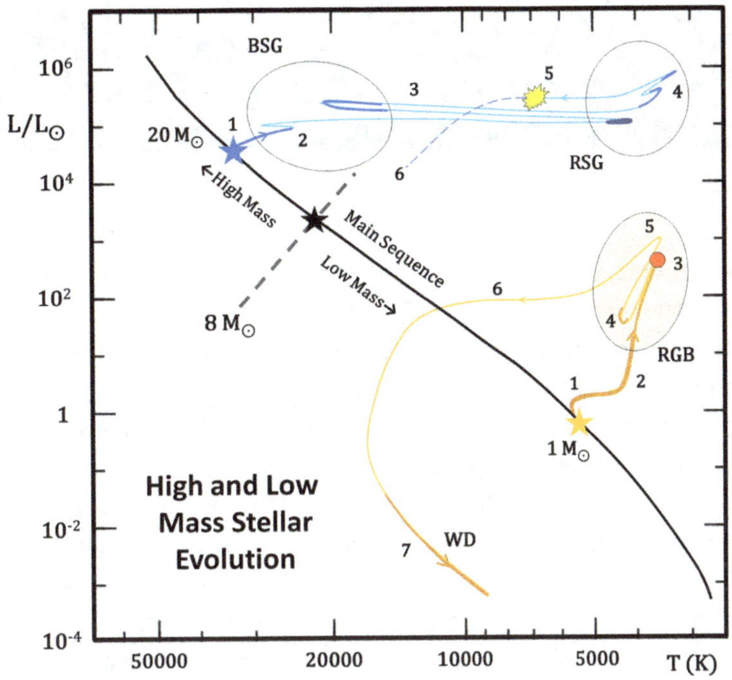

Fig 12.8

An H-R Diagram summarizing the main stages of evolution of a $20M_\odot$ (blue curve) and $1M_\odot$ star (orange curve). In both cases, the stars begin their evolution on the Main Sequence (the black diagonal curve). The approximate demarcation between High-Mass Stellar Evolution (illustrated here by the $20M_\odot$ star) and Low-Mass Stellar Evolution (illustrated by the $1M_\odot$ star) is shown by the gray line crossing the Main Sequence at around $8M_\odot$. In both evolutionary tracks, thicker lines reflect where the speed of evolution is slowest and so where stars spend more of their time. For the $1M_\odot$ star, evolution begins as H becomes depleted in the core (location 1), then slowly moves toward the Red Giant Branch (RGB; 2). At the top of the RGB, the star experiences the He Flash in the core (3, denoted by a red dot), then spends a fairly stable period as a star in which He is being fused into heavier elements in the core (4). As He runs out, the star again evolves up to the red giant region (5), but never ignites C or O fusion. It then begins to shed its envelope, becoming hotter and smaller as the core becomes more exposed (6). Finally, the now-naked core becomes a white dwarf star that slowly, eternally, cools down (7). Evolution of the $20M_\odot$ star begins in a similar way when H is exhausted in the

Therefore, stars with masses below this limit participate in the process known as *Low-Mass Stellar Evolution*, all essentially following the basic evolutionary path we described at Destination 11 (and summarized in Figure 12.8). But Main Sequence stars more massive than about $8M_\odot$—stars like those currently evolving away from the Main Sequence in NGC 1850—go through the process known as *High-Mass Stellar Evolution* that we've described here for the case of a $20M_\odot$ star (also summarized in Figure 12.8). The result of high-mass evolution is generally some sort of supernova explosion. Sometimes this event destroys the star and sometimes it leaves behind a remnant, either a neutron star or a black hole. And in nearly all cases, gas enriched in newly created heavy elements is expelled into the ISM at high velocity, driven out either at the time of the supernova explosion, or earlier by violent stellar winds at the star's surface.

Well, at least that's the story. And what a story! Byzantine, lots of plot twists, quite violent. And it even includes some unexpected characters. But does this *really* happen?

12.4 WITNESSING HIGH-MASS STELLAR EVOLUTION

As we saw earlier, a significant portion of the ionized gas in the LMC is distributed in distinctive shells or arcs (Figure 12.4). Doppler measurements of spectra of this gas confirm that some of these structures are indeed expanding, some at thousands of *km/s*. Normal stellar winds cannot do this, but stellar explosions can. This is important—though circumstantial—evidence that stars do explode. Can we add more to our case?

As it happens, one star in the LMC has exploded in historical times. This star—known at the time of the explosion as SN1987A to designate it as the first supernova discovered in 1987—was not only the most luminous star in the LMC for a time, it was, for a few weeks, almost as luminous as *all* of the stars in the LMC combined (see Figure 12.9). The energy release was similar to what you'd expect from the collapse of a core of a few solar masses.[21] But most amazing of all, the star that exploded to become SN1987A was actually known *before* it blew up! It even had a name: SK-69 202 and many of its basic properties were known. For example, SK-69 202 was a $20M_\odot$ star with a luminosity of about $10^5 L_\odot$. It was a blue supergiant, located in a position in the H-R Diagram that confirmed it had already evolved *off* the Main Sequence (see Figure 12.8).

If we look carefully, we discover unusual rings of gas surrounding the location where SK-69 202 had been—the location where SN1987A exploded (see Figure 12.10). These rings could not have expanded to their present location since the explosion occurred; that would require that they have expanded at much greater than the speed of light. The rings must be related to material ejected over a period of a few thousand years by the intense stellar winds of the doomed star *before* the explosion. The supernova itself—the light from the actual explosion of the star—faded to near invisibility within a couple of years.[22] Only then did it

21 Which is comparable to the total gravitational energy of the star. From Destination 8 we know that's given approximately as $E_{graw} = GM^2/R$, where G is the Gravitational Constant, and M and R are the mass and radius of the star just before the explosion. For a $10M_\odot$ star with a radius a bit over $3R_\odot$, this corresponds to about 10% of the total energy the Sun will emit over its entire 10 Gyr lifetime. The difference, of course, is that an exploding star releases this energy in just a few seconds!

22 This statement is somewhat more complicated than it seems. The actual release of energy in a supernova occurs, as we have seen, near the core of the star, buried far below the exploding envelope. The radiation from the explosion helps to lift up the outer layers of the star, but it does take time to see this effect. As the envelope expands, however, it also starts immediately to try to cool down, just as any expanding gas would. If this were the entire story, a supernova would actually remain bright for a few days, maybe a couple of weeks, tops. But there is something else going on. Among the many elements

core (1) until He fusion begins without a Helium Flash (2). Afterward, the star follows a complex set of loops in the upper H-R Diagram (3) as various new types of thermonuclear fusion ignite in the core and outer shells (see Figure 12.7 for a schematic view of the star's structure during this time). The star can be seen as a red supergiant (4; denoted by the reddish region) or a blue supergiant (the bluish region). Finally, as the core becomes dominated by Fe, the star explodes as a supernova (5) leaving behind either nothing or a remnant object such as a neutron star or black hole (6). Notice the very different temperature and luminosity regimes that the high-mass $20M_\odot$ star and the low-mass $1M_\odot$ star inhabit during their respective evolutionary phases. The full evolution of $20M_\odot$ star shown here from the Main Sequence to the supernova takes about 1–2 *Myr*. The evolution of the $1M_\odot$ star from Main Sequence to the planetary nebula stage is about 1 *Gyr*. In both cases, these timescales correspond to about 10% of the stars' lifetimes on the Main Sequence, about 10 Myr and 10 Gyr, respectively.

Fig 12.9

Two images of the region in the LMC containing SN1987A, one obtained before (*right*) and one obtained during (*left*) the explosion. Can you spot the supernova?! The arrow in the pre-supernova image at *right* identifies the star that exploded *before* it blew up. This remains the only case where we have such detailed information about the progenitor of a supernova.

Images from D. Malin/AAO.

become apparent that the massive star that had existed at that location was gone (see Figure 12.10)! This is direct proof that young, massive stars do explode. But there is still a glaring loose end. Try as we might, we do not see at the location of SN 1987A—the location of the exploded star SK-69 202—any direct evidence of a central neutron star or black hole that we predicted. Was this an example of a star that completely destroyed itself? Or did SN1987A leave behind a neutron star or black hole remnant that, for whatever reason, we have not yet been able to spot? As we ponder this, it raises an even more fundamental question. Just how does one find a neutron star or a black hole at the location where a star has exploded? It dawns on us that maybe this isn't so easy.

Before we can address that last question, we have to recall an important detail that occurs during the collapse of an object. When we visited the Orion Nebula (Destination 6), we saw how molecular clouds collapsed into protostars, and ultimately into mature, Main Sequence stars. The collapse was accompanied by increased rotation because the material must conserve angular momentum, which we defined as $J = MV_{rot}R$. Thus, as the size of an object, R, decreases, its rotation velocity, V_{rot}, must increase to keep the angular momentum,

---

formed during the explosion, one particular nucleus, ^{56}Co (cobalt with 27 protons and 29 neutrons, which add up to 56) is radioactive (see the discussion in the previous section of this destination). As this element decays on a typical timescale of about 10 weeks, it emits x-rays that are absorbed by and heat the expanding shell of gas. This keeps the gas sufficiently hot to radiate visible EM radiation for many weeks, sometimes even a few years. Eventually, the ^{56}Co decays away and the expanding shell becomes too tenuous to absorb much radiation, and so the supernova fades from view.

J, constant.[23] Even though the circumstances are radically different, the same *must* be true of collapsing stellar cores. If they had any angular momentum to start with—and there is good reason to believe that the cores of Main Sequence stars do rotate[24]—they must be spinning very rapidly by the time they reach the size of a white dwarf. Even more rapidly if they collapse down to the size of a more compact neutron star or black hole.

Fig 12.10

(*Left*) Images of SN1987A many years after the star SK-69 202 exploded. The large image shows the general region around the supernova (visible at the center). There are many isolated massive stars here, in addition to a lot of gas of the Hot and Warm ISM. The *upper right* image shows the region around SN1987A in more detail. The location of the star SK-69 202 matches the pinkish dot at the center of the inner elliptical ring. Before the explosion, this star had been as bright as the other two nearby stars. The very rapidly expanding envelope of the supernova is still buried in that small central region. The three rings visible around the core are gas that had been ejected via stellar winds by the star long before it exploded.

Wide-field photo from A. Burrows/STScI/NASA. Inset photo from STScI/NASA.

Rotation is not the only thing enhanced as the core collapses. Magnetic fields also become more intense during the contraction of the core. These fields—already likely pretty strong before the core collapsed[25]—can become so intense in a white dwarf, in a neutron star, or around a black hole that they radically alter the motions of charged particles, such as ionized atoms or free electrons, that happen to pass near the compact star's surface (see a less extreme but still

23 Of course, this argument assumes there's little or no mass loss, a questionable assumption for a supernova. Our reasoning still applies, however, since it is only the angular momentum of the core that matters to our present discussion. In many cases, the core mass of a star does not decrease that much due to the explosion.

24 The surface rotation of Main Sequence stars can be measured from the Doppler effect through careful analyses of their spectra. It seems reasonable to suppose that if the surface of a star rotates—and all types of Main Sequence stars do appear to rotate—their cores do too. We can actually measure the rotation inside the Sun from careful analysis of seismic waves on its surface. These are measured from long-term studies of minuscule changes in the Sun's luminosity on timescales of seconds to months. Interestingly, it appears that—in the Sun at least—the core rotates even faster than its surface.

25 The surface magnetic field of the Sun (see Figure 12.11) is likely generated from motions of its core. Though we do not have as detailed information for other stars, they too have surface magnetic fields that can be detected in many ways.

impressive example near the Sun's surface; Figure 12.11). Due to the nature of the magnetic force, these particles will generally be funneled onto the surface of a compact star near its magnetic poles, producing two unusual effects. First, the regions where the particles impact the star tend to heat up, producing bright—and aptly named—*Hot Spots* on the compact star's surface. Second, as charged particles travel toward the surface along paths dictated by the magnetic fields, they emit radiation that is directed preferentially toward or away from the surface above the hot spots. Such *Synchrotron Radiation*—which we encountered earlier near the Galactic Center at Destination 10—is produced by energetic charged particles as they travel within a strong magnetic field. Unlike thermal emission, which peaks at wavelengths defined by the temperature of the source (Wien's Law; Destination 6), synchrotron radiation is emitted with a distinctively different spectrum that extends without a peak from radio waves to x-rays.[26]

If this picture reflects what really happens to a collapsed core after the envelope has been blown away in a supernova explosion, we might expect to observe some distinctive effects. First, the magnetic poles of compact stars should emit light with a synchrotron spectrum that is directed along narrow beams pointing outward the star's surface. This is often called the *Lighthouse Effect* since it is reminiscent of how those nautical beacons use specialized optics to emit light in narrow beams. Second, if the magnetic poles of the star are not coincident with the rotation poles[27], the 'lighthouse beam' and the underlying hot spot will spin with the surface of the star. If the beam points at us while we observe the rotating star, we would see it 'flash' periodically, at a very stable rate corresponding to the star's rotation period. Nothing like this has been spotted where SN1987A exploded, but a number of other stars in the LMC do behave this way. One of these—with the rather clinical-sounding name *PSR B0540-69* (see Figure 12.3)—exhibits a sequence of flashes that repeats every 50 *milliseconds* (abbreviated as *msec*), or about 20 times per second! What is this bizarre object?

One thing it *cannot* be is a rotating white dwarf. At Destination 11 we saw that these stars have radii of around 5000–7000 *km*. If a white dwarf were rotating at a period of 50 *msec*, the surface would have to travel at three times the speed of light! In fact, due to structural limitations, white dwarfs can't rotate faster than about once per minute. That's still pretty fast (!), but not as fast as PSR B0540-69. On the other hand, a neutron star—which has a radius about 1000 times smaller than a typical white dwarf—*can* rotate at this rate. Indeed,

26 This cannot be strictly true or synchrotron radiation would emit an infinite amount of energy. Indeed, the spectrum does have a sharp peak—quite distinctive and very different from the broad, smooth peak of a thermal spectrum (see Figure 6.15)—near the peak kinetic energy of the particles producing the radiation. For electrons near neutron stars, this can be in the x-ray portion of the EM spectrum.

27 As they are not on Earth. Our planet's North Pole, where the imaginary rotation axis passes through the surface, is in the center of the Arctic Ocean. But the North Magnetic Pole is hundreds of miles away in Northern Canada, and has actually moved over 1000 miles closer toward the rotation pole since the early 20th century!

Fig 12.11

These arcs above the surface of the Sun—known as *Solar Prominences*—illustrate matter flowing along magnetic field lines. The fields at the surface of a neutron star are over 1 million times stronger than those at the Sun's surface.

Photo from TRACE/NASA.

neutron stars exert such strong gravitational force at their surface that they can hold themselves together even if they were to rotate with periods as short at 0.5 *msec*, or 2000 revolutions per *second*! At such a rapid rotation rate, the star's surface would be moving at over 50% the speed of light.

A side trip over to PSR B0540-69 in the LMC (Figure 12.3) confirms our suspicions.[28] The object we find here is far too small to be a white dwarf. Instead, what we find is a neutron star—one that is a mere 20 *km* across!—spinning at the phenomenal rate of 20 times a second, a blur to our eyes. If we take rapid images to 'freeze' the spin of the star, we discover it has two bright hot spots associated with its magnetic poles. Visible above both poles are pale shafts of eerie blue light, the synchrotron radiation from charged particles traveling along the star's magnetic field. Because the field is anchored to the surface of the star, the hot spots and the synchrotron emission regions also rotate with the star's surface, spinning at 20 times per second. With care, we can place ourselves at a location that lets us periodically look straight down one of the beams as it spins with the star. From that vantage point, we first see the flash of one of the poles as it faces us. We then may see a second, weaker flash as the beam and hot spot from the other pole pass by further out of our line of sight (Figure 12.12).[29] Then, 50 *msec* after the first flash, the first pole returns to

28 When we visit this pulsar, we have to balance safety with our desire to see the object clearly. Keep in mind, a 20 *km* diameter neutron star is tiny, comparable in size to Manhattan. That island is visible from Earth orbit, 300 miles up, but from one *AU* it would be impossible to discern by eye. Even from a distance equal to the Sun's radius, 700000 *km*, a neutron star would appear very small and it would be difficult to see much surface detail. We could try to get closer, but eventually all the high-energy particles accelerated in the intense magnetic fields of the star would become dangerous. If we avoid the synchrotron beams, we might be able to venture to within a few tens of thousands of *km* from such a star. From that distance, the star would only be about 7 *arcmin* in diameter in our sky, or 1/4 the Moon's diameter as seen from Earth. Not great, but probably about the best we can do.

29 Because the magnetic poles are offset from the rotation axis, even if we are aligned to see down one of the beams, in general we will not be positioned to see down the other beam. For this reason, as the star flashes, invariably one of the flashes is very bright, the other dimmer. To get a sense of this effect, get a globe and identify some location (not

view and the cycle repeats. And repeats. And repeats. Precisely the lighthouse effect we anticipated. Since the timing of the flashes is set by the rotation period of the star, it is extraordinarily stable.[30] There are many such stars—dubbed *Pulsars*—that have been discovered in the LMC and our Galaxy (and even beyond). But here at PSR B0540-69, we see the name is a misnomer. Pulsars do not vary in brightness because they pulsate. They flash because they—and their strong magnetic fields—rotate rapidly.

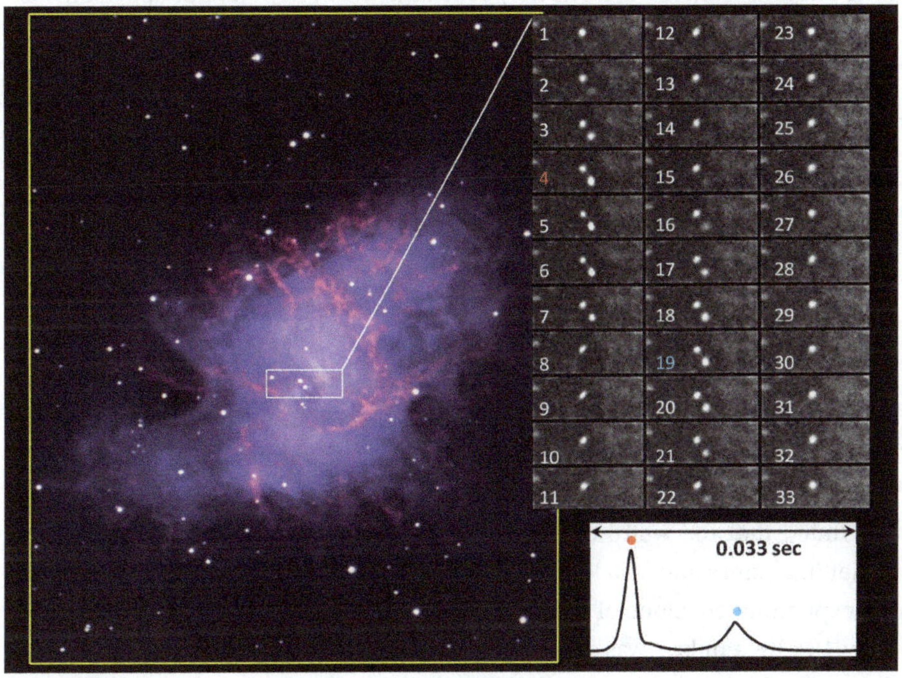

Fig 12.12

(*Left*) The image shows the famous supernova remnant M 1, commonly known as the Crab Nebula. The gas that makes up the nebula is the expanding debris from the envelope of the star that exploded as a bright supernova of 1054. The set of 33 images that make up the inset (*upper right*) make of a sort of movie of the central region of the nebula. It is evident that one of the stars in the field fluctuates significantly in brightness. The *Light Curve*—a plot of how brightness changes with time—of the *Variable Star* is shown (*lower right*). Two distinct peaks are visible in both the series of images and the light curve. These are denoted by red and blue numerals in the inset images, and by red and blue dots in the light curve. The astonishing thing is that the full duration of the light variations is 0.033 *seconds*! The only thing that can fluctuate in brightness this rapidly and regularly is a rotating neutron star. When one is seen to vary as this one does, it is known as a *Pulsar*. This image shows how the Crab Pulsar varies in optical light, but it is also seen as a pulsar in every part of the EM Spectrum, from x-rays to radio waves. The pulsar PSR B0540-69 that we are visiting in the LMC has a period of about 0.050 seconds, only a bit longer than the Crab Pulsar shown here.

Image montage by N. A. Sharp/NOAO/NSF.

The view of the pulsar in PSR B0540-69 is so alien, so fantastic, we almost failed to notice the rest of the sky here. We soon appreciate that a bluish-gray glow surrounds us, somewhat reminiscent of the sky at the Orion Nebula. But there are no young, hot, massive stars here to ionize the gas. So why does the sky glow? A spectrum reveals that some of the light is synchrotron radiation coming from a vast region around the pulsar. Evidently, magnetic fields and charged particles—both likely carried out by the expanding envelope of the

on the poles or on the equator) and the point directly on the opposite side (the so-called antipode of the first location). As you spin the globe, you'll note that you may get a good look at one of those positions but not the other as the surface passes by.

30 Normally this is true, but occasionally the periods of these flashes change quite suddenly. This can be understood if the 'crust' of the star—a neutron star is practically solid—suffers a small change, essentially a 'starquake'! That changes the distribution of mass in the star, which can alter its rotation period. The only reason we can see this is that even a tiny quake—in some cases, motion of under 1 *cm* at the star's surface—can create a noticeable change in the period. There's also a longer term increase of the periods of all pulsars, but this effect is the subject of one of the questions at the end of this destination.

exploded star—surround the pulsar, extending up to a few *pc* away. Some of the radiation is in emission lines—synchrotron radiation produces only continuous radiation—which exhibit quite different relative brightnesses compared to the emission lines we saw in the Orion Nebula, particularly for emission lines arising from Oxygen (O) and Silicon (Si). The differences imply that the atoms in the nebula here are ionized primarily by collisions with energetic particles accelerated by the pulsar's rotating magnetic field, not by absorption of energetic photons from hot stars as they are in the Orion Nebula. Our spectrum also reveals that most of the gas surrounding the pulsar is expanding outward at thousands of *km/s*. This gas is clearly from the envelope of the star that exploded thousands (perhaps just hundreds) of years ago and gave birth to the pulsar. Remarkably, some of the gas is falling back into the vicinity of the central neutron star and fuels the synchrotron beams of the pulsar.[31] It is a fantastic sight. And it fits neatly into our basic picture of how massive stars evolve.

All this is impressive, but it would still be nice to actually see a supernova, then discover the neutron star—probably a pulsar—where the original star had been located. Unfortunately, nobody recorded the supernova that produced PSR B0540-69. However, in 1054 AD, Chinese astronomers did note the appearance of a 'guest star' in the constellation Taurus.[32] Easily visible even during daytime, the star eventually faded from view in about two years, then, after a few more years, from memory. But in the 20th century, astronomical sleuths concluded that the weird object *M 1*—the first object in Messier's catalog of comet imposters and also known as the *Crab Nebula* (Figure 12.12)—is in fact the expanding envelope of the star that produced the 1054 supernova. In the early 1970's, a pulsar was found to be located in the very center of M 1, one of the first pulsars ever discovered (Figure 12.12). For many years, its period of 33 *msec* was the shortest known, though some more rapidly spinning pulsars have subsequently been found. Since pulsars slowly spin down as they age[33], the rapid rotation of the Crab Pulsar is a sign that it formed only a few hundred years ago! So here's our proof. In M 1 we have a case where a known supernova produced a known pulsar. It is a spectacular confirmation of every aspect of our story of how high-mass stars evolve.

---

31 If the gas is expanding from the pulsar, why would any fall back? In fact, the expanding envelope of the star would have exhibited a range of expansion velocities. Some of the material was moving slower than the escape speed from the neutron star and eventually falls back into the pulsar's vicinity.

32 While the best records of this event are from Chinese astronomers, Arab, European, and even North American astronomers saw it too. In the case of the Europeans, at least, the heavens were believed to be 'perfect' by the ruling classes, so there was probably a lot of social and religious reasons to ignore or downplay even as spectacular event as this one.

33 This slowdown of pulsar rotation has been directly observed. But why does it occur? You can explore the reasoning in a question at the end of this destination.

12.5 VICTORY OF A RELENTLESS FORCE

Except for one thing. Black holes. Up to now, we've accumulated pretty convincing evidence that white dwarfs and neutron stars exist, and that neutron stars form during supernova explosions. We've also seen that a Supermassive Black Hole likely exists near the Galactic Center. But what about smaller black holes? Ones that form during the deaths of massive stars? Do these objects *really* exist and, if they do, can we convincingly tie them to the evolution of high-mass stars?

To track down these objects and to get some sense of how they form, we must, paradoxically, visit one of the most luminous stars in the LMC. Known as *LMC X-3* (Figure 12.3), this star appears from afar to be a single, fairly normal-looking B-type Main Sequence star. Similar, in fact, to Alcyone, the most luminous star in the Pleiades (Destination 8). But the odd name of this star—in particular the 'X-3' designation—means that it is a luminous source of x-ray radiation, one of the most powerful x-ray emitters in the LMC. This is peculiar. Even the hottest stars—which normally have thermal spectra—emit very little x-ray radiation. A typical B-type star like this one should emit virtually no x-rays at all. And there is another oddity. Although the spectrum of this star shows evidence of only a single B-type star, its absorption lines periodically shift back and forth. An application of the Doppler Equation (Destination 6) tells us that the visible star's velocity changes by several hundred *km/s* on a timescale of 1.7 days. We've seen this behavior before. The star's velocity is changing because it is orbiting another body. But this body can't be seen at optical wavelengths. And, apparently, it emits copious a quantity of x-rays.

Wait a minute! A B-type star such as the one here has a mass of about $6.6M_{\odot}$. It's not easy to push around a star this massive, particular at a relatively high velocity of hundreds of *km/s*. Its companion—even though we see no optical trace of it—must also be quite massive. To figure out just how massive, we write the orbital radius, a (in *AU*), of the visible star to be $a = PV$, where P is the orbital period in *years*, and V is the star's orbital velocity relative to its companion in units of the Earth's orbital velocity, 29.8 *km/s*. If we plug this into Newton's version of Kepler's Third Law (Destination 4) and then solve for the combined masses of the stars, we get

$$M_1 + M_2 = PV^3,$$

Fig 12.13

An artist's impression of a binary star system like LMC X-3 composed of a massive, hot main-sequence star (right) and a black hole (left) at the center of an accretion disk. This disk—as the gravitational energy of infalling material heats it up through friction—can become very bright from thermal radiation as matter spirals down onto the event horizon of the black hole at its core. Since the orbital speeds increase inward in the disk—due to Kepler's Third Law—the frictional heating increases at smaller radii, producing a strong temperature and color gradient in the disk. Our eyes see the disk only over an intermediate range of radii since the outer, comparatively cooler regions emit mostly in the infrared, while the inner, hottest portions of the disk nearest the black hole emit predominantly UV and x-ray EM radiation. A steady flow of gas from the outer layers of the B-type star (at right) feeds the disk through the L1 Lagrangian Point (see footnote 36) and leads to a hot spot where the stream impacts the disk. A pair of 'jets' are produced from material accelerated from near the black hole and in a direction perpendicular to the disk. The mechanisms that produce such jets are discussed at Destination 21.

Artwork by M. Weiss/ CXC/NASA.

where M_1 is the mass of the visible star, and M_2 is the mass of the unseen companion.[34] If we plug in the values for LMC X-3, we find that $(M_1 + M_2) \approx 18M_\odot$. Less than half of this mass can be accounted for by the visible star (only $6.6M_\odot$), implying that the unseen companion in LMC X-3 has a mass of about $11M_\odot$. If this were a normal Main Sequence star or an evolved star—a red or blue supergiant, say—with a mass this large, we'd easily see it since it would dominate the light of the binary system. But we don't. Instead, we see only the light from a single B-type Main Sequence star. And a lot of x-rays.

As we approach LMC X-3 to get a better view of what's going on, we find a spectacle unlike any we've seen so far (Figure 12.13). Even the bizarre view of the pulsar PSR B0540-69 did not prepare us for this. The B-type star we saw from far away is certainly here, luminous and shining intensely with a blue-white color characteristic of its 15000 K surface temperature. But we now see the star is quite odd in one obvious respect. Though we've become used to stars being spherical, this one isn't. Instead, it is shaped almost like a teardrop, one end round, the opposite side pointy. Even more bizarre, from this pointy

34 The equation given here is a simplification of a complex analysis. Normally, we measure the motions of stars about their center of mass, but this equation assumes the velocity, V, is measured relative to one of the bodies. It is a somewhat subtle but critical distinction. In practice, one needs to measure the velocities of both stars, in which case $V = V_1 + V_2$. If that isn't possible—and it's not in the case of LMC X-3—we must make some assumptions about the mass of one of the stars. In the case of LMC X-3, we can estimate the mass of the visible B-type star since it certainly seems reasonably similar to a normal Main Sequence star (Destination 8). We can then use a variant of the equation given here to determine the mass ratio of the two stars in the system. Since we assume we know the mass of the one of the stars—the visible one—we can then estimate the mass of the invisible star.

end of the star, a thin stream of luminous gas extends outward toward an even weirder spectacle. There, beside the B-type star, is a gargantuan disk, quite thin and pancake-like in shape, and about as large in diameter as the B-type star. The outer parts of the disk glow with a faint reddish tint, similar to a stove burner at low temperature.[35] As we peer closer toward its center, the disk's color transitions smoothly from orange, then yellow, then bluer, all the while becoming intensely brighter to our eyes. As we peer even further inward within the disk, our eyes reveal that the light from the disk steadily increases in brightness toward the disk center, welder-blue in color. Our instruments reveal more. The central regions of the disk emit intense ultraviolet and x-ray radiation, the origin of the emission that defines LMC X-3 as a powerful x-ray source! At the very heart of the disk, we finally see the explanation for all the strange features of LMC X-3. There, buried in the glare of the disk itself, is a deep inky blackness. Not just the absence of light, but the banishment of the idea that light can even exist. At the center of this disk is an $11M_\odot$ black hole.

As we look at the spectacle, we soon come to realize that the matter in the disk around the black hole arises from material being drawn out from the envelope of the B-type star from the intense tidal forces that the black hole exerts on its companion. This material cannot fall directly into the black hole because it has too much angular momentum. Instead, the matter from the B-star settles into orbit around the black hole, forming what is known as an *Accretion Disk.*[36] Due to Kepler's Third Law, the orbital velocity of the material in the disk steadily increases inward as does the density of matter in the disk. This change in speed generates friction between material at different distances from the black hole, heating the disk in the process. Close to the black hole—near the center of the accretion disk where the density and orbital velocities are highest—the friction becomes so great that the matter heats up to temperatures reaching millions of degrees *K*, high enough to emit thermal radiation in the form of x-rays. There is a curious irony here. Though black holes are the darkest objects we can imagine, when matter is dumped onto them, they give rise to accretion disks that can attain very high luminosities.

With what we know about high-mass stellar evolution, we can imagine a plausible scenario to explain the existence and properties of LMC X-3. A few million years ago, two massive stars formed a binary system with a very short period of

35 Looking further out in the disk, we would be able to detect infrared radiation, implying from Wien's Law (Destination 6) that the disk is even cooler in that region.

36 This process, though reasonably well understood, is still rather bizarre. In effect, the outer layers of the B-type star stream toward the black hole to a special point known as the *L1 Lagrangian Point*. Any material at that location feels a balanced gravitational force from the star and black hole. Matter flows away from the star because it is trying to expand—that is, the star has begun its evolution off the Main Sequence—and matter is being pushed through the L1 point. It's the large quantity of material flowing from the star that produces the accretion disk that ultimately feeds the central object. We'll explore the properties of accretion disks in more detail as we encounter them again at Destinations 17 and 21.

about one day; such binary systems are known to occur in active star-forming regions like the Orion Nebula (Destination 6). One of these stars may have been as massive as 30 $M_\odot$, while the other was considerably less massive, say around 18$M_\odot$. After about 3–4 million years, the more massive star evolved through its late stages, ultimately exploding as a supernova, its core collapsing into an 8$M_\odot$ black hole. Amazingly, the explosion did not destroy the binary system, though it may have significantly altered the orbital period and semi-major axis. Some of the mass of that had been the lower-mass companion star—the B-type star we see today—was probably blown away during the explosion, but the companion still retained around 12$M_\odot$. Within a few million years, tidal forces[37] brought the stars back into compact, circular orbits, though with a somewhat longer period than before the first star exploded as a supernova. As the surviving 'normal' star began expanding due to its own evolution, it grew large enough that its envelope started to pour onto the black hole. Not directly, of course. As we have seen, the matter needs to conserve angular momentum and so an accretion disk formed around the black hole (see Figure 12.13). Energy losses in the disk, due to friction and radiation, allowed a steady flow of mass onto the black hole. Over time, the black hole gained 2–3$M_\odot$ from its stellar companion, while about 1–2$M_\odot$ was radiated away as energy or lost through a wind emanating from the binary system. It is at this point that we see it as LMC X-3. If the B-type companion loses enough mass, it might end up as a white dwarf orbiting the black hole. Due to its unusual evolution next to its massive companion, the B star might even explode as a supernova, leaving a neutron star or another black hole. Regardless of the details, the result is that both stars will have become inert remnants that—if far enough apart—will orbit each other indefinitely.[38]

It's a shame that we can't actually see individual stars go through their life cycles completely. But observations of enough stars—and the stars here in the LMC have been particularly helpful—reveal insights into what must be happening. What we've seen is that massive stars, rare as they are, live their lives in the fast lane: At high luminosity, high speed, and they go out with bang! Good thing, too. Without these stars, there would be no heavy elements in the Universe, nor would these elements have gotten back into the ISM from which later generations of stars could form. Stars like the Sun. Without massive stars, we would not exist.

---

37 The tides that we are familiar with on Earth arise due to the difference in the gravitational force from the Moon on the near and far sides of our planet. The same sort of forces can distort the shapes of individual stars in a close binary system, causing them to move closer together in the process.

38 If they are not 'far enough apart', the remnants may eventually merge. Depending on the nature of the remnants that coalesce in this process, the result could be a explosive event—a supernova—possibly accompanied by the production of powerful gravitational waves similar to the sort first detected in 2015.

DESTINATION 12 QUESTIONS

12.1. The pulsar of SN1987A has not yet been discovered. Give at least two reasons why this may be the case. Saying "We don't know where it is" won't work! We know *precisely* where the pulsar should be if it exists: right where SK-69 202, the star that exploded, had been.

12.2. If the nebula around a pulsar shines in part from the energy from particles accelerated by the pulsar's magnetic field, what does this imply for the pulsar itself? That is, what is the only source of energy that the pulsar can tap? Remember, it has *no* fusion and it is *not* collapsing. Our usual suspects are not available. Read the notes about the Crab Nebula and its pulsar quite carefully to get a hint.

12.3. Find some basic information—name, age, mass, H-R Diagram, etc.—for an LMC or SMC star cluster other than NGC 1818 and NGC 1850. If possible, provide an image of the cluster or a link to an image of it on the web.

12.4. One of the reasons we believe pulsars to be neutron stars is that their rapid, stable periods can be understood only in terms of rotation, and only very dense objects can rotate stably. To illustrate this, calculate the maximum rotation speed of the following objects:

a) The Earth.
b) The Sun.
c) A white dwarf (assume $R = 7000\ km$, $M = 0.7M_{\odot}$).
d) A neutron star (assume $R = 10\ km$, $M = 2.0M_{\odot}$).

HINT: The maximum rotation is roughly when the centripetal acceleration (V_{rot}^2/R) equals the gravitational acceleration at the surface (GM/R^2). From that velocity, determine the minimum period each object could rotate. Compare that to periods of 0.001 *sec* that some pulsars are known to exhibit. Which, if any, of these objects can conceivably spin that fast?

12.5. What is the Schwarzschild Radius (Destination 10) of the black hole in LMC X-3? Describe something you are familiar with that is of comparable size.

12.6. How fast are the following objects moving at their equator?

a) The Earth.
b) A white dwarf, assuming it has a rotation period of 1 minute.
c) The surface of the Crab Pulsar. Assume the star has a radius of 10 *km* and is rotating at the rate given in the text.

12.7. What is the density of a neutron star? Justify the values you use to calculate this density from information in the text of this destination.

12.8. For the scenario described for the binary system that led to the formation of LMC X-3 (and described at the end of this destination), use the

Main-Sequence Mass-Age relation from Destination 8 to estimate the time after the formation of the binary when the key events described in the scenario would have occurred.

12.9. On the web, find information about a specific pulsar other than the Crab Pulsar and PSR B0540-69 of the LMC. List the distance, period, and at least one other bit of information about the pulsar.

12.10. The atomic weights of ^{28}Al, ^{56}Fe, and ^{112}Cd (those are Aluminum, Iron, and Cadmium) are 27.982, 55.935, and 111.903 atomic mass units, respectively. Imagine fusing two of these Al nuclei into one Fe nucleus. Then imagine fusing two of these Fe nuclei to make one Cd nucleus. Which of these two reactions is exothermic (produces energy) and which endothermic (consumes energy)? Given what you observe from these comparisons, explain under what conditions Al, Fe, and Cd might be able to form in stars.

12.11. How fast would the Sun rotate if it collapsed, with no mass loss, to a radius of 10 *km*? You will need to look up or calculate the rotation speed of the Sun at its current radius.

12.12. A careful analysis of the Crab Nebula, M 1, reveals that it is expanding outward from the central Crab Pulsar.

a) How long ago did the nebula begin its expansion, assuming that it was formed during the supernova explosion of 1054?

b) Assuming the nebula is expanding at a speed of 5000 *km/s*, how far has it expanded since the supernova exploded? Express your answer in *km/s* and *pc*.

c) If the nebula is observed today to be two *arcmin* in radius, how far away is the nebula? Express your answer in *pc*.

12.13. Let's explore the properties of an accretion disk around a black hole. We assume the disk has a Keplerian rotation curve (see Destination 9), so that the orbital velocity, V, of material in the disk varies as $V \propto 1/\sqrt{R}$, where R is the distance from the center of the disk (that is, the location of the black hole).

a) Draw a schematic plot showing how orbital velocity, V, varies with radius, R, in such a disk.

b) The temperature of different locations within an accretion disk, which we'll denote as T, depends on how rapidly the orbital speed varies. With a little calculus it is possible to show that this means $T \propto R^{-3/2}$. What is the relative temperature of the disk at $R = 2$ and $R = 5$ if we suppose that $T = 20000K$ at $R = 1$? (Notice that the specific units of R do not matter.)

c) At what wavelength, in *nm*, does the disk emit its peak thermal radiation at $R = 1$, 2, and 5? What portions of the EM Spectrum do these peak wavelengths correspond to (see Destination 5, Figure 5. 6). Estimate the rough range over which the disk emission peaks in the optical portion of the EM Spectrum. Compare this to the full range of the disk that, in the units we are using, might realistically extend from about $R = 0.01$ to $R = 30$ or so.

Destination 13

THROUGH THE HEART OF DARKNESS: THE GALACTIC HALO

13.1 THE LONELY OUTSKIRTS OF THE GALAXY

It's not the first time we've chosen to travel somewhere that doesn't appear to contain anything (recall Destination 5). But this is different. Here, some 80 *kpc* from the Galactic Center—and essentially the same distance from the Sun—we have arrived at a place that seems to be the epitome of emptiness. If we strain, we can, at best, see *one or two* stars in the entire sky! The only oasis for our parched eyes are the pale glows of our Galaxy, the Magellanic Clouds and two other faint smudges known as *M 31* and *M 33* (Destination 14). Since there are no stars here, there are unlikely to be any planets either. And so, lacking a solid base on which to park, we simply float here. In a vast, dark desert.

The view of the Galaxy—our home!—is humbling. The Galactic Disk shines weakly, a dim gray patch about 10–15 *deg* across, noticeably smaller and more nearly circular in appearance than it appeared from near LMC (Destination 12). The Galactic Bulge is evident, as are broad spiral features that extend out through the Disk itself. But we see none of the trillion or so stars that we know inhabit our Galaxy. Not even individual globular clusters—each composed of hundreds of thousands, sometimes millions, of stars—are visible to our eyes from this place. And our Sun is now hopelessly lost, a speck of light in the Galaxy that is virtually impossible to distinguish from other stars in Galactic Disk, even with the aid of a powerful telescope. Distance has reduced them all to invisibility in this Alien Sky. The combined light of all the stars, all the clusters, all the nebulae that we know reside in our Galaxy manages only to produce a feeble glow in the sky of this lonely place.

The emptiness of the sky here makes a deep impression on us. We are accustomed to thinking that any place in the Universe, no matter how alien it might be, would always have stars in its sky. But our own Earth illustrates how wrong our perceptions can be. If we were randomly dropped onto the surface of the Earth from outer space, three times out of four we'd end up in a body of water. Yet most of us think of land when we think of Earth.[1] Here, in this lonely place far from the center of our Galaxy, we get a first inkling that the same may be true of the Universe as a whole: Most of its volume is so distant from any planet, star, or galaxy that typically none—or very few—would be easily seen from random locations in space. Our comfortable mental image of star-filled skies everywhere in the Universe is fundamentally wrong. The desolate view

1 A Planet which, as far as its surface is concerned, really should be called 'Water.'

we see here is far more typical. Our expectation that the sky, any sky, should be full of stars reflects a bias developed from our confinement deep within the Disk of our crowded Galaxy.[2]

As bleak as the view appears from this spot—or, perhaps, because of this—we soon come to appreciate some features of the few things we *can* see in this sky. The near-circular outline of the Galaxy reflects the path we took to get here. As in Destination 9, we have traveled relentlessly in the direction defined by the location of the South Galactic Pole as seen from Earth (see Figure 9.5), but this time continuing for 80 *kpc*, a much greater distance than we have covered on any previous journey. The view from this location reveals the Galactic Disk from 'above'—that is, perpendicular to the Galactic Plane—that is similar to an aerial view from directly above a stadium. Astronomers refer to this as the *Face-on* view of a galaxy. From here, the Magellanic Clouds are about as distant as they are from Earth, though the LMC is now the more remote of the two, the opposite of what we see from Earth. Our new location also offers a fresh perspective of these familiar galaxies. The LMC—a thin disk that's tilted about 30–35 *deg* from our viewing angle from Earth (Destination 12)—is now tilted about 75 *deg* from our new line of sight, providing a nearly *Edge-on* view of the galaxy that causes our familiar neighbor to appear very different from how it looks from home (see Figure 13.1 to see how perspective alters the galaxy's appearance). The SMC's elongated appearance—it looks like a stubby cigar from here—comes about because we are now viewing it from the side, rather than nearly along its long axis as seen from Earth. Since we've moved 80 *kpc* from Earth—a distance comparable to the distances of the Magellanic Clouds from Earth (55 and 65 *kpc* for the LMC and SMC, respectively)—it's not surprising these galaxies can look significantly different from our new location (see the discussion from Destination 3 on parallax).

But the two other patches of light—M 31 and M 33—are different. They look virtually unchanged in shape, angular size, and brightness compared to their appearance from Earth. On earlier voyages, when we traversed what we thought at that time were vast distances—the Earth to Mars (Destination 2), the Sun to α Cen (Destination 4)—it was disconcerting to realize that the locations and appearances of stars that are even more remote had hardly changed. In the same way, it's clear to us now that M 31 and M 33—whatever they may be—are far more distant than the 80 *kpc* we've traveled to reach this lonely

2 A totally empty sky is not likely something that anyone would be familiar with, where 'anyone' would refer to all possible living beings that may exist in the Universe. It seems a safe bet that life needs the energy from a star, the stability of a planet, and the chemistry of heavy elements to develop and flourish. That's only likely to happen in galaxies such as ours, and, even then, likely only in a galactic disk where plenty of star formation and chemical evolution—driven by the life cycles of stars (as we saw in Destinations 11 and 12)—can occur. To experience a truly empty sky such as the one we see here would require that we seek it out. As we can now attest, that would require a *very* long journey to the outermost reaches of a galaxy like ours.

Fig 13.1

The LMC from Earth (*lower right*) is viewed nearly face-on. This gives us a nearly unobstructed view of the entire central bar and disk regions of the galaxy. But from our new location, about 80 *kpc* south of the Galactic Plane, we get to see the LMC from a very different perspective, causing it to look more like the galaxy shown the image shown on the *left*. This particular galaxy—NGC 55, a near twin of the LMC but about 50 times more distant—shows how the LMC might appear if viewed edge-on. For all of us trapped on Earth, the perspective of any given galaxy is not going to change much over a human lifetime, or even over millions of years. But the different appearances of galaxies such as the LMC and NGC 55—galaxies that are very similar—indicate that the perspective from which galaxies are viewed strongly affects their appearance. We'll need to keep this in mind as we try to compare and understand the basic properties of other galaxies throughout the Universe.

Photos from ESO (NGC 55) and W.-H. Wang (LMC).

spot. Once again, we've conquered a vast distance only to discover that it is tiny compared to what awaits us beyond our current destination.

Another thing we've learned on our travels is that appearances, especially when we restrict ourselves to visual observations, can sometimes produce a distorted view of our environs. At Anti-Cen (Destination 5) we found 'empty' space to be full of neutral gas that comprises the Warm ISM. The 'dark' clouds of the Orion Nebula and in the Galactic Disk—seemingly devoid of matter to our eyes—turned out on more careful inspection to be filled with molecular gas, and, sometimes, regions of ongoing star formation (Destinations 6 and 10). But as we look around us with radio and microwave telescopes, we realize there is no hidden reservoir of interstellar gas here. Apart from the ever-present, ultra-low-density gas of the Coronal ISM (Table 6.2 of Destination 6) and a few isolated wisps of neutral and ionized gas, even the ISM shuns this place.

But there *is* something here. Something relatively close by, though very hard to see. With the aid of a small telescope, or even binoculars, we can just make it out. Faint stars! Hundreds, perhaps even thousands of them. Individually, they are all too dim to see with the naked eye. That's true of the Galaxy too, yet we see it easily, if faintly, in this sky. These nearby stars were hard to spot because they are spread out over a wide swath of this sky. Their collective light is not concentrated enough to produce the smooth glow of the Milky Way we are used to from Earth and that we see here as we look back toward the Galaxy.

But this is not some random collection of stars. They are definitely related to one another. As we map their distribution carefully, we discover that these stars congregate into a single broad structure that spans over 10 *deg* in this sky, a structure that is definitely, but mildly, centrally concentrated

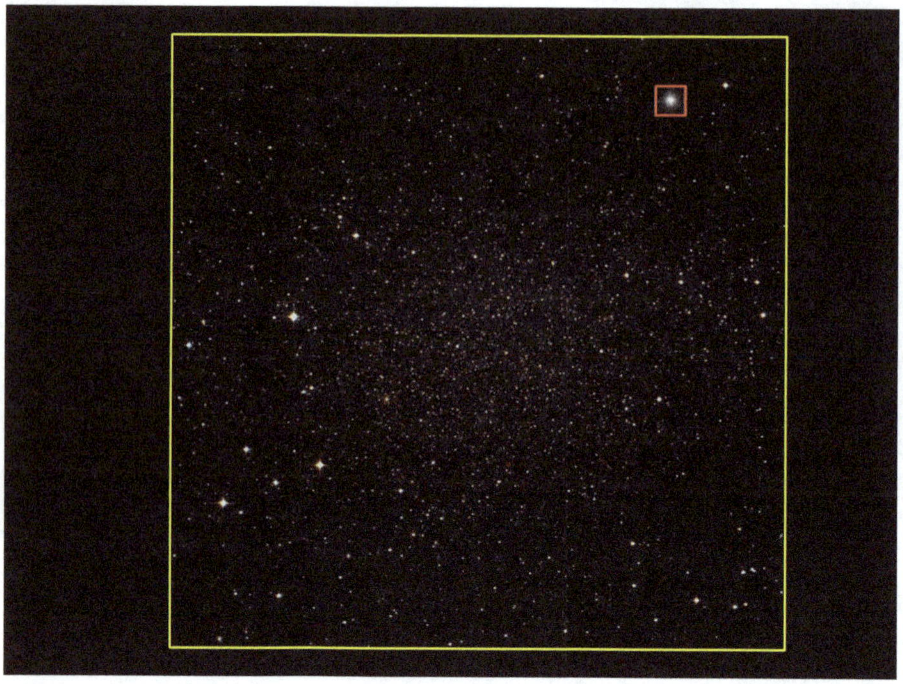

Fig 13.2

This view covers the central region of the Sculptor Dwarf Spheroidal Galaxy, spanning about 750 *pc* on a side. The globular cluster M 92 is shown to the same scale within the red square in the upper right (compare this to the much more detailed images of globular clusters in Figures 11.2, 11.4 and 11.11). Though the mixture of stars in the two objects is quite similar—they are both composed of mostly ancient stars with low chemical abundances—their distributions are clearly very different. While there are about 5-10 times more stars in Sculptor than in M 92, the stars of the galaxy are spread out through a volume that's about one billion times larger than the volume taken up by the cluster.

Sculptor photo from NASA.

(see Figure 13.2). Intrigued, we study the stars in this object further using some of the tools we've developed at previous destinations. For example, the stars exhibit an H-R Diagram remarkably similar to the one of M 92 (see Figure 11.5). Like that cluster, the stars in this object are also predominantly very old—around 10–13 *Gyr*—and there are lots of them. While M 92 has a couple of thousand red giants (Destination 11), the H-R Diagram reveals this object to contain about ten times as many. Spectra of these red giants reveal that—also reminiscent of M 92—the stars of this system exhibit very low abundances of heavy elements, ranging from about 1% to 3% the abundance of such elements in the Sun.

But this is no globular cluster (see Figure 13.2). With a little effort, we determine the average distance to this grouping of stars to be about 10 *kpc* from our current vantage point, about 79 *kpc* from the Sun.[3] If M 92 were 10 *kpc* away—only slightly further than it is from Earth—it would appear as a compact, fuzzy ball of light concentrated within about 10 *arcmin* in the sky that would be easy to spot with a pair of modest binoculars. A globular cluster ten times brighter—which is consistent with the total number of stars we see in this object compared to M 92—would be easily visible to the naked eye even from a distance of 10 *kpc* (you can get a sense of this from Figure 13.2). But this congregation of stars located far from the Galactic Disk is something

3 We've hinted at the existence of techniques that can be used to get distances to objects like these at Destination 11 (see the footnotes in that destination), which we'll explore in detail starting at Destination 14. For now, we can just accept that getting the distance to these stars is something we can do with reasonable precision.

Fig 13.3

A comparison of a typical Dwarf Spheroidal Galaxy (*upper left*; this one is known as Leo I and is similar in size to Sculptor) and a large galaxy similar to ours (*lower right*; the actual galaxy shown is in this image is NGC 4414). The approximate location of the Sun is also shown for reference (open circle). Recall that the distance from the Galactic Center to the Sun is about 8.5 *kpc*, which is also about the diameter of the LMC (see Figure 12.3). It is clear from this comparison why galaxies like Sculptor and Leo I are considered 'dwarfs'!

Photos from D. Malin (Leo I) and STScI/NASA (NGC 4414).

different. Not a cluster at all, it is an example of a type of object known as a *Dwarf Spheroidal Galaxy.* This particular one is the *Sculptor* Dwarf Spheroidal Galaxy, because it lies within the confines of the constellation Sculptor as seen from Earth (Figures 13.3)[4], the same constellation in which the South Galactic Pole is located (Destinations 9 and 10; Figure 10.2).

Although much larger than a globular cluster (see Figure 13.2), galaxies like this one are tiny by comparison to systems like our Galaxy. The stars in Sculptor, for example, extend at most a couple of *kpc*, much smaller than our Galaxy or even the LMC (see Figure 13.3). And though it contains a respectable number of stars—around 5–10 million, which is significantly more than M 92—it is still very much a small system when compared to the population of a trillion or so stars found in the Galaxy. The shape of Sculptor—nearly circular in the sky—also suggests that it is more or less spherical in shape. These types of galaxies certainly are 'dwarfs' and they are definitely 'spheroidal' in shape, but as we shall see later, determining whether or not they deserve to be called 'galaxies' is a bit more complicated.

Sculptor's presence so far away from the Galaxy begs the question: What's it doing out here, seemingly in the middle of nowhere? Just a coincidence? In fact,

4 Historically, these galaxies are often named after the constellations in which they are found. This naming system is somewhat clumsy—what do you do when there is more than one found in a single constellation? (see Figure 13.3 for the answer)—and, especially, of late, not always followed. But it will do for our present discussion. To add to the confusion, and because saying 'dwarf spheroidal galaxy' is a mouthful, these galaxies often referred to by just the name of their constellation. So, the Sculptor Dwarf Spheroidal Galaxy (see Figure 13.2) becomes just 'Sculptor,' and so on.

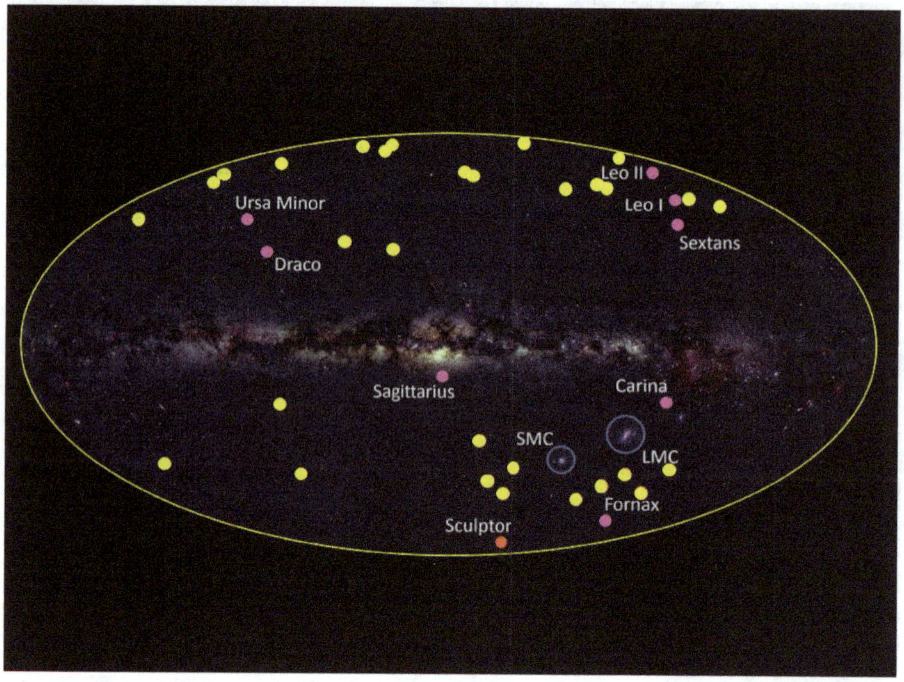

Fig 13.4

An image of the sky from Earth showing the locations of all known dwarf satellite galaxies of our Galaxy (as of mid-2015). The LMC and SMC are centered within the blue circles. The pink dots (and the one red dot) denote the largest of the dwarf spheroidal satellite galaxies, labeled by the constellation in which they are found (and, in the case of the Leo galaxies, appended by a roman numeral). These systems were all discovered before 1995. The yellow dots show the locations of dwarf galaxies discovered since then. The somewhat uneven distribution of the yellow dots—there are more dwarf galaxies known in the North Galactic Hemisphere corresponding to the upper half of this image—reflects the fact that sensitive searches for dwarf galaxies have only recently begun in the Southern sky. There is also an apparent concentration of dwarfs near the Magellanic Clouds, perhaps a hint that these galaxies—like our home Galaxy—have their own retinue of satellite systems. Our current Destination—Sculptor—is denoted by the red dot near the South Galactic Pole at the bottom of the map.

Image by Axel Mellinger.

a total of about 40 galaxies[5] similar to Sculptor are found within 250 *kpc* of the Galaxy. Not a large number, particularly considering how vast a volume this represents, but certainly not zero. Along with the Magellanic Clouds, these objects constitute a population of *Satellite Galaxies* associated with our Galaxy (Figure 13.4). Which tells us something rather surprising. This location, so barren we reasonably wondered if there was *anything* here, is not empty after all. Here, far outside the Galactic Disk and the Galactic Bulge, we find ourselves within a third major component of the Galaxy. Known as the *Galactic Halo*, this component is the largest and, in many ways, the most enigmatic of the Galaxy (see Table 13.1 for a comparison of the three principal components of the Galaxy).

5 It may seem odd that there is uncertainty with such a small number. But there are two reasons for this. First, as recently as the mid-1990's, a total of nine of dwarf spheroidal galaxies similar to Sculptor were known to be associated with our Galaxy. In recent years new surveys of the sky have uncovered more systems at a rate that now exceeds one per year! The census of these objects is changing all the time, quite possibly different when you read this footnote from when it was written (see Figure 13.4). The second reason is a bit more subtle. Although Sculptor is clearly more luminous than M 92, some recent discoveries are fainter than globular clusters such as M 92. One galaxy is even less luminous than the Pleiades! As a result, it is starting to become tricky to know what's a dwarf galaxy and what's a cluster. Some systems currently believed to be dwarf galaxies may be reclassified as clusters, and vice versa. If this seems strange, consider Pluto, which was recently 'demoted' to dwarf-planet status, consistent with recent discoveries of other planetoids in the outer Solar System. Of course, the bodies involved—clusters, dwarf galaxies, planets, or dwarf planets—do not change in nature simply because they are reclassified by humans! But the distinctions implied by these classifications may still be important, as they may help us understand how different types of objects formed and evolved. We shall see this is the case when it comes to more general schemes of galaxy classification (Destination 16).

13.2 DARK MATTER IN THE GALACTIC HALO

Even out here, 80 *kpc* from the Galactic Center, most of the satellite galaxies of the Galaxy are far away from us. Only Sculptor can be said to be 'nearby', and yet even it is 10 kpc from us, further than the distance from the Sun to the Galactic Center. Moreover, the only reason Sculptor is even that close is that we *chose* to travel to a destination that placed us relatively near it. It turns out that these satellites seem to be fairly randomly distributed within a huge spherical volume surrounding the Galaxy, ranging from about 20 *kpc* to 250 *kpc* from the Galactic Center (see Figure 13.5). Filling such a vast volume with about 100 small satellite galaxies—a number that aims to account roughly for the many dwarf systems likely remaining to be discovered—means that any given dwarf will typically be located about 50–100 *kpc* from its nearest neighbor. That's similar to our current distance from the Galaxy.[6]

From the distribution of satellite galaxies, it's clear they do not comprise an extension of the Galactic Disk. Nor do they define some other disk-like system within the Halo. This means that—reminiscent of what we saw in M 92 (Destination 11)—we can't assume the orbits of these satellites around the Galaxy are circular, oriented in a plane, or even traveling in the same direction. That's too bad, because the dwarf satellites probe the most distant regions of the Galaxy. They can, in theory, provide the best means of determining the total mass of the Galaxy. We *could* try to do this by measuring a complete orbit of even one satellite system, but these galaxies have orbital periods of around 2–3 *Gyr* or longer! That's probably more time than we'd like to wait.[7] Fortunately, there is an alternative. We can be clever and measure things that are comparatively easy to determine—for example, the radial velocities, positions, and distances of the individual galaxies—then use that information to determine the total mass of the Galaxy. But to do this, we need some new tools.

---

6 The true distribution of these galaxies is not really random. Dwarfs are comparatively rare within about 70 *kpc* from the Galactic Center, and their distribution falls off outside that distance with increasing radius. But the fact remains that a typical dwarf satellite is many tens of *kpc* away from its nearest neighbor. It really is lonely out here in the Halo.

7 Actually, It's not nearly as bad as this. As we'll see shortly, we can measure radial velocities for individual galaxies. And, if we are reasonably patient—a few years, not *Gyr*—we can also measure proper motions for some of these galaxies (the motion across our line of sight; see Destination 9). This lets us estimate their orbital paths. The problem is that the measurements are still very difficult, and restricted to only the closest galaxies. The resulting orbits we derive are not very precisely determined. Still, it's a start, and this approach can produce a pretty good idea of a galaxy's orbit after 'just' 10–20 years of effort. The technique we explore here, however, is also applicable to cases where we have no hope of *ever* measuring the even a portion of a galaxy's orbit.

When we estimated the Galaxy's mass from the Sun's orbital motion around the Galactic Center (Destination 9), we resorted to our workhorse, Kepler's Third Law. We would have gotten exactly the same result had we used a relation known as the *Virial Theorem*. To appreciate how this approach works, let's consider some particles moving under the influence of their mutual gravity. Perhaps these are stars within a cluster, or maybe dwarf galaxies in the Galactic Halo. The Virial Theorem tells us that a simple relationship holds between the gravitational energy of these particles and the energy of their motion,

$$-U = 2K.$$

The term U stands for the gravitational potential energy of the particles, while K represents their total kinetic energy. This equation tells us that these two forms of energy—gravitational and kinetic—are intimately related in a self-gravitating system in a remarkably simple manner. But, there is a catch. For the Virial Theorem to be valid, the particles must be in *equilibrium*. That is, the particles must be in stable orbits, not all contracting inwards toward the center of mass of the system, nor all expanding outward from the center. A star cluster such as M 92 (Destination 11) is in gravitational equilibrium. Although every particle (star) in the cluster is in motion, the cluster will appear essentially unchanged millions, even billions, of years from now (see Figure 11.11). A collapsing protostar (Destination 6) or a supernova (Destination 12)—both rapidly changing their configurations—are not in gravitational equilibrium. We can legitimately analyze the cluster with the Virial Theorem, but not the protostar or the supernova.

To see more clearly how the Virial Theorem works, let's considering a minimal gravitational system consisting of just two objects. We can prove from Newton's Universal Law of Gravity that the gravitational potential energy of the two bodies—one of mass m, the other of mass M, and separated by a distance R—is $U = -GmM/R$. If we further assume that $M >> m$, then the bigger body is essentially stationary.[8] Consequently, the kinetic energy can be written as $K = \frac{1}{2}mV^2$, where V is the velocity of the smaller body relative to the larger body (which we have assumed is located very close to the system's center of mass). This is similar to the cases of a Planet orbiting the Sun (Destinations 2 and 7), or the Sun orbiting the Galaxy (Destination 9). Combining these results for U and K, we can rewrite the Virial Theorem as

$$M = \frac{RV^2}{G}.$$

8 This is by no means a critical assumption, but it does greatly simplify the analysis.

This is *precisely* the same relation for the velocity of an object like the Sun orbiting the Galaxy (Destination 9). But there's a subtle difference. So long as the objects are in stable orbits—and regardless of the shape of the orbits—we can use the Virial Theorem even if all we know is the separation between the bodies and their relative velocity at any moment.

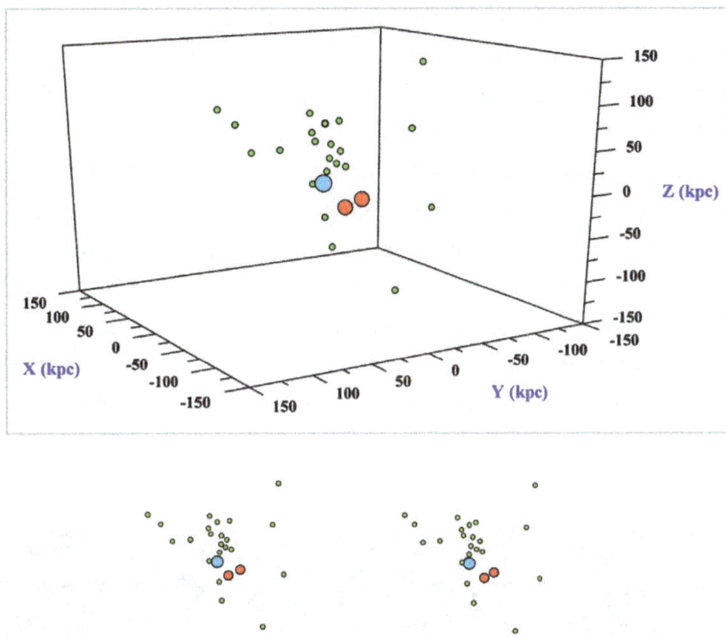

Fig 13.5

(*Top*) A 3-D plot of the satellite galaxies surrounding the Milky Way. The MW itself is the central blue dot, while the two red dots are the Magellanic Clouds. The smaller green dots show the locations of the known dwarf satellites as of October, 2008. Notice the scale, which ranges from −150 to +150 *kpc* in each axis. This is a *big* volume. (*Bottom*) These two images form a stereo pair of the galaxy distribution shown in the upper figure. Together, these figures provide a three-dimensional 'view' of the distribution of dwarf galaxies around the Milky Way. To see this, place your eyes about 6–9 inches from the page, then let them relax so that the two lower images merge. When they do, the galaxies will pop out from the page! The effect is not subtle. When you see it, you'll know!

This sounds perfect. By measuring the distance of a dwarf galaxy from the Galactic Center and measuring its velocity, we can use the Virial Theorem to estimate the Galaxy's mass. Unfortunately, we can't really do that. We rarely measure the total velocity, *V*, for a dwarf galaxy, but rather only its radial velocity, its motion toward or away from us. The problem is that interpreting the radial velocity of a single dwarf galaxy is ambiguous. A small radial velocity may mean the dwarf is moving slowly around a relatively low-mass parent. Or, it might mean the dwarf just happens to be moving across our line of sight, possibly at high velocity (see Figure 13.6). Unless we know the total velocity of the object, we can't distinguish between these cases, even though one implies the Galaxy's mass is low, while the other that the mass is high. But, there is a way to get around this problem. A statistical approach.

To get a sense of how this works, imagine a bag of rocks. We can count how many rocks there are in the bag as we weigh each one, after which we can easily calculate the average weight of rocks in the bag as simply the total mass divided by the number of rocks. Now, the average weight may be quite useful information, but there's more that we can extract from these measurements. To understand this, let's consider now a more specific example consisting of *two* bags of rocks, each containing four rocks with the following individual weights: (4,4,4,4) and (1,7,3,5). It's apparent that the *average* weight of the rocks, 4.0, is the same for the two bags. But it is also clear that the samples are not identical. In the first sequence, every number equals four, implying that the variation in the weights within the sample is zero. In the second sequence, the spread of values is clearly much larger. The variation or spread in the values can be defined mathematically as a quantity known as the *Dispersion* of the sample. The first sample has no variation in the weights, so its dispersion is zero. But the second sample clearly exhibits a spread in the weights and has a non-zero dispersion.[9]

We can apply the same analysis to radial velocity measurements to determine the *Velocity Dispersion* of a set of objects (see Figure 13.6). The beauty of the Virial Theorem is that it can be applied to any number of bodies if we use statistical estimates of their velocities and separations by replacing the velocity, *V*, with the velocity dispersion—which we'll denote as σ_V—and the orbital radius, *R*, with some sort of average separation—which we'll denote as $\bar{R}$. This is truly remarkable: By taking a snapshot of the random motions of particles in *any* sort of self-gravitating system that's in equilibrium, we can estimate the system's total mass. And without measuring orbits.

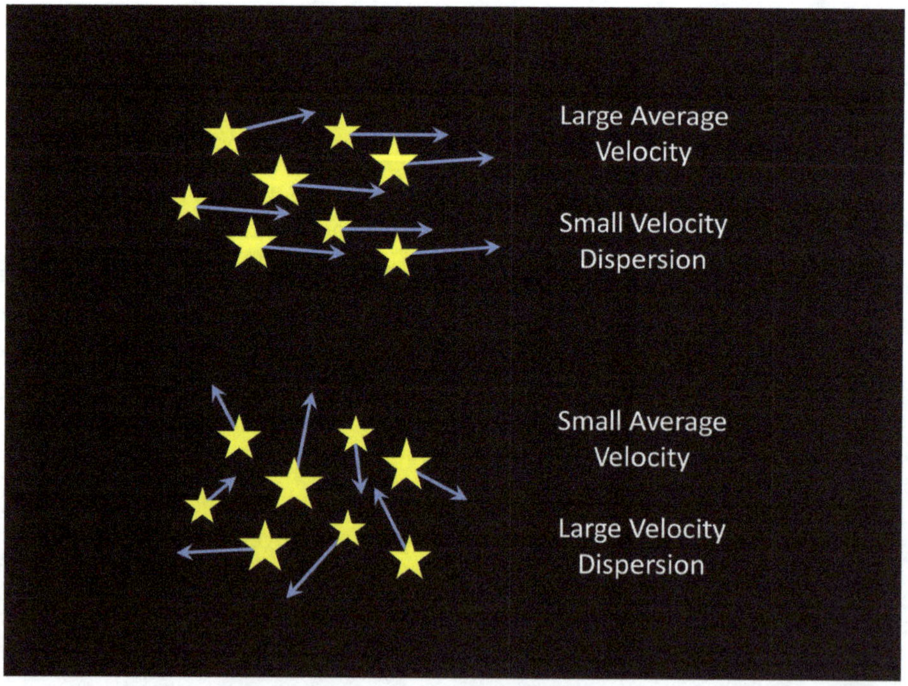

Fig 13.6

The spatial distribution of stars in the *top* and *bottom* groups in this diagram is identical. However, since the velocities of the stars in the *top* group—denoted by the direction and length of the blue lines—are very similar, the velocity dispersion of the group is low even though the average velocities of the stars may be rather high. This represents a situation similar to the Sun and neighboring stars moving along nearly circular in the Galactic Disk. If we used the velocity dispersion of these stars in the Virial Theorem, we'd grossly underestimate the mass of the Galaxy in which they are moving. Instead, we use the mean velocity of the stars—that is, their orbital velocity about the Galactic Center—to estimate the Galaxy's mass (see Destination 9). For the *bottom* group of stars, the velocities are both random in direction and variable in size. For this group, the velocity dispersion is large even though the average velocity may be small, or even zero. As long as this group of stars is gravitationally bound, we can calculate its collective mass using the group's velocity dispersion in the Virial Theorem.

This lower image could represent the motions of individual stars in a cluster or galaxy, or of individual dwarf satellite galaxies within the Galactic Halo.

9 Mathematically, there are a number of ways to define a dispersion. For one pretty standard definition (as described in a problem at the end of this destination), the dispersion of the weights of the second sample of rocks is 1.3. By the same token, since all the values of the first sequence equal the mean of the sequence, the dispersion in that case is exactly zero by any mathematical definition of dispersion. That just confirms that there is no spread in the weights of the rocks in that case.

This approach can work for stars moving within a globular cluster or a galaxy. And it can work for dwarf galaxies moving within the Galactic Halo. It turns out that we can measure the radial velocities of dwarf galaxies fairly easily; when we do, we find that their velocity dispersion is $\sigma_V \sim 220$ *km/s.* The symbol, '~,' which means 'approximately equal to,' reminds us that this is a statistical estimate that will have some inherent uncertainty no matter how precise the individual radial velocity measurements may be.[10] If we take the average distance of these galaxies from the Galactic Center to be around $\bar{R} \sim 150$ *kpc*, we can use the Virial Theorem (careful with the units!) to solve for the Galaxy's mass

$$M = \frac{\bar{R}\sigma_V^2}{G} \geq 1.6 \times 10^{12}\ M_\odot .$$

That's nearly two trillion solar masses! Keep in mind what this result refers to. The dwarf galaxy satellites are found out to about 250 *kpc* from the Galactic Center, so this result tells us the mass of our Galaxy out to a significant fraction of that distance. But not beyond. Consequently, the mass we derive from the Virial Theorem represents a *lower limit* to the Galaxy's total mass. That's why there is a greater-than-or-equal sign, '≥,' in the our estimate of the Galaxy's mass. It might weigh even more. We just don't yet have any way of measuring velocities further out.[11]

In one curious way, the mass we derive for the Galaxy is highly peculiar. As we look at the Galaxy from our current vantage point, it is hard to deny that it appears to be mostly *over there*. That is, all the stuff that emits radiation is localized in a comparatively small part of the sky that includes the stars and ionized gas that we see visually, and also all the gas and dust that we can detect through radio waves, microwaves, x-rays, or any other form of EM radiation. In contrast, such stuff appears to be essentially absent from this location.[12] Yet

10 There is a technical detail here. The actual dispersion we'd measure from radial velocities is more like 125 *km/s*. But if we assume the orbits are randomly oriented, we can expect that the radial velocity dispersion gives us a reasonable estimate of one of the three total velocity components of the galaxies. That is, a given galaxy could also be moving N–S and E–W, neither of which would affect the radial velocity. To correct for this statistically, we multiply the observed radial velocity dispersion by $\sqrt{3}$. Thus, our best estimate $\sigma_V = \sqrt{3}\sigma_{RV}$, is or about 220 *km/s*. This detail does not alter our basic results significantly, so it is safe to ignore it as you probably now wish to do!

11 Not quite. We'll see at Destination 14 that there is a way to estimate our Galaxy's mass, indirectly, beyond 250 *kpc*. That result is entirely consistent with what we are find from analyzing the motions of dwarf galaxies in the Halo.

12 The Coronal ISM—which we have seen is present out here—cannot by itself account for the Galaxy's huge mass. Despite the enormous volume that this gas fills, its very low density means that the total mass in this component of the ISM—perhaps as much as $3 \times 10^8\ M_\odot$—is a tiny fraction of the Galaxy's total mass. But the gas is not entirely irrelevant to this discussion. Recall that temperature reflects the speeds of particles in matter (Destination 5). If we convert the temperature of the Coronal ISM—around 1–2 million *K*—to a velocity dispersion for the atoms in the gas, we get around 150 *km/s*. That's very similar to the velocity dispersion for the dwarf satellite galaxies, meaning

TABLE 13.1 Principal Components of Our Galaxy

GALACTIC COMPONENT	LUMINOSITY ($L_\odot$) (1) / NUMBER OF STARS (1)	VISIBLE MASS ($M_\odot$) (1) / VISIBLE+DARK MASS ($M_\odot$) (1)	AGE RANGE	HEAVY ELEMENT ABUNDANCE	DENSITY LAW (2)	COMMENTS
Disk *Dest. 9*	*Luminosity:* 6.2×10^9 52%	*Visible:* 2.5×10^{10} 51%	0–10 *Gyr*. Still forming stars today at rate of $1 M_\odot/yr$ over entire Disk.	Similar to Sun to 1/3 of Sun's heavy-element abundance.	**Exponential** $H = 3.5\ kpc$ $h = 50 - 350\ pc$ $\rho_0 = 0.8\ M_\odot/pc^3$	• Sun located near Galactic Plane 8.5 *kpc* from Center • Most ISM of Galaxy is in Disk • 10% of Disk mass is in ISM • Width/thickness ratio ~20:1 • Seen as Milky Way in sky • Dominated by circular motions of gas/stars with flat rotation curve with $V_{circ} \sim 220\ km/s$
	Number: 5.0×10^{10} 45%	*Visible+Dark:* 2.5×10^{10} 1%				
Bulge *Dest. 10*	*Luminosity:* 4.6×10^9 38%	*Visible:* 1.8×10^{10} 38%	Mostly older than 10 *Gyr*. No current star formation except near the Galactic Center.	Ranging from 1/10 to over 5 times higher than Sun's heavy-element abundance.	**Power Law** $R_c = 1.5\ kpc$ $\rho_0 = 1.5\ M_\odot/pc^3$ *Power* ~ –4	• Dominant near Galactic Center • Contains stars with high chemical abundances • Ellipsoid shape • Mostly devoid of ISM, apart from near Galactic Center • Some rotation, but random stellar motions most common
	Number: 4.6×10^{10} 41%	*Visible+Dark:* 1.8×10^{10} 1%				
Halo *Dest. 13*	*Luminosity:* 1.1×10^9 10%	*Visible:* 5.7×10^9 11%	Ancient; stars possibly all older than 12 *Gyr*. No star formation today apart from some associated with dwarf galaxies.	Very low on average, ranging from 20% to much less than 1% of the Sun's heavy-element abundance.	**Power Law** $R_c = 5\ kpc$ $\rho_0 = 0.005 M_\odot/pc^3$ *Power* ~ –3	• Most extended component of Galaxy, out to at least 250 *kpc* • Approx spherical in shape • Dominated by DM • Globular clusters mostly belong to the Halo • Coronal ISM pervades Halo • Little rotation; dominated by random stellar motions • Much less luminous but far more massive than Disk+Bulge • Where dwarf galaxies and tidal streams reside
	Number: 1.6×10^{10} 14%	*Visible+Dark:* 2.6×10^{12} 98%				

Notes to Table 13.1:

1. The cumulative values for luminosity, numbers of stars, and visible mass are restricted to the inner 30 kpc of the Galaxy. The total mass (Visible+Dark), however, refers to an outer limit of 250 kpc. The percentages list how much each component contributes to the total (either mass or luminosity, depending on the column).

2. All the structural parameters for the Galaxy are uncertain, but these values describe fairly well the relatively contributions of the three principal components. The density parameter ρ_0 used here refers to the Galactic Center; the same symbol is used in Destination 9 but in that case it refers to the density of stars at the location of the Sun.

the motions of dwarf satellite galaxies—isolated, tiny islands in this otherwise empty ocean of darkness—are telling us that there's still something here. Something that contributes a lot of mass but very little radiation. We've encountered the need for *Dark Matter* before to explain the flat rotation curve of our Galaxy (Destination 9). But that was within the Galactic Disk. In that region of the Galaxy we could account for at least half of the mass with stuff we could see, leaving open the possibility that Dark Matter was just something we had overlooked. Something like very low-luminosity stars, or planets, or perhaps unmatched socks.[13] Stuff we understand but just happens to be hard to see.

Out here, that argument fails. The visible Galaxy amounts to no more than about 5% of the total mass we have measured here in the Halo. The motions of distant dwarf galaxies imply that vast majority of the Galaxy's mass must be in the form of Dark Matter. Sadly, having traveled here to the Galactic Halo doesn't really help us figure out the nature of this enigmatic stuff. We look around us. There's no evidence here of some vast horde of low-luminosity stars or planets that might contribute to the Galaxy's mass.[14] Black holes with masses up to $10^6 M_\odot$—similar to the one at the Galactic Center—are not present out here. Really massive black holes could do the trick in theory, but they would significantly distort the Galactic Disk in a way we do not observe. Out here in the Halo, far from the visible portions of the Galaxy, it is no longer

---

that the gas is also gravitationally bound to the Galaxy, though probably more confined closer to the center. If we were to apply the Virial Theorem to the gas particles, we'd get a very similar mass as before, about $10^{12} M_\odot$. The Coronal ISM is not Dark Matter. But its existence tells us Dark Matter is present.

13 OK, probably not socks. But they must go somewhere!

14 This is not a mere conjecture; there are direct measurements that rule out many of these sorts of objects as the source of Dark Matter in the Galactic Halo. One approach relies on gravitational lensing—the bending of light as it passes by massive objects—to identify unseen objects as they pass between a background star and Earth. Such *microlensing*, as this phenomenon is called, has been detected (see Destination 7 to see how microlensing can be used to find extrasolar planets). But the detailed results do not reveal any additional star-like or even planet-like objects in the Halo than what we already know about. Microlensing doesn't rule out unmatched socks or some other form of 'normal' matter on a tiny mass or size scale as the source of Dark Matter. But, as we will see at Destination 24, observations from very early times in the evolution of the Universe do.

reasonable to think we have missed that much 'normal' matter.[15] Perhaps Dark Matter isn't 'normal' stuff at all but composed of so-called *Weakly Interacting Massive Particles* (WIMPS) that effectively do not interact with normal matter except through gravity. Despite intense effort, experiments to detect such particles on or near Earth have, as of 2015, failed to detect WIMPS or any other form of matter that could conceivably comprise the Galaxy's Dark Matter.

Even if we can't figure out what it is, we can still broadly characterize the properties of Dark Matter in the Galaxy from our observations. One common way to do this is to introduce the *Mass-to-Light Ratio*, typically denoted as M/L (pronounced 'M over L'). This is defined as the total mass of something divided by its total luminosity. It is usually most convenient to provide this ratio in 'solar units.' That is, we measure mass in solar masses ($M_\odot = 2.0 \times 10^{33} g$), and luminosity in terms of solar luminosities ($L_\odot = 3.9 \times 10^{33}\ erg/s$). For the Sun, that makes it simple: $(M/L) = 1M_\odot/1L_\odot = 1$. For the inner components of the Galaxy—that is, the Galactic Disk and Bulge—we found a mass of around $10^{11}M_\odot$ (Destination 9, Table 13.1). The total luminosity of these components from all the EM radiation they emit (see Table 13.1) is about $0.9 \times 10^{11}L_\odot$. The resulting mass-to-light ratio of the inner Galaxy is therefore about 1.1, very similar to that of the Sun. But when we include the Galactic Halo—which we have seen is predominantly in the form of Dark Matter—the Galaxy's mass is nearly 10 times larger with virtually no change in the total luminosity (see Table 13.1). Out to our present location at about 80 *kpc*, the Galaxy's mass-to-light ratio is around 10. If we consider the total mass of the Galaxy out to 250 *kpc*, we infer a higher mass-to-light ratio of around 25–30. Mass-to-light ratios of around one or less are consistent with 'normal' matter—that is matter composed of stuff similar to what the Earth, Sun, and ISM are made of. But significantly larger values of M/L—values of ten or higher—imply the presence of considerable Dark Matter. In this sense, the M/L ratio offers a convenient way to state what we already concluded: The Galactic Halo is dominated by Dark Matter. We'll use this often as we explore other galaxies, galaxy groupings, and even the Universe as a whole.

There's one other interesting feature of Dark Matter we can explore here. Just as in the Halo, we can measure the radial velocities of stars within the Sculptor Dwarf Spheroidal Galaxy—the galaxy we've parked beside here in the Halo—and use the Virial Theorem to estimate its mass. We find that Sculptor

---

15 That is, matter made up of protons, neutrons, and electrons. In this sense, even black holes can claim to be 'normal' since they probably formed as the result of the evolution of massive stars that were themselves composed of normal stuff. Since protons, neutrons, and electrons are examples of subatomic particles known as *Baryons*, 'normal' matter is often called *Baryonic Matter* by astronomers and physicists. For a variety of reasons (see previous footnote), it is widely assumed that dark matter is composed of some sort of *non*-baryonic matter.

has a mass-to-light ratio of around 10–15, a value that suggests that that this insignificant little satellite galaxy also contains Dark Matter! But if we carry out the same measurements for a globular cluster—for example, M 92 (Destination 11)—we find a mass-to-light ratio of about 1.5, consistent with no Dark Matter. Is the presence of Dark Matter what *defines* a galaxy? Possibly. As we'll see at future destinations, all dwarf galaxies similar to Sculptor and all systems such as our Galaxy appear to possess Dark Matter. *Every single one, with no known exceptions.* And it does not end there. All structures larger than individual galaxies—including the Universe itself—appear to also contain a lot of Dark Matter. For now, we must live with this mystery. But make no mistake. Dark Matter is not some minor curiosity of our Universe. We'll discover that the existence of our Galaxy—the pale patch of light we see glowing faintly in this Alien Sky—would likely not have occurred in the first place without the help of Dark Matter.

13.3 THE STRUCTURE OF THE GALACTIC HALO

We've commented that there are very few stars out here in the Galactic Halo. But there are some. In fact, having kept a record of how many stars we saw as we traveled to our present location, we find that the Halo—like the Galactic Bulge (Destination 10)—obeys a Power-Law Density Law given by

$$\rho_{Halo}(R) = \rho_0\left(1 + \left[\frac{R}{R_{Core}}\right]\right)^{-3}.$$

This relation is 'shallower' than for the Bulge since the power-law exponent for the Halo law, −3, is more positive than for the Bulge (which you may recall was equal to −4; see Destination 10). The core radius, R_{Core} of the Halo—about 5 *kpc*—is also considerably larger than for the Bulge, while the central density, ρ_0—approximately $0.005M_\odot/pc^3$—is much lower (see Table 13.1). These structural parameters tell us that stellar content of the Halo, tenuous as it is, extends much further out from the Galactic Center than either the Galactic Disk or the Galactic Bulge (see Figure 13.7).

But it's very important to appreciate that this density law describes the distribution of the *visible* component of the Halo. The *mass* in the Halo must be distributed differently. Since the effective rotation curve of the Galaxy remains flat out to at least 80 *kpc* (where we're now located) and probably well beyond[16], Newton's Universal Law of Gravity (Destination 9) tells us that the mass density in the Halo must also vary as a power law with the form $\rho_{Halo}(R) \propto 1/R^2$. Since the stellar density falls off faster than this (as $1/R^3$), the Galaxy becomes progressively 'darker' the further out we travel from its center (see Figure 13.9). We've already seen that this is the case by noting that the mass-to-light ratio of the Galactic Halo increases steadily with increasing distance from the Galactic Center. This seems to be a fundamental characteristic of all galaxies, even tiny galaxies like Sculptor: When Dark Matter is present in a system, it seems always to extend over a far larger volume than the visible matter.[17]

Fig 13.7

A schematic view of the Galactic Halo with an image of a galaxy similar to ours at the center. The red dot shows a location about 80 *kpc* from the Galactic Center, about equal to our location at our present destination near the dwarf galaxy Sculptor. This image illustrates that the Halo (the white-gray area) is far more extended than the Galactic Bulge and the Galactic Disk, and that the Halo's density drops off steadily outward (see Table 13.1). Since the Halo is essentially a spherical structure, it would appear practically the same as we see here no matter from what angle we were to view it. There is also a glaring error in this diagram. Because the Halo has such a low density of stars, it has to be depicted here about 1000 times brighter than it actually is. Otherwise, we'd not be able to see it in this figure. Finally, it is important to realize that the central, brightest portion of the Halo has been suppressed in this figure so we can see the central Galactic Disk. In reality, the density distribution of the Halo peaks at the Galactic Center.

If we were to survey the Halo more fully, we would discover that the power-law stellar density law we've described is too simplistic. In some places we find that there are relatively more stars than this relation predicts. We saw a similar effect in the Galactic Disk (Destination 9) where variations in stellar density—particularly among the most massive, youngest stars—revealed the presence of long, coherent features that comprise spiral arms (see too Destination 14). The 'overdense' regions in the Halo are different. Rather than spiral features, they appear to align themselves along extended *Streams* that crisscross the

16 We did not actually measure the Galaxy's rotation out here in the Halo, but the velocity dispersions we did measure can be reasonably 'translated' to an equivalent rotation velocity. When we do this, the result is consistent with the flat rotation curve that we measured directly in the Disk, extending out into the Halo.

17 Again, 'visible' here refers to any radiation from any part of the EM Spectrum.

Halo in a complicated and seemingly random way (see Figure 13.9). Some Halo streams lie perpendicular to the Disk. Some appear more closely aligned with the Galactic Plane. Some follow broad, elongated loops. Some are more nearly circular in shape. In a few cases, Halo streams are clearly associated with individual satellite galaxies. One example is the *Sagittarius Stream*, which clearly originates from the Sagittarius Dwarf Spheroidal Galaxy, the closest external system to our Galaxy. Though this particular dwarf galaxy is small and compact, the stream that extends from it stretches more than once across the entire sky (see Figures 13.9 and 13.10)![18]

Fig 13.8

This schematic image that shows the known extent of the Dark Matter Halo around our Galaxy. The (tiny!) inner visible region contains the Galactic Bulge, the Galactic Disk and, more extended, the visible central parts of the Galactic Halo (see Table 13.1). The red dot denotes a location about 80 *kpc* from the Galactic Center, the distance we are from the Center at the present destination. The Dark Matter is depicted here as the large black region surrounding the visible Galaxy. As with the visible Galactic Halo, the Galaxy's Dark Matter Halo appears to be essentially spherical in shape. It would appear the same regardless of our viewing angle—*if* we could see it! Since the mass density of the Dark Matter in the Halo falls off from the Galactic Center less steeply than the visible matter, the Galaxy becomes progressively darker—with an increasing mass-to-light ratio—further out from the Center. Beyond about 250 *kpc*—the region depicted in pink in this image—we have little information on the Dark Matter distribution in our Galaxy, though there's no reason to doubt the Dark Matter Halo extends beyond that distance. There are no galaxies known—not even galaxies as small as Sculptor—that lack Dark Matter Halos.

The dwarf galaxy/stream relationship is an important clue to the origin of the extended Halo structures. Since they are so small in mass compared to the Galaxy, dwarf satellite galaxies experience strong tidal forces as they pass through the inner parts of the Halo and pass near the Galactic Disk and Bulge. These tides cause stars to be shed from the satellite to produce the coherent streams we see in the Halo (as illustrated in Figure 13.10).[19] This process

18 Another famous example is the *Magellanic Stream*, which, as the name implies, is associated with the Magellanic Clouds. Unlike the Sagittarius Stream, the Magellanic Stream does not appear to contain any stars, but is delineated instead by emission from clouds of neutral H emitting 21-*cm* radiation (see Destination 5). The origin of the Magellanic Stream is probably more complex than the Sagittarius Stream, but it is no less impressive, extending over 120 *deg* over the sky.

19 Tides arise from the difference in the gravitational force from one side of a body to the other. This means that tides become more severe as the distance between bodies starts to become comparable to the sizes of the bodies. Thus, as a dwarf galaxy gets very close to the Milky Way, the difference in the gravitational force on the near and far sides of the satellite relative to the Galaxy can becomes stronger than the internal

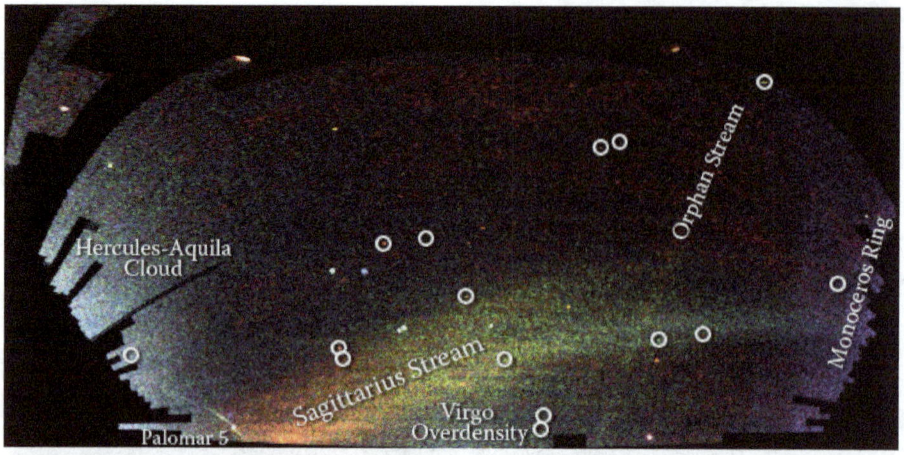

Fig 13.9

A map of the stellar density in the Galactic Halo over about 10% of the sky near the North Galactic Pole as seen from Earth. The lighter regions represent areas in the sky where there are relatively more stars in the Halo (though still very few). At least five streams are noted in the image, including the Sagittarius *Stream* that spans the entire sky, and the Palomar 5 stream that is associated with a globular cluster. The origins of the stars that populate the Monoceros, Orphan, and GD-1 Streams remain unknown. Toss in the Virgo Overdensity and the Hercules-Aquila Cloud and the many dwarf galaxies discovered in this region of the sky (white dots in the image) and it becomes apparent that the Halo, seemingly nearly empty and simple in form, is actually a pretty complicated structure.

Image from V. Belokurov/ Cambridge/SDSS.

also builds up the stellar component of the outer Galaxy as stars lost from the dwarfs are assimilated into the Halo itself. We will discover later that the way dwarfs appear to be merging into the Galaxy's Halo today resembles the processes by which all galaxies formed during the early history of the Universe (Destination 20).

For a structure that appeared at first so empty, the Galactic Halo has turned out to be filled with surprises. It harbors numerous satellite galaxies orbiting the Galaxy, some being shredded by gravitational tidal forces. The Halo is where most of the mass of the Galaxy seems to reside but very little of its light. And the Halo is telling us something about how galaxies form. We will have to travel quite far away and back into time to explore the details of this process. But one thing is clear. Even apparently barren locations can reveal profound features of the Universe. The Galactic Halo won't be the last example.

---

gravitational force of the satellite that is trying to hold the smaller galaxy together. When this happens, stars become 'unbound' from the satellite and start to follow independent orbits about the Galaxy. When enough stars are lost in this manner, they can form long streams that extend both in front and behind the main body of the original satellite. The Sagittarius Stream is the most spectacular example of this effect in the Galactic Halo.

Fig 13.10

An artist's impression of the structure of Sagittarius Stream as it winds around the Galactic Disk and Bulge. Note that the actual stream is shown much brighter than it actually is—otherwise we'd not see it in this picture—but in other respects the image is roughly to scale. The complex structure of this stream is the result of tidal forces that have relentlessly torn the Sagittarius Dwarf Spheroidal Galaxy apart over the past few billion years, steadily adding new stars to the Halo.

Artwork by D. Martínez-Delgado/MPIA and G. Peréz/IAC.

DESTINATION 13 QUESTIONS

13.1. At 220 *km/s*, how long would it take a typical halo satellite to complete one orbit with a radius of 80 *kpc*? If this is considerably shorter than the age of the Galaxy—about 14 *Gyr*—then it is reasonable to use the Virial Theorem to estimate the Galaxy's mass using the satellite dwarf galaxies. Do you find that it *is* reasonable to do so?

13.2. Let's say that the mass of the Galaxy is 'only' $10^{11}M_{\odot}$. What velocity dispersion would we have expected to find for the dwarf galaxies? Assume that the average distance of these galaxies from the Galactic Center is 150 *kpc.*

13.3. Use the Main-Sequence Mass-Luminosity relation (Destination 8) to determine the mass-to-light ratio of a $20M_{\odot}$ star and for a $0.1M_{\odot}$ star. Assume both stars are on the Main Sequence. Which of these could conceivably be a source of dark matter?

13.4. The very hot (Coronal) gas of the Galactic Halo supposedly comes from supernovae explosions. Let's see if this is reasonable.

a) Assume that the Coronal ISM is present out to 100 *kpc* from the Galactic Center at an average density of 10^{-7} H atoms per cubic *cm*. What is the total mass of gas in this very hot phase?

b) Now, assume that a supernova goes off every 100 *yr* in our Galaxy. Let's further suppose that each explosion ejects $1M_{\odot}$ of material into space that eventually becomes part of the Coronal ISM. If the Galaxy has been producing supernovae at this rate for the past 10 *Gyr*, what is the total mass that they could have injected into the Coronal ISM?

c) Have supernovae been able to produce enough material to account for the Coronal ISM?

d) What other tests might you envision that could address the question of whether the Coronal ISM comes mostly or entirely from supernovae?

13.5. Imagine that a galaxy is composed of an exponential disk (Destination 9), a bulge (Destination 10), and a halo (Destination 13). Calculate the relative densities at the following radial distances. In each case, assume you are in the plane of the disk and that the relevant structural parameters are the same as we found for our Galaxy. At each radius, clearly indicate which component dominates the local stellar density. You can answer this problem with a plot if you prefer.

a) $R = 0$ *kpc.*

b) $R = 1$ *kpc.*

c) $R = 8$ *kpc.*

d) $R = 20$ *kpc.*

e) $R = 150$ *kpc.*

13.6. The Crab Nebula, the remnant of a supernova (see Destination 12), is expanding outward from its origin at about 3000 *km/s*. If the radius of the nebula is about 2 *pc*, calculate its mass using the Virial Theorem. Given the likely origin of the nebula as the explosion of a high-mass star, explain why the result you determined from the Virial Theorem provides or does not provide a good estimate of the nebula's mass.

13.7. Estimate your M/L ratio. **HINT:** Estimate your surface area and assume you radiate like a thermal body as described by Planck's Law (Destination 6). That information allows you to calculate your luminosity and you know your mass from your weight. Express the result in Solar units. Based on your calculations, are live humans likely a good candidate for dark matter?

13.8. Look up on the web descriptions of the properties of any one of the named galaxies in Figure 13.4. Report on some of the basic information for the galaxy you chose. Be sure to include relevant data for the galaxy's distance from the Sun, its luminosity, its total stellar content, its metal content, its size, or any other of its fundamental characteristics.

13.9. Empty as it is, the Halo is by far the largest component of the Galaxy.

a) Compare the volumes of the Galactic Disk and the Galactic Halo. A reasonable assumption for the disk is that it is a thin cylinder, while the Halo is a sphere. Justify any values for the dimensions of either structure that you might need to calculate their volumes.

b) If the 10^{11} stars of the Disk were uniformly distributed in the volume of the Halo, how far, on average, would the stars be located apart from one another?

13.10. Convert the 220 *km/s* velocity dispersion of the Halo to a gas temperature using the relation in Destination 5 that relates the gas kinetic energy to temperature. Assume that the Halo gas is composed of Hydrogen atoms only. What does this result say about the coronal gas of the Halo and, in particular, whether it is gravitationally bound to the Galaxy or not?

13.11. For this problem, refer to the attached table of the radial velocities and distances (as measured from the Galactic Center) for some of the dwarf galaxies in the Galactic Halo.

a) What is the mean velocity, $\langle V \rangle$, and the velocity dispersion, σ_V, of this sample? You can determine the dispersion in various ways, but one common approach is to determine for each galaxy the square of the velocity difference between its radial velocity, V_i, and the mean velocity of the sample, then sum these up and divide by the number of galaxies, N. That is,

$$\sigma_V^2 = \frac{3}{N} \sum (V_i - \langle V \rangle)^2 .$$

In case it is new to you, the symbol Σ means 'take the sum of what follows.' The factor of three accounts for the fact that we measure the radial velocity dispersion only from Earth. Express the dispersion, σ_V, in units of *km/s* and *cm/s*.

b) Calculate the mean distance of these galaxies from the Galactic Center in units of *kpc* and *cm*.

c) Use this information to estimate the Galaxy's mass out to this mean distance you found in part (b). Express your results in units of *g* and $M_{\odot}$.

d) Compare your result with the estimate of the mass of the Galaxy given in the text of Destination 13. Explain any differences between the two mass estimates.

DATA FOR Question 13.11

SATELLITE GALAXY	RADIAL VELOCITY (*km/s*)	DISTANCE FROM GALACTIC CENTER (*kpc*)
Sculptor	84.3	79
Leo I	187.1	254
Carina	32.5	103
Fornax	–19.6	141
Ursa Minor	–104.1	68
Draco	–120.7	76
Sextans	88.0	94
Leo II	23.5	225

13.12. Prove that the Virial Theorem works for the Earth. You can do this in a general way using symbolic analysis, or you can plug in values for the orbital properties of the Earth, its mass, and the mass of the Sun to show that the gravitational and kinetic energy of the Earth are equal in the manner specified by the Virial Theorem.

Destination 14

THE NEARBY SPIRAL GALAXIES M 31 AND M 33

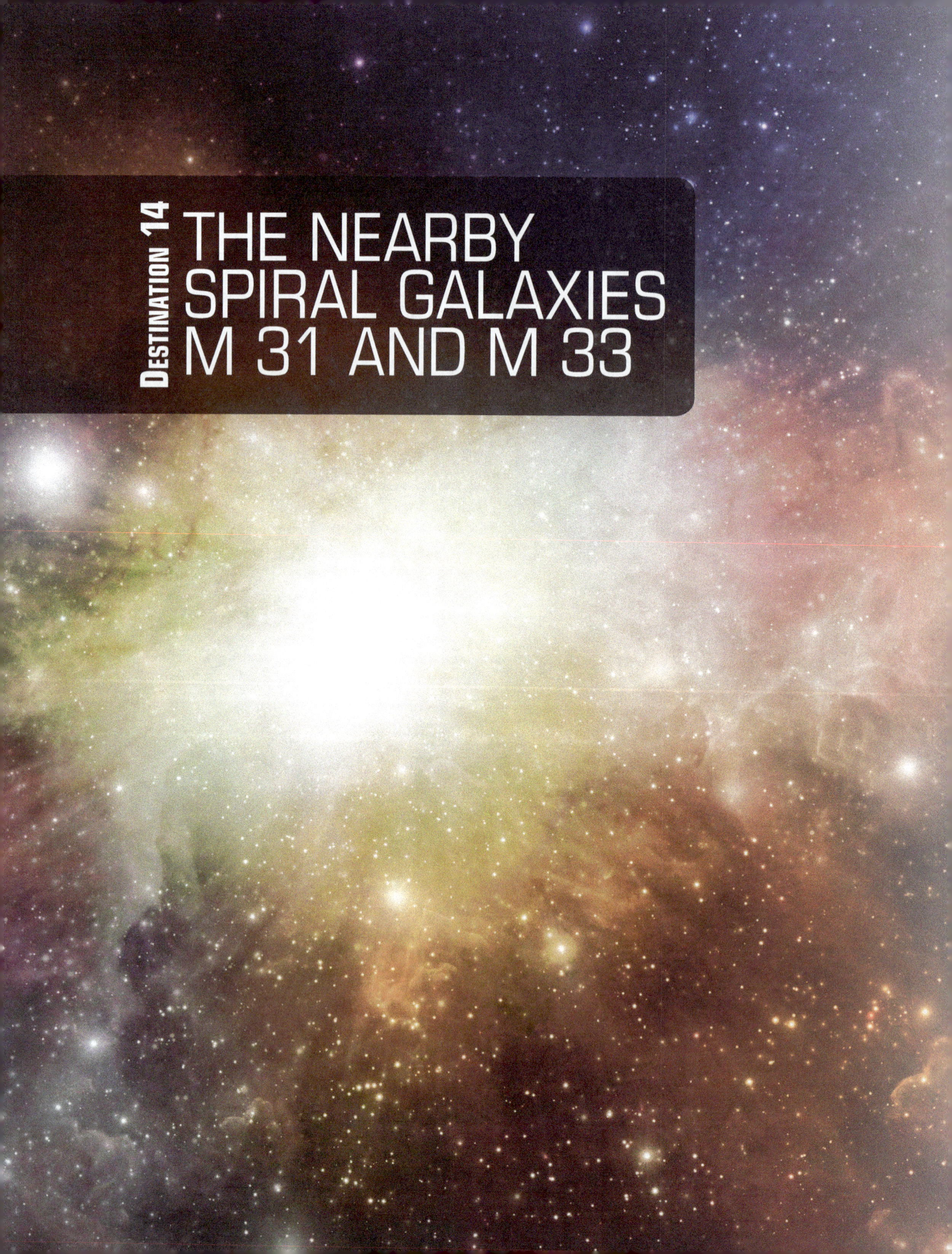

14.1 LEAVING OUR GALAXY BEHIND

Very strange. Unsettling, really. Having just left our lonely outpost in the Galactic Halo (Destination 13), we now find ourselves about 100 *kpc* from the center of our Galaxy on the way to our next destinations, M 31 and M 33. The choice of where to go on this journey was pretty obvious. Of the five places easily visible in the sky from the Halo, we had already explored the Galaxy, the LMC, and had taken a pass on the SMC. Only M 31 and M 33—whatever they are—were left for us to explore. And yet, experienced as we have become to visiting new, *very* distant places, we feel more than the typical trepidation and excitement of a voyage into the unknown. There is something about this particular trip that makes us more nervous than usual: The distance we have to travel. We already discovered in Destination 13—from our vantage point in the Galactic Halo—that M 31 and M 33 must be extremely far away. And now, at a distance some 100 *kpc* from the Sun, both objects have hardly changed from how they appear from Earth (see Figure 14.1). Essentially the same shape. The same size. The same brightness. From experience (see Destinations 3 and 4), we know this means that both M 31 and M 33 must be much more distant from the Sun than 100 *kpc*. But how much more distant? That's what we still don't know. *That's* what's so unnerving.

As we can guess from their names, both M 31 and M 33 are included in Charles Messier's catalog of comet imposters. But it's unfair—and incorrect—to give credit of their discovery to Messier. Both objects are visible by eye from Earth (Figure 14.1) and both have been known since antiquity. M 31, the brighter of the two, was long referred to as the *Andromeda Nebula* for the constellation[1] in which it's located, and is relatively easy to see with the naked eye from any reasonably dark site. M 33, noticeably dimmer, was also known as the *Triangulum Nebula*.[2] It can just be seen without optical aid by people with good vision and under very dark skies. But we can excuse Messier's enthusiasm. Even in a small telescope, M 31 is clearly elongated in shape, reminiscent of a comet with a short tail, while M 33 looks somewhat like some eruptive comets (see Figure 14.1). Like all objects in Messier's catalog—and no doubt, much to

1 We have already described the celestial soap opera that Andromeda—daughter of King Perseus and Queen Cassiopeia—is entangled in back at Destination 3. Of course, at our present location far from the Galactic Disk, even the notion of 'constellations' makes little sense. There aren't *any* easily visible stars here, let alone patterns of stars!

2 Some constellations do not have particularly creative names. Triangulum is the Celestial Triangle.

Messier's disappointment—neither M 31 nor M 33 moved relative to the stars (see Figures 11.1 and 14.1). Not after a day. Not after many years. And not even by very much after we have traveled far into the Galactic Halo![3]

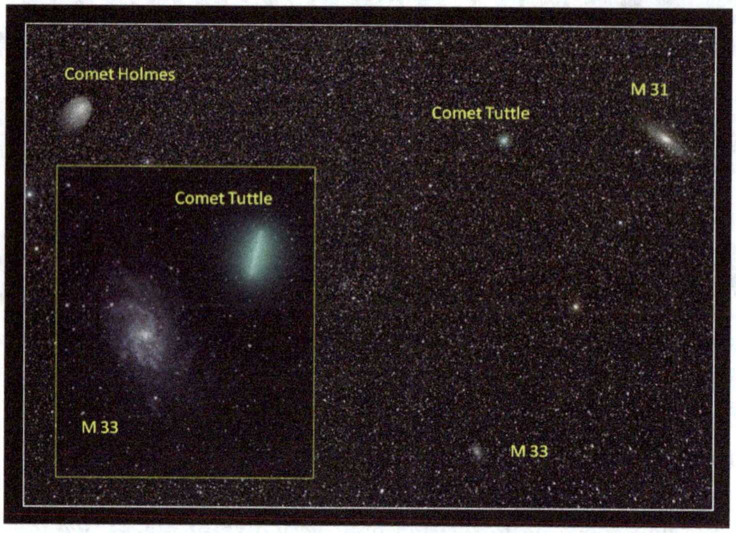

Fig 14.1

(Background) A wide-field view of the sky showing M 31 and M 33, along with two comets, Comet Holmes and Comet Tuttle, in a photo taken in Dec 2007. Messier would have enjoyed finding the comets, but he considered the other two objects to be nuisances. We can see from this figure that they really do mimic the appearances of comets. There is also a star cluster near the center of this image, NGC 752. *(Inset)* Comet Tuttle—a few days after the background photo was taken—passing by M 33. The camera tracked the stars, causing the comet to appear as an elongated blur as it moved relative to the stars during the exposure. Compare this image to Figure 11.1, which was taken at about the same time, but with the camera tracking the comet. In that image, it's the background stars and M 33 that appear smeared.

Inset photo by E. Kempen. Background photo by R. Ligustri.

About the only good thing to be said of a long journey like this is that we have a lot of time to sightsee. And since there's so little else to look at in this sky—and since they are our current destinations—we focus our attention exclusively on M 31 and M 33. As we train our various telescopes at them—telescopes that let us study these objects over the entire EM Spectrum (Destination 5)—we come to realize that both M 31 and M 33 are complex objects. Both are filled with stars (Figure 14.2), the brightest of which are typically very blue and often associated with emission nebulae similar to the Orion Nebula (Destination 6; Figure 14.2). These are obviously associated with star-forming regions, some of which are evidently immense in extent. We can also spot compact star clusters that remind us of M 92 (Destination 11; Figure 14.2), but which must be far more remote since they look so small from here and because their stars are so faint. M 31 possesses a centrally bright region mostly devoid of young, blue stars, but that contains many red stars, reminiscent of the Bulge of our Galaxy (Destinations 9 and 10). Both M 31 and M 33 contain extensive distributions of neutral gas that we can detect from 21-*cm* radiation (Destination 5; see Figures 14.7 and 14.9). We also can plainly see dark molecular clouds that obscure some of the light of the galaxies in the optical, and which emit far-infrared and microwave radiation (Destinations 6 and 10; see Figure 14.7). As we identify more of these details—features we've seen before in our Galaxy and in the Large Magellanic

3 Given how far away M 31 and M 33 are—which we'll determine a bit later—it would in fact be easy to estimate their distances after having traveled 100 *kpc*. You can explore some methods that would allow you to do so in some of the questions at the end of this destination.

Cloud—it becomes clear that both M 31 and M 33 are independent *galaxies*, distinct islands of stars, gas, and, as we'll see, dark matter, that inhabit the vast, empty space far beyond the limits of our Galaxy (Figure 14.3). Not just mere 'nebulae': M 31 is the Andromeda *Galaxy*, and M 33 is the Triangulum *Galaxy*.

For us—at our present location far from our Galaxy—this conclusion may seem obvious. But from Earth, and before telescopes could make out the details we've observed in both M 31 and M 33, it was not at all clear that these objects—or others like them found spread over the sky—were located beyond our Galaxy. As astronomers of the late 19th and early 20th centuries started to discern some of the details we've seen, they too began to suspect that M 31 and M 33 had to be very distant. If we assume—as the astronomers from those eras did—that some of the features we've identified in M 31 and M 33 are similar to objects found in our Galaxy and the Magellanic Clouds, we can use them to estimate the distances to their parent galaxies. For example, if some of the star clusters in M 31 are the same physical size as M 92, we can compare their apparent sizes and, knowing the distance and size of M 92, derive the distance to M 31 (where the clusters reside) by using the Small-Angle Formula from Destination 4. Likewise, we can identify the bluest, brightest stars in M 33 and compare those to what we believe are similar stars in our Galaxy or the Magellanic Clouds—say, from their colors or spectra—assuming them to have the same luminosity. The differences in brightness would then reflect differences in distance (see the Inverse-Square Law for Brightness from Destination 7). These methods were useful to reveal over a century ago that M 31 and M 33 must be many *hundreds* of *kpc* from Earth. That also helped to explain why more direct methods of obtaining distances—for example, parallax measurements (Destination 3)—had utterly failed when applied to M 31 and M 33. Both objects are so remote that their parallaxes are far too small to measure, a conclusion that's still true today.

Fig 14.2

As we study M 31 and M 33 in detail, we discover they both contain examples of all the sorts of places we have visited in our Galaxy and the LMC, including prominent exponential disks composed of myriad individual stars (NGC 206 is a particularly bright region of the disk of M 31), spectacular globular clusters (G1, located in the halo of M 31, has about 50 times more stars than M 92), and star-forming regions (the giant emission nebula NGC 604 in M 33 is 100 times more luminous than the Orion Nebula). M 31 and M 33 are full-fledged galactic systems, distant islands independent of our Galaxy.

Photos by B. Mochejska-Hedberg/U. Warsaw (NGC 206) and STScI/NASA (G1 and NGC 604).

These methods of estimating the distances to M 31 and M 33 are certainly useful; they provide us with a sense of the true scale of the journey we have undertaken. But the result causes us to take pause. At our present location—about 100 *kpc* from the Galactic Center—it appears that we have traversed only about 10–20% of the distance to M 31 and M 33! But can we get a more precise estimate of their distances? After all, we can't be certain that the clusters in M 31 really are identical to the ones—such as M 92—in our Galaxy. Or that the brightest blue stars of M 33 have the same luminosity as their counterparts in the Magellanic Clouds.[4] Our distance estimates for both galaxies might be off by a factor of two or more. Not so great from a navigational standpoint. How can we do better?

As we continue our detailed study of M 31 and M 33—what else is there to do on this very long voyage?—we soon spot something quite astonishing. The galaxies are *changing* with time. And not merely due to our travel toward them. Long before the galaxies appear to increase in apparent size or brightness as we move closer, we notice something more subtle. Some of the individual stars in M 31 and M 33 are changing their brightness (Figure 14.4). We've encountered examples of *Variable Stars* before. The most spectacular are supernovae, which change brightness quickly and radically, signaling the explosive destruction of a high-mass star (Destination 12).[5] But the stars we've spotted in the two galaxies are not examples of such *Eruptive Variable Stars.*[6] Instead, they vary in a highly

4 Such methods of getting distances to galaxies are sometimes called *Secondary* methods. They depend on our knowing the distances to some fiducial (reference) objects—for example M 92 or the luminous stars in the Magellanic Clouds—and assuming that the objects we know the distances to are intrinsically the same as the one we want to get the distances of. What makes these approaches secondary is that we can't get the distances to the fiducial objects using direct methods. Trigonometric parallax—the best example of a *direct* method of getting distances—does not work for objects as remote as M 31 and M 33 (see Destination 3), or even for M 92 or the Magellanic Clouds. *Primary* distance methods also use fiducial objects, but in these cases the distances to those are obtained with direct methods. We'll summarize a full range of distance estimators—ranging from direct, primary, secondary, and even tertiary methods—at Destination 18.

5 Not all supernovae result from the explosion of a high-mass star. Some are believed to occur when white dwarfs in binary systems merge, in the process explosively becoming a single neutron star. We'll explore these types of supernovae—known as Type Ia Supernovae (pronounced 'One-A')—later on. They turn out to be particularly useful for determining distances to very remote galaxies.

6 A large number of different types of so-called *Eruptive Variable stars* are known. There are supernovae—which we have already discussed, see the previous footnote and Destination 12—but also novae and dwarf novae plus some other rather unusual stars that undergo large-scale brightness variations. The mechanisms that cause these stars to vary in brightness include instabilities in accretion disks, unstable thermonuclear fusion at the surfaces of stars, violent ejection of mass, stellar mergers of stars, and even the formation of soot (!) to name a few. Many of these systems are visible through their emission of x-ray radiation, an indication of the very high temperatures and extreme conditions these phenomena can produce. It would be fascinating to visit some of these variable stars from up close, but we already have a pretty full itinerary. Too many destinations, too little time!

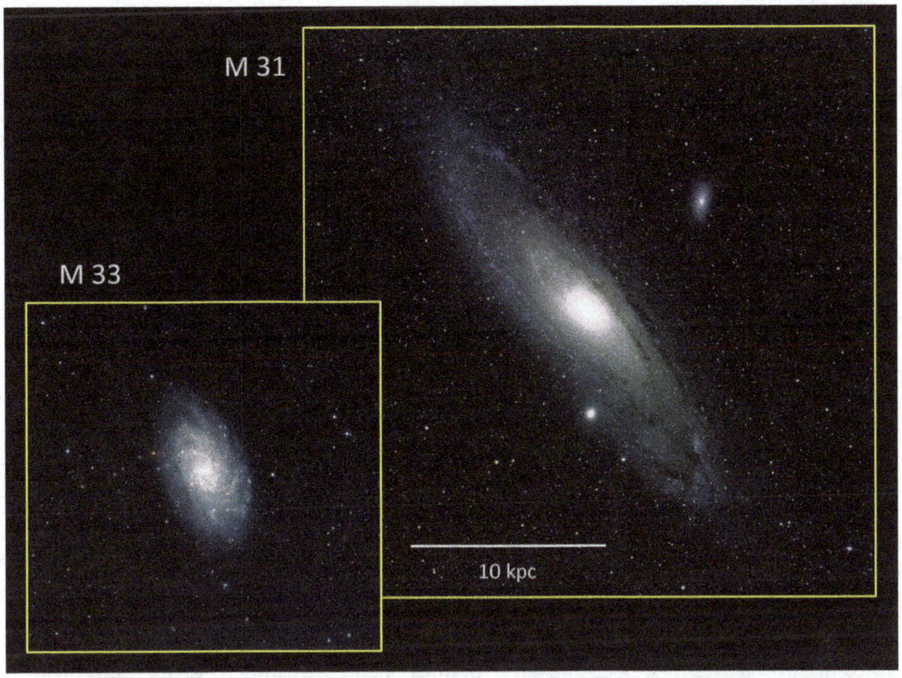

Fig 14.3

Magnificent optical views of M 31 and M 33. The two images reproduced here are shown on the same scale to illustrate the true relative sizes of the two galaxies. There are two obvious satellite galaxies near M 31: NGC 205—located at the 1 o'clock position relative to the center of M 31—and, M 32—located almost directly below the center of M 31, near the edge of the galaxy's disk. Interestingly, M 33 has few, if any, satellite galaxies of its own. Certainly none as evident as NGC 205 or M 32.

Photos from D. de Martin, DSS/Palomar Observatory (M 33) and NOAO/AURA/NSF (M 31).

regular manner, periodically brightening and dimming on timescales of days, weeks, even years (see Figure 14.3). We've seen regular brightness variations before in pulsars, but those occur on timescales of seconds or less (Destination 12). Eclipsing binaries (Destination 8) can exhibit brightness variations on timescales ranging from days to years as these stars do, and we certainly can spot some eclipsing systems in both M 31 and M 33. But the stars that are currently commanding our attention in the two galaxies are different. And they are the key to getting the precise distance estimates for M 31 and M 33.

14.2 CEPHEID VARIABLE STARS: A NEW WAY TO ESTIMATE EXTRAGALACTIC DISTANCES

As we study these variable stars, we see that they display some remarkable features. Their *Light Curves*—that is, plots of how their brightness vary with time—typically exhibit a characteristic sawtooth pattern (Figure 14.5). This contrasts with most eclipsing binaries, which have constant brightness when

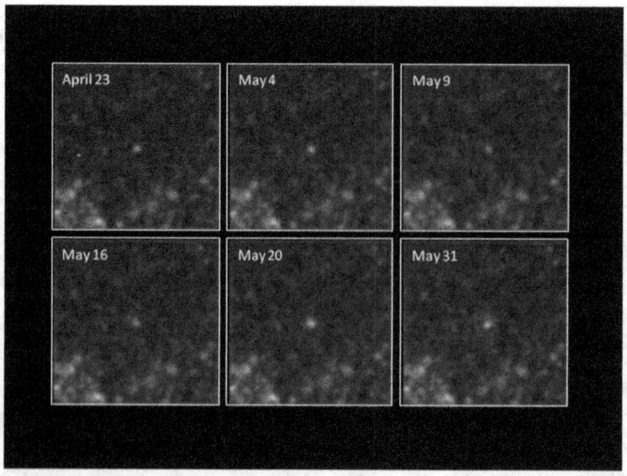

Fig 14.4

Six images of a Cepheid in a nearby galaxy (in this case, M 100). Notice how the Cepheid, the isolated star at the center of each image, has obviously changed in brightness over a period of a bit more than one month. A careful analysis of these changes reveals that these stars vary periodically in brightness over well-defined periods. The star shown here has a period of about 50 days. These variations are considerably more sedate than the spectacular pyrotechnics of a supernova (Figure 12.9, Destination 12), and not as rapid as the periodic variations in brightness seen in pulsars (Destination 12).

Image mosaic from STScI/NASA.

the stars are not eclipsing one another.[7] The intervals between successive times when the stars are brightest—the *Period* of their brightness variations—are typically quite stable for any given star. Many of these variable stars, particularly ones with yellow and orange colors, exhibit periods ranging from as short as 2–3 *days* up to around 100 *days*. Because their properties are similar to those of a famous bright variable star located near Earth—the naked-eye star known as δ Cephei—these kinds of variable stars are commonly referred to as *Cepheids*.[8] What's most remarkable about them, however, is that for the Cepheids in M 31 and M 33, there seems to be a tight relationship between the stars' periods and their average brightnesses. The longer the period, the brighter the Cepheid appears.

Now, this seems very odd. Brightness, after all, is something we perceive as the result of combining a star's distance and its intrinsic luminosity, a fundamental point we discovered when we formulated the Flux-Luminosity-Distance relation at Destination 7. Since the distance of a given star can be anything, even if all stars had the same luminosity—and we know that's not the case—they would still have different brightnesses. So how can the period of a Cepheid possibly correlate with its brightness? The period is something *intrinsic* to the star—similar

7 There is a class of eclipsing binaries—known as *Contact Binaries*—in which the stars are in physical contact. The proximity of the stars to one another produces strong tidal effects that distort their shapes in a manner similar to what we saw in LMC X-3 (Destination 12). These stars aren't even remotely spherical, and as they orbit one another, they appear to change shape over time from our vantage point. These stars thus vary continuously in brightness, unlike most eclipsing binaries. The light curves of contact binaries—which are known to exist in M 31 and M 33—are fairly regular in appearance—not the sawtoothed pattern we see in the variables we are currently studying—nor do the colors of contact binaries vary in the same way. It's easy to tell them apart.

8 *δ* Cephei was one of the earliest stars to be discovered to be variable. Like some of the other variable stars—such as Mira, the 'Marvelous' star, and Algol, the 'Demon' Star—*δ* Cephei is visible to the naked eye. The British astronomer John Goodricke—a man who was profoundly deaf, but who had, uncharacteristically for that era, received an excellent private education—discovered its variability in 1785 when he was 21.

to the luminosity—and would be the same regardless of how far away the star is located. But brightness does depend on distance. It should vary randomly.

Even from our crude distance estimates we've used so far, we know that we are still far from both M 31 and M 33. *Really* far away. If we make an educated guess based on what we know about the sizes of our Galaxy and the Magellanic Clouds, we can assume that both galaxies are 10-30 *kpc* in size. Since both are hundreds of *kpc* distant, this means that all the stars in each galaxy are essentially at the *same* distance.[9] This is equivalent to saying that the differences in distance to the Eiffel Tower and the Cathedral of Notre Dame in Paris are irrelevant if you live in San Francisco. Only when you get to Paris do the actual distances between these locations start to matter. In a similar way, from the Sun—or even from our current location 100 *kpc* from the Sun—all the Cepheids in M 31 are effectively at the *same* distance, equal to the average distance of the galaxy. Ditto for M 33.

This has an important consequence that helps us reveal a fundamental—and very useful—characteristic of the Cepheids we are monitoring in M 31 and M 33. We know from the Flux-Luminosity-Distance Relation (Destination 7) that the flux, or brightness, of some star—let's call it 'star A'—is given by $F_A = L_A/4\pi D_A^2$. If we consider a second star—we'll call it 'star B'—then its flux can be written in a similar way as $F_B = L_B/4\pi D_B^2$. We can combine these results—by dividing the equation for star A by the one for star B—to calculate the ratio of the brightnesses of the two stars as

$$\frac{F_A}{F_B} = \frac{L_A}{L_B}\left(\frac{D_B}{D_A}\right)^2.$$

If both stars are in a distant galaxy, then for all practical purposes, $D_A = D_B$, and so we can rewrite this formula as

$$\frac{F_A}{F_B} = \frac{L_A}{L_B} \quad \text{if } D_A = D_B.$$

That is, the ratio of the brightnesses equals the ratio of the luminosities *if the stars are located at a common distance.*

The variable stars we see in M 31 and M 33 now make more sense. Since they are all at essentially the same, large distance for a given galaxy, the fact that brighter Cepheids have longer periods must be because they have higher luminosities. Likewise, the stars with the shorter periods—the ones that appear faintest—must have the lowest luminosities. This relationship is known as the *Cepheid Period-Luminosity Relation*, or just the *Cepheid PL Relation* for short (Figure 14.5). It was discovered by American astronomer Henrietta Leavitt (Figure 14.5) who, in the early 20th century, found a clear relationship between the periods and brightnesses of Cepheids in the Small Magellanic Cloud (SMC;

9 This point is so important and so pervasive in astronomy we discussed it before in footnotes in Destinations 6 and 11.

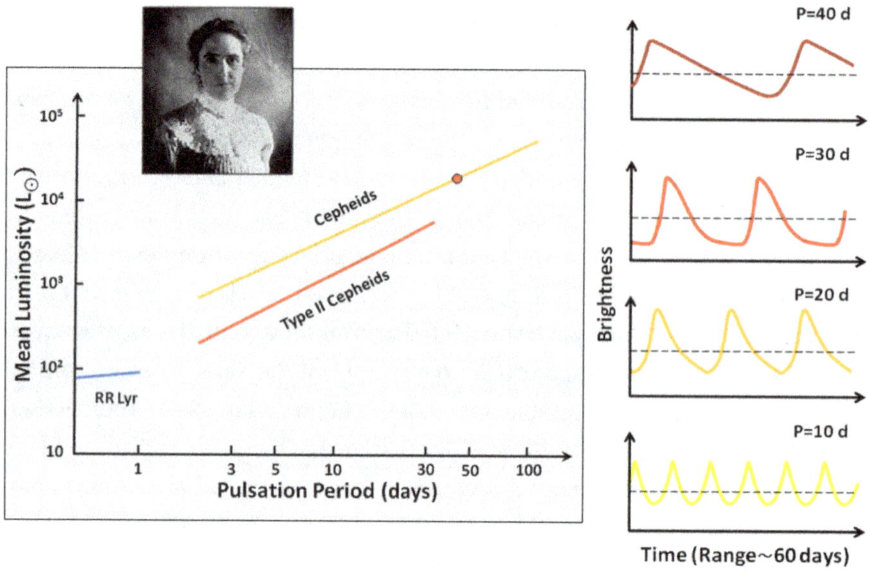

Fig 14.5

(*Right*) Examples of four light curves of Cepheids. In each plot, time advances to the right and brightness toward the top. The width of each plot is about 60 *days*. Typical brightness variations for Cepheids can be as large as a factor of four over a full pulsation cycle. Though Cepheid periods range from about 3 to 100 *days*—similar to the range seen in other classes of variable stars—these characteristic 'sawtoothed' light curves make Cepheids fairly easy to identify. The black dashed lines correspond to the brightness of each star averaged over their light variations. (*Left*) Period-Luminosity (PL) Relations for three types of pulsational variable stars, including Cepheids. The red dot shows the location of a 40-*day* Cepheid on its appropriate PL relation (with period shown along the *x*-axis of the plot). By simply measuring the period of the Cepheid, we can use the PL Relation to determine the star's luminosity (the *y*-axis scale). The other two PL Relations are for Type II Cepheids and RR Lyrae stars, pulsating stars commonly found in globular clusters such as M 92 (Destination 11). The inset photo shows Henrietta Leavitt, the astronomer who discovered the PL Relation for Cepheids in the early 20th century from her systematic studies of these stars in the SMC.

Destination 12). She realized—as we have for M 31 and M 33—that all the Cepheids in the SMC are practically at the same distance from the Sun. That is, the relation she found between period and brightness among SMC Cepheids revealed a relationship between period and *luminosity* for these stars (Figure 14.5).

The luminosity variations that Cepheids exhibit come about from physical pulsations of their atmospheres. These stars—typically yellow or red supergiants that are 3–30 times more massive than the Sun—literally swell up then shrink down in size during each pulsation cycle. In some cases, their radii vary by more than 50% in just a few days.[10] Cepheids are not the only stars that pulsate in this manner. So-called Type II Cepheids and RR Lyrae stars also pulsate. Each type of star—all much larger than the Sun but similar in mass—obeys a distinct PL Relation (see Figure 14.5). Virtually all types of pulsating stars[11]

10 Imagine the Sun doing that! It would be astonishing to see a star vary significantly in size so rapidly.

11 And there are other pulsating stars than just Cepheids (sometimes known as *Classical Cepheids* or *Type I Cepheids*), Type II Cepheids (sometimes known as *W Virginis Stars*), and RR Lyrae Stars. For example, *Dwarf Cepheids* are Main Sequence stars that pulsate on timescales of a few hours, while *ZZ Ceti Stars* are White Dwarfs that pulsate on timescales of minutes (!). A somewhat different class of pulsating stars known as *Long-Period Variables* consists of red giants and red supergiants that pulsate on timescales of months to years. Each of these types of pulsating variable stars—including their various subclasses—obey their own distinct PL Relations. Ironically, the one type of variable star with 'pulsate' in its name—pulsars—do not actually pulsate at all (see Destination 12)! As you've noticed, astronomical naming systems can be confusing!!

vary in brightness due to instabilities in their outer atmospheres, not because the energy they produce in their cores varies.[12]

Fascinating as Cepheids are, how do they provide precise distance estimates for M 31 and M 33? To make progress on this, we have to make the key assumption that Cepheids in one galaxy—such as ours—are fundamentally the same as Cepheids in another—such as M 31 or M 33. That is, we assume that Cepheids are what astronomers refer to as *Standard Candles*, objects that behave in the same way regardless of where they are located. As long as that's true—and it's a reasonable assumption given the physical structures of stars—then if we can determine the period of even one Cepheid in a distant galaxy, we can estimate its luminosity from the PL Relation (Figure 14.5). If we can also measure the star's average brightness (Figure 14.5), the Flux-Luminosity-Distance relation gives us the distance to the Cepheid. And to the galaxy in which it resides.

A practical problem with this approach is that we must have a good determination of the distance of at least one Cepheid before we can use the PL Relation to get distances to other Cepheids. And we must get the distance and luminosity of such a 'calibrating' Cepheid *without* using the PL Relation or else our reasoning would be circular. Historically, this step has proven tricky[13] and only fairly recently has it become possible to measure reliable trigonometric parallaxes—and therefore distances—for even the closest Cepheids. In addition, a few tricks for determining the Cepheid distances have been developed in recent years.[14] Taking all this into account, the best modern calibration of the Cepheid PL Relation can be written as

$$\log\left(L_{Ceph}\right) = 1.1\log\left(P_{Ceph}\right) + 2.5,$$

12 That would happen on timescales of thousands of years or longer, not days.

13 Of course, that did not stop people from using early version of the PL Relation, even when they suspected—or even knew—that the calibration of the relation was questionable. For example, early in the 20th-century, Edwin Hubble used a crude calibration of the PL Relation to determine that M 31 and M 33 are hundreds of *kpc* away. But, because of significant calibration errors in the PL Relation he adopted, his distances turned out to be about a factor of two smaller than our modern estimates. His results proved definitively that M 31 and M 33 are distinct galaxies from our own, but they led to some serious confusion when people tried to compare in detail the properties of those galaxies with ours.

14 One *extremely* clever example is the use of *light echoes* from a nebula surrounding the bright Cepheid RS Puppis. As this Cepheid pulsates, its light reflects off dust and gas that the star itself has ejected through a wind over the past tens of thousands of years. The result is a ripple-like expansion of bright rings out from the Cepheids as seen from Earth, an effect that can be seen spectacularly in movies of the star obtained over many months. The rate at which those ripples expand—combined with knowledge of the speed of light—allows one to estimate the star's distance to high precision. In the case of RS Puppis, its distance is 1993 *pc*, with an uncertainty of only 28 *pc*. It is currently the Cepheid with the most precisely known distance from the Sun.

where a Cepheid's luminosity, L_{Ceph}, is given in solar luminosities, $L_{\odot}$, and the pulsational period, P_{Ceph}, is in *days*. As we've seen, Cepheid periods range from about 3 to 100 days, so the PL Relation tells us the average luminosities of these stars range from about 1000 $L_{\odot}$ to 50000 $L_{\odot}$, implying they are quite luminous stars. That's a nice feature because it means Cepheids can be seen from far away, making them useful to determine distances to even comparatively remote galaxies.[15] Using the PL Relation, we can use our observations of Cepheids in M 31 to determine that the galaxy is about 780 *kpc* from the Sun, or about 2.5 *million* light-years. Because the calibration of the PL Relation is now quite good, we know this result is accurate to about 3%, or about ± 25 *kpc*.

So now we know. Long and arduous as our voyages within our Galaxy have been, traveling to M 31—or to M 33, which Cepheids reveal is 860 *kpc* from the Sun—is significantly more difficult. This is going to be a very *loooooong* journey.

14.3 THE GALAXY, M 31, AND M 33: THE CLOSEST SPIRAL GALAXIES

But, eventually, we do arrive. And our first stop is a spot in the inner halo of M 31, about 10 *kpc* from the galaxy's center. Stretching over nearly half of the entire sky, the huge glowing bulk of M 31 outlines a roughly circular shape, the light of its central regions considerably brighter and centrally concentrated. We are looking 'down' at the galactic disk and the galactic bulge of M 31, structures that are similar in appearance to the Disk and Bulge of our Galaxy when seen from a similar perspective (Destination 11). We discover that the disk of M 31 has an exponential light profile similar the Disk of our Galaxy (Destination 9)[16],

15 Quite generally, having a high luminosity is a particularly good characteristic for any standard candle.

16 You may have noticed what appear to be a few inconsistencies here. The exponential density law we defined at Destination 9 provides a way to estimate the mass distribution in the disk of our Galaxy. But, here at M 31, we are measuring the *light* distribution. We can get away with this if we assume that the mass-to-light ratio (Destination 13) of the stars that make up the disk do not change too much over the face of the disk. This turns out to be a reasonable assumption and so our use of the light from the M 31 disk to tell us about its mass distribution is not a bad one. The other inconsistency is that the exponential density law tells us about how much matter there is in a given volume (that's the definition of density). But we are not measuring the mass density of stars in M 31. Instead we are observing what's known as the *Surface Density* of stars in the galaxy. That is, we see all the stars along the line of sight as we look *through* the disk. What may not seem obvious—but is true—is that if you look

though with a larger scale length of about 5.5 *kpc*, compared to 3.5 *kpc* for our Galaxy (Destination 9). The light profile of the bulge of M 31 is also similar in form to what we see in our Galaxy but with a central density, ρ_0, that's two times higher: $3M_\odot/pc^3$ in M 31 compared to $1.5M_\odot/pc^3$ for our Galaxy (Destination 10). We notice too that M 31 has many globular clusters, about twice as many as our Galaxy. In short, we find the same major components in M 31—an exponential disk, a central bulge, and an extended halo—as in our Galaxy (Figures 14.3 and 14.6). But they are all a bit larger and a bit more luminous in M 31 than in our Galaxy.

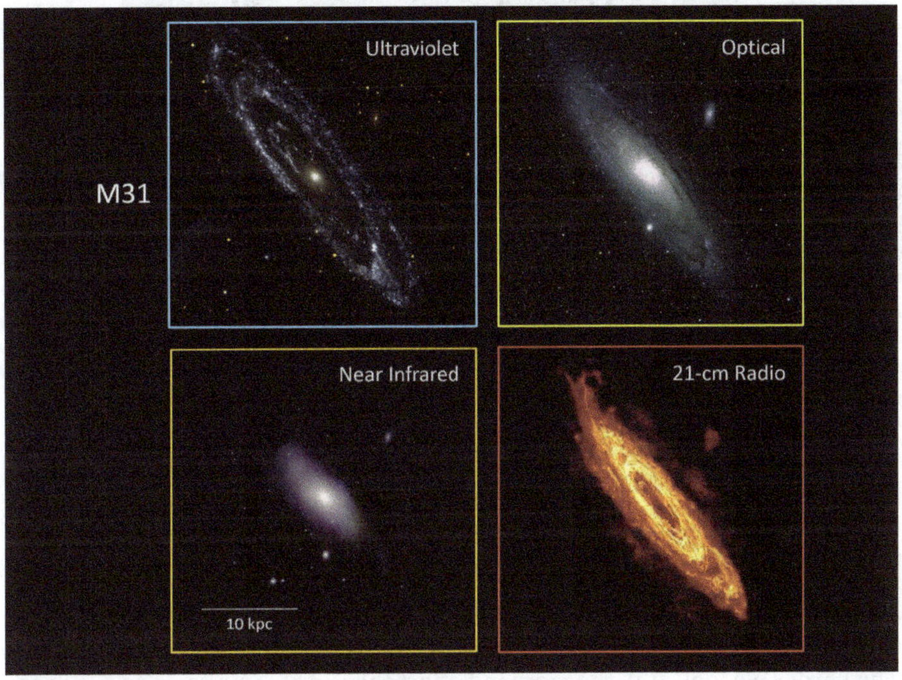

Fig 14.6

Images of M 31 in the light of radiation from four different portions of the EM Spectrum. All images are shown to the same scale as indicated in the lower left panel. (*Upper Left*) Ultraviolet radiation reveals the locations of the hottest, youngest, and most massive stars forming in M 31. These stars clearly trace immense, arced structures—spiral arms—within the galaxy's disk. (*Upper Right*) An optical image of M 31 (reproduced from Figure 14.3) reveals many components of M 31, including its bulge, disk, dark molecular clouds, and star-forming regions. (*Lower Left*) M 31 as seen in the near-infrared. This light comes mostly from red giants in the bulge of M 31. Few of the young stars and spiral features visible in the UV and optical images are apparent here. (*Lower Right*) This image shows the 21-*cm* radiation from M 31, effectively tracking out the distribution of the brightest parts of the Warm ISM in the galaxy's disk. This gas is clearly more spread out than the young stars in the UV image, though both are distinctly associated with the disk. The central 'hole' in the neutral H is easily seen here, a feature that M 31 shares with our Galaxy.

Images from GALEX/JPL/NASA (UV), NOAO/AURA/NSF (Optical), 2MASS/JPL/NASA (IR), and R. Braun et al./Westerbork (21-*cm*).

The differences between the two galaxies are not merely of comparative size. As we study the magnificent disk of M 31, we notice that it has pronounced spiral structures that extend gracefully outward from near the center of the galaxy. We saw evidence of similar *Spiral Arms* as we traveled through the Disk of our Galaxy (Destinations 9 and 10). But the arms in the M 31 disk seem longer, somewhat more distinct than in our Galaxy. This is ironic because, from Earth, the arms in M 31 are hard to see. But the disk of M 31 is tilted by about 77 *deg* from our line of sight from Earth, or nearly edge-on. That orientation makes it difficult to trace its arms clearly as seen from home. But it is not only the arms that make the disk of M 31 seem different to us. We notice that M 31's

---

through an exponential disk with a somewhat constant scale height of visible stars (Destination 9), the surface density will also obey an exponential law. And that's what we seen in the disk component of M 31 and M 33. We can be pretty confident they both possess exponential disks, just like our Galaxy.

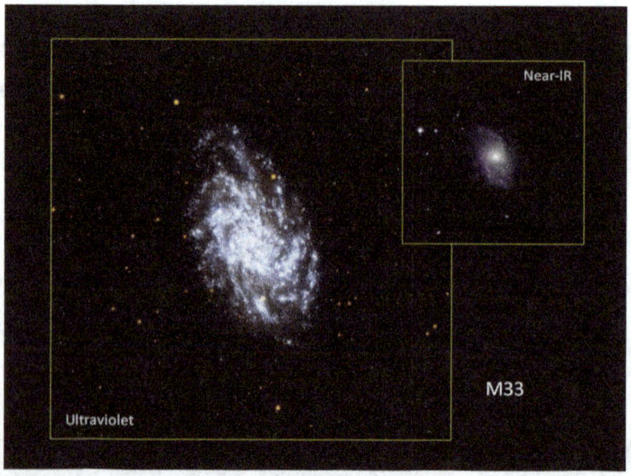

Fig 14.7

M 33 as seen in ultraviolet light (*left*) and in the near-infrared. Both images are shown to the *same scale*. It is clear that while M 33 has a well-defined disk with obvious spiral features traced out by its young stars, it has a minuscule bulge.

Image from GALEX/JPL/NASA (UV) and 2MASS/JPL/NASA (IR).

bulge—though similar in form to the Bulge of our Galaxy—is a comparatively more prominent feature in M 31. In our Galaxy, the total luminosity of the Bulge is about half that of the Disk. But in M 31, the bulge and disk have nearly equal total luminosities. This tells us is that M 31 is not simply a scaled-up version of our Galaxy (the details are summarized in Table 14.1). None of this should mask the key point: M 31 and our Galaxy are clearly similar in overall form. But they do exhibit definite differences.

To get a different perspective on the range of properties that large disk galaxies may exhibit, we shift our attention to M 33. Even from near M 31, we had already seen that M 33 is also a disk galaxy (see Figure 13.1). But we have now shifted to a spot about 5 *kpc* 'above' the center of M 33 where we see the full extent of M 33's disk face-on. As seen from Earth—where we have a similar, nearly face-on view of M 33—it's obvious why M 33 is often also referred to as the *Pinwheel Galaxy.* But there are some odd things about this galactic pinwheel. The harder we look at it, the more indistinct the spiral arms seem to be. These are not the long graceful arms of M 31, but more like arcs or segments of arms, broken up and detached from one another and from the center of M 33. The center! Suddenly, we realize the other strange thing about M 33. The spiral pattern of the arms shows clearly where the galaxy's center lies. But unlike M 31 and our Galaxy, there's no obvious bulge located at the center of M 33 (Figures 14.3 and 14.7).

What M 33 does certainly possess is an exponential disk. Its scale length is approximately 1.4 *kpc*, much smaller than the Galaxy (3.5 *kpc*) or M 31 (5.5 *kpc*). That's consistent with the overall size of M 33; it is clearly the smallest of these three galaxies. And if we look carefully, particularly taking advantage of the infrared radiation that luminous red giants emit, we do detect a tiny bulge-like structure at the center of M 33. But it seems almost an afterthought in this galaxy. The bulge of M 33—if that's really what it is—has

TABLE 14.1 A Comparison of the Largest Disk Galaxies of the Local Group

GALAXY PROPERTY	M 31	THE GALAXY	M 33	LMC
Total Luminosity	$2.0 \times 10^{10} L_\odot$	$1.2 \times 10^{10} L_\odot$	$4.1 \times 10^{9} L_\odot$	$9.1 \times 10^{8} L_\odot$
Disk Scale Length	5.5 *kpc*	3.5 *kpc*	1.4 *kpc*	1.4 *kpc*
Visible Radius	20 *kpc*	15 *kpc*	6 *kpc*	5 *kpc*
'Full' Extent (includes DM halo)	200 *kpc*	150 *kpc*	50 *kpc*	25 *kpc*
Disk Rotation Velocity	250 *km/s*	220 *km/s*	140 *km/s*	80 *km/s*
Disk Luminosity	$7.5 \times 10^{9} L_\odot$	$6.2 \times 10^{9} L_\odot$	$4 \times 10^{9} L_\odot$	$8.3 \times 10^{8} L_\odot$
Disk Visible Mass	$3.0 \times 10^{10} M_\odot$	$2.5 \times 10^{10} M_\odot$	$8 \times 10^{10} M_\odot$	$2.7 \times 10^{9} M_\odot$
Bulge Luminosity	$7.0 \times 10^{9} L_\odot$	$3.5 \times 10^{9} L_\odot$	$7 \times 10^{7} L_\odot$	$8.3 \times 10^{7} L_\odot$
Bulge Visible Mass	$3.5 \times 10^{10} M_\odot$	$1.8 \times 10^{10} M_\odot$	$3.5 \times 10^{8} M_\odot$	$4.2 \times 10^{8} M_\odot$
Bulge/Disk Luminosity Ratio	0.95	0.56	0.02	0.10
Total Gas Mass	$2.2 \times 10^{9} M_\odot$	$2.8 \times 10^{9} M_\odot$	$1.7 \times 10^{9} M_\odot$	$5.7 \times 10^{8} M_\odot$
Gas/Vis Mass Ratio	0.03	0.07	0.20	0.18
Total Mass to Vis Radius	$2.9 \times 10^{11} M_\odot$	$1.7 \times 10^{11} M_\odot$	$2.7 \times 10^{10} M_\odot$	$7.5 \times 10^{9} M_\odot$
M/L to Vis Radius	15	14	7	8
Total Mass to 'Full' Extent	$2.9 \times 10^{12} M_\odot$	$1.7 \times 10^{12} M_\odot$	$2.3 \times 10^{11} M_\odot$	$3.7 \times 10^{10} M_\odot$
M/L to 'Full' Extent	150	140	56	41

Notes to Table 14.1:

The dark matter halos of both M 31 and the Galaxy extend beyond M 33 and the LMC, respectively, so it is unclear how to define the total extents of the smaller galaxies. The values chosen are somewhat arbitrary but they illustrate the general trends.

a total luminosity at least 50 times fainter than the total luminosity of the disk (Table 14.1).

There can be little doubt that our Galaxy, M 31, and M 33 are all galaxies of a similar basic type. In fact, they are all examples of what astronomers refer to as *Spiral Galaxies* in acknowledgment of the prominence of the most obvious optical features in the disks of these galaxies (Figures 14.3, 14.6, and 14.7).[17] Even though we have visited only three spiral galaxies so far, we already see that such galaxies can exhibit real differences in size, luminosity, bulge-to-disk luminosity, and the prominence of the spiral arms. As it happens, these three galaxies—the closest spiral galaxies to the Sun—embody much of the range of properties that virtually all spiral galaxies exhibit.

14.4 WHAT ARE SPIRAL ARMS?

As we study the spiral arms in M 31, M 33, and our Galaxy, we come to the realization that these arms are, in some sense, an illusion. They certainly appear prominent to our eyes. But that's largely because the arms always seem to be closely associated with luminous, blue stars, themselves often associated with star-forming regions like the Orion Nebula (Destination 6). These stars act like track lights that highlight the arms' locations (Destination 8), particularly in the ultraviolet and optical regions of the EM spectrum where they copiously emit radiation (Figures 14.2, 14.6, and 14.7). The space between arms is *not* empty, however. There definitely are stars there, lots of them, but mostly in the form of redder Main Sequence stars—including middle-aged stars like the Sun—as well as red giants associated with the late stages of the evolution of low-mass stars (Destination 11). We can see these stars particularly well when we look at galaxies in near-infrared radiation where they emit much of their energy (see Figures 14.6 and 14.7 for M 31 and M 33, and Figure 10.5 for our Galaxy).

But it is clearly not just the bright blue stars that reveal spiral arms. The ISM, particularly the dark molecular clouds of the Cold ISM (Destination 6), often also beautifully trace out the arms in many spiral galaxies (Figures 14.2 and 14.8). The neutral gas—the dominant component of the Warm ISM—also commonly traces out arm-like features to some degree (Figure 14.6). And as

17 It is becoming increasingly common to refer to galaxies such as M 31, M 33, or our Galaxy as simply *Disk Galaxies*. This acknowledges the fundamental structure in which most of the stars and gas are typically found in these systems. It also reflects the fact that young 'spiral' galaxies may have recognizable disks before they form prominent arms (see Destination 20).

we've already noted, ionized gas from the Hot ISM is often found around the luminous blue stars that help define the optical extent of the arms. None of this obviously explains why the spiral arms exist, but at least there's some consistency here. To form young stars, warm gas must be compressed into denser cold clouds that then collapse to produce stellar nurseries such as the Orion Nebula (Destination 6). In some way, the arms in spiral galaxies appear to be triggering this process. Finally, there's an additional important feature of spiral arms worth mentioning, and that's their scale. In M 31, tracers such as molecular clouds (Figure 14.2) and young stars (Figure 14.6) can be tracked along arms that can be followed over distances in excess of 20–30 *kpc* end to end! Any sensible explanation for how spiral arms form—and why they are so common in so many disk galaxies—must account for all these observed features.

Fig 14.8

(*Left*) A beautiful example of bioluminescence in a lake in Australia. Microscopic organisms in the water emit light when they are agitated by waves that break along the shore. The brightest regions of light form long, coherent structures due to the wave action. (*Right*) The central regions of a nearby spiral galaxy (M 51) show beautifully the structure of spiral arms. In both arms visible here, it is apparent that the dark molecular clouds are distributed primarily along the inside—concave—edges of the arms. The pinkish H II regions (similar to the Orion Nebula, Destination 6), young blue stars, and older redder stars dominate the structure of the arms in sequence as one looks progressively from the inner edges of the arms toward the outer—convex—sides. The density wave in the ISM of the galaxy—the wave that, in analogy to the bioluminescence, instigates star formation—is moving more slowly than the stars and gas orbiting within the disk. As a result, the stars and gas of the disk 'catch up' to the wave, causing the observed progression from younger to older objects from the concave inner side of the arms to convex outer side.

Photo by P. Hart/www.philhart.com (*left*) and STScI/AURA/NASA (*right*).

There's a phenomenon back on Earth that might provide some useful insight to the problem of spiral arms. When a boat passes through warm waters, its wake can agitate tiny organisms causing them to emit light in a process known as *Bioluminescence* (Figure 14.8). At night, it is an eerie sight to see the faint glow of bluish-green light trailing behind the ship. Since the organisms can maintain the glow for only a short while, the light gradually fades behind the ship until the sea reverts to its inky blackness even though the plankton are still present in the water. In an analogous manner, a coherent wave moving through the ISM of a galactic disk could compress the gas as the wave passes by. This could then lead to the formation of high-density molecular clouds that then begin to collapse under their own gravity, ultimately forming stars. As in any star-formation process, the most massive stars will form first and shine brightly due to their high luminosities. But they will also be first to die. Like the bioluminescent trail, a galactic disk can light up in response to a passing compression wave, but only near the disturbance. And, again like the organisms in the water, though there is still plenty of gas left behind after the wave passes, there is no disturbance to initiate further star formation. As the wave passes on, only progressively older, fainter stars are seen. Is this how spiral arms behave in galaxies?

There are two difficulties with this idea. The first has to do with the shapes of spiral arms. To appreciate this problem, imagine three stars in our Galaxy positioned along a line extending out from the Galactic Center. Let's also imagine that one of the stars is the Sun so that its orbital velocity is about 220 *km/s* and its orbital radius is about 8.5 *kpc*. The second star is located halfway to the Galactic Center from the Sun, while the third is two times farther out from the Galactic Center than the Sun. Now, let's trace the motions of these stars for a while. After about 240 *Myr*, the Sun will have returned close to where it started. Because the Galaxy's rotation curve is flat (see Destination 9), the other two stars are also moving at about 220 *km/s*. Consequently, they will have traveled the same total distance over the same period of time. But they'll no longer be lined up. Because the circumference of its orbit is two times smaller than the Sun's orbit, the inner star will have completed *two* full revolutions about the Galactic Center. The outer star, on the other hand, has an orbital circumference that's twice the Sun's orbital distance, so it will have completed only half of its longer orbit. A long, very elastic cord, connecting the stars would start to trace out a beautiful spiral structure, just as we observe in spiral galaxies. The problem is that the process would not stop. Real galaxies like our Galaxy, M 31, and M 33 are much older—about 40 times older—than the time the Sun requires to complete an orbit of the Galactic Center. Thus, if spiral features in real galaxies were tied to the disk rotation, they would 'soon,' comparatively speaking, produce spiral arms that wrap around dozens of times around the galaxy. That's *not* what we see. Instead, the arms in spiral galaxies usually wrap at most twice, and often less, around the galaxy centers.

A second problem has to do with the direction in which spiral arms rotate in galaxies. We know from playing with various spinning toys—for example, whirligigs and pinwheels—that objects with flexible arms tend to rotate so that the arms trail behind the direction of rotation. We can measure the direction of rotation of spiral galaxies using the Doppler effect (Figure 14.9). In some cases this lets us confirm that the spiral arms in galaxies do indeed generally appear to trail behind their direction of rotation. Whew! At least that's reasonable. But there's still a subtle problem. If our bioluminescence analogy is correct—that is, if a wave passing through the ISM is what generates the star formation that traces out spiral arms—then the gas and molecular clouds should be located on the leading edge, or the convex side, of the arms. It is this side that is the likely first point of contact if the wave that generates the arms is moving through and affecting the gas in the ISM of a galaxy's disk. But that's not what's observed.

The molecular gas and dust associated with spiral arms invariably appears on the *inside*, or concave, edges of the arms (see Figure 14.8).

Astronomers have scratched their heads over this apparent paradox for some time. The best answer anyone's come up with is known as *Spiral Density Wave Theory*. Though somewhat complex in detail, this model does solve the problems we noted earlier. For example, as we suspected from the winding problem, the arms are not material objects or standing waves rotating with the galaxy disks. Instead, analagous to the wake of the ship, they represent the locations of wave-like disturbances—density waves—that are created by gravitational instabilities within galactic disks. But these waves don't pass through the disk in the way the wake passes through the sea. It's more accurate to say that the disk passes through the wave. That is, the wave rotates about the galaxy more slowly than the disk, causing the stars and gas in the disk to overtake it.

The tricky part of this idea involves producing and maintaining gravitational instabilities that result in long-lived, spiral-shaped density waves in galaxy disks.[18] We already know something about gravity instabilities. Dense clouds of gas collapse under their own gravity in a classic type of instability that allows star formation to occur in the Cold ISM (see Destination 6). But instabilities in disks are much more complex. For example, a rotating bulge near the center of a disk can change its shape due to gravitational interactions with the disk. This can cause the bulge to become more elongated in shape, reminiscent of the bar-shaped center of the LMC (Destination 12) and the asymmetric structure of the Galactic Bulge (Destination 10). As this process progresses, the asymmetric bulge can start to affect the disk in return, producing instabilities that lead to variations in the gravitational force within the disk in the form of spiral-shaped density waves.

Since the centers of galaxies rotate more slowly than their outer regions (see Figures 9.13 and 14.9) and since density waves are tied to the rotation speed of the central bulge or bar, a nice feature of this idea is that the density wave rotates more slowly through the disk than do most of its stars and gas. Consequently, matter in the disk can 'catch up' to the wave at a speed dictated by how fast the inner part of the galaxy rotates and the maximum rotation speed of the disk. This implies that the first point of contact between gas in the disk and the density wave is *from behind* the wave. Contrary to our earlier guess, we'd therefore expect to see molecular clouds on the inside (trailing) edges of arms, with young stars and H II regions, then only older stars dominating along a sequence extending out to the leading (outside) edge of the arm. Amazingly, this is just what we observe (Figure 14.8). Because the arms are produced in

---

18 If spiral arms survived for, say, only one or two disk rotation periods—around 0.5 *Gyr*—it would be a monumental coincidence that we see so many spiral galaxies in the Universe. Much easier to imagine that spiral arms become established early—and that they last a long time—in typical disk galaxies.

response to the passage of a wave that passes through the entire disk, this also explains why star formation turns on in such a highly coordinated fashion over the full extent of the arms (Figure 14.8).[19] Though this idea of density waves seems to explain the key features of spiral arms, it reveals a strange, unexpected aspect of galactic disks. And serves to indicate just how complicated galaxies really are.

14.5 THE PECULIAR ROLE OF DARK MATTER

Our discussion of the nature of spiral arms relies on two key assumptions: That spiral galaxies rotate and they exhibit flat rotation curves. We can confirm the validity of the first assumption—that disk galaxies rotate—through Doppler measurements (see Destination 6) of their stars and/or gas. Such observations invariably reveal that the stars and gas on one side of spiral galaxies are blueshifted—therefore approaching us—while the stars and gas on the opposite side are redshifted, and hence moving away from us (Figure 14.9).[20] The simplest interpretation that is consistent with the obvious longevity of disk galaxies[21] is that all disk galaxies rotate in the plane in which most of their stars and gas reside. Just like our Galaxy (Destination 9). By measuring the magnitude of the Doppler Shifts, we can determine the rotational velocities of the galaxies directly.[22] In many cases, these techniques can be used to produce

19 It's the same with the bioluminescence. Organisms on one side of the wake—or at different ends of a long wave (see Figure 14.8)—will 'light up' at the same time as the ones on the other side. Not because they communicated with one another, but because the wake—or wave—coming from a single source, passed by them at the same time.

20 As we will see later, most galaxies exhibit a systematic redshift as if they are moving away from us, sometimes at astonishing speeds approaching the speed of light. Nonetheless, one side of such galaxies may be less redshifted than the other side, indicating the systems are rotating. In practice, one first determines the mean redshift of the galaxy, then subtracts that off before determining the redshifts and blueshifts associated with the galaxy's internal rotation.

21 It would be an astonishing coincidence for the Sun to find itself in the Galactic Disk along with all the other stars, clusters, and gas in the Disk if this arrangement were some sort of short-lived, random configuration. The same reasoning holds for other spiral galaxies. The obvious order of galaxy disks implies they have been around for many rotation periods.

22 Well, almost. We have to account for the tilt of the galaxy relative to our view. If we see a galaxy edge-on, these measurements give us the rotation speed of the galaxy pretty directly. But if the galaxy's disk is tilted, what we measure is lower than the true rotation speed of the disk. One common way to estimate the tilt is to assume the real disk of a

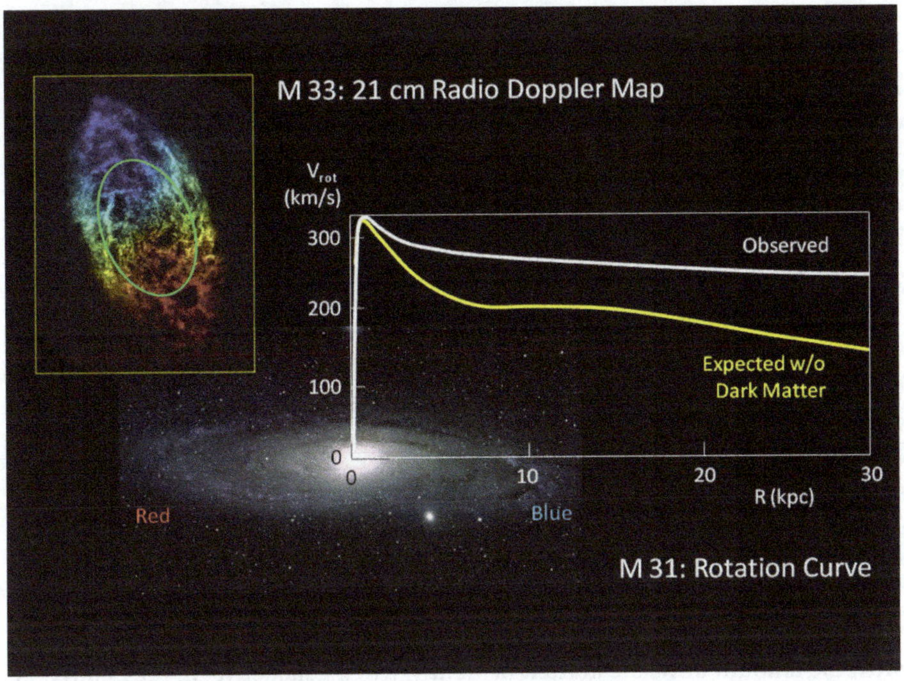

Fig 14.9

Two views of galactic rotation. At *upper left* is a 'Doppler Map' of M 33 obtained using 21-*cm* radiation from the Warm ISM in the galaxy. The colors correspond to the Doppler Shift seen in the 21-*cm* line over the face of the galaxy. Red regions are 'redshifted,' meaning they are moving away from us, while blue regions show where regions are approaching us, causing the lines to be 'blueshifted.' Intermediate colors show intermediate velocity variations. Green, in particular, corresponds to regions that are neither approaching nor receding from us. In those parts of the galaxy, the gas is moving perpendicular to our line of sight and so the Doppler Shift is zero. The green ellipse superimposed on the map is the rough outline of the optically visible portions of M 33 (for example, as seen in Figure 14.3). Notice that the neutral gas extends far beyond the brightest portions of the galaxy's disk. (*Right*) The rotation curve of M 31 is superimposed, to scale, on an image of that galaxy. The observed rotation is given as the white curve, while the yellow curve shows the velocities we'd see in the absence of Dark Matter. The M 31 rotation curve flattens out at about 250 *km/s*, while the rotation curve for M 33—which can be derived from the Doppler Map shown here—also flattens out at large distance from the center of the galaxy, but at a

rotation curves for galaxies, allowing us to test the validity of our second assumption that disk galaxies exhibit flat rotation curves.

When we use these techniques on M 31 and M 33—and also, incidentally, on the LMC—we find they all exhibit flat rotation curves that extend to large distances from their centers (Figure 14.9). In fact, such flat rotation curves are seen in all normal disk galaxies, *with no known exceptions*. In our Galaxy (Destination 9), the flat part of its rotation curve settles to a velocity of about 220 *km/s* at large radius (see Figure 9.13). We'll refer to the velocity where the rotation curve of a galaxy flattens out at large radius simply as *the* rotation velocity for that galaxy. Values for the rotation velocities of M 31, M 33, the LMC, and the Galaxy are listed in Table 14.1. What's immediately apparent from these results—admittedly for a small sample of just four systems—is that the more luminous galaxies rotate faster. As it happens, this appears to be a universal feature of disk galaxies: When we expand our sample to include the rotation velocities and luminosities of galactic disks for hundreds of galaxies that that have been adequately measured, we confirm our hunch based on the four galaxies listed in Table 14.1. There is indeed a strong correlation between the rotation velocities and luminosities of disk galaxies that has the form

$$L_{Disk} = 6.2 \times 10^9 L_\odot \left(\frac{V_{rot}}{220\ km/s} \right)^4 .$$

galaxy is circular and then calculate the angle of tilt from the shape (usually elliptical) of the galaxy in the sky.

This is an amazing result. It implies that we can predict the luminosity of the disk component of a galaxy by knowing how fast the disk rotates! Known as the *Tully-Fisher Relation* (or just *TF Relation*) in honor of the astronomers who discovered it, this result implies a remarkable uniformity in the rotational properties of all disk galaxies. And it works amazingly well. The TF Relation lets us 'predict' the disk luminosities of M 31, M 33, and the LMC to within 5% of their observed values simply by knowing how fast they rotate (see Table 14.1).[23]

rotation velocity of about 140 *km/s*. The image of M 31 is labeled to show which side is approaching us (the 'Blue'shifted end of the galaxy), and which is receding from us (the 'Red'shifted side) relative to the center of the galaxy.

Image of M 31 from D. Thilker/NRAO.

Now, if this were the end of the story, that would be impressive enough. But there's more. In all galaxies where we can measure rotations curves out to large distances from the galaxy centers—for example, our Galaxy, M 31, and M 33, but also dozens of others—we discover that the flat rotation curves extend far out into their halos where the motions become dominated by the distribution of dark matter (Destination 13). If this is generally true—again, there are *no known exceptions*—the TF Relation tells us that the relative amounts of Dark Matter and visible matter in galaxies is always present in precisely the 'right' proportion to ensure that the galaxy's rotation velocity strongly correlates with the amount of *visible* matter. Evidently, when galaxies form they converge somehow on a universal way of doling out the relative amounts of Dark Matter and visible matter they possess. Regardless of how different the galaxies appear. Regardless of how different their total masses are. We'll have to keep this astonishing result in mind as we venture back in time to the era when disk galaxies first formed in the early Universe (Destinations 20–22).

14.6 THE LOCAL GROUP

Before we leave the vicinity of M 31 and M 33, we decide to do one final reconnaissance of the galaxies from a location midway between them. From here, the two galaxies are about 100 *kpc* away, by far the brightest patches of light in the sky. M 31 appears nearly face-on from this vantage point, while M 33 is seen nearly edge-on, precisely opposite of how the two appear from Earth. Otherwise, this is a very empty sky. As in the Halo of our Galaxy

23 If you think about this for a bit, you'll realize that this means we can use the Tully-Fisher Relation to estimate the distances to disk galaxies. All we have to do is measure their rotation velocities—sometimes easy, sometimes not—and then calculate the luminosity of the disk from the Tully-Fisher Relation. If we then measure the disk's brightness, we can use the Flux-Luminosity-Distance Relation (Destination 7) to get the galaxy's distance. Of course, it's not that simple. You have to correct for inclination (see the previous footnote), and, if the galaxy has a particularly big bulge, you have to remove its contribution to the galaxy's rotation and brightness.

(Destination 13), at our present location far out in the halo of M 31[24], there are very few stars. Very little of anything in fact.

But this region is not quite as empty as it first appears. As we look carefully around us, we find that M 31, just like our Galaxy, has a collection of dwarf satellite galaxies. Two of these—M 32 and NGC 205—are located quite close to the center of M 31 (see Figure 14.3), both just barely visible to the naked eye from our current vantage point for people with very good vision. But most of the other companions of M 31 that we detect around us are quite distant from us and from M 31, all located well out into the galaxy's halo. As for our Galaxy, many of the new dwarf companions of M 31 have been discovered relatively recently and it's almost certain that many more of these tiny galaxies will be found in coming years. Even so, there are no fewer than 29 satellite galaxies known to be surrounding M 31, a number comparable to what we see around our Galaxy. If we look carefully, we detect some stellar streams surrounding M 31, reminiscent of the Sagittarius Stream (and others) encircling our Galaxy (see Figures 13.9 and 13.10).

Aren't we are overlooking something? And then we see it! We had to search carefully, but eventually we find our Galaxy in this bleak, empty sky. Our home now shines as a small, pale patch of light, barely visible to the naked eye. From here, near M 31 and M 33, our home Galaxy appears somewhat harder to see than M 31 appears from Earth. On reflection, that's not surprising; after all, our Galaxy is not as luminous as M 31 (see Table 14.1). The LMC and SMC—both easily visible to the naked eye from Earth—are now invisible without at least modest optical aid. All the destinations we have visited on our voyages so far, all our experiences, everything we've ever known, is near that small smudge: Our Galaxy. And the Sun? Finding our home star would be nearly impossible from our current position even with the aid of a large telescope. Not long ago, the very idea of spanning the distance to the nearest star seemed unimaginably difficult. And it is. But now we know that reaching even the nearest large galaxies beyond ours requires a journey that's nearly *1 million times longer.*[25]

---

24 Although we are located halfway between the two galaxies, we are still deep inside the gravitational 'realm' of M 31. As we've discovered, M 31 is much more massive than M 33, so the gravitational forces acting on any objects midway between the two come mostly from the mass associated with the larger galaxy. In fact, we can estimate where the gravitational forces of the two galaxies are equal (in a homework problem at the end of this destination). It's much closer to M 33.

25 Ponder that for a moment. Think of the shortest journey you can imagine that actually gets you from one useful place to another. Say, the distance from your bed to the bathroom. Twenty feet? Then multiply by one million. That's about 3800 miles! More than once across the United States. You'd certainly not want to have to travel that far to brush your teeth every day! We may be getting somewhat inured to huge numbers, but a factor of 1 million in distance is still a big deal.

As we look around us at this dark, empty sky, we realize with some unease that we see *only* the Galaxy, M 31, and M 33 from this place. Just three galaxies. We might have expected that after traveling the vast distance from home to reach this place near M 31 and M 33 that we'd start to see other bright galaxies located beyond M 31 and M 33 that are too faint to be seen by eye from Earth. It makes us wonder: Is this all there is? We know the answer is 'no'; after all, we've already referred to the existence of other galaxies. But what's clear from the view of this Alien Sky is that there are no other large galaxies *nearby*. The Galaxy and M 31 are, by far, the two dominant galaxies in a region of space containing the closest 70+ galaxies from the Sun and known collectively as the *Local Group* (see Figure 14.10). Most of the galaxies of the Local Group—in particular the dwarf spheroidal galaxies (see Destination 13)—are concentrated around the two giant spirals, M 31 and our Galaxy. But about 15 small galaxies, mostly of a type known as *Dwarf Irregular Galaxies*, are spread through a much larger volume that defines the full extent of the Local Group as a roughly spherical region with a radius of about 750 *kpc* (see Figure 14.10).

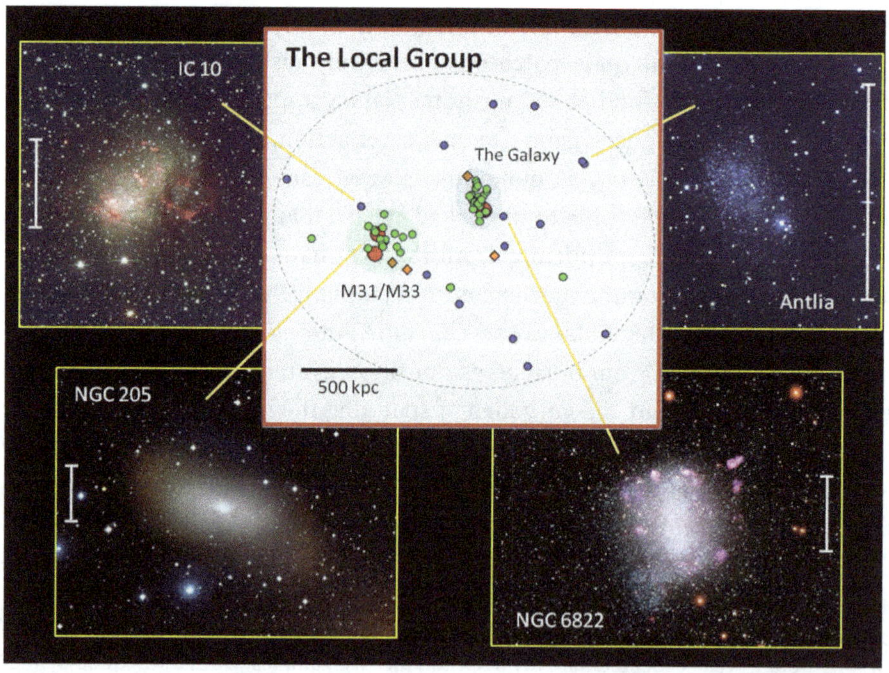

Fig 14.10

(*Upper center*) A plot of the distribution of galaxies in the Local Group as seen from a location far away. Our Galaxy, M 31, and M 33 are shown as large red dots. Green dots are dwarf spheroidal galaxies similar to Sculptor (Destination 13). These are very strongly concentrated around the main galaxies of the Local Group, our Galaxy (light blue ellipse), and the M 31/M 33 pair (light green ellipses). The blue dots show the locations of *dwarf irregular galaxies*. These are predominantly spread over the volume of the entire Local Group, generally not associated with any of the individual big galaxies in the region (M 31, M 33, and our Galaxy). The orange diamonds show the locations of galaxies with properties that are a mixture of what is seen in dwarf spheroidal and dwarf irregular galaxies. A scale of 500 *kpc* is shown and a rough outer boundary of the Local Group is denoted as a dashed ellipse. The montage surrounding the plot displays optical images of four of the 70+ individual dwarf galaxies of the Local Group: IC 10, NGC 205 (see also Figure 14.2), NGC 6822, and Antlia. In each image, the vertical bars denote a length of 1 *kpc* at the distance of each galaxy. Yellow lines point to the locations of each galaxy in the map of the distribution of the galaxies in the Local Group.

Images from P. Massey/ Lowell Observatory/NOAO (IC 10), M. Mateo/LCO/OCIW (Antlia), J.C. Cuillandre/CFHT and G. Anselmi/Coelum (NGC 205), and NOAO/AURA/NSF (NGC 6822).

The clumpy nature of the distribution of galaxies within the Local Group, and, indeed, even the existence of the Local Group as a distinct set of galaxies, is not some oddity of our neighborhood. As we shall see at our next few destinations, this is a common feature of the distribution of galaxies in the Universe. A feature that's telling us that galaxies are not distributed randomly through extragalactic space. *There is structure in the Universe on scales far larger than individual galaxies.* And that's why the sky here is so bleak. We see no other bright galaxies because the next distance ranges we must traverse to reach them are considerably larger than the 800 *kpc* journey that brought us here to the vicinity of M 31 and M 33. We have barely begun to probe extragalactic space.

DESTINATION 14 QUESTIONS

14.1. Oddly enough, Cepheids (and other pulsating variable stars) appear most luminous when they are *smallest* in size. How can this be? **HINT:** Look back at the factors that control a star's luminosity (Destination 8). Size is one of these. What's the other?

14.2. Let's say that a density wave that forms a spiral arm in the Galaxy travels at 150 *km/s* at the location of the Sun. How rapidly does the Sun approach the wave? Assume that there are two symmetric waves in the Galaxy—that is, there are two spiral arms that are precisely 180 *deg* apart in the disk. If the Sun is currently inside one of the arms (as far as we can tell, this is not actually the case) how long will it take to get to the other arm? **HINT:** Use the information you know about the Sun's orbital velocity and orbital radius (Destination 9).

14.3. Draw the situation described in Destination 14 regarding the winding problem in spiral galaxies (see Section 14.4). Demonstrate in your drawing(s) how the situation described there produces overly wound-up arms, quite different than the ones we see in real spiral galaxies.

14.4. Let's say that Cepheids obey the following period-luminosity relation (repeated from the text):

$$log_{10}(L) = 1.1\, log_{10}(P) + 2.5,$$

where P is the Cepheid period in *days*, and L is the Cepheid mean luminosity in solar luminosities, $L_\odot$ (the notion in the equation above emphasizes that base-10 logarithms are being used). Let's say that from Earth we observe that the Cepheids in the LMC with periods of 100 *days* are 6000 times brighter than the Cepheids in M 31 with 5.2-*day* periods. How much further away is M 31 than the LMC? If the LMC distance is approximately 50 *kpc*, how far is M 31? **HINT:** You can rewrite the period-luminosity relation as

$$L = 316\, P^{1.1}$$

where P is in days and L is in units of solar luminosities, $L_\odot$. You may find this form of the PL relation to be easier to manipulate for this problem.

14.5. As described in Destination 14, our Galaxy, M 31, M 33, and the Magellanic Clouds are the brightest members of the 'Local Group,' a, well, local group of galaxies around the Galaxy. Write a paragraph or two describing some key features of one member of the Local Group *other than* any of the five listed here. Basic information such as distance, age(s), size, luminosity, mass, etc., would be good to include in your description.

14.6. Based on the rotation curve of M 31 shown in Figure 14.9, what is the mass of that galaxy within a radius of 30 *kpc*?

14.7. Analogs of all the sorts of objects found in our Galaxy are almost certainly present in M 31. Yet, because of the great distance to the galaxy (D_{M31} = 780 *kpc*), these objects would appear much smaller and much fainter than the cases we know about near the Sun. For the following objects—all destinations we have visited—estimate how large they would appear in the sky from Earth if they were located in M 31 and also by what factor would they appear fainter. For each object, the angular diameter, θ, and the distance, D, from Earth is provided.

a) The Orion Nebula: $\theta = 60$ *arcmin*, $D = 500$ *pc* (Destination 6).

b) The Pleiades: $\theta = 90$ *arcmin*, $D = 140$ *pc* (Destination 8).

c) M 92: $\theta = 20$ *arcmin*, $D = 8.5$ *kpc* (Destination 11).

14.8. Look up on the web information about a specific pulsating star that is *not* a Cepheid. What is the period, P, and luminosity, L, of the star? Where is it located in the sky? What galaxy or what part of our Galaxy is it located in? Report on any other interesting information you may run across involving the star.

14.9. Given the information in Table 14.1, how well do M 31, M 33, the LMC, and our Galaxy conform to the Tully-Fisher relation described in this destination? What does this imply regarding the use of the Tully-Fisher relation as a means of obtaining galaxy distances?

14.10. Use the Period-Luminosity relations in Figure 14.5 to answer the following questions:

a) What are the approximate mean luminosities of the four Cepheids whose light curves are shown schematically in Figure 14.5?

b) If we discover that the star with the 10-day period is actually a Type II Cepheid (such stars are found in some globular clusters), how luminous is the star?

c) If you discovered a pulsating variable in a galaxy, how could you be certain that the star is or is not likely an RR Lyrae star?

14.11. Let's say that we discover two 30-*day* Cepheids in M 31. One of these appears to be on the near side of the galaxy's disk (as seen from Earth), while the other is on the far side. If the two stars are, in reality, about 30 *kpc* apart, what is the ratio of their brightnesses, assuming they each have the same intrinsic luminosity? If we can measure brightness differences to about one part in 100, would such stars allow us to measure directly which side of M 31 is closest to Earth?

14.12. A curious feature of M 31 is that it is approaching the Galaxy. Why this is curious will be more evident when we explore the motions of more distant galaxies (they are essentially all apparently moving *away* from us), but, for now, let's explore an interesting implication of M 31's motion.

a) If M 31 is 780 *kpc* from our Galaxy and is approaching us at 100 *km/s*, how long will it take the two galaxies to collide? Assume that the closing speed stays constant (this is not true, but adequate for our purposes here).

b) Now suppose that the two galaxies started moving apart from one another when the Universe was born, some 14 *Gyr* ago (we will explore this later) and they have never since been close together. Imagine that this initial motion apart represents the start of an 'orbit' the two galaxies have since followed. What is the approximate orbital period of the Milky Way-M 31 binary system?

c) Take the current separation of the two galaxies as the 'semi-major axis' of this orbit. What is the combined mass of M 31 and our Galaxy? If we assume that M 31 is about twice as massive as our Galaxy, what are the individual masses of each galaxy? Compare this to the result we found for the Galaxy (Destination 13). **HINT:** Use Kepler's Third Law along with the period you derived in part (b).

14.13. In this problem we explore the relative numbers of dwarf galaxies in M 31 and M 33 based on the relative gravitational influences of the two galaxies on dwarfs located in their vicinity.

a) Calculate where the center of mass of M 31 and M 33 is located (you may want to review from Destination 4 how this is done in binary stars). You can assume the two galaxies are 200 *kpc* apart. Use the data on the masses of M 31 and M 33 in Table 14.1. For each galaxy, multiple masses are listed; justify which values you used for the calculation of the center of mass location.

b) Now, estimate the position along the line between the two galaxies where the gravitational forces from each galaxy on a small mass (like a dwarf galaxy) are equal. You may want to look through Destination 9 for a refresher on Newton's Universal Law of Gravity. How far is this force-balancing point from each galaxy? **NOTE:** There are actually two answers to this calculation! Only one, however, refers to a location between the galaxies; that's the one you should use for part (c). What does the other answer mean?

c) What is the ratio of the respective 'spheres of influence' of the two galaxies? **HINT:** Take the distances you calculated in part (b) (see NOTE), calculate a volume around each galaxy, then take the ratio.

d) On the basis of this calculation, explain why it is reasonable that M 31 has over 30 satellite galaxies while M 33 has at most one or two.

Destination 15

THE GIANT ELLIPTICAL GALAXY MAFFEI 1

15.1 BEYOND THE LOCAL GROUP

Each journey we've undertaken so far has kept us in close proximity of spiral or dwarf galaxies. What we've discovered is that the Galaxy and M 31 represent classic examples of 'giant' spiral galaxies, spectacular structures characterized by vast rotating disks of stars and gas spanning tens of *kpc* (Destinations 9 and 14). We also discovered that dwarfs vary considerably in form and size, ranging from systems like the Magellanic Clouds—fairly large 'dwarfs' that possess some ordered large-scale structure—to much smaller dwarf irregular galaxies such as NGC 6822 and IC 10 (see Figure 14.10). We've also encountered dwarf spheroidal galaxies such as Sculptor (Destination 13), comparatively tiny systems that seem to congregate strongly around large parent galaxies such as our Galaxy and M 31 (see Figure 14.10). At least this is what we see in the Local Group. But is this a representative sample? Do the galaxies of the Local Group—the large spirals, the dwarf irregular galaxies, and the inconspicuous dwarf spheroidal galaxies—give us a complete picture of the full range of galaxy types in the Universe?

Based on experience, we can guess that the answer is 'no.' If our explorations have taught us anything, it is that we need to be very careful about generalizing from what we see in our immediate neighborhood.[1] The stars around the Sun—though typical of the Galactic Disk—are not at all representative of stars in the Galactic Bulge, or of stars located in globular clusters or the Galactic Halo. We should therefore be somewhat skeptical that the Local Group provides a fair census of all types of galaxies we are likely to encounter elsewhere in the Universe. After all, the Local Group is about 1500 *kpc* in diameter—or, to introduce a new unit that we'll find increasingly useful—about 1.5 *Megaparsec*, or 1.5 *Mpc*. That turns out to be less than one trillionth the volume of the observable Universe![2] It would be like concluding that you have seen all the flora

1 This reasoning applies to the Earth too. You'd not get a very good sense of the nature of the physical structures of the Earth's surface—canyons, deserts, volcanoes, rain forests, etc.—from any single location. Nor would you experience all the different human cultures from just one place. Which is why we explore. Not merely for the excitement of visiting new places, but to truly appreciate the full range of what's possible. And that's usually far more than what's imaginable.

2 Uh-oh. There's one of those big numbers again. In this case, this large number implies that every human on Earth could have—to themselves—a volume about 200 times that of the Local Group, or about 9 *Mpc* in diameter. It's also approximately equal to the ratio of the volume of a red giant—a star with a radius about equal to one *AU*—to the volume of the Earth. Does that help? Probably not. But the point is simple: As big as the Local Group is, it encompasses a virtually negligible fraction of the total volume of the Universe. The only reason we care about it is it just happens to be *our* volume!

of the world by studying only the plants found in one's backyard.[3] We need to explore beyond the Local Group.

Fig 15.1

Two luminous members of the Sculptor Group of galaxies are shown to approximately the same relative scale. (*Left*) NGC 300 is shown as it appears in the ultraviolet (*upper left*) and in optical light (*lower left*). (*Right*) NGC 253 is illustrated as it appears in optical light (*upper right*) and the near-infrared (*lower right*). As we saw in the Local Group, galaxies can appear quite different depending on which part of the EM Spectrum they are observed. The only elliptical galaxies in the Sculptor Group or M 81 groups (Figure 15.2) are dwarf systems, similar to the Sculptor Dwarf Spheroidal Galaxy in the Local Group (Destination 13).

Photos from ESO (optical images), GALEX/JPL/NASA (UV), and 2MASS/IPAC/NASA (IR).

As it happens, there are two other comparatively 'nearby' groups of galaxies beyond the Local Group. The closest is the *Sculptor Group*—named after the constellation in which it is located[4]—while the other is known as the *M 81 Group*—named after its most prominent member, the large spiral galaxy Messier 81. Both groups are located about 2–3 *Mpc* from our Galaxy, a rather typical separation between galaxy groups.[5] As we explore these two nearby groups, we discover some interesting new galaxies that certainly differ in detail from those found in the Local Group (Figures 15.1 and 15.2). But, somewhat to our surprise, the Sculptor and M 81 groups contain a remarkably similar mixture of galaxies compared to what we see locally. Both groups contain a handful of dominant spiral galaxies (Figures 15.1 and 15.2), a few large irregular

3 My backyard—which is not too large—is almost exactly one trillionth the surface area of all the land on Earth. There *are* a lot of weeds there, but I suspect some plant species are not present!

4 This is the same constellation in which the South Galactic Pole is located as seen from Earth (see Figure 9.5 in Destination 9).

5 There's an important implication to this statement. If a typical group is about 1.5 *Mpc* in diameter and groups are typically 2.5 *Mpc* apart, then groups comprise only about 20% of all space if groups are distributed more or less uniformly through the Universe. Of course, the Local Group itself is pretty empty. The galaxies that comprise it fill up less than 1% of the group's volume. And, as we have seen as we journeyed within our Galaxy, galaxies themselves are mostly filled with empty space. The bottom line is that there is just a whole lot of nothing in the Local Universe. And it's even worse than that. As we shall discover (Destination 18), galaxy groups and clusters (Destination 16) are not randomly distributed in space. There are vast regions that are, by comparison, almost entirely devoid of galaxies.

Fig 15.2

Two luminous members of the M 81 Group of galaxies shown to approximately the same relative scale. (*Left*) M 81 is shown as it appears in far-infrared radiation (*upper left*) and in the optical (*lower left*). In the IR image, shorter-wavelength radiation traces the locations of red stars and is shown as blue in this image. Longer-wavelength radiation, shown in red, traces the location of warm dust heated by the radiation of young, hot stars. In the optical image, the colors are reversed: Redder parts of the image show the locations of older stars, while the young stars, here seen directly, show up in blue. (*Right*) NGC 2403, seen here as it appears in 21-*cm* radio emission (*upper right*) and in the optical (*lower right*). Notice how much more extended the radio emission from neutral H atoms of the Warm ISM appears compared to the optical image.

Photos from NOAO/AURA/NSF (M 81/NGC 2403 optical), NRAO (NGC 2403 radio), and JPL/NASA (M 81 far-IR).

systems similar to the Magellanic Clouds, many dwarf galaxies similar to the Sculptor dwarf spheroidal galaxy (Destination 13), and numerous dwarf irregular galaxies (Destination 14). Not only do these resemble galaxies found in the Local Group, but they also exhibit similar tendencies regarding how they are distributed. In all three groups, the dwarfs are generally clustered around the large spiral galaxies. The mixture of galaxy types and the relative numbers of different types of galaxies is not very different from what's seen in the Local Group. It begins to dawn on us that perhaps the Local Group *is* representative of all galaxies after all. Could spirals, dwarf irregulars, and dwarf spheroidals be the only types of galaxies in the Universe?

Actually, even before leaving the Local Group we could have surmised the answer to this question is 'no.' There is one oddball galaxy in the Local Group. One we've noted before but largely ignored up to now. From Earth, this galaxy appears very compact, its faint outer 'edge' nearly perfectly round.[6] And it is located so close to M 31 that it appears to be part of the larger galaxy's disk, which just emphasizes how comparatively tiny it is. This galaxy—known as M 32—is an example of a *Dwarf Elliptical Galaxy* (Figure 15.3). Though less luminous than the LMC, it is more luminous than virtually all the dwarf spheroidal galaxies it superficially resembles. And its central concentration—the density of stars at its core—is far larger than that of any of the dwarf galaxies in the Local Group.[7] Even higher than the central densities of our Galaxy or M 31!

6 Galaxies do not have sharp outer edges (a point we made at Destination 9), but there does come a point where it becomes impossible to trace a galaxy's stellar content beyond a certain distance. The place where this happens is often referred to as a galaxy's 'edge.'
7 These distinctions may not seem sufficient to justify calling M 32 a different type of galaxy. But, we shall soon see that true elliptical galaxies share many more properties with M 32 than M 32 shares with dwarf spheroidal galaxies like Sculptor (Destination 13). For example, the high central density of M 32 implies it contains a supermassive central

Fig 15.3

M 32 (*left*) is shown in detail and relative to its parent galaxy, M 31 (*right*). It is clear from these images that M 32 indeed has an elliptical outline, but also that it is a true dwarf system, considerably smaller than even the bulge of M 31. What is not so evident from this image is the comparatively bright but very compact—almost star-like—nucleus of M 32.

Photos from W. Promper (*left*) and NOAO/AURA/NSF.

In many respects, M 32 is unique. And not just in the Local Group. Neither the Sculptor nor the M 81 groups contain anything like it. So what do we make of M 32? We have already seen that tidal forces can affect galaxies, and that smaller galaxies can be severely altered due to these forces. Perhaps the strong gravitation field of M 31 has severely transformed M 32 structurally. At some distant time in the past, M 32 may have been more like a spiral galaxy or large irregular system that has been transformed by tidal effects into the strange object we see today. This is not farfetched: As we'll see at Destinations 16 and 20, galaxies can change from one type to another as they interact gravitationally.

Fig 15.4

A wide-field view of Maffei 1 (*right*) and Maffei 2 (*left*) as seen in Earth's sky. The large number of stars visible in this image is to be expected in a field located only half a degree from the Galactic Equator (see Destination 9). There is a considerable amount of dust and gas from the Galactic Disk between us and the two galaxies, making them appear much fainter, redder and smaller than they would if we could see them without any obscuration. This is why these galaxies escaped detection by astronomers until 1968.

Photo by S. Faworski and S. Walker.

black hole, similar to, but smaller than the one in the Galactic Center (Destination 10). Dwarf spheroidal galaxies like Sculptor exhibit very low central stellar densities with no obvious sign of a central black hole. Despite cosmetic similarities, elliptical galaxies and dwarf spheroidals do really seem represent distinct types of galaxies.

But it's unclear if that's really what has happened to M 32 and it is tempting to simply treat it as an oddball case among the nearly 200 galaxies that we've identified in the Local Group and the Sculptor and M 81 groups. But we can say one thing: M 32 is unquestionably a *dwarf* galaxy. As such, it doesn't alter our thinking that perhaps all large galaxies are spiral galaxies, while all smaller systems are dwarf irregular or dwarf spheroidal galaxies, with the occasional dwarf elliptical galaxy—like M 32—tossed in. Yes, based on what we've seen so far, we'd be justified in reaching this conclusion. But it would still be wrong.

There is another galaxy group just beyond the Local Group that demonstrates this. Known as the *Maffei Group* after the Italian astronomer who discovered the group's principal galaxies in the late 1960's[8], it is located about 3100 *kpc*, or 3.1 *Mpc* from Earth. That's comparable to the distance from our Galaxy to the Sculptor and M 81 groups, or about four times more than the distance between our Galaxy and M 31. It seems odd that this group went undiscovered for so long. The problem is that Maffei Group is situated (mostly) within the constellation Cassiopeia in Earth's sky, quite close to the Galactic Equator defined by the Milky Way (see Figure 15.4, Destination 9). Because of this, the galaxies of the Maffei Group—it has about ten currently known members—are heavily obscured by gas and dust in our own Galactic Disk. This makes the group very difficult to see and study optically, and it made it hard to discover in the first place. The two brightest galaxies of the group are known as *Maffei 1* and *Maffei*

Fig 15.5

Maffei 2 (*left*) and Maffei 1 (*right*). Though Maffei 2 is clearly some sort of spiral galaxy, Maffei 1 is not. Obscuration by dust and gas in our Galaxy renders both galaxies much fainter and smaller in appearance than they really are; in effect, we can easily see only the central regions of the galaxies. Although it is not apparent from this image, Maffei 1 is considerably larger, more massive, and more luminous than Maffei 2. It is *not* a dwarf galaxy!

Photos from 2MASS/IPAC/NASA.

8 The Maffei Group has had a series of identity crises since its discovery. It is located fairly close to another set of galaxies known as the *IC 342 Group*. Some astronomers consider the two groups to be one entity, but some consider them as separate, as we do here. For a time, the Maffei Group was even considered a possible outer component of the Local Group. The key point is that galaxy groups are amorphous, their 'boundaries' ill defined.

2 (Figure 15.5).[9] Maffei 2 is yet another example of a large spiral galaxy in a nearby group, broadly similar in appearance to our Galaxy or M 31.

It's Maffei 1—our next destination—that interests us now. Plainly not a spiral system (see Figure 15.5), this galaxy bears a much closer resemblance to the dwarf elliptical galaxy M 32 (Destination 14), and, in some respects, with dwarf spheroidal galaxies such as Sculptor (Destination 13). Like those galaxies, the outline of Maffei 1 in Earth's sky can be traced out well as an ellipse, in this case a somewhat elongated ellipse. The overall light distribution—a bright center fading gradually into the dark sky—also broadly resembles what we see in M 32 and Sculptor. But Maffei 1 is no dwarf galaxy. Maffei 1 is the closest example of a *Giant Elliptical Galaxy.*[10]

15.2 MAFFEI 1: A GIANT ELLIPTICAL GALAXY

Now that we are here, it's obvious. We've never seen a sky like this before. True, we've had spectacular views of other galaxies—our Galaxy, M 31, and M 33 (Destinations 13 and 14). But at those previous vantage points out in the halos of those galaxies, we'd always been struck by the lack of stars in the sky at locations far from the galaxy's disks. Not here. At a distance of about 10 *kpc* from the center of Maffei 1, we see about 500 stars in this Alien Sky. Not a huge number—particularly compared to the skies within our Galaxy or the LMC (Destinations 9, 10, and 12)—but far more stars than we've seen in any of the skies we have visited since we began our explorations in the Galactic Halo and extragalactic space (Destinations 13 and 14). Most of the stars we see here are distinctly red in appearance with perhaps at most 2 or 3 bluish stars shining

9 It turns out that Maffei 1 and Maffei 2 comprise the *entire* catalog of galaxies that Maffei discovered. By comparison, Charles Messier's catalog contains 110 objects, a few of which are bogus (this lends 'charm' and a little mystery to his catalog). Some modern astronomical catalogs contain hundreds of millions of objects. To be fair, Maffei was not looking for galaxies—or comets—when he found Maffei 1 and 2. He was searching for young stars in star-forming regions, a far more sensible thing to do close to the Galactic Plane, a region where we know most star formation occurs (Destinations 9 and 10).

10 Maffei 1 is actually on the small side of the size distribution of elliptical galaxies. Some ellipticals are known that are considerably bigger (Destination 16). But, as we shall see shortly, Maffei 1 is significantly larger than our Galaxy and M 31. Or, for that matter, than any other galaxy in the Local Group, Sculptor, and M 81 groups, and among the remainder of the galaxies of the Maffei group. Maffei 1 is certainly the largest galaxy we've encountered so far on our journeys, justifying, for now at least, the 'giant' label we've assigned to it.

feebly in the sky. The stars here appear fairly uniformly spread out over this sky, typically about 10 *deg* apart. But these stars are only part of what's visible here. Something else in this Alien Sky that makes it unique compared to any other we have beheld.

An immense glow spreads over a quarter—maybe more—of the visible sky. At its outer extremes, this patch of light imperceptibly merges into a dark sky punctuated by the sparse smattering of red stars. The indistinctness of the 'edge' makes it tricky to gauge the system's full extent. Obviously wider than 90 *deg*, the glow would appear up to 1.5 times larger to people with particularly sensitive eyesight. But it is not faint everywhere. Where we can see it easily, the texture of this roughly circular glow reminds us of the silky appearance of the Milky Way seen from Earth. And, as it turns out, for good reason. From our location 10 *kpc* from the center of Maffei 1 we can easily confirm—even with a pair of binoculars—that the smooth glow of light spread above us in this sky arises from the contribution of myriad faint stars (see Figure 15.6). Just as the faint glow of the Milky Way is the sum of the light of distant stars within the Galactic Disk, the glow of Maffei 1 arises from a vast number of stars too distant and too faint to be seen individually by the naked eye.

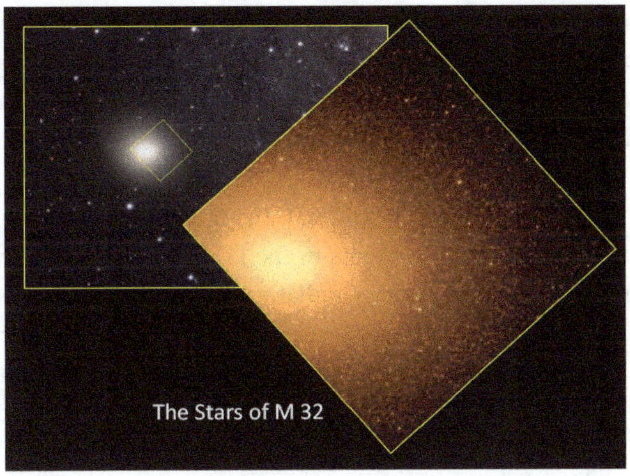

Fig 15.6

A detailed view of M 32 (*right*) showing the galaxy is composed of stars with a strongly concentrated density distribution, peaking at a very high central stellar density. Notice too how smooth the light distribution is; there is no sign of extensive quantities of gas and dust in M 32. Nor are there any extremely luminous, young blue stars that would show up clearly as distinct, bright sources in this view of the galaxy. The appearance of Maffei 1 from a comparable distance would be similar in most respects to this view of M 32, though with a few notable differences. The total number of stars in Maffei 1 is about 100 times what's seen in M 32. Also, Maffei 1 is more elongated as seen from Earth. Finally, although the central density of Maffei 1 is high, M 32's central density is even higher, by about a factor of 10–20.

Photos from W. Promper (*left*) and STScI/NASA.

But there are some important differences between the appearance of the Milky Way from Earth and that of Maffei 1 from our current location. The roundish outline of Maffei 1 (Figure 15.5) is nothing like the extended, narrow band of light that shines from the Milky Way as it encircles Earth's sky. What we see here is more reminiscent of the round outlines of the Galactic Bulge or the central bulge of M 31 when we viewed them from outside the disks of those galaxies (Destinations 9, 10, and 14). But there is no disk here. We would see essentially the same view from *any* location 10 *kpc* from the center of Maffei 1. The reason is that Maffei 1 is nearly spherical in form—though obviously not perfectly spherical as evident from Figure 15.5—reminiscent of stellar systems such as the Galactic Bulge (Destination 10) and M 92 (Destination 11). But Maffei 1 is far larger than either of those stellar systems. Though about 75% of the stellar content of the Maffei 1 is concentrated within 10 *kpc* of the galaxy's center, the fact that we see individual stars around us is proof that the galaxy extends well beyond our location 10 *kpc* from its core.

It also strikes us that the glow of Maffei 1 appears very smooth just about everywhere we look (see Figures 15.5 and 15.6). There are none of the 'holes,' 'spurs,' or 'rifts' that characterize the appearance of the Milky Way as seen from Earth (Destination 9). Since the extended glow of Maffei 1 arises from the light of distant stars and not from gas emission as in the Orion Nebula (Destination 6), this simple observation suggests the galaxy contains very little dust and molecular gas of the Cold ISM. It turns out that this is no illusion. As we explore using telescopes that detect infrared and microwave radiation, we confirm what our eyes tell us. There's very little molecular gas in Maffei 1, very little dust. Nor is there much neutral gas from a warm ISM. Using radio telescopes sensitive to 21-*cm* radiation from neutral H atoms in the ISM (Destination 5), we discover that Maffei 1 emits very little of this radiation, a sure sign the galaxy is nearly devoid of neutral interstellar gas. The lack of blue stars or obvious nebular regions like the Orion Nebula (Destination 6) suggests there is very little star formation going on in this galaxy and very little gas in the hot phase of the ISM. Just about the only gas we see around us in any significant quantity is the very hot, low-density gas of the Coronal ISM, which reveals itself through its x-ray emission (Figure 15.7). Interestingly, the Coronal gas of Maffei 1 is hotter than in the Galaxy or M 31, exhibiting a temperature of about 10–15 million *K*. This very hot gas is also comparatively more common in Maffei 1, amounting to about 10% the mass of stars in the galaxy. Nonetheless, it is still fair to say that Maffei 1 is a galaxy that—as far as its visible component is concerned[11]—appears to be composed almost exclusively of stars. If we add up all the light from these stars, we discover that Maffei 1 has a total luminosity about 2.4 times higher than the Milky Way, or about 1.5 times higher than M 31 (see Tables 14.1 and 15.1).[12] Compared to these galaxies, Maffei 1 is truly a 'giant' system.

And then there's the center of Maffei 1. We've lamented (Destination 9) how it is not immediately obvious where the center of our Galaxy is located as we view the Milky Way from Earth. The dust and gas in molecular clouds effectively

11 The word 'visible' here refers to the detection of *any* sort of EM Radiation from the galaxy, not just optical radiation.

12 If Maffei 1 is so luminous, why didn't we see it from past destinations like the Galactic Halo (Destination 13) or from near M 31 and M 33 (Destination 14) where the obscuration by the Galactic Disk would not have been an issue? As it happens, even from such locations, Maffei 1 would (barely) remain invisible to everyone except those few with uncommonly sensitive vision. However, even a small pair of binoculars would readily have revealed Maffei 1 as a bright, large galaxy, almost spanning the diameter of the Moon as seen from Earth. If our Galaxy was tilted at just a slightly more favorable angle, Maffei 1 would be a spectacular sight—and a famous object—in Earth's sky!

TABLE 15.1 Properties of Elliptical/Spheroid Galaxies

GALAXY PROPERTY	MAFFEI 1	GALACTIC BULGE	M 32	SCULPTOR
Total Luminosity	$2.8 \times 10^{10}\ L_\odot$	$3.5 \times 10^{9}\ L_\odot$	$4.0 \times 10^{8}\ L_\odot$	$2.2 \times 10^{6}\ L_\odot$
Size of Reference Radius (95%)	20 *kpc*	7.0 *kpc*	2.0 *kpc*	1.0 *kpc*
Velocity Dispersion	330 *km/s*	190 *km/s*	110 *km/s*	17 *km/s*
Total Mass Inside of Reference Radius	$3.8 \times 10^{12}\ M_\odot$ (150 *kpc*)	$5.9 \times 10^{10}\ M_\odot$ (10 *kpc*)	$5.9 \times 10^{9}\ M_\odot$ (5.0 *kpc*)	$7.0 \times 10^{7}\ M_\odot$ (2.0 *kpc*)
M/L Ratio inside of Reference Radius (Solar Units)	18	17	15	25

Notes to Table 15.1:

1. Maffei 1 and Sculptor show little internal rotation. The Galactic Bulge and M 32 exhibit some rotation, though always less than the velocity dispersion.
2. The total masses and mass-to-light ratios (M/L) are calculated for mass and light interior to the reference radii noted in the Table.

mask it from view, rendering the Galactic Center invisible to the naked eye. But not here. We can plainly see a fuzzy, cream-colored 'star' shining at the center of the extended glow from Maffei 1 that dominates this sky. As bright as all but the brightest few stars in Earth's sky, we have no doubt of the special nature of this nearly point-like object: This is the *nucleus* of Maffei 1. The view reminds us more of the compact nuclear core in M 32 (Figure 15.6), not the diffuse central concentration of stars that we saw at the center of dwarf spheroidal galaxies such as Sculptor (Figure 13.2). And it really is a remarkable place. One of the few H II regions in Maffei 1—a small one, considerably fainter than the Orion Nebula (Destination 6)—is located in its nucleus. And so are a lot of stars! The light from the nucleus of Maffei 1 alone shines with a luminosity about 100 times greater than M 92 (!), equivalent to the light of about 100 million Sun-like stars packed within the central 10 *pc* of the galaxy.[13]

There are other features that signify the nucleus of Maffei 1 is a special region in the galaxy. A few dark molecular clouds are visible in this region, some of the few blemishes in the otherwise smooth glow of starlight from Maffei 1. Evidently, some process allows molecular clouds to form and survive near the nucleus of Maffei 1 but not throughout most of the rest of the galaxy. We'll discover (Destinations 16, 19, and 21) that this is not uncommon in elliptical galaxies, a clue to how they interact with other galaxies. And then there are the globular clusters (Destination 11). Surrounding the nucleus of Maffei 1 is a host of approximately 600 of these massive, ancient star clusters. About half of the clusters lie within 10 *kpc* of the galaxy's center, with the rest extending to beyond 100 *kpc* from the galaxy's center. This is reminiscent of how globular clusters are distributed around the Galactic Center, but Maffei 1 contains three times more of these old clusters than our Galaxy does. From our vantage point, most of the clusters are spread out over the same part of the sky that contains the extended glow from the main body of Maffei 1. And most are easily visible with even modest optical aid since there's little dust and gas blocking our view. A few—maybe 10 to 15—are even visible to the naked eye, standing out prominently in this sky. Apart from the nucleus of Maffei 1 and the few dark clouds near the center, the clusters provide about the only visual respite within this relentlessly star-filled galaxy.

Our proximity to Maffei 1 lets us measure its basic properties with some precision. We can determine luminosities and temperatures (colors) of the stars around us to produce an H-R Diagram for the galaxy (Destination 8), confirming what our eyes are telling us. The most luminous stars in Maffei 1 are almost exclusively red giants, similar to the ones we first encountered in M 92 (see Figure 11. 5 of Destination 11). And we discover why we see so few blue stars around us within the galaxy. The brightest Main Sequence stars that we see near us in Maffei 1 have masses of around 1 $M_{\odot}$, considerably redder and less luminous than the young blue stars we found in young star clusters such as the Pleiades (Destination 8), or the ones in the LMC (Destination 12).[14] Maffei 1 is old. The galaxy has essentially formed no stars in nearly 10 billion years.

---

13 The nucleus of Maffei 1 would be yet another spectacular place to visit if we only had the time. From near the center of the nucleus of the galaxy we'd see over 1 million stars in the sky! But we'd also see something even more impressive: An isolated Supermassive Black Hole (Destination 10). At Destinations 17 and 21, we'll describe what *that* might look like, and what the likely properties of black holes at the centers of Maffei 1 and other massive galaxies would be like.

14 The few blue stars we can see in the sky here are not young stars, but examples of *Blue Horizontal Branch Stars* (BHB) similar to the ones seen in some ancient globular clusters such as M 92 (Destination 11; see Figure 11.5).

Is this representative of the galaxy as a whole? In our Galaxy, the mixture of the ages of stars in the Bulge is very different than, say, the outer Disk. Consequently, we cannot simply assume that a local measurement of the stellar mixture near our current location applies to the entire system. Unfortunately, producing an H-R Diagram of all 100 billion stars in Maffei 1 would be extremely tedious, if not impossible. But there's another approach. By taking spectra of the smooth light of Maffei 1, we can collect light from all its stars simultaneously. Analyzed with care[15], such spectra can be used to extract the underlying mixture of ages and chemical abundances of the stars producing the smooth light from the galaxy. When we do this for Maffei 1, we confirm that the bulk of the stars in Maffei 1—well over 95%—is similar to what we see in the immediate volume around our present location: Full of red giants and mostly very old. But, to our surprise, these old stars of Maffei 1 seem to contain the same fraction of heavy elements seen in the Sun, or about 2% by mass. The last thing we encountered as old as this was M 92 (Destination 11), which we found has an extremely low heavy-element abundance by mass, about 100 times lower than the Sun's. The old stars of the Galactic Bulge (Destination 10) exhibit a spread of heavy-element abundance, ranging from 10% of the Sun's abundance to ten times higher. Obviously, chemical evolution progresses at different rates and in quite different ways in various environments.[16] There is an important lesson here. Just because M 92, the Galactic Bulge, and Maffei 1 all appear superficially similar and have comparably old ages, we can't simply assume they all evolved the same way.

The spectra we have obtained to determine the stellar content of Maffei 1 can be used to study the stellar motions within the galaxy. From analysis of Doppler Shifts in these spectra, we discover that Maffei 1 is *not* rotating. Instead—just like M 92 and the Galactic Halo (Destinations 11 and 13)—the stars in Maffei 1 appear to be moving randomly through the galaxy. From inside the galaxy, the impression is a bit unnerving. Some stars, exhibiting blueshifts in their spectra, are moving toward us at high speed, while others, redshifted, are moving away. If we measured the proper motions of these stars, we'd find

---

15 Astronomers refer to spectra that include all (or much of) the light from a complicated source as *Integrated Spectra*. The trick in using integrated spectra is to disentangle the contributions from different types of stars and nebulae that might be present in the object. For example, red giants are much more luminous than low-mass Main Sequence stars are. But Main Sequence stars are much more numerous (this has to do with the much longer lifetimes of stars on the Main Sequence compared to the Red Giant phase described in Destination 11), so it is not immediately obvious which type of star contributes the most light to an integrated spectrum. These kinds of details—and many others—can greatly complicate the interpretation of integrated spectra. But it is usually worth it. These spectra provide information about distant galaxies that we often can't obtain any other way.

16 We won't have a chance to make any sense of this until we travel back to a time when galaxies such as Maffei 1 and globular clusters like M 92 were forming. That was many *Gyr* in the past (Destination 21).

that some are moving north, some south, and as many east and west.[17] This is not like the Galactic Disk where stars orbit the Galactic Center in an orderly

Fig 15.7

Elliptical galaxies as they appear in x-ray radiation. The two sets of smaller images show how NGC 4697 (*lower left*) and NGC 1553 (*upper right*) are composed of both bright x-ray point sources (likely x-ray binaries similar to LMC X-3 from Destination 12) and the diffuse emission of x-rays from the Coronal ISM. It is clear that the x-ray gas can take on complex forms, very different from the elliptical form that the stars trace out (*left, upper* for NGC 4697; *right, lower* for NGC 1553, both shown at the same scale as the x-ray images). In the center is a composite optical/x-ray image of the elliptical galaxy NGC 1132 (the x-ray brightness is denoted by the purple-blue glow across the image). The x-rays are roughly centrally concentrated like the stars, but seem to extend out further from the center.

X-ray images from Chandra/NASA. X-ray+optical image by M. West/Chandra/NASA. Optical images by D. Verschatse/Observatorio Antihue, Chile (NGC 4697) and from DSS/STScI/NASA (NGC 1553).

way, all at about the same speed, all in about the same direction. Somewhat like cars on a freeway. The stellar motions within Maffei 1 are more reminiscent of a demolition derby, with stars moving in all directions. And, very quickly. The velocity dispersion[18] of the stars around our location 10 *kpc* from the center of Maffei 1 is about 330 *km/s*, considerably higher than the 220 *km/s* velocity dispersion that we observed in the Galactic Halo (Destination 13).

Just as we did for our Galaxy at Destination 13, we can use this information to estimate the mass of Maffei 1 with the Virial Theorem:

$$M = \sigma_V^2 \overline{R} / G.$$

In this relation, σ_V is the velocity dispersion measured at some radius, $\overline{R}$, from the galaxy center. When we insert the values we observe for Maffei 1, this relation tells us that the mass of the galaxy is about $5.0 \times 10^{11} M_\odot$ out to a radius of $\overline{R} = 10$ *kpc*. That's approximately 12 times more than the mass observed in the galaxy's stars and its minuscule amount of gas. This suggests that Maffei 1—just like our Galaxy and M 31 (Destinations 9, 13, and 14)—contains Dark Matter. But to determine how much, it would be nice to have a better estimate of the total mass of the galaxy, not just the inner 10 *kpc*. Fortunately, there is a way to

17 Absent a planet orbiting a star, exactly how we'd define the cardinal directions here—north, south, east, and west—is anyone's guess. The stars around us would still appear to be moving in random directions.

18 This result takes into account the technical detail described in a footnote at Destination 13. In summary, velocity dispersions are typically measured along one direction, so we have to multiply by that value by $\sqrt{3}$ (about 1.7) to get the 'true' three-dimensional dispersion if we also assume the stars in the galaxy have no preferred direction of motion. In the case of Maffei 1, the actual measured radial velocity dispersion is around 190 *km/s* from which we infer the full dispersion is 330 *km/s*.

do this. By measuring the velocities of the outermost globular clusters of Maffei 1—some of which are found 100 *kpc* from the galaxy's nucleus—and of its most distant satellite galaxies, we can determine the galaxy's velocity dispersion far from its center in a manner similar to the method we used in the Galactic Halo (Destination 13). When we do this for Maffei 1[19], we discover that the dispersion is still around 330 *km/s,* even though we've measured it out to a radius of $\overline{R}$ = 150 *kpc* from the galaxy's center. Now the mass that we derive for Maffei 1 is $3.8 \times 10^{12} M_{\odot}$ for Maffei 1, over three times our Galaxy's mass at a comparable distance from the Galactic Center (Destination 13), and about twice as massive as M 31 (Destination 14). There's no doubt now. At a distance of about 150 *kpc* from its center, the mass-to-light ratio of Maffei 1 is around 130, a clear indication that the galaxy's total mass is dominated by Dark Matter. It's the same conclusion we've reached for *every* galaxy that we've studied in comparable detail.

Results likes these regarding the velocity dispersions in other elliptical galaxies we've studied seem to reveal a trend. At Destination 10 we found that the Galactic Bulge—essentially a small (though not a dwarf) elliptical galaxy that happens to be at the center of our Galaxy—has an internal velocity dispersion of around 190 *km/s.* M 32—the dwarf elliptical companion of M 31 (Destination 14)—has a dispersion of around 110 *km/s* (see Table 15.1). This trend is reminiscent of the Tully-Fisher Relation for spiral galaxies (Destination 14), which revealed a similar correlation between luminosity and disk rotation velocity. If we consider a big sample of elliptical galaxies—one that includes many more than the three galaxies we've studied so far—we find a similar relation of the form

$$L_E = 5.9 \times 10^9 L_{\odot} \left(\frac{\sigma_V}{220\, km/s} \right)^4 .$$

Here, L_E is the luminosity of the elliptical component of a galaxy[20] in $L_{\odot}$, and σ_V is its velocity dispersion measured in *km/s.* This is known as the *Faber-Jackson Relation* after the astronomers who discovered it. From a practical standpoint, this relation is very useful since we can estimate an elliptical galaxy's luminosity by simply measuring its velocity dispersion, a quantity that does not depend on distance. We can then use the Flux-Distance-Luminosity Relation (Destination 7) to determine the galaxy's distance.[21] But this relation—just like the Tully-

19 This has never actually been done for Maffei 1. The obscuration by dust and gas in our Galaxy makes it very difficult to find its more remote globular clusters or dwarf galaxies. But these sorts of measurements have been carried out on other, similar elliptical galaxies and we're assuming that Maffei 1 behaves the same way.

20 What this means is that for, say, the Galactic Bulge, you'd measure only the light from the Bulge, while ignoring the light from the rest of the Galaxy.

21 As you may have noticed, the predominantly old stellar content of Maffei 1 implies that it does not contain any Cepheid variables. We therefore cannot use the PL Relation (Destination 14) to estimate this galaxy's distance. But we can measure its velocity

Fisher Relation for spiral galaxies (Destination 14)—tells us something more fundamental and somewhat mysterious. Here again we see a close relationship between the Dark Matter content of a galaxy—which controls the value of the velocity dispersion, σ_V—and the galaxy's luminosity. It seems that galaxies of all types and all sizes display a strong relationship between their Dark Matter content and how much 'normal' matter—like stars and gas and any other stuff we can detect through its EM Radiation—they contain.[22]

When we left the Local Group, we were lulled into thinking that the largest galaxies are always spiral systems. In contrast, it seemed that elliptical galaxies are always dwarf systems. The properties of Maffei 1—in particular its size, luminosity, and mass—reveal a more complicated situation. Elliptical galaxies span an enormous range of properties (see Table 15.1), even broader than what we see among spiral galaxies (see Tables 14.1 and 15.2). We shall discover that some the largest, most luminous, and most massive galaxies known are elliptical systems (Destinations 16 and 21). And yet, at the other extreme, we've seen that dwarf spheroidal galaxies—different but clearly closely related to normal elliptical systems—are the smallest, dimmest, and least massive galaxies in the Universe. Systems like the Galactic Bulge and M 32 are examples of galaxies with properties between these extremes.

---

dispersion and use the Faber-Jackson Relation to estimate its luminosity. Combined with its apparent brightness, we then obtain its distance. There are other ways to get the distance to Maffei 1—methods that we will not attempt to describe in detail here—and, in general, they agree pretty well with each other and the Faber-Jackson Relation. For Maffei 1, the biggest problem is estimating precisely how much obscuration dims the galaxy as seen from Earth. That's why distance estimates for this galaxy have yo-yo'ed dramatically since its discovery in the late 1960's. The distance we've used—3.1 *Mpc*—places Maffei 1 well outside the Local Group but still implies that it is relatively 'nearby.'

22 Well, almost. A question at the end of this destination will reveal that at least some dwarf spheroidal galaxies (Destination 13)—in particular, Sculptor—appear not to obey the Faber-Jackson Relation. These galaxies differ from most elliptical/spheroid systems in that they are believed to be almost entirely composed of Dark Matter and so the normal matter they contain hardly contributes to their total masses. Precisely why this causes them to deviate from the Faber-Jackson Relation remains unclear.

15.3 CLASSIFYING GALAXIES: THE HUBBLE SEQUENCE

A common feature of virtually all fields of science is the need to *classify* things.[23] Such bookkeeping has, of course, practical benefits as it helps condense a huge amount of information into a more compact form. But a truly useful classification scheme does more than just place items into pigeonholes. It also helps us to better *understand* the nature of the things we classify. What we have learned from exploring Maffei 1 is that there is real variety among galaxies beyond what we inferred from the Local Group alone. Can we come up with a classification scheme that helps summarize the complex variety of galaxies while simultaneously revealing something about their underlying nature?

There are some things we already know that might be helpful in formulating such a scheme. We've seen that there are at least two fundamental forms of 'giant' galaxies: *Elliptical* (or, to be more general, *Spheroidal*) galaxies similar in form to Maffei 1 and M 32, and also *Spiral* (or, more generally, *Disk*) galaxies similar to M 31, our Galaxy, and M 33. As we travel further afield in the nearby Universe, we shall confirm that these two types of galaxies do indeed represent the most common families of large, luminous galaxies in the Universe today. But

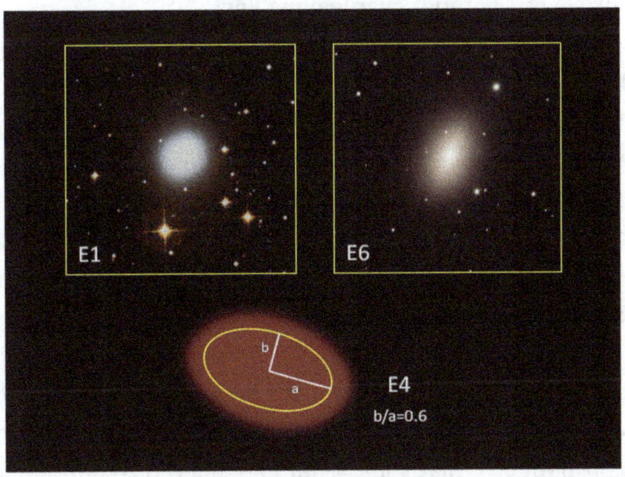

Fig 15.8

An example of an E1 elliptical galaxy (*left*) and an E6 elliptical galaxy (*right*). The drawing shows how an ellipse can be fit to the apparent shape of an elliptical galaxy (in red). The ellipse can be described by its semi-major axis, a, and its semi-minor axis b. For an elliptical galaxy of type En, n is defined as $10[1-(b/a)]$. For the case shown here, $(b/a) = 0.6$ and so $n = 4$. The example 'galaxy' is thus of Hubble type E4, intermediate between the two galaxies in the top images.

Photos from DSS/STScI/NASA (*left*) and by D. Verschatse/Observatorio Antihue, Chile.

23 And it is hardly just a trait of scientists. It seems that everyone—musicians, artists, doctors, linguists, car enthusiasts, and even screwdriver manufacturers—classify stuff into groups. Obviously, this is a distinctly human trait. But it also indicates that most areas that humans are interested in quickly become sufficiently complicated that it becomes necessary to introduce the simplification that classification systems provide.

we've also seen here at Maffei 1 that not all the elliptical/spheroid galaxies are the same (Table 15.1). Nor are all spiral/disk galaxies the same. That's apparent from comparing M 31, our Galaxy, and M 33 (Table 14.1, Destination 14).

Let's consider elliptical galaxies first, mostly because they superficially seem most similar to one another. They should be easiest to classify. Indeed, direct photos of these galaxies typically reveal systems that do, broadly speaking, closely resemble each other (Figure 15.8). Often, the only feature that readily distinguishes them is their apparent shapes. Some of these galaxies—for example, M 32 (see Figure 15.6)—appear nearly perfectly circular in the sky. Others, like Maffei 1, are far more flattened in appearance. Round elliptical galaxies are classified as E0 (that's a 'zero,' not the letter 'O') systems. At the other extreme, the flattest elliptical galaxies are denoted E7 with a smooth transition between zero and seven between these two types (see Figure 15.8 for details on how the shape index—the number used to specify the galaxy type for ellipticals—is calculated). In practice, what this means is that a galaxy classified as E7 appears more elliptical than an E6 galaxy, which is more elliptical than an E5, and so on.

This classification system is very easy to apply (Figure 15.8), but it does beg a question: How does the actual three-dimensional shape of the galaxy relate to its apparent shape? We've already seen that when we view spiral galaxies from various angles, they can appear anywhere from very elongated to nearly circular (Destination 14). Yet, we know that all spirals are—in reality—essentially thin, circular disks. Since we are now adept at extragalactic travel, we could just move to different locations to convince ourselves that elliptical galaxies exhibit genuine variations in their three-dimensional shapes. But we don't have to. We can actually prove this without ever leaving Earth.

The reasoning is statistical in nature, but straightforward. Imagine that *all* elliptical galaxies are intrinsically of type E7. That is, when seen from the side, they appear to have the shape of a flattened ball like a football or rugby ball. If that's true, we'd obviously never see any galaxies flatter looking than about E7. But we would, by chance, occasionally see some of the galaxies end-on, making them appear essentially round in the sky. We'd classify those systems as E0 galaxies. It turns out—this is where the statistics enter the argument—such misclassifications should be comparatively rare *if* all galaxies really are E7 in shape. In the same sense, we should typically expect to see these galaxies from a perspective that makes them appear pretty flattened, and so we should see many more elongated galaxies (types E5 to E7) than round galaxies (types E0 to E2). As it happens, that's *not* what is observed. When we identify and study large samples of elliptical galaxies, it becomes clear that there are far too many E0 systems to explain by chance if all the galaxies are in reality E7 systems (see Figure 10.6 for a graphic example of this). Consequently, at least some elliptical galaxies must be nearly spherical in shape. On the other hand,

Fig 15.9

A montage of spiral galaxies illustrating a range of Hubble types. Distinguishing features include the contrast, openness, and smoothness of the arms, and also the prominence and shape of the central bulges. These images are displayed here to enhance the relative brightnesses of the spiral arms and the central bulges of the galaxies.

Photos by A. Block/Mt Lemmon SkyCenter/U Ariz (NGC 2841), R. Gendler and A. Hornstrup/ESO (NGC 1232), R. Gendler, J.-E. Ovaldsen, C. Thöne, and C. Feron/ESO (NGC 1365), and P. Massey and K. Neugent/Lowell Obs (NGC 6946).

we do see flattened elliptical systems, so some must truly be ellipsoidal in shape. Thus, there is real variety in the three-dimensional shapes of elliptical galaxies. This is a remarkably important conclusion, one that we reached merely by classifying elliptical galaxies by reference to their shape on the sky. This is a promising start to our efforts at galaxy classification.

What about the spiral galaxies? We have already noted that these show systematic variations exemplified by M 31, our Galaxy, and M 33 (Destination 14, see Table 14.1). Similar trends are seen among spiral galaxies outside the Local Group (Figures 15.1 and 15.2). Some spirals have majestic spiral arms. Typically, these galaxies also seem to possess prominent central bulges similar or even larger than observed in M 31. Such galaxies are classified as *Sa Galaxies*. Other spirals appear to have less-distinct arms and less-prominent bulges. They more closely resemble our Galaxy and are designated *Sb Galaxies*. Other galaxies have indistinct arms that are clearly spiral in form when first glanced at, but are often very difficult to trace for long distances within a given galaxy. Like M 33 (Destination 14), these galaxies rarely possess much of a central bulge and are known as *Sc Galaxies*. Some spiral galaxies have particularly indistinct arms and no obvious bulge component at all; these are known as *Sd Galaxies* (Figure 15.9).

When present, the bulges of spiral galaxies appear remarkably similar to elliptical galaxies—usually of type E5 or flatter. The only real difference is that, unlike a normal elliptical, bulges are surrounded by the exponential disks that define spiral/disk galaxies. A significant fraction of all spiral galaxies—maybe 20–30%—are different. While they still have central bulges—sometimes

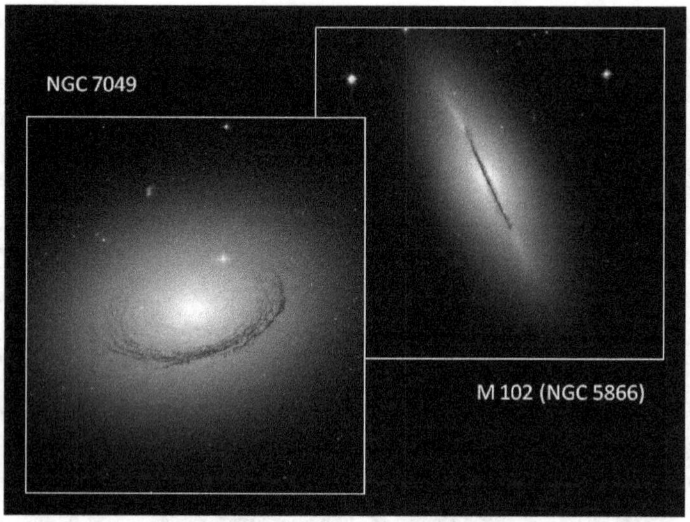

Fig 15.10

Two examples of S0 galaxies. Notice how each system shows evidence of disk-like features, both in the presence of gas and dust and, particularly in the case of M 102 (*right*) a weak stellar disk. Although S0 galaxies reveal their hybrid nature most clearly, many elliptical galaxies—when studied in detail—also reveal evidence of gas and/or stellar disks buried within them. And the reverse is true too: Most spiral galaxies with bulges effectively have an elliptical galaxy at their centers. This is not too surprising. Virtually all galaxies tend to reveal at some level the dual characteristics of spheroidal and disk systems.

Photos from STScI/NASA.

prominent, sometimes not—these tend to be notably elongated in appearance. Instead of looking like an elliptical galaxy trapped in a disk, these are often reminiscent of the shape of a candy bar or a bar of gold. We saw an example of such a *Central Bar* (or, simply *Bar*) in the LMC (Destination 12).[24] Spectra of these bars or, in some cases, optical or infrared images confirm they are composed of stars. When present, bars are typically centered in the galaxy's exponential disk.[25] The spiral arms in such galaxies—known as *Barred Spiral Galaxies*—usually originate from the ends of the bars. As such, they can be classified in broadly the same way as other, unbarred spirals. The key difference is that a capital 'B' is added to the designation. Consequently, *Barred Spirals* are designated as *SBa*, *SBb*, *SBc*, and *SBd Galaxies*.

Though we've considered spiral and elliptical galaxies as distinct types of galaxies, there's one unusual class of galaxy that seems to bridge the two classes. Many of these hybrid galaxies mimic comic-book representations of flying saucers: A disk containing molecular gas, dust, and stars embedded within an otherwise fairly normal-looking elliptical galaxy. There is usually little evidence of spiral arms in the disk, and, even compared to Sa galaxies, the central bulges of these systems are dominant compared to the disk (Figure 15.10). These galaxies are known as *S0 Galaxies* (as before, that's a 'zero,' not the letter 'O').[26] Anywhere from 1% to 5% of all moderately luminous galaxies

---

24 We also noted that the Bulge of our Galaxy may be bar-like in structure (Destination 10).

25 Though not always. The bar in the LMC, for example, is significantly offset from the center of the galaxy's disk and the galaxy's center of rotation.

26 In Hubble's classification scheme, S0 galaxies have always seemed out of place, as if they are weird transition systems between spiral and elliptical galaxies. But, as people have studied these galaxies in greater detail, it has become clear that systems with S0-like characteristics are widespread. We've already seen in detail that spiral galaxies often exhibit bulges. S0 galaxies seem to just be the extreme case of a disk that has a dominant bulge. At the other extreme, it has become clear in recent years that many elliptical galaxies also contain disks—often very small in both size and mass—but rotating

TABLE 15.2 Comparison of Different Galaxy Types

PROPERTY	ELLIPTICAL GALAXIES (INCLUDING BULGES)	SPIRAL GALAXIES	IRREGULAR GALAXIES	DWARF SPHEROIDAL GALAXIES
Mass	10^7 to 10^{14} $M_\odot$	10^{10} to 5×10^{12} $M_\odot$	10^7 to 10^{10} $M_\odot$	10^7 to 5×10^9 $M_\odot$
Luminosity	10^6 to 10^{12} $L_\odot$	10^9 to 5×10^{11} $L_\odot$	10^5 to 10^9 $L_\odot$	10^3 to 10^8 $L_\odot$
Rotation Velocity	50 to 100 *km/s*. Never more than velocity dispersion, usually much less.	50–350 *km/s*	15–100 *km/s*	0 to 30 *km/s*
Velocity Dispersion	50 to 500 *km/s*. Never less than rotation velocity and usually much larger.	Disk: <10% of Rotation Velocity Halo: 80–300 *km/s*	Low, usually <10 *km/s*. Irregular galaxies have minimal or no halos.	5 to 30 *km/s*
Gas Fraction	Very low, often 0%	Sa: <10% Sb: 5% to 20% Sc, Sd: 15% to 50%	5% to >50%	Usually 0%. Rare exceptions may have up to 5%.
Current Star Formation Rate	Typically very low, though with some notable and unusual exceptions.	Sa: <0.1 $M_\odot/yr$ Sb: 0.1 to 2 $M_\odot/yr$ Sc, Sd: 0.5 to >5 $M_\odot/yr$	10^{-4} to > 1 $M_\odot/yr$	Typically none or very low.
Examples (Destination)	Maffei 1 (D15) M 87 (D16) Centaurus A (D21)	The Galaxy (D9–D13) M 31, M 33 (D14) M 81, NGC 253 (D15)	Magellanic Clouds (D12) IC 10, NGC 6822 (D14) Antlia (D14)	Sculptor (D13) Sagittarius (D13) NGC 205 (D14)

in the nearby Universe are classified as S0 galaxies, depending on classification details.[27]

All of these galaxy types—Ellipticals ranging from E0 to E7, Spirals ranging from Sa to Sd (or SBa to SBd), and S0 Galaxies—exhibit a certain amount of symmetry in their structure (Figures 15.7, 15.8, and 15.9). But there are many galaxies that appear to have much more haphazard, often highly asymmetric forms. The official designation for these sorts of systems is the catch-all

disks nonetheless. Thus, the two characteristics that appear about equally prominent in S0 galaxies—their disk and bulge components (see Figure 15.10)—are simultaneously present to some extent within most large galaxies.

27 Some of these details? S0 galaxies with central bars are known as SB0 galaxies. There are also S0 subclasses that are based on the gas/dust content of a given galaxy.

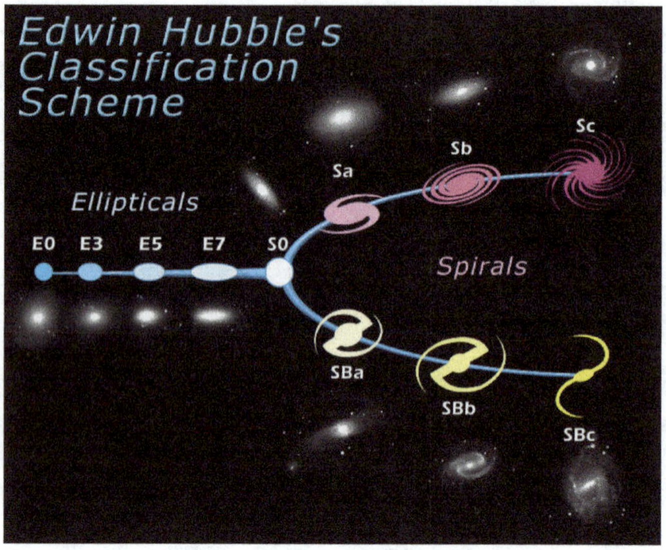

Fig 15.11

A summary of the Hubble Classification Scheme shown as his classical *Tuning Fork Diagram*. Hubble's classification scheme—based only on how galaxies appear in the sky—organizes a wide range of physical characteristics in galaxies (see Table 15.2). The ability to classify many galaxies in this way points to some important underlying processes of how galaxies form and evolve, though these processes are not yet fully understood.

term *Irregular Galaxies*. Many of the dwarf galaxies of the Local Group—for example, the Magellanic Clouds (Destination 12)—and many galaxies in other nearby groups are technically known as irregular galaxies.[28] Although most irregulars are dwarfs, they come in all sizes. When we travel to regions of particularly high galaxy density (Destinations 16 and 20), we'll discover that sometimes even large, luminous galaxies can appear highly irregular in appearance.

This classification scheme we've described here was first introduced by Edwin Hubble in the 1920's, and has remained widely used in astronomy since that time (see Figure 15.11). As a result, this specific classifications of galaxies used in this scheme are known as *Hubble Types*. Hubble believed that his scheme revealed a smooth evolution of galaxies from one type to another, perhaps similar to how low-mass Main Sequence stars gradually evolve into Red Giants (see Destination 11). And he was partly correct. As we'll soon discover, galaxies can indeed change their type, usually as a result of fairly catastrophic processes. More significantly (Destinations 20 and 21), had Hubble lived some 5–10 billion years earlier, he could not have even come up with the same classification scheme *because galaxies were totally different in appearance in the past* (as we'll discover at Destinations 21 and 22). This suggests that Hubble's classification scheme has a rather complicated applicability to real galaxies, one that does not fully describe how galaxies evolved into the forms we see today, but that nonetheless describes those forms fairly well.

And this is why Hubble's scheme remains very useful. From a practical standpoint, it's much easier to say that M 31 is an Sa galaxy than having to always repeat its fundamental characteristics, particularly when we want to

28 The Magellanic Clouds exhibit some features characteristic of Sc or Sd galaxies, yet retain an overall irregular appearance.

compare it to other galaxies. But much more importantly, Hubble's classification scheme, based solely on the appearances of galaxies in the present-day Universe, seems to provide some insight into fundamental differences between galaxies of various types (see Table 15.2). Differences we did not consider as we classified the galaxies in the first place.

To illustrate this point, we can consider a number of galaxy characteristics that are not used explicitly in Hubble's classification scheme. For example, spiral galaxies all exhibit clear signs of rotation, just as we see in M 31 and our Galaxy, while ellipticals generally do not. Like Maffei 1 and M 92, their stars travel along randomly oriented orbits that aren't even approximately circular. As we have seen, we can analyze these motions—rotational or random—to estimate the masses of typical spiral and elliptical galaxies. It's striking that ellipticals come in a wide range of luminosities and masses ranging from giant systems—some ellipticals are the biggest single galaxies known—to tiny dwarf spheroidals. Spirals span a far smaller range in properties. There are, for example, no true dwarf spirals, although some dwarf irregular galaxies exhibit exponential disks that simply lack obvious spiral features (for example, the LMC). Spirals also generally contain considerable amounts of dust and gas, often closely correlated to their subtype. Sa galaxies, for example, have the least gas and dust among spirals, while Sc and Sd systems have the most. What's

Fig 15.12

An image of a large elliptical galaxy (M105 at *right*), an S0/E7 galaxy (NGC 3384 at *top*) and a spiral galaxy (NGC 3389 at *lower left*), all at about the same distance. This illustrates the true relative sizes and colors of such galaxies, a good approximation of what our Galaxy would look like next to a giant elliptical galaxy such as Maffei 1. Although the spiral galaxy in this image is the smallest of the three galaxies shown here, it still takes light over 60000 *yr* to travel from end to end in that 'little' spiral galaxy! We must not forget that these galaxies—all galaxies—are truly gargantuan structures.

Photo by J. Misti.

clear is that virtually all spirals contain far more gas and dust than most normal elliptical galaxies. Not surprisingly, spirals also exhibit evidence of recent or even ongoing star formation—usually throughout their disks. In contrast, most ellipticals appear virtually devoid of recently formed stars. Table 15.2 summarizes various physical features such as these and others that distinguish spiral and elliptical galaxies. The key point is that none of these factors—internal motions, mass/size ranges, gas content, star formation, etc.—play a direct role in how individual galaxies are classified within Hubble's scheme.[29] And yet they all strongly correlate with Hubble type.

The fact that we can tie so many physical properties of galaxies through their Hubble classification is impressive for a system that's essentially based on what galaxies look like. This seems as surprising as discovering that all blonde people have the same weight and age. Or, like saying that all red stars have the same mass, even though we know some might be supergiants much more massive than the Sun, while others might reside on the Main Sequence with masses only a fraction of the Sun's. Hubble's classification scheme may be simple, but it clearly points to something rather profound. There is an underlying order in the process of galaxy formation—even though the process is undoubtedly complex—that results in producing the full range of magnificent galactic systems we see around us today (see Figure 15.12). Systems that, despite all their complexity, we can reliably classify with Hubble's simple scheme.

29 Make no mistake: These properties indirectly relate to the classification system. After all, rotationally supported disks form arms (Destination 14) that are then used to classify spiral galaxies. Dust, gas, and star formation can make arms more or less prominent in individual systems and thus affect how they would be classified. But Hubble certainly did not appreciate all this at the time he developed his classification scheme.

DESTINATION 15 QUESTIONS

15.1. Let's say that you find an elliptical galaxy and measure its long axis to be 20 *arcmin*, and its short axis to be 10 *arcmin*. What is the Hubble type for this galaxy? Assuming that the long axis remains equal to 20 *arcmin*, what would the short axis length be if the galaxy had a Hubble type of E0? What if it had a Hubble type of E7? You'll find the information you need to solve this in Figure 15.8 of this destination.

15.2. This is a web question. Provide pictures of two elliptical galaxies of different Hubble types and at least three other non-elliptical galaxies all of different Hubble types. Print the galaxy images, and include their names and Hubble types.

15.3. Classify the four spiral galaxies in Figures 15.1 and 15.2. Specify at least two characteristics of each galaxy that you used to help determine what type of spiral galaxy each one is. Be as specific about this as you can, given the detailed discussion about the Hubble classification scheme in this destination as your guide.

15.4. The dwarf spheroidal satellite galaxies surrounding our Galaxy are among the least luminous elliptical galaxies known. Centaurus A (often abbreviated to 'Cen A') is one of the most luminous normal elliptical galaxies (this is a different elliptical galaxy than the one we visited: Maffei 1).

a) If Cen A is 50 times further away than one of these dwarf spheroidal galaxies but Cen A appears 10 times *brighter* as seen from Earth, how much more luminous is Cen A? Use the same relations between distance, brightness, and luminosity that we introduced initially for stars. With care, these relations can also apply to galaxies as demonstrated here.

b) Assume that this luminosity ratio is equal to the mass ratio of these two types of galaxies; that is, assume the mass-to-light ratios are equal. Find an example among one of the classes of the animal kingdom—for example, birds, or mammals, or reptiles, etc.—of a pair of species that exhibits a similar mass ratio among adults.

15.5. Given the information in Table 15.1, how well do Maffei 1, the Galactic Bulge, M 32, and Sculptor conform to the Faber-Jackson Relation described in this destination? What does this imply regarding the use of the Faber-Jackson Relation as a means of obtaining galaxy distances?

15.6. Elliptical galaxies appear rather simple in structure today, but they must have had pretty exciting early histories. This question explores what we can learn about how star formation proceeded in typical massive elliptical and spiral galaxies.

a) How much mass is in stars in Maffei 1 if we assume the M/L ratio of the galaxy's visible stars is about 2.0?

b) If we assume all the stars formed over a period of 1 *Gyr*, what was the average star formation rate in the galaxy?

c) Compare this to our Galaxy by assuming that it has formed its stars over a 10 *Gyr* period. **HINT:** Follow the recipe to estimate first the mass in stars and keeping in mind that for a spiral galaxy it is better to assume M/L = 1.

15.7. Based on your knowledge of their stellar and ISM contents, describe how the spectra of Elliptical and Spiral galaxies would likely differ. Be as specific as you can and also extend the comparison over as wide a range of the EM Spectrum as you are able.

15.8. For this question, we will try to simulate how we observe elliptical galaxies appear in the sky as seen from Earth.

a) Find an oblong ball—a football or rugby ball would work fine—and determine its intrinsic ellipticity by measuring its long and short axes and then using the Hubble ellipticity formula given in Figure 15.8.

b) Have a friend toss the ball up and down while you take ten (or more) digital pictures of the ball while it is in the air. Be sure the ball is tossed as randomly as possible each time so that its orientation relative to the camera will not necessarily be the same.

c) From the images you took, roughly estimate the type of 'elliptical galaxy' the ball appears to be—either perfectly round, and hence an E0, or as oblong as it can appear—and make a table summarizing your results.

d) In a similar manner, show the results of such an experiment for a perfectly round ball such as a baseball or tennis ball. In this case, you can take another set of photos, or you can ponder a simple mental experiment to get the result.

e) Explain how your results help to confirm the discussion in the text regarding the fact that the distribution of elliptical galaxy shapes implies that not all of these galaxies are oblong systems that we just happen to see end-on.

15.9. For the maximum velocity dispersions listed for each type of galaxy listed in Table 15.2, calculate the temperature of the hottest gas that could be bound to the most massive Elliptical, Spiral, Irregular, and Spheroidal galaxies. Explain why we see x-ray emission from the Coronal ISM only from fairly massive elliptical and spiral galaxies.

15.10. Imagine that Maffei 1 was located where the Andromeda Galaxy (M 31) exists (that is, 780 *kpc* from the Sun). For the following questions, refer to Tables 14.1 and 15.1.

a) How much brighter would Maffei 1 appear from this location than M 31 does in Earth's sky?

b) How much stronger would the gravitational force between our Galaxy and Maffei 1 be compared to the force between our Galaxy and M 31?

c) How large would Maffei 1 appear in our sky, assuming we can trace its structure out to a radial distance of at least 30 *kpc*?

15.11. Although we have largely ignored it in our discussion at this destination, look up basic information on Maffei 2 including (a) its Hubble type, (b) its distance from the Sun, (c) its radial velocity as measured from the Sun, and (d) any other key physical properties of the galaxy such as its mass, luminosity, size, etc.

Destination 16

THE VIRGO CLUSTER: A SKY OF GALAXIES

16.1 THE VIRGO CLUSTER OF GALAXIES

Surrounding us is a sky of galaxies! We are near the center of the *Virgo Cluster*[1], some 18 *Mpc* from the Sun (see Figure 16.1). This is easily our longest journey so far—about six times more distant from the Sun than Maffei 1—but one worth the rigors of the trip. The sky around us is utterly devoid of individual stars, but it is far from empty. There are so many galaxies here—about one hundred are easily visible to the naked eye—that, with a glance, we can see a dozen simultaneously[2]. It is a sublimely beautiful sight. But also a bit disturbing. We have learned enough about galaxies that the

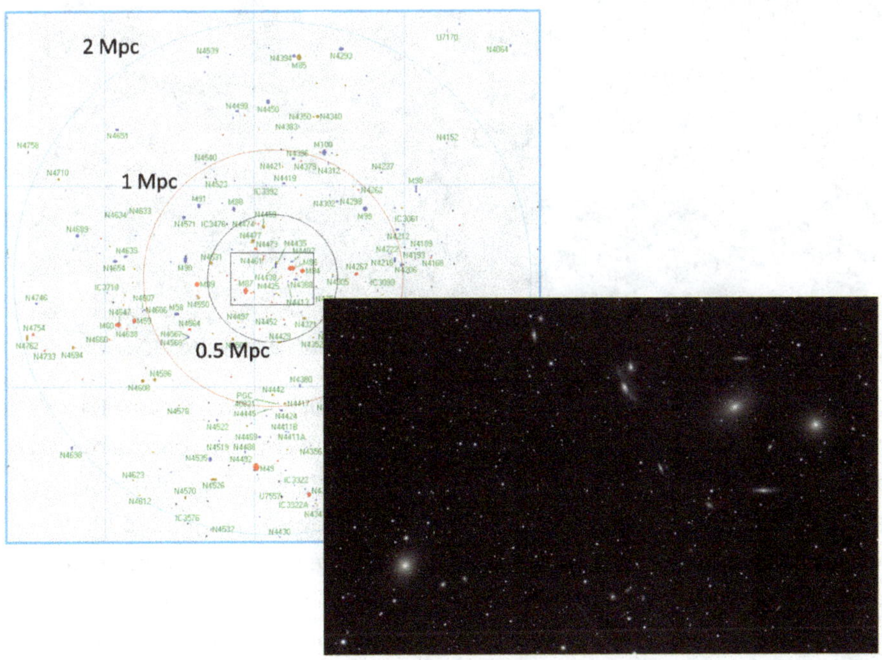

Fig 16.1

(*Lower right*) A visual image of the central region of the Virgo Cluster as seen from Earth. Three of the largest elliptical galaxies of the cluster, M 87 (lower left), and M 86 and M 84 (left and right, respectively, of the pair in the upper right) are clearly visible. They dominate the view from our vantage point roughly midway between M 87 and M 86. (*Upper left*) A map of the sky from Earth showing the principal galaxies of the Virgo Cluster. The concentric circles are at 0.5 *Mpc* (black), 1.0 *Mpc* (red), and 2.0 *Mpc* (blue) from the cluster center. The field in the image at *lower right* is shown as a black rectangle in the map near the cluster center. In this map, red symbols represent elliptical galaxies (giants and dwarfs), brown are S0 systems, and blue/purple symbols denote spiral and irregular galaxies.

Photo by M. Theusner. Map based on data compiled by R. Powell.

1 Many bright galaxy clusters are named after the constellation in which they are located. At least where their *centers* are located. The Virgo Cluster—the closest to Earth—is quite large in the sky (see Figure 16.1); it spills well outside the boundaries of its namesake constellation.

2 If we are willing to go to the minimal trouble of using even a modest pair of binoculars, we would see nearly 600 galaxies in this sky! Some are so close that they would reveal detailed features such as H II regions, spiral arms, globular clusters, and galactic nuclei. From Earth, there are about 6–8 galaxies that can be seen with modest binoculars, but only the Magellanic Clouds are close enough that we can easily see some of their internal details with such modest optical aid (Destination 12).

profound implications of what we are seeing in this Alien Sky slowly dawns on us. These pale, fuzzy patches of light are actually vast structures extending tens of thousands of *parsecs* across, each one comprised of billions—in some cases, trillions—of stars. And here in the heart of the Virgo Cluster, we discover that galaxies can congregate together to produce an even bigger structure! And, as a bonus, a wondrous view.

One side of the sky is dominated by M 87 (Figure 16.2), a truly giant elliptical galaxy that is about four times more luminous than Maffei 1 (Destination 15) and nearly ten times more luminous than our Galaxy (Destinations 13 and 14). From our vantage point near the cluster center, this behemoth shines

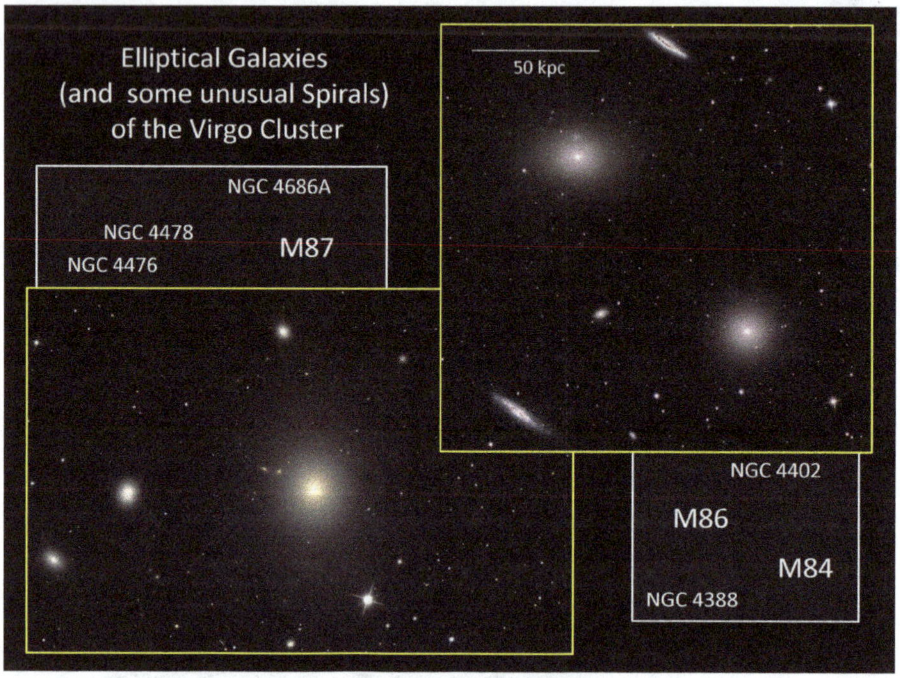

Fig 16.2

Elliptical galaxies of the Virgo Cluster. (*Lower left*) The giant elliptical M 87 dominates one side of our sky from the cluster center. It is such a large galaxy that its 'satellites' NGC 4686A, NGC 4478, and NGC 4476 (visible in the image) are comparable in luminosity to our Galaxy! (*Upper right*) The other side of our sky contains M 86 (left in the image) and M 84 (right), both a little less luminous than M 87. All three of these giant elliptical galaxies are much more luminous than Maffei 1 (Destination 14), the first 'giant' elliptical galaxy we encountered on our journeys.

Photos by A. Block/Mt. Lemmon SkyCenter/UA (*left*) and G. Morgan (*right*).

with a total brightness equivalent to some of the brightest stars in Earth's sky. The complex structure associated with its unusually bright core is readily apparent; we'll explore the detailed nature of the enigmatic nuclei of massive galaxies at Destinations 17 and 21. For now, it is the soft, extended glow of the main body of M 87 that fascinates us. Here before us is the combined light of over one trillion stars, stretching some 3.5 *deg* in our sky, equivalent to seven times the diameter of the full Moon as seen from Earth (Destination 1)[3]. Nearby are three 'satellite' galaxies. But what satellites! One of these, the elliptical galaxy NGC 4478, is almost as luminous as our Galaxy (Figure 16.2).[4] By themselves, this small grouping of galaxies makes an amazing spectacle,

3 Next time you see the full Moon rising, imagine it as seven times wider, and as a galaxy containing well over 1 trillion stars!

4 That really is some satellite!

yet there's much more to see here. As we turn to view the opposite side of the sky[5], we see two other giant elliptical galaxies—M 84 and M 86—shining nearly as brightly as M 87, and each extending over an even larger area in the sky. Off to the side we can easily spot M 49, the most luminous, largest galaxy of the Virgo Cluster. Though M 49 is almost twice as distant from our location near the cluster center as M 84, M 86, and M 87, it still manages to be one of the brightest, largest galaxies in our sky. It is clearly a giant among giants. As we ponder the view, it seems that most of the bright galaxies we see near the center of the Virgo Cluster are elliptical galaxies. And, indeed, it is not an illusion: Eight of the ten brightest galaxies in this sky are spheroid systems that we would classify as E (elliptical) or S0 galaxies using Hubble classification scheme (Destination 15).

One of the bright galaxies in this sky that's not a spheroid is M 91 (Figure 16.3), a classic example of a barred spiral galaxy of type SBb in Hubble's classification system (Destination 15). Hanging over us, well separated in the sky from any of the brighter elliptical galaxies in our sky, the central bar of M 91 is plainly visible, centered in a pale gray disk that stretches about 2.5 *deg* in the sky. As we study other galaxies around us, we notice that spiral galaxies become relatively more common among the fainter systems in this sky. Some of these—for example, M 58 (Figure 16.3)—shine brightly enough that we can discern their central bulges. Even their delicate spiral arms—snaking through disks that visually span 4–5 times the diameter of the Moon—are faintly visible. Oddly, some of the dimmer—and as it turns out, a bit more distant—spiral systems such as the majestic galaxies M 88 and M 99 (Figure 16.3) often exhibit more pronounced spiral structure than the brighter, typically closer spiral galaxies visible in the sky near the cluster center.[6]

Other Hubble types are also well represented among the galaxies of the Virgo Cluster. No fewer than ten S0 galaxies—systems that are rare outside of clusters—are readily visible from here with the unaided eye. As we noted

---

5 There are no stars at this location—or, more precisely, too few to reasonably expect that one might be located nearby. In such a place the chance of finding a planetary base is close to zero. We are just floating in space, holding our position close to the cluster center. We don't have to wait for a planet to rotate to reveal new parts of the sky. We can inspect the entire sky from one spot, at one time, just by looking around. We should get used to this. There are few planetary systems we can expect to encounter at any of our future destinations.

6 If it seems there are a lot of Messier objects in the Virgo Cluster, it's no illusion. Thanks largely to the presence of the cluster, Messier discovered 14 objects within a region 10 *deg* in diameter in this part of the sky. One of these, M 49, was the first Virgo member galaxy ever found. In this regard, Messier—through his heirs of course—can technically lay claim to discovering the cluster itself! All of the Messier objects associated with Virgo are excellent comet imposters when viewed through a small telescope. It's quite possible Messier disliked springtime—when Virgo is most easily visible in the sky—even more than summer when globular clusters are common (Destination 11).

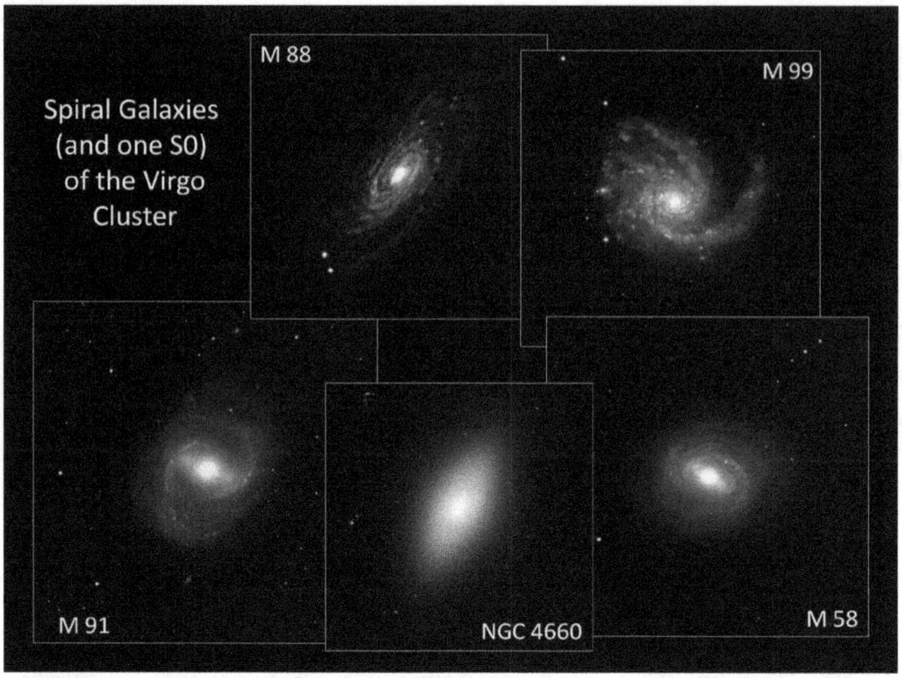

Fig 16.3

Four spiral galaxies and one S0 galaxy of the Virgo Cluster, all visible to the naked eye from our current location at the center of the Virgo Cluster. All of these galaxies are located outside the core region of the cluster (Figure 16.1), though NGC 4660 is quite close to M 60, a large elliptical galaxy in the Virgo Cluster. We shall find this is a general trend. Spiral galaxies favor the lower-density regions of clusters like Virgo, while S0 and elliptical galaxies tend to be found near the higher-density regions closest to cluster centers. These trends define what is known as the *Density-Morphology Relation*.

Photos from STScI/NASA (NGC 4460) and NOAO/NSF (all others).

earlier, a few of these are among the brightest galaxies in our sky (Figure 16.3). Many fainter galaxies we see here appear considerably more indistinct, often just faint puddles of light to our unaided eyes. Most of these are plainly larger than the extent of the full Moon from Earth, implying that they are comparatively nearby. Some dwarf galaxies that can be seen in this sky—for example, NGC 4431 and IC 3363—illustrate this point. Though low in luminosity, they're visible to the naked eye from the center of the cluster because they happen to be nearby. Unless we've been uncommonly lucky, this suggests that Virgo contains many dwarf systems.

This whirlwind tour of the sky near the center of the Virgo Cluster reveals many fundamental characteristics of galaxies and clusters in general. Perhaps the most obvious is that galaxies seem to prefer to hang out together, a characteristic we first noted the moment we left our Galaxy (Destination 13). Dwarf galaxies tend to congregate around larger galaxies—such as our Galaxy and M 31—in groupings that typically span hundreds of *kpc* (Destinations 13 and 14). Big galaxies—along with their satellites—themselves tend to band together to form groups such as the Local Group and the Maffei Group (Destinations 14 and 15). Such groups have typical sizes of around 1–2 *Mpc*. *Galaxy Clusters* such as the Virgo Cluster represent the next step—but not the last one!—in a hierarchy of what astronomers typically refer to simply as *Structure*. Instead of a few dozen dwarfs surrounding a larger parent galaxy, or a few large galaxies comprising a single group, clusters typically contain thousands—sometimes tens of thousands—of galaxies. Though Virgo is generally regarded as a comparatively small galaxy cluster, it still contains about 30 galaxies more luminous than our Galaxy, 16 that are more luminous than Maffei 1. In total, the cluster contains over four thousand galaxies[7], all crammed in a volume that's not that much bigger than the Local Group.

And that's the key. What makes a cluster such as Virgo stand out is not merely the number of galaxies it contains, but rather the density. The cluster itself is around 4–4.5 *Mpc* in diameter, roughly eight times larger in volume than the Local Group. Since the cluster conservatively contains about 80 times more galaxies[8], the galaxy density is ten times greater than in the Local Group. Another way to state this is that the average separation between luminous galaxies in Virgo is a few hundred *kpc*, not 0.5–1 *Mpc* as is typical in groups. It is no surprise that we see so many galaxies in the sky here: There really *are* a lot of galaxies jammed into a relatively small volume in the Virgo Cluster. As a result, many of them are close enough to the cluster center and sufficiently luminous to appear comparatively bright to us as we view them in this spectacular Alien Sky.

But not only beautiful. Having so many galaxies in one place is also very convenient. By itself, the Virgo Cluster acts as a sort of showcase of the broad range of galaxy types defined in Hubble's classification scheme (Destination 15). There are elliptical galaxies such as M 49, M 84, M 86, M 87, and NGC 4478 (Figure 16.2). Many of these are considerably more luminous than Maffei 1, but there are many dwarf ellipticals and dwarf spheroidals here as well. Some of these galaxies appear nearly circular in form; for instance, luminous M 87 is clearly an E0 galaxy. Others are more elongated; M 86 is usually classified as an E3 or E4 galaxy.[9] The numerous spiral galaxies in Virgo—we've already seen M

7 As of 2012, the Virgo Cluster has about 1600 cataloged member galaxies. However, this census is incomplete. Due to its large distance (about 18 *Mpc*), it isn't easy to find low-luminosity galaxies in Virgo, and certainly impossible to find galaxies as low in luminosity as the faintest systems known to exist in the Local Group (Destination 13). If the Virgo Cluster has a similar proportion of faint galaxies, its total galaxy content could easily exceed four thousand, and perhaps even reach 6–10 thousand. This inability to count everything, sometimes even roughly, is a common problem in astronomy and is known as *Incompleteness*.

8 As noted in the previous footnote, it is a little tricky to make this comparison because the completeness of galaxy counts in the Local Group and the Virgo Cluster are quite different. One way to deal with this is to restrict the galaxy count in the Local Group to about the same luminosity limit that we can reach at the distance of the Virgo cluster. If we do that, we find that only about 20 of the 70 or so galaxies of the Local Group would be visible if they were as distant as the Virgo Cluster is from Earth. For the sake of comparison, we'll assume there are about 1600 galaxies in Virgo, and just 20 in the Local Group to the same minimum luminosity.

9 As we saw at Destination 15, the Hubble classification scheme for elliptical galaxies depends on our viewing angle for individual galaxies. What the shapes of the elliptical galaxies in Virgo would be when seen from the cluster center is difficult to guess. This illustrates a key limitation to Hubble's classification scheme for elliptical galaxies. Some astronomers have suggested different approaches such as classifying ellipticals by the degree to which they exhibit disk-like features (see Destination 15). But that's generally much harder to determine than the apparent shape of a galaxy on the sky. As a result, Hubble's classification scheme as described in Destination 15 remains in common use.

58, M 88, M 91, and M 99 (Figure 16.3)—span the full range of possible types, barred and unbarred, and ranging from subtypes Sa (or SBa) through Sd (or SBd). And, as we've seen, even S0 galaxies (Figure 16.3)—uncommon in most environments—are relatively numerous in this cluster. Hubble's basic classification scheme also covers the most common types of galaxies in Virgo, the dwarf irregular and dwarf spheroidal galaxies. About 1200 of these are known in the cluster, comprising roughly 80% of all cataloged galaxies in Virgo.[10] It seems that Hubble's system (Destination 15) does a very good job of characterizing the galaxies of the Virgo Cluster.

16.2 GALAXIES IN COLLISION

But not a perfect job. There are some galaxies in Virgo that fail to fit neatly into Hubble's scheme. One good example happens to be one of the brightest galaxies in our sky: NGC 4438 (Figure 16.4). Though comparable to M 31 in size and luminosity, NGC 4438 is no 'normal' spiral galaxy and certainly not any sort of elliptical or S0 system. It is unlike any galaxy we've encountered at any earlier destination or elsewhere within the Virgo Cluster. And yet, strange as it is, NGC 4438 is not entirely unrecognizable. It possesses some familiar features—ones we've seen before in normal galaxies—but that are weirdly distorted. For example, we see dark lanes and patches in its light distribution, revealing the presence of molecular clouds of the Cold ISM, reminiscent of what we might expect to see in a disk galaxy (Destinations 6 and 9). Radio telescopes reveal that NGC 4438 emits 21-*cm* radiation from neutral H atoms in its Warm ISM (Destination 5), another common feature of disk galaxies (Destination 14). Its gas content has a mass of about $10^9 M_\odot$, not too unusual and approximately half the mass of gas seen in M 31 (see Table 14.1 in Destination 14). NGC 4438 also contains a distinct stellar component that has the appearance of a small elliptical galaxy or perhaps a galactic bulge. But surrounding this bulge-like component is no normal disk. Though full of young stars and H II regions like many disks in spiral galaxies, this component in NGC 4438 has a highly asymmetric shape more like that of a potato chip than a normal, flat disk. And then there are the arms. The long structures we see in NGC 4438 are weirdly angular

---

10 Though dwarf galaxies certainly outnumber the larger galaxies in all clusters and groups, even when added together, they invariably amount to less than the total mass and luminosity contained in the most luminous 10% of cluster or group members.

in appearance, not at all like the graceful spiral arms we've grown accustomed to seeing in the disks of typical spiral galaxies.

Try as we might, we can't really fit this galaxy into any of the basic types defined in Hubble's classification scheme. Is the central bulge-like structure an elliptical galaxy, or even an S0? Are the unusual arms and distorted, asymmetric disk somehow related to what we see in normal spirals? Also, it seems too luminous and, somewhat ironically, too strange and asymmetric in structure to be classified as an irregular system. Galaxies like NGC 4438 that defy simple classification with Hubble's scheme are often generically referred to as *Peculiar Galaxies.*[11] Some peculiar galaxies resemble NGC 4438 by having oddly juxtaposed or moderately distorted features that would appear familiar if seen in 'normal' galaxies. Many peculiar galaxies can only be described as bizarre; some exhibit internal structures unlike any we've seen before (see also Figure 16.6). One

Fig 16.4

(*Right*) A close-up view of the peculiar galaxy NGC 4438 near the center of the Virgo Cluster (the galaxy is visible in Figure 16.1). The highly peculiar structure of the galaxy is evident, including its unusual bulge-like component, its highly distorted 'disk,' and its long, angular 'arms.' But it also contains some 'normal' features typical of spiral/disk galaxies, including clouds of dust and molecules comprising its Cold ISM, neutral hydrogen in its Warm ISM, young stars in the 'disk' region, and older, redder stars in its bulge-like component. (*Left*) A composite image of NGC 4438—and its companion NGC 4435—and the large elliptical galaxy M 86. The red filaments trace the distribution of ionized gas distributed along a 'bridge' between the two galaxies. The gravitational interaction between NGC 4438 and M 86 has likely produced the strange structure of NGC 4438, while gas surrounding M 86 (as in Maffei 1, see Destination 15) has helped to strip out and ionize the gas from NGC 4438 to produce the long filaments.

Photos by K. Crawford (*right*) and T. Tal and J. Kenney/NOAO/NSF (*left*).

11 This is obviously a catch-all sort of classification that, in effect, admits that some galaxies cannot be classified! However, one should be careful not to assume that all peculiar galaxies are peculiar for the same reasons. While it might be reasonable to expect that all normal spiral galaxies have had similar histories that led to their distinctive appearances, we'll see that there seem to be many ways to make a galaxy 'peculiar.' On top of this, as we learn more about galaxies in general, it becomes easier to spot oddities in just about any system. In that respect, virtually all galaxies are peculiar in some way. As a general rule, astronomers tend to assign the term 'peculiar' in cases where it is suspected that external effects may be strongly affecting a galaxy's morphology or the galaxy is particularly luminous. In contrast, galaxies are considered 'irregular' if their forms are likely dictated by internal processes. Usually, these tend to be comparatively lower luminosity systems.

common feature about peculiar galaxies is that they're often found in compact galaxy groups or near the centers of galaxy clusters (see Destination 19). This must be telling us something important about how such peculiar galaxies come about. But what is it about these environments that favor the presence of such unusual galaxies?

We've already seen that the Virgo Cluster is considerably more densely populated with galaxies than the Local Group by about a factor of ten. But this is only part of the story. Galaxy groups and clusters—including Virgo and the Local Group—are not static. The galaxies are in motion relative to one another. As it happens, the two large galaxies of the Local Group—our Galaxy and M 31—are currently approaching one another at about 150 *km/s*, implying that the two may 'collide' in about 5–6 *Gyr*. In Virgo, typical closing speeds between galaxies are much higher, easily exceeding 1000 *km/s* (we'll explore the details of the motions of galaxies in Virgo a bit later in this destination). Larger velocities? Smaller separations? If galaxy collisions can conceivably occur in low-density environments such as the Local Group (see Destination 13), it seems clear that collisions between galaxies should occur considerably more frequently in crowded regions such as the Virgo Cluster.

And they do. When we analyze the motions of galaxies in a dense environment such as Virgo in a little detail[12], we infer that a pair of galaxies is likely to collide somewhere within the central region of the Virgo Cluster every 0.5–1 *Gyr*. That sounds like a long time, but in fact it's approximately equal to the time a single galaxy requires to travel across the central region of the cluster. What this means is that the odds are good of seeing colliding galaxies—or recently collided systems—near the center of a cluster such like Virgo. Does this explain the peculiarities of galaxies such as NGC 4438? To answer that, we need to explore what we mean when we say that galaxies 'collide.'

As it happens, galaxies don't run into one another in the same way that things collide in everyday experience. For instance, when cars run into one another, the problems that arise stem from the fact that the components of the individual cars involved in the collision are trying to share the same volume of space at the same time. At the atomic level, the individual particles in the cars—the atoms and molecules that comprise them—can't easily pass by one another due to the strong electrostatic repulsion of their electrons. Those forces guarantee that the cars come into what we refer to as 'contact.' The severity of such collisions depends on other factors, such as the relative speeds of the cars, their directions of motion, and also the sizes and masses of the cars. Consequently, 'car interactions' can range from minor fender benders to far more serious incidents that may induce considerable damage.

Galaxies are different. Because they contain mostly empty space, certain components of galaxies can pass right through each other without suffering

12 And you can do this for yourself in a problem at the end of this destination.

what we would call direct physical contact. In addition, galaxies can strongly affect one another even before they 'collide'—and even sometimes even when they don't collide at all—through their mutual gravitational influences. These differences make galaxy 'collisions' far more varied in character than the comparatively 'simple' range of collisions between things such as cars or balls. Astronomers tend to use a broader term—*Galaxy Interactions*—to describe the range of phenomena associated with galaxy collisions and near-misses. Systems that are involved in such events are, quite reasonably, referred to as *Interacting Galaxies*.

Broadly speaking, there are three major types of galaxy interactions. The tamest cases occur when galaxies pass sufficiently close to one another to induce mild effects without radically altering their structures. Some outcomes of mild interactions might be to amplify spiral arms in galaxy disks, or to bend, or warp galactic disks (this may be happening in M 31; see the 21-*cm* map of the galaxy in Figure 14.6). The next level of interactions occur when massive galaxies pass only a few diameters apart, close enough to produce strong *Tidal Effects* in their neighbor. In such cases, the gravitational forces experienced by stars that lie between the galaxies can differ greatly from the forces at the op-

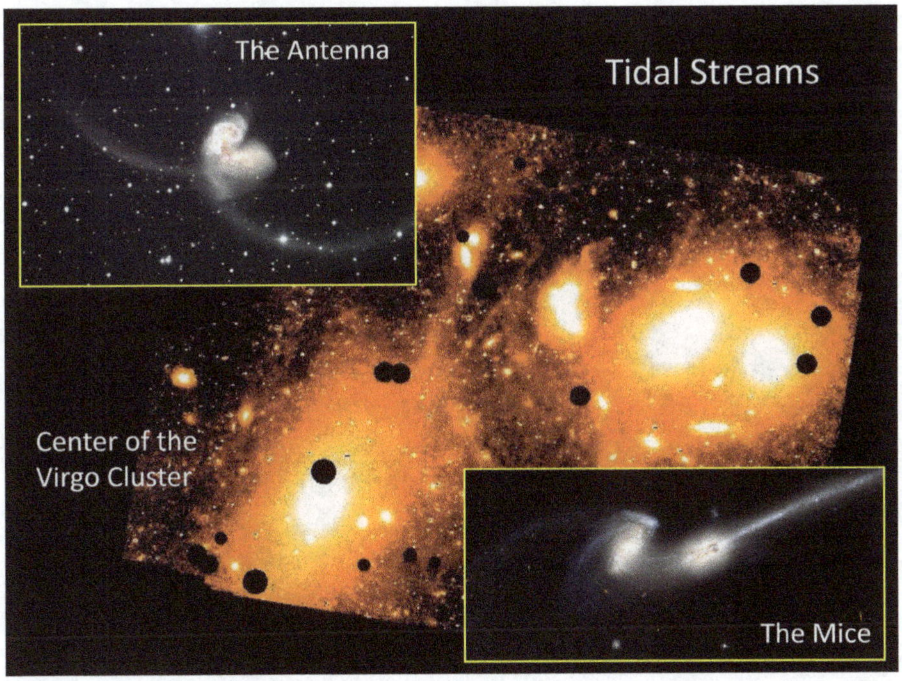

Fig 16.5

Examples of tidal streams associated with galaxy interactions. (*Upper left*) *The Antenna* is a collision between two disk galaxies similar in size to our Galaxy. Although the main bodies of the galaxies are only now coming into close proximity, tidal streams were generated even before the main collision began. (*Lower right*) *The Mice* is another example of how tidal streams are generated as galaxies interact. (*Center, in orange*) A very 'deep' image of the center of the Virgo Cluster—this image covers almost the same region shown in the lower right image in Figure 16.1—shows linear features and plumes all likely associated with moderate to strong galaxy interactions near the cluster center. The round 'bullet holes' in the image correspond to locations of bright foreground stars that were removed during the processing of the image. A large fraction of all the features visible in this image are galaxies, most members of the cluster.

Photos from ESA (Antenna), STScI/NASA (The Mice), and C. Mihos/CWRU (Virgo center).

posite ends of the galaxies. If the galaxies involved are disk galaxies, stars that may have been traveling along circular orbits—such as the Sun in the Galactic Disk (Destination 9)—will now follow different orbits due to the change in the gravitational field. This process can lead to the formation of large, potentially complex features known as *Tidal Arms* or *Tidal Streams* (Figure 16.5).[13] Because the internal motions in galaxies can be quite large, tidal streams can take on remarkable forms. Some are long, thin loops. Others may appear as straight or angular streams. Sometimes, tidal encounters create features that extend off in multiple directions, and even, at times, in the direction of motion—that is, *ahead* of—a galaxy in an interacting system (Figure 16.5)!

But the real fireworks arise in the third class of galaxy interactions. In these cases, galaxies really do 'collide' in the sense that they share the same volume of space at the same time. As strongly interacting galaxies approach one another, they induce noticeable gravitational tides even before they meet. The tidal features produced in these instances tend to be quite bright (see Figure 16.5), pulling out many stars and sometimes much of the gas within the individual interacting galaxies. The presence—or lack—of tidal features can help us decide if two galaxies are truly interacting, or if they just happen to be close to one another in the sky. The most bizarre—and often spectacular—effects occur just before, during, or just after direct collisions (Figure 16.6). To begin to appreciate the details, we have to expand on an earlier point: Some components of galaxies—specifically, those composed of stars—can pass through one another during galactic collisions. But others—in particular, the ISM—cannot.

Fig 16.6

A gallery of galaxy interactions. Most of these galaxy pairs would be hard to classify in Hubble's classification scheme because they are in the process of being transformed in a major interaction. At upper left, for example, one disk galaxy has passed through the middle of another. The gas in the 'bullet' appears to be partly stripped, partly in the throes of intense star formation, while the target galaxy has become a ring of star formation! Other panels show galaxies being shredded or distorted, tidal arms being produced, and normal galaxies being transformed into bizarre shapes. Some of these interacting systems closely resemble different stages in the process of major mergers as calculated using modern computer simulations (see Figure 16.8).

Photo mosaic by A. Evans/STScI/NASA.

13 If this process seems odd or hard to envision, imagine that a second star passes through the inner Solar System. As the star approaches the Sun, planets such as Earth start to feel the sum of the gravitational forces of the two stars, causing the planet to follow a new, possibly radically different orbital path. The same is true of stars within interacting galaxies, except that, for galaxies, there are so many stars following similar orbits that coherent streams can be produced (see Figures 16.6 and 16.8).

Let's consider first what happens to the stars. Unlike galaxies, individual stars are minuscule compared to the typical distances separating them. The Sun, for example, is about 40 million Solar radii from its nearest neighbor, Proxima Cen (Destination 4). The Milky Way, on the other hand, is about 50 times the radius of its optically visible disk away from M 31. Given the typical sizes of galaxies and the total number of stars they contain, it is easy to show that stellar collisions are extremely rare. That's not to say stars in interacting galaxies are entirely unaffected. We've already seen that tidal effects can cause some stars to be pulled out of their galaxies to form gigantic tidal streams and arms. But if we were to look closely within tidal features or even the centers of interacting galaxies when they are colliding, the stars themselves only very rarely run into one another.

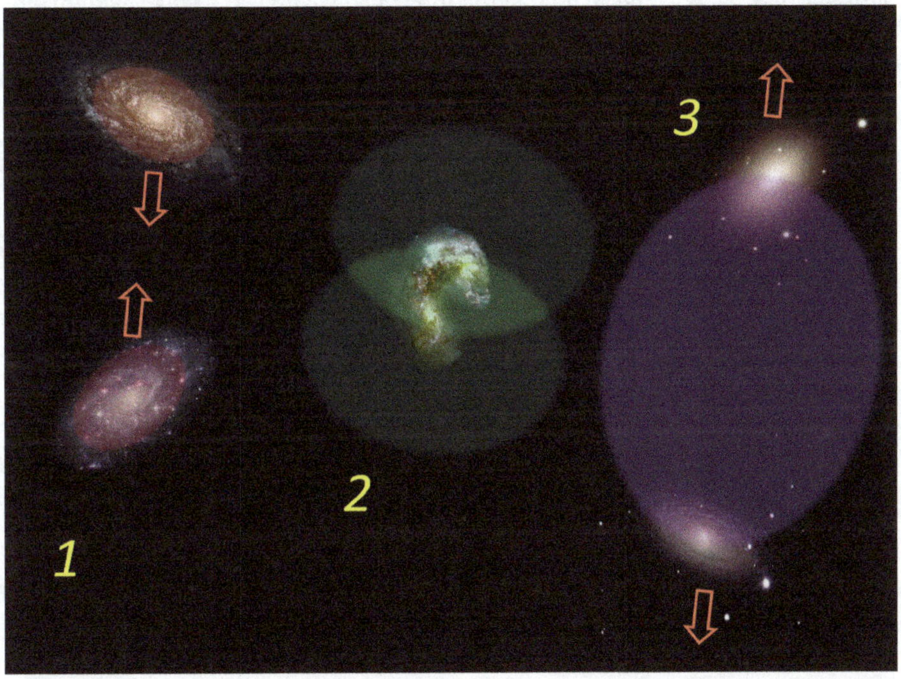

Fig 16.7

A schematic illustration of how gas stripping can remove the gas from interacting galaxies. At stage 1 (*left*), two normal gas-rich spiral galaxies approach one another. Their comparatively cool gas is shown in red within the disks. At stage 2 (*center*), the galaxies collide, causing the gas to compress and form stars. The image shown here is of the center of the Antenna galaxy (see Figure 16.5) where star formation is occurring at an exceptionally high rate. This also heats up all gas not involved in star formation, as shown by the green regions surrounding the star-forming core. During this phase, the interacting galaxies may emit copious ultraviolet and infrared radiation associated with young stars and heated dust in star forming regions. At stage 3 (*right*), the gas-poor galaxies continue on, leaving behind a significant fraction of their gas as a very hot IGM component—shown as a purple cloud—that begins to orbit within the cluster independently of its parent galaxies. The efficiency and heating of this process depend sensitively on the details of the collision, but the stripped gas in the galaxies can now have temperatures of many million *K*. This figure suggests the galaxies are not significantly altered due to the interaction (stage 3), but in fact both systems would likely become distorted and form tidal arms after an interaction like this.

Galaxy images from STScI/NASA., and Adam Block/Mount Lemmon SkyCenter/University of Arizona

The ISM behaves quite differently. This may seem strange at first. After all, compared to their sizes, atoms and molecules in the ISM are even more widely separated from their nearest neighbors than stars are.[14] The difference is that the total number of atoms and molecules in the ISM—over 10^{60} in our Galaxy[15]—is vastly larger than the number of stars—about 10^{11}. As a result, atoms and molecules in the ISM of interacting galaxies do collide efficiently as the

14 Yet another claim you are invited to explore in a question at the end of this destination.

15 In recent years, numbers such as 'billions' or even 'trillions' have become common to consider in the context of computer storage capacities or speeds, and even in terms of the costs of various loans, bailouts, and debts in national economies. But the number 10^{60} is in a different category. That is one trillion, trillion, trillion, trillion, trillion.

ISMs of galaxies interact.[16] This causes the gas to compress and become denser. Under favorable circumstances this can initiate star formation, sometimes in spectacular bursts along the collision front or in tidal streams (Figure 16.6). For this reason, many interacting galaxies emit intense ultraviolet and blue optical radiation arising from the numerous hot, young stars being formed in the compressed gas, as well as optical emission lines from ionized nebulae (Destination 6). They also tend to be quite luminous in the far-infrared as large amounts of dust—leftovers from the star-formation process (Destination 6)—are heated up by young stars. Interactions between gas-rich galaxies can also remove their gas as the stars move on, leaving their ISM behind, a process known as *Gas Stripping* (see Figure 16.7).

The properties of NGC 4438 are best understood in terms of a direct collision with another galaxy. The strong distortions in the galaxy, the clear evidence of stripped gas flowing behind it (see Figure 16.4), strongly suggest that it collided with M 86, one of the giant elliptical galaxies so prominent in the sky from the center of the Virgo Cluster. Given the typical relative speeds of galaxies at the cluster core, this collision must have occurred some 150 *Myr* ago. Though NGC 4438 was severely affected by this encounter, M 86 was comparatively unscathed due to the fact that it is about ten times more massive than NGC 4438. Such cases where one of the interacting galaxies is considerably more massive than the other represent examples of *Minor Interactions.*[17] These usually result in comparatively small changes in the structures of one, and sometimes both galaxies involved. In contrast, *Major Interactions* typically involve galaxies of more nearly equal mass that pass close to one another. In such cases, both galaxies can suffer transformative changes that radically alter their structure and stellar/gas content; Figure 16.6 show a number of spectacular examples of major interactions.

Single encounters between galaxies are rarely the end of the story. When a galaxy such as NGC 4438 interacts with another galaxy like M 86 (Figure16.4), the systems tend to lose some of their bulk orbital energy. This in turn causes them to slow down and shrink their orbits, virtually guaranteeing they will interact again ... and again ... each time passing closer than before. Eventually, this process can lead to a *Galaxy Merger* (Figure 16.8). In the case of NGC 4438

16 Recall from Destination 5 that the particle density in the Warm ISM is typically around 1 atom per cm^3. At the typical temperature of the gas—100 K to 2000 K—the time for atoms to collide is usually about 1500 years (see Destination 5). However, colliding galaxies require a few million years to fully pass through one another, so this is plenty of time for atoms to collide many times and allow the gas in the two galaxies to mix together, compress, and even commence star formation.

17 Another example is the interaction between the Large Magellanic Cloud (LMC) and our Galaxy. The LMC, apparently starting to pass the Galaxy for the first time, will likely be strongly distorted, perhaps destroyed by the encounter. On the other hand, our Galaxy may develop a warped disk, but likely nothing more severe from this encounter.

and M 86, larger M 86 will eventually subsume the smaller galaxy in a process known as a *Minor Merger.*[18] Such mergers can be transformative to the smaller galaxy—NGC 4438 will completely dissolve into M 86!—but they do not so strongly affect the larger system, apart from making it a bit more massive. *Major Mergers*—ones that involve galaxies of more nearly equal masses—are different (Figure 16.8). In these cases, the merger process tends to radically alter both galaxies. The Local Group is a good example. As we noted earlier, the Galaxy and M 31 are heading toward each other and will almost certainly undergo a major merger some 6–8 *Gyr* in the future. The Local Group will then contain only one principal galaxy (see Figures 16.8), and it won't be a spiral. Our Galaxy and M 31 are destined to merge into a single elliptical galaxy (Figure 16.8), one similar in mass and luminosity to Maffei 1 (Destination 15).

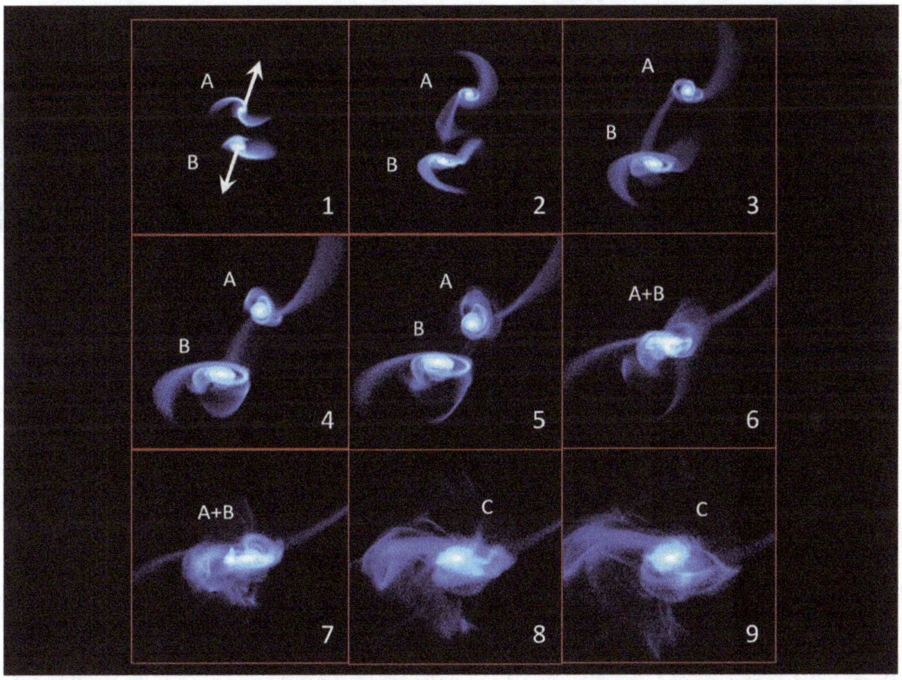

Fig 16.8

Snapshots from a movie simulating the (future) major merger of our Galaxy and M 31, but which can serve as an illustration of any similar major merger between any two galaxies that we'll denote 'A' and 'B.' The first frame (*upper left*) shows the galaxies after their first close passage, about 5–6 *Gyr* from now. Within about 1–2 *Gyr*, the two separate, then re-collide (frames 2–6), eventually merging into a new galaxy—denoted 'C' in the figure. When this happens in the Local Group, its two big spirals will have become a single elliptical galaxy, probably quite similar to Maffei 1.

Image from simulations by J. Dubinski/CITA.

An interesting implication of this discussion is that galaxy interactions and mergers should be most common in the densest parts of clusters and in dense groups. This helps explain why so many elliptical galaxies concentrate near the center of many clusters, including Virgo (see Figure 16.1). These galaxies likely represent the products of repeated mergers near the cluster core, mergers that are still occurring today and will continue to occur far into the future. But the flip side of this argument is that while this activity is happening in the cluster

18 Another example of a minor merger is the Sagittarius Dwarf, which is being torn apart by tidal forces as it falls into our Galaxy (Destination 13). From the dwarf's perspective, this is a pretty severe interaction. But for our Galaxy—about 100 times more massive than Sagittarius—it's not such a big deal.

core, the outer regions of clusters should be considerably more sedate. In the cluster periphery, the density of galaxies is considerably lower and so collisions or even near-misses between galaxies will occur much less frequently. This implies the outer regions of clusters should contain a 'normal' mixture of galaxies, more similar to what's seen in low-density regions such as the Local Group. And indeed, this is very often the case. Outside the dense cores of many clusters—including the Virgo Cluster—we find mostly normal, gas-rich spirals, many dwarf irregulars, dwarf ellipticals, and spheroids (though fewer than in cluster cores), but very few giant ellipticals or S0 galaxies. The striking manner in which galaxy types and gas contents depend on location within a cluster is known as the *Density-Morphology Relation*. It reflects the different relative frequency of galaxy interactions at the centers of clusters compared to their peripheries. And, indirectly, the profound effects such interactions can have on the structures of individual galaxies.

16.3 THE HUBBLE LAW

Now that we have surveyed the Virgo Cluster in some detail—and more out of curiosity than anything else—we decide to look for our home. The Galaxy. This turns out to be tricky, and when we ultimately find it, we understand why it was so hard. At a distance of 18 *Mpc*—the distance to the Virgo Cluster—our Galaxy is not very impressive. It is now just one of many faint galaxies located beyond the cluster visible from here. A system that is now so faint that we require a moderately large telescope to see it at all. As for the Sun, no telescope known—or even contemplated in the near future—could identify our home star at this distance. It is much too faint and too crowded by myriad other stars in our Galaxy for us to spot it from so far away. It's a startling realization, one that produces an odd sensation. We are thoroughly cut off from home. Though we know it is there, we can't even see the Sun anymore from this destination.

But we *can* see our Galaxy, and we feel strangely compelled to study it in some detail. Perhaps because it *is* our home. Or perhaps because we realize that we know so much about our Galaxy that we can study it from a great distance with some degree of confidence. One simple way to do this—though by no means the only way—is to obtain its optical spectrum (we used this approach at Destination 15). As we would have guessed, the combined light of our Galaxy shows evidence of young and old stars (Destinations 8–11), emission lines from star-forming regions like the Orion Nebula (Destination 6), and even evidence of rotation (Destinations 9 and 14). But the spectrum of our Galaxy reveals something else. Something strange that we certainly did not expect. Our Galaxy seems to be moving away from us. And rather rapidly. We can discern this by comparing the observed wavelengths of absorption and emission lines of well-known elements with the wavelengths we measure in a laboratory for the same lines from the same elements. What we find is that the Galaxy's spectrum is shifted to the red. That is, to longer wavelengths (see Figure 16.9).

The simplest interpretation of this result is that the Galaxy—relative to our location at the center of the Virgo Cluster—displays a Doppler Shift (Destination 6 and 7) arising from its motion relative to our current location. And as we did at Destination 6, we can define the shift as a wavelength offset, $\Delta\lambda = \lambda_{obs} - \lambda_0$, where λ_{obs} is the observed wavelength and λ_0 is the 'true,' or 'rest,' wavelength of the emission or absorption line measured in a laboratory (the subscript on λ_0 is 'zero,' not the vowel in the word 'GO'). The Doppler Equation (Destination 6) can be used to determine the velocity producing the shift as

$$V_r = c\left(\frac{\Delta\lambda}{\lambda_0}\right) = cz\ .$$

In this relation, V_r is the radial velocity we infer for the Galaxy (or any galaxy when we apply this equation more generally), and c is the speed of light. Notice too that we've also defined a new term, z, known as the *Redshift*[19], as $z \equiv \Delta\lambda/\lambda_0$ (the symbol '$\equiv$' means that this is a definition, in this case for the symbol z). If we apply this equation to our observation of the Galaxy, we discover its velocity relative to us is about 1300 *km/s*! Since positive velocities are defined as motions away from us, this appears to be telling us that the Galaxy is moving away from us—away from the Virgo Cluster—at this moderately high speed. Of course, the opposite is true. As seen from Earth, the Virgo Cluster (on average) appears to be moving at this speed away from the Galaxy. What's peculiar about this result is that we have no idea where this large speed came from. As far as the instruments in our spacecraft are concerned, we began our journey at rest relative to the Local Group. And everywhere we stopped along the way, we always came to rest.[20] Yet, somehow, now that we've arrived at Virgo, we find that we are actually moving very rapidly relative to our starting point! How did this motion arise?

---

19 As noted earlier at Destination 6, the Doppler Equation also allows for negative—that is, approaching—velocities. Such spectra are said to be blueshifted. However, it will become apparent soon why it is reasonable to refer to $z = \Delta\lambda/\lambda_{True}$ exclusively as a redshift. In practice, very few galaxies—and no galaxies more distant than about 30–40 *Mpc*—exhibit blueshifts.

20 By 'came to rest,' we are referring to our motions relative to local galaxies. For example, at the center of mass of the Local Group, we would certainly see the galaxies around us in motion, but their mean speeds would average to zero. Likewise in the Virgo Cluster: The galaxies around us are certainly moving, but their average speed is zero. We are at rest relative to the average motions of the entire cluster.

Intrigued, we study this a bit further. From our current location near the center of the Virgo Cluster, our Galaxy happens to be located in the sky near another galaxy cluster, the *Fornax Cluster*. From Earth, this cluster is also about 18 *Mpc* away, but it is located at a position that is nearly at the opposite point in the sky from where the Virgo Cluster is situated. Now that we are in Virgo, this means is that the Fornax Cluster is about 36 *Mpc* away from our current location. When we measure the Doppler shifts from the spectra of galaxies in the Fornax Cluster, we discover that those galaxies are also shifted to the red, to longer wavelengths. But, much to our surprise, this shift is now equivalent to 2600 *km/s*!! This seems weird, but not as strange as what we discover next. As we obtain spectra of moderately distant galaxies all around us in this sky—galaxies that are located well beyond the outer limit of the Virgo Cluster—we soon realize that *every* galaxy seems to be moving away from us. All exhibit spectra that are shifted to longer wavelengths. Shifted to the red.

These observations of the redshifts of distant galaxies reveal an important clue that might help us interpret what's going on. In particular, it is interesting that when we obtained spectra of galaxies twice as distant as the Galaxy from Virgo, the Doppler Shift also doubled. As we observe other galaxies, galaxies for which we can estimate their distances independently using Cepheids or other types of distance indicators that work over *Mpc* scales, we soon appreciate that this is a widespread result. Everywhere we look, all galaxies exhibit spectra that reveal recessional velocities as estimated from their Doppler Shifts. And, invariably, these shifts appear to be proportional to distance. That is,

$$V_r = H_0 D,$$

where V_r is the observed recessional (or redshifted) velocity of a galaxy, D is the galaxy's distance, and H_0—pronounced 'H-naught'—is the very special proportionality constant for this equation. There is no additional constant since the average velocities of galaxies near us (essentially at zero distance) are also zero. This simple relation is known as the *Hubble Law* and H_0 is commonly referred to as the *Hubble Constant* in honor of the same person—Edwin Hubble—who came up with the galaxy classification scheme we first encountered at Destination 15 (Figure 16.9).[21] Because the Hubble Law also implies that galaxies can only be moving away from us—the distance D and Hubble Constant H_0 are both positive, so that means V_r must be too—it is taken as proof of *Universal Expansion*.

21 The name 'Hubble Constant' is unfortunate because this parameter, as we shall see as we travel to other epochs in the Universe, it is not a constant! Still, if we only consider how the Universe expands 'locally' (see Destination 18 for more on what this means), then H_0 is indeed—or at least should be—a fairly constant value.

Noting that $H_0 = V_r/D$, we realize that our observations of the Galaxy and the Fornax Cluster from our current location in the Virgo Cluster suggests that the value for the Hubble Constant is about 72 *km/s/Mpc*. The strange units are chosen for convenience (to astronomers at least!) since we generally measure velocities in units of *km/s*, and galaxy distances in *Mpc*. But don't be fooled. The units of the Hubble Constant are really inverse time—that is, one over time—since both *km* and *Mpc* are units of distance. The remarkable significance of this seemingly esoteric detail will become apparent shortly.

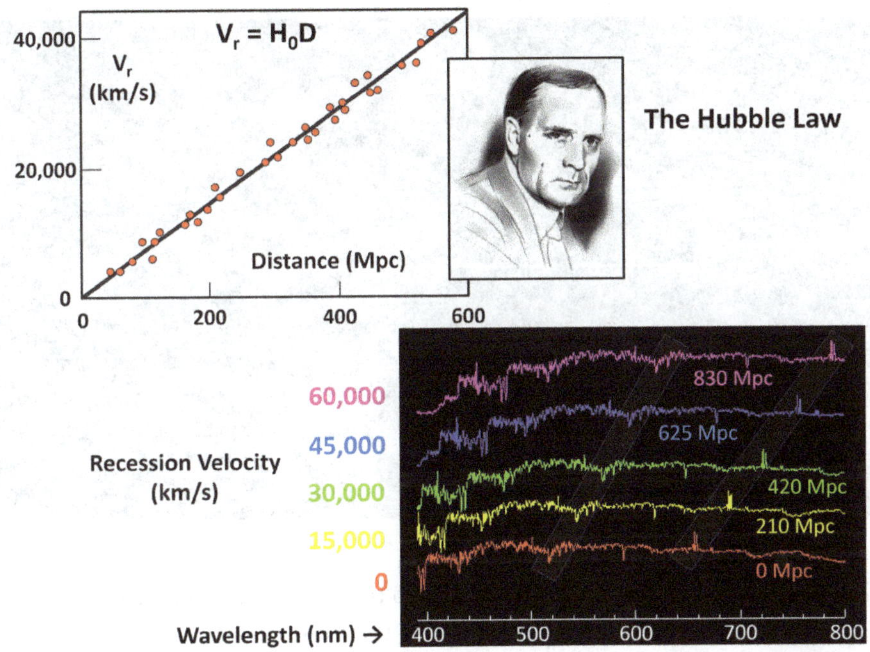

Fig 16.9

The Hubble Law (*upper left*) reflects the observed relationship between the distances of galaxies, *D* (along the *x*-axis), and their recession velocity, V_r (the *y*-axis). The best current estimate of the Hubble Constant, the proportionality constant of the Hubble Law, is H_0 = 72 *km/s/Mpc*. A portrait of Hubble is shown in the *inset*. The spectra (*lower right*) show how the recession velocities are measured. Note that as distance increases, the entire spectrum is shifted and stretched to the red—that is, to longer wavelength—consistent with the Doppler Equation, $V_r = c\,(\Delta\lambda/\lambda_0)$. The two shaded regions show how certain features (a Mg absorption line at left, and a H emission line at right) change their wavelength with increasing distance.

Image of galaxy spectra by M. Subbarao/SDSS.

The Hubble Law could not be simpler, yet its implications could not be more profound. One reason is that there is no sense of *direction* in the Hubble Law, only distance. That is, we observe the same Hubble Law with the same Hubble Constant no matter where we look in the sky. There's also no sense of *location* specified in the Hubble Law, implying that the Hubble Law is the same everywhere. Of course, Hubble discovered his law from Earth, but we would measure precisely the same Hubble Law from here at Virgo. From either location, galaxies would appear to be moving away from the observer at velocities proportional to their distance. That is, they would obey the Hubble Law and use the same value of the Hubble Constant. But how can this be? How can galaxies all appear to move away from two places simultaneously?

To understand this apparent paradox, we can do a simple thought experiment. Imagine four galaxies arranged in space so they all lie on a line, each separated from its nearest neighbor by a distance *D* (see Figure 16.10). Let's name these galaxies A(xel), B(etty), C(harlene), and D(ave) and let's further assume that we happen to be located in galaxy Betty (denoted '*B*' in Figure 16.10). From Betty, we can go to the trouble of measuring distances and Doppler Shifts of other galaxies to determine that the Hubble Law applies from that location. As a result, the two closest galaxies, Axel on one side and Charlene (galaxies '*A*' and '*C*,' respectively, in Figure 16.10) on the other are both a distance D away, so their recession velocity is $V_A = V_C = H_0D$. Galaxy Dave (galaxy '*D*' in Figure 16.10), on the other hand, is at a distance of $2D$, so its velocity is $V_D = 2H_0D$. This is precisely what the Hubble Law predicts: Galaxies appear to move away from us at velocities proportional to their distance. And we know this can happen:

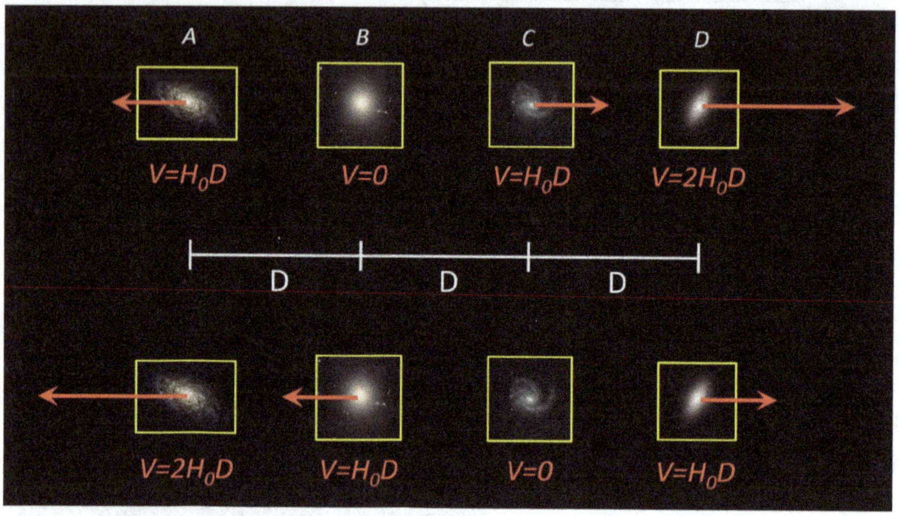

Fig 16.10

(*Top*) The velocities one would measure from galaxy B(etty) for galaxies A(xel), C(harlene), and D(ave) are shown as red arrows. These velocities are chosen to ensure they obey the Hubble Law. The velocities measured for the same galaxies from C(harlene) (*lower images*) were not estimated from a Hubble Law but simply calculated by zeroing galaxy Charlene's velocity and applying the same velocity shift to the other three galaxies. However, the *same* Hubble Law is seen in this case as was observed for the upper case. This illustrates that the simplicity of Hubble Law's implies one will see the same behavior of the recession velocities of galaxies from *any* location in the Universe. That is, there is no preferred 'center' or any preferred 'direction' in the Universe.

Galaxy images from STScI/NASA.

The Hubble Law is valid from *our* Galaxy so the situation we envision for Betty certainly does occur in at least one place in the Universe.

Now, imagine that we move to galaxy Charlene (lower section of Figure 16.10). Once there, we come to rest with respect to that galaxy and so we measure its velocity, V_C, to be zero. If we wish to predict the motions of other galaxies, we have to first compensate for galaxy Charlene's motion as illustrated in Figure 16.10. This correction makes galaxy Betty—where we used to be located—appear to be moving away from us at a velocity equal to H_0D. Galaxy Dave also now seems to be moving away from us at half its speed as seen from Betty, or H_0D. Relative to Charlene, Betty and Dave are both a distance D away. Galaxy Axel, on the other hand, is now moving at a velocity of $2H_0D$, and its distance is now $2D$. Notice that we did not impose the Hubble Law on the galaxies we observe from Charlene. We simply corrected for the motion of Charlene as seen from Betty (see Figure 16.1). Yet it's easy to see that the *same* Hubble Law we observed from Betty must also be valid at Charlene. This is amazing. The simple form of the Hubble Law ensures that if the Hubble Law is valid *anywhere*, it is valid *everywhere*. In more technical terms, the expansion of the Universe implied by the Hubble Law is *Uniform*—the same from all locations. Since our analysis (Figure 16.10) remains valid regardless of which directions the galaxies we considered are located means that the Hubble Law is also *Isotropic*—the same in all directions from a given location.

As amazing as this seems, there's a second, more astonishing implication about the Universe that's built into the Hubble Law. Notice that if a galaxy is moving away from us at some velocity V_r and its current distance is D, then it must have been very close to us at some time in the past equal to $T = D/V_r$. It turns out that this is true for *any* galaxy we might consider. To see this, let's return to our example with the four galaxies (Figure 16.10). If we imagine that we are again in galaxy Betty, it is clear from the Hubble Law that galaxies Axel and Charlene were near us a time $T = D/V_r = 1/H_0$ ago (recall the Hubble Law in its standard form states that $V_r = H_0D$, so all we've done is rearrange the equation). If we consider galaxy Dave, it was near galaxy Betty at a time $T = 2D/V_r = 1/H_0$ in the past. The reason for this is that for Dave, $V_r = 2H_0D$ since it is located at a distance $2D$ from Betty. You can prove to yourself that it doesn't matter which galaxy we use to calculate T. As long as the Hubble Law holds, *all galaxies* were close to one another at time $T = 1/H_0$ in the past! As it happens, T corresponds to the very special time when all galaxies—indeed, all matter and all of space—were close to one another. It corresponds to what astronomers often refer to—rather grandly—as the *Age of the Universe.*[22]

There's another profound implication of the Hubble Law, one we can envision if we consider a bunch of dots imprinted on the surface of an expanding balloon. No matter which dot we choose on the balloon's surface, all the others will appear to move away from it. Most importantly, as long as we restrict ourselves to the surface of the balloon, it is apparent that there's no obvious 'center' to the expansion. When the balloon is small and the dots all appear close together, they do not congregate about any single predefined point on the balloon's surface. The dots are *all* close to one another simultaneously. As the balloon expands, it is also clear that every dot is moving away from every other dot. As a result, *every location* appears to be at the center of expansion, which can only make sense if *no place at all* is the center of expansion!

The balloon analogy illustrates a final strange, but central feature of Universal Expansion implied by the Hubble Law. Imagine that we are two-dimensional beings who are restricted to living only on the surface of our expanding balloon. Since we can't move off the surface, we may not even recognize that a third dimension exists. That is, the balloon's surface represents our entire 'universe.' In this case, though we can see the expansion going on, we do not see the universe expanding *into* anything. Space is simply getting larger and larger. It is similar

22 Well, not quite. This argument implies that the velocities of all galaxies remain constant with time. But that's certainly not the case since the mutual gravitational forces from all the galaxies are slowing down the Universal Expansion. At this stage, that's really just a detail. The time $T = 1/H_0$ nevertheless remains a measure of the Age of the Universe to within a factor of two or three.

for us in our three-dimensional Universe. It is definitely expanding—the Hubble Law tells us that this is the case—but at the same time, it is not expanding *into* anything. Certainly not into a normal three-dimensional volume that we can easily envision or potentially identify. The only way to reconcile this is to appreciate that from our perspective, *space itself is expanding*!

This is a key point because it is quite distinct from saying that galaxies are 'moving away from one another.' That simpler interpretation of Universal Expansion envisions galaxies moving around in some fixed grid of space, somewhat like a crowd of people dispersing from a single location. But that is not what is happening with galaxies. An observer in any galaxy anywhere in the Universe would be convinced he/she is standing still relative to local space and that it's all the other galaxies that are in motion.[23] Since that's inherently inconsistent, the only way around this is to adopt our earlier conclusion that space itself is expanding. This also means that the Doppler Shifts we observe in galaxies do not really reflect motions in the usual sense. Rather, as radiation travels to us from distant galaxies, its wavelength increases as the space in which it is traveling steadily expands. By the time we receive the radiation, the increased wavelength generates a Doppler Shift that we *interpret* as a velocity. But it's an illusion. If we could somehow see galaxies 'instantaneously'—and avoid all the inconsistencies *that* implies—we'd see that galaxies are, in fact, all standing still relative to one another.[24] Confusing? Strange? Absolutely. On our upcoming future journeys into the past (Destinations 18–25)[25], we will try to develop a clearer sense of what it means for space to 'change size.'

16.4 THE DARK MATTER OF THE VIRGO CLUSTER

There's one last thing to note before we depart the Virgo Cluster. Because we are near the center of mass of the cluster, we appear to be standing still with respect to the average motions of the cluster's galaxies. But as we measure the Doppler Shifts—and hence velocities—of galaxies located within the cluster, we notice that some are moving rapidly toward us, while others are moving

---

23 Of course, one first has to remove any 'local' motions due to gravitational effects of nearby stars, galaxies, or clusters. We'll explore this in more detail at Destination 18.

24 Though, in fact, that's not quite right since then the galaxies would start to move due to their mutual gravity. The expansion of the Universe is a critical and fundamentally unavoidable features of our Cosmos. If you remove it ... well, that's not our Universe anymore.

25 The grammar of time travel can be confusing!

rapidly away from us. That is, the galaxies of the cluster itself exhibit a large range of blueshifts and redshifts.[26]

There are a number of important conclusions to draw from this. The first is that these motions reflect the internal velocity dispersion of the cluster. We should now be familiar that this information—along with some idea of the cluster size—can give us an estimate of the *mass* of the cluster (see Destinations 13 and 15). In the case of the Virgo Cluster, the dispersion[27] is quite large, $\sigma_v = 1330$ km/s. If we take the radius, $\overline{R}$, of the cluster to be around 2.0 *Mpc*, this implies a total cluster mass of

$$M_{Virgo} = \sigma_V^2 \overline{R} / G = 6.2 \times 10^{14}\, M_\odot.$$

We can also measure the total luminosity of all the galaxies in the cluster to be $L_{Virgo} = 2.3 \times 10^{12}\, L_\odot$. The cluster's mass-to-light ratio from these two values turns out to be

$$(M/L)_{Virgo} = 270$$

in solar units. This is much higher than the mass-to-light ratios—around 10 to 20—that we found for individual galaxies such as M 31, Maffei 1, and our Galaxy.[28] The Virgo Cluster is *full* of dark matter, containing even more of this enigmatic stuff than the individual galaxies within the cluster. Amazingly, the largest known galaxy clusters have masses approaching $2 \times 10^{15}\, M_\odot$ and mass-to-light ratios of 500 or higher. Whatever dark matter is, clusters contain a lot of it. Yet sadly, even from inside the Virgo Cluster—a place known to be full of dark matter—we still can't discern the true nature of this stuff.

There is another way to probe the dark matter in galaxy clusters that generally is not possible in individual galaxies. We've seen that as galaxies interact they can shed gas via tidal effects or stripping (see Section 16.2). Much of the gas stripped from galaxies during interactions (Figure 16.7) will steadily contribute

26 The internal motions of Virgo are so large that, from Earth, about 30 of its galaxies exhibit blueshifts despite a mean cluster recession velocity of about 1300 *km/s*. These rare exceptions to the Hubble Law comprise the majority of *all* galaxies in the sky that are known to exhibit blueshifts.

27 As before (Destinations 13 and 15), we've corrected the observed dispersion for the fact that radial velocities give only one (of three) velocity components of galaxies observed in the cluster. The *observed* radial velocity dispersion of Virgo is around 770 *km/s*.

28 This comparison is a bit tricky. The total masses for the individual galaxies—these are listed in Tables 14.1 and 15.1—include their full dark matter halos (see Destinations 11, 14, and 15). However, in Virgo, we can't easily determine the mass to a comparable distance from the cluster center. For our present purposes, we'll compare Virgo's mass with the masses of galaxies measured to the limits of their 'observable' mass (these values are also listed in Tables 14.1 and 15.1). Regardless of this detail, the cluster certainly has a higher mass-to-light ratio than the individual galaxies we've explored so far.

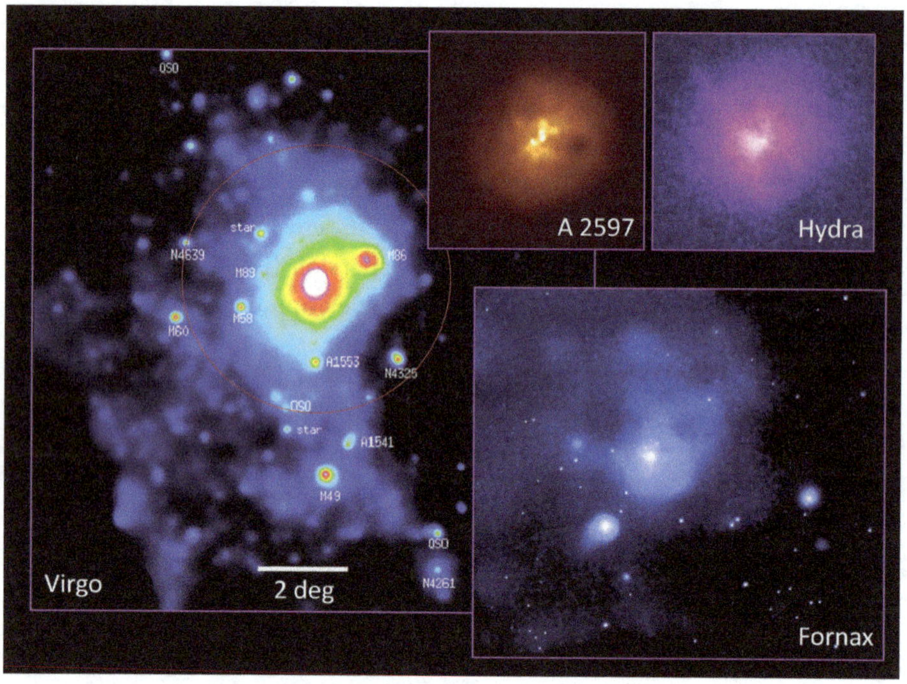

Fig 16.11

X-ray images of galaxy clusters. Virgo (*left*) and Fornax (*lower right*) are the two closest galaxy clusters to our Galaxy and so reveal the most detail. Notice how there are bright patches around many individual galaxies as well as more diffuse—but still quite structured—radiation from a general IGM. The circle in the Virgo map is centered on the brightest part of the x-ray emission and has a radius of 1 *Mpc*. The upper right images show the x-ray images of two more distant clusters, again revealing the existence of patchy, hot gas of the IGM of these clusters. This hot gas likely results from the heating of stripped gas from galaxies (Destinations 14 and 15) and the gas lost from their stars during their evolution (see Destinations 11 and 12).

Images from Rosat/MPE (Virgo) and Chandra/NASA (all others).

to a cluster's *Intergalactic Medium*, or *IGM*. Since the stripped gas is well mixed, the gas particles will tend toward speeds similar to those of the galaxies—that is, thousands of *km/s*. When we convert these speeds to a temperature (see Destination 5), we find that the IGM should have temperatures of tens of millions *K* (Figure 16.11)! Galaxies passing through this hot IGM—even ones that avoid strong interactions with other galaxies—can lose their gas through *Ram Pressure Stripping*, adding even more gas to the IGM.[29] From the perspective of understanding the nature of dark matter, the IGM is interesting because it provides another means of measuring cluster masses. Assuming that the gas has had time to mix throughout the entire cluster—and this is likely true near cluster cores, but at best only marginally true at the cluster peripheries—we can use x-ray observations that reveal the gas temperatures to measure the masses of clusters. The results are substantially in agreement with what the galaxies tell us. But the hot IGM in clusters reveals something more subtle. The total quantity of hot gas in galaxy clusters (Figure 16.11), though abundant, is still far too little to account for the total cluster masses we observe. And yet the mere existence of the very hot gas in the IGM confirms that there must be a lot of dark matter in clusters. If there weren't, the hot gas would diffuse away. But

29 This helps explain why we see many anemic (gas-poor) spiral galaxies and S0 galaxies—likely normal spirals that have been stripped of their gas but not yet tidally disturbed—near the center of Virgo and many other clusters (this is related to the Density-Morphology Relation we identified in Section 16.2). It also implies that once galaxy interactions start in earnest within a cluster—particularly as gas gets injected into the IGM—the process can accelerate rapidly.

despite all this, the bottom line remains the same: Though we have figured out way to trace it better and better, we still do not know what the dark matter is.

There is an additional conclusion we draw from the motions of galaxies in clusters that is more profound than 'merely' measuring cluster masses. It's based on the obvious fact that galaxies in clusters exhibit both redshifts *and* blueshifts relative to the cluster centers. It's as if the Hubble Law does not apply inside galaxy clusters! But this is not really news. Within the Solar System, or within our Galaxy, or even within the Local Group, both redshifts and blueshifts are commonly observed, and all with no obvious favoritism toward redshifts. The Hubble Law, strictly speaking, does not apply in any of those cases either. But this does not mean the Hubble Law is somehow 'invalid.' Instead, it tells us that whenever there is enough mass locally for gravity to overcome the expansion of space, it does. On Earth, or within our Galaxy or from anywhere in the Local Group, or within the Virgo Cluster, there is enough mass concentrated into a small enough volume that the motions of people, planets, stars, and galaxies are dictated by gravitational forces arising from local matter rather than from the Universal Expansion implied by the Hubble Law.[30] But in the depths of intergalactic space far beyond the boundaries of clusters and groups—regions where there is virtually no matter and no strong gravitational fields—there are no local sources of gravity that could halt or slow down the expansion embodied in the Hubble Law. It's only in vast volumes of nothingness where space itself can expand.

This is actually a good thing. If the Universal Expansion were significantly more vigorous, there might be no regions where matter could collapse due to its own gravitational attraction anywhere in the Universe. And without the collapse of matter, there'd be no galaxies, no star formation, no planets. And no 'us' to study Universal Expansion.

---

30 That's also why you should not bother trying to measure the expansion of the Universe in your house! Not only are the gravitational forces of the Earth, Sun, and the Galaxy sufficient to keep the Universal Expansion in check, even the materials holding your house together can overcome the expansion on that scale. Only where the density of mass is extremely low can Universal Expansion overcome local gravitational forces. We can therefore expect to see direct evidence that the Universe is expanding only when there is a lot of intervening empty space.

DESTINATION 16 QUESTIONS

16.1. Imagine a typical Virgo Cluster galaxy is a sphere 30 *kpc* in diameter and that 200 of such galaxies are confined to the inner 1 *Mpc* of the cluster. What is the typical separation of the galaxies in this part of the cluster? If the typical relative velocity of galaxies in the cluster core is 700 *km/s*, how long would it take one galaxy to reach its closest neighbor (on average) if they headed straight toward one another? Does this result imply that collisions were common or uncommon in the approximately 7 *Gyr* lifetime of the cluster?

16.2. When Hubble first discovered the expansion of the Universe, he estimated the value of H_0 to be about 500 *km/s/Mpc*. How much younger was his Universe compared to the age we derive for the Universe today for a Hubble Constant of 72*km/s/Mpc*?

16.3. Recall from Destination 5 that gas temperature is directly related to the typical speed of the particles that make up the gas. As we saw, the Virgo Cluster as a whole contains a very hot intergalactic medium, similar to the very hot Coronal ISM of the Galaxy. For this to be stable, the gas in Virgo must have a typical velocity equal to the velocity dispersion of the cluster. What is the temperature of this very hot gas in the Virgo Cluster?

16.4. Use the Hubble Law for the expansion of the Universe for the following questions.

a) What is the distance to a galaxy with a recession velocity of 10000 *km/s* for a Hubble Constant of 80 *km/s/Mpc*?

b) You measure the distance to a galaxy to be 120 *Mpc* and its recession velocity 6600 *km/s*. What Hubble Constant would you derive from this galaxy?

c) What is the distance to a galaxy for which you measure the Hydrogen-Beta line to be at 505 *nm*? The normal wavelength of this line is 486 *nm* and you can assume a Hubble Constant of 100 *km/s/Mpc*. For this, you will need to use the Doppler Shift Equation, $\Delta\lambda/\lambda_0 = V/c$, where $\Delta\lambda = \lambda_{obs} - \lambda_0 = V/c$ is the difference between the observed wavelength λ_{obs}, and the true wavelength, λ_0, of the emission line.

16.5. As we've seen at this destination, the number $1/H_0$ is intimately related to the Age of the Universe—or at least the time when all galaxies, all matter, all space were compressed closely together. If the Hubble Constant, H_0, has the value 72 *km/s/Mpc*, what is the value of $1/H_0$ in seconds and in years? We noted that the globular cluster M 92 may be as much as 14 *Gyr* old. Is that age compatible with this value of the Hubble Constant? If you see a problem when comparing these ages, speculate on how it might be resolved.

16.6. Maffei 1 (from Destination 15) is located 3.1 *Mpc* from the Sun and it exhibits a mean radial velocity as measured from the Sun of 66 *km/s*.

a) From this information, what Hubble Constant do you derive for Maffei 1?

b) Compare this to the accepted value of the Hubble Constant, H_0 = 72 *km/s/Mpc*. Explain why the result for Maffei 1 appears to be very different from this more accepted value.

c) Given the basic information for Maffei 1 listed in Table 15.1, explain why you would or would not expect to see some of the components of that galaxy to exhibit negative—that is, blueshifted—velocities relative to the Sun.

16.7. Astronomers often refer to redshifts rather than recession velocities or distances, implicitly assuming the validity of the Hubble Law and the conversion of redshift to distance. The questions below illustrate the intimate relation between velocity, redshift, and distance.

a) From the information on its recession velocity given in the text, what is the redshift z for the Virgo Cluster? From the Hubble Law, how distant is the cluster?

b) The Coma Cluster of galaxies is about 80 *Mpc* from the Sun. What is its recession velocity and redshift?

c) The Shapley Supercluster (which we will discuss at Destination 18) has a redshift of about 0.065. What is its recession velocity and distance?

16.8. What is the ratio of the gravitational force exerted by the Moon on the near and far side of the Earth? This force difference is the source of the tidal forces that act on our planet. For this problem, you will need to look up the diameter of the Earth and the separation of the centers of the Earth and Moon, but *not* the masses of either body. You will also need to use Newton's Universal Law of Gravity (Destination 9), keeping in mind that as far as the Earth is concerned, the Moon can be reasonably approximated as a point mass.

16.9. Assume for the moment that the mass of the Virgo Cluster is concentrated in a single point at the cluster center. Calculate the orbital times for particles in circular orbits about this mass at distances of

a) 0.5 *Mpc* (in the cluster core)

b) 1.0 *Mpc* (outside the cluster core)

c) 2.0 *Mpc* (near the outer edge of the cluster)

Given that the cluster is probably around 7 *Gyr* in age, what do these timescales tell you regarding the likelihood that a given part of the cluster is in Virial Equilibrium? **HINT:** Recall that to use the Virial Theorem (Destination 13)—and, hence, to be able to assume that a body is close to Virial Equilibrium—requires that the orbital times are short compared to the age of the system.

16.10. Look up and identify a galaxy cluster other than Virgo or Fornax that is located within a distances of about 100 *Mpc* from the Sun. Report on the basic properties of the cluster (distance, mass, size, etc.) and on one or two of its noteworthy features.

16.11. Look up detailed descriptions of five of the galaxies in the Virgo Cluster. You may choose any galaxy in the cluster, including ones we have discussed in this destination, but be sure to provide some new information about the system(s) beyond a listing of their basic data (size, mass, Hubble type, etc.) for at least one of the galaxies.

16.12. Based on our discussion in Destination 16 on the way galaxies interact in clusters, discuss what else, other than hot gas, might be expected to reside within the volume of 'empty' space between cluster galaxies. **HINT:** Consider what resides in galaxies other than the ISM.

16.13. Some of the galaxies that are certainly physically located within the Virgo Cluster exhibit radial velocities that we measure to be negative as seen from the Sun. That is, these galaxies are not receding from us, but in fact appear to be moving toward us! The Hubble Law would imply that this is not possible. However, what specific features of the cluster make this curious result understandable? **HINT:** Consider what else, other than the Hubble Law, may affect the velocities of galaxies in the Virgo Cluster. Compare the magnitude of this effect with that expected from the Hubble Law.

Destination 17

THERE'S SOMETHING STRANGE HERE: THE SEYFERT GALAXY M 77

17.1 THE MONSTER IN THE MIDDLE

It's impressive—not to mention somewhat surprising—how so much of our understanding of galaxies can be gleaned from their optical appearances. Classification systems such as Hubble's (Destination 15) that are based on the visual forms of galaxies are remarkably effective at isolating key properties that let us infer other galactic characteristics we don't observe directly. It's as if we would need to know only what a galaxy looks like—and perhaps also its luminosity—to fully characterize its properties[1]. But galaxies are not obligated to be easy or convenient to classify and there is no guarantee any optical classification scheme will capture the full range of features galaxies may exhibit. Our next destination—the spiral galaxy M 77[2]—will illustrate this point by introducing us to a broad range of extraordinary new phenomena that galaxies can exhibit but that play no role in Hubble's classification system.

Located in the constellation Cetus, the Whale, M 77 is about 14.4 *Mpc* from the Sun. Although somewhat closer than the Virgo Cluster, it's still a very long journey (!), leaving us plenty of time to study the galaxy before our arrival. From a few *Mpc* away, we can almost convince ourselves that M 77 is just a normal spiral (see Figure 17.1). The well-defined arms in an exponential disk, the prominent bulge, the clear evidence of star formation, gas, and dust, all indicate that this is an Sb spiral galaxy (Destination 15), not drastically different from our Galaxy or M 31 (see Destination 14). When we take spectra of the main body of the galaxy[3], we find it is composed of normal stars (Destinations 8 and 10–12), star-forming regions (Destination 6), a complex ISM reminiscent

---

1 The extent to which this is true already tells us something remarkable about how galaxies form and evolve. As we'll see (Destinations 20 and 21), galaxy formation is a rather messy process, more reminiscent of galaxy interactions (Destination 16) than some clean, orderly process. And yet, from apparent 'chaos,' galaxies emerge that fit into Hubble's Classification System (Destination 15). The mere fact we can classify galaxies at all is amazing, a critical clue to understanding how galaxies have developed over the history of the Universe. It's just not the complete story.

2 Yet another Messier object! And, as we'll soon see, a unique one. It is intriguing how—without knowing anything about the true nature of the objects he discovered—Messier identified numerous fundamentally interesting objects. Many of them are the brightest or most striking examples of their class, a big reason the names he gave them are still used to identify these objects. Alas, M 77 will be the last Messier object we shall visit on our journeys.

3 And not just optical spectra. As we know from our past explorations, observations covering the full EM Spectrum from radio waves to x-rays are needed to fully explore the components that make up the galaxy.

of what we see in our Galaxy (Destinations 5 and 6), and even a number of exotic binary stars similar to LMC X-3 (Destination 12). Measurements of the Doppler Shifts of emission lines (mostly from the ISM) and absorption lines (mostly from stars) reveal that M 77 has a flat rotation curve with $V_{rot} \sim 250 km/s$ (Destination 9), properties consistent with the presence of an

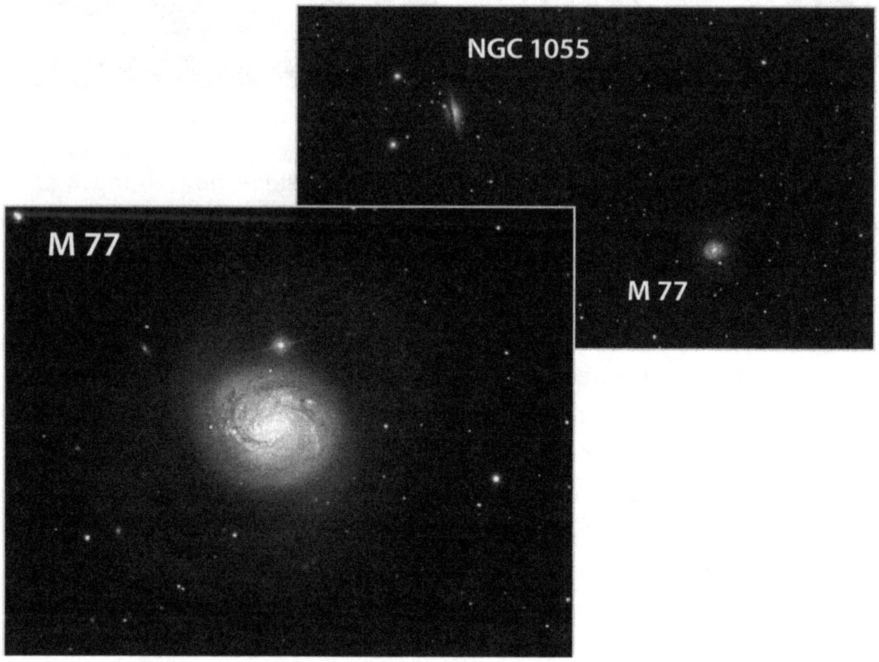

Fig 17.1

(Lower Left) The Seyfert galaxy M 77, a fairly normal spiral galaxy—though with an unusual, faint extended outer disk—hosting an Active Galactic Nucleus (AGN) in its center. From this vantage point, the AGN, buried at the center of the galaxy's bulge, is about 10% as bright as the rest of the galaxy. As we shall see, much of the AGN's luminosity is not directly visible from Earth, and the AGN has a total luminosity about equal to the rest of M 77. (*Upper Right*) M 77 has one large nearby companion, the edge-on spiral galaxy NGC 1055, located a few hundred *kpc* away and about four times less luminous than M 77 itself.

Photos by S. Rosen (wide-field image) and NOAO/AURA/NSF (M 77 detail).

extended halo of Dark Matter (Destination 13).[4] In short, M 77 seems to be a normal Sb spiral galaxy (Destination 14), one that is slightly more massive than M 31 (Destination 14).

Well, almost normal. There's one prominent feature of M 77 that sets it apart from any spiral galaxies we've explored so far: The galaxy's *nucleus*. Not the bulge. That component is easily visible (see Figures 17.1 and 17.4) and seems fairly normal for a spiral galaxy of M 77's type (Destinations 14 and 15). But the nucleus is different. Located precisely at the center of the galaxy's light distribution and at the location about which the galaxy's disk rotates, its most noteworthy feature is its remarkable brightness compared to the rest of the galaxy: Over 10% of the *entire* luminosity of M 77 is produced in its nucleus.[5] And yet, even from a moderate distance, we can see clearly the nucleus is compact, taking up

4 But that's not news. Dark Matter has been inferred to exist in every galaxy ever studied that has adequate measurements. Finding a galaxy with *no* Dark Matter? That would be *big* news.

5 As we'll soon appreciate, galaxy nuclei can emit radiation over a very wide range of the EM Spectrum. Consequently, when we refer to the 'luminosity' of the nucleus, we are implicitly including the energy output from all forms of EM Radiation, from radio waves to high-energy γ-rays (see Figure 5.6).

less than 0.01% of the volume of the galaxy's disk. To put that into perspective, the luminosity-to-volume ratio of M 77's nucleus is equivalent to having an engine smaller than a walnut powering a high-performance automobile. From a practical standpoint, however, the compactness of the nucleus of M 77 poses a problem. This region is so small—it's essentially star-like in appearance from our vantage point a few *Mpc* away (Figure 17.2)—it seems impossible to discern in detail what's going on in this obviously extraordinary region.

We've run into this problem before with binary stars (Destinations 8 and 12), many of which are so distant from Earth they appear as single stars. Despite that, it's still possible to explore the properties of such binary stars in two ways.

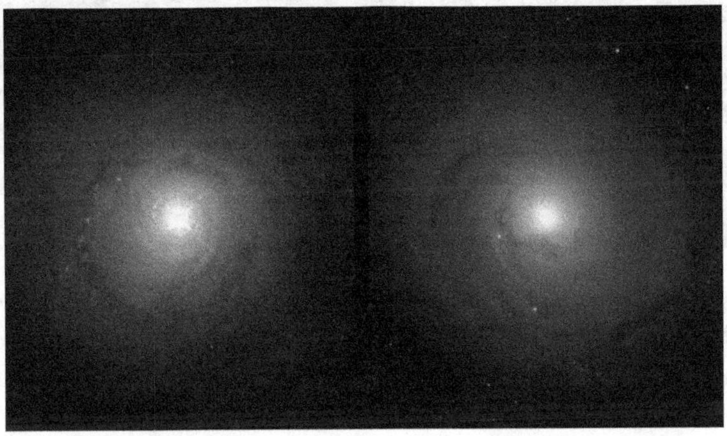

Fig 17.2

Images of two similar spiral galaxies. The galaxy on the *right*, NGC 3277, has a small central bulge. The galaxy on the left is NGC 5548—a Seyfert galaxy very similar to M 77—which contains a very bright central AGN clearly visible in the image. The spikes extending from the AGN are produced in the telescope, but they indicate that the bright source is point-like at the resolution of the telescope. That is, it appears smaller than the smallest feature the telescope can discern.

Image montage by W. Keel/UA with images from STScI/NASA.

First, we have seen that spectra can reveal individual stars in binary systems directly (Destination 8), or uncover unseen companions from periodic Doppler shifts of specific spectral features (as in LMC X-3, Destination 12). Second, changes in brightness—due to eclipses or changes in viewing aspect—can also reveal the presence of individual stars in binary systems (Destination 8). But these techniques do not merely identify binary stars. Spectra and brightness variations can provide basic information about these systems—for example, orbital periods and velocities, inclinations, stellar sizes—that allow us to investigate their properties in detail. Can we explore the nucleus of M 77 in similar ways? The short answer is 'yes,' but we shall soon discover we are dealing with a situation that's more complex, more bizarre, and far more difficult to interpret than comparatively 'simple' binary stars.

The spectrum of the nuclear region of M 77 (Figures 17.3 and 17.4) immediately tells us that something strange is going on here. It's unlike anything we've encountered before, revealing a number of extraordinary features. Perhaps most striking is the wavelength/frequency range over which the emission is seen. As shown in Figure 17.3, the nuclear region of M 77 emits radiation over essentially the entire EM Spectrum, from radio waves to x-rays, even extending to the most energetic γ-rays (see Figure 5.6). In contrast, stars emit radiation essentially as

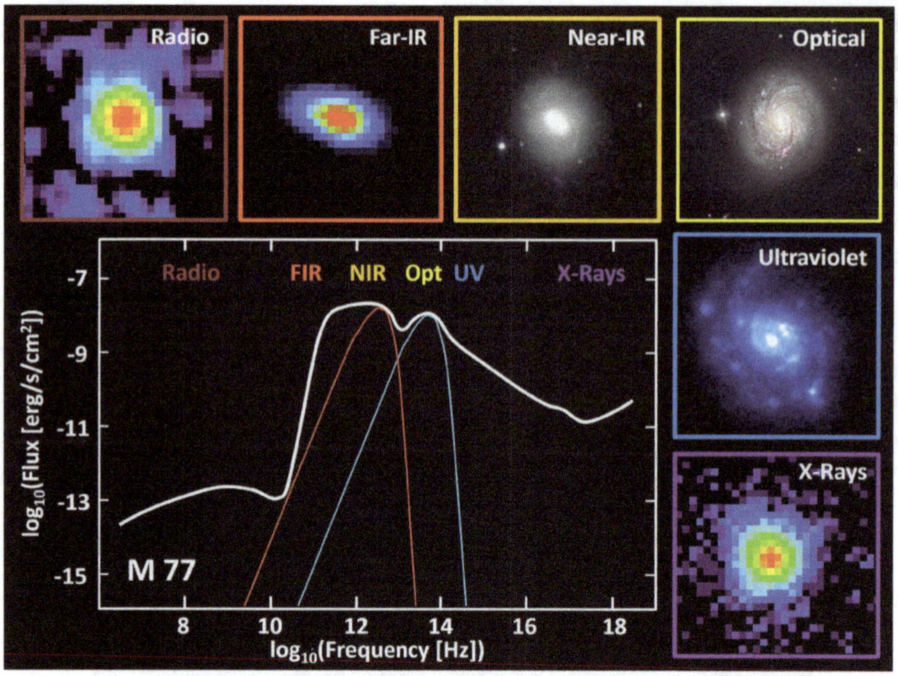

Fig 17.3

(*Upper left, clockwise to lower right*) M 77 is easily visible over the full range of the EM Spectrum. These images—obtained at the radio, far-infrared, near-infrared, optical, ultraviolet, and x-ray regions of the EM Spectrum—reveal a range of emission processes in the galaxy. The radio and x-ray images are not resolved (see Figure 17.2), so it implies that for M 77 most of the radiation from those spectral regions is emitted from the nuclear region only. (*Lower left*) A schematic representation of the spectrum of the central AGN of M 77, plotted from the radio to the x-ray portions of the EM Spectrum (white curve). Notice that the galaxy emits significant flux (the *y*-axis) over a factor of 10^{12} in frequency (the *x*-axis). The red and blue curves show the shape of spectra that we'd expect to see from 100 *K* and 10000 *K* thermal sources. It's clear that no single thermal source—or even a combination of such sources—can account for the full spectrum of M 77's nucleus. The radio emission (corresponding to frequencies lower than about 10^{10} *Hz*) is best understood as synchrotron emission generated by energetic electrons interacting with an extended, but relatively weak, magnetic field (Destination 12).

Images and spectral data from IPAC/Caltech/NASA.

single-temperature thermal sources (Destinations 8), their spectra fading to invisibility away from their peak flux defined by Wien's Law (Destination 6; see also Figure 17.3 where some thermal spectra are shown). Even a cluster composed of stars with different temperatures can't explain the broad spectrum we see in the nucleus of M 77. To do so would require stars ranging in temperature from a few degrees above absolute zero to a few million *K*. But since the coolest stars have temperatures of at least 1000 *K*, while the hottest stars cannot exceed about 100000 *K* (Destinations 8 and 10), we can be quite certain that the nucleus of M 77 does not shine (exclusively) with starlight. The radio emission from the nucleus (Figure 17.3)—also 'non-thermal' in nature—has the characteristic spectrum that we'd expect from synchrotron emission, a form of radiation we encountered earlier in the Galactic Center and in pulsars (Destinations 9 and 12). The existence of this radiation implies the presence of energetic electrons interacting with strong magnetic fields in a low-density gas, and apparently confined within a tiny volume buried in the nucleus of the galaxy.

And then there are the emission lines. The optical spectrum of the nucleus of M 77 exhibits a number of strong emission features (Figure 17.4). Many of these arise from moderately ionized atoms such as Oxygen (O), Neon (Ne), and Helium (He). Some of these emission lines can form only in fairly low-density gas (about 10^4 particles per cm^3) that is being bathed with energetic, ultraviolet photons. This is reminiscent of conditions associated with the Hot ISM that we encountered at the Orion Nebula (Destinations 6). However, unlike the emission lines in Orion, many of the lines in the spectrum of M 77's nucleus are 'broad'; that is, they span a relatively wide wavelength range. Assuming that this is due to the Doppler

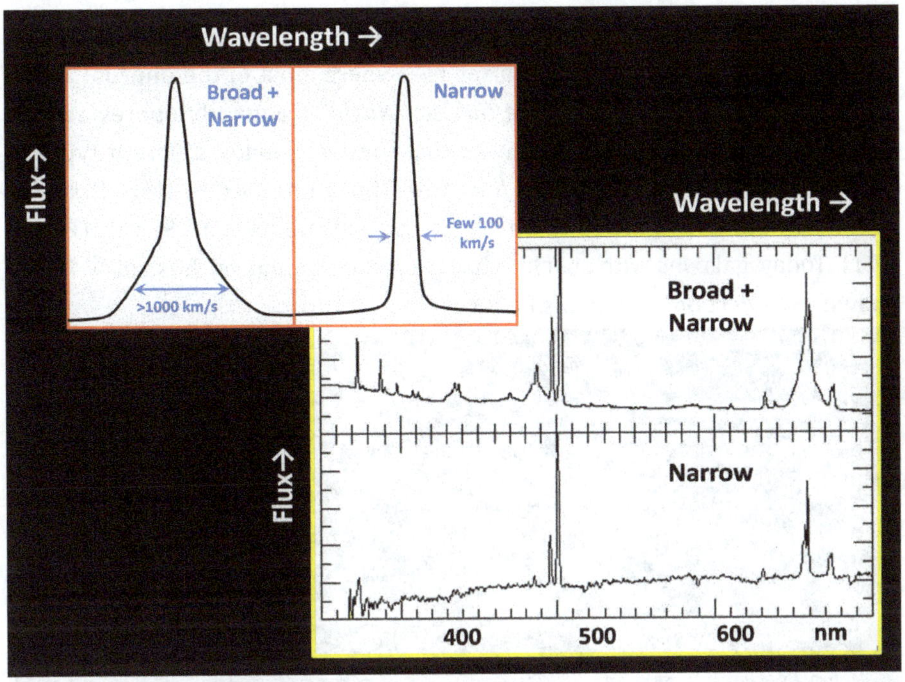

Fig 17.4

(*Lower right*) Optical spectra of two Seyfert galaxies shown as plots of the flux (*y*-axis) and the wavelength (*x*-axis). The upper spectrum—labeled 'Broad+Narrow'—exhibits a range of emission lines from ionized H to highly ionized Fe (Iron). Note that some of the lines appear very broad at their bases. The lower spectrum—which happens to be of M 77 and is labeled 'Narrow'—exhibits fewer lines, all of which appear much narrower than the broad emission lines of the Broad+Narrow spectrum. The Broad+Narrow spectrum represents a so-called Seyfert 1 galaxy, while the Narrow spectrum is characteristic of a so-called Seyfert 2 galaxy (see footnote 10 of this destination for more details). (*Upper left*) A detailed view of the broad+narrow (*left*) and the narrow (*right*) emission lines seen in different Seyfert nuclei. The widths of the lines reveal large motions of the gas producing the emission. As we will soon see, the narrow emission lines arise from emission of radiation from gas located far from the central supermassive black hole of an AGN, while the broad line components are produced from gas located closer to the central black hole. Both regions are heated and ionized by copious ultraviolet and x-ray radiation produced deep within the core of the AGN.

effect (Destination 6), the large breadth of these lines implies fairly speedy gas motions. In fact, speedier than we can account for from just the thermal particle speeds we'd expect from gas associated with the Hot ISM at a typical temperature of about 10000 *K* (Destinations 5 and 6). We saw this behavior before in the dark molecular clouds of the Cold ISM around the Orion Nebula (Destination 6) where we interpreted the broad emission lines of those clouds as signs of internal rotation. The same seems to be true of the gas in the nucleus of M 77, but with a striking difference: The lines from the nuclear region of M 77 reveal rotation velocities of up to 500 *km/s*, far faster than the few *km/s* seen in the cold clouds in the Orion Nebula (Destination 6).

These remarkable spectral features in M 77 and in other galaxies with similar nuclei were first noted with curiosity—but little comprehension—early in the 20th Century. As it happens, approximately 3% of all fairly nearby galaxies[6] more luminous than the LMC (which is about 10% as luminous as our Galaxy; see Table 14.1) contain nuclei that exhibit spectral peculiarities similar to what we see in M 77. Since it is already clear from their spectra that these nuclei

6 We have to restrict ourselves to only nearby galaxies because if we consider systems that are too distant, the finite speed of light implies that we begin to look back a significant distance in *time*. It turns out that nuclear activity in galaxies has *evolved* such that it was far more vigorous in the past (see Destination 21). Consequently, it's not fair to compare distant (younger) galaxies with comparatively nearby (older) galaxies of the *Present-Day Universe*. Even with this restriction, we still have a pretty large chunk of space available to explore. For most situations, galaxies closer than about 300–400 *Mpc* can be considered to be sufficiently 'local' that we can compare their properties without worrying too much about evolutionary effects. We'll explore all this in more detail at Destination 18.

encompass extreme environments, such galaxies are generically referred to as *Active Galaxies.*[7] The nuclei themselves—where most of the unusual action takes place—are known as *Active Galactic Nuclei,* frequently abbreviated as *AGN.*[8] We'll discover at Destination 21 that there are many different types of AGN in the Universe. The closest few AGN—including M 77—were first systematically studied in the 1940's by American astronomer Carl Seyfert (Figure 17.1). Today, galaxies with specific characteristics similar to those of M 77 are known as *Seyfert Galaxies* in his honor.

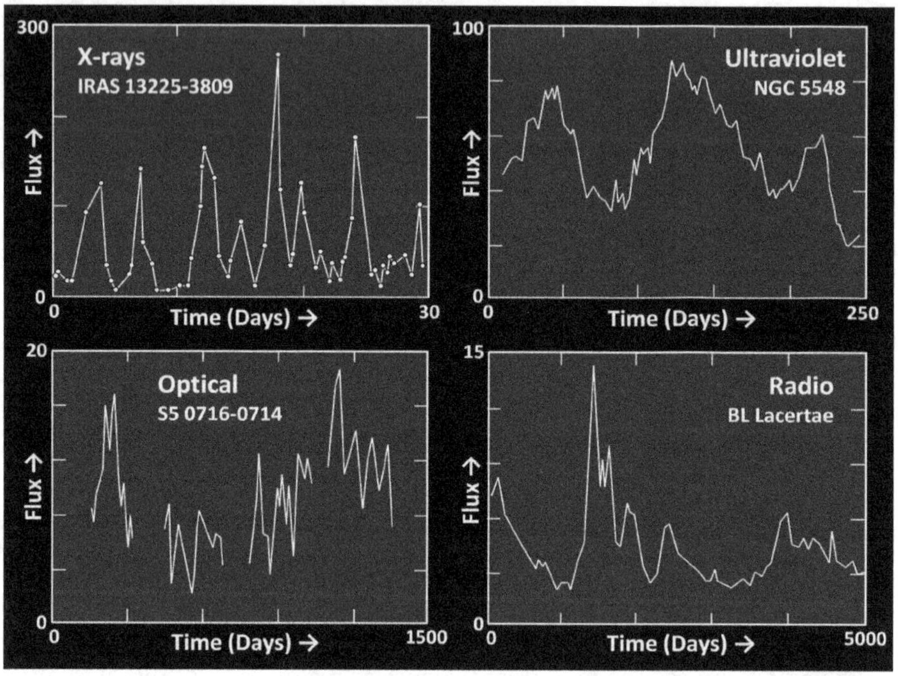

Fig 17.5

AGN such as the nuclear region of M 77 often vary significantly in brightness as illustrated in these *light curves*, or plots of the flux from a source (y-axis) as a function of time (x-axis) (see Destination 14). These variations are usually visible at all wavelengths, ranging from the x-ray (*upper left*), Ultraviolet (*upper right*), optical (*lower left*), and radio (*lower right*) portions of the EM Spectrum. In general, the timescales of these variations—evident from the x-axis of each plot—become shorter at shorter wavelengths, implying that the higher-energy radiation is typically produced in more compact regions. The arcane names of the individual galaxies associated with each light curve are specified in each frame. Virtually all AGN show some variability akin to what's illustrated here.

Although the spectral characteristics of the nuclear region of M 77 are noteworthy, we soon realize something even more amazing about this region: It varies in brightness (Figure 17.5)! Now, we *have* seen plenty of examples of variability in astronomical sources before. Planets in the Solar System vary in brightness as their distance from Earth changes over time (Destination 2). Stars such as Cepheid variables (Destination 14) change their luminosity as their atmospheres pulsate. Exploding stars such as supernovae (Destination 12) vary in brightness in a particularly extreme manner. And, as we've already noted, binaries can vary in brightness if their individual stellar components interact or eclipse one another. But this is different. We are now seeing something on

7 These are distinct from the class of Interacting Galaxies we encountered at Destination 16. It turns out that many Active Galaxies are also Interacting Galaxies. But there are also many Active Galaxies that do not appear to be interacting, and also Interacting Galaxies that are not active in the way M 77 is.

8 Grammar warning! AGN can stand for 'Active Galactic Nucleus' or 'Active Galactic Nuclei.' That is, AGN can be singular or plural. Kind of like the words 'fish' and 'moose'. In all these cases, you have to go by context to know which is being used.

a vastly larger scale: The *entire* nuclear region of a galaxy is changing brightness (Figure 17.5). Not a single star, but a small *region* that is emitting as much energy as an entire galaxy!

To fully appreciate why this is so strange, let's consider the implications of the variability of the nucleus of M 77. A good analogy is a large stadium full of people attending a sporting event. Imagine that we are located somewhere outside the stadium where we can easily hear the roar of the crowd. Imagine too that, unknown to us, cheerleaders inside the stadium decide to try to support their side by holding up signs directing the crowd to chant 'Go Team'. Everyone is fired up and wants to pitch in on the cheer, and so the entire crowd in the stadium starts the chant simultaneously when they see the signs flashed at them. As they do, the sound propagates outward at about 700 mph—the speed of sound—from every person. If the stadium is fairly large, say 300 *m* in diameter (about 1000 *ft*), it takes sound about 1 *sec* to travel from one end to the other. From the outside, this causes the cheer—which is perfectly recognizable inside the stadium—to become indistinct. Instead of hearing a crisp 'Go Team, Go Team, Go Team ...,' we first hear the sound from the nearest end of the stadium, followed about a second later by the sound from the far end of the stadium. Since the sounds overlap, what we end up hearing is so indistinct that it sounds more like a steady roar than a clear message: 'GToeaGmGToeGmG ... .' That is, it sounds more like noise than a message. If the stadium were smaller, the arrival times from the near and far sides of the structure would be closer together and we'd be more likely to hear the 'Go Team' chant a bit more distinctly. Less like noise. More like a cheer.

This example suggests is that if we want to generate a *coherent* signal—for example, coherent variability of a galaxy's nucleus—the size of the emitting region must be relatively small. How small? We found that if the time it takes to say 'Go Team' is comparable to the time it takes sound to cross a stadium, the signal will be perceived as noise from outside the stadium. That is, just a steady, indistinct roar. On the other hand, if the stadium is small compared to the time it takes sound to travel across it, we'd hear a more coherent signal (a clearer 'Go Team'), not just an unintelligible clamor. The same logic holds for a galactic nucleus, only now the relevant speed—the fastest speed at which a signal can be transmitted—is the speed of light. What this means is that the size of any region that varies in brightness in a coherent way can't be larger than the time it takes light to travel across the region. If we denote the size of the emitting region as R_{AGN} this means that

$$R_{AGN} \leq c\ \Delta t$$

where Δt is the timescale of variability observed in the galaxy nucleus. For many Seyfert galaxies, the variability timescale, Δt, can be as short as a *few days* (see Figure 17.5), meaning the emission region must be less than a few light-days across. If significantly larger, luminosity variations would be washed out and the light output would appear steady.

This result is astounding. The 'diameter' of the Solar System is about half a light-day, only a bit smaller than the emitting region of the AGN in M 77. Somehow, this comparatively tiny volume at the center of this remarkable galaxy generates a luminosity equivalent to about 4 billion Suns! We simply must go in for a closer look.

17.2 THE TWO FACES OF THE MONSTER

After a *very* long journey, we finally arrive at our first stop about 50 *kpc* short of our ultimate destination, the nucleus of M 77. As we look up at the majestic view of the galaxy in this Alien Sky, our attention is irresistibly drawn to the galaxy's nucleus. Even at this moderate distance, it shines as brilliantly as the brightest nighttime star in Earth's sky, Sirius, and with a steely blue color that reminds us of the hottest stars of the Orion Nebula (Destination 6). But the nucleus of M 77 is no star. And it is not close. We are almost 19000 times more distant from the center of M 77 than Sirius is from the Sun (about 2.6 *pc*). The startling brightness of the nucleus at such a comparatively large distance illustrates the gargantuan luminosity that an AGN can produce.

Though we still have a way to go to reach the nucleus itself, we are close enough to discern details that were difficult to see from farther away. Surrounding the nucleus itself and spanning about one degree in our sky—about twice the diameter of the Moon as seen from Earth—is a complicated tangle of clumpy cloud-like structures (see Figure 17.6). There is no evident order or pattern to how these structures are distributed in space. No spiral arms, no obvious disk-like shapes. Spectra of these regions reveal that they are the source of some of the emission lines we observed earlier. Ionized, fairly dense, and moving with a typical velocity of hundreds of *km/s*, these nebulae are distributed through a volume a few hundred *pc* in diameter surrounding the nucleus. We can readily make out patterns that look a lot like the dark clouds in Orion (Destination 6) and in the Milky Way as seen from Earth (Destination 12). This indicates there may be some dense molecular clouds near the nucleus of M 77. Microwave and radio observations confirm the presence of molecular gas, and also reveal an oddly linear feature cutting across the region[9] (Figure 17.6). The overall complexity is reminiscent of the Galactic Center (Destination 10), but with two big differences. The total extent and quantity of the gas near the center of M 77 is considerably greater than what's found near the comparatively anemic center of our Galaxy. Second, of course, is the AGN itself. If located in the Galactic Center (Destination 10), this nucleus would have appeared brighter than the brilliance of the full Moon as seen from Earth (see Figure 10.7). We could not have missed it. The center of our Galaxy is certainly unusual, but it does not contain an AGN like the one we see here.

Given what we've learned about the nucleus of M 77—not to mention the mystery that still surrounds our understanding of what's going on there—it seems prudent to scout the region out a bit more from our current, comparatively safe distance. With that in mind, we move to different locations in the halo of M 77, constantly maintaining a distance of 50 *kpc* from the nucleus. In effect, we are scouting the nucleus from a large circular orbit around it. As a result, we get to view the nuclear region from different vantage points.

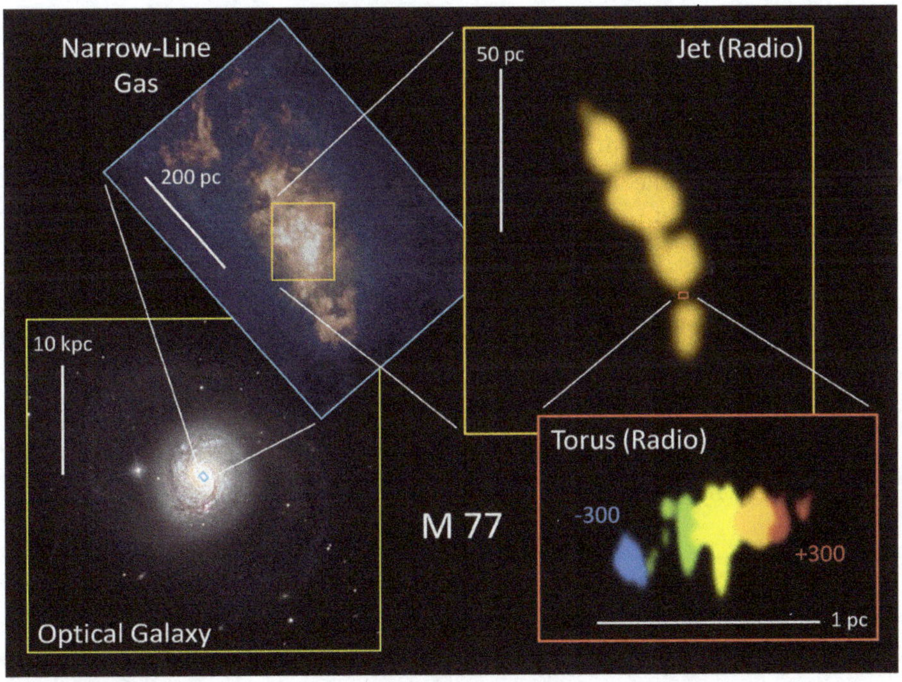

Fig 17.6

(*Lower left*) An overall optical view of M 77 (see also Figure 17.1). (*Upper left*) A detailed optical view of the central region of M 77 surrounding its AGN. The clouds of gas seen in this view span the central few hundred *pc* of the galaxy and are illuminated and heated by a source of radiation much hotter than any star and with a total luminosity similar to the rest of the galaxy. (*Upper right*) A radio image showing the structure of an elongated, clumpy structure about 50 *pc* long that appears to have been ejected from the central nuclear region of M 77 (in the small red rectangle). This feature is likely a *jet* of gas recently ejected by the central AGN (see footnote 9; we'll learn more about jets at Destination 21). (*Lower right*) A highly detailed image of the radio emission from the true nuclear region of M 77. We can see clouds of dense gas that exhibit Doppler shifts (denoted by colors in this image) ranging from a blueshift of about −300 *Km* (on the *left*) to a redshift of about +300 *km* (*right*). These motions almost certainly reflect rotation of the clouds around some massive central object that is hidden from view by the clouds themselves. These thick clouds encircling the AGN are often collectively referred to as the *Torus Region* (see footnote 14), though it is likely quite clumpy in structure. Photos from NOAO/NSF (M 77 overview) and STScI/NASA (core of M 77).

Artwork adapted from data from A. Gallimore et al. and J. Roy.

To our surprise, as our viewing angle of the nucleus of M 77 changes, we see the nucleus become brighter. Remember, we're not moving closer to the nucleus,

9 Astronomers like to refer to such structures as *Jets* in analogy to water jets spewing from a fountain. As it turns out, that's precisely what these structures are doing. Radio observations from Earth can detect the motions of bright blobs in these jets moving steadily outward from the nucleus at speeds approaching the speed of light. We'll return to AGN jets at Destination 21 where we will encounter far more impressive cases than the puny—but still impressive—jet we see here in M 77.

so this is not due to the Inverse-Square Law for Brightness (Destination 4). Intrigued, we continue to move around until we find a location where the nucleus appears brightest, almost ten times brighter than it appeared from our initial vantage point. Though we are *still* 50 *kpc* from the galaxy's nucleus, it now appears comparably bright as the rest of M 77 combined! Much brighter than any nighttime star in Earth's sky. And yet, as before, virtually all of this light seems to arise from the nearly point-like nuclear region of the galaxy.

Even more amazing, the spectrum of the nucleus has transformed as we changed our viewing angle! Though the emission lines we had seen earlier are still apparent, many additional lines are now visible. As we identify these, we realize that many arise from extreme ionization states—for example, Ne (Neon) and Ca (Calcium) that have lost up to four electrons, and Fe (Iron) that has lost up to nine (!) electrons. These ions can form only if there is a nearby source of extremely energetic photons, a source much hotter than any star. But it's not only the specific emission lines that have changed. Some of the lines we now see in the spectrum of the M 77 nucleus—including some of the ones we had identified earlier—are considerably broader than any lines we saw from our initial viewing point. In a few cases, these lines exhibit velocity widths of up to 2000 *km/s* (see Figure 17.4).

The changes are so extreme, so striking, that it is as if we are looking at a different galaxy. A different AGN. But we're not. Both the narrow-lined, low-ionization spectrum of the fainter nucleus and this broader-lined, high-ionization spectrum of this brighter nucleus arise from the *same* nucleus in the *same* galaxy.[10] Somehow, the fundamental character of the galaxy's core has been altered simply because the angle at which we view the nucleus has changed. The lesson here is clear. Some astronomical objects differ from one another because they really are fundamentally different. But sometimes, differences between objects can be illusions. In this case, our two views of the nucleus in M 77 suggest that the inferred nature of a Seyfert galaxy's spectrum, and even the luminosity we estimate for its AGN, appears to depend on the viewing angle.[11] This represents

---

10 Of course, from Earth we can't do this experiment since galaxies are oriented at an angle that changes only very slowly with time. Yet the existence of some 'narrow-lined' Seyfert galaxies—known as *Seyfert 2* galaxies—and other 'broad-lined' Seyfert galaxies—dubbed *Seyfert 1* systems—was apparent soon after the discovery of the this class of galaxy. The likely reason for the existence of two types of Seyfert galaxies was eventually suspected to arise from differences in the orientations of the emitting regions in two types of galaxies. This suspicion has been confirmed through very clever observations of the reflected light in AGN that—similar to how a mirror can let us see both the side and front of something simultaneously—allow us to view some AGN from two different orientations simultaneously.

11 Technically, we would say that the radiation from an AGN—or a flashlight, or a pulsar (Destination 12)—is not emitted uniformly in all directions. Such radiation is said to be emitted in an *anisotropic* manner. In contrast, the Sun—or a standard light bulb, or any other normal star—emits about the same total radiation in all directions. The emission in such cases is said to be *isotropic*.

a crucial hint to understanding what's going on in the unusual nuclei of galaxies like M 77.

17.3 THE NAME OF THE MONSTER

The details of what we've seen in the nucleus of M 77 are certainly fascinating but unquestionably complicated. One might even be tempted to say 'chaotic.' But it's important not to let the details mask the key characteristics that all AGN exhibit: An object comparable in size to the solar system and containing matter moving at extraordinarily high speeds generates as much energy as the amount produced by an entire galaxy. How can we explain the existence of such seemingly incompatible features in AGN?

We've already tentatively interpreted the motions that broaden the emission lines we see in AGN spectra as rotation.[12] If that is correct, we can use techniques we've developed earlier to get an idea of the mass of the emitting region (for example at Destinations 6 and 9). To see how we can do this, recall that Seyfert galaxies exhibit optical emission lines with widths that correspond to a velocity range—denoted as ΔV—of up to $2000 km/s$, and that these nuclei can vary in brightness on timescales as short as 5 *days*. As before, we'll denote the timescale of variability as Δt. If we take the rotation speed to be equal to half the line width (that is, $V_{rot} = \Delta V/2$) and we estimate the orbital size as $R_{AGN} = c\Delta t$, we can use the techniques introduced in Destinations 6 and 9 to calculate an 'upper limit' for the mass within the emitting region as

$$M_{AGN} = \frac{V_{rot}^2 R_{AGN}}{G} \approx \frac{(\Delta V/2)^2 c\Delta t}{G} = 10^6 M_\odot.$$

Though we have measured the rotation velocities and radii in very different ways, this result is remarkably similar to the mass we derived for the massive object at the heart of the star cluster IRS 16 in the Center of our Galaxy (Destination 10). And our conclusion now is the same: The only way to stably

12 If the motions were due exclusively to expansion, we'd expect to see AGN of widely different sizes. However, surveys show that they are all tiny. In a few cases, including M 77, the nuclear rotation has been observed directly (see Figure 17.6) in a manner similar to how we inferred that the dark clouds in Orion are rotating (Destination 6).

cram so much mass—at least one million Solar masses—into a volume comparable in size to the Solar System is in the form of a supermassive Black Hole (Destination 10). Certainly, the volume of the nuclear region is big enough for it to fit. At Destination 10 we learned how to estimate the 'size' of a black hole (or, just 'BH' for short) by calculating its Schwarzschild Radius,

$$R_{BH} = \frac{2GM_{BH}}{c^2} .$$

For BHs typical of Seyfert galaxies—around $10^6\ M_{\odot}$—this turns out to be about four times the radius of the Sun. About 10 *light-seconds*! That's small enough to easily fit into the emitting region of any known AGN, ranging from the comparatively mild nucleus of M 77 to other, more extreme types of AGN we'll encounter later (Destination 21).

Unfortunately, there seems to be a fundamental paradox here, one we did not have to contend with for the supermassive BH at the Galactic Center (Destination 10). How can the epitome of darkness—a black hole—generate the enormous luminosities we see in AGN such as M 77? We have one important clue. At Destination 12, we discovered that LMC X-3—a binary star containing a low-mass BH—is also very luminous, particularly in the ultraviolet and x-ray portions of the EM Spectrum. In that instance, the high luminosity arises from an accretion disk surrounding a BH. Is that what's happening here in M 77? The only way to answer this is to push on and complete our journey to the very heart of the galaxy. And so, off we go. Our initial course takes us directly along the line of sight from which the nucleus of M 77 is seen from Earth.

As we near the nucleus—now less than 1 *pc* away—the bright clouds of gas we saw from afar now surround us (Figures 17.6 and 17.7). Their spectra indicate they are composed of ionized gas similar to the Orion Nebula (Destination 6), but it is also apparent that much of their glow comes from reflected light. That's surprising because these clouds are tens, even hundreds of *pc* away. To appear so bright, the source of radiation they are reflecting must be intensely luminous. And yet, where is it? Ahead of us, we see the outlines of immense dark clouds (Figure 17.6), their bulks revealed by the glow of their bright edges. Beautiful as the view is, we are stunned when we realize that these clouds *are* the outer edges of what must be a vast accretion 'disk' at the center of M 77. The outskirts of the M 77 disk are not composed of optically bright gas like we saw in the disk surrounding the BH in LMC X-3 (Destination 12), but instead is composed of large, dense molecular clouds (see Figures 17.6 and 17.7). Radio, microwave, and infrared spectra confirm that the clouds are rotating around something that lies ahead of us but is frustratingly blocked from view by the clouds themselves (see Figure 17.6); this hidden region must be the source of the light we see reflected off the clouds around us. The Doppler shifts of molecular and atomic emission lines from the clouds ahead of us reveal that one side of the complex is approaching us at about 300 *km/s*, while the opposite side is moving away from us at a similar velocity (see Figure 17.6), perfectly consistent with rotation around something quite massive.

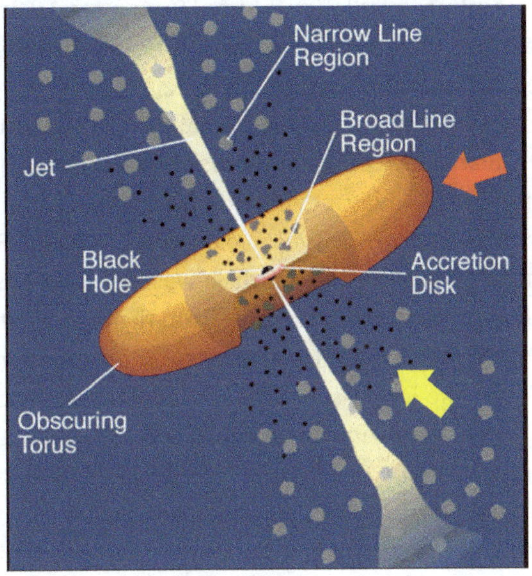

Fig 17.7

A schematic drawing of the central regions of a generic AGN. From the perspective of an observer viewing this region along the line of sight denoted by the red arrow, the thick outer disk (labeled as the 'Obscuring Torus') can block the view of the hottest, fastest-moving portions of the inner accretion disk; in the figure we can see the inner disk because the torus has been conveniently cut away in the drawing. However, when viewed from the line of sight denoted by the yellow arrow, the central, hotter, faster-moving regions of the AGN can be seen directly. It is the unusual geometry of the disk—really a fat torus—that allows a single AGN to appear different depending on the angle at which it is observed. We view M 77 from Earth along a line of sight similar to the red arrow. Many other Seyfert galaxies—ones that exhibit emission lines that are broader and of higher ionization than the emission lines seen in M 77—are viewed along lines of sight more similar to the yellow arrow. Yet they are basically the same sorts of objects. Other characteristic regions of AGN—the Narrow-Line and Broad-Line regions, the inner Accretion Disk and Jets, and of course the central supermassive Black Hole—are also highlighted in this diagram.

Adapted from original image by C. M. Urry and P. Padovani.

The presence of cold molecular gas is not the only thing that makes this disk different from the one in LMC X-3. This disk is a monster! Something *only* a few light-days across may be small by galactic standards, but it's mighty big for us! And it is also very different in shape. This is no thin pancake (see Figure 12.13), but, based on the sizes of these outermost molecular clouds, much fatter, more reminiscent of a wide automobile tire. If we want to move forward—and we certainly do!—we could try to punch through these clouds. But that approach seems rather risky since we are not entirely sure what lies on the other side. So we decide on a different strategy. Taking a cue from our earlier reconnaissance of the AGN from our distant (and safe!) vantage point in the halo of M 77, we decide to travel 'up and over' the edge of the molecular 'disk' so that we can peer into it while maintaining our 'safe' distance 1 *pc* from the center itself.[13]

As we rise over the edge of the disk we are greeted by a dazzling view. Before us lies the inner extent of the accretion disk at the heart of the AGN in M 77. A portion of it—a broad, annular region—shines intensely brightly in the optical, ranging from a deep reddish color at larger radius, to a dazzling blue-white glare nearer the disk's center. Further in, ultraviolet radiation, x-rays,

13 Actually, it would not be very safe here at all. Even 1 *pc* away, the radiation from the center of the AGN would be intense over virtually the entire EM spectrum. Unless we have some very sophisticated protection, we'd be fried by this radiation in short order. Of course, we'll assume we do have such protection so we can carry on to the heart of the AGN without this becoming a suicide mission.

and even γ-rays—forms of radiation our eyes cannot see—are being emitted copiously along with optical light making the inner portions of the disk appear dazzlingly bright. Throughout the regions where optical and higher-energy radiation is produced, the disk appears fairly thin, roughly similar to the structure we saw at LMC X-3 (Destination 12). Beyond the optically visible annulus, however, we detect infrared radiation emitted from material. In this region, the disk gradually thickens, clumps up, and transitions into the large, fat molecular clouds we viewed earlier from 'outside' the disk.[14]

We can now see why Seyfert galaxies can have such different observed properties. If the molecular clouds of the outer disk block the inner disk, we can see only low-ionization, fairly slowly moving clouds located far from the disk center. This zone is known as the *Narrow-Line Region* of an AGN (see Figure 17.7). But if our viewing angle reveals the central disk, we see hotter gas, more extreme conditions, and higher rotational velocities from the disk itself as well as from clouds that are located closer to the center of rotation and therefore moving at a faster speed. This is the *Broad-Line Region* of an AGN and is the source of the wider emission, higher-ionization lines seen in some AGN spectra (Figures 17.4 and 17.7). Of course, the more direct view of all this activity also makes the AGN appears *much* brighter (Figure 17.2). Luminous enough in energetic ultraviolet and x-ray photons to ionize and illuminate clouds that are hundreds of *pc* away.

If we focus on only a small section of the disk, we find that its spectrum is essentially thermal in nature. But, unlike the surface of a star, the disk as a whole is not characterized by a single temperature. Instead, the disk exhibits a *Temperature Gradient*—a general characteristic of all accretion disks—in which the disk temperature ranges from a few hundred *K* at its outer edge to many million *K* near the center. That's why we see mostly microwave and far-infrared radiation from the outer parts of the disk, but higher-energy radiation as we look closer to the center of the disk. The temperature gradient arises because the orbital speed of the material in the disk changes with distance from the center according to Kepler's Third Law. The friction—technically known as *Viscosity*—within the differentially rotating disk heats the gas most effectively near the center of the disk where the orbital velocities vary rapidly, and less efficiently in the outer regions where the orbital speeds taper off more slowly with radius thanks to Kepler's Third Law. Since AGN disks are (comparatively) large, they can attain immense luminosities, much higher than the considerably smaller disks found around stellar-sized black holes as in LMC X-3.[15] But

---

14 From this location we can see why the complex of molecular clouds surrounding the 'true' accretion disk is sometimes referred to as a *Torus* (see Figure 17.7) But the torus is rotating too (see Figure 17.6), and the material in it is destined to flow into the inner accretion disk.

15 This is the same reasoning from Destination 8 that allowed us to estimate stellar sizes. Basically, the more surface area, the higher the luminosity of any thermal source. Since the disk of an AGN is millions of times larger in radius than disks surrounding

in all cases, it's the heating within the accretion disk that can transform black holes—the darkest objects imaginable—into the engines that produce some of the most luminous objects in the Universe!

As we ponder this, we sense that this explanation of how disks shine does not really explain how the luminosity is maintained for long periods of time. Nor is it clear why some supermassive BHs—like the one here in M 77—have large, healthy accretion disks surrounding them, while others—such as the one in the center of our Galaxy—do not.[16] One way to address these issues is to write the luminosity of an accretion disk in the most general manner imaginable:

$$L_{Disk} = \frac{Energy}{time} = \frac{\eta M c^2}{\tau}.$$

This result uses Einstein's Equation, $E = Mc^2$, along with a new term, the *Efficiency Factor* denoted by η (the Greek letter 'eta'), and a timescale, τ (the Greek letter 'tau'). This equation can explain *any* energy source since there is always a tradeoff between mass and energy (or vice versa) whenever power/luminosity (energy per time) is generated (Destination 8). For typical energetic chemical processes—such as burning gasoline—the efficiency factor is minuscule, $\eta \approx 10^{-10}$. In such cases, the conversion of mass to energy is not readily apparent even to very careful measurements. But for disks surrounding BHs, the efficiency can become much higher, more like $\eta \approx 0.1$ to 0.5! Since η cannot exceed 1.0, such values are astoundingly large, implying that 10–50% of the mass in the disk will eventually be converted into energy that is emitted as thermal radiation.[17] If we solve the previous equation for the quantity M/τ—which we will refer to as the *Mass Accretion Rate*—we can determine how rapidly mass must travel through the disk on its way to the central black hole in order to account for the observed luminosity of an AGN. In the case of M 77, the nuclear luminosity is about $4 \times 10^{10} L_{\odot}$, three times higher than the luminosity of our entire Galaxy! If we assume $\eta = 0.1$, the mass accretion rate turns out to be $M/\tau \sim 0.03\ M_{\odot}/yr$.

There's an interesting implication of this result. If 10% of the accreted matter in the nucleus of M 77 is converted to energy, the other 90% must eventually reach the supermassive BH, steadily increasing its mass. At the current rate of

---

smaller black holes, such as the one in LMC X-3 (Destination 12), it has much more surface area. It will therefore exhibit a larger luminosity for a given temperature range.

16 This problem does not arise in LMC X-3 since we know that there is a companion star that provides the mass that is transferred to the black hole. It's not so clear what the source of mass is in an AGN such as M 77.

17 According to current ideas, reaching such fantastically high efficiencies is possible only if the black hole generates a strong magnetic field that permeates the accretion disk. Such fields are expected to exist on general grounds, and, as we've seen from the presence of radio synchrotron radiation (Figure 17.3), magnetic fields certainly do appear to be present in AGN.

mass accretion in M 77, it would take 'only' about 30 *Myr* to produce the $10^6 M_\odot$ BH that we infer exists at the center of the galaxy[18]. But we know from the ages of old stars clusters such as M 92—about 14 *Gyr*, Destination 11—that galaxies like M 77 are *much* older than this. They simply can't have been accreting mass at this rate this over the entire age of the Universe (Destinations 15 and 19) or else the central BH would be much more massive than it is. Perhaps the BH formed only fairly recently. As we'll see in Destination 21, it seems that most—and quite possibly all—elliptical galaxies and bulges of spiral galaxies contain supermassive BHs. It would be an impossible coincidence for all of these to have formed within the past few ten million years, just in time for us to see them.

A more reasonable explanation is that the nuclear activity of M 77 and, by implication, of all AGN, is fundamentally episodic in nature. This implies that we just happen to be around at the right time to see M 77 in an active, but short-lived state.[19] And it helps explain why so many galaxies are *not* strongly active. The BH at the center of our Galaxy, currently dormant, must have at some point in the past gained its mass, perhaps quickly. As it did so, it would have had a luminous accretion disk feeding its central supermassive BH. The Galaxy would have been host to an AGN. But not for long. For a mass accretion rate similar to what's seen in M 77, it would take only 100–200 *Myr* to build the $5 \times 10^6 M_\odot$ at the Galactic Center, a small fraction of our Galaxy's age.[20] It seems that accretion disks associated with supermassive BHs *can* generate immense luminosities in AGN. But not indefinitely. Certainly not for the entire age of the Universe.

---

18 There's plenty of uncertainty in the masses of black holes in most AGN. However, even if the mass of the BH in M 77 were ten times higher—that is, about 10 million $M_\odot$—the basic argument we are making here would remain valid.

19 Similar reasoning applies to volcanoes. Some of these mountains are forming today, venting magma to the surface of the Earth where the volcano forms. If any volcano were to do this for the entire history of the Earth—nearly 5 *Gyr*—it would produce a far larger mountain than any we see. Even if we had only a snapshot of the Earth's surface, we could therefore be quite certain that volcanoes—though occasionally spectacularly active—must lie dormant most of the time. And so it is for AGN as well.

20 We might imagine altering these timescales by changing the efficiency factor, η. However, if we choose to make this parameter much smaller than 10%, higher mass accretion rates would be required to produce a typical Seyfert luminosity, implying the central BH would gain mass even more quickly. On the other hand, we can increase η, but only to a maximum of about 0.5 (50% efficiency). Even in this case, plenty of mass makes it to the BH to increase its mass, so the time to build a supermassive BH is still a small fraction of any galaxy's age. The unavoidable conclusion is that the central BHs in galaxies simply cannot be active all the time (we'll return to this at Destination 21).

17.4 EINSTEIN'S LEGACY: AN INTRODUCTION

We are not quite there. After traveling so far, it would be impossible for us to resist the urge to press on to the very center of M 77, as close to its central supermassive BH as we dare. But the view that awaits us is so bizarre that we cannot easily interpret it until we understand how light behaves in the presence of the intense gravitational fields near BHs. The essence of this behavior is embodied in Einstein's *Theory of General Relativity* or simply *General Relativity* (or even more simply, GR). The implications of GR are so profound that they challenge many everyday perceptions that we have of the real world. Perceptions we 'know' to be true. For example, the notion that, in empty space, light travels in a straight line from where it is emitted to where it is detected. Or, that time always flows at the same, constant rate everywhere in the Universe. General Relativity tells us that neither of these 'obvious' statements are correct.

The reason involves a fundamental, overriding feature of Nature that we can illustrate with a simple example. Imagine standing near a railroad track as a train passes by. Also imagine there is someone on the back of the train who, as she passes, throws a ball to us at a speed that is similar to that of the train, but in the opposite direction of the train's motion. It is evident that the person on the train and us, standing on the ground beside the tracks, will have very different perceptions of the speed of the ball. Though the person on the train threw the ball at high speed to ensure that it reaches us, from our perspective we'd see the ball was tossed softly with essentially little net speed relative to the ground where we stand. The reason is that the speed of the ball that we measure from our stationary position next to the tracks equals the speed of the ball as thrown by the person on the train *minus* the speed of the train itself. This behavior is called *Galilean Relativity* in honor of Galileo (see Destination 9), the first person to elucidate these ideas systematically.

Einstein was the first person to realize that light *cannot* behave this way. To demonstrate why, let's consider an amped-up case of our earlier example in which we now pretend the train is moving at half the speed of light. If the person on the train shines a light source at us, we might expect to see the light traveling at a different speed—in fact, about half the speed of light—as specified by Galilean Relativity. But we don't. In this case, *both* sets of observers—us on the ground and the person on the train—independently measure the light emitted from the train to travel at precisely the *same* speed: The speed of light!

Einstein realized that if this were *not* true, basic laws of physics observed on the train would differ from the same laws observed on the ground beside the tracks. But that's impossible. If the laws of physics varied depending on where one is located and how fast one is moving, we'd expect to see very strange effects. Although we may not have handy any trains that move at a significant fraction of the speed of light, we do have distant galaxies. And, as far as we can tell, the same basic laws of physics hold in remote galaxies as in our Galaxy, even though they can appear to be traveling at high speeds relative to us due to the Hubble Law (Destination 16).[21] This fundamental modification to Galilean Relativity was named *Special Relativity* by Einstein. When applied to cases where accelerations can occur due to a strong gravitational field—say, near a black hole—one must use General Relativity instead.

The concept that the speed of light is always the same regardless of where you are or how fast you are moving has some profound but unusual implications. To see this, let's imagine that we've discovered a 'naked' BH, one lacking surrounding material in the form of a disk or anything else.[22] We'll assume we know the mass of this BH to be similar to that of the supermassive BH in M 77, or about $10^6 M_\odot$. Knowing this, we can determine the BH's Schwarzschild radius using the relation $R_s = 2GM_{BH}/c^2$ (Destination 10). Recall that nothing located closer than R_s from the center of mass of a BH can escape since it would have to travel faster than the speed of light to do so. That's why the spherical 'surface' defined by the Schwarzschild radius is known as the *Event Horizon.*[23] This is a good, descriptive name since, as we'll soon see, any events that occur inside the 'horizon' defined by the Schwarzschild radius are forever hidden from the Universe that lies beyond R_s.

21 Definitive tests of whether or not all the same laws of physics hold in remote galaxies as on Earth are tricky, but some similarities are obvious. Emission/absorption lines of atoms in remote galaxies appear precisely the same as on Earth, implying that the atoms from which these lines arise (see Destination 6) are also the same. And since the formation of these lines involve a range of physical laws that govern subatomic particles and radiation, we can infer these laws are the same as on Earth.

22 It's been claimed that some supernovae—ones arising from massive stars in binary systems—might impart a 'kick' to the black hole produced in the process or to a preexisting black hole such as in LMC X-3 (Destination 12). Such BHs—likely naked since they would have no source of mass to accrete onto them—can be flung rapidly from the site of the explosion, possibly orbiting in the halos of galaxies such as M 77, or even in the Halo of *our* Galaxy. These are unlikely to be supermassive BHs, but some of those might exist outside of galaxy nuclei as well, flung out as a result of galaxy collisions (Destination 16). In any case, we'll presume we found one so we can study it in (comparative) safety compared to an active BH in an AGN.

23 This is strictly true only for a comparatively 'simple' BH, one that is not rotating nor has electric charge. In more general cases, the location and shape of the Event Horizon can be quite complex and it may even be possible to receive information from 'within' these horizons under special circumstances.

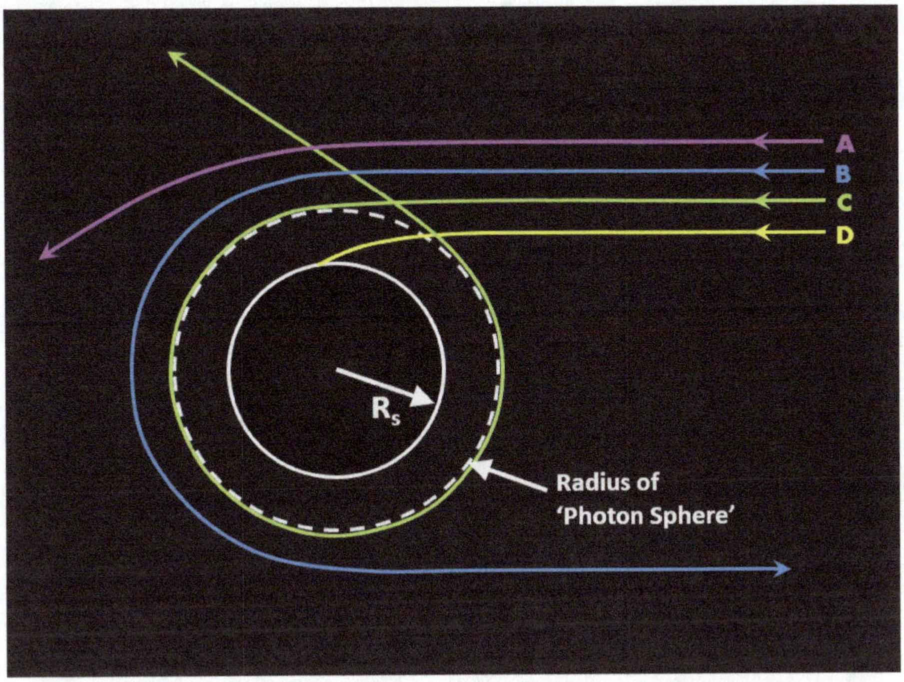

Fig 17.8

Light does not travel in what we perceive to be straight lines in the presence of a gravitational field. In this image, light rays produced at various sources (denoted A, B, C and D) travel along paths that are curved as they pass the vicinity of a black hole. The light originating from source A passes far enough from the event horizon (denoted at the solid white circle) and photon sphere (dashed white circle) so that it is merely deflected by a few degrees from its original path. Light from source B passes closer to the black hole and is deflected by 180 *deg* causing the light to return to where it was produced! The light originated at C passes just outside the photon sphere, making nearly one complete orbit of the black hole. Finally, the light from source D passes within the photon sphere and, inevitably, to the event horizon where it travels 'into' the black hole.

We dare not approach the event horizon ourselves, at least not for such a comparatively 'low-mass' BH.[24] But we cannot resist playing with it from a distance. In this spirit, we decide to toss a lit flashlight into our BH. We watch in fascination as the gravity of the BH accelerates the flashlight toward the surface—that is, the event horizon. As the flashlight nears the event horizon, its speed is approaching the speed of light. At the same time, the radiation we observe from the light grows redder and dimmer even though the lamp was designed not to change the color or brightness of the light in any way. And through the entire process, the speed we measure for the light we receive from the flashlight remains steadfastly equal to the speed of light even as the flashlight itself hurtles toward the BH.

It turns out that the only way for this strange situation to occur is if time *slows down* for the receding flashlight compared to our (apparently stationary) location! This strange effect is known as *Time Dilation* since it causes moving clocks to appear to advance more slowly relative to clocks that are stationary relative to the observer. One practical consequence of this is that the frequency of light from a source moving away from us—like our flashlight falling into the BH—will be reduced. After all, the time interval between crests of light from, say, an emission line, is basically a very precise clock with a natural time interval equal to the inverse of the frequency, or $1/\nu$. Since we see the moving 'clock' appear to go slower due to time dilation, the frequency of the light will be smaller.

24 For relatively 'light' BHs, the tidal forces on a human body—the difference in the gravitational force from ones head to toes—would tear it apart long before reaching the event horizon.

Of course, if the frequency is reduced, the light from the moving source will have a longer wavelength, λ, and a lower energy, E, since these properties of EM radiation are always related to one another by $\nu = c/\lambda = E/h$ (Destination 5; c is the speed of light, and h is the Planck constant). Combined with the Doppler effect (Destination 6)[25], this explains why the light from our flashlight has been redshifted; that is, why the light becomes redder and dimmer as the flashlight falls into the BH.[26] Strictly speaking, these unusual effects would arise just from Special Relativity if the flashlight was moving at a high, steady speed. But since the flashlight is also nearing a black hole where the gravitational field is strong, its light must also 'climb out' of the field to reach us. That produces an additional redshift to the light that arises from General Relativistic effects.[27]

Another important result of General Relativity has to do with the path that light follows. It is common sense to think of light traveling along the shortest distance between two points. But this isn't strictly correct. The more precise thing to say is that light travels along the path that requires the *least time* for it to complete a trip between two points. When gravitational fields are weak—as they are on and near the Earth—the distinction between these shortest distance and shortest time is not noticeable. But in strong gravitational fields—such as near a BH—this is no longer true. For two reasons. First, according to General Relativity, space itself is 'warped' by the presence of any source of gravity—that is, mass. As a result, 'straight lines' may appear curved due to what is known as the *Curvature of Space* (we'll return to this concept at Destination 19). Second, since the rate at which time advances can change depending on location near a strong gravitational source, the path light travels to minimize its travel time is also distorted. In these circumstances, light will follow what appear to an outside

---

25 When the speed of a receding object becomes comparable to the speed of light, the simple Doppler Equation we derived at Destination 6, $V/c = \Delta\lambda/\lambda_0$, must be replaced with its relativistic version

$$\frac{V}{c} = \frac{(1+z)^2 - 1}{(1+z)^2 + 1},$$

where $z = \Delta\lambda/\lambda_0$. You can prove for yourself that this turns into the classical Doppler Equation when V is very small compared to the speed of light. As you'd expect, this relativistic form of the Doppler Equation already accounts for time dilation.

26 If the flashlight had been shot toward us at very high speed, the same arguments apply. Time would *still* appear to run slower at the flashlight as seen by a stationary observer, but the Doppler effect would be reversed. The net effect is to make the frequency of the light from the flashlight appear higher to the stationary observer (and the wavelength shorter, and the energy higher). The light from the moving source—the flashlight—would therefore appear bluer and brighter, the exact opposite of what we see if the source of light is receding from us.

27 The General Relativistic redshift would occur even if the flashlight was 'hovering' near the BH's event horizon.

observer to be highly curved—even circular—paths near BHs (see Figure 17.8)! We can see this for ourselves with our naked BH. If we were to shine light at it, we'd see a faint ring of light around the event horizon from radiation that wraps entirely around the BH and comes back toward us! At a distance of 1.5 Schwarzschild radii from the center of our BH (that is, for $R \leq 3GM_{BH}/c^2$), the paths can become so curved that light can be trapped in orbit around a BH. This unusual region is known as the *Photon Sphere* of a BH.

Fig 17.9

Figure 17.9: The satellites of the *Global Positioning System* (*GPS*) are all in motion relative to one another, as well as with respect to users on Earth. These satellites—one of which is shown here in an artist's impression—would be unable to provide precise positional information without the corrections to the flow of time prescribed by the Theory of General Relativity. Many of the extreme effects that General Relatively predicts near BHs—for example, gravitational waves—have been observed directly. Along with the practical operation of the GPS, many other experiments on and near Earth and others observable in astronomical sources demonstrate directly that many of the bizarre phenomena—time dilation and curved space—that arise from General Relatively really do occur. Einstein's 'theory' is correct!

Artwork from NASA.

This is all starting to sound a bit crazy. And, in any case, these results are based on the *theory* of General Relativity.[28] But the term 'theory' here does not imply uncertainty. It refers to the specific collection of concepts embodied in the overall picture we call General Relativity and which we can use to expand our thinking. For example the concepts in GR can be used to predict how light and matter should behave near strong gravitational sources. These predictions, in turn, can be tested experimentally and used as a springboard to the identification and understanding of new concepts that extend, or perhaps even falsify, GR. But 'mere' theories such as General Relativity are not just academic constructions. They can be of practical use. The *Global Positioning System* (GPS) can pinpoint one's location on Earth to a precision of better than one meter (Figure 17.9). It is now used routinely for navigation by hikers, and in cars, boats, and aircraft. It's even used by waiters to find tables in restaurants! But none of this would work without the 'Theory' of General Relativity. The GPS consists of dozens of satellites orbiting the Earth, each one is moving at a significant velocity relative to the others and relative to the surface of the Earth. Consequently, time advances at a different rate for each satellite, precisely as predicted by General Relativity. Without accounting for time dilation effects, the GPS wouldn't function at anywhere near the precision it does. And that's just one

28 All the results from Special Relativity are 'bundled' into General Relativity, so we are tacitly referring to both arms of Einsteinian relativity.

example. There are many other more esoteric experiments and applications that fully verify fundamental tenets of General Relativity.[29] But the GPS nicely illustrates that the 'Theory' of General Relativity is used in everyday applications, largely by people who are utterly unaware of the fact. Einstein was right.

17.5 EINSTEIN'S LEGACY: GETTING PERSONAL

With the help of General Relativity, we are getting closer to being able to 'view' directly an accreting BH like the one in M 77. But there are still some important effects to demonstrate that can account for predictions GR makes regarding what we'd see near an accreting BH. So, let's perform one last experiment. A somewhat more ambitious one. We'll have a volunteer fall into a BH! To avoid an overly gruesome result, we need to use a 'naked' supermassive BH[30] with a mass of at least $10^9\ M_\odot$ (one billion solar masses!) and a Schwarzschild radius of about 20 *AU*. Make no mistake. What we are proposing is certainly a suicide mission, though perhaps not for the reason we'd think. The average 'density' of the mass within this particular BH is only 0.02g/cm^3, considerably less than the density of water![31] Our volunteer will easily survive the passage through event horizon since the resulting tidal forces on him and his ship will be minuscule.[32] But then? Well, let's not think about that just yet.

29 One of the earliest involved a long-known anomaly of the motion of the Planet Mercury around the Sun, a problem that Einstein wasn't even aware of as he developed his ideas. When the discrepancies in Mercury's orbit were fully accounted for by General Relativity, Einstein proclaimed the notification of the result one of the sublimely happiest moments of his professional life.

30 One relatively nearby candidate would be the supermassive black hole in the galaxy NGC 3115. This BH has a mass of about 2 billion $M_\odot$ and appears to be quite dormant at the present time. The precise BH we choose to explore for this part of our experiment is not critical. We've chosen a BH with a mass of 1 billion $M_\odot$ in order to make the duration of the volunteer's voyage reasonable (for us!) and his passage through the event horizon 'survivable' ... at least initially.

31 This calculation assumes the mass of a BH is uniformly distributed within the Schwarzschild Radius. However, as far as modern physics is concerned, there is *nothing* that can stop the collapse of the mass within a BH. Thus, all the mass of the BH should all be concentrated at its center in a point-like structure of nearly infinite density known as a *singularity*. As we'll see later, the tidal forces as the singularity become extreme, but only well inside the event horizon.

32 This is not quite true. The tidal forces *would* be survivable, but there is another danger. If our volunteer free-falls into the BH, the acceleration will become sufficiently high—10–20 times the force of gravity on Earth—to make the descent quite unpleasant,

After some solemn farewells, we are ready. Our volunteer floats in his spacecraft beside our mother ship, both orbiting about $10R_S$ from the center of the BH—a comparatively safe distance of about 200 *AU*. Our volunteer's destination is the BH itself, which appears as an inky blot about 1 *deg* across. The primary indication that something bizarre awaits our explorer is the grotesquely distorted view of the sky in that direction, a clear sign of the extreme space curvature near the BH (see Figure 17.10). After we assure that the computers on both spacecraft have synchronized their clocks, we watch the TV signal beamed to us from our explorer's vessel as he starts his final descent.[33]

After falling for thirty hours as measured by the clock on the scout ship, our volunteer is now just three Schwarzschild radii from the center of the BH. For him, things still seem pretty normal. To the front is a view of an ever-larger BH (though perhaps a bit too large?), while, to the back, he can still make out the mother ship floating serenely in a field of stars. From the mother ship, however, things are beginning to appear odd. For one, the explorer is now moving at a sizable fraction of the speed of light—about 0.05*c*. As a result, we have to alter the receiver to a lower frequency to pick up the TV signal from the scout ship. Since there is no mechanism on the volunteer's ship to alter the signal's frequency, we realize we are seeing the Doppler effect at work (see Destination 6). More alarming is what we see in the video feed. The clock on the infalling ship appears to have slowed down. Our clocks on the mother ship tell us that 30 *hr* 8 *min* have elapsed since the volunteer began his journey, but the clock we see on the TV clearly says that only 30 *hr* have elapsed! The clocks *should* be far more precise than that.[34]

Five hours later according to the scout ship's clock, our intrepid explorer reaches a distance $1.5R_S$ from the center of the BH. Now, there's no mistaking that things have become quite odd. The forward window of the volunteer's ship shows nothing—a deep inky blackness that seems to have swallowed the very idea of light. Certainly, he could expect the BH to appear larger as he approaches, but why is the *entire* front view now appear black? The explorer is still well

---

possibly fatal. To avoid this, our volunteer could use rockets to slow the fall, or find a more massive BH. For a $10^{10}\,M_\odot$ BH, for example, all the same events we describe here would occur about 10 times more slowly and with (survivable) forces about 10 times weaker. We've chosen a $10^9\,M_\odot$ BH here simply for convenience: For this lower-mass BH, the experiment takes only a couple of days, not weeks. Plus, we can just assume our volunteer will be tough enough to survive! *We're* safe and sound back in the mother ship in any case!!

33 The details of this descent are actually rather complex. Our volunteer intends to fall straight into the BH. To do so requires that he first get rid of his orbital angular momentum (Destination 6). The mother ship needs to do the same to remain in communication with our explorer's ship, effectively 'hovering' above it. Both maneuvers require a lot of energy.

34 The time delays for the signal to reach us from our volunteer's ship—which have become considerable—have been accounted for before comparing of the clock times on the two spacecraft.

above the horizon, yet half the sky is filled by the BH itself. This is analogous to what an ant sees on the Earth's surface: Half of its field of vision is filled by the Earth, half by sky. But far *above* the Earth's surface, we'd expect to see the planet shrink away, producing a view filled with relatively more empty sky. Apparently, that's not what happens near a BH. Well above its event horizon, it looks like we are already there.

To the rear, our explorer still sees stars, but they too look unusual. For one thing, all appear noticeably redder than before. And fainter. But why? Compared to the distances to the stars, the explorer's ship has hardly moved at all. Consequently, according to the Inverse-Squared Law for Brightness (Destination 3), the brightnesses of the stars should *not* have changed by any discernible amount. More alarming, however, is that some of the stars that had been *in front* of him are now visible *in the rear viewport!* This is a consequence of the extremely curved paths that light must follow in the vicinity of the BH to reach the ship. From its present location, the only way light from stars located to the front of the ship can reach the explorer's ship is for it to follow highly distorted trajectories (see Figure 17.8). Stars that had been visible in front of the ship now appear to be located to the side—or even slightly behind—the spacecraft! As it happens, our volunteer is now at the photon sphere. There are no light paths that can reach him from any location closer to the BH. Any light that does reach him must come from *behind* his infalling ship. Ahead looms utter darkness so complete that no light—no radiation of any sort—can penetrate.

Fig 17.10

Figure 17.10: (*Lower right*) The view of a 'naked' BH from a safe distance. The distorted view of the Milky Way of our Galaxy results from the extreme bending of light around the Schwarzschild radius of the BH (see Figure 17.8). The object depicted in the two red circles is M 31. It is visible at two locations since some of its light bends around the far side of the BH into our line of sight. (*Upper left*) Although the view of the BH is hypothetical, the images of *Lensed Galaxies* (in blue) are not. In each case shown, the mass of a foreground galaxy has 'bent' light sufficiently so that the image of a background galaxy (shown in blue) is distorted into arcs or even complete circles (known as *Einstein Rings*).

Photos by STScI/NASA (*upper left*) and A. Hamilton/CU and A. Mellinger.

From the mother ship, things seem even weirder. The TV signal from the scout ship has shifted to very low frequencies—larger wavelengths—and is becoming very weak and hard to detect despite the relative proximity of the transmitter. This effect is partly due to the high speed of the scout ship—which now appears to be traveling at over 70% the speed of light—but also by an effect known as the *Gravitational Redshift.* Even if the scout ship could hover above the BH and remain motionless relative to the mother ship—a *very* energy-intensive maneuver—any radiation emitted by the explorer's craft would appear redder (lower frequency, longer wavelength) and less energetic. The reason is that even light has to work—that is, use energy—to escape from the vicinity of the BH. This effect acts like Doppler shift, but it is not due to motion. It results from the strong gravitational field of the BH.[35] Together, the Doppler effect and the gravitational redshift combine to make the view of the scout ship grow rapidly much fainter as it approaches the event horizon. As we try to hold the signal by changing the frequency of our receiver, we can still—barely—detect our volunteer's TV signal. What we see is downright alarming. The clocks on the scout ship indicate that only 35 *hr* have elapsed since the start of our volunteer's journey. Yet our clocks on the mother ship tell us that *36 hr* 7 *min* have elapsed!

Finally, as the explorer reaches a point just above the event horizon, his view has become bizarre. Ahead—blackness. But now, even the view *behind* him is mostly black. As if he were near the bottom of a deep well, only a small circle of sky remains visible. With a telescope, he can see (some) stars, all extremely faint, all very red in appearance. And now there is no doubt about it. Some of the stars and galaxies that had been directly in front of him when he set out were now visible almost directly at his rear. This is his last chance. The only way out now is to accelerate instantly to nearly the speed of light and head straight up in the direction where the stars are visible.[36] This is the only direction in which light—or anything else—can escape the BH. Any less speed, any other direction, will inexorably take the ship on a path to the event horizon. And then ... Too late! Poof! He passes the event horizon. Yet the seconds continue ticking by. Pinching himself, our explorer realizes he is still alive. And 'inside' a black hole!

From the mother ship, we watch with a mixture of horror and amazement. Before our eyes, the infalling ship starts to grow fainter, redder, its clocks continuously slowing down as the ship nears the Schwarzschild radius. After just over forty hours by the local clock—40 *hr* 17 *min* to be more precise—the

35 Gravitational redshift occurs when light travels away from any source of gravitation, including, say, the Earth. The effect is just much, much more pronounced when light originates near a BH.

36 As noted earlier, rapid acceleration to near the speed of light would be pretty painful, and at some point impossible to do without killing our volunteer, no matter how capable his spacecraft. This point is often ignored by movies and television shows where accelerating to the speed of light seems to be as easy—and as fast—as saying 'Make it so.'

explorer's ship appears to just vanish. Not because it entered the BH—as far as we could tell, it was last visible frozen just above the event horizon—but because it ... vanished! But the lasting mystery was the clock. The last video frames we received from the volunteer's ship showed his clock reading 37 *hr* 30 *min*. And grinding to a complete stop as the signal disappeared.

What happened? The most startling effect is the apparent slowing down of time on the scout ship as perceived from the vantage point of the mother ship. Since time seemed to change its rate, everything associated with time also changed. That includes how light behaves. Since we can characterize all EM Radiation by its frequency—that is, how often the waves go up and down every second (Destination 5)—once time slows down, then the rate at which the waves oscillate will also appear to decrease. Consequently, the frequency of the light—and of the TV signal—from the infalling ship decreased. Since the basic relation between wavelength and frequency, $c = v\lambda$, still holds (that is, the frequency of EM Radiation times its wavelength still equals the *unchanging* speed of light, c; Destination 5) the wavelength, λ, increases and the radiation appears redder. But light also has energy according to the relation $E = hv = hc/\lambda$ where h is Planck's constant (Destination 5). Thus, as the frequency, v, decreases and the wavelength, λ, increases, the energy, E, of the light decreases. That's why our view of the scout ship and the strength of the TV signal dropped off so suddenly that it appeared that the ship vanished just as it reached the event horizon: The explorer's ship became so faint and the light shifted so far to the red, that there is no way to observe it. As the wavelength of the radiation from the scout ship became longer and longer, the ship—and our volunteer—were cut off from our Universe.

And so now, finally, we are ready. We can look at the regions directly around the BH in M 77 and have some hope of understanding what we see!

17.6 THE MAW OF THE MONSTER

With some trepidation, we head to the very center of the accretion disk of M 77. Keep in mind, the disk is *light-days* in size, a few tens of billions of *km*. The 10^6 $M_\odot$ black hole—about 6 million *km* in diameter—is just a tiny speck at the center of this immense structure. The optically bright portion of the disk appears as a gargantuan ring of light extending some 7 billion *km* from the BH, considerably bigger than the Solar System. This is where much of the optical radiation is produced—especially the radiation exhibiting the 2000 *km/s* rotational velocities that we observe from the broad emission lines in the spectra of some Seyfert galaxies (see Figure 17.7). Even though we see nothing obvious, we know the disk extends well beyond the optically visible region. In the outer regions of the disk, where it is comparatively cool, it emits most of its energy in the form of infrared radiation. The inner parts of the disk—which are also optically faint—emit most of its energy in the form of ultraviolet and x-ray radiation. As we gaze upon this breathtaking structure, we begin to wonder. Will we see *any*thing near the black hole? If we extrapolate from what we see in the disk, it seems that the material there will be so hot that most of the energy will be emitted as extremely short wavelength radiation far outside the range our eyes can see.[37]

But it is close to the BH itself where the odd relativistic effects we've described earlier begin to alter our view. As we peer into the very center of the disk—well inside the domain of the BH—we realize that we are seeing material falling through the intense gravitational field just above the event horizon. The material *is* hot, with temperatures reaching many tens of millions *K*. But the redshift produced by the strong gravitational field near the BH allows us to see this radiation—originally emitted as x-rays—as optical photons! And what we see is astounding. Matter in the final phases of falling into the BH tries, in vain, to remain in orbit, madly spiraling down to the event horizon.[38]

Fig 17.11

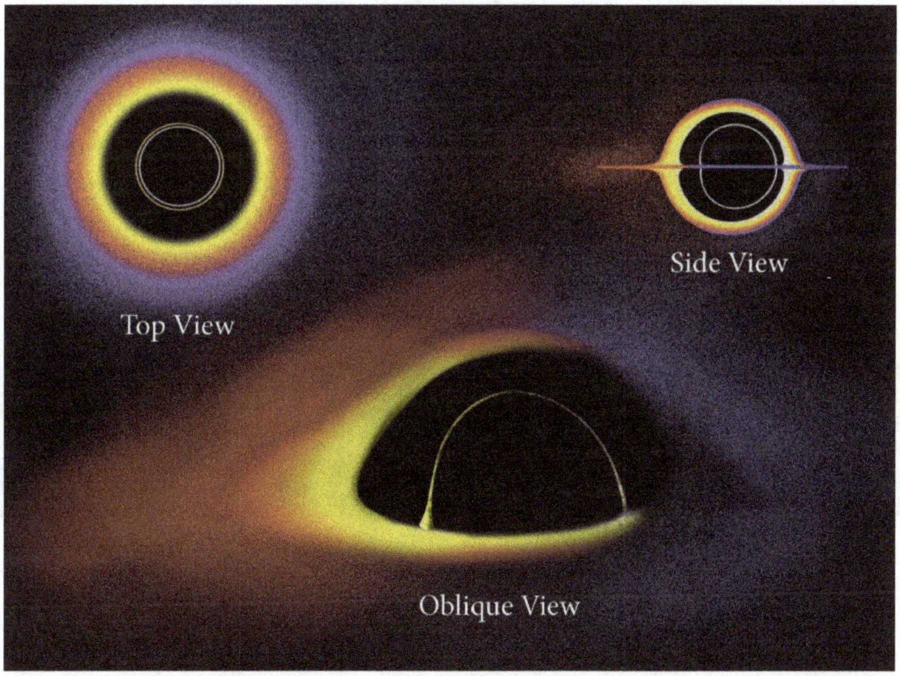

Figure 17.11: Three views of the inner accretion disk near a BH. In all cases, the location of the event horizon is depicted as a white curve. At *upper left* is a top view of the disk, showing it to be circular in shape. At *upper right* is a side view that reveals the disk to be very thin. Notice that the side view reveals a view of the disk similar to the top view (*upper left*). This comes about from light that is sufficiently diverted by the intense gravity of the BH so that we can see the 'far side' of the disk directly both above and below the location of the BH (if you look carefully, you can even see a similar 'secondary image' of the disk in the top view at upper left). The central image shows an oblique view of the disk; the left side of the disk is strongly redshifted, while the left side is blueshifted due to the disk rotation. This boosts the brightness of the blueshifted side, making the disk appear even more asymmetric than this image implies. The oblique view also shows how we can see both the top and the bottom sides of the far portion of the disk as the light is diverted as it passes by the BH. The

37 There is a subtle, technical point here (and in our description of the accretion disk at LMC X-3 back at Destination 12). The specific appearance of an accretion disk is actually quite sensitive to how fast the temperature rises inward toward the center and how the mass is distributed in the inner parts of the disk.

38 Indeed, inside the photon sphere, there are no stable orbits around the BH. None of the matter and few photons have any hope of escaping from this region. The future for

The result is mind-bendingly bizarre (Figure 17.11). One side of the BH is exceedingly bright where material falling toward the event horizon spins in our direction. The gravity causes a tremendous redshift that allows this super-hot material to become visible in the optical while the rotation of this matter within the disk at nearly the speed of light induces a powerful *blueshift* that intensifies the light. The other side—where the material is moving away from us—suffers the same intense gravitational redshift plus a velocity redshift, making the radiation from that side of the BH so unenergetic that we see almost nothing at all at any wavelength. And in the center of it all is the exquisite darkness that marks the event horizon of the BH itself (Figures 17.7 and 17.11), serene within the maelstrom surrounding it. Due to the highly curved paths light take near the BH, we can even see portions of the disk 'above' and 'below' the event horizon, causing it to appear grotesquely distorted. It looks nothing like the flat, circular disk we know it must be.

Here at the center of M 77 we see more plainly than ever before that a black hole consumes matter insatiably as long as new 'food' is delivered to it via the accretion disk. And that's the key. The disks around BHs 'shine' only as long as they accrete matter (at Destination 21, we'll explore where that matter comes from). Eventually, the accretion disk here at the center of M 77 will dry up when the source of infalling matter is expended. Of course, the BH will continue to exist, but *incognito* since it will not have any mass accretion to generate a luminous disk. Our Galaxy contains one such 'dormant' BH (Destination 10), similar in most respects—apart from mass accretion—to the 'active' supermassive BH in M 77. Do most large galaxies contain such 'dead' BHs? We'll discover at Destination 21 that the answer is very likely 'yes.' But the story of how supermassive BHs come about in the first place in so many galaxies is something we will have to address later on as we explore how galaxies themselves begin to form in the young Universe (Destinations 20–22).

asymmetry in the shape of the event horizon arises from distortions in space as it is dragged around the BH, a feature of strong gravitational fields that further distorts the appearance of objects near BHs. What's important to appreciate is that what we would ***never*** see is a flat, thin disk, even though that's the actual form of the inner accretion disk just above the event horizon of the BH.

Images from calculations by N. Quien, R. Wehrse, and C. Kindl.

17.7 IN MEMORIAM

By the way, whatever happened to our volunteer? The one who passed through the event horizon in that $10^9 M_\odot$ black hole? In one sense, he never did make it through. From our vantage point on the mother ship, time stopped on his ship as he reached the event horizon. It would take him an infinite time, as measured on the mother ship's clock, to pass into the BH. But he became effectively invisible on

virtually everything in this region is on the other side of the event horizon.

a much shorter, entirely finite time.[39] Even more confusing, from the volunteer's perspective, the clock on the scout ship did not slow down at all. As far as he was concerned, he really *did* pass through the event horizon. Into the black hole.

What would he see? We'll never really know for sure. His trip not only took him forever outside of the realm of our Universe, but also beyond the realm of well-understood modern physics. We could guess that he probably got to see the environs of the central *Singularity*, the ultra-dense core where the all the mass of the BH resides. What this would look like is unclear since it is not obvious how, or if, light travels from the singularity to our observer's eyes. Nor is it clear how so much compressed mass—now effectively subatomic in size—would behave.[40] But one thing is certain. Our explorer ultimately could not avoid being sucked toward the singularity itself. Inside the Schwarzschild radius, space starts to act more like time, flowing steadily toward the center of the BH where all its mass is concentrated. Whatever he does, however much energy he exerts to try to avoid it, space within the event horizon drifts continuously, mercilessly, toward the singularity at the center of the BH. It may take a few hours on the infalling spacecraft's clock—or perhaps a near-infinite time—but the flow of space ensures that the singularity will eventually have the scout ship, along with our brave but foolhardy explorer, for lunch.

---

39 What would that time be? One definition would be when the wavelength of radiation from the scout ship matches the physical extent of the Universe. But it hardly matters. Any definition based on a finite wavelength or frequency of radiation would occur essentially simultaneously from the perspective of the mother ship during the very short interval when our volunteer's ship reaches the event horizon.

40 We have already seen that small things behave in unusual ways; remember the Bohr atom from Destination 6? In much the same way, the singularity of the BH might start to have quantum mechanical properties (Destination 11) that lead to fantastical behavior. No one has worked out how to merge quantum and gravitation effects, so it is hard to predict precisely what would happen. Worse, as our volunteer has demonstrated, it would be impossible to broadcast any direct observations of the singularity from inside the event horizon to the outside Universe. It would be a purely personal—and suicidal—experiment!

DESTINATION 17 QUESTIONS

17.1. Describe how the active nucleus of a Seyfert galaxy can exhibit rotation velocities of over 1000 *km/s* while the outer stellar disk of the galaxy itself rotates at only about 200–300 *km/s*. This is a somewhat subtle question, so think carefully.

17.2. Let's say internal combustion engines have an efficiency factor, η, of 50%. How far could you drive on one *kilogram* of gas? Assume that you travel continuously at 100 *km/hr* and that your car consumes energy at a rate of 10^{11}*erg/sec* to maintain this speed. **HINT:** Use Einstein's Equation with η like we did for the nucleus of the Seyfert galaxies.

17.3. Explain why the 'mean density' of a black hole gets smaller as its mass increases. **HINT:** Use the equation for the Schwarzschild radius and the definition of density. Assume—even though this is incorrect—that the mass within the event horizon of a black hole is distributed uniformly through the entire volume within one Schwarzschild Radius.

17.4. Let's consider how long a massive black hole in an AGN can produce energy.

a) The mass of a black hole in a luminous active galactic nucleus is believed to be about $10^9\,M_\odot$. If this mass was completely converted into energy, how much total energy could you extract based on Einstein's Equation, $E = Mc^2$?

b) The luminosity of a bright active nucleus is about $10^{14}\,L_\odot$. Let's say that the black hole in such a nucleus converts only 10% of the mass that ultimately goes into the black hole as energy. That is, 10% of the answer to part (a). How long can the quasar have produced light at this luminosity? Remember that Luminosity = Energy/time.

c) Let's turn this around. Assume that a typical active nucleus remains active for the entire age of the Universe, or about 14 *Gyr*. How much mass must the black hole in such a nucleus consume to produce the luminosity in part (b) at 10% efficiency?

d) Which result—part (b) or (c)—seems more reasonable to you? Why?

17.5. Look up detailed information about a Seyfert galaxy other than M 77 = NGC 1068. List the Hubble class, distance, and luminosity of the host galaxy, and at least two physical properties of the AGN (e.g., luminosity, rotational velocity (specify for which emission line), age, mass-accretion rate, size, variability, flux at any wavelength, etc.).

17.6. Our volunteer traveled 180 *AU* during his descent to the event horizon of the naked supermassive BH we studied at this destination. From our perspective, his journey was 37.5 *hr* long. How does this compare to the time light would travel from the event horizon to us?

17.7. The 'narrow' emission lines in AGN can range from 100–500 km/s in width. This gas typically has a temperature of about 10000 K. How does the velocity we infer from the line widths compare to the average speed of the gas particles in gas of this temperature? (See Destination 5.) You can assume the gas is composed exclusively of ionized H. Are we justified in interpreting the line widths as an indication of either rotation or bulk random motions of the gas in an AGN?

17.8. Based on the information in the lower right panel of Figure 17.6, what is the inferred mass of the supermassive BH in M 77?

17.9. How wide would the Earth appear in the sky in degrees if you were suspended 1.5 times its radius from the Earth's center? Compare this to the apparent size of a BH at the photon sphere located at a distance of 1.5 Schwarzschild Radii from its center.

Destination 18

THE BOÖTES VOID

18.1 MAPPING THE DISTRIBUTION OF GALAXIES

Never once during our travels so far have we strayed far from individual galaxies. We started out our voyage exclusively inside or in the immediate proximity of our Galaxy (Destinations 1–13). As we explored galaxy groups (Destinations 14 and 15), and clusters (Destination 16), we invariably found ourselves within 1–2 *Mpc* from at least one galaxy comparable in mass and luminosity to systems such as our Galaxy or Maffei 1. Maybe that's normal. Maybe it's simply not possible to roam further from big galaxies. Maybe galaxies are distributed within a more-or-less uniform 'sea' of galaxies, galaxy groups, and galaxy clusters. If so, and if there are enough galaxies in this 'sea', then it could be that no matter where we go, we'd always find ourselves relatively close to some big, luminous system.

But it turns out that it's *not* the case that galaxies are distributed uniformly in space. And we can tell simply by looking at where galaxies appear in the sky. From Earth, a careful mapping of galaxies reveals a complex, curiously interconnected distribution (Figure 18.1), nothing at all like the smooth distribution we'd expect to see if galaxies, groups, and clusters are distributed uniformly throughout the Universe.

However, saying that galaxies are not uniformly spread out in space is not the same as knowing their true, three-dimensional distribution. Is there a way to determine this? If we have to rely solely on galaxy positions in the sky, it's quite hard to do. There are many complicated ways that galaxies could be distributed in three-dimensional space yet produce essentially the same distribution we see in the two-dimensional sky. The key to breaking this uncertainty is to measure galaxy distances. This is no surprise: We've seen before at earlier destinations the fundamental importance of determining distances. It was only after we discovered how to measure the distances to stars (Destinations 3 and 7) that we could map their complicated distribution within our Galaxy (Destinations 9 and 10). And it was the Cepheid Period-Luminosity Relation (Destination 14) that provided the means for us to obtain distances of galaxies up to 20 *Mpc* away, allowing us to probe the detailed structures of galaxy groups and clusters. But, vast as these distances to local galactic systems are, the scale of the Universe is far larger. Indeed, just about all of the *Observable Universe*—which we will define as the spherical volume surrounding us that has a radius equal to the distance light has traveled over the age of the Universe (Destinations 16 and 19)—lies *far* beyond 20 *Mpc*. Far beyond the practical range over which we can measure parallaxes. Far beyond the range over which we can use Cepheids to measure distances. If we hope to determine the distribution of galaxies over a large fraction of the volume of the Universe[1], we need a tool that lets us measure galaxy distances out to hundreds—even thousands—of *Mpc*.

It turns out we already have that tool. During our visit to the Virgo Cluster (Destination 16), we explored the profound implications of the Hubble Law, including the realization the Universe has a finite age, and its role in providing a first glimpse into the nature of Universal expansion. But there's also a very practical aspect to the Hubble Law. One we can exploit to map the distribution of galaxies in the Universe and is absurdly easy compared to measuring stellar distances via parallax measurements (Destination 3) or determining distances to nearby galaxies using Cepheids (Destination 14). By simply obtaining a galaxy's spectrum that contains emission or absorption lines—for example, 21-*cm* lines in the radio portion of the EM Spectrum, molecular lines in the microwave or infrared regions of the EM Spectrum, optical lines from stars or H II regions

Fig 18.1

The distribution of galaxies in the sky. (*Upper left*) An all-sky map of galaxies obtained from an infrared survey. Each detected galaxy is shown as a small dot, and over 1.6 million galaxies are depicted in this map. The uneven, filamentary-like distribution of galaxies is particularly evident. This survey detects galaxies efficiently to a redshift of about $z = 0.1$. The faint blue band in this map shows the location of the Milky Way. (*Lower right*) A map of galaxy positions in the Southern Hemisphere that corresponds to a portion of the right side of the all-sky map. Here, galaxies are plotted with a color code that gives a rough idea of their distances. White points are nearby, red-brown points more distant. Though this map covers only about 10% of the sky, it goes much 'deeper'—that is, revealing fainter, more distant galaxies—than the all-sky map and includes about 2 million galaxies. The filamentary distribution of galaxies is evident in this map as well.

All-sky map by T. H. Jarrett, J. Carpenter, and R. Hurt/2MASS/IPAC/NASA; Galaxy count image from 2DF/APM surveys.

1 Actually, there are some tricks we can play with galaxies—tricks that do not work for stars—that allow us to estimate their distances beyond the range where Cepheids are useful. Unlike stars that, apart from the Sun, appear as point sources from Earth, galaxies are generally *resolved*. That is, we can often discern their structure in enough detail to measure their apparent sizes in the sky. If we assume all galaxies of a certain type—for example, all Sc galaxies (Destination 15)—are approximately the same diameter, we can guess their distances based on their apparent sizes using the Small-Angle Formula (Destination 4). This can be rather crude, and can fail for distant galaxies for which we cannot measure reliable sizes. It also assumes that galaxies—even galaxies of a given Hubble type—are more similar to one another than they really are. This was how Hubble first estimated galaxy distances to discover the Hubble Law, but it would be good to devise a better, more precise method.

(Destination 6), or x-ray emission lines from hot, highly ionized gas—we can measure a galaxy's *Redshift*[2] as

$$z = \Delta\lambda / \lambda_0,$$

where $\Delta\lambda = \lambda_{obs} - \lambda_0$ and λ_{obs} is the observed wavelength of a spectral feature that has a true, or rest, wavelength of λ_0. Using the Doppler Equation (Destination 6), we can determine a galaxy's recession velocity[3] to be $V_r = cz$. We can then use the Hubble Law to estimate the galaxy's distance, $D = V_r/H_0 = cz/H_0$, as long as we known the value of the Hubble Constant, H_0 (Destination 16). It really does seem pretty simple. All we need is one observation—a galaxy's spectrum—and one measurement—a galaxy's redshift, z. Then, thanks to the Hubble Law, we can derive one important result: The galaxy's distance.

But there are complications.[4] Although identifying the Period-Luminosity Relation for Cepheids was relatively easy (we could see brighter Cepheids have longer periods in M 31; see Destination 14), calibrating the relation so it can be used to measure distances reliably was far more complicated. People had to somehow determine the distance to a galaxy, such as M 31, that contains Cepheids but without using the Cepheids to get the distances (the process would be circular otherwise). The same is true of the Hubble Law. Though Hubble himself had sufficient information years ago to see the overall trend of increasing recession velocity with distance (see footnote 1), determining the precise value of H_0 turned out to be much more difficult. For the past 80 years, astronomers remained quite busy—and, occasionally, at each other's throats[5]—trying to figure out the value of H_0, This long effort seems to finally be reaching a consensus, and for our purposes, we'll adopt the value of the Hubble Constant from Destination 16: $H_0 = 72\ km/s/Mpc$. It seems unlikely that future work will change this by more than about 5% (or $\pm\ 4\ km/s/Mpc$).[6]

Another subtle complication with using the Hubble Law to determine galaxy distances is that the method assumes a galaxy's observed redshift comes about only from the expansion of the Universe. We've seen, however, that

---

2 We defined redshift back at the Virgo Cluster (Destination 16), but it is so important that it bears repeating.

3 Remember (Destination 16), the 'recession velocity' that we measure arises from the change in the scale of space that light traverses as it reaches us from distant galaxies.

4 There are *always* complications!

5 If you want to get some sense of this, read *Lonely Hearts of the Cosmos* by Dennis Overbye. You can learn a bit about the efforts, passions, failures, and triumphs of astronomers who worked to determine the distance scale of the Universe and, ultimately, the value of H_0. Though the drama associated with determining H_0 is mostly over, it is being replaced by new efforts to determine other aspects of Universal expansion. We'll explore that starting at Destination 19.

6 This modern estimate of H_0 is far more accurate than Hubble's original guess of $500 km/s/Mpc$. You can explore the implications of such a large Hubble Constant in a question at the end of this destination.

galaxies in clusters move at significant speeds relative to one another due to their mutual gravitational attraction. From Earth, for example, at least a dozen galaxies in the Virgo Cluster (Destination 16) have negative 'redshifts.' That is, the sum of their recession velocity due to the expansion of the Universe and their motions internal to the cluster produce a net blueshift for these galaxies, implying that they are approaching us.[7] This would lead to an absurd negative distance from a strict application of the Hubble Law, yet we know that these galaxies must be at the same distance as the other members of the cluster. Fortunately, this problem shows up only for galaxies in dense regions such as clusters, and is really problematic only for regions relatively near our Galaxy (to understand why, see the questions at the end of this destination). But it does illustrate that some care is needed when using the Hubble Law to interpret redshifts as distances.

Because it's so easy to use, the Hubble Law provides a powerful way to survey the distribution of galaxies in the Universe. The problem is that in order to produce a comprehensive map of the true distribution of galaxies in the Universe, many galaxies need to be observed. Over the past 30 years, ambitious *Redshift Surveys* have been carried out to produce galaxy maps that extend to ever-greater redshifts, ever-greater distances. The first such surveys obtained spectra of hundreds of galaxies in narrow regions of the sky. More recent surveys include redshift measurements of hundreds of thousands—even millions—of galaxies, often spread over quite large sections of the sky. These surveys have reached the point where they provide a good sense of how galaxies are distributed in the *Local Universe*, a volume that extends out to a radius of around 500 *Mpc* from our Galaxy.[8] More importantly, it appears that they also give us a *representative* map of the distribution of galaxies. That is, although redshift surveys have by necessity so far all been carried out from Earth, the general features of the galaxy distribution over a volume comparable in size to the Local Universe—which is essentially just a random region of the Universe

---

7 You may recall that M 31 (Destination 14) also has a negative velocity relative to Earth, yet the Hubble Law predicts its velocity to be about 50 *km/s* and moving away. The mutual gravitational attraction between the two galaxies has overcome the expansion of the relatively little amount of space that separates them. In the case of Virgo, the large internal velocities within the cluster (Destination 16) and the cluster's proximity to us (so that the average Hubble velocity, $V_r = H_0 D$ of its galaxies is relatively small) explain why some of its galaxies exhibit blueshifts as seen from Earth.

8 The 'Local Universe' is a vague, somewhat subjective concept. As we'll see starting at Destination 19, as we consider objects farther and farther away, we cannot escape the fact that we see them as they appeared in the past. This simply reflects that fact that the EM Radiation we receive from remote objects—even traveling at the large, but finite speed of light—can take a lot of time to reach us. For our purposes, the *Local Universe* encompasses a volume over which the time it takes light to reach us is never more than about 10% the age of the Universe (see Destination 14). In terms of round numbers, that's about 500 *Mpc*.

about 500 *Mpc* in radius—would appear statistically the same no matter where we are in the Observable Universe.[9]

Out to about 20 *Mpc*, galaxy redshift surveys have confirmed and refined what we have already gleaned from our journeys from the Galaxy to the Virgo Cluster (Destinations 14–16): The majority of galaxies reside in well-defined groups or clusters. But there are also many systems—perhaps up to 25% of all galaxies—that appear to be *free-floaters*: Galaxies that are unassociated with any specific group or cluster (Destination 14). As we've seen while traveling around this volume, there is just about always at least one easily visible galaxy in the sky, and often two or more (Destinations 14 and 15). But this is a minuscule portion of the Observable Universe and not even a very significant fraction of the Local Universe. Galaxies within 20 *Mpc* of Earth exhibit recession velocities of under 1500 *km/s*, equivalent to a redshift of $z = 0.005$ or less. Since modern redshift surveys probe far beyond this distance—out to many hundreds of *Mpc* in some cases—we can infer distances to galaxies that have redshifts as large as $z = 0.2$, equivalent to nearly 20% the speed of light.[10] That's much larger than the 20 *Mpc* volume we have explored so far.

By expanding our scope to the full extent of the Local Universe revealed by redshift surveys—out to 500 *Mpc* (see Figure 18.2)—we find that the

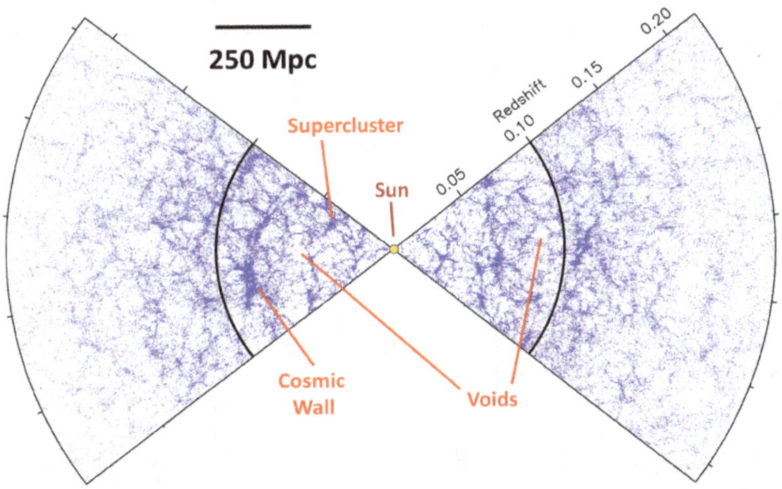

Fig 18.2

A so-called *Pie Diagram* from a deep redshift survey of galaxies. The Sun is located at the center of the plot, at the apex of the two pie 'slices.' Individual galaxies are plotted as blue dots corresponding to their angular location within the slice (measured in a circular arc around the Sun) and their redshift (measured outward from the Sun with the scale shown in the figure). The resulting image shown here is analogous to how doctors slice tissue to be able to view its detailed content with a microscope. The black arcs at about 500 *Mpc* from the Sun represent the 'completeness' range of this survey. The distribution of galaxies inside those arcs is approximately 'true' in that every galaxy that could be measured was measured. Beyond the arcs, galaxies become too faint to have their redshifts measured. Those regions are said to be 'incomplete.' Because of this, the apparent drop in the average number of galaxies beyond $z \approx 0.1$ is an artifact of the survey. This region within 500 *Mpc* from the Sun represents what we are calling the 'Local Universe', the volume in which the Universe has changed comparatively little over the time it takes light to reach us, or about 1.5 *Gyr*. Some examples of large structures—a cosmic wall, a supercluster and two voids—are noted (see Table 18.1).

Image from M. Colless/2DF survey.

9 This turns out to be a very important point. We will return to this later in this destination and at Destination 19.

10 Why not precisely 20%? After all, according to the Doppler Equation (Destination 6), a galaxy's recession velocity, V_r, is equal to *cz*. The problem is that the classic Doppler Equation—the one we've been using so far—works only when the recession velocity is significantly smaller than the speed of light. It is clear from the Hubble Law (Destination 14) that this assumption begins to fail as we consider more and more distant galaxies: Recession velocities should eventually reach then exceed the speed of light! But Special Relativity (Destination 17) tells us that this can't happen. Instead, both the Hubble Law and the Doppler Equation change form as $V_r \rightarrow c$. We'll return to this at Destination 19 as we are forced to confront what happens as galaxy redshifts approach—and even exceed—a value of 1.0.

TABLE 18.1 Summary of Large-Scale Structures in the Universe

STRUCTURES (*EXAMPLE*, DESTINATIONS)	EXTENT (*Mpc*)	OPTICAL LUMINOSITY ($L_{\odot}$)	TOTAL MASS ($M_{\odot}$)	M/L (SOLAR UNITS)	NUMBER IN OBSERVABLE UNIVERSE	TOTAL OPTICAL LUMINOSITY ($L_{\odot}$) [FRACTION]	TOTAL MASS ($M_{\odot}$) [FRACTION]
Dwarf Galaxies (*Sculptor*, **D11**, **D14**)	0.003	2.0×10^{6}	1.4×10^{8}	70	1.0×10^{13}	2.0×10^{19} [12%]	1.4×10^{21} [11%]
Large Spiral Galaxies (*The Galaxy*, **D9**, **D10**, **D14**)	0.5	1.0×10^{10}	4.0×10^{11}	40	1.4×10^{10}	1.4×10^{20} [83%]	5.7×10^{21} [44%]
Giant Elliptical Galaxies (*Maffei 1*, **D15**)	0.8	3.0×10^{10}	1.0×10^{12}	33	2.9×10^{8}	8.6×10^{18} [5%]	2.9×10^{20} [2%]
Galaxy Groups (*The Local Group*, **D14, D15**)	2	4.0×10^{10}	2.0×10^{12}	50	7.0×10^{8}	2.8×10^{19} [16%]	1.4×10^{21} [11%]
Galaxy Clusters (*The Virgo Cluster*, **D16**)	4	2.3×10^{12}	6.0×10^{14}	260	7.0×10^{6}	1.6×10^{19} [9%]	4.2×10^{21} [33%]
Rich Galaxy Clusters (*MS1054-03*, **D19**)	5	6.9×10^{12}	1.8×10^{15}	260	1.2×10^{6}	8.1×10^{18} [5%]	2.1×10^{21} [16%]
Superclusters (*The Shapley Supercluster*, **D18**)	30	4.1×10^{13}	5.4×10^{15}	130	1.0×10^{6}	3.9×10^{19} [23%]	5.1×10^{21} [40%]
Voids (*The Boötes Void*, **D18**)	90	1.1×10^{14}	5.5×10^{15}	50	7.1×10^{5}	7.7×10^{19} [45%]	3.9×10^{21} [30%]
Walls (*The Great SDSS Wall*, **D18**)	100	2.1×10^{14}	2.7×10^{16}	130	1.0×10^{5}	2.0×10^{19} [11%]	2.6×10^{21} [20%]
The Observable Universe (see Notes 1 and 2)	8600	1.7×10^{20}	1.3×10^{22}	75	1	1.7×10^{20} [100%]	1.3×10^{22} [100%]

Notes to Table 18.1:

1. The Observable Universe is defined as the volume corresponding to the distance light can travel over the 13.5 *Gyr* age of the Universe (see Destinations 15 and 19).
2. The total mass of the Universe was obtained by calculating the present-day critical density (see Destination 19) for a Hubble Constant of $H_0 = 72 km/s/Mpc$ and a mass fraction of 0.27.

distribution of galaxies takes on a very different character than what we see within 20 *Mpc* of our Galaxy (Figure 18.2). There is now no illusion that galaxy groups and clusters are distributed uniformly in space. On this larger scale we still see that such groupings exist, but it is apparent that galaxy clusters themselves form clusters, grouping together into even larger coherent structures known as *Superclusters* (see Figure 18.2 and Table 18.1). One of these—the *Shapley Supercluster* (see Figure 18.4)—by itself encompasses a volume about 30 *Mpc* in diameter and includes at least 15 major clusters. We can also see (Figure 18.2) that galaxy clusters and superclusters often align along filamentary-like structures that snake through much of the volume of the Local Universe (Figure 18.2). Some of these can easily be traced over 100 *Mpc*—seven times the distance from our Galaxy to the Virgo Cluster! These gigantic entities are often referred to as *Cosmic Walls* and are among the largest coherent structures known to exist in the Universe (see Figure 18.2 and Table 18.1). Taken together, these vast filaments and walls define structures that somewhat resemble surfaces of soap bubbles (Figure 18.5). *Gargantuan* soap bubbles! Some of these encompass volumes that are well over 100 *Mpc* in diameter.

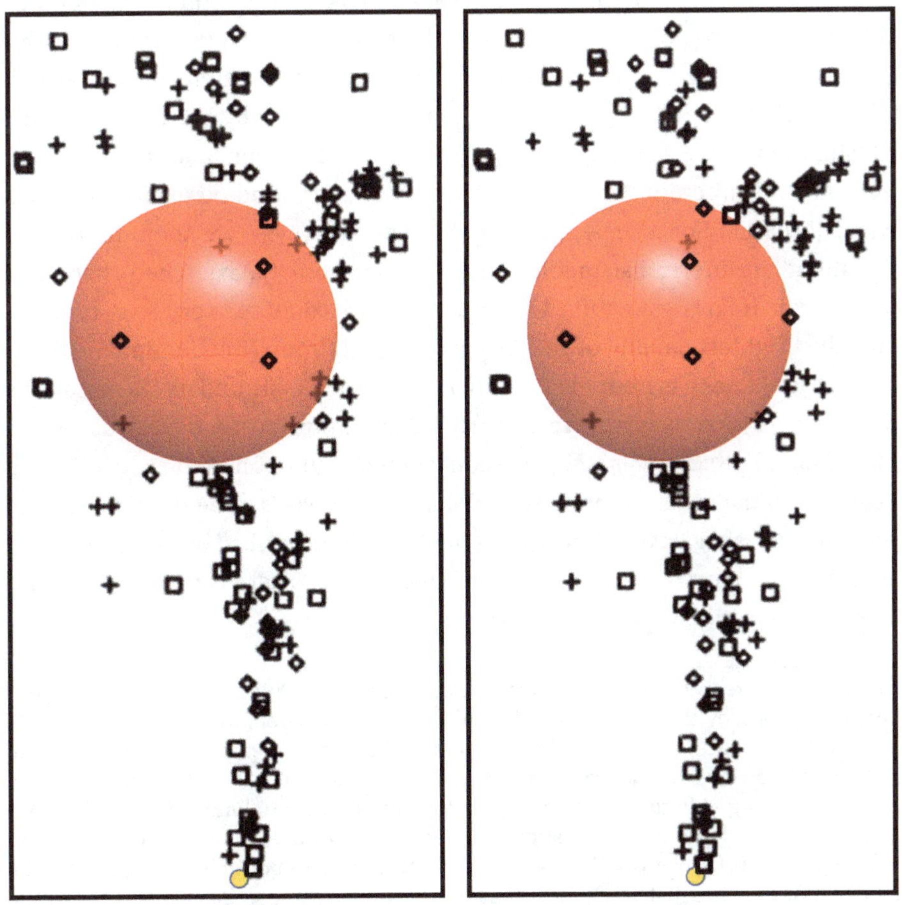

Fig 18.3

A stereoscopic view of the Boötes Void. The basic data here—the small black points—are redshift data similar to the results shown in Figure 18.2 except that a number of slices are shown simultaneously on top of one another. The Sun is visible at the bottom of each panel as an orange dot. From the perspective of the viewer, the diamonds represent closer galaxies, squares denote galaxies at a middle-distance, and crosses galaxies that are furthest away. The red sphere represents a region where very few galaxies are found—the Boötes Void. To see this in stereo, relax your eyes and stare at the image from about 4 to 6 inches. As the separate images begin to merge, you will suddenly see the points appear to pop out of the page; it is not subtle! The 3-D view makes it clear the Void is comparatively galaxy-free; the few galaxies that do appear to be within it are actually either in the foreground or in the background from this perspective.

Stereo image by M. Mateo based on data from R. Kirshner et al.

Here, finally, we find our next destination. Not the surfaces of these cosmic bubbles. Those are the regions where most galaxies reside and correspond to places we have already explored during our earlier journeys. Our goal now is to visit the immense, seemingly empty, space that lies *within* one of these bubbles. These vast zones—which astronomers refer to as *Cosmic Voids*—fill about 95% of the volume of the Local Universe. Yet they contain a much smaller fraction—at most 50%—of all galaxies (see Table 18.1). One of the first such voids to be discovered spans a broad region in Earth's sky centered on the constellation *Boötes*, the Herdsman (see Figure 18.3). The *Boötes Void* is our next stop.

18.2 THE DARK, LUMPY UNIVERSE

As we look around us at this utterly black Alien Sky at the center of the Boötes Void, the incomprehensible emptiness of this place presses upon us like a weight. We console ourselves by recalling that this is not the first time we've traveled to somewhere that there appeared to be nothing nearby. At 'Anti'-Centauri (Destination 5), we visited a random location of interstellar space near the Sun. But that was different. We could still see stars there. And by exploiting other parts of the EM Spectrum—from radio waves to x-rays—we soon discovered the abundant interstellar medium that pervades our Galaxy. Then there was the Galactic Halo (Destination 13). From there we could see very few stars and only the brightest handful of galaxies of the Local Group (Destination 14), with just a trace of hot gas from the Coronal ISM detectable around us (Destinations 5, 11, and 15). Yet, empty as those earlier destinations were, there was always *something* to be seen, *something* we could detect. But not here it seems (Figure 18.4). From the center of the Boötes Void, there appears to be nothing at all to see. No stars. No galaxies. No gaseous medium. To the naked eye we gaze upon an unbroken, oppressive darkness surrounding us.[11] There is no lonelier, more isolated place in the entire Universe.

11 As we'll see, voids do contain galaxies. But voids are very, very large, and the numbers of galaxies, by the standards of the Galaxy's neighborhood, rather low. As a result, we could, by chance, see a galaxy in our sky, even here at the center of the Boötes Void. But it would be rare. In practice, there's about a 1%—or about 1 in 100—chance of our *just barely* being able to see one galaxy in the sky from a randomly selected location inside a void's volume. More realistically, the odds of being able to *easily* see a galaxy are more like 1 in 1000. For our discussion, we'll assume we're not that lucky (or, unlucky if you want to see it that way). This Alien Sky really is devoid of galaxies visible to the naked eye.

We are now 210 *Mpc* from our Galaxy, a vast distance that adds to our profound sense of isolation. Our majestic home has been reduced to a faint, inconspicuous smudge, so dim we require a fairly large telescope to see it at all.[12] But it is not merely the distance that adds to our sense of separation from home. From here at the center of the Boötes Void, the radiation we detect from our Galaxy requires about 700 *Myr* to reach us. Even if we take the trouble to find it, what we see from here is not really our Galaxy anymore, but an image of what it was like hundreds of millions of years in the past.

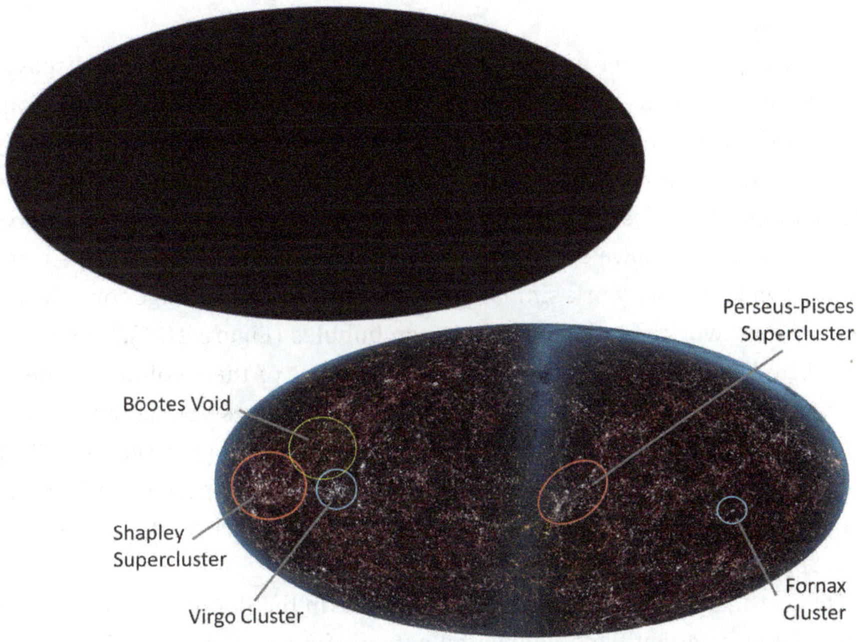

Fig 18.4

(*Upper left*) To the naked eye, we see nothing in the sky of the Boötes Void. But galaxies are visible; they are just faint and require telescopic aid to see. The galaxy distribution from Earth (*Lower right*; this is a repeat of the all-sky map shown in Figure 18.1) is likely a good approximation of what we would see with telescopic aid from the center of the void with a few exceptions. Since there are no galaxy clusters in the void itself, we would not see clusters like Virgo and Fornax (blue circles) The Shapley Supercluster (the left-hand red circle) is located near the edge of the Boötes Void, about four times closer to the center of the void than it is to Earth. It would appear four times wider in the sky from a location at the center of Void. Finally, unless we chose to do so, the odds are minuscule that we'd randomly find ourselves *inside* a galaxy within the void; the pale blue band that represents the Milky Way would not be seen either.

All-sky galaxy map by T. H. Jarrett, J. Carpenter, and R. Hurt/2MASS/IPAC/NASA.

But it's not just our Galaxy that requires a large telescope to be seen from this place. From most places near the center of the Boötes Void, we would require some sort of optical aid—often a fairly large telescope—to see *any* galaxies at all (Figure 18.4). The closest few galaxies that we spot turn out to be located inside the Void, typically about 3–4 *Mpc* away. Some of these are about as bright as galaxies located in the neighboring of our Galaxy—for example, the Sculptor, M 81, and Maffei Groups (Destination 16)—appear from Earth. These galaxies tell us the Boötes Void is not completely empty. But their scarcity also reveals that galaxies are certainly considerably rarer here than anywhere we've visited before. To be more precise, the average number density within the Boötes Void is about 5–10 times lower than the density of galaxies in a volume containing the Local Group and its neighboring groups. All told, there are about 10 thousand galaxies as luminous as our Galaxy spread throughout

12 The Sun is utterly undetectable using any known, planned, or even dreamed-of telescope. We are utterly cut off from home.

the Void, around one million if we count dwarf galaxies. Many of the galaxies in the Void are isolated free-floaters, while the rest are members of the relatively few small groups that permeate the Void's volume. There are no galaxy clusters in the Boötes Void, and certainly no superclusters or cosmic walls (Figure 18.4). In contrast, a similar-sized volume centered on our Galaxy contains about a dozen galaxy clusters, and even a significant fraction of a supercluster.[13] The Boötes Void is a vast galactic desert, a large volume of the Universe where the average galaxy density is low and where galaxies are generally found only in isolation or in small galaxy groups. It is like this throughout all voids. It is what *defines* voids.

As we explore further out from the center of the Boötes Void, we discover that the density of galaxies rises fairly suddenly some 40–50 *Mpc* from our location. Here we encounter a region where galaxies are distributed mostly within clusters, superclusters, and cosmic walls, structures that we now realize define the *edges* of the Boötes Void.[14] These regions of higher galaxy density—where we have spent most of our time at past destinations—trace an interconnected network similar to the structure of a sponge, or a bird's bone, or a network of interconnected soap bubbles (Figure 18.5), all characterized by large empty cavities that make up most of their volumes. One of the most prominent features along the edge of the Boötes Void is the Shapley Supercluster, which we identified from redshift surveys (see section 12.1). From our current location at the center of the void, this supercluster is about four times closer to us than it is from Earth, spanning over 60 *deg* across this Alien Sky (see Figure 18.4).

The fact that there is so very little to see here in the Boötes Void offers two important clues about the fundamental nature of the distribution of matter in the Universe. The first has to do with how galaxies are spread throughout space. More than any other place we have visited, the Boötes Void underscores just how *lumpy* the Universe is. We've now encountered regions—galaxy clusters—that contain thousands of galaxies within volumes spanning a few *Mpc,* and others—voids—that contain comparatively very few galaxies within volumes spanning tens, sometimes hundreds of *Mpc*. What this tells us is that galaxies exhibit a *Hierarchy of Structure*, ranging from individual luminous galaxies and their satellites all the way to superclusters and cosmic walls (Table 18.1).

---

13 The Virgo cluster is often taken to be the nucleus of a rather anemic supercluster known (aptly) as the *Virgo Supercluster* or sometimes the *Local Supercluster*. The (vague) boundaries of the Virgo Supercluster generally encompass five clusters (including the Virgo and Fornax clusters; Figure 18.4, Destination 16), as well as many hundred groups, including, notably, the Local Group.

14 Voids are generally complex structures, often with intricate three-dimensional shapes. Nevertheless, it is customary to try to approximate their extents with spherical or ellipsoidal volumes. In our case, we can reasonably pretend that the Boötes Void is spherical. The distribution of galaxies around us tells us that the Void's radius is about 45 *Mpc*.

Fig 18.5

The Local Universe (*right*) is roughly similar in form to other cavity-filled structures such as interconnected soap bubbles (*left*), though vastly different in scale: The view of the Local Universe shown here spans about 300 *Mpc*, but only a few inches for the soap bubbles. Yet both are characterized by similar topologies comprised of large empty volumes surrounded by thin shells of matter.

Galaxy distribution image from M. Colless/2DF survey.

The largest of these structures define the outer edges of voids.[15] If we consider regions at least a few *Mpc* in diameter, the *Density Contrast* between the centers of rich galaxy clusters and the center of a void is about 10000:1, about the same density contrast of lead to air. We can get some sense of what this means by comparing the spectacular sky of galaxies that we saw at the Virgo Cluster (Destination 16)—not even a particularly rich cluster—and the oppressively barren sky we see here at the center of the Boötes Void.

The second observation is particularly evident because it is the most obvious thing about this place: The sky is dark here. Really dark. There are no stars to be seen at all. No galaxies close enough to be seen by the naked eye. Nothing to interrupt the darkness. Is this a surprise? Back on Earth, any child would happily inform us that it is dark at night because the Sun is no longer in the sky (Figure 18.6).[16] There's no Sun near our current location. No stars at all. Isn't it rather obvious why the sky is so dark here? But we are not children anymore, and we've learned enough on our voyages to realize that perhaps the obvious

---

15 One can also legitimately think of the voids as part of the hierarchy of structure, a sort of 'anti-structure' that defines where galaxies are *not* found. Table 18.1 includes voids as part of the census of structure in the Universe.

16 The night sky of Earth, even when the Moon is not up, is quite far from being completely dark. Apart from man-made lighting (Destination 9), the sky itself naturally glows from scattered sunlight and emission from excited atoms (as in the Orion Nebula, Destination 6). There is also a significant amount of light that comes from sunlight reflected off dust particles in the Solar System. These sources add up to quite a bit of light in Earth's sky. But here—and in fact at many of our destinations—there are no similar sources of light in our vicinity. To our eyes, this Alien Sky is as dark as any in the Universe.

answers to questions like these are not always the correct ones.[17] So as we ponder the bleak, lonely view around us, we slowly come to realize that perhaps the sky should not be dark after all. Not here. Not anywhere in the Universe.

Fig 18.6

From Earth, the sky is pretty dark at night. From the center of the Boötes Void, the sky is *really* dark. There are no stars, no galaxies, no sky glow as we see in Earth's sky. But why is the sky dark at all? From anywhere? At any time?

To understand this peculiar claim, let's first consider our newfound knowledge on the distribution of galaxies in space. We've already noted that galaxies are not distributed perfectly uniformly on scales of a few *Mpc*. For example, the density of galaxies in the Local Group (Destination 14) is quite different from the center of the Virgo Cluster (Destination 16). But even when we consider larger chunks of the Universe—regions a few tens of *Mpc* in size—the Universe can appear strikingly different from one place to another (Figure 18.7). On this scale, we might be entirely in a void, or possibly in the much more populous edge of a void. But redshift surveys tell us that this trend does not continue indefinitely. As we noted earlier, when we consider much bigger chunks of the Universe—say, regions 250–500 *Mpc* in diameter—the distribution of galaxies begins to exhibit much greater uniformity. On this scale, one region of the

17 Though it is worth remembering that the obvious answers *are* sometimes the right ones!

Universe appears, statistically, to be pretty much identical to any other. That is, the Universe appears to be *Homogeneous* on scales greater than a few hundred *Mpc*, even though it is decidedly *not* homogeneous on smaller scales.

This seems like a strange concept, but we encounter it all the time. A good example—and one that mimics the structure of the Universe a bit—is a beehive. From close up, it is easy to see the individual cells that comprise the hive—structures that reasonably approximate how galaxies are distributed in voids and filaments and other structures surrounding the voids. However, from a distance, the same hive would appear to us as a much more uniform structure. Another way of saying this is that the individual cells of the hive are small compared to the overall size of the hive. Over a large enough scale, the uneven detailed structure of the hive—or of the Universe—begins to appear uniform. And *that's* the problem. If the distribution of galaxies in the Universe really is uniform on any scale, we can argue that the sky should not be dark. In fact, it should appear everywhere to be about as bright as the surface of a star!

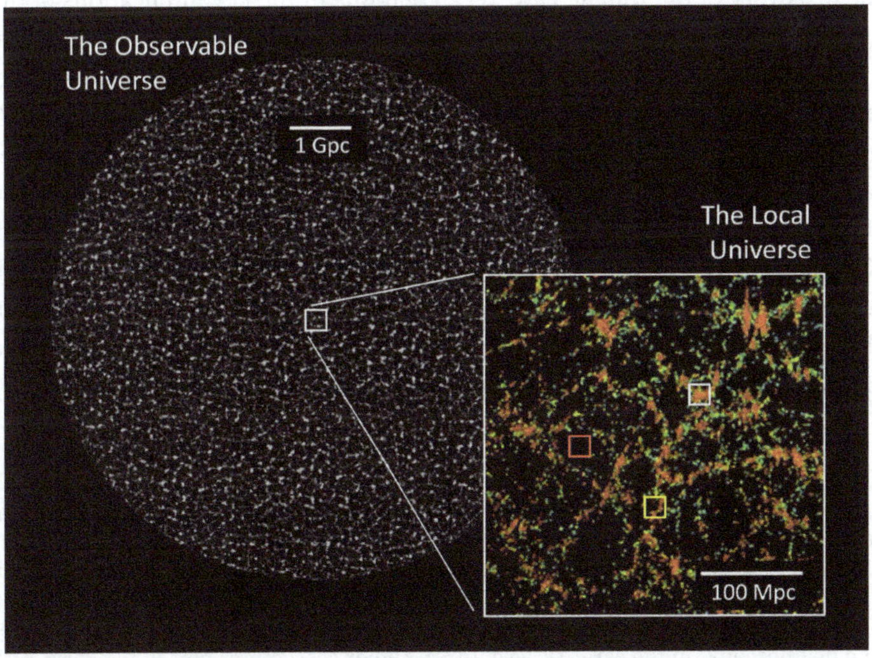

Fig 18.7

A section of the Local Universe (lower right) as observed in a redshift survey exhibits a wide range of structures. The distribution of galaxies in this region includes superclusters (inside the small white square), normal-density regions similar to what we see around our Galaxy (yellow square), and voids (red square). The key point is that the distribution of galaxies varies considerably on scales of about 20 *Mpc*—the sizes of the 'small' squares. (*Left*) This map shows the distribution of galaxies in a thin slice of the Observable Universe where we pretend that we can see this entire volume simultaneously (for reasons we'll start to explore at the next destination, we can't). Notice that wherever we might place the reference square—now 350 *Mpc* on a side—in this larger map, the appearance of the galaxy distribution would appear statistically the same. The Universe is clearly highly lumpy on scales of tens of *Mpc*, but becomes much more uniform—effectively homogeneous—on scales of hundreds of *Mpc*.

Wide-field image from R. Powell. Inset by M. Blanton/SDSS.

To understand this, consider being at the center of a vast, transparent onion-like structure, one with a number of concentric layers and in which galaxies are located in the same distribution we see in our Universe. If the layers are larger than about 200–300 *Mpc* in thickness, the argument above suggests that we can fairly assume each layer to have the same average density of galaxies. We'll assume the average luminosity of the galaxies, L_{gal}, are constant throughout every layer of this cosmic onion.[18] If we define each layer to have an average distance,

18 This is not an essential assumption, but it does simplify the analysis. Most importantly, it doesn't invalidate our final conclusions.

D, from the center of the onion, it is easy to prove that the number of galaxies in each layer is proportional to D^2 times the layer thickness, a quantity that is essentially equal to the volume of each layer. Now, if all the layers have the same thickness, and the thickness is small compared to the size of the onion, the volume varies by D^2 for each successive layer.[19] But we also know from the Inverse-Square Law for Brightness (Destination 4) that the average brightness of galaxies in any given layer is proportional to the inverse of D^2, that is $1/D^2$. We can calculate the total light from any layer by multiplying the number of galaxies within it—this is proportional to D^2—by the factor that describes how the light falls off with distance, which is proportional to $1/D^2$. Since the product of D^2 and $1/D^2$ is a constant, what we see is that *each layer contributes the same total brightness as any other layer!* If there are an infinite number of layers, then an infinite amount of light would reach us. The sky would not be dark!

Of course, we know the sky *is* dark, so some aspect of our argument is incorrect. People grappled with this for *centuries*[20], but the final resolution required the discovery of the Hubble Law. Recall that one of the key implications of the Hubble Law is that at some time—approximately equal to $1/H_0$ (Destination 14)—all galaxies were closer together. This was, in some fundamental sense, the time when the Universe was 'born.' For our present purposes, it isn't important to know precisely how long ago that was (we'll explore this in detail at Destination 19). But what does matter to us now is that the age of the Universe—approximately$1/H_0$—is finite. Because light travels at a finite speed, this means we can see only a finite number of layers in our cosmic onion from its center, only a finite amount of light. So, despite the large-scale uniformity of the Universe, the sky is dark.[21] And the fact the sky is dark tells us we live in a Universe of finite age. Pretty profound for such a simple observation!

---

19 We did some stealth calculus here. Although the volume of a sphere is proportional to D^3, the volume of a thin shell around a sphere changes as D^2.

20 The problem of the dark night sky is often referred to as *Olbers' Paradox* after an early-19th century astronomer who formulated the problem. The paradox, however, has a longer recorded history, going back at least to the time of Halley (of comet fame) and de Chéseaux (in the early 18th century), Kepler (yes, our Kepler in the early 17th century) and the English astronomer/mathematician Digges as early as the late 16th century. The resolution of the problem—the finite age of the Universe—was proposed by Lord Kelvin (we met him at Destination 8) and, oddly enough, Edgar Allan Poe. However, before the Hubble Law was discovered, no one could justify why the Universe might have a finite age.

21 There is also a more subtle contribution to the sky's darkness. The Hubble Law tells us that the recession velocities of galaxies increase with distance, ultimately approaching the speed of light as we consider more and more distant systems. According to Einstein's Theory of Special Relativity (Destination 17), as objects approach the speed of light they become fainter (we saw this effect—partly due to the effects of gravity, partly due to near light-speed motions—in the vicinity of the supermassive black hole we visited in Destination 17). Thus, not only are there a finite number of layers in our cosmic onion, but the combined light from galaxies in each layer decreases as the distance increases,

18.3 'LIGHT' IN THE DARKNESS: THE COSMIC MICROWAVE BACKGROUND

Having run out of conventional things to observe from this lonely, dark vantage point at the center of the Boötes Void—and being thoroughly bored by the lack of optically visible objects in the sky—we turn to telescopes sensitive to radiation from other portions of the EM Spectrum. When we look around us in x-rays, we discover sources spread all over the sky that are clearly associated with individual galaxies of some sort. Most of these are AGN, some similar to the Seyfert galaxy M 77 that we visited at Destination 17, others are associated with more luminous types of objects known as *Quasars* that we'll explore in detail at Destination 21. When we look around us at infrared or radio wavelengths, we again see considerable radiation, but virtually all of this is clearly coming from individual, faint galaxies. It seems that for all forms of EM radiation, the space *between* these individual galaxies—the sky itself—is essentially dark.

Well, almost all. There is one exception. When we look around us using microwave detectors—EM radiation with wavelengths around 1 *mm*—we see something very different. Everywhere we look, we see that the sky is aglow in microwave radiation (Figure 18.8). Unlike the optical, infrared, radio, or x-rays—or any other part of the EM Spectrum where the radiation comes mostly from individual galaxies, groups, and clusters—this microwave radiation fills every bit of the sky. There are essentially no individual sources such as galaxies, stars, or emission nebulae.[22] It is the *sky itself* that is shining in microwave radiation! And it is not peculiar to our current location: We'd see the same microwave radiation no matter where we observe from. Earth. The Virgo Cluster. Or here at the center of the Boötes Void. There is not an infinite amount of microwave radiation—we'd be fried if there were—so this radiation has nothing to do with our cosmic onion argument that suggested the sky should be bright. And yet, this microwave radiation is unmistakably present. In every direction. From every location in the Observable Universe.[23]

---

contrary to our expectation from the Inverse-Square Law. The sky would still be dark without this effect, just not quite as dark.

22 Such sources can and do emit microwave radiation, but the dominant microwave flux from our location at the center of the Boötes Void comes from the sky itself.

23 And probably even beyond. For reasons we'll begin to explore at Destination 22 and try to explain at Destination 25, we can surmise that we'd see the same sort of all-sky microwave radiation were we to observe the sky from locations beyond the edge of what we are calling the Observable Universe (see Figure 18.7). And, quite possibly, *far* beyond.

As we try to grapple the meaning of this, we notice another feature of this radiation. Something even stranger. By observing at a range of microwave frequencies, we discover that the spectrum of this radiation is that of a thermal source characterized by a single temperature (see Destinations 6 and 8 and Figure 18.8). This *Microwave Spectrum*—the spectrum of the sky!—is in fact one of the most perfect examples of a thermal spectrum known. This also means we can use Wien's Law (Destination 6) to estimate the sky's temperature by simply measuring the wavelength at which the spectrum is brightest (around $\lambda_{max} = 1.1\ mm$).

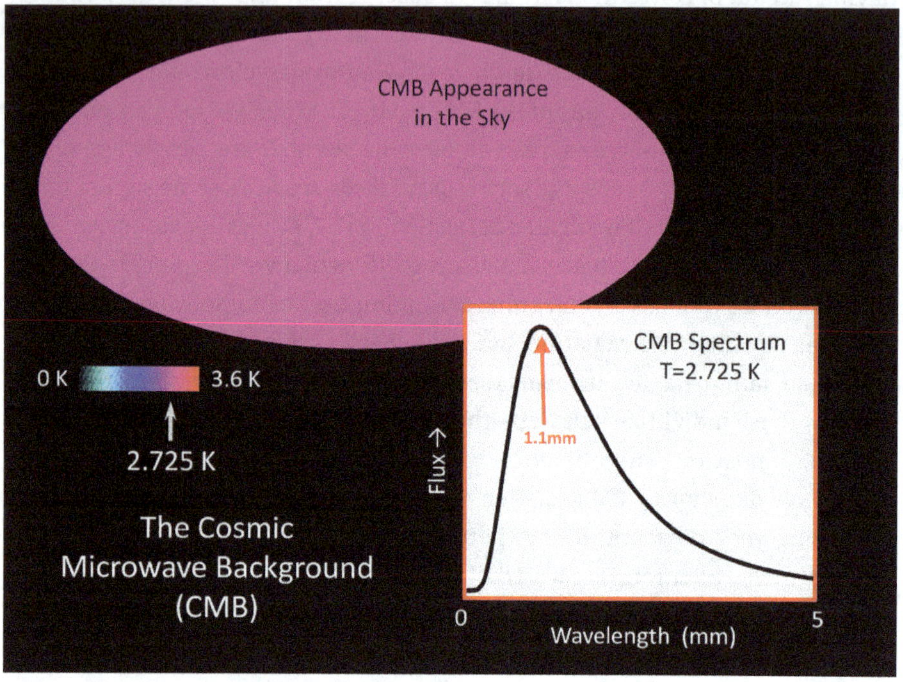

Fig 18.8

(*Upper left*) An all-sky map showing the appearance of the Cosmic Microwave Background (CMB)—a view that would appear essentially the same from Earth as from the center of the Boötes Void. The uniformity of the color represents the fact that the entire CMB radiates as a thermal source at a single, well-defined temperature—2.725 *K*—to better than one part in one thousand (see the temperature scale in the figure). (*Lower right*) A schematic plot of the spectrum of the CMB. The curve shows the expected spectrum of a thermal source with $T = 2.725\ K$ (Destination 6). Observations of the spectrum of the CMB agree with this curve to incredibly high precision.

When we do this, we find that $T_{sky} = 2.725\ K$, a temperature barely above absolute zero! What's especially remarkable about this radiation is how uniform the temperature appears to be. Regardless where in the sky we look, the temperature is always the same to better than one part in a thousand. Not only are we seeing that the sky emits microwave radiation as a nearly perfect thermal source, but also that the spectrum is incredibly uniform over the entire sky. This point and the pervasiveness of the radiation over the sky imply that the source of this microwave radiation must be very distant.[24]

24 This is not strictly true. We could—by some incredible cosmic coincidence—just happen to be located smack-dab in the center of some gigantic, warm shell, one that's smaller than the observable Universe but far from us. Perfectly tuned, this arrangement could produce radiation with the properties that we've observed. However, observations of very distant galaxy clusters and AGN (Destinations 19 and 21) reveal that the source of the all-sky microwave radiation we observe must originate beyond these

Much further away than any stars, galaxies, superclusters that we see in the Observable Universe.

This radiation—*Cosmic* in origin, *Microwave* in type, and originating from a source that is in the *Background* of all other objects we can detect in the Observable Universe—is of singular importance. As we'll explore in detail at Destination 22, the *Cosmic Microwave Background* radiation turns out to be a relic of the early Universe. The *CMB* (as it is commonly abbreviated) reveals conditions from the era we will refer to as the *Early Universe*, the time soon after the Universal expansion revealed by the Hubble Law commenced (Destination 14).

Is there a pattern here? We've already inferred from the Hubble Law that the Universe was very dense in the past when all galaxies—indeed, all matter in the Universe—were much closer together (Destination 16). Since expansion causes things to cool down, the Universe had to have been 'hotter' in the past (at Destination 23 we will explore what it means for the Universe to have a 'temperature'). But if it was hotter, that implies it was acting like a thermal source and it would have radiated energy as any thermal source must. Since the emission of this radiation was a long time ago, the radiation has had to travel through a vast distance to reach us today. If it's far away, the Hubble Law indicates its recession velocity—its redshift—should be very high. That means the radiation would appear much redder to us than when it was emitted. In fact, it would appear as if it arose from a much cooler thermal source. A thermal source with a temperature of only 2.7 *K*. Ouch! This is a pretty complex line of reasoning, and we'll return to some of these steps at later destinations. But the main point is simple. And profound. The CMB radiation arrives to us from a time when the Universe was much younger, hotter and denser than it is today. The Universe has evolved!

The high redshift of the source of the CMB may explain its temperature, but it does not explain why the CMB appears so smooth over the sky. As we look around us from our current location near the center of the Boötes Void, we are struck by the 'lumpiness' of the Local Universe. The network of structures defined by cosmic walls, voids, superclusters, clusters, groups, and even individual galaxies (see Figure 18.2) stands in stark contrast to the incredible uniformity of the CMB. Somehow the complex hierarchy of the colossal structures that exist in the Local Universe (Table 18.1) has arisen from the incredibly smooth Early Universe revealed to us by the CMB. To start to resolve this fundamental dilemma—forming complexity from uniformity—we need to travel back in time. Back to when the Universe was significantly younger. But before we can confidently do that, we must first learn how to tell time in an expanding Universe of finite age.

---

objects. Indeed, it arises from beyond any objects we've ever observed, or, just about, ever *can* observe.

DESTINATION 18 QUESTIONS

18.1. Calculate the 'total luminosity' of the CMB. Express your answer in solar luminosities and in terms of the luminosity of our Galaxy. **HINT:** Use the relation for the surface flux of a thermal radiation source from Destinations 6 and 8. Then calculate the radius of the observable Universe, assuming that its age is $1/H_0$. Use the speed of light to convert this age to a distance.

18.2. Let's say there are 1000 galaxies in the Virgo Cluster confined in a volume of diameter 3 *Mpc*. If there are 10000 galaxies in the entire Boötes Void, which has a diameter of 50 *Mpc*, what is the density ratio of visible galaxies between the two regions?

18.3. If the CMB radiation peaked instead in the infrared, say at a wavelength of 5 *microns* (one *micron* equals 10^{-4} *cm*), would the Universe be younger or older than we believe it to be today? Explain your reasoning.

18.4. This problem tries to give you a sense of the density contrast between a void and a galaxy cluster.

a) Assume there are 10 million people in a city with a radius of about 10 *km* (somewhat like New York City). What is the surface area covered by this city in units of km^2? What is the surface density of people in this city in units of number of people per km^2?

b) Now, how much larger in radius would an area containing 10 million people have to be to have a surface density that is 1/10000 the value you got in part (a)? What is the surface area of this region? Look up the surface areas of *countries* and identify the country with a comparable surface area. That is how much larger the area in part (a) has to be to become 10000 lower in surface density!

c) There is an important way in which this argument differs from what is actually happening when we compare the density of galaxies in voids and clusters. Can you identify this difference?

18.5. From the center of the Boötes Void, the void's edge is about 40 *Mpc* away and about 10 *Mpc* thick. What recessional velocity and redshift would galaxies have that are located at the near and far sides of the edge of the Void, as measured from the Void center? Be sure to describe any assumptions or values of constants that you may have adopted to answer this. Let's say that you carry out a redshift survey from the void center and that you observe galaxies that are located in three distinct void edges: The edge of the Boötes Void and two other void edges at 65 and 105 *Mpc*, respectively. If the edges of these voids are similar to the edge of the Boötes Void, draw a plot that schematically shows the number of galaxies with increasing redshift. Be sure to show fairly precisely the redshifts of any features that might arise in the plot.

18.6. Redshift surveys can be affected by so-called *Fingers of God* that cause remote cosmic structures to seemingly point back to Earth. This problem explores how these arise and the problems they can cause.

a) Imagine we observe the redshifts of galaxies in a cluster located at 35 *Mpc* from Earth. Use the Hubble Law with a Hubble Constant of $H_0 = 72\ km/s/Mpc$ to determine the mean redshift of the cluster.

b) Since the cluster galaxies are moving within the cluster, it happens that some galaxies are moving toward us at a velocity equal to the internal velocity dispersion of the cluster, and some are moving away from us at that same velocity. If the particular cluster has a velocity dispersion of $1500 km/s$, what range of velocities would we see in the cluster? Give this answer in terms of the minimum and maximum velocities we could expect to see for cluster members.

c) Now, take the minimum and maximum values from part (b) and use the Hubble Law to estimate the galaxy distances, ignoring the fact that you know them to be cluster members. How does the distance range from the 'closest' and 'furthest' galaxy compare to the actual size of the cluster, likely around 3–4 *Mpc*?

d) Can you explain why this leads to Fingers of God? **HINT:** You may want to make a pie-slice diagram similar to Figure 18.2, but use distance instead of redshift.

18.7. Imagine a set of nested shells of central radii $R = 1, 5, 10$, and 100 *Mpc*, all sharing the same center.

a) Calculate the volume of each shell, assuming it is 0.1 *Mpc* thick.

b) Plot this volume against R^2 and use that to prove that the shell volumes are indeed proportional to R^2 as claimed in the text. This is the heart of Olbers' Paradox.

18.8. Use Wien's Law (Destination 6) and your knowledge of the EM Spectrum (see Destination 5 for a refresher) to estimate reasonable temperatures if the CMB were instead:

a) The Cosmic 'Radio' Background.

b) The Cosmic 'Optical' Background.

c) Relative to the present day, when might it be (or when was it possible) that the CMB had these temperatures as seen by an observer in the Universe at that time? You can try to be precise about this, but simply answering 'earlier,' 'same time,' or 'later' than the present—and an explanation of your reasoning—is sufficient.

18.9. We will explore some tiny, but crucial, non-uniformities of the CMB at Destination 22. In this problem, we will investigate what is known as the *dipole asymmetry* of the CMB.

a) Assuming a mean CMB temperature of $2.73K$, use Wien's Law (Destination 6) to determine the wavelength of the peak of the CMB spectrum.
b) As measured from Earth, one side of the CMB is about 0.1% hotter than the mean temperature of the CMB, while the opposite side is about 0.1% cooler. Use Wien's Law to estimate the change in the wavelength of the peak of the CMB spectrum that this temperature difference corresponds to.
c) Now, use the Doppler Equation (Destination 6) to translate the wavelength shift from part (b) into a velocity.
d) Given your results, can you provide a possible explanation for this so-called dipole asymmetry of the CMB?

18.10. As we saw at Destination 6, cold clouds of the ISM can also emit microwave radiation. Without leaving the Earth, how can we be sure that the CMB does not originate from nearby clouds rather than from a very distant source as we have inferred in this destination?

18.11. In the text, we noted that some common structures (sponges, bird bones, soap bubbles) have topologies similar to that of visible galaxies at the edges of voids. Look up and identify at least one other object that has the same sort of topology (that is, three-dimensional distribution) seen in the distribution of galaxies. Provide a photograph or drawing that demonstrates the similarity.

18.12. This question explores the relative 'crowdedness' of the Earth's surface with humans compared to the Observable Universe with galaxies.

a) Use the information in Table 18.1 to estimate the 'filling factor' of galaxies in the Universe. For our purposes, this factor corresponds to the fraction of the volume that galaxies take up relative to the total volume of the Observable Universe. Explain your reasoning in coming up with this number.
b) Now, calculate a similar 'covering' factor for humans. In this case, estimate the surface area that a person takes up, then all people, then divide that total area by the surface area of the Earth.
c) Based on this calculation, which is more 'crowded': The surface of the Earth with humans or the Observable Universe with galaxies?

Destination 19

BACK IN TIME TO A DISTANT GALAXY CLUSTER

19.1 NAVIGATING SPACE AND TIME

As much as we have tried to ignore it, we can no longer fool ourselves: Our trips are getting too long. By traveling ever-larger distances in the Universe, it is becoming clear that our voyages require so much *time* that our destinations may change significantly before we arrive. The Virgo cluster, the Seyfert galaxy M 77, and even the Boötes Void (Destinations 16–18) were all located within a few hundred *Mpc*—less than one billion *light-years* distant. The fundamental characteristics of most galaxies—including our Galaxy with an age of at least 10–13 *Gyr* (Destination 13)—typically haven't changed that much over the past few hundred *Myr* (Figure 19.1). But what about longer timescales? During the time light travels 1 *Gpc*—about 3.3 *Gyr*—the Sun can circle the Galactic Center nearly 13 times (Destination 9). Stars with masses as low as $1.5M_{\odot}$ can complete their Main Sequence lifetimes (Destination 8). Billions of new stars can be born within our Galaxy (Destinations 6 and 7). Three billion years ago, the Magellanic Clouds—currently among the closest dwarf satellite companions of our Galaxy (Destination 12)—were located far in the outskirts of the Local Group (Destination 14), and the Sagittarius dwarf galaxy—today, almost completely disrupted by its interactions with the Galaxy—was a fairly normal dwarf spheroidal system in the Galactic Halo (Destination 13). At that distant time in the past, immense structures like the Virgo Cluster (Destination 16) were just beginning to assimilate into coherent, recognizable galaxy groupings resembling the ones we see around us today. Three billion years ago, simple single-celled organisms were just getting established on a planet their descendants would later call Earth.

The point is that planets, stars, galaxies, groups, clusters, and even larger cosmic structures (Destination 18) are *evolving* with time[1]. On galactic scales, this evolution is partly driven by the steady changes in the quantity, ages, chemistry, and distribution of stars and gas that comprise individual galaxies,

---

1 When applied to individual galaxies by astronomers, the term 'evolution' is somewhat similar to the way biologists use the same term to describe how species adapt over multiple generations to changing conditions and mutations. A major part of any galaxy's evolution results from the steady changes that occur in multiple generations of stars and gas that comprise galaxies, as well as by changes that are driven by interactions with other galaxies (Destination 16). This contrasts with the use of the term 'evolution' when applied to stars (Destination 8) where 'evolution' refers to the changes that occur in a *single* star. If astronomers were consistent with their nomenclature—and they rarely are!—this process would be more accurately referred to as stellar 'aging.'

perhaps sometimes modified by interactions with other galaxies (Destinations 13 and 16). Over a few hundred *Myr*—perhaps even a few *Gyr*—these changes are comparatively slow. But eventually, at some point in the past, our Galaxy—*any* galaxy—had to have been significantly different that it is today. On even longer timescales more comparable to the Age of the Universe (Destination 16), the appearance and distribution of the matter that ultimately evolves into galaxies such as ours may have been unrecognizable compared to what we see in the Local Universe. And through it all, the Universe itself would have evolved

Fig 19.1

The galaxy M 83 is about 3 *Mpc* away. We see this galaxy not as it exists today, but as it was about 10 *Myr* ago. Although the details of its appearance will have changed over that period of time, we can be quite certain that the large-scale characteristics of the galaxy are essentially the same now as they were 10 *Myr* ago. If this same galaxy were placed 500 times further away, however, we would see it as it existed at a time so far in the past that it might well be unrecognizable, probably utterly unlike the beautiful spiral galaxy we see here.

Photo by D. de Martin/ ESO.

in its own way as it steadily expands over cosmic time (Destination 16).

For our purposes, this causes a bit of a conundrum. Since it takes time for any EM Radiation emitted from a distant source to reach us, we unavoidably see things as they appeared at the time they emitted the radiation we subsequently detect at Earth.[2] For astronomers, this offers the opportunity for a sort of time

2 We introduced this concept at Destination 18, but in a somewhat parenthetical way. But now, as we start to consider the locations of our destinations in space *and* in time, this concept that we see distant objects as they were in the past becomes centrally important.

travel that allows us to view the Universe and its contents at earlier times. But there is a price we pay for this opportunity. Our view from Earth of the early Universe is necessarily limited by the gargantuan distances associated with viewing things from the far past. As a result, objects that we view directly in the young Universe invariably appear faint and small. It is virtually impossible to discern their structure at the level of detail we enjoy as we study nearby galaxies (Destinations 14–16).[3] Even so, observations of distant galaxies, imperfect as they may be, reveal without doubt that the constituents of the Universe—even the Universe itself—have fundamentally evolved over time. These observations of distant structures whet our appetites for a closer look. It would be so much more informative to *go* to these places. To see them as they were in the past.

With this in mind, it's clearly time to change the fundamental nature of our journeys. Although we have mastered travel over vast dimensions in *space*, we now need to travel in *time*. Only then can we hope to visit remote destinations to see them firsthand at the time they emitted the radiation we receive from them today on Earth. But this introduces a new set of problems.[4] Perhaps the most practical of these is that we now have to figure out how far in time we need to travel, not merely how far in space. Over the course of our past journeys, we've learned about many ways to estimate distances to remote objects (Destinations 3, 8, 9, 14, and 16; these methods are summarized in Table 19.2). Now we must learn how to navigate in *time*.

19.2 TELLING TIME IN THE UNIVERSE

To understand how this is possible[5], we need to return to our discussion regarding the implications of the Hubble Law at Destination 16. There we found that because the Universe is expanding uniformly—a fundamental implication of the Hubble Law—it had to have had a *beginning*. To be more specific, there must have been a time when all matter and all space was compacted together.[6]

3 One clever way that holds promise is to use galaxy clusters as gravitational lenses that can magnify distant background galaxies and make them appear brighter. Of course, we have no real control of such 'lenses,' but it turns out that, by chance, many clusters do in fact magnify background galaxies sufficiently that we can discern details and even obtain good-quality spectra to analyze their content.

4 Not the least of which is that there's no known way to carry out time travel!

5 Not time travel, but being able to determine how far back in time we have to travel.

6 This state of near-infinite density of the Early Universe is often referred to as the *Big Bang*, though this is a somewhat misleading term (see Destination 25).

In this sense, that means that there was a time when the Universe's clock read zero.[7] But that's not all. The Hubble Law also gives us some idea of the current *rate* of the expansion of the Universe through the value of the Hubble Constant. Among other things, this lets us estimate that the time since the expansion commenced must be around $T_0 \oplus 1/H_0$ (Destination 16). But we need more detail. Knowing the current expansion rate of the Universe—or, by implication, its approximate age—is not sufficient to determine how far back we must travel in time to visit a specific distant destination. It would be like trying to reconstruct a car's journey—including all its decelerations, accelerations, stops, turns, etc.—from knowing only its current speed and how long it's been on the road. The problem seems like a real stumper.

But, there is a way. To illustrate, let's consider the motion of a ball. Not just any ball, but one that we encounter in seemingly empty space that's moving at some velocity, V_b (see Figure 19.2). For the moment, let's assume the ball's motion is affected only by gravitational forces (Destination 9) that arise from both seen and (possibly) unseen masses along its trajectory. Let's also imagine that we somehow know the initial velocity of the ball, V_i—perhaps it was imprinted on the ball by whoever (or whatever) put the ball in motion in the first place

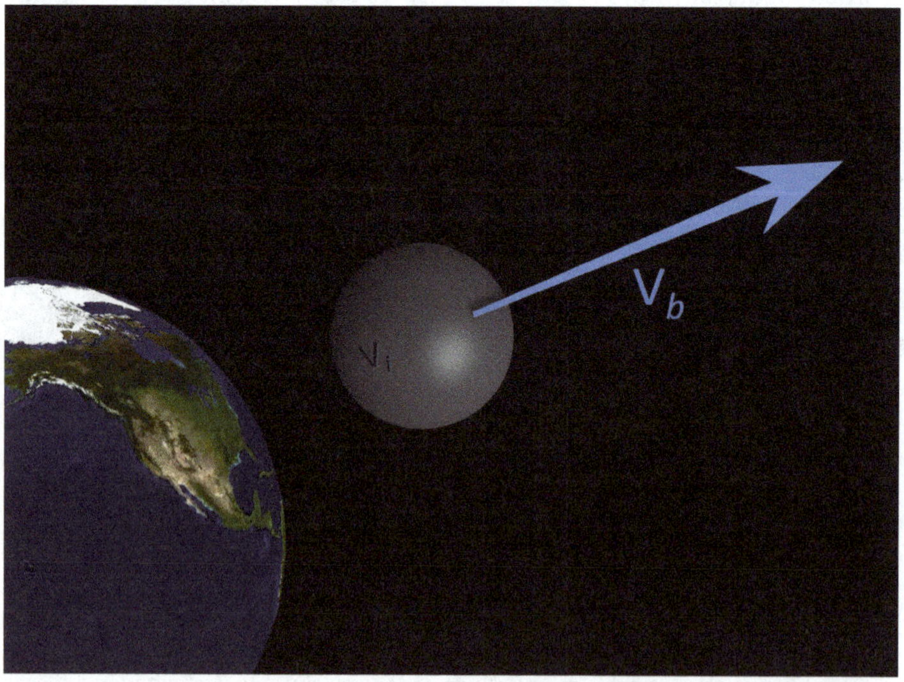

Fig 19.2

We can get some sense of the various ways that the Universe can evolve by consider a free-floating ball moving at a velocity V_b and initially propelled at a velocity V_i (imprinted on the ball itself). We assume that the direction of the ball's motion has not changed during its travels (apart from possibly turning around). We also assume—at least initially—that the ball's speed has been affected only by gravitational forces arising from visible or unseen matter (denoted here by a pale Earth).

Image of Earth from K. Swartz/Great Circle Mapper and mapping data from Blue Marble/NASA.

7 Or, at least, very nearly zero. As we'll see at Destination 25 there was a time—about 10^{-44} *sec* after the onset of the expansion of the Universe—before which the known laws of physics break down, making it impossible for us to explore the very earliest moments of the Universe. Prior to then, our standard concepts such as time and space may not have had any recognizable meaning.

(see Figure 19.2).[8] Assuming the ball has followed a straight course, we can appreciate there are only three ways that the current and initial velocities might be related. First, maybe they are equal so that $V_b = V_i$. Though the ball could have experienced some combination of accelerations and decelerations that have caused its velocity to precisely equal V_i, it seems more reasonable—and certainly a lot simpler—to suppose the ball has *always* been moving at a velocity equal to V_b. This is equivalent to saying that the ball has experienced no forces of any kind after it began moving.

A second possibility is that the ball—still moving in the same direction or possibly in the *opposite* direction from which it was launched—has slowed down so that $V_b < V_i$. We can guess that its deceleration probably arose from gravitational forces arising from masses located in the region the ball passed through. We may even be able to see these masses in the form of nearby stars or planets. But even if we don't, the change in the ball's velocity implies that something—visible or not—exerted a net gravitational force that slowed it down. If the force is strong enough, the deceleration may have even caused the ball to stop its motion and turn around. In this case, V_b is not merely less than V_i, it is less than zero, meaning it has changed direction.[9] The third possibility is that the ball is now moving faster than its initial velocity so that $V_b > V_i$. In this case, the ball has accelerated since it began its journey. Perhaps it is falling toward a massive object, one we may or may not see. Or perhaps something other than gravity gave it a push. If that's the case—that there are non-gravitational forces that may have affected the ball's motion—then analyzing its motion may prove to be even more complicated than we had originally suspected.

The Universe can behave in any of the same ways we have described for this hypothetical ball.[10] The big difference is that while the ball is affected by the distribution of mass in the space through which it travels, the gravitational force—and any other force for that matter—that acts on the Universe arises from the Universe itself.[11] To analyze the motion of the ball, we must at least

8 This implies we also know how we are moving relative to the source of the ball. This goes back to our discussion of Galilean Relativity back at Destination 17.

9 If we were very lucky, we may have caught the ball just as it came to a stop. But if so, all we have to do is wait a while and it will start to move again, this time in the opposite direction from where it came.

10 Keep in mind that there is one important way in which the expansion of the Universe and the motion of the ball are fundamentally different. The ball moves *through* space and its motion is defined relative to some sort of reference. A planet, the cosmic microwave background (Destination 18), a spacecraft. Something. In the case of the Universe, its expansion is not the same as something that changes size within some container. The expansion of the Universe involves space itself. Nevertheless, the basic ways in which the expansion of the Universe can evolve are reasonably illustrated by the motion of our free-floating ball.

11 Which has the curious implication that the Universe must either expand or contract. To see this, imagine the Universe were perfectly static—that is, neither expanding nor contracting—but finite in size. The net gravitational force that it exerts on itself would

know how mass is distributed *around* it. To analyze the expansion of the Universe, we must at least know how mass is distributed *within* it.

As an extreme example, let's consider a model describing an expanding Universe that contains no mass. Such a Universe can be said to be expanding only if space itself changes its 'size' or, to use the more technically correct term, its *Scale* (Destination 16). To describe this evolution, we can define a variable—denoted as *R*—that represents the scale of the Universe and that changes value from 0.0 at the time when the expansion of the Universe began, to a value of 1.0 at the present time.[12] The lack of mass in this model of the Universe means it

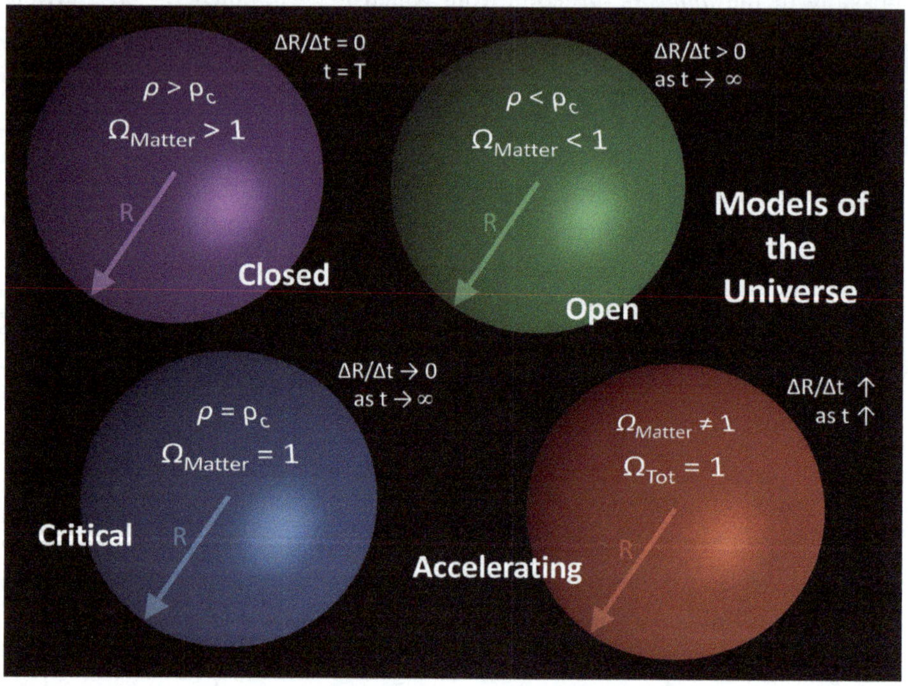

Fig 19.3

The four major classes of Universe models. Each ball in this Figure represents the Universe at some stage of its expansion characterized by a scale of *R*. For each case, the rate of expansion, $\Delta R/\Delta t$ at some time *t* after the expansion began is denoted. (*Upper left; purple*) A *Closed Universe* has a mean density, ρ, larger than the critical density, ρ_c so that $\Omega_{Matter} > 1$. This Universe stops expanding after a finite time, *T*, after which it contracts back on itself. (*Lower left; blue*) The *Critical Universe* has a mean mass density that equals the critical density at all times ($\Omega_{Matter} = 1$). It will continue to expand forever, getting closer and closer to zero expansion as time advances toward infinity. (*Upper right; green*) An *Open Universe* has a lower mean density than the critical density and so $\Omega_{Matter} < 1$. It will continue to expand forever, approaching a constant expansion rate as time advances toward infinity. The *Wide-Open Universe* is the special case of an Open Universe in which the density is zero. Its expansion rate remains constant for all time. (*Lower Right, red*) An *Accelerating Universe* has something in it that causes the expansion rate to increase relative to the deceleration caused by any mass that it may contain. Relevant models for an Accelerating Universe

---

soon cause it to begin to collapse and it would cease to be static. The only way around this is if the Universe is also infinite in extent in both space and time so that the net gravitational force everywhere is zero. But the dark night sky (Destination 18) assures us that that is not the case. As a result, no plausible model for the Universe allows it to be static for more than an instant.

12 This concept of the 'scale' of the Universe is important. If it seems unclear, imagine a balloon that is steadily filled with air. Before it begins to expand, we would say that the scale, or size, of the balloon is $R = 0$. However, once it reaches its final size, we would then say its scale, or size, has reached a value of $R = 1$. If we happen to know the separation between two points on the surface of the balloon when it reaches its maximum size—let's call that separation *x*—then the separation between these points was *xR* as the balloon expanded. In the case of a spherical balloon, this scale factor is simply $R = r/r_{max}$, where *r* is the radius of the balloon at any time as it expands, and r_{max} is its maximum radius. The same basic idea can be applied to three-dimensional space to generate mathematical models of how the Universe evolves. And, just like the balloon example, the scale—or size—of space can be defined in terms of a scale factor *R* that varies from 0 (at the beginning of the expansion) to 1 (today).

have $\Omega = \Omega_{Matter} + \Omega_\Lambda = 1.0$, where Ω_Λ is the contribution to the mass density by a mysterious effect—dubbed *Dark Energy*—that causes the the acceleration. After initial deceleration due to gravity, the expansion rate of an Accelerating Universe increases steadily with time. Both the Accelerating Universe and the Critical Universe—by virtue of having $\Omega = 1$—are also examples of *Flat Universes*.

also lacks gravitational forces that can slow down its expansion. Consequently, the *Expansion Rate*, $\Delta R / \Delta t$, remains constant, where ΔR represents a change in the scale of the Universe over a time interval equal to Δt. What this simply means is that the scale of the Universe is proportional to the amount of time that has elapsed since the expansion commenced—that is, $R \propto t$ (recall that the symbol '$\propto$' means 'proportional to'). The current age of this massless Universe—which we will denote as T_0—is precisely the value we encountered when we first discussed the Hubble Law: $T_0 = 1/H_0$ (Destination 16).

Of course, unless we are badly fooled, there *is* mass in our Universe.[13] That implies that the Universe must be slowing down due to the gravitational force it exerts on itself. If we assume that the matter in the Universe is distributed homogeneously in all directions—and, on scales of the Local Universe or larger, this seems to be reasonably true (Destination 18)—then the evolution of the Universe should depend only on how much mass it contains (refer to Figure 19.3 and Table 19.1). In one special case—the *Critical Universe*—the mass content of the Universe is just sufficient to cause it to stop expanding after an *infinite* time has elapsed. If there's less mass in the Universe than the critical case, then the expansion of the Universe slows down steadily, though it never stops, not even after an infinite time has passed. This is known as an *Open Universe*. Our earlier model of a massless Universe is a special sort of Open Universe that we'll call the *Wide-Open Universe*. If the Universe has more mass than the critical case, then gravity will cause the expansion to stop and reverse itself in a finite time. This is known as a *Closed Universe*. All of these models assume that gravity alone acts on the expansion of the Universe. But there is another model—the *Accelerating Universe*—in which the expansion rate increases with time relative to how it would evolve given the amount of mass the Universe contains. This model stands apart from the others because it implies that something other than gravity affects Universal expansion (Figure 19.3).

Like the wide-open case we discussed earlier, the Critical Universe is fairly simple to analyze. In particular, the scale, R, of the Critical Universe varies with time as $R \propto t^{2/3}$, and its age is given by $T_0 = 2/3H_0$. This is considerably shorter than the age we derived for the Wide-Open Universe (see Table 19.1). But that makes sense. After all, if there is mass slowing the expansion of the Universe, then the expansion rate was, on average, higher in the past than it is today. It therefore takes less time for the Critical Universe to attain a given size, R, and Hubble Constant, H_0, than for the Wide-Open Universe expanding at a constant rate.[14] In general, any Universe that contains mass—apart from possibly an

---

13 If you doubt this, just pinch yourself.

14 We can see this clearly if we return to our earlier car analogy. Imagine we see two cars pass us at the same speed. We are told that car 1 has traveled a distance, D, of 500 *miles* to reach us, always traveling at a constant speed, V, of 50 *mph*. Since $VT = D$, we know that the car has been on the road for $T = 10$ *hrs*. All we know about the second car is that it has been continuously slowing down since it started, also from a location 500

TABLE 19.1 Summary of Parameters for Different Universe Models

PARAMETERS[1]	APPLICABLE REDSHIFT, Z	WIDE OPEN	OPEN	CRITICAL	CLOSED	ACCELERATING
$\Omega_{Matter,0}$ (Matter)	0.0	0.0	0.27	1.0	1.5	0.27
$\Omega_{\Lambda,0}$ (Dark Energy)	0.0	0.0	0.0	0.0	0.0	0.73
$\Omega = \Omega_{Matter,0} + \Omega_{\Lambda,0}$ (Total)	0.0	0.0	0.27	1.0	1.5	1.0
Present-Day Age of Universe, T_0 (*Gyr*)	0.0	13.5	11.1	9.1	8.3	13.5
ρ_0 (g/cm^3)	0.0	0.0	2.6×10^{-30}	9.7×10^{-30}	1.5×10^{-29}	2.6×10^{-30}
Lookback Times, τ in *Gyr* for $z = 0.8$ [D19] $z = 2.0$ [D20] $z = 4.0$ [D21]	0.8	6.0	5.8	5.3	5.1	6.7
	2.0	9.1	8.4	7.3	6.8	10.2
	4.0	10.9	9.8	8.2	7.6	11.9
Age of Universe, T in *Myr* at $z = 100$ [D22] $z = 1200$ [D23]	100	59.6	16.4	8.8	7.2	16.6
	1200	0.52	0.32	0.20	0.17	0.32

Note to Table 19.1:

1. The notation [DNN] refers to the destination for which the given redshift applies. For example, the lookback times for Destination 19 (the young galaxy cluster MS 1054) are based on its redshift of $z = 0.8$. In all cases, a present-day Hubble Constant of $H_0 = 72$ *km/s/Mpc* is assumed.

Accelerating Universe—requires less time to reach a given size and expansion rate than the Wide-Open Case (Table 19.1).

---

miles away. We may not know its precise average speed, but we do know it is higher than its current speed of 50 *mph*. As a result, its time on the road, *T*, must be less than 10 *hrs*. In the case of the Universe, the distance that it has 'traveled' is the scale factor *R*, which we know equals 1.0 at the present time. The 'speed' of the expansion is given by the Hubble Constant, H_0.

This is progress. After all, having a handle on the relative ages of different models of the Universe is useful, akin to a navigator knowing the size of the Earth.[15] But simply knowing how big the Earth is doesn't tell us the distance between two locations. We need more than that to navigate. We also require *coordinates* for the two locations and we have to specify *how* we intend to travel between them (see Figure 19.4). In the same way, we need to define some sort of coordinate that tells us how far back in time we must travel to reach remote destinations.

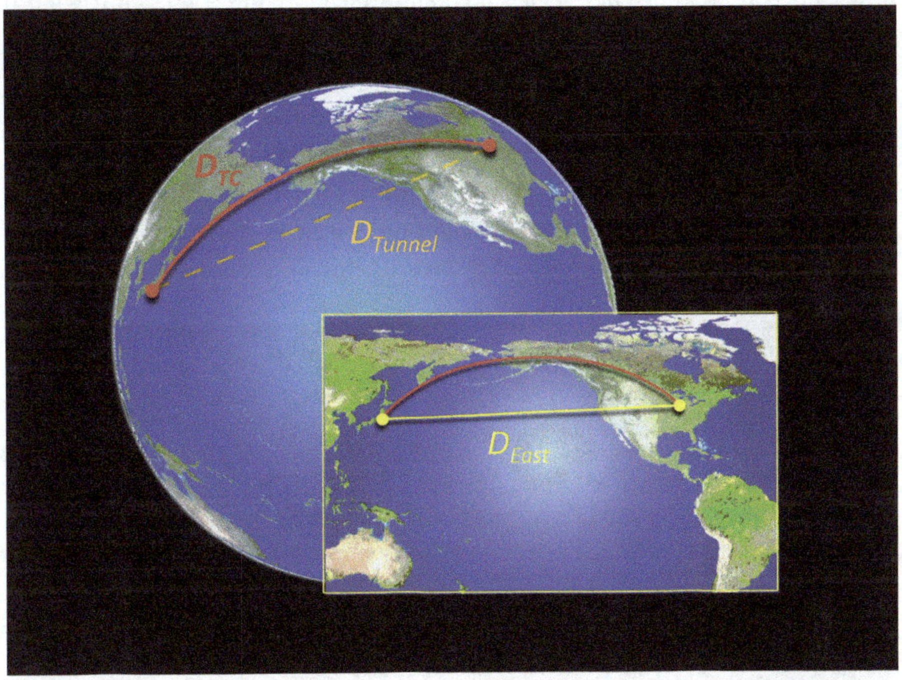

Fig 19.4

The 'shortest' distance by air between Tokyo and Chicago, D_{TC}, is shown by the red path on the globe of Earth (*upper*). This distance is considerably shorter than the 'straight line' distance between the two cities, D_{East}, depicted as a yellow line on the flat map (*lower*), which corresponds the path one follows by maintaining a constant compass course for the entire journey. Since the Earth is a three-dimensional body, the 'true' minimal distance between Tokyo and Chicago, D_{Tunnel}, is the path through a straight tunnel between them (shown as dashed orange line on the globe). Imagine what these paths would be like if, at the same time we tried to travel between the two cities, the surface of the Earth was uniformly expanding. This describes the nature of space that light traverses as it reaches us from distant objects in an expanding, and possibly non-flat Universe.

Earth images produced with K. Swartz/Great Circle Mapper and Blue Marble/NASA.

We start by defining the *Light-Travel Distance*, D_L, as the distance light traverses to reach us from a distant object in the Universe (Figure 19.5). The *Lookback Time*—denoted with the lowercase Greek letter τ (tau)—is that distance divided by the speed of light, or

$$\tau = \frac{D_L}{c}.$$

The tricky part of this process is determining D_L. The reason partly reflects the fact that different models of the Universe imply different geometries of space, some of which can allow multiple 'shortest' paths between two points (the Earth provides an analogy as shown in Figure 19.4 and described below). More

15 Perhaps the biggest practical blunder related to miscalculating the size of the Earth was made by Columbus. To his dying day, he was convinced he had traveled most of the way around the globe to reach the East Indies in Asia.

importantly, calculating the light-travel distance, D_L, is greatly complicated by the expansion of the Universe.

To understand why, let's define the shortest distance along Earth's surface between two locations on Earth—say, Tokyo and Chicago (Figure 19.4)—as D_{TC}. Now, imagine that the radius of the Earth were suddenly doubled, with its surface features stretching uniformly to maintain their shapes. At the end of this expansion, distances between locations on the Earth's surface will have doubled compared to the time before the expansion. In particular, the new 'shortest' distance between Tokyo and Chicago will be $2D_{TC}$. But what if we happened to be traveling between the two cities while the expansion occurred? It's clear that the distance we'd travel is somewhere between D_{TC} and $2D_{TC}$. However, the precise distance that we would have covered depends on exactly how the Earth expanded and not just by what factor. This means that to determine lookback times we need to know by what factor the Universe has changed its scale, R, during the time light has taken to reach us from distant sources. And we need to know what *kind* of Universe we live in so we can

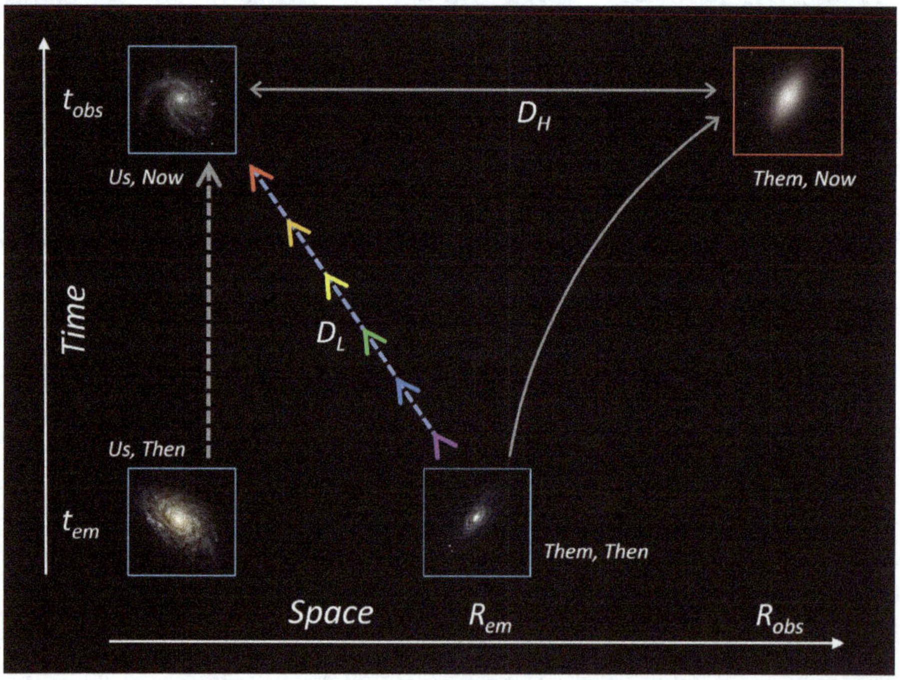

Fig 19.5

A *Space–Time Diagram* that describes how we see distant objects in an expanding Universe. Time increases upward in this plot, and distance from the observer—which we assume is us—increases to the right (as denoted by the white arrowed lines). The upper left image represents our Galaxy (labeled 'Us,Now') at the present time, t_{obs}, when we receive radiation from a distant galaxy (labeled 'Them,Then'). That radiation was emitted at the time t_{em} in the past, and its path from its source to us is depicted by the curved line to emphasize its passage through expanding space. The colored arrowheads denote how the radiation becomes redder as it travels through the expanding Universe (see Figure 19.6). This expansion, shown schematically by the dashed blue lines, which, over time, carries the remote galaxy to its current location labeled 'Them,Now' and us from our past location ('Us,Then') to our current location ('Us,Now'). The fact these lines are straight is consistent with the simple form of the Hubble Law we discovered at Destination 16. Notice that at the time the remote galaxy emitted its radiation, t_{em}, it was a distance R_{em} away from us (the horizontal separation between 'Us,Then' and 'Them,Then' in the Figure). But when we observed that radiation at the time t_{obs}, it was at a distance R_{obs} away, equivalent to

determine how the expansion evolved before we can calculate the light-travel distance, D_L.

It turns out that determining the change of the scale of the Universe is relatively easy. Recall from Destination 14 that the wavelength of EM Radiation lengthens as it passes through empty space (Figure 19.6). This stretching results directly from the change in the scale, R, of the expanding Universe during the light's journey to us. If the emitted wavelength from a specific emission line produced in a distant

its Hubble Distance, D_H (the horizontal separation between 'Us,Now' and 'Them,Now' denoted by the gray, double-arrow line). We see the remote galaxy as it existed its *lookback time*, τ, in the past. This is defined as the time it takes light to traverse its path from its origin to where we observe it (that is, $\tau = D_L/c = t_{obs} - t_{em}$). The lookback time is what's required to figure out how far back in time we must travel to visit our distant destinations as they existed in the past. Galaxy images from STScI/NASA.

galaxy is λ_{em} and the observed wavelength of the same line is λ_{obs}, then we know that the Universe was $R = \lambda_{em} / \lambda_{obs}$ times smaller when the light was emitted that it is today.[16] Since the redshift of a distant source is defined as $z = (\lambda_{obs} - \lambda_{em}) / \lambda_{em}$ (see Figure19.6), it's easy to prove that $R = 1/(1 + z)$. Since R varies from a value of 0 at the start of the expansion of the Universe to 1 today, this tells us that the redshift varies from $z = 0$ (today) to $z = \infty$ (at the edge of the Observable Universe; see Destination 18). This remarkable result tells us that the redshift—something we can easily measure from the spectra of astronomical objects—allows us to track the evolution of the expansion of the Universe (Figure 19.5).

Lookback times can be calculated for any of the models of the Universe that we have considered. For some of the cases the results are relatively simple. For example,

$$\tau = \frac{z}{H_0(1+z)} \quad (\textit{Wide-Open Universe}),$$

while for a Critical Universe, the lookback time is

$$\tau = \frac{2\left(1-(1+z)^{-3/2}\right)}{3H_0} \quad (\textit{Critical Universe}).$$

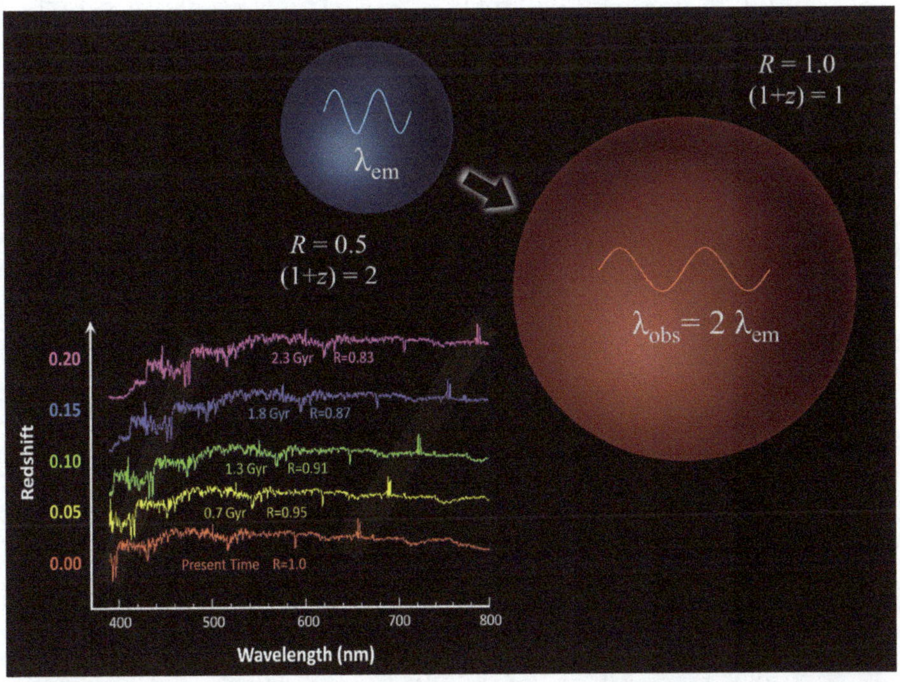

Fig 19.6

The redshift, z, is defined as $\Delta\lambda / \lambda_{em}$, where λ_{em} is the emitted (rest) wavelength of a spectral feature. Some representative spectra of galaxies at different redshifts ranging from 0.0 to 0.2 redshifts are plotted at *lower left*. These are the same spectra from Figure 16.9, but now labeled by redshift instead of recession velocity ($V_r = cz$). The shifts in absorption lines from Calcium (along the left sides of the spectra) and an emission line of Hydrogen (along the right sides of the spectra) are highlighted by shaded rectangles. The spectra are labeled by their lookback time in an Accelerating Universe with $\Omega = 1.0$ and $\Omega_{Matter,0} = 0.27$, and by the scale of the Universe, R, at the time when the radiation was emitted. (*Upper Right*). These balls represent schematic views of a section of the Universe at the present time (red; $R = 1$) and

16 To be a bit more precise, what this means is that the average separations of galaxies was $R = \lambda_{em} / \lambda_{obs}$ times smaller in the past. Both in the past and now, the Universe could still be infinite in extent.

For the other cases we've considered—the Open, Closed, and Accelerating Universes—the equations for the lookback time are more complicated, but they can be readily plotted. Some specific examples that illustrate how lookback time varies with redshift are shown in Figures 19.7 and 19.8.

We are almost there. By measuring the redshift, $z = \Delta\lambda / \lambda_{em}$, of our next destinations we can determine by what factor the Universe has expanded since light left those remote locations. And, by knowing the redshift and the Hubble Constant, H_0, we can determine the lookback time ... *if we also know what kind of Universe we live in.*

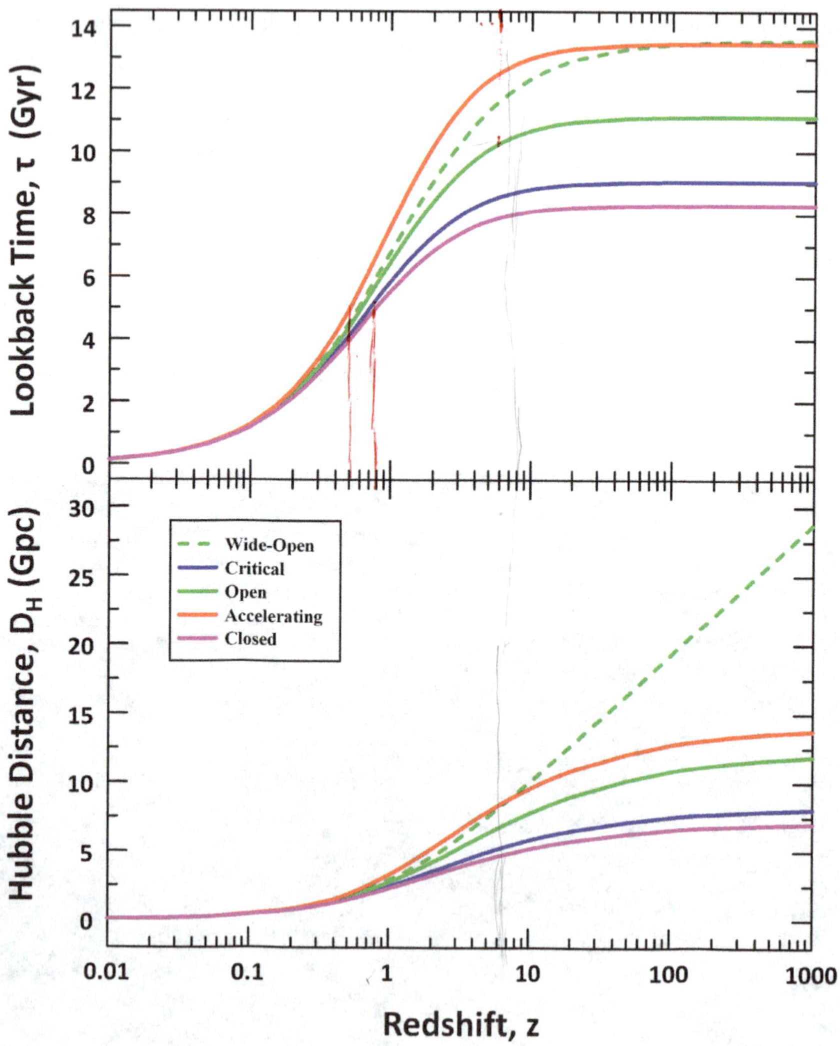

when the scale of the Universe was half its present value (blue; $R = 0.5$). Radiation emitted with a wavelength λ_{em} (shown as the blue waveform) will stretch along with the expanding Universe until it is detected with a wavelength λ_{obs} (shown as the red waveform that has twice the wavelength of the emitted radiation shown in blue). The wavelength provides an 'imprint' the scale of the Universe that can be written as $R = 1/(1 + z)$. Since the redshift, z, equals $(\lambda_{obs} - \lambda_{em}) / \lambda_{em}$, it is easy to see that if $\lambda_{obs} = 2\lambda_{em}$, then $(1 + z) = 2$. Consequently, R was equal to 0.5 when the radiation was emitted in the example shown here.

Fig 19.7

Plots of how lookback time (τ; *Upper*) and Hubble Distance (D_H; *Lower*; see Figure 19.5) vary over a wide range of redshift for different models of the Universe identified in the figure legend. The Open model plotted here adopts $\Omega = \Omega_{Matter,O} = 0.27$ while the Closed model assumes $\Omega = \Omega_{Matter,O} = 1.5$. The Accelerating model has $\Omega_{Tot} = 1.0$ and $\Omega_{Matter,O} = 0.27$. Both the Critical and Accelerating models are examples of Flat Universes with $\Omega_{Tot} = 1.0$, the difference being that in the Critical model the density is fully in the form of matter (both ordinary and dark). By definition, $\Omega_{Tot} = 0.0$ in the (massless) Wide-Open model.

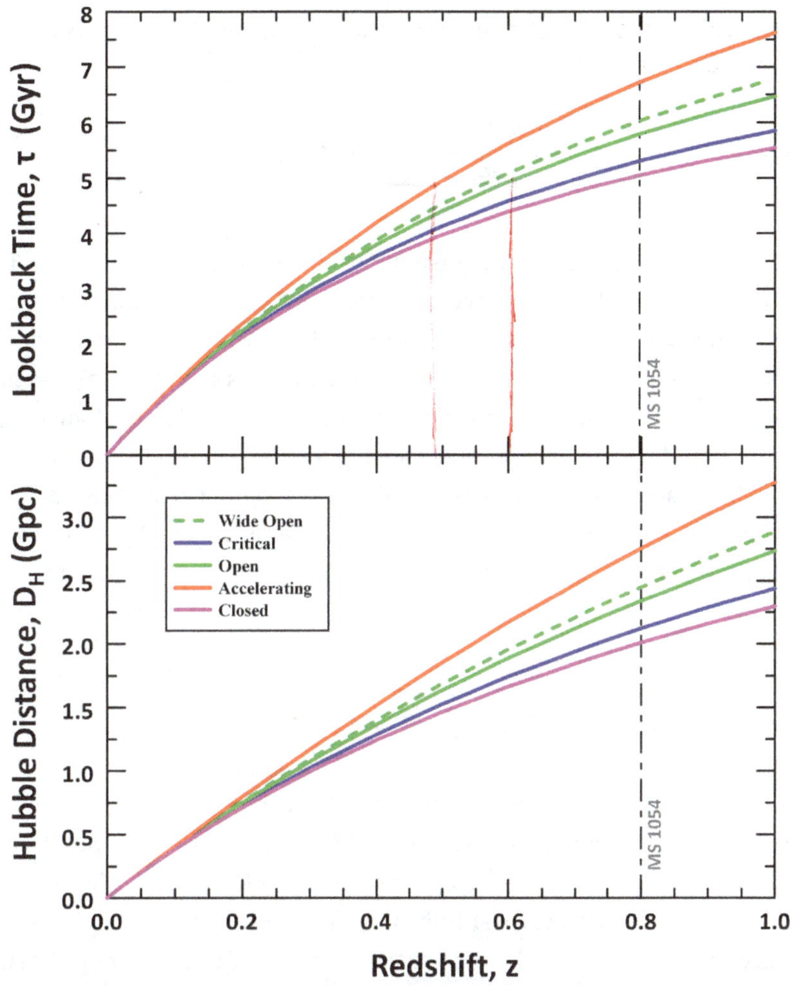

Fig 19.8:

Plots of the lookback time (τ; *Upper*) and the Hubble Distance (D_H; *Lower*; see Figure 19.5) for the same models shown in Figure 19.7. In these plots, the *x*-axis has been modified to illustrate how different the models evolve at low redshifts ranging from $z = 0$ to $z = 1$. The redshift of MS 1054 ($z = 0.8$) is noted at by a dashed line in both plots.

19.3 WHAT KIND OF UNIVERSE DO WE INHABIT?

As we saw earlier, a key feature that distinguishes different models of the Universe is the mass content. It's the mass that establishes the gravitational force that the Universe can exert on itself, which, in turn, dictates the rate at which the

Universe evolves. We can define a quantity known as the *Critical Density*[17], $\rho_{c,0}$, as

$$\rho_{c,0} = \frac{3H_0^2}{8\pi G} = 9.7 \times 10^{-30} \; g/cm^3,$$

which lets us conveniently distinguish different types of Universes (see Table 19.1). If the actual—that is, observed—total mass density of the present-day Universe equals $\rho_{c,0}$, then we live in a Critical Universe. If the mass density exceeds $\rho_{c,0}$, then we live in a Closed Universe that is destined to eventually 'collapse back' on itself in the future. If the mass density is lower than $\rho_{c,0}$, we live in an Open Universe of some sort (see Figure 19.3) that will never cease to expand. Since the relationship between the actual and critical mass densities of the Universe can help us determine which type of Universe we live in, it is convenient to define a new parameter, $|_0$ (pronounced 'Omega-naught') as

$$\Omega_0 = \frac{\rho_0}{\rho_{c,0}},$$

where ρ_0 is the current mass density of the Universe. For a Closed Universe, $\Omega_0 > 1$, since the actual mass density exceeds the critical density. Similarly, an Open Universe has $\Omega_0 < 1$, the special case of the Wide-Open Universe corresponding to $\Omega_0 = 0$. For a Critical Universe, $\Omega_0 = 1.0$ since this case is defined to be the one where the actual mass density precisely equals the critical density (see Figure 19.3).

This suggests that if we carry out a census of matter in the Universe to determine Ω_0, we can figure out what sort of Universe we live in. For this to work, our census must satisfy two important rules. First, the mean mass density we measure must come from a volume of space that's big enough to provide a fair sample of the entire Universe. From our observations at Destination 18, the smallest possible volume that's suitable would be comparable in size to the Local Universe, a region extending to about 300 *Mpc* around our Galaxy. Second, we have to account for *everything* that might contribute to the mass density of the Universe. Some of the stuff we have to include is comparatively obvious: Normal and Dark Matter. Some is surprising: The 'mass' associated with radiation. And some of it is downright strange: The 'mass' of empty space. A nice feature of the Omega parameter is that we can calculate its value for individual contributors, then add up these individual quantities to obtain the total value of Ω_0.

17 The subscript '0' (naught) refers to the values of quantities at the present time. This distinction is important because just about all the parameters that describe the Universe vary with time. That includes the Hubble 'Constant,' the mean density of the Universe, and the Critical Density.

If we consider all the visible mass in galaxies spread throughout the Local Universe[18], we find that mean density of visible matter corresponds to $\Omega_{Vis,0} \approx 0.02$, only 2% of the current value of the critical density, $\rho_{c,0}$. But we know that visible galaxies are always accompanied by Dark Matter (Destinations 13–16). If we add up that mass, we find that $\Omega_{DM,0} \approx 0.25$ within the volume of the Local Universe, over ten times the density of visible matter! We can define a new parameter, $\Omega_{Matter,0}$ (pronounced 'Omega-Matter-Naught') as the sum of these two values. $\Omega_{Matter,0} = \Omega_{Vis,0} + \Omega_{DM,0} = 0.27$. This seems to tell us that there's only enough matter in the Universe to account for about 27% of the present-day critical density, $\rho_{c,0}$. If this were the end of the story, it would imply that we live in an Open Universe. But there are a few things we need to consider before we jump to this conclusion.

First, have we truly accounted for *everything* that might contribute to the mass density? One obvious thing we've missed is radiation. As we discovered at Destination 18, the entire sky emits radiation in the form of the CMB (the Cosmic Microwave Background). Since Einstein's Equation, $E = mc^2$, tells us that mass and energy are fundamentally equivalent[19], this implies that we need to include the contribution of the CMB radiation to our estimate of Ω_0. It turns out that the equivalent mass density of the CMB radiation—which dwarfs the total light emitted by stars and galaxies—can be calculated as

$$\rho_{rad,0} = \frac{4\sigma T^4_{CMB,0}}{c^3},$$

where σ is the Stefan-Boltzmann constant (Destination 6), and T_{CMB} the temperature of the CMB radiation we observe today (2.725*K*; Destination 18).[20] Using this equation to calculate $\rho_{rad,0}$, we find that $\Omega_{rad,0} \approx 2 \times 10^{-4}$, a minuscule fraction of the present-day critical mass density of the Universe. Including the contribution of radiation to Ω_0 does not alter our conclusion that we appear to live in an Open Universe.[21]

---

18 Of course, this has to be done statistically since it is impossible to carry out such a census perfectly. Nonetheless, we do know the amount of matter in the form of visible material pretty well throughout the Local Universe.

19 In all respects. If we had a ball of photons from EM Radiation, they would exert a gravitational force equal to their equivalent mass $M = E/c^2 = hv/c^2$, where $E = hv$ is the energy of the photons of frequency v (see Destination 5).

20 Yes, we are using T to represent the age of the Universe and temperature! Be sure you understand from context (or from the units) whether T refers to time or temperature!

21 So why did we even bother to consider radiation? As we'll see, the Universe was hotter in the past than it is today. Far in the past, it was much hotter. Since the radiation density is such a strong function of temperature, T^4, it used to be a more significant component of Ω when the Universe was hotter. Thus, while the present-day Universe is said to be *Matter Dominated*, at earlier times it was *Radiation Dominated* and the mass density of radiation, ρ_{rad}, actually exceeded the density of matter (dark and visible). Partly because of this, the Universe behaved very differently back then than it does now (Destinations 24 and 25).

TABLE 19.2 Astronomical Distance Determination Techniques

NAME OF METHOD	METHOD TYPE[1]	APPLICABLE RANGE	APPLICABLE DISTANCE RANGE[2]	COMMENTS [DESTINATION(S) TO FIND DETAILS]
Radar Ranging	Geometric	Solar System	<1 *AU*	Establishes scale of Solar System and value of the Astronomical Unit [D1–3].
Parallax	Geometric	Solar Neighborhood	0–500 *pc*	First step to get distances outside Solar System [D3].
Main Sequence Fit	Stellar Astrophysics	Galaxy; Nearby Galaxies	Within 0.5 *Mpc*	Relies on uniformity of Main Sequence properties in star clusters [D8].
RR Lyr Stars	PL Relation Standard Candle	Galaxy; Local Group	Up to 1 *Mpc*	Most found in old stellar systems such as Globular Clusters [D11].
Binary Stars	Geometric Stellar Astrophysics	Galaxy; Local Group; Nearby Groups	Up to 3 *Mpc*	Typically relies on analysis of orbital sizes and velocities of stars in binary systems [D4, D8].
Cepheid Variables	PL Relation Standard Candle	Galaxy; Local Group; Nearby Groups; Nearby Clusters	Up to 20 *Mpc*	Historically, one of the first known extragalactic distance methods. Well calibrated and extensively studied [D14].
Tully-Fisher and Faber-Jackson Relations	Galactic Astrophysics Scaling Relation	Galaxy Clusters; Local Universe	10–300 *Mpc*	Relies on underlying uniformity of disk galaxies (TF Relation) and elliptical galaxies (FJ Relation). [D14, D15].
Type Ia Supernovae	Standard Candle	Local Universe; Moderate Redshifts	From 10 *Mpc* to $z \approx 1$	A powerful technique to explore distances where different Universe models make significantly different predictions [D19].
Gravitational Lensing	Geometric Galactic Astrophysics	Moderate to High Redshifts	From $z \approx 0.1$ to 2.0	Based on the analysis of gravitationally deflected light from distant sources [D16].
Sunyaev-Z'eldovich Effect	Galactic Astrophysics	Moderate to High Redshifts	From $z \approx 0.1$ to 1.0	Based on how CMB radiation [D18] interacts with the hot gas in galaxy clusters [D16, D19].
Redshift	Cosmological	Entire Universe beyond 30 *Mpc*	All Redshifts beyond about 0.01	Converts redshift to distance for a given model for the Universe. Local motions can distort distances for close sources [D18, D19].

Notes to Table 19.2:

1. The various methods are defined as
   - Geometric: Requires only geometric concepts to determine distances.
   - Stellar Astrophysics: Relies on known or derived properties of stars to estimate distances. This method requires that distances are independently determined for at least some calibrating stars.
   - PL Relation: Exploits the existence of a relationship between periods and luminosities of specific classes of variable stars.

(*CONTINUED*)

- Galactic Astrophysics: Relies on known or derived properties of galaxies to estimate distances. This method requires that distances to at least some calibrating galaxies are independently known.
- Scaling Relation: Based on comparisons of galaxy properties with known global relations that tie these properties together.
- Standard Candle: These indicators are assumed or known to be identical in properties wherever they are found, usually spanning a limited range in luminosity. They serve as standard beacons.
- Cosmological: Uses models of the expansion and geometry of the Universe to convert redshifts to distances.

2. Some methods are rapidly evolving, and their applicability expanding. Parallax measurements, for example, will be attainable to far greater distances than currently possible with the successful completion of the GAIA mission in the 2010's.

But there still is a reason we may want to rethink our apparent conclusion that we live in an Open Universe. As it turns out, every model of the Universe that we have considered (see Figure 19.3 and Table 19.1) can be evaluated to provide a measure of the current distance of any object—a distance we'll refer to as the *Hubble Distance*, D_H, and which is not the same as the light-travel distance, D_L (see Figure 19.5). The problem with the Open Universe is that distant galaxies and galaxy clusters do not appear to obey the relation we'd expect between redshift and Hubble Distance (see Figure 19.8).

How can we know this? The key is to determine the relation between the distance to remote galaxies and their redshifts by using distance estimates obtained with methods that do not involve the Hubble Law (or else the reasoning becomes circular). Although none of the distance methods we've discussed so far are up to the task (see Table 19.2), methods developed and refined over the past few years have steadily improved our ability to measure Hubble distances to remote objects. One particularly powerful approach involves the use of variable stars known as *Type Ia Supernovae*. Like the supernovae we encountered at Destination 12, Type Ia supernovae (pronounced 'Type One-A') temporarily become very luminous when they explode. So luminous, in fact, that they can be detected out to distances of many hundreds of *Mpc*. But Type Ia supernovae do not arise from the explosions of dying, massive stars (Destination 12). Instead, they appear to occur when white dwarfs (Destination 11) in binary systems catastrophically merge. Fortunately, their precise origin is not critical for our current purposes. The key feature that interests us now is that Type Ia supernovae appear to all be remarkably alike. No matter where—or when—they occur within the Universe, these stars all appear to reach nearly the same peak luminosity. As such, they represent a particularly useful class of *Standard Candles*—objects with well-defined and uniform luminosities—that we can use to map the expansion of the Universe to extraordinarily large distances and, hence, redshifts (Figure 19.9 and Table 19.2).

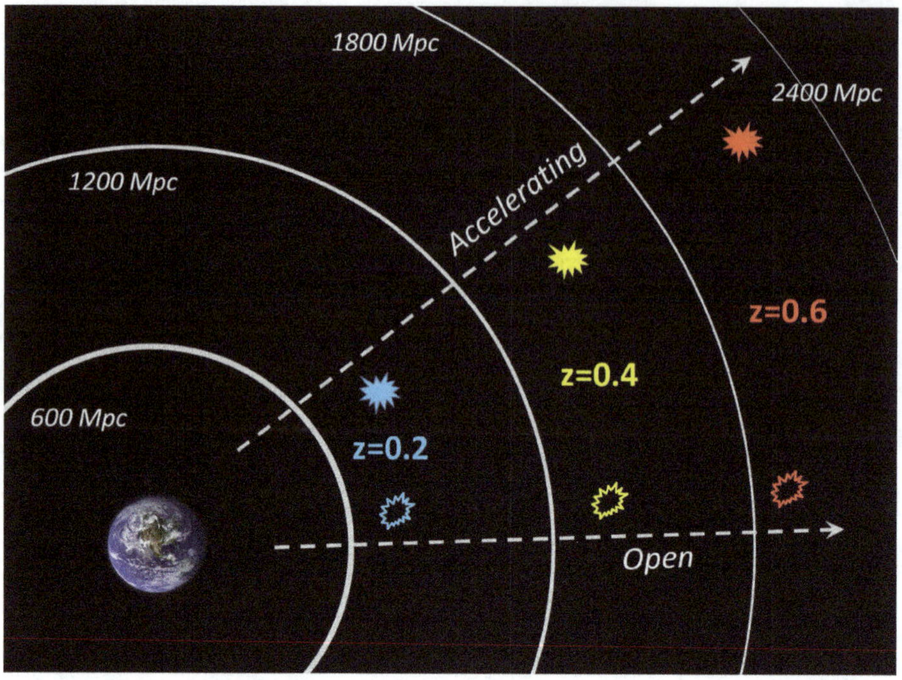

Fig 19.9

Distances of Type Ia supernovae at redshifts of 0.2 (blue), 0.4 (yellow), and 0.6 (red) are shown to scale (apart from the size of the Earth!) for an Open Universe (open symbols) and an Accelerating Universe (filled symbols). By independently measuring distances and redshifts of these exploding stars, we can compare their redshift-distance relation with these model predictions (see also Figures 19.7 and 19.8) to determine which of these two Universe models—or any other—best agrees with what we observe. The Accelerating Universe predicts that Type Ia supernovae will be further away and fainter at a given redshift than they would be in an Open Universe. Modern observations of Type Ia supernovae agree much better with the predictions from the model of the Accelerating Universe with $\Omega_{Tot} = 1.0$ and $\Omega_{Matter,0} = 0.27$ than with any model for an Open Universe.

Earth image from JAXA.

As studies of distant Type Ia supernovae have progressed, it has become increasingly clear that the Universe does not appear to be evolving as we'd expect for an Open Universe with $\Omega_0 = 0.27$. Instead, distant galaxies appear to be further away from us given their redshift than we'd have expected for this model (see Figure 19.9). This result suggests that empty space not only partakes in the Universal expansion revealed by the Hubble Law (Destination 14), but it seems to also contain something that causes the expansion to accelerate. This 'something' is referred to as *Dark Energy*, but its nature is utterly mysterious. Models of an Accelerating Universe that best fit the observations suggest that the 'push' from Dark Energy is becoming relatively more important with time. Its effects have been significant only relatively 'recently,' that is, during the past 6–8 *Gyr* of the history of the Universe. The possibility that the Universe could exhibit this strange behavior was first realized by Einstein over 80 years ago, but he rejected the idea as absurd.

We can calculate the contribution of Dark Energy to Ω_0 if we treat the force that it exerts on empty space as a strange 'negative' pressure. Normally, when a gas expands its pressure *decreases*, lessening any tendency toward further expansion.[22] But the empty space of the Universe appears to behave in precisely the opposite way. The more it expands—and therefore the emptier it gets—the *higher* the pressure, accelerating the expansion! As we have for matter and radiation, this lets us calculate an effective mass density associated with Dark

22 Recall that this behavior plays a key role in stellar evolution as we saw at Destinations 8, 11, and 12.

Energy through a parameter known as Ω_Λ (the subscript is the Greek capital letter Lambda so this parameter is typically pronounced 'Omega-Lambda'). Various observations—including Type Ia supernovae and careful analysis of the CMB (Destinations 18 and 23)—suggest that $\Omega_{\Lambda,0} \approx 0.73$. Astonishingly, this value makes the total density parameter of the Universe, $\Omega = \Omega_{Matter} + \Omega_\Lambda$, precisely equal to 1.0.[23] It is important to appreciate that this does not mean that the Accelerating Universe is the same as the Critical Universe, even though both have $\Omega = 1.0$. In particular, in a Critical Universe the density parameter arises exclusively from the amount of matter—and radiation—in the Universe. The gravitational forces of that matter on itself cause the expansion to stop after an infinite time has elapsed. In the case of the Accelerating Universe, even though $\Omega = 1.0$, the expansion will *not* stop after an infinite time has passed. Just the opposite: It expands even more rapidly with time.

The Accelerating and Critical models of Universal expansion share one very important feature. Although we have defined Ω in terms of the density of the Universe, it turns out that this parameter also describes the *geometry* of space. For any Universe where $\Omega = 1.0$—no matter how it arrives at this value—the geometry of space is said to be *Flat*, equivalent to the three-dimensional geometry of a flat sheet of paper. In a *Flat Universe*, standard *Euclidean Geometry*—the kind we all first learn by the time we get to high school—applies. For a Universe with $\Omega \neq 1$, the geometry of the Universe is said to be *curved*. This would be similar to the geometry one observes on a curved surface such as a sphere (an example of a closed geometry) or a saddle (an open geometry). In a curved geometry, standard Euclidean geometric rules do not apply. For example, parallel lines can cross. Or diverge. The sum of the angles within triangles may add up to more than 180 *deg*. Or less. For this reason, the term *Flat Universe* is a more appropriate name for any Universe—as ours appears to be—that has $\Omega = 1.0$.

And so we have what we came for ... finally! Our understanding of the ways that the Universe can evolve and our observations of how it appears actually to evolve give us the information we need to determine how far back in time we need to go to reach a remote destination. If we have a target's redshift, and we know the key parameters that describe the Universe in which we live—$H_0 = 72\ km/s/Mpc$, $\Omega_{Matter,0} = 0.27$, and $\Omega_{\Lambda,0} = 0.73$—we can use plots such as Figures 19.8 and 19.9 to determine the lookback times and Hubble distances to even the most remote destinations we might want to visit. We now know how to navigate in both time and space anywhere in the Universe, a significant achievement. But it may seem somewhat hollow. We still have no idea why the Universe is expanding in the first place. Or why it is accelerating, let alone why the acceleration is precisely what's

23 You may notice that there are no subscripts '0' on these values. The reason is that if the total value of Ω is equal to 1.0 at some time, it will be equal to 1.0 at all times. What does change are the individual values of Ω_{Matter} and Ω_Λ, so we do have to specify the time in the history of the Universe when we measure those. But since the sum of $\Omega_{Matter,0}$ and $\Omega_{\Lambda,0}$ appears to equal 1.0 now, it always will and always has.

needed to make the geometry of the Universe perfectly flat. Or what Dark Energy or Dark Matter are, both critical components of any models that aim to describe the evolution of our Universe. These remain open questions. Ones that address some of the most fundamental mysteries in modern astronomy.

19.4 MS1054-0321: A GALAXY CLUSTER AT Z = 0.8

Now that we know how to calculate lookback times in the Universe, let's start our travels in time with a comparatively timid first step. Our goal is to visit a galaxy cluster with a redshift of $z = 0.8$. Its formal name is MS 1054-0321, but we'll shorten that a bit to 'MS 1054' (see Figure 19.10). This unpoetic designation refers to the cluster's position in Earth's sky, which incorporates the basic information regarding to *where* we have to go. The cluster's redshift means that the radiation we observe from its galaxies has shifted to wavelengths 80% longer than when the radiation was emitted. For example, ultraviolet radiation emitted from the cluster is perceived at Earth to be optical radiation; optical radiation is shifted to the infrared. And so on. This means that from Earth, optical observations reveal different portions of the spectra of galaxies in this cluster than we are accustomed to observing from nearby galaxies with much smaller redshifts. Most importantly for our purposes, however, the cluster's redshift tells us how far back in time we view the cluster. It tells us to *when* we have to travel.

So how far have we had to travel in space and time to reach MS 1054? Assuming we live in an Accelerating Universe with $\Omega_{Matter,0} = 0.27$, $\Omega_{\Lambda,0} = 0.73$, and $H_0 = 72\ km/s/Mpc$ (as we concluded in the last section of this destination), we can inspect Figure 19.8 (also Table 19.2) to see that the lookback time to MS 1054 is $\tau = 6.7\ Gyr$ corresponding to a light-travel distance of $D_L = 2.1$ *Gpc*.[24] Since the Accelerating Universe with these parameters has an age—as measured from present-day Earth—of 13.5 *Gyr*, this means that we have had to travel nearly halfway back in time from the instant when the expansion of

24 For reference, the Hubble distance to MS 1054 is $D_L = 2.8\ Gpc$, about 20% higher than the light-travel distance. The Hubble distance is how far MS 1054 is located from us today—in the 21st century from Earth—not the separation of our young Galaxy and the cluster 6.7 *Gyr* ago, which was about 1.5 *Gpc* (see Figure 19.5 to review the various distances involved). This discussion should make it apparent that determining time and distances in an expanding Universe can be confusing!

the Universe commenced in order to visit this cluster so that we can see it as it existed as seen from Earth today (see Figure 19.5). This can be confusing, but as time travelers we'll need to get used to this sort of reasoning. Nor should we fail to put the scales of the timespans we are traversing into broader perspective. The oldest stars in the Galactic Disk have ages of about 10 *Gyr* (Destination 9), while the Sun itself is only about 5 *Gyr* old (Destination 8). This last point means that even if we could somehow identify our incredibly faint Galaxy from the remote vantage point of MS 1054, we'd see it as the comparatively 'young' Galaxy that existed at a time *before* the Sun and the Earth were born.[25] The sense of homesickness that we feel as we travel back in time to visit destinations in the early history of the Universe will be a positively strange emotion. We will find ourselves yearning for a home that has yet to come into existence!

But that's not a problem for us now. As we look around us, the view of this Alien Sky—a sky from far away and long ago—dispels any thoughts of home. Here in the core of the cluster MS 1054 (Figure 19.10) we behold a sky of galaxies unlike any we've seen before (Destinations 14 and 16). Just with the naked eye we can immediately see that this cluster is denser than the Virgo Cluster (Destination 16): Surrounding us are nearly 400 galaxies, all easily visible in this sky, nearly four times as many as we could easily spot by eye from the core of Virgo. Many of the galaxies are recognizable as specimens that could be classified within Hubble Classification System (Destination 15). There are many spiral galaxies here—nearly one third of the galaxies in and near the cluster core are of this type—as well as numerous disk-like galaxies best described as gas-poor Sa or S0 galaxies in Hubble's classification system (Destination 15). Many of the spirals have unusual structures or blue colors, suggestive of recent interactions and active star formation (Figure 19.10). Perhaps most striking, however, is the comparative lack of large elliptical galaxies here at MS 1054. In Virgo, we found that nearly all of the most luminous galaxies in the cluster core are large elliptical systems. But here, though the most luminous galaxy in the cluster is a large elliptical galaxy, the vast majority (80%) of galaxies in the core of MS 1054 are not (Figure 19.10). It is a strikingly obvious difference: The center of the Virgo Cluster is dominated by S0 and large elliptical galaxies (see Figure 16.1), but here at the center of MS 1054, there seem to be relatively many more disk galaxies.

Not all the galaxies we see here are easy to classify, however. Back in Virgo, we found one example of an interacting galaxy—NGC 4438 (see Figure 16.4)—but here at the center of MS 1024, a much larger fraction of the luminous galaxies appear to be interacting systems, perhaps as many as 20–30% of all galaxies in the region. In some cases, the interactions appear to be comparatively tame

25 Of course, we'd see the Galaxy as it appeared at a time even *earlier* than the time we are visiting to view MS 1054. Just as it takes time for light from the cluster to reach us today on Earth, it took time for light from our Galaxy to reach the cluster in the past (see Figure 19.5, reversing 'Us' and 'Them' in the plot).

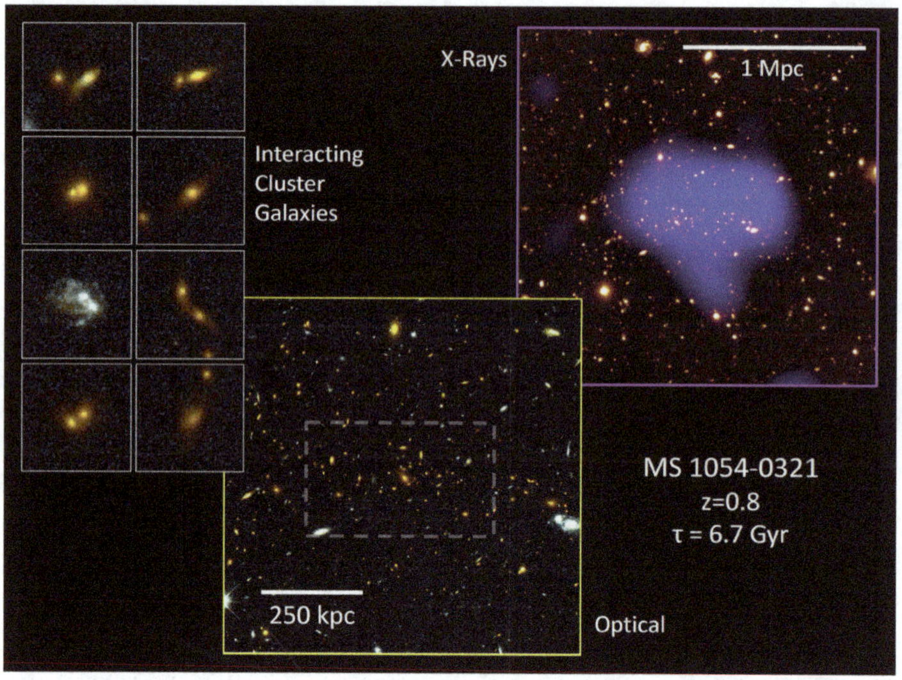

Fig 19.10

(*Lower center*) An optical image of MS 1054-0321, a galaxy cluster with a redshift of $z = 0.8$ and a lookback time of 6.7 *Gyr* (see Table 19.1). Many of the brightest cluster galaxies are disk-like or interacting systems—some examples are shown in more detail (*left*). Unlike nearby rich clusters which are typically dominated by luminous elliptical systems, there is only one prominent elliptical galaxy at the center of this rich cluster. The pale dashed rectangle represents the size of the Virgo cluster depicted in Figure 16.1 (*Upper right)* The x-ray emission from MS 1054—at a somewhat different scale than the optical image—illustrates that the gas in the cluster has evolved significantly despite the relative youth of the cluster.

Optical images from STScI/NASA. X-ray image from Chandra/NASA.

in the sense that they involve relatively small S0 or E galaxies with little internal gas, and therefore not much induced star formation. But quite a few of these galaxies are spectacular, exhibiting the kinds of fireworks we see in interacting systems near Earth (Destination 16). These galaxies are also often very blue and luminous compared to normal spiral or irregular galaxies (Destination 15), and many display tortured, complex morphologies that make them difficult to identify as anything other than 'irregular' galaxies within Hubble's classification system (Figure 19.10).

We've already noted that MS 1024 is a rich cluster with a considerably higher density of galaxies at its core than the Virgo Cluster. Based on its velocity dispersion of nearly 2000 *km/s*, it is also considerably more massive. All that is just a matter of scale; MS 1024 is a larger, denser cluster than Virgo. But there is another subtle feature apparent in this sky at the core of MS 1024 that distinguishes it from Virgo in a more fundamental way. As we look around us, we come to realize that the distribution of high- and low-luminosity galaxies in MS 1024 appears to be different from what we saw in Virgo. By carefully counting the number of galaxies in the cluster and noting their luminosities, we find that there are relatively more dwarf galaxies in MS 1024 for the number of larger, more luminous galaxies compared to what's seen in Virgo or in other galaxy clusters and groups located near Earth. Many of these dwarfs are irregular systems, full of gas and some forming stars vigorously (Destinations 12 and 14). Some are comparatively gas-free dwarf ellipticals and dwarf spheroidals (Destinations 13 and 15). Like the more luminous galaxies, a large fraction of the dwarfs appear to be associated with interacting systems (Destination 16). In the Local Group, there is a strong segregation of irregular dwarfs to the outskirts of the group, while dwarf elliptical and spheroids are found almost exclusively close to the largest galaxies, M 31 and our Galaxy (Destination 13).[26] This effect is noticeably less pronounced here in MS 1054 where dwarf galaxies of all types are spread throughout the cluster. There is only a weak tendency for dwarf irregular galaxies to avoid the cluster core; they do not avoid the cluster center to the extent that they do in Virgo.

What can we make of all this? The properties of the spiral and disk-like galaxies in MS 1054 seem easiest to understand. Having traveled back in time, we can appreciate that some of these galaxies must be young. After all, the Universe itself is only half as old—6.7 *Gyr*—as it was when we left Earth (see Table 19.1). Not only are the stars in these galaxies younger than in the older galaxies of the Local Universe, but it appears that many of the spiral galaxies in MS 1054 formed many stars early in their lifetimes. Even the gas-poor Sa and S0 galaxies that are common here are consistent with this. They reveal galaxies that formed large numbers of their stars within the first few *Gyr* of the history of the Universe, but which now lack gas due to stripping processes—for example, interactions and tidal effects (Destination 16)—that operate efficiently in the core of this dense cluster. This stripped gas has incidentally helped to produce the very hot intergalactic medium that we detect as *x*-ray radiation from the cluster (Figure 19.10, see Destination 16), similar to what we see in many clusters near the Earth in the Present-day Universe.

The other characteristic feature of MS 1024—the lack of large elliptical galaxies—is a bit harder to understand. But we can glean some nuggets of information about individual galaxies and galaxy clusters that might help. Recall that the large elliptical galaxy Maffei 1 (Destination 15) seems to contain a lot of mostly very old stars. These must have formed over a short time period compared to the stars in a typical spiral galaxy where star formation occurs over much more extended time periods. We also know the internal motions of elliptical and disk galaxies are fundamentally different. Most of the stars in spiral/disk galaxies follow orderly, nearly circular orbits confined to a thin disk (Destinations 9 and 14). But in elliptical galaxies (Destinations 13 and 15), stars move along more randomly distributed orbits. We've seen too that young and old clusters have different mixtures of galaxies. The relatively young cluster MS 1024 contains more spirals, fewer luminous elliptical galaxies, and more dwarf galaxies than do older clusters such as Virgo. Could it be spiral/disk and irregular galaxies progressively merge through random collisions to produce larger systems that look more like elliptical galaxies? It is hard to know for sure

26 A similar segregation exists in the Virgo Cluster, but we did not stress it when we visited (Destination 16). Nevertheless, the dwarf galaxy content near the center of Virgo is dominated by dwarf elliptical and dwarf spheroidal galaxies (see Destination 15), while dwarf irregular galaxies are much more common toward the outer regions of the cluster.

Fig 19.11

Examples of galaxy cannibalism. NGC 1316 (*Left*) and NGC 1275 (*Right*) both exhibit clear signs of molecular and ionized gas that likely came from the ISM of recently consumed gas-rich galaxies. Both of these unusual elliptical galaxies reside at the centers of nearby galaxy clusters. Both are very large and massive. The gas we see in these galaxies today likely represent the remnants of gas-rich galaxies that have merged over time to produce these large systems.

Images from STScI/NASA.

based on one visit to this lone cluster, but the circumstantial evidence we've summarized certainly supports this idea (see Figure 19.11).

Although we cannot wait around to see this process in action, it is possible to generate computer models that describe how galaxies interact and merge with one another. These simulations show that multiple random collisions of spirals can indeed lead to the formation of larger systems that can look awfully similar to giant elliptical galaxies (Figure 16.8). This is probably not the *only* way to form elliptical galaxies—small elliptical galaxies (Destinations 13 and 15), for example, probably did not form from collisions if only because the odds of dwarf-dwarf mergers is considerably lower. But galaxy collisions—galaxy cannibalism, really (Figure 19.11)—might be the best (only?) way to form *big* elliptical galaxies. And what better place to promote collisions between galaxies than in the high-density cores of clusters?

This makes sense in another way. Although we did not really stress it at the time, some of the Virgo cluster elliptical galaxies—especially the giant elliptical galaxy M 87 (Destination 16)—are also active galaxies. As galaxies collide, interactions of the gas and dust in the original galaxies cause this material to heat up and lose much of its angular momentum (Destination 6). Once this gas cools down—possibly through tidal interactions or from emission of x-ray radiation like what we see from MS 1054 and other clusters (Destination 16)—it has no choice but to fall toward the centers of the galaxies where, very likely (as we shall see at Destination 20), a supermassive black hole awaits (Destination 17).

Though we have only scratched the surface of what we could explore in MS 1054, our trip back in time has demonstrated one very important thing to us.

Galaxies evolve. Some of that evolution—for example, the way that young spiral galaxies appear bluer than older disk galaxies—is ultimately related to the fact that the stars and gas within galaxies change over time, a process known as *Passive Galaxy Evolution*. But some of the evolution—such as the transformation of spiral/disk and irregular galaxies into larger elliptical galaxies or the stripping of gas from disk galaxies represent *Active Galaxy Evolution*—are related to the environments of galaxies and how galaxies interact. It seems fitting—indeed, inevitable—that galaxies evolve in our evolving Universe. Yet it is also noteworthy that the galaxies in MS 1054, which exist in the Universe when it was half as old as it is today (in the 21st century), have already become recognizable entities. This begs the obvious question: What did these galaxies evolve from? We shall have to press even further back in time to address that fundamental question.

DESTINATION 19 QUESTIONS

19.1. Let's consider a Wide-Open Universe in which the expansion rate, $\varnothing R / \varnothing t$ is constant, where R represents the scale of the Universe as it varies with time, t. Let's say also that at some time the distance between two objects is D and the recessional velocity that each measures for the other is V. What is the Hubble Constant? The idea here is not to supply a value, but rather an expression using only symbols. Now, say that the Universe doubles in age. By what factor has it changed in scale? What is the new value of the Hubble Constant in terms of the value you derived originally?

19.2. We will discover later on that the Cosmic Microwave Background (CMB; see Destination 18) originates at a source that has a redshift, z, of about 1200. What is the lookback time to this era from the early history of the Universe? Use the relevant equations to determine this for both the Wide-Open and the Critical cases for a Hubble Constant of 72 $km/s/Mpc$. What fraction of the age of the Universe do these times represent for each case?

19.3. Many of the galaxies in MS1054-0321 emit Hα line radiation from the transition of electrons in Hydrogen atoms (see Destination 6). That line has a rest wavelength—the wavelength at which it was emitted—of 653 nm. Assuming that the cluster has a redshift of $z = 0.8$, at what wavelength will the line be observed in the spectra of the cluster's galaxies?

19.4. Calculate the lookback times for the following objects, assuming that we live in a Critical Universe. Express your answer as a fraction of the age of the Universe.

a) A galaxy cluster with a redshift of $z = 0.5$.

b) A quasar with a redshift of $z = 4.5$.

c) A galaxy where the Lyman-alpha emission line of Hydrogen (this is the transition between orbital levels $n = 2$ and $n = 1$) is observed at 450 nm. The rest wavelength of this line is 121 nm.

d) For each case (a)–(c) above, indicate if the Sun was already born when the light that we see from each of these three sources was emitted.

19.5. As noted in the text, the scale factor of the Universe, R, is defined as $R = \lambda_{em} / \lambda_{obs}$, where λ_{em} is the emitted wavelength of a spectral feature, and λ_{obs} is the observed wavelength of that feature. Combine this definition of R with the Doppler Equation to prove that $R = 1/(1 + z)$.

19.6. Refer to Table 19.1 for this question.

a) Which of the five models imply the largest present-day age for the Universe? What age is that?

b) What is the lookback time for an object observed at $z = 0.8$ as a fraction of the age of the Universe for each model?

c) What is the critical density in units of H atoms per cubic centimeter for each model?

19.7. Use Figures 19.7 and/or 19.8 to answer the following questions:

a) What is the redshift of a galaxy that has a lookback time of 5 *Gyr* in the Accelerating and Critical Universes? This corresponds to the era when the Sun was born.

b) If the Galactic Disk formed about 10 *Gyr* ago, at what redshift would we observe other galaxies that existed at this same era in the history of the Universe for the cases of a Wide-Open and Closed Universe?

c) Only two Universe models can accommodate a globular cluster such as M 92, which has an observed age of about 13 *Gyr*. Which models are these? Which of these two models can be reasonably ruled out on general grounds? Explain why.

19.8. The x-ray and optical images in Figure 19.10 are on approximately the same scale. Use the information in that figure to estimate the mass of the cluster, assuming that the x-ray gas is bound and that its temperature is about 17 million *K*. **HINT:** You will need to convert the gas temperature into a characteristic velocity (Destination 5), which you can then use to estimate the cluster mass with the Virial Theorem (Destinations 15 and 16). Compare this to the mass of the cluster that you'd derive from the 2000 *km/s* velocity dispersion obtained from radial velocities measured for its galaxies.

19.9. Choose one of the distance methods listed in Table 19.2 and find a specific example of the use of that method to get the distance to an astronomical object. Explain what that object is, its derived distance, and at least two of its intrinsic properties such as luminosity, mass, etc., that were derived once the distance was known.

19.10. Look up information on some cluster other than MS 1054-0321 that has a redshift, *z*, greater than 0.5. Report on the key properties of the cluster, including its richness, mass, size, luminosity, etc.

Destination 20

HIGH REDSHIFT GALAXIES: THE YOUNG AND RESTLESS UNIVERSE

20.1 THE GALAXIES OF THE EARLY UNIVERSE

Like the view from within a rich cluster, we are awestruck by the sight of so many galaxies here! Over one hundred are easily visible in this sky, many appearing larger that the Moon does from Earth (Destination 1). And as we peer more intently at this Alien Sky, we discern the ghostly glow of nearly 1000 more galaxies, most dimly apparent at the very limit of our eyesight. It is as if the gossamer band of the Milky Way has been chopped up and scattered across the sky to produce a blotchy pattern of ghostly puddles of light everywhere we look. We have not encountered a sky so filled with distinct objects—there are nearly 1/6th as many *galaxies* in this sky as there are *stars* in Earth's sky!—since leaving the vicinity of our Galaxy (Destinations 9, 10 and 12). Indeed, there are more galaxies in this sky than we could see from the center of the Virgo cluster of galaxies (Destination 16). Even more than we saw from within the young galaxy cluster MS 1054 (Destination 19).

But this is no cluster. We have traveled back in time to a location in the Early Universe that corresponds to a redshift of $z = 2$. The galaxies we see around us are so distant from the Present-Day Earth[1] that their light requires 76% of the age of the Universe—over 10 *Gyr*—to reach our home based on our adopted Accelerating Universe model with $H_0 = 72$ km/s/Mpc, $\Omega_{Matter,0} = 0.27$, and $\Omega_{\Lambda,0} = 0.73$ (Destination 19). And though this sky reminds us of the view from the core of a galaxy cluster, we are in fact at some random location in space in the Early Universe. This is the sky we would see anywhere in the Universe at $z = 2$. A sky of galaxies.

And *what* galaxies! The Hubble classification system (Destination 15) that has served us so well during our journeys into extragalactic space (Destinations 14–18)—and even at the distant galaxy cluster MS 1054

1 Even thinking about time travel can be quite confusing and our journeys to the early Universe are no exception. It helps to be a bit careful about how we refer to things that happen(ed) now (then), compared to how they happen(ed) then (now). See? It really *is* confusing! To be clear and consistent about all this, we'll explicitly refer to features of the *Present-Day Universe*—that is, from the era we refer to as the 21st century of the Holocene Epoch on Earth, some 13.5 *Gyr* after the Universe began its expansion (see Table 19.1 of Destination 19)—with the adjective 'Present-Day.' Likewise, we'll continue to use the term *Early Universe* (first introduced at Destination 18) to refer to features of the Universe here at $z = 2$. This wording should help us distinguish features of the old Universe—that is, the Present-Day Universe—from analogous or contrasting features that we identify in the younger Early Universe at our current destination at $z = 2$.

Fig 20.1

A sky of galaxies! Virtually every object in this image of a small region of the sky known as the Hubble Ultra-Deep Field is a galaxy (you can spot the four foreground stars of our Galaxy from their 'diffraction spikes,' artifacts produced by the Hubble Space Telescope with which this image was obtained). Many of the fainter galaxies and a few of the brighter ones in this image are seen as they would appear from our current destination: The Early Universe at $z = 2$, when the Universe was considerably younger and more compact than it is today. The galaxies we see here are very different from ones near Earth in the Present-Day Universe.

Hubble Ultra-Deep Field image from STScI/NASA.

(Destination 19)—utterly fails us here. Even a cursory inspection of the galaxies in this sky at $z = 2$ reveals that few of them conform to the Hubble scheme (Figure 20.1). Where are the classical spiral galaxies with their graceful arms, distinct central bulges, and exponential disks? Among the brighter galaxies in this sky we see a handful that resemble inclined disk systems, vaguely similar in form to our Galaxy or M 31 (Destinations 9 and 14). Some even appear to possess central bulge-like structures (Destination 10). But all of these galaxies lack well-defined spiral arms and are virtually all asymmetric in appearance, only vaguely reminiscent of the beautiful spiral galaxies we see in the Present-Day Universe (Destinations 14 and 15). Yet these galaxies contain far too much gas and dust to be classified as S0 galaxies (Destination 15). Perhaps stranger, disk-like galaxies seem to be comparatively rare here. From Earth, the five galaxies we can readily see with the naked eye are all classical spiral and irregular systems. The most luminous galaxies of the closest groups that lie beyond the Local Group in the Present-Day Universe are also generally spirals, the nearest exception being Maffei 1 (Destination 15). Even within Present-Day galaxy clusters—for example, the Virgo and Fornax clusters—spiral galaxies are common outside the dense cluster cores (Destination 16). But not here.

Instead, *most* of the galaxies in this sky are irregular systems, some quite unlike any non-interacting galaxies we've encountered before in the Present-Day Universe. The most prominent examples exhibit internal structures that are more complex and chaotic in appearance than we see in typical, irregular galaxies of the Present-Day Universe such as the Magellanic Clouds (Destination 12). Some of these are as luminous as our Galaxy or M 31, making them in effect 'giant' irregular galaxies. Many of these strange systems somewhat resemble classical interacting galaxies seen in the Present-Day Universe, often looking quite similar to examples of 'major mergers' involving galaxies of comparable size (Destination 16). But, if these unusual galaxies of the Early Universe are also merging systems—and we shall soon see that they probably are—then they are clearly much more chaotic. Rather than involving pairs of large galaxies as is typical of major mergers in the Present-Day Universe (see Destination 16), these 'giant irregular' galaxies of the Early Universe appear typically to involve mergers of numerous small galaxies—each of which are also generally unusual in appearance—that seem to be coalescing simultaneously (Figure 20.2).

Fig 20.2

Galaxies of the Early Universe are typically bizarre, often displaying unusual shapes rarely seen in galaxies of the Present-Day Universe. It is apparent that the Hubble sequence of galaxy classes (Destination 15) had not become established by $z = 2$ when the Universe was about 24% the age of the Present-Day Universe. We can also appreciate from these images that the galaxies of the Early Universe were typically much smaller, lower in luminosity and mass, and bluer on average than galaxies today. And there were many more of them. Yet, strange as these young galaxies appear, somehow over the next 2–3 *Gyr* they will evolve into the sorts of normal galaxies that we see inhabiting the Present-Day Universe (Destinations 14 through 19).

Hubble Ultra-Deep Field image from STScI/NASA.

Although spectacular to look at, it is important to appreciate that these 'giant irregular' galaxies—and the handful of disk-like systems—are comparatively rare here in the Early Universe. By far, the majority of the galaxies in this sky appear to be fairly normal dwarf irregular systems. Some appear unusually bright to us, but that simply reflects their proximity.[2] Most of these galaxies

2 This is similar to how the Magellanic Clouds are readily visible from Earth (Destination 12), but are too faint to be seen by the naked eye from the vicinity of M 31

appear extremely faint in this sky, making them hard to see with the naked eye. And yet such irregular galaxies are one of the few familiar things in this sky at $z = 2$, superficially similar to dwarf irregular galaxies seen in the Present-Day Universe. But there is something missing here. As we complete our inventory of galaxies in this sky, it dawns on us that there are no large elliptical galaxies—at least, none that we can readily see near us. Nor are there many dwarf elliptical or dwarf spheroidal galaxies. This seems odd given how extremely common they are in the Local Group (Destinations 13 and 14). Indeed, just about every galaxy here contains significant amounts of gas and dust, making it difficult to imagine classifying any of them as gas-poor elliptical or spheroidal systems regardless of their overall shapes or forms (Destination 15).

This is clearly a topsy-turvy place. Galaxies *are* abundant here. But many of the prominent galaxy types defined in Hubble's classification scheme (Destination 15) are rare (spirals), abnormal (irregulars), or absent (ellipticals, S0's). So varied are the shapes, sizes, luminosities, and structures of the galaxies at $z = 2$ that it is not obvious that galaxies *can be* successfully classified in this sky (Figure 20.2)! And yet, we must not forget: This is the *same* Universe we've been exploring at earlier—that is, later!—destinations.[3] The bizarre galaxies we see here in the Early Universe at $z = 2$ must somehow evolve into systems so regular that they can serve as the basis of Hubble's classification system in the Present-Day Universe (Destination 15). And they must do so relatively soon. In 'just' a few billion years—about when the Universe is twice its age here at $z = 2$—we already know that fairly normal-looking galaxies will exist and will be inhabiting galaxy clusters such as MS 1054 (Destination 19).

The structural potpourri that galaxies exhibit at $z = 2$ implies that their *Morphologies*—that is, their structural appearances—may be of limited use to examine their basic properties in detail. So we turn to other methods we've developed to measure the sizes, colors, luminosities, internal motions, and even chemical compositions of these young galaxies. What we discover is that the distributions of many of the observed properties of galaxies in the Early Universe differ significantly from those of galaxies that exist in the Present-Day Universe. For example, the young galaxies of the Early Universe are, on average, significantly bluer than galaxies of the Present-Day Universe. In one obvious respect, this isn't too surprising. After all, the Early Universe itself is only 25% as old as the Present-Day Universe, placing a limit on how old any stars here can be. Since younger stars tend to be bluer than older ones (Destinations 6, 8, and 12) it makes sense that younger galaxies would appear bluer, too. On top of this, many of the galaxies here are actively forming stars as we observe them. They exhibit

---

(Destination 14). However, the fact that there are so many nearby galaxies is significant, a point that we'll return to shortly.

3 The grammar of time travel really can be confusing! For a hilarious take on this, see Douglas Adams' *The Restaurant at the End of the Universe.*

strong emission arising in from star-forming regions such as the Orion Nebula (Destination 6), or colors so blue that their light must be dominated by stars that are only tens of millions years old (Destinations 6 and 8). Again, this makes sense; galaxies have just not had enough time to use up all their gas to form stars.

Perhaps what's oddest about these galaxies is how their sizes, luminosities, and masses compare with galaxies of the Present-Day Universe. Few of these young galaxies have diameters larger than about 10 *kpc*, considerably smaller than the optical diameters of our Galaxy or M 31 (30–40 *kpc*; see Table 14.1 of Destination 14). And because they are smaller, these galaxies are also generally dimmer than galaxies at home, on average about 5–10 times less luminous than typical galaxies in the Present-Day Universe.[4] We can also obtain spectra to estimate the masses of these galaxies, either from measurements of their rotational speeds (Destination 9) or from the random motions of their stars and gas (Destination 13). These observations tell us that galaxies here in the Early Universe are also typically 3–5 times less massive than galaxies of similar luminosity that exist in the Present-Day Universe.

The fact that galaxies of the Early Universe are smaller does not mean there's less matter associated with galaxies here compared to the amount of mass associated with galaxies in the Present-Day Universe. The reason for this is simple: There are *many more* galaxies here at $z = 2$. On average, the density of galaxies in Early Universe is about 10–20 times higher than in the Present-Day Universe[5], and their distribution in space is a lot smoother. In particular, there are no galaxy clusters here, no cosmic walls, no voids (Destination 18). Wherever we choose to travel within the Early Universe at $z = 2$, our sky would generally appear about the same. Sometimes we'd see about twice as many galaxies as we see in the sky from our present location, sometimes about half as many. This is in striking contrast to the Present-Day Universe where we can travel from the dense cores of galaxy clusters—where the sky is filled with galaxies (Destination 16)—to the profound emptiness of voids where it is rare to see any galaxies in the sky at all (Destination 18). In the process, we traverse regions that exhibit a range of nearly a factor of 10000 in galaxy number density. Here at $z = 2$ the range in galaxy density is considerably more modest, about 10:1. Not only do individual galaxies evolve, but so does their distribution in space.

---

4 This is true despite the fact that these galaxies generally contain many young—and therefore comparatively luminous—stars.

5 The expansion of the Universe means that any volume containing a fixed region of space will also expand as the Universe evolves. To be meaningful, comparisons of the densities of matter or galaxies at some redshift and in the Present-Day Universe must account for this change. Astronomers refer to a fixed region of space that expands with the Universe as a *Co-Moving Volume.* It is possible to calculate how such a volume evolves once the parameters for a model of the Universe are specified (Destination 19). All the comparisons we carry out—say, of the number density of galaxies at $z = 2$ and in the Present-Day Universe—always refer to co-moving volumes.

20.2 THE GOLDEN ERA OF GALAXY FORMATION

A combination of factors lie behind the unusual characteristics of the galaxies we see here in the Early Universe. But the fundamental factor is time. And it is why we have traveled far back to a specific *time* to reach this otherwise random place. That is, '*where* we are' is far less important than '*when* we are.' It is also why we have begun to refer to our location in terms of a redshift—which is equivalent to a lookback time (Destination 19)—rather than with a more standard spatial coordinate like distance. But the fact remains that we have traveled in *both* time and space to get to this destination in order to position ourselves near galaxies that have a redshift of $z = 2$ as seen from Present-Day Earth. How far have we come?

For the Accelerating Model of the Universe—characterized by $H_0 = 72$ $km/s/Mpc$, $\Omega_{Matter,0} = 0.27$, and $\Omega_{\Lambda,0} = 0.73$ (Destination 19; Table 19.1)—the Universe is 3.3 *Gyr* old here at $z = 2$, about 24% of the age of the Present-Day Universe (see Figure 19.7). Consequently, we've traveled back in time about 10.2 *Gyr*, the equivalent of a light-travel distance, D_L, of 3.1 *Gpc* (see Figure 19.5 and Destination 19).[6] This vast distance in time and space has one unsettling, though somewhat surreal, consequence. At all previous destinations we have visited it was always possible—at least in principle—to spot the Earth, the Sun, or our Galaxy. That's no longer true from here. From our current location far from home in both space and time, we *cannot* see our Galaxy under any circumstances. Not because it is too faint, but, rather, because the Universe here at $z = 2$ is too young for EM Radiation from our Galaxy to have reached us at this location and time. The technical way of saying this is that our Galaxy has not yet crossed over the *Horizon* that defines the totality of what we can see from this location at $z = 2$. And it won't ... for a few *billion* years! Home is no longer merely far away. From this location and at this time in the Early Universe, our Galaxy has not yet entered into the Observable Universe.[7]

6 This makes the added assumption that we have traveled to a region near specific galaxies that we can see at $z = 2$ from Earth in the Present-Day Universe. In fact, if we had just stayed near the Galaxy as we traveled back in time, the sky we would see around us would not be that different from what we describe. Perhaps the weirdest aspect of such a time-only journey is that the stuff we'd see around us would be the matter from which our own Galaxy formed! And yet our Galaxy, at least in any recognizable form, would be nowhere to be seen.

7 The concept of a horizon, or limit, to the Observable Universe is very important. We will return to it at Destination 22 when we consider the implications of the incredible uniformity of the Cosmic Microwave Background (Destination 18).

Even though our Galaxy may not yet be part of our Universe here at $z = 2$, we can still appreciate that this era—when the Universe was only 3.3 *Gyr* old—must have been a special time in the development of galaxies such as ours. As we saw at Destinations 8, 10, and 11, analyses of the oldest stars in our Galaxy show that they have ages ranging from about 10 to 13 *Gyr*. That's similar to how far back in time we have had to travel to reach our current location. Our Galaxy is certainly not unusual in this respect; it appears that all galaxies contain at least some stars this old. Many galaxies consist almost entirely of ancient stars older than 10 *Gyr* as measured in the Present-Day Universe (for example, Maffei 1; see Destination 15). This tells us that by traveling back in time to $z = 2$, we have reached an era when all galaxies were actively forming stars for the first time or had just recently finished producing most of their stars during a comparatively short, intense period of star formation. Here at $z = 2$ we appear to find ourselves near the peak of the *Era of Galaxy Formation*, a broad span of time in the Early Universe during which galaxies were, on average, actively forming their first generations of stars (see Figure 20.3).

Does this interpretation explain the appearance of the sky here in the Early Universe? Does it account for the striking properties of the galaxies that inhabit this place? To get a start on answering these questions, recall from Destination 19 that the scale of the Universe, R, is changing as the Universe expands such that $R = 1/(1 + z)$. At our present location at $z = 2$, this means that the average

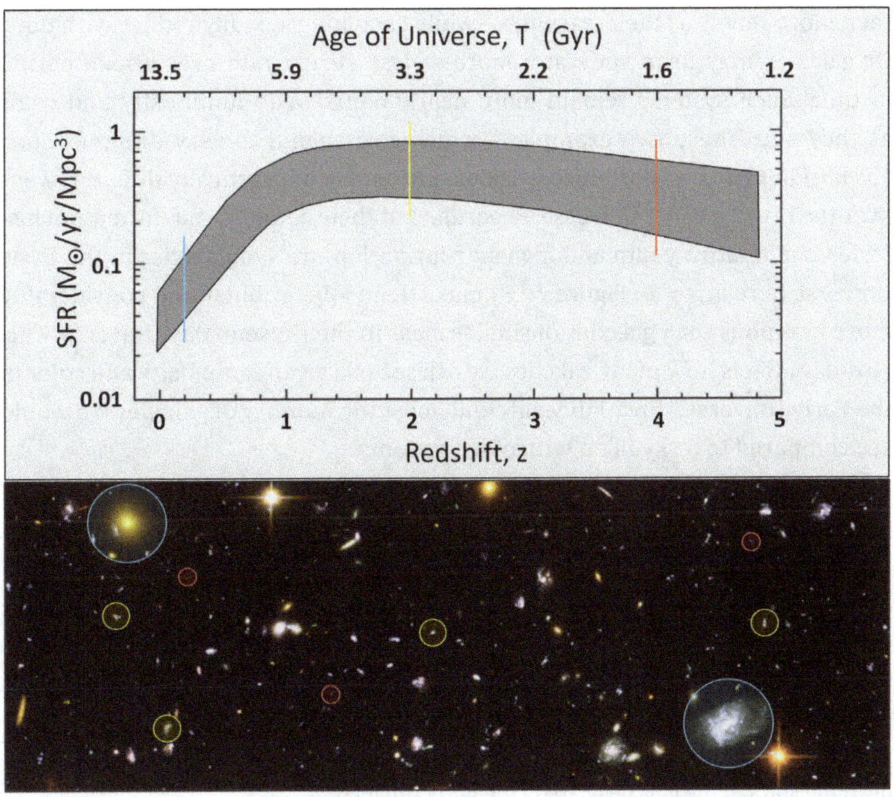

Fig 20.3

(*Upper*) The darker gray band shows the range in the average star-formation rate, in 'co-moving' (see footnote 5 of this destination) units of $M_{\odot}/yr/Mpc^3$, for galaxies over a range of redshift from $z = 0$ to $z = 5$. The distribution of the star-formation rate peaks broadly between redshifts of about 1–3. Notice too how the average star-formation rate of the Universe has steadily declined over the past 60% of the age of the Universe since $z = 1$. (Lower) A small portion of the Hubble Ultra-Deep Field (Figure 20.1) illustrates what galaxies of comparatively low redshift ($z \approx 0.2$ in blue), moderate redshift ($z \approx 2$ in yellow and corresponding to the current destination), and high redshift ($z \approx 4$ in red and corresponding to galaxies we shall visit at Destination 21) look like as seen from Earth. These redshifts are also noted on the upper plot with the same color coding. Notice that while some low-redshift galaxies are still forming stars (the spiral at lower right in the large blue circle), many other low-redshift galaxies are 'red and dead' (such as the elliptical galaxy at upper left in the blue circle). In contrast, just about all galaxies at redshifts higher than around $z = 1.5$ appear to be supporting some star formation, sometimes at extremely high rates. The presence of galaxies that have effectively ceased star formation in the

separation between galaxies is only $1/(1 + 2) = 1/3$ what it is in the Present-Day Universe.[8] But as the linear scale, R, of the Universe becomes smaller as z increases, then space must also be getting considerably more crowded. That's because the volume of the Universe varies as R^3, or as $1/(1 + z)^3$. An immediate consequence of this is that all galaxies within a given volume of the Present-Day Universe would have been packed into a volume that is $1/(1 + z)^3 = 1/27$ as large at $z = 2$ even if the galaxies do not change at all between the two epochs. To put it in perspective, this effect alone would be equivalent to seeing 27 times more easily visible galaxies in Earth's sky—over 100 in total—rather than just the five we do see (our Galaxy, M 31, M 33, and the Magellanic Clouds; Destination 13). Simply due to the expansion of the Universe, galaxies at $z = 2$ will appear closer together compared to the Present-Day Universe even if the galaxies or galaxy numbers do not change. This represents a big reason we see so many galaxies in this Alien Sky.

Present-Day Universe is a large part of the reason the overall rate (as seen in the upper plot) is lower today than in the past.

Hubble Ultra-Deep Field image from STScI/NASA.

But we've seen that individual galaxies do change, if only because the stars that comprise them evolve. Consequently, galaxies that contain recently formed stars tend to be bluer and more luminous than older galaxies of the same mass simply because young stars are generally bluer and more luminous than older stars. These stellar evolutionary effects can drastically alter how a galaxy's appearance changes over time. At one extreme, systems that experience a single 'burst' of star formation start off as very blue, intensely luminous systems (for their mass), which then fade in luminosity—quite quickly at first, then more slowly as their stars age—while becoming steadily redder with time. Or galaxies may form stars at a more-or-less steady rate over a long period of time. Such systems remain more nearly constant in luminosity and color as they age. These two examples—which correspond to very different *Star-Formation Histories*—produce galaxies that evolve in drastically different ways over the history of the Universe. Regardless of their detailed star-formation histories, the relative youth and high star-formation rates of galaxies in the Early Universe here at $z = 2$ (Figure 20.3) make them appear bluer and considerably more luminous than galaxies of similar mass in the Present-Day Universe. This strongly affects how many galaxies we can see of a given luminosity and color in the Early Universe.[9] Invariably, this will boost the number of galaxies we would see compared to a sky filled with older systems.

8 This is strictly true only for relatively nearby galaxies. If we were to consider galaxies that are comparatively far apart, their separations would vary with redshift at a rate that differs from $1/(1 + z)$. The reason is that, just like in the Present-Day Universe, the Universe at $z = 2$ is expanding. Thus, distant galaxies will exhibit a redshift from galaxies in the Early Universe just as they do in the Universe today. We'll assume that most of the galaxies visible in this sky here at $z = 2$ are comparatively nearby and so we do not have to worry about how the volume of space changes with redshift as measured at $z = 2$.

9 We have already noted that galaxies here at $z = 2$ are 3–5 times more luminous than Present-Day galaxies of similar mass, a difference directly related to the preponderance of young stars in the galaxies here in the Early Universe.

Finally, recall that the average number of galaxies is considerably higher in the Early Universe than it is in the Present-Day Universe. Not merely due to the evolution of space in the expanding Universe, but also because there really are more individual galaxies in a given volume. On the other hand, these are generally smaller, less luminous, and less massive galaxies compared to the ones we see in the Present-Day Universe, so there are clearly two trends at work here to affect how many galaxies we would see here at $z = 2$. The increased number of galaxies will obviously boost the total number we can see, while the fact that each galaxy will have a lower average luminosity works in the opposite direction, reducing the number we can see.

Another factor affecting how galaxies appear and evolve here at $z = 2$ has to do with their distribution in space. We've already noted that the Early Universe is much more homogeneous (smoother) than the Present-Day Universe. But it is not perfectly smooth. Some regions are slightly *overdense* than average.[10] When galaxies that happen to be located in such regions become gravitationally bound, they may begin to interact—perhaps merge—with their neighbors. Due to the nature of gravity, this process always begins in the smallest, densest regions (Destination 23), implying that small galaxies generally form first.[11] Once formed, these galaxies can then merge with other small galaxies to form larger systems that can, in turn, merge to form even larger galaxies and so on,

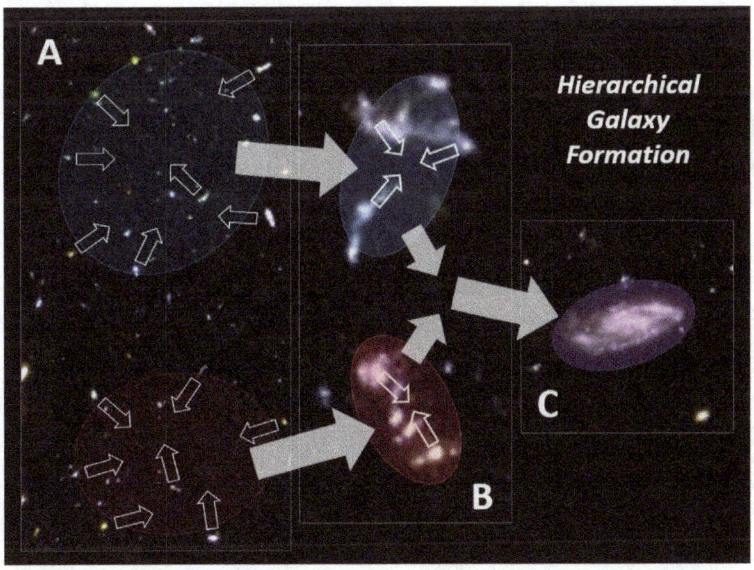

Fig 20.4

A schematic picture illustrating the basics of Hierarchical Galaxy Formation. Small systems form first (left, panel A), then, through successive mergers (middle, panel B), build up into progressively larger galaxies (right, panel C). The existence of interacting galaxies in the Local Universe (Destination 16) and dwarf galaxies accreting onto larger galaxies such as ours (Destination 13) tells us that this basic process is still going on in the Present-Day Universe, though at a much lower rate than in the early Universe at redshifts of $z = 1$ and higher.

10 An example of this is the Local Group where the mutual gravitational attraction of M 31 and our Galaxy has caused these two systems to approach one another (Destination 14) despite the opposing effect of Universal Expansion.

11 How small? We'll return to that interesting question when we probe conditions when this process of galaxy formation was truly commencing (Destination 22). But, to get to the punch line, the smallest galaxies that could form in the Universe had masses similar to those of the smallest dwarf systems we see in the Present-Day Universe (Destinations 13 and 14).

a process known as *Hierarchical Galaxy Formation* (Figure 20.4). We've already seen that there is strong circumstantial evidence from the morphologies of galaxies in the Early Universe that small systems are indeed merging into larger ones (Figure 20.2).[12] And, of course, as smaller galaxies merge into larger ones, the total number of galaxies will decrease while the average sizes, luminosities, and masses of galaxies will increase. This process drives the evolution of galaxies like the ones we see in the Early Universe to become more like galaxies we see in the Present-Day Universe.

Taken together, these effects—a smaller, younger Universe, more active star formation, more numerous but lower luminosity galaxies, variations in the space density of galaxies—help to explain the number, colors, and sizes of galaxies we see here in the Early Universe. They imply that this era in the Early Universe corresponds to a particularly active period of star formation within just about all galaxies. It is the era when most galaxies made most of their stars. An era that can produce the beautiful sky of galaxies we see around us.

Before we press on, there's one final thing to note about the galaxies here at $z = 2$. The spectra we obtained to study the internal motions of these galaxies—and that we used initially to determine the galaxies' masses—also provide information regarding their chemical compositions (Destinations 8 and 11). What we find is that these galaxies of the Early Universe are generally quite deficient in heavy elements compared to the Sun. In one respect, this makes sense. During the next few *Gyr*, stellar evolution will produce heavy elements in the cores of both low- and high-mass stars that form, live, and die in galaxies. Some of this gas will start to form the Coronal ISM of galaxies. Some will be recycled into the ISM of individual galaxies, from which subsequent generations of stars will arise (Destinations 11 and 12). And throughout the process, the chemical abundances of stars and the galaxies in which they reside will steadily rise over time. If we run this argument in reverse, it implies that as we travel to even higher redshift—that is, back in time to an era *before* galaxies or stars had formed (Destinations 22–25)—the abundance of heavy elements must steadily decrease. This raises an interesting question: What was the chemical composition of the Universe like before the first stars existed? If heavy elements were present at that time, where did they come from if not from stars? The answers to these questions are to be found further back in time, in the first few *minutes* in the history of our Universe. We still have a long way to go.

---

12 Not to mention the presence of streams and disrupted dwarfs around our Galaxy (Destination 13), clear relics of past and ongoing mergers of dwarfs.

DESTINATION 20 QUESTIONS

20.1. Let's pretend for the moment that M 31 and the Milky Way do not attract each other gravitationally. If the distance between the two is increasing due to the expansion of the Universe, how far apart were the two galaxies at a time corresponding to $z = 3$ compared to today? You may want to look back to Destination14 to get the Present-Day distance between the Milky Way and M 31.

20.2. We can define something called the free-fall time as $\tau_{ff} = 1/\sqrt{G\rho}$. This is an approximate measure of how long it would take something to collapse if the only force involved were gravity. Imagine a section of the Universe today that has a mean density two times the Present-Day critical density (see Destination 19). How long will it take for such a region to collapse? How long will it take a region that is 100 times denser to collapse? What do you think that these numbers tell you about galaxy formation?

20.3. List two things in everyday life that form from a hierarchical process similar to the one described in this destination to explain galaxy formation. Note in what way their formation is indeed hierarchical. List two things that do *not* form hierarchically. Describe briefly in what manner their formation is *not* hierarchical. The things you list could be made directly in Nature or they may be manmade.

20.4. How many galaxies similar to the Sculptor Dwarf Spheroidal Galaxy (Destination 13) would be required to build up a galaxy like M 31? Calculate this number based on the relative luminosities of the galaxies and on their relative masses; you may want to use the information in Tables 14.1 and 15.1. Does the number of Sculptors that are required to account for the luminosity of M 31 agree with number required to account for the larger galaxy's mass? If not, why are they different?

20.5. This question explores how the current rate of star formation locally compares with the rate during the peak of the era of galaxy formation in the Early Universe.

a) Assume that the Local Group is a spherical volume with a radius of 2 *Mpc*. What is the volume of the Local Group in Mpc^3?

b) If we assume that the Galaxy has produced approximately 5×10^9 $M_\odot$ of stars over its lifetime (see Destinations 9 and 10), what is the total mass of stars produced in the Local Group over the history of the Universe? You may assume that M 31 has produced twice as many stars as the Galaxy, and that all the other galaxies of the Local Group combined have produced about 20% as many stars as the Galaxy over the lifetime of the Universe.

c) If the Universe is assumed to be 13.5 *Gyr* old, what is the average star-formation rate of the Local Group over its lifetime in the same units shown in Figure 20.3?

d) How does the star-formation rate in the Local Group compare with the peak average star-formation rate in the Universe as read from Figure 20.3? You should find a significant difference between the star-formation rate of the Local Group and the average rate in the Early Universe. Provide a simple explanation for this difference. **HINT:** Is the Local Group a 'typical' volume of the Universe? You may want to study Table 18.1.

20.6. Let's assume we are comparing two galaxies that we see near us in the Present-Day Universe. One is a spiral galaxy that has formed a total of $10^9\ M_\odot$ in stars steadily over its 12 *Gyr* lifetime. The other is an elliptical galaxy that produced the same total amount of stars but did so in the first 1 *Gyr* of its life.

a) What are the average star-formation rates in the two galaxies, assuming both were formed about 12 *Gyr* ago?

b) What are the peak star-formation rates in the two galaxies in units of $M_\odot/yr$?

c) Describe the differences in the stellar content of the two galaxies based on what you know about their star-formation histories.

d) Which galaxy is likely more luminous today? Which galaxy was more luminous 11.5 *Gyr* ago? Explain your reasoning for your answers to both questions.

20.7. The text makes the odd claim that our Galaxy would not be visible from galaxies that we see today from Earth at $z = 2$. This problem explores the reasoning for this claim.

a) Assume that the light-travel distance, D_L between these remote galaxies and our Galaxy varies in proportion to the scale of the Universe, R. This means that the light-travel *distance* from the $z = 2$ galaxies we have visited and our(younger) Galaxy is $1/(1+z) = 1/3$ as large as the light-travel distance from our Galaxy to those remote systems as seen in the Present-Day Universe (3.1 *Gpc* as noted in the text). What is the light-travel *distance* between these remote galaxies and our Galaxy at the epoch corresponding to $z = 2$?

b) What is the light-travel *time* between those remote galaxies and our Galaxy as calculated at the $z = 2$ galaxies?

c) Explain why these results confirm the conclusion that light from our Galaxy has not reached these remote galaxies when we consider what is in their Observable Universe at $z = 2$.

20.8. Remote galaxies from the Early Universe tend to make a big splash in the news when they are first found. Look up the current 'most distant galaxy' and write up a short description of its properties. Using the plots in Destination 19 (Figures 19.7 and 19.8), what is the lookback time and Hubble distance to this galaxy? Be sure to specify what cosmological model you assumed to calculate these quantities.

Destination 21

QUASARS AT THE DAWN OF GALAXY FORMATION

21.1 YOUNG GALAXIES AND 'FALSE STARS'

Looking around us, we are startled to see ... Stars! Not very bright ones. And not many. But there is no question about it: Distinctly visible in this sky are about two dozen stars—or objects that certainly look like stars to our eyes. Most of these shine dimly near the limit of our vision. But a handful—perhaps 3–5—could charitably be described as 'bright'. Many stars of comparable brightness are (or will be!) easily visible in Earth's sky on clear, dark nights. But so are hundreds of others that appear much brighter than any of the 'stars' we see here. The view around us is not overwhelming, but remarkable nonetheless. We've not been able to see easily *any* star-like objects in the skies we have visited ever since leaving the vicinity of M 31 (Destination 14). These tiny points of light—feeble as they are—remind us of the star-filled sky of Earth. A sky that is nearly inconceivably distant from us in space and time from our current vantage point in the Early Universe.

But there's more to see here than just these faint 'stars'. Dim galaxies similar to the ones we saw at $z = 2$ (Destination 20) are also visible here, though in significantly smaller numbers. In all, perhaps 200 galaxies shine with a pale ghost-like light in this sky, many hovering near the limit of our vision. At most, about 30 of these galaxies are comparatively 'easy' to see with our unaided eyes. Some of these brighter systems appear noticeably 'clumpy' in appearance, as if we can *almost* make out individual stars. The overall effect is a sparser-looking sky compared to the galaxy-rich skies of our last two Destinations (MS 1054 at Destination 19 and the $z = 2$ Universe at Destination 20), but with the notable addition of those enigmatic 'stars' shining faintly around us.

We have arrived at a time in the Universe corresponding to a redshift of $z = 4$ as measured from Present-Day Earth. Our preferred model for an Accelerating Universe—the one characterized by $H_0 = 72\ km/s/Mpc$, $\Omega_{Matter,0} = 0.27$, and $\Omega_{\Lambda,0} = 0.73$, (Destinations 19 and 20)—tells us that we've traveled 88% of the way back to the time when the Universe began its expansion. From the Present-Day Universe, that corresponds to a lookback time of nearly 11.9 *Gyr*. But this is not only a younger Universe. It is also considerably more compact. The scale of the Universe here at $z = 4$ is $R = 1/(1 + z) = 1/5$—or just 20%—the scale of the Present-Day Universe. A fixed region of space defined in the Present-Day Universe has a volume that is only $1/(1 + z)^3 = 1/125$ as large here in the $z = 4$ Universe (see Figure 21.1).

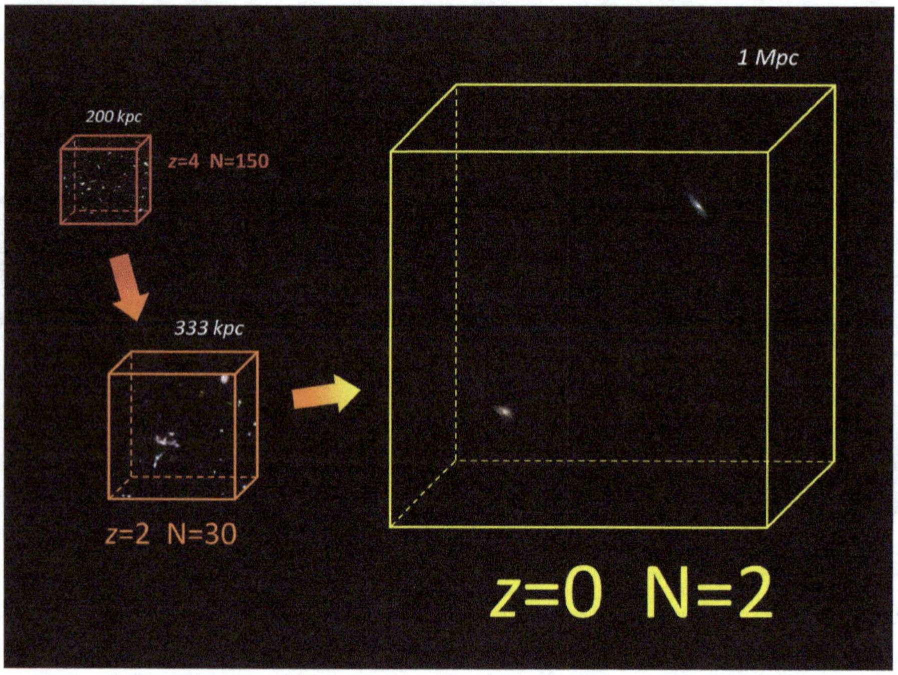

Fig. 21.1

The evolution of a fixed, or co-moving (see footnote 5 at Destination 20), volume over the expansion history of the Universe. Since the scale of the Universe varies as $R = 1/(1 + z)$, the volume evolves as $1/(1 + z)^3$. The cubes in this figure illustrate the evolution of a cubic volume that has 1 *Mpc* sides in the Present-Day Universe. As shown, at $z = 2$ the sides of this volume have shrunk by a factor of three, and at $z = 4$ by a factor of five. The corresponding volumes are $3^3 = 9$ times smaller for $z = 2$, and $5^3 = 125$ times smaller for $z = 4$. In addition to these changes in the scale of the Universe, the galaxies within the volume are also evolving. At the present time—corresponding to a redshift of $z = 0$—a typical one Mpc^3 volume of the Universe such as the Local Group contains typically one or two large galaxies comparable to the Milky Way or M 31 (Destination 14). At $z = 2$, the same (but smaller) volume contains 15 times as many galaxies, while at $z = 2$ more like 75 times as many galaxies. Consequently, the total number of galaxies is decreasing as the Universe ages, while the individual galaxies are on average steadily becoming bigger and more luminous.

As we consider the implications of all this, it starts to seem surprising that we see such a sparse sky here. After all, if objects in the Universe are considerably closer together on average here in the compressed space of the Early Universe, they should appear correspondingly brighter (due to the Inverse-Square Law for brightness, Destination 4). And indeed they do. But this simple expectation is complicated by the fact that the objects in the Universe—the galaxies and the enigmatic 'stars' we see here at $z = 4$—are evolving. As far as the galaxies are concerned, we've already discovered the hierarchical nature of their evolution (Destination 20), meaning there are likely more galaxies in a given volume here at $z = 4$ than in the Present-Day Universe, and even more than at $z = 2$. This boosts the number-density of galaxies—that is, how many galaxies are present in a given volume—in a manner that amplifies the change in the volume of the Universe as it expands. But the galaxies here at $z = 4$, though about five times more abundant in number than galaxies at $z = 2$, are typically smaller and have lower average star-formation rates (see Figure 20.4). The net effect is that we actually detect fewer galaxies in this sky than in the sky at $z = 2$. But the galaxies we do see here tend to be younger and bluer than systems at $z = 2$ (Destination 20). *Much* younger and bluer than galaxies of the Present-Day Universe (Destinations 14–16). Also, the galaxies here at $z = 4$ are generally closer together—after all, the Universe is more compact (Figure 21.1)—which is why they appear clumpier in appearance. The galaxies are so close together here that we can nearly make out individual luminous stars and star-forming regions even with the naked eye[1].

1 We certainly could if we had even a modest pair of binoculars.

Galaxies here at $z = 4$ are also typically much more gas-rich than the galaxies we know will exist later in the Present-Day Universe (Destinations 14–16). On the surface, this isn't hard to understand. After all, we already know that stars form out of gas in the ISM (Destination 6) and these galaxies are forming a lot of stars. But the implication here at $z = 4$ is important. The galaxies we see around us in the Early Universe are so young that they must contain some of the first stars[2] to have formed. Ever. In the Universe. That, in turn, means that at even earlier times—before *any* stars had formed—galaxies must have been entirely gaseous structures (Destination 22), not the mixture of gas *and* stars we are used to seeing here at $z = 4$ or at any later epoch in the history of the Universe.[3]

It's also important to emphasize a point we noted at our last destination. The view here at $z = 4$ depends primarily on *when* we are in the Universe rather than *where* we are. As we travel back in time, the distribution of matter—both dark and visible—becomes steadily more homogeneous[4], making it increasingly unnecessary to specify our *location* in space. That is, all locations in the Early Universe start to all look essentially alike. This has become so pronounced that we can imagine that for this journey, we have traveled *only* through time to reach this destination. That is, we have ended up in a region of space surrounded by the material that will eventually become the Local Group some 12 *Gyr* in the future.[5] In this sense, we might be 'near' our Galaxy ... though

2 The stars referred to here are not the 'stars' we can see with our naked eyes in this sky. On average, we are still too distant from individual galaxies here at $z = 4$ to be able to see their constituent stars, apart from the occasional supernova (Destination 12). But we can reasonably expect to sometimes be able to make out individual star-forming *regions* within galaxies. These are often so compact—perhaps related to the formation of globular clusters (Destination 11)—that they could appear as faint, star-like objects embedded in the pale light of individual galaxies. By contrast, the enigmatic 'stars' we can see in this sky tend to be loners, typically unassociated with the light of any readily visible galaxy.

3 Although people have looked hard for them, there are no known cases in the Present-Day Universe of galaxies that contain absolutely no stars. There aren't even any known cases of galaxies that lack stars older than about 10 *Gyr*. Perhaps the closest counter-examples are 'tidal dwarfs,' which are formed during galaxy interactions (see Destination 16). It is unclear, however, if these systems typically remain gravitationally bound as real galaxies are. It seems that forming stars—and doing so very early in the history of the Universe—represents a fundamental requirement for an object to become what we would call a galaxy in the Present-Day Universe.

4 We can already appreciate that this has to happen: Recall the incredible uniformity of the Cosmic Microwave Background (CMB; Destination 18) compared to the pronounced clumpiness of the Present-Day Universe. To the extent that the CMB reflects the way matter was distributed in the Early Universe (which we'll explore in detail at Destination 23), this tells us that the Universe must become smoother—with matter more uniformly distributed through space—as we travel back in time. We've begun to see this directly on our voyages. Already at $z = 2$ (Destination 20), we see that the Early Universe lacks the kinds of large-scale structures such as cosmic walls, voids, galaxy clusters that characterize the distribution of galaxies in the Present-Day Universe (Destination 18).

5 It's important to be clear about this, even at the risk of reiterating a point we've already made at Destination 20. Had we traveled to the vicinity of a specific galaxy that

we'd never guess it. The gas and stars that will eventually coalesce to form our Galaxy, M 31, and the Magellanic Clouds are still too widely dispersed into smaller individual galaxies here at $z = 4$ to be recognizable as the building blocks of those future systems.[6] But we know that, eventually, some of the dwarf-like galaxies here at $z = 4$ will ultimately begin to merge, starting the hierarchical process that will result in the formation of the larger galaxies that inhabit the Local Group in the Present-Day Universe. Including our Galaxy. Here, at the dawn of galaxy formation in the Universe, these dwarf systems have only just begun to collapse toward one another as their mutual gravity locally overcomes the expansion of the Universe. As we ponder this Alien Sky, it is strange to contemplate that the matter that will eventually form our Sun, our Earth—even us—is here, all around us. Even stranger that we cannot recognize it.

21.2 QUASARS: EXTREME ACTIVE GALACTIC NUCLEI

But what about the enigmatic 'stars' we see in the sky here at $z = 4$? As we study these objects in more detail, we soon come to suspect they are not stars at all. At least not like any we have encountered before. The first hint comes from studying their variability. Though we know of many sorts of variable stars—pulsating stars such as Cepheids (Destination 14), eclipsing binaries (Destination 8), or supernovae (Destination 12) to name a few—the stellar-like objects in this sky don't behave like any known types of true variable stars we have encountered before. Many brighten or fade for months or years at a time, only to reverse the trend or remain constant in brightness for equally protracted periods. Others vary rapidly and violently in brightness—sometimes by factors of ten or more—on timescales as short as weeks or days.[7] There's no

---

exhibits a redshift of $z = 4$ as seen from Present-Day Earth, we would indeed have had to travel a long way in both time *and* in distance to reach it as it existed when its radiation reached us. But it is no longer *necessary* to make large moves in position to reach this destination—or any of the ones we have yet to visit—because the Universe is becoming steadily more homogeneous as we travel back in time. Our destinations are no longer *locations*, but rather *time periods* in the history of the Universe.

6 Just as a pile of bricks, even if they are located at a building site, would not alone give us a clear idea of the form of the final structure that will be built with the bricks.

7 Since we are looking up in the sky with our eyes, we first notice this variability in the form of optical radiation. But the variability of these 'stars' isn't restricted to visible light. They change brightness in all portions of their spectra, from radio waves to x-rays, and even γ-rays (see Destination 17).

obvious periodicity to these variations, certainly nothing similar to the strict periodicity we see in the light curves of Cepheids or binaries. And rarely do these 'stars' rapidly rise in brightness only to steadily fade to invisibility as we observe in supernovae.

The second hint of a non-stellar interpretation for these objects comes from their unusually blue colors. Since stars are essentially thermal bodies, they emit spectra similar in form to what Planck predicted for such sources (Destinations 6 and 18). But there's a limit to how blue a thermal source can appear, and for temperatures higher than about 12000 *K*, their optical colors remain unchanged regardless of the temperature.[8] This behavior—which astronomers refer to as *Saturation*—effectively rules out ultra-blue colors for normal stars even if their temperatures were arbitrarily high. The problem is that many of the strange 'stars' in this sky at $z = 4$ are bluer than the thermal saturation limit; they simply can't be single-temperature thermal sources. They cannot be stars, variable or otherwise.

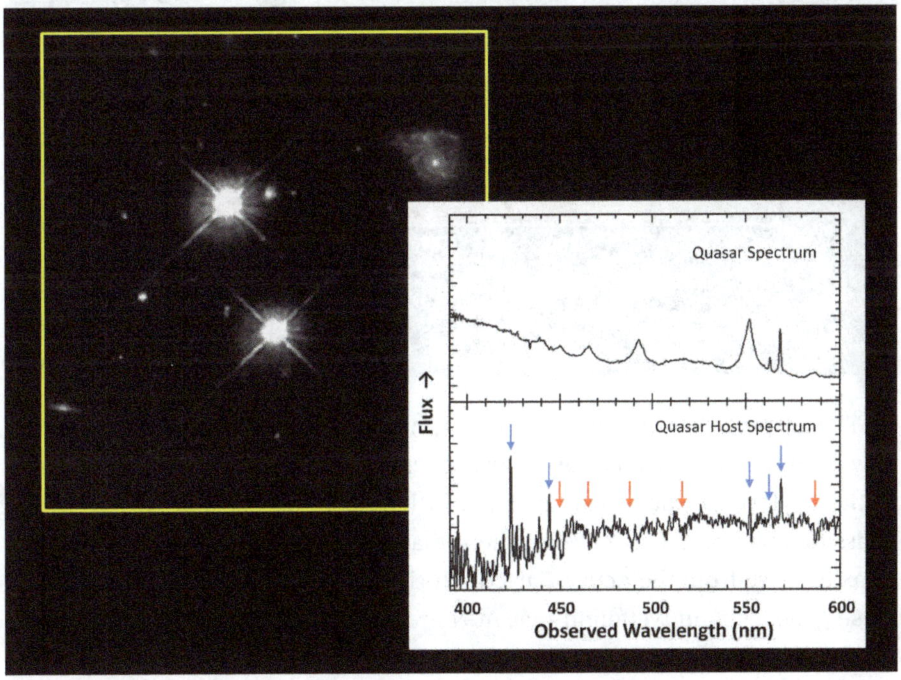

Fig. 21.2

(*Upper left*) An image of a star—which is a few hundred *pc* away—and a quasar that is so distant the light we see was produced when the Universe was only 20% its present age. To appear so similar yet be so far away, the quasar's luminosity must be over one trillion times larger than that of the star. The star is the upper of the two. (*Lower right*) A spectrum of a quasar (top) and its underlying host galaxy. The quasar's spectrum shows mostly broad emission lines and an underlying blue continuous spectrum. The galaxy—essentially composed of stars and star-forming regions—shows a redder continuous spectrum punctuated by absorption lines (red arrows) and narrow emission lines from H II star-forming regions (blue arrows; Destination 6).

Photo from STScI/NASA.

But it's the spectra of these objects that clinch the case (Figure 21.2). Many of these point-like sources typically exhibit strong, often very broad emission lines, a blue continuum increasing in brightness toward the ultraviolet, and an almost complete lack of stellar absorption lines. They are certainly unlike the

8 This occurs because the peak of the EM Radiation emitted by a thermal body continues to move to shorter wavelengths as its temperature increases (Wien's Law; Destination 6). Due to the detailed nature of the spectrum of a thermal source, in the optical part of the EM spectrum, there is a maximally blue color we can observe for very hot sources.

spectra of any types of stars we've encountered before.[9] When these strange objects were first identified in the early 1960's, these unusual properties and remarkably star-like appearances (Figure 21.2) led to their being dubbed **Quas***i-Stell***ar** *Objects*, or *Quasars* for short.[10] The name reflects the fact that, at the time of their discovery, people understood that quasars could not possibly be normal stars. But what they actually were remained a mystery.

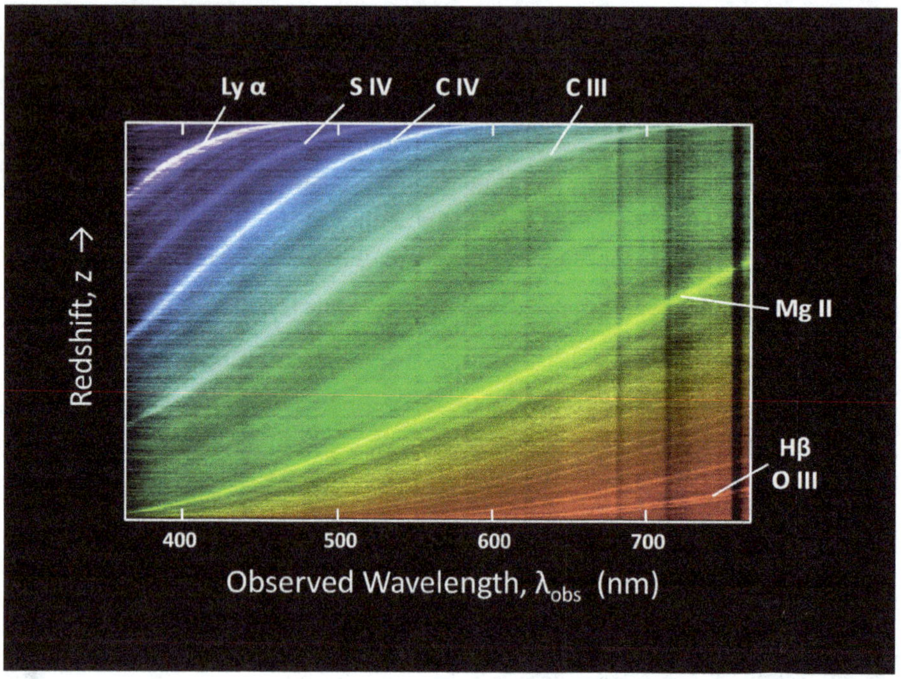

Fig. 21.3

This colorful image is a 'stack' of individual spectra of over 10000 quasars ordered by increasing redshift (toward the top) and *observed* wavelength, λ_{obs}. Each horizontal line in the plot—far too many to be seen individually—is one spectrum. The dark vertical lines—mostly at right—correspond to absorption features in Earth's atmosphere. The graceful curves show how the emission lines arising from specific elements and arising in the quasars vary with redshift. Many of these lines arise from high-ionized elements as labeled (see Destination 5 to review the notation used for ions). Each emission line starts near its true, or rest, wavelength at the bottom of the image, and progressively moves toward longer wavelength (to the right in the plot) as higher-redshift quasars are plotted. The colors indicate the region of the EM spectrum where the radiation originated at the quasar. Red+Orange corresponds to the optical portion of the spectrum (about 360–760 *nm*); Yellow+Green, the near ultraviolet (200–360 *nm*); and Teal+Blue, the far ultraviolet (100–200 *nm*).

Image of quasar spectra from 2QZ survey team.

The properties of quasars are certainly not star-like, but they are not entirely new to us. Back at Destination 17 we observed a similar range of phenomena in the Seyfert galaxies such as M 77, including rapid variability and strong emission spectra. We discovered there that the unusual properties of Seyfert galaxies arise from the Active Galactic Nuclei (AGN) that reside in the cores of these galaxies. And we found that AGN are driven by accreting Supermassive Black Holes (SMBH). Are quasars also related to this AGN phenomenon? As we'll see, the short answer is 'yes,' but only in the same way that a garden lizard and a Tyrannosaurus Rex are both reptiles. As this comparison suggests, quasars differ from Seyferts partly in scale; as we'll soon see, quasars are the

9 In fact, when the first spectra of quasars were obtained, it was clear they were unlike any known types of stars. But that did not stop some people at the time from trying to explain them as stars. Once people properly interpreted the spectra in terms of high redshift, it became evident to all that these objects were extragalactic in nature and absolutely not stars.

10 Astronomers also commonly refer to these objects as *QSOs*. But there are probably enough acronyms floating around from past destinations (for example, ISM, AGN, BH, CMB) to keep us happy (or confused?). It seems preferable to avoid adding any more unless absolutely necessary.

most extreme forms of AGN in just about all key respects (see Table 21.1 for a review of AGN properties). But this analogy also points to a more fundamental difference between quasars and Seyferts, one that has to do with *when* we typically find these different types of AGN in the history of the Universe. The supermassive black holes that power Seyfert AGN reside in the Present-Day Universe, some 13 *Gyr* after galaxy formation commenced. But we are now at $z = 4$ and the Universe is only 1.6 *Gyr* old at the dawn of galaxy formation. And we can see that this era is also the *Age of Quasars.*

To understand why this is so, we need to delve more deeply into the detailed properties of quasars. A good place to start is in studying their spectra. As we've already seen, most quasar spectra exhibit broad, bright emission lines, many of which arise from ionized atoms of elements such as Hydrogen (H), Carbon (C), Magnesium (Mg), Oxygen (O), and Silicon (Si), among others (Figures 21.3 and 21.4). These emission lines, especially the ones from multiply-ionized atoms, reveal generally more extreme conditions—higher gas temperatures and densities—than we see in Seyfert nuclei (Destination 17). Quasar lines are also typically much broader, implying even higher velocities and hence even higher BH masses (Destination 17). Like Seyferts, quasars also emit strongly over the entire EM Spectrum. What's noticeably different is the relative power of radio and x-ray radiation that quasars can produce. While Seyferts typically emit most of their radiation between far-infrared and ultraviolet wavelengths (see Figure 17. 4), quasars often emit their energy more uniformly over the full range (Figure 21.4), from radio waves to x-rays, and, in some cases, all the way out to high-energy γ -rays (Destination 5).

TABLE 21.1 Summary of AGN Types and Properties

TYPE OF AGN (1) [REDSHIFT RANGE]	LUMINOSITY RANGE (2) [MASS-ACCRETION RATES]	SUBTYPES (1)	SPECTRAL FEATURES	COMMENTS [DESTINATION(S) FOR MORE DETAILS]
Seyfert Galaxies [$z < 0.1$]	$10^{10} - 10^{11} L_\odot$ [$0.01 - 0.1\ M_\odot/yr$]	Type 1	Broad and narrow emission lines including highly ionized gas. Weak radio and x-ray emission.	Associated with spiral galaxies. AGN often show optical variability in brightness and/or spectra [D17].
		Type 2	Narrow emission lines only, mostly moderately ionized gas. Weak radio and x-ray emission.	Associated with spiral galaxies, but rarely exhibit much variability [D17].

(CONTINUED)

TABLE 21.1 *(CONTINUED)*

Quasars [z = 0.2–6.4]	$10^{11} - 10^{14} L_\odot$ [0.1–500 $M_\odot$/yr]	Radio Loud	Broad and narrow emission lines + powerful radio emission.	Mostly associated with luminous Elliptical or interacting galaxies. Extended radio lobes. Typically variable in brightness [D21].
		Radio Quiet	Broad and narrow emission lines. Weak radio emission though often x-ray luminous.	Mostly associated with luminous Elliptical or interacting galaxies. Typically variable in brightness [D21].
		Ultra-Luminous IR Galaxies	Optically invisible, but very luminous far-IR sources. Sometimes radio loud or quiet.	Possibly highly dust-obscured quasars. Some may not be AGN at all but rather luminous galaxies during intense periods of star formation.
Radio Galaxies [z = 0–5.2]	Up to – $10^{13} L_\odot$ [Low to10 $M_\odot$/yr]	Broad Line	Broad and narrow emission lines with strong radio emission. Often weak in optical and x-rays.	Associated with luminous Elliptical galaxies with widely separated but bright radio lobes. Typically variable in radio and optical brightness [D21].
		Narrow Line	Narrow emission lines only with powerful radio emission. Weak in optical and x-rays typically.	Associated with luminous Elliptical galaxies. Exhibiting a wide range of radio morphologies. Often variable in radio and optical brightness [D21].
Blazars [z = 0.2–3]	10^{11} to $10^{13} L_\odot$ (Relativistic boost[3]) [10 to 500 $M_\odot$/yr]	BL Lacertae Objects	Bright but nearly featureless continuous optical spectra. Powerful radio sources.	Mostly associated with Elliptical galaxies. Highly variable from radio to X-rays. Exhibit strong magnetic fields [D21].
		Optically Violently Variable Quasars	Broad+narrow emission lines. Extremely luminous in radio, optical, and sometimes x-rays.	Highly variable at all wavelengths. Associated with elliptical and interacting galaxies.
Low Ionization Nuclear Emission Region (LINER) [z < 0.1]	< 10^6 to $10^9 L_\odot$ [<< 0.01 $M_\odot$/yr]		Typically narrow emission lines of low-ionization gas.	The least luminous of all AGN. Some (many) may arise from star formation and other non-AGN activity. LINER activity seen at some level in most spiral galaxies [D10].

Notes to Table 21.1:

1. Some of these AGN types have not been discussed explicitly in the text. See the final column for references to the text for those that have.
2. These values refer to the total luminosities over all wavelengths. The ranges indicate the typical range seen among members of each sub-type of AGN. Radio bright quasars and Radio galaxies exhibit radio luminosities of 10^6 times or higher than observed for normal giant elliptical galaxies.
3. Blazars are believed to be examples of quasars where, by chance, we happen to be viewing them almost directly down the angle of their jet. Due to relativistic effects, this can cause the brightness to be boosted by large factors of well over 100 at times. Thus, it is difficult to determine the true luminosities of the underlying AGN. However, the power of the jets implies most blazars are typically powerful quasars.

There is another intriguing difference between quasars and Seyferts. The AGN in Seyferts are found at the centers of spiral galaxies (Figure 17.1, Destination 17). In contrast, quasars appear at first glance to be more point-like in structure with no evident extended structure. This characteristic is so common it is even built into the name: *Quasi-stellar* objects. However, if we look very carefully at the regions surrounding quasars, we discover that *every one* is accompanied by some extended distribution of light. Unlike what we see in Seyferts, the 'fuzzy' light surrounding quasars is comparatively dim relative to the quasar itself. Sometimes the light surrounding a quasar is fairly regular in shape, sometimes it is highly structured, almost chaotic in appearance (Figure 21.5). Taking our cue from Seyferts, we can reasonably conclude that the light surrounding quasars arises from the *Host Galaxies* that harbor these luminous AGN.

Why is this light from the host galaxies of quasars so difficult to detect? It's not because these galaxies are low-luminosity systems. The internal velocities of quasar hosts are typically hundreds of *km/s*, similar to what we see in large galaxies such as ours, M 31, or Maffei 1 (Destinations 9, 14 and 15), and similar to the kinds of galaxies that surround Seyfert AGN (Destination 17). But if that's true, then quasars—even the closest ones—must be phenomenally further away than Seyferts or else we'd see the quasar hosts much more easily. As it happens, quasars are not located at 'mere' tens of *Mpc* from Earth (like M 77; Destination 17), but at least many hundreds of *Mpc* away from Earth in the Present-Day Universe. To be as bright as they appear at such monumental distances, quasars must be extraordinarily luminous. Indeed, they are far more luminous—at all wavelengths—than Seyfert galaxies (Figure 21.4), some ten times to well over one thousand times more luminous than our entire Galaxy (see Table 21.1). It's no wonder, then, why it's so difficult to detect the host galaxies of quasars. Not only are the hosts small in appearance—due to the typically large distances to quasars—but the AGN's immense luminosity typically swamps the host's light[11] (Figure 21.5). Even here at $z = 4$, the nearest quasars are still about 15–20 times more distant from us than M 31 is from Earth in the Present-Day Universe (Destination 14), making it hard to spot the host galaxies

11 Of course, we could have determined the luminosities of quasars from measuring their redshifts, then using the Hubble Law to estimate their distances (Destinations 16 and 18). Either in the Present-Day Universe or at $z = 4$ so long as we use the correct value of the Hubble Constant and Ω's appropriate for each era. Historically, this simple method of getting quasar distances was slow to develop because no one initially believed that such gargantuan redshifts—the closest quasar to Earth, an object known as 3C 273, has a redshift of 0.17—were even possible. As we noted in the previous footnote, this made it difficult to recognize the spectra for what they were. But there was also resistance because astronomers back then appreciated that the large distances implied by quasar redshifts implied that these objects had to have astoundingly large luminosities (see a question at the end of this destination).

of these AGN. As we look around this Alien Sky, we begin to marvel that we can see such distant objects with our naked eyes![12]

These quantitative differences between quasars and Seyfert galaxies are certainly striking, but we mustn't forget that quasars and Seyferts are ultimately the same sort of object. Both are examples of AGN, but representing two of the distinct known classes of AGN (Table 21.1). Like all AGN, quasars and Seyfert galaxies owe their extreme properties to properties of a gargantuan accretion disk surrounding a supermassive black hole (SMBH) located at the centers of their host galaxies. The compact nuclei, rapid motions, and extreme range of thermal emission are all features that support this interpretation. But the quantitative differences among AGN *are* significant. The luminosities, host galaxy properties, and the way different types of AGN are distributed in the Universe (Table 21.1) tell us something about how AGN and normal galaxies evolve. They may even help reveal the origin of the SMBHs that power AGN.

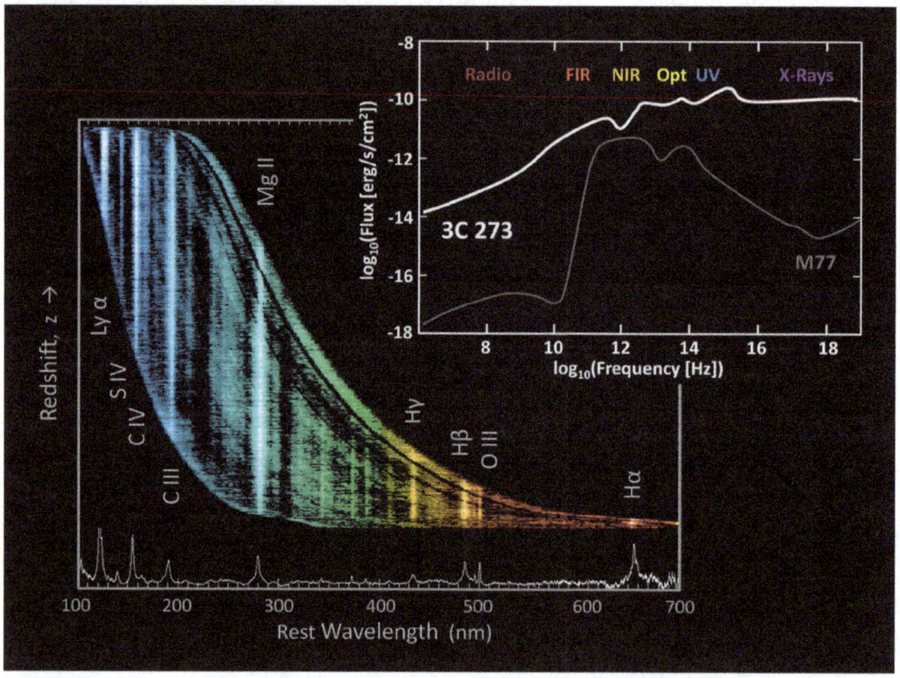

Fig. 21.4

(*Lower left*) This image shows the same spectra from Figure 21.3 but now plotted in terms of their rest wavelengths. The curved edges of the colored region corresponds to the edges of the observed wavelength limits—at about 360 and 770 *nm*—at the left and right sides of the image in Figure 21.3. The plot at the bottom shows the composite spectrum obtained by adding up the individual 20000 spectra represented in the colored swath. (*Upper right*) The EM Spectrum of the quasar closest to Earth—an object known as *3C 273*—is shown as a white curve extending from radio waves to x-rays. The dim gray line is the spectrum of the Seyfert galaxy M 77 (from Figure 17.4) shifted to the same distance. The comparatively much higher luminosity of the quasar at all wavelengths is evident. In the radio and x-rays, 3C 273 is up to 10000 times more luminous than the Seyfert. And this is not a particularly luminous quasar!

Image of quasar spectra from 2QZ survey team.

To start to address all this, recall (Destination 17) that the luminosity from an AGN accretion disk can be given by

$$L_{AGN} = \eta c^2 \left(\frac{\Delta M}{\Delta t} \right),$$

12 For comparison, in the Present-Day Universe, 3C 273 is about 900 times more distant from us than M 31, about 40 times more distant than the Virgo Cluster. Yet, remarkably, it can be seen with a small telescope, despite being slightly over 600 *Mpc* away!

where $(\varnothing M / \varnothing t)$ is the *Mass-Accretion Rate* onto the accretion disk surrounding the BH at the AGN's core (see Destination 17 for the slightly different, original version of this equation and Figure 17.7 to review the basic AGN model). The parameter η specifies the efficiency with which the BH can convert mass to energy and is defined to have a value between 0.0 and 1.0. Detailed models of accretion disks imply that $\eta \approx 0.1$ for most types of AGN, so it is clear from this equation that the only practical way to attain high luminosities in AGN is to increase the mass-accretion rate.

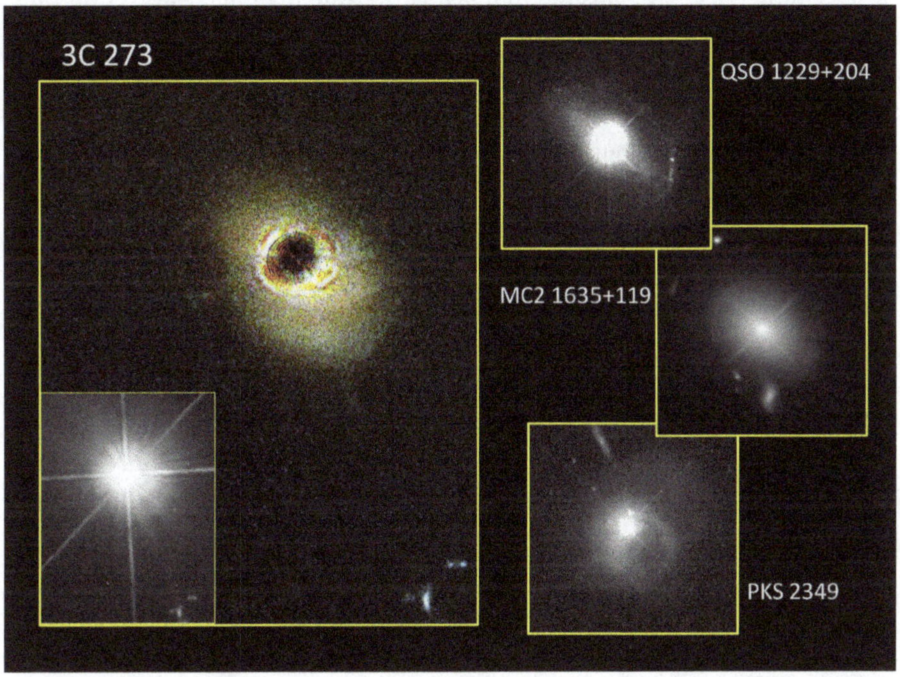

Fig. 21.5

When studied carefully, quasars invariably reveal their underlying *Host Galaxies*. In some cases such as 3C 273, the light from the AGN so swamps the underlying galaxy that only fairly sophisticated techniques can definitively detect the host. In the image (*left*), the bright image of the quasar (*lower left*) has been digitally subtracted. In other cases (*right*), the host galaxies are more readily visible despite the light of the quasar. The host galaxies of most quasars are typically either elliptical galaxies (3C 273 and MC2 1635 + 119) or likely interacting systems (QSO 1229 + 204 and PKS 2349). The spikey features extending from the quasars are artifacts produced by the optics of the telescopes used to obtain these images.

Images from STScI/NASA.

How high a rate is needed? The most luminous quasars we see here at $z = 4$ have luminosities of about $10^{14} L_\odot$, equivalent to about 10000 times the luminosity of our Galaxy! If the accretion disks surrounding SMBHs in AGNs are really able to convert mass into energy with an efficiency of 10%, then the previous equation tell us that the most extreme quasars require a mass-accretion rate onto their central BHs of about 50–100 $M_\odot/yr$ to keep shining at their observed luminosities. Even the least luminous quasars we see here at $z = 4$ require from 0.1–5 $M_\odot/yr$ to maintain their observed luminosities![13] It's these extreme luminosities—and the correspondingly very high mass-accretion rates—that set quasars apart from all other types of AGN (see Figure 21.4 and Table 21.1). Quasars are the most luminous, steadily shining single objects in the entire Universe.

What do we make of such high mass-accretion rates? In one sense, these rates are *not* absurd ... at least not for the Universe at $z = 4$. In the comparatively

13 Recall that M 77 required 0.03 $M_\odot/yr$ to shine at its observed luminosity (Destination 17).

cramped quarters of this early epoch in the history of the Universe (Figure 21.1), galaxies in overdense regions have a good chance to collide and interact on fairly short timescales as their mutual gravity binds them together against the expansion of the Universe.[14] Should a BH already exist in the densest parts of such regions, it can accrete matter—particularly material that has little angular momentum (Destination 6)—that makes its way close to where the BH resides. This not only helps build up galaxies, especially massive galaxies, but also feeds the black hole with the material required to produce an accretion disk. This line of reasoning makes it plausible to imagine that the high mass-accretion rates necessary to explain quasar luminosities can occur in environments where mergers and interactions are especially active. In short, the sort of environment that appears to be common here in the Early Universe. In this respect, quasars likely act as very luminous beacons of the initial stages of hierarchical galaxy formation (Destination 20).

But there seems to be some basic problems with this interpretation. First, when—and how—do the black holes that evolve into AGN form in the first place? The evolution that we have observed from the Present Day to $z = 2$ and now to $z = 4$ implies that the Universe is becoming on average more uniform and more gaseous as we move further back in time (recall the discussion of this important point at Destination 20). These don't seem to be the kinds of trends that would promote BH formation at the earliest times in the history of the Universe. But we've seen that there may be a way—at least a way to form stellar-mass BHs—that is driven by processes that govern star formation in the Early Universe (Destinations 6 and 14). We'll return to this issue later in this destination when we explore how the first generation of stars may have formed in the Universe, and what their role is in producing the 'seed' black holes that could eventually evolve into AGN.

The other problem with our interpretation explaining how quasars shine is associated with the masses of the BHs typically seen in AGN. At first, this may seem odd. The equation accounting for the luminosity of an AGN does not actually require that we specify the mass of a BH. A small BH can shine with the same luminosity as a much more massive one as long as the mass-accretion rates and efficiency factor, η, are the same.[15] But the mass of the BH does matter in three important, but somewhat indirect ways. First, the extreme luminosities

---

14 And the Universe is expanding vigorously here at $z = 4$ The Hubble 'Constant' is 422 *km/s/Mpc* here, nearly six times higher than in the Present-Day Universe. This means that regions must be even denser here to be able to collapse under the influence of their own gravity. To be specific, the average mass density of the Universe here at $z = 4$ is now 3.3 × 10–28 g/cm^3, about 130 times higher than in the Present-Day Universe. Many large regions in the Local Universe that are gravitationally bound today—including regions like the Local Group—were still expanding at z = 4.

15 Though, of course, the lower-mass BH will gain mass at a proportionally much faster rate than the more massive one. Meaning the small BH will become a large BH pretty quickly, an issue we'll discuss in more detail shortly.

of quasars introduces a problem—one we encountered when discussing the upper limit on the masses of stable stars (Destination 8)—related to the *Radiation Pressure* that EM Radiation can exert on matter. In particular, if the luminosity of a source is too high, the outward force from its radiation pressure can exceed the inward pull of gravity. The luminosity when this occurs—which is known as the *Eddington Luminosity*—does depend on the mass of the source, so long as the source is hot enough to be mostly ionized. This is a safe assumption for the emitting regions of quasars given the strong evidence of highly ionized atoms seen in their spectra (see Figures 21.2–21.4). Due to this effect, we can infer that the most extreme quasars at $z = 4$ require masses of at least 2–4 billion $M_{\odot}$ to shine stably at their fantastic luminosities. Any lower mass and their own radiation would blow the accretion disks in quasars apart.

Which leads to the second problem with the BH masses in quasars. If we know the total mass of a BH, M_{BH}, and its mass-accretion rate, $(\varnothing M / \varnothing t)$, we can calculate how long it would take to produce a BH of that mass (assuming the accretion rate is constant) as

$$\tau_{Accretion} = \left(\frac{1}{1-\eta}\right)\frac{M_{BH}}{\left(\Delta M / \Delta t\right)}.$$

For the most extreme quasars we see here at $z = 4$, the Eddington analysis implies that the BH masses cannot be smaller than about $M_{BH} = 4 \times 10^{9} M_{\odot}$, but the luminosities demand mass-accretion rates of at least $\Delta M / \Delta t = 100\ M_{\odot} / yr$. Together, these imply accretion timescales of at least 440 *Myr* (assuming $\eta = 0.1$) for the most massive BHs in the most luminous quasars. For the Present-Day Universe, this is a short timescale when compared to the Age of the Universe (13.5 *Gyr*; see Table 19.1). But the Universe here at $z = 4$ is only 1.6 *Gyr* old, leaving only a bit over 1 *Gyr* for all the processes—locally halting the expansion of the Universe, having interactions and mergers between young galaxies, producing 'seed' BHs—that *must* precede the formation of a quasar's central SMBH. But the problem is worse than this. The most distant quasar known has a redshift of over 7.0, meaning that it existed when the Universe was less than about 800 Myr old. The processes that led to the formation of this remarkable object had at most a few hundred *Myr* to do its job. Obviously, Nature figured out a way to achieve this since remote, highly luminous quasars do exist. But it remains a difficult problem for us to explain how she did it.

We can take this discussion and basically flip it around to explore the third and final problem our model for quasar formation implies for BH masses. Suppose that a SMBH starts off with a mass of $M_{BH,0}$ and grows in mass at a rate $\varnothing M / \varnothing t$. Then, after some time T, its mass will have grown to

$$M_{BH} = T\left(\frac{\Delta M}{\Delta t}\right)(1-\eta) + M_{BH,0}\ .$$

Don't' be intimidated by this equation: it is just saying that if we add mass to a quasar, it will get more massive with time! More specifically, if we take the mass and mass-accretion rate for our extreme quasar from the earlier example, and if we also assume it shines at the same luminosity over the 11.9 *Gyr* that separate the $z = 4$ Universe from the Present-Day Universe, the BH would have a mass of $10^{12}M_{\odot}$ today. That's comparable to the entire mass—visible *and* dark matter—of our Galaxy. But it's not merely the fantastic magnitude of this mass that's the problem here. The real dilemma is that no known SMBHs of this mass exist in the Present-Day Universe. Moreover, unlike the $z = 4$ Universe, there are (essentially) no quasars in the Present-Day Universe. Yet, if they were growing in mass at the rate we are describing here, we would easily see such quasars in the vicinity of the Sun today. But we don't.

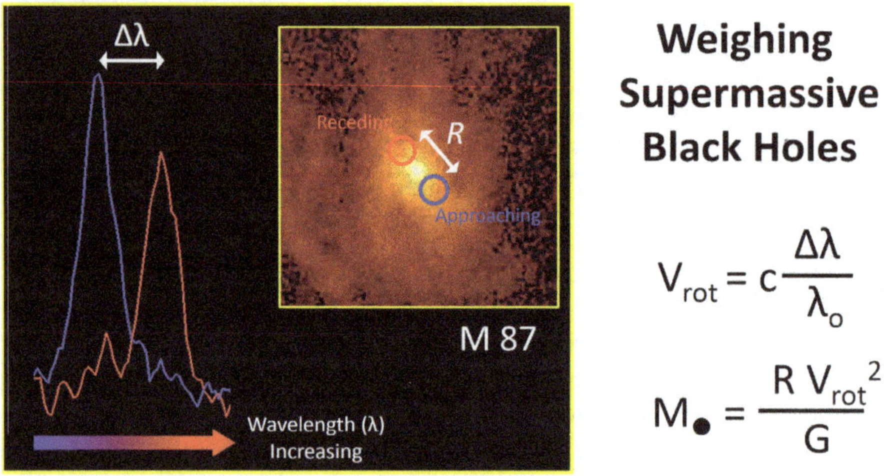

Fig. 21.6

For nearby massive galaxies such as M 87 of the Virgo Cluster (see Destination 16), it is possible to determine the masses of its central supermassive Black Hole. Spectra obtained of the light in the two circular regions on each side of the galaxy's bright nucleus (*inset picture*) reveal shifted emission lines. The lines from one side of the nucleus are clearly blueshifted, implying that material is moving toward us relative to the nucleus. The other side is shifted to the red, implying motion away from us. Using the Doppler Equation (Destination 6), we can use the observed wavelength shift, $\Delta\lambda$, to estimate the rotation velocity near the nucleus, V_{rot}. Since we know how far apart the regions are in the galaxy where the spectra were obtained, denoted as *R*, we can use the rotation velocity to estimate the mass of whatever is in the nuclear region.

Images and data from STScI/NASA.

One way to prove these points is to consider the supermassive black hole at the center of our Galaxy (Destination 10). If BHs like this one are common in other galaxies, that means that the nuclei of other galaxies can also harbor BHs that are not producing an AGN. Surveys to hunt for these BHs and to measure their masses (the technique to do this is described in Figure 21.6) reveal that perhaps all spheroidal systems more massive than the smallest dwarf galaxies (Destinations 13 and 15)—and even some globular clusters (Destination 11)—may contain nuclear BHs. Based on the motions of gas in the cores of these galaxies, none of these SMBHs appear to exceed $2 \times 10^{10}M_{\odot}$ in mass (Figure 21.9), consistent with the upper mass limit implied by the most extreme quasars known in the Early Universe. In addition, surveys of AGN in the Local Universe (as defined at Destination 18), reveal that there are no quasars in the Present-Day Universe that are even 10% as luminous as the extreme quasars seen at higher redshifts ranging from about $z = 1$ to 6.

Can we reconcile all these results? One way that does *not* work is to alter the efficiency factor, η, in the equation to calculate AGN luminosities. Ironically, lowering this value would cause BH masses to grow even faster since less of the accreted mass is converted into energy and so more falls into the BH. But raising η to its maximum value of 1.0 is also no good, as that would imply that no mass at all makes its way to the BH, in clear contradiction to the known existence of a BH at the center of our Galaxy (Destination 10) and other galaxies (Figure 21.6). Our estimates of how fast BH masses grow can't be off by more than a factor of 2–4. The only reasonable alternative—and one that is similar to the conclusion we reached for Seyfert galaxies at Destination 17—is that any given quasar can maintain its high luminosity for only a comparatively short time. We've already seen that in the most luminous quasars, it takes about 1 *Gyr* to 'build' a BH of mass $10^{10}M_{\odot}$, a value close to the maximum mass we see that any SMBHs can achieve. Lower-luminosity quasars would take somewhat longer to produce a comparable mass or they may end up generating lower-mass BHs. Regardless, what this interpretation demands is that mass accretion onto SMBHs must eventually cease. And it must do so on timescales much shorter than the age of the Present-Day Universe or we'd see many luminous quasars comparatively near the Sun today. Of course, without accretion, such 'dormant' BHs can't produce an accretion disk and therefore can't generate any luminosity. These non-accreting Black Holes can finally live up to their name and become terminally dark objects, transformed into massive fossils hidden in plain view at the centers of massive galaxies.

It is interesting to consider the implications of this idea for the SMBH at the center of our Galaxy (Destination 10). The presence of this object implies that at some time in the past our Galaxy almost certainly resembled an active galaxy of some sort. If it accreted mass quickly—but for a short time—its luminosity may have rivaled those of quasars. But if it accreted mass more slowly—and thus for a longer time—it may have resembled a Seyfert galaxy with a typical mass-accretion rate of about 0.01–0.1 $M_{\odot}/yr$. The second scenario is more likely, in part because we see so many normal spiral galaxies associated with Seyfert AGN, but also because ultra-luminous quasars are rare at all times in the Universe.[16] The fact that there may be so many 'inert' SMBHs in the Present-Day Universe also suggests that these objects could be 'rejuvenated' into AGN

16 This is not inconsistent with the fact that we can see a few of these AGN in our sky here at z = 4. The Universe here is more compact, and luminous quasars are considerably more common than in the Present-Day Universe. This combination makes a few of them bright enough to be seen by the naked eye from a random location here in the Early Universe. But this doesn't mean they are 'common.' Here at z = 4 there is only about one quasar per 10^4 Mpc^3, a volume that contains millions of what pass as 'normal' galaxies at this epoch.

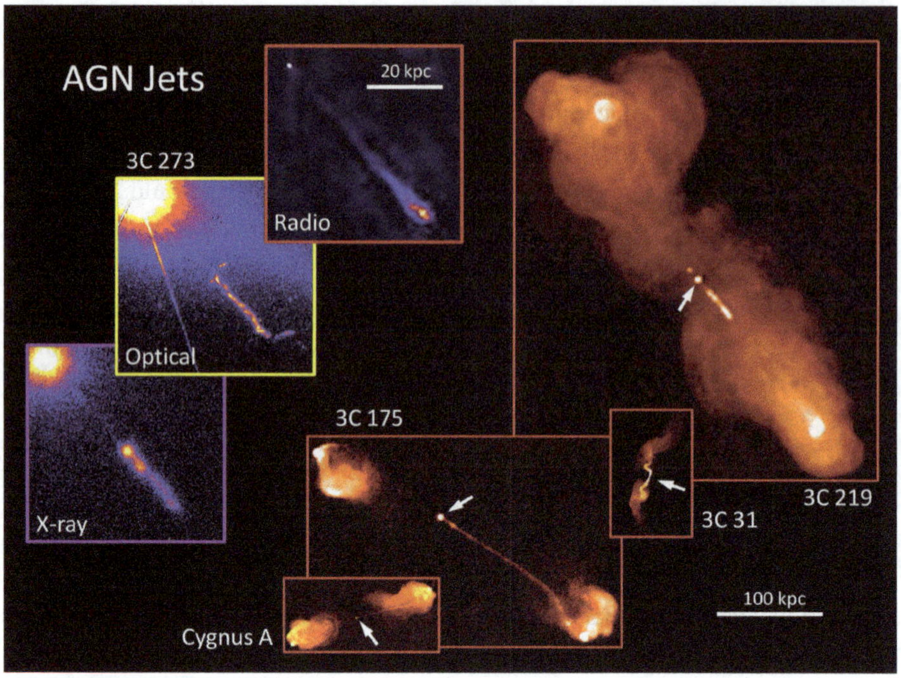

Fig. 21.7

(*Left*) Images of the jet associated with the nearby quasar 3C 273. This jet is readily visible from radio waves to x-rays. Notice that the jet's form varies at different wavelengths due to variations in the conditions along the jet. All three images are shown at the same scale as indicated by the 20 *kpc* bar in the radio image. The quasar itself is the bright-point source at the upper left in each image. (*Right*) Images of four other jets associated with AGN. These images are shown at a common scale defined by the 100 *kpc* bar at the lower right. In each image an arrow points to the galaxy nucleus. 3C 273 and 3C 175 are quasars. 3C 31, 3C 219, and Cygnus A are all examples of radio galaxies (see Table 21.1). In many cases shown here, the radio power from the jet far exceeds that of the rest of the galaxy.

Images of 3C273 from Chandra/NASA, NOAO/AURA/NSF and Jodrell Bank/U. Manchester/PPARC. All other images from NRAO/AUI.

by restarting accretion onto the BHs. Indeed, there are cases in which nearby galaxies appear to be doing precisely this (Figure 21.7).

Quasars appear to be very good at producing more than just immense luminosities. Many exhibit prominent 'jets' of materials that appear to have been emitted at high speeds from their cores.[17] It is quite likely that some of the quasars we can see in our sky here at $z = 4$ have a jet, but we'd not be able to easily see them. For one, as we will see, jets seem to originate from very close to the inner accretion disks in quasars and other AGN (see Figure 17.7). Odds are that a typical jet is simply too small and too close to the central BH to be easily visible in most AGN here in the young Universe at $z = 4$. More fundamentally, many jets—particularly larger ones—tend to emit most of their energy in the form of radio synchrotron radiation (see Destinations 10 and 17) that arises when ionized matter moves rapidly through a magnetic field. As we study quasars at radio wavelengths, we appreciate that many of them do indeed radiate intense radio emission from jet-like structures (Figure 21.7). But jets are not only seen to be associated with quasars. Another class of AGN—systems known as *Radio Galaxies*—often do not have particularly bright or otherwise noteworthy nuclei, yet may possess truly spectacular radio jets.[18] These structures can extend over hundreds of *kpc*, and some range up to a few *Mpc* in length (Figure 21.7)!

17 As noted back in Footnote 17.9, for astronomers, a 'jet' is any fairly linear feature that appears to emanate from a compact source. The term derives from the concept of 'water jets' from hoses or fountains.

18 It seems inappropriate to refer to these galaxies as AGN. In many cases, their nuclei—the 'N' in 'AGN'—are not particularly spectacular, sometimes invisible. But

How do the jets form? Even from our vantage here at $z = 4$ it is hard to tell. But one thing is clear. Specialized observations, which can discern jet details very close to the centers of some AGN, reveal directly that the matter in jets is moving away from the nuclear regions at very high speeds, often approaching the speed of light (Figure 21.7). This suggests that jets must form very close to the central BH in AGN, probably in response to how matter traveling through the accretion disk reacts to the radiation and intense magnetic fields in the region (Figure 17.7). These conditions appear to promote a way to 'launch' charged particles from near the central BH in a direction most likely perpendicular to the surrounding accretion disk, while accelerating the matter that comprises the jet to nearly the speed of light. Although the details remain unclear, this process has to be powerful in order to accelerate the particles that comprise jets sufficiently to overcome the gravitational field of the central BH and still have enough energy to be moving so fast as they travel far beyond the BH's effective gravitational range.

Powerful AGN jets do not exist in isolation: They are always associated with galaxies, and often with galaxies that are located in groups or clusters. The material spewed out from the central black hole and its accretion disk must pass through the gas of its host galaxy or cluster. What happens in such interactions is poorly understood, but there are known examples where star formation appears to be spawned by the passage of these jets through the ISM located far from where the jets were produced. This sort of behavior is an example of what astronomers refer to as *Feedback*, where one process induces another, seemingly unrelated one. As we look around us and appreciate the sheer power of the jets we see in some of the quasars and radio galaxies here in the Early Universe, it seems reasonable to imagine that AGN feedback may play a significant role in producing stars in young galaxies (Figure 21.8). In other instances, the jets can propel matter far from the host galaxy, perhaps reaching as far as other galaxies or the inter-galactic medium (IGM; Destination 16) of the cluster the galaxy may reside in (Figure 21.7). When this happens, gigantic radio lobes are formed at the locations where jets collide with gas located far beyond the confines of the host galaxy itself. This too is a form of feedback in which jets can inject energy and matter far from their source and cause the IGM to heat up and emit radiation as a result of the interaction.

---

jets almost certainly originate in the vicinity of SMBHs, so the fact that we do not see a bright nucleus means the BH has either stopped accreting since launching the jet, or it is obscured and not directly visible. In any case, the existence of jets in such galaxies almost certainly indicates fairly recent nuclear activity, justifying their inclusion among more traditional AGN (Table 21.1).

Fig. 21.8

A rejuvenated AGN? These images of the neaby Elliptical galaxy Centaurus A (or just Cen A for short) in x-rays (*upper left*), ultraviolet light (*lower*), and optical (*upper right*) illustrate that dormant supermassive BHs can become AGN again if a new source of mass accretion is initiated. The optical images show clear evidence of a recent merger, particularly apparent in the disturbed distribution of the dark molecular gas bisecting the galaxy. The x-ray image shows clear evidence of energetic processes, including hot gas (in red), a huge expanding bubble of gas, and an obvious jet emanating from the galaxy's nucleus (light blue). The ultraviolet image shows a number of strange blue stellar associations located well outside the main body of the galaxy but roughly aligned with the x-ray jet. This appears to be an example of *Feedback* where the jet seems to have somehow induced star formation far from the main body of the galaxy.

Optical image from S. Binneweis and J. Pöpsel/ Capella Observatory, X-ray image from Chandra/ NASA, and UV image from GALEX/JPL/NASA.

21.3 QUASARS IN THE EARLY UNIVERSE: IMPLICATIONS FOR GALAXY FORMATION

Although the basic picture of how AGN form and evolve seems consistent with the general features of these galaxies—both in the Early Universe and in the Present-Day Universe—we have been largely avoiding a fundamental question. If quasars are indeed powered by SMBHs, how did the black holes form in the first place? Up to now, the only way we've seen to actually *make* black holes is through the evolution of stars. And not just any stars, but specifically *massive* stars with initial Main Sequence masses greater than about $8M_\odot$ (Destination 12). Could the BHs in the AGN we see here at $z = 4$ be somehow related to the remnants of the very *first* massive stars to ever exist in the Universe? To explore this question, we need to consider the details of how star formation might have proceeded in the Early Universe.

An important way to start is to recognize a clear trend that is now unmistakably present here at $z = 4$. As we study the stars and gas in typical galaxies here, it becomes quite apparent that they are noticeably deficient in heavy elements—that is, all elements other than H and He—compared to galaxies of the Present-Day Universe, and even when compared to galaxies located 'only' 1.7 *Gyr* in the future at $z = 2$ (Destination 20). The change in the chemical composition is now so striking that it can affect key stages of the star-formation process. For example, gas that lacks significant quantities of heavy elements such as Carbon (C), Oxygen (O), and Nitrogen (N) cannot easily form dark molecular clouds. One way to appreciate this is to study the list of molecular species found in such clouds back at Destination 6 (Table 6.2). Many of these molecules contain C, N, or O atoms, and often two or even all three. Yet we identified these clouds as the birth sites of stars in our Galaxy and presumably all other galaxies (Destination 6). Right away, this tells us that star formation here at $z = 4$—or even earlier in the history of the Universe—must have been a fundamentally different process than later in the history of the Universe when heavy elements were more abundant.

One consequence of the lower chemical abundance in the Early Universe that affects how stars are born is a possible change in the Initial Mass Function (see Destination 11; we'll refer to this as the IMF), the distribution of masses of stars born in a single star-forming event. In our Galaxy, the IMF dictates that there are typically large numbers of low-mass stars produced in a star-forming region for each massive star produced. The basic form of the IMF appears to be a universal feature of star formation, at least when heavy elements are sufficiently abundant in the gas that stars are forming out of. But here, at $z = 4$ we find ourselves at a time when there are far fewer heavy elements around. Precisely the elements that most help to regulate the star-formation process. Under such conditions, it may be the case that *only* very massive stars can form. That is, stars with masses in excess of $100M_{\odot}$. Such stars would follow the basic outline of high-mass stellar evolution we explored at Destinations 8 and 12, living on the Main Sequence for very short times, perhaps only a few hundred thousand years. And, just like present-day massive stars, they are doomed from birth (Destination 12). After a smorgasbord of nuclear fusion processes that forms every element from H to Fe, they eventually explode as supernovae where heavier elements are produced in the intense furnace of the supernova itself (Destination 12).

This evolution has two highly significant effects here in the Early Universe. First, it starts the ball rolling on the creation of heavy elements. Stars that form out of the gas ejected from these first stars—even those that formed very early in the history of the Universe—can now go through a formation process that is far closer to what's seen in our Galaxy (Destination 6). In other words, the first stars, if they are mostly massive, can rapidly 'pollute' the Early Universe with freshly made heavy elements. After that, all subsequent stars can begin to form and evolve 'normally' as described at Destinations 8 and 12. The second beneficial consequence is that the compact remnants of many of these massive stars will likely be black holes. If these stars happened to also form in clusters—and we know that's possible since massive globular clusters (Destination 11) are among the oldest known objects in the Present-Day Universe—then many massive BHs may form at roughly the same time and at roughly the same place. Such 'black-hole clusters' are inherently unstable according to General Relativity, and each individual black hole in the cluster—indeed, *everything* in the cluster—will rapidly merge into one supermassive black hole.[19] Mix in some mass accretion from galaxy mergers and interactions—events that were comparatively common in the Early Universe (see Destination 20)—and all the key ingredients to form a luminous AGN are in place. Voilà! A quasar is born.

Is this the right description of how stars and active galaxies first came onto the scene in the Early Universe? It is hard to be sure. Even as time travelers, we can spend only a limited time at any given epoch of the Early Universe. From Earth, the era when the first stars may have formed is so distant that it is very difficult to see these processes in any detail. The story of how the first stars form—and how or even if they are related to the formation of quasars and other active galaxies—remains unsettled.

19 Which, as recently observed for the first time, would result in the formation of gravitational waves. As the ability to detect such waves improves in the future, it may become possible to directly 'observe' events like the ones described here as comparatively small BHs merge to form large-mass BHs at the centers of galaxies.

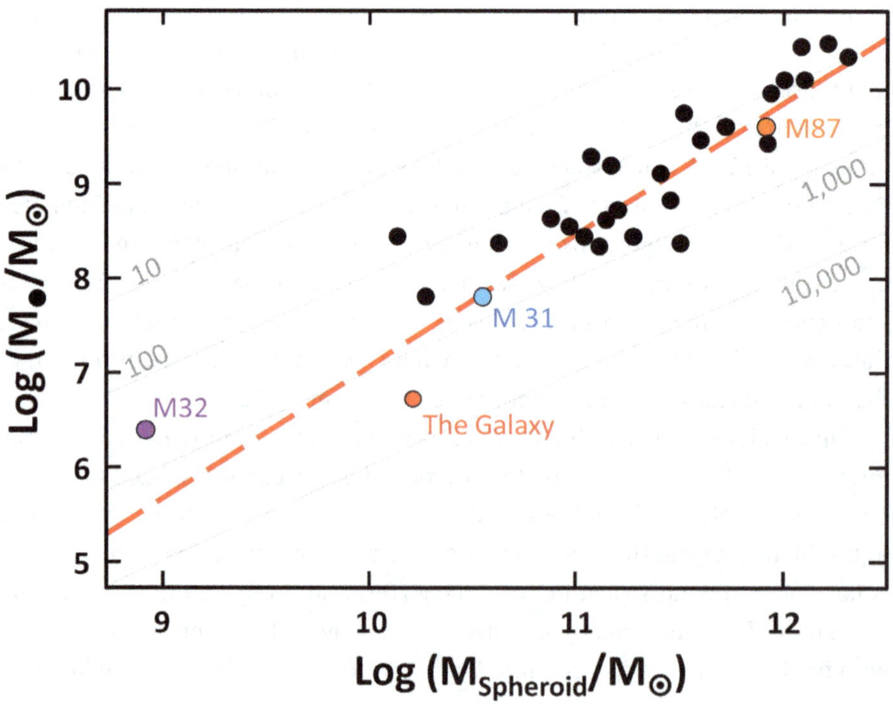

Fig. 21.9

This plot shows the masses of SMBHs (*y*-axis) plotted against the masses of the corresponding galaxies (*x*-axis). The red dashed line shows the best fit to these data, indicating a remarkable correlation between the two masses. The locations of four galaxies—M 31, M 32, and our Galaxy in the Local Group (Destination 14), plus M 87 from the Virgo Cluster (Destination 16; see also Figure 21.6)—are highlighted. The thin gray lines show lines of constant mass ratio of the spheroid and BHs. Thus, all galaxies on the line labeled '100' have central SMBHs that have 1% of the mass of their spheroids. In this plot, spheroids are defined as either the entire galaxy (for E and S0 systems; see Destination 15), or just the central bulge for spirals (Destinations 10 and 14).

There is one hint that suggests quasar formation and star/galaxy formation are intimately related. We've already seen that from studies of the motions of stars and gas around BHs in the Present-Day Universe, not only can we find these relic AGN engines, but also measure their masses (Figure 21.6). The fact that BHs exist in the centers of galaxies in the Present-Day Universe is beautifully consistent with our picture of how AGN evolve. That's gratifying, but it is not a real surprise. What is astonishing is that the masses of the black holes in galaxies of the Present-Day Universe are strongly related to the mass of the visible *spheroid* component of the galaxies (Figure 21.9). For a spiral galaxy, the spheroid component is the galaxy's bulge, while for an elliptical galaxy, it's the entire galaxy. In large elliptical galaxies, the SMBH is roughly 1% the mass of the stellar component of the galaxy, while in our Galaxy, the central BH is only about 0.05% of the mass of the Galactic Bulge (Destinations 10 and 14). What the trend in Figure 21.9 tells us is that in some fundamental way, the central SMBH of a galaxy is intimately related to the total stellar content of the spheroid that surrounds it. How this correlation came about remains a mystery. But whatever the answer, it tells us that the central SMBHs of galaxies are probably not oddities. Instead, they appear to be intimately related to the process of galaxy formation. And it probably tells us that all massive spheroids, at least down to the level of dwarf elliptical galaxies, probably contain supermassive black holes.

The Alien Sky here at $z = 4$—a sky of 'false stars' that we have discovered to be quasars—has opened up new windows to the basic process of galaxy formation, revealing some of earliest steps involved in forming galaxies in the Universe. But not the *first* step. Had there been no variations in mass density within the Early Universe (Destination 19), there would have been no collapsing clouds of gas from which to form the first stars. No first black holes. No first galaxies. To understand how and when those original density variations arose—in effect, the seeds of galaxy formation—we must travel even further back in time. Before any stars or galaxies had formed. Even before most of the chemical elements that make up our Universe had been created.

DESTINATION 21 QUESTIONS

21.1. If the supermassive black hole in the center of the M 31 has a mass of $10^7 M_\odot$, how long would it have taken to form if it shined over some period of time as a (small) quasar with a luminosity of $10^{13} L_\odot$? Assume an efficiency factor, η , of 0.25. Notice that these numbers are a bit different from the ones for the Milky Way given in the text, so you need to calculate the result anew. But the results in the text do give you an idea of the likely scale of your final answer.

21.2. Some people believe that the first stars may have had mass of up to $250 M_\odot$. Use the mass-luminosity relation and the mass-age relations for Main Sequence stars (Destination 8) to estimate the luminosities and lifetimes of such stars. Is it reasonable to imagine that such stars could have formed before quasars became active in the Early Universe? Explain your reasoning.

21.3. The largest radio galaxy is about 10 *Mpc* end to end. How long would it take a jet emitted at a velocity of $0.5c$ (half the speed of light) to extend to this distance? Given your result, is it reasonable to suppose that a jet emitted from the nucleus of an active galaxy could make such a large structure within the age of the Universe?

21.4. On the web, look up information on a quasar with a redshift, z, greater than 1.5. Describe the basic properties of this quasar, including things such as its name, its luminosity, its redshift, the properties of its host galaxy, its age, the mass of its central black hole, etc.

21.5. Imagine that the two peaks of the emission lines from a quasar, shown at the lower left in Figure 21.6, are separated by a wavelength that corresponds to a velocity difference of 1000 *km/s*.

a) If the emission line shown in the figure is the Lyman α line but is observed at an average wavelength of 525.0 *nm*, what is the redshift of this quasar? **HINT:** You will have to look up the wavelength of the Lyman α line. This particular line corresponds to the transition from $n = 2$ to $n = 1$ in a H atom (see Destination 6 to review how emission lines are formed).

b) Given the information in this problem what is the separation of the blue and red emission lines in the quasar spectrum in *nm*?

c) If the two locations noted at the upper right of Figure 21.6 are 10 light-days apart, what is the mass of the central black hole?

21.6. From Figure 21.9, read off the masses of the black holes in M 87 and M 31. If it took 10^8 *yr* for both black holes to reach their final masses, what are the respective mass-accretion rates of each black hole? What was the ratio of their peak luminosities?

21.7. One definition of the 'Era of Quasars' is the span in redshift from $z = 7$ to $z = 1.0$. Use the equations and plots in Destination 19 to calculate the duration of this span in years for

a) A Wide-Open Universe.

b) A Critical Universe.

c) An Accelerating Universe with the standard parameters we have adopted for this model: $H_0 = 72\ km/s/Mpc$, $\Omega_{Matter,0} = 0.27$, and $\Omega_{\Lambda,0} = 0.73$.

21.8. Let's imagine that a region in the Early Universe where a globular cluster (Destination 11) is forming. That region contains a total of $10^5 M_\odot$ of 'pristine' gas that will turn into stars. Here, 'pristine' means that the gas is composed essentially of H and He.

a) If a 'first star' of mass $150M_\odot$ forms in this region, how long will it live before (likely) exploding as a supernova? **HINT:** Use the age-mass relation from Destination 8 and make some assumption about the lifetime of the star after it leaves the Main Sequence. You may find some useful guidance for that last part in Destination 12.

b) If, when the star explodes, it pollutes the gas with 10% of its mass in the form of heavy elements, what is the average 'metallicity' of the stars that form out of the polluted gas? Here, 'metallicity' is defined as the fraction of a star's mass that is in the form of heavy elements (that is, elements other than H and He).

c) Compare the metallicity of this gas to the metallicity of the Sun (0.02) and to the metallicity of the globular cluster M 92 (0.0001). What does this say about the nature of the types of stars that form right after the first stars in the Early Universe?

21.9. Assume a quasar is accreting matter at the rate of $1M_\odot/yr$. How far away would this quasar have to be located to appear as bright from Earth as Sirius, the brightest star in Earth's sky? **HINT:** You will need to find plausible values for the distance of Sirius and its luminosity in solar units.

21.10. When their true distances were first understood, the luminosities of quasars turned out to be so high that it was hard to imagine what could power them. One idea was that perhaps they are regions where supernovae are occurring at such a high rate that the combined light accounts for the quasar luminosity.

a) What is the luminosity of a quasar with a mass-accretion rate of $1M_\odot/yr$? Adopt a reasonable efficiency factor, η.

b) If a single supernova has a luminosity of $10^{10}L_\odot$, how many supernovae would be required to account for the luminosity of such a quasar?

c) Just on the basis of parts (a) and (b), does this supernova-driven explanation for a quasar seem reasonable? Use your knowledge of supernova properties (Destination 12) to help answer this question.

d) What other comparisons between quasars and supernovae could be used to test the validity of this idea?

Destination 22

THE DARK AGES

22.1 A SIMPLER UNIVERSE

Our eyes strain to see something in this sky ... anything. But there is only darkness here. Is this a cosmic void similar to the one in Boötes that we visited at Destination 18? If so, then there should be galaxies surrounding us, though so distant that they are beyond the range of our unaided eyesight. So we look around us with optical telescopes. First with small ones, then larger telescopes. Still nothing. Gradually—and somewhat alarmingly—it dawns on us that there is essentially no optical radiation here at all. Not from stars, galaxies, AGN, or even from the sky itself as in the Orion Nebula (Destination 6). We have now traveled to the time in the Early Universe corresponding to a redshift of $z = 100$. Back to the era astronomers aptly refer to as the *Dark Ages*. At this redshift, we have traveled to a time that is 99.9% of the way from the Present-Day Universe to the moment the Universe began its expansion. A time when the Universe was only 17 *million* years old.[1] As we approach the origin of time itself, we seem to have encountered our most desolate destination yet.

But is this reasonable? Can the Universe really be empty here at $z = 100$? We think of all that we know will exist later on in the Universe. That is, the places in the future that we have visited in our past destinations[2]: Planets. Stars. Star Clusters. Galaxies. Galaxy Clusters. Black Holes. Quasars. We are now at an early epoch of the same Universe where we know all these things will exist at some later time. The raw material to produce all these things *must* be here. Somewhere. In some form. But where?

Taking our cue from past destinations—particularly Destinations 5 and 18—we decide to look for this matter by observing the sky around us with detectors sensitive to radiation other than visible light (Figure 5.8). Starting first with higher-energy radiation, we try our luck with X-ray detectors. But we the view is the same: Nothing. It's no different when we look around us with ultraviolet or even higher-energy γ-ray detectors. The sky here is eerily dark in all forms of radiation that have wavelengths shorter than optical light.

---

1 You may wonder why, after the lengthy discussion at Destination 19, we are not specifying the *kind* of Universe this age estimate assumes. One reason is that we have concluded it is a safe bet that we live in an Accelerating Universe with $H_0 = 72\ km/s/Mpc$, $\Omega_{Matter,\,0} = 0.27$, and $\Omega_{\Lambda,0} = 0.73$ (Destination 19, Table 19.1). But the real reason is that it doesn't matter very much anymore what kind of Universe we reside in. As we travel to earlier and earlier eras in the Universe, the difference between the various models becomes smaller and smaller. We'll explore this point further at Destinations 23–25.

2 There's that confusing time-traveler grammar again!

So, we turn to the opposite extreme of the EM spectrum characterized by low-energy, long-wavelength radio waves to carry out our search for the material we know must be here somewhere. Initially, this too proves to be disappointing ... until we tune our radio receivers to receive radiation with a wavelength of about 21 *cm* (equivalent to a frequency of about 1.4 *GHz*).[3] At that wavelength, we immediately discover that the entire sky here at $z = 100$ is blazingly bright! Thinking back to one of the first destinations we visited—another seemingly empty region near the Sun that we dubbed 'Anti-Centauri' (Destination 5)—we immediately recall we have encountered this radiation before. It arises from neutral Hydrogen atoms when their electrons 'flip' their spin to become 'anti-parallel' to the spin of the protons comprising their nuclei. The change in energy causes the atom to emit 21-*cm* radio radiation, the same radiation that we now see all around us here at $z = 100$ (see Destination 5 to review the details of this emission process).[4] There is indeed a lot of matter here—we knew there had to be. And most of it is in the form of neutral Hydrogen gas.

Our past voyages have taught us that if we see emission from neutral Hydrogen—which, if it were in our Galaxy or other galaxies, we'd refer to it as the Warm ISM[5]—we are likely to also see emission from other elements in the ISM and possibly detect radiation from molecular gas here as well (the Cold ISM). In our Galaxy—which from our vantage point here at $z = 100$ does not yet exist in any sort of recognizable form—individual molecules emit radiation at wavelengths corresponding to unique sets of emission lines, many of them detectable with microwave receivers (Destination 6). So we turn our attention to observations that will allow us to detect Carbon Monoxide (CO) by tuning our microwave receivers to the wavelength corresponding to CO's emission lines. And we see ... nothing. OK, well, another common molecule in the ISM of our galaxy is H_2O. Water. So we look for

---

3 That's *giga*-Hertz, or 10^9 cycles per second. For comparison, typical radio and television signals are transmitted at about 100 *mega*-Hertz (*MHz*).

4 It is a bit more complicated than this. Since the Universe is quite vigorously expanding here at $z = 100$—the Hubble 'Constant' has a value here of nearly 38000 *km/s/Mpc*!—even nearby neutral H clouds exhibit significant redshifts from our current location. Consequently, the '21-*cm*' radiation we see here in the Dark Ages will actually be observed over a wide range or wavelengths extending from 21 *cm* to about 250 *cm*. More highly redshifted neutral H radiation (it seems inappropriate to call it '21-*cm*' radiation anymore) would not be seen. As we'll see at our next destination, such gas would come from an earlier time when there was no neutral H at all and therefore no neutral H radiation.

5 This name—the Inter*stellar* Medium—is no longer really appropriate since there are no stars in the Universe here at $z = 100$. But it is not the Inter*galactic* Medium since there aren't any galaxies around either. Rather than try to come up with a clever new name, we'll use the one we know—the ISM—and just keep in mind that here in the Dark Ages, the ISM has had a very different history than the gas we see within galaxies in the Present-Day Universe.

its emission lines. And again: Nothing. Ammonia? No. And so it goes. No formaldehyde. No ethanol. No ammonia. In fact, *none* of the molecules we see in the Cold ISM of our Galaxy (see Table 6.1) exist here. Not even (much) molecular *Hydrogen*—H_2—is present in the gas surrounding us in this place. How can this Universe—so obviously full of atomic H—be utterly devoid of molecular gas?

As we probe the gas here at $z = 100$—say, by collecting it in a manner similar to what we did at Destination 5—we realize that the answer to this questions lies partly in the makeup of the gas itself. By analyzing the composition of the ISM here at $z = 100$ carefully, we confirm that most of it consists of simple Hydrogen atoms. More precisely, about 76% of the gas, by mass, is in the form of H atoms.[6] The next most common component of this gas is Helium (He), comprising about 24% of the total mass of the gas here in the Dark Ages. Which adds up to 100%. Does this mean there are no other elements here? As we analyze the gas of the Dark Ages carefully, we find that there are indeed atoms of elements other than H and He present in the ISM. But not many. Only with difficulty do we detect trace amounts of Lithium (Li) and even smaller quantities of Beryllium (Be): About one Li atom for every 1 billion H atoms, and no more than one Be atom per 100 billion H atoms. And that's it. Just four elements. Try as we might, we cannot detect any other elements in this gas. No Carbon. No Oxygen. No Iron. No Gold. No Lead. Only H, He, and these trace quantities of Li and Be.

But these are not just any four elements. They are the first four elements of the periodic table. The four lightest elements. The four elements whose atoms contain the fewest number of protons. There is one proton in H, two in He, three in Li, and four in Be. This can't be a coincidence.

One consequence of our survey of the composition of the gas at the Dark Ages is that it helps explain for the most part why there are no molecules here: There are no chemical elements to form molecules. Helium, being a noble gas, rarely combines with other atoms to form molecules of any sort. And there is so little Li and Be that any molecules they could form would be extremely uncommon. In short, there simply are few molecules that could form easily out of this particular mixture of elements. Well, except for one. There's plenty of H in the ISM here in the Dark Ages, so why don't we detect much molecular Hydrogen here at $z = 100$? Something must be inhibiting the formation of this simple molecule at this early time in the Universe despite the presence of so much atomic Hydrogen in the gas here.

---

6 Though not all of these H atoms are the same. A small fraction consists of Deuterium atoms, H *isotopes* that contain one proton and one neutron in their nuclei. We'll return to these atoms and isotopes at Destination 24 when we encounter the origin of the first elements.

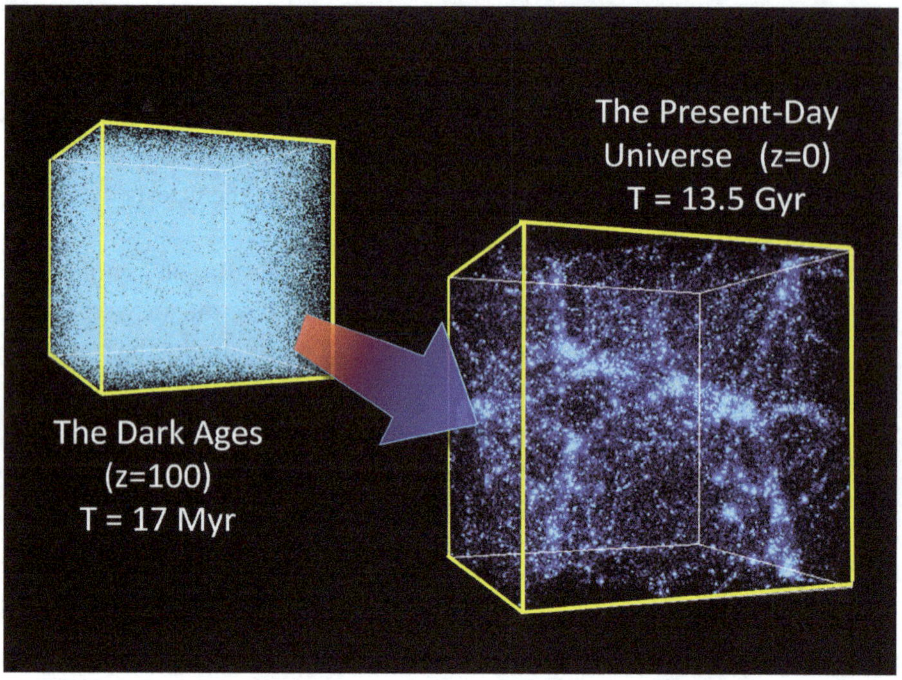

Fig. 22.1

(*Left*) A model representation of the distribution of *dark matter* in the early Universe (in this case at $z = 100$). Notice how uniformly distributed the material is in the volume. (*Right*) The same volume of space (corrected for the expansion of the Universe) at the present time. From the smooth beginnings at $z = 100$, all the detailed and complex structure at $z = 0$ arises due to the role of gravity in an expanding space. A key reason for this dramatic transformation of how the material in the Universe is distributed is dark matter, which helps ensure that visible matter can collapse gravitationally fast enough to form the structures we see in today's Universe.

Model images by A. Klypin and A. Kravtsov.

As we look around us more carefully, we come to appreciate that the neutral atomic H we can see with 21-*cm* radiation is not uniformly distributed in this Alien Sky. There are distinct clumps of gas here, some barely denser than the average mass density[7], but some considerably more dense. These regions—particularly the densest ones—must be where the first stars and galaxies will form. But even the densest of the clumps are not particularly extreme in their properties compared to what we've seen at other destinations. In particular, the *Density Contrast*—that is, the ratio of the maximum and minimum gas densities—is pretty small, at most around 10:1 or so (Figure 22.1). This is far lower than the mass density contrast in the Present-Day Universe, which we found is closer to 10000:1 (Figure 22.1; Destination 18). Along with its striking lack of nearly all of the chemical elements, the distribution of visible matter here at the Dark Ages is a lot smoother than we know it will be in the distant future. What do these features tell us about our strongly evolving Universe?

7 The average mass density here at $z = 100$ is 2.7×10^{-24} g/cm^3. Though this is still fairly low, it is over 1 million times higher than the mean mass density of the Present-Day Universe. This density happens to be comparable to the mean density of the ISM in our Galaxy.

22.2 THE PROBLEM WITH BEING SMOOTH, PART 1: WHY SO LUMPY NOW?

As we ponder these evolutionary changes in the Universe, we begin to understand that the remarkable smoothness of matter here at $z = 100$ is a problem. To appreciate why, it's useful to stop thinking in terms of lookback time (or redshift) and start thinking in terms of the elapsed time since the expansion of the Universe commenced. That is, in terms of the Age of the Universe as defined at each destination. Since we are traveling back in time as we visit epochs corresponding to higher and higher redshifts (as measured from the Present-Day Earth), the Age of the Universe is becoming progressively smaller. Here at $z = 100$, the Age of the Universe is about 17.2 *Myr*, almost 0.1% the 13.5 *Gyr* age of the Present-Day Universe. Although 17.2 *Myr* is a comparatively short time, a lot can still happen in that time: Stars with masses in excess of about 10 $M_{\odot}$ can form, live out their entire lives on the Main Sequence, die, and eject almost all their mass back into space (Destinations 6, 8, and 12). Why haven't they? Part of the answer comes from our discovery at our last destination that star formation might be strongly suppressed when there is no molecular gas (Destination 21). But, easy or not, we *know* the Universe figured out some way to form a first generation of stars. Yet here in the Dark Ages there are no stars at all; in the not-too-distant future, stars will be common. The problem we face is that the timescales involved—the expansion of the Universe on one hand, and stellar evolution on the other—seem inconsistent. From what we know about stellar evolution and how stars might form in gas with little or no heavy elements (Destinations 8 and 21), the Universe seems to have had plenty of time to have formed some stars by now. Yet there are none. What's inhibiting star formation here in the Dark Ages?

The answer is tied to the fundamental feature of Universe: It's expanding! And, here at $z = 100$, quite vigorously (see footnote 4 of this Destination). But forming stars requires *collapse*, not expansion. Consequently, it takes time for stars to form in the early Universe because the formation process, whatever the details might be, first must overcome Universal Expansion. This causes the initial process of cloud collapse to take quite a bit longer than we'd expect based on how normal molecular clouds collapse to form stars in star-forming regions in our Present-Day Galaxy (Destination 6). The basic reason for this is fairly simple, even if the details can be complicated.

Imagine a volume of matter inside an infinite expanding Universe. Imagine too that the mass density in this volume started off perfectly uniformly, apart from one subregion where the density is initially just a bit higher than average. As this Universe expands, the mass of the matter in this region attracts itself gravitationally. At first, not sufficiently to start to collapse, but enough to steadily slow down the expansion within the subregion.[8] As the overall expansion causes the matter in the Universe to become more diffuse, the self-gravity of the overdense subregion can ultimately stop the expansion locally. At that point, the subregion becomes gravitationally distinct from the other matter in the Universe and it can start to collapse onto itself, in effect evolving independently of the rest of the expanding Universe. The slow but steady increase in the mass density for overdense subregions in an expanding Universe is known as the phase of *Linear Growth.* The behavior of a subregion can be calculated pretty simply during this phase. In contrast, *Non-Linear Growth* occurs when overdense regions start to collapse on themselves more vigorously. This stage in the evolution of a subregion is much harder to model. But it's a critical stage during which the actual formation of structures such as stars, galaxies, galaxy clusters, and even larger components of the Universe occurs (see Destination 18).[9]

Simple calculations based on these ideas that include only the visible matter of the Universe while mimicking the conditions of the early Universe—in particular, how its expansion and temperature evolve with time—suggest there's simply not enough time for quasars to form by a redshift of $z = 4$. Yet, we've been there! We've seen firsthand that quasars do indeed exist by $z = 4$ (Destination 21).[10] The same calculations of how structures grow in the Early Universe also suggest that the dawn of galaxy formation should have commenced no earlier than around a redshift of $z = 2$ or less. But we've been there too, and discovered

---

8 What about the effect of the rest of the matter in the Universe on this subregion? It turns out that the gravitation effects of the Universe cancel out, so long as the initial mean density was uniform and the Universe as a whole is a lot larger than the subregion. This implies that all subregions similar to the one we are considering will behave the same way. As far as these overdense subregions are concerned, their initial expansion and subsequent collapse means that they behave much like isolated, little versions of the Closed Universe we considered at Destination 19.

9 A good analogy for all this would be a tomato tossed up in the air from the surface of the Earth. While the tomato is moving through its arc in the air—even while it is still moving upward—we can readily calculate its motion as long as we know from where it started and how rapidly it was moving at that time. And we can readily predict where it will be in the future. That corresponds to the linear growth phase of substructure in the Universe. But when the tomato hits the ground—splat!—the result is hard to predict in detail. Yes, we know it will be messy, but we do not know, say, where any one of the seeds in the tomato will end up after impact. This 'messy' process—which in our case involves formation of stars and galaxies—comprise the non-linear growth phase.

10 Indeed, as of 2012, quasars are known to exist out to redshift of at least $z = 7$. This just makes the issues we are discussing here even harder to understand.

that galaxy formation was in full swing by that time (Destination 20). Indeed, normal galaxies exist at even earlier times, and certainly by $z = 4$ where we saw them firsthand (Destination 21). And galaxy clusters? These same models of the growth of structure in the Early Universe imply that they should just start forming around 10–12 *Gyr* after the expansion of the Universe began. We know that too is wrong. The mature galaxy cluster we visited at a redshift of $z = 0.8$ (MS 1054 at Destination 19) was already present in the Universe after about half that time had elapsed. It's as if the Universe cannot make up its mind. We see extraordinary smoothness at the earliest times and dramatic clumpiness soon afterwards. Seemingly, *too* soon afterwards.

One possible resolution of this enigma comes from dark matter (DM). We've already seen that in individual galaxies (Destinations 14 and 15), DM appears to extend far beyond the extents of their visible matter and appears to dominate the total mass contents of all galaxies. What if this is a general feature of dark matter? That is, visible matter may be just the tip of the iceberg of matter in the Universe—the part we can see. But what if for each galaxy we can see—and perhaps even for some we cannot detect because they have too few stars—there is an immensely larger reservoir of dark matter? People who worry about how galaxies form argue that it is the distribution of DM—not the visible matter—that leads to the relatively rapid formation of structures such as stars, galaxies, and clusters during the evolution of the Universe. Once we let the dark matter of the Universe be distributed differently than the visible stuff, it can be 'tailored' to ensure that the structures we see in the Universe form fast enough to explain what we see at high redshift. As an illustration, DM may exhibit much larger variations in density than visible matter exhibits. This could produce regions of high mass-density concentrations where matter—both dark and visible—could begin to congregate, ultimately forming stars and galaxies. If the density variations of the far more abundant DM are sufficiently pronounced, these overdense regions can promote more rapid collapse in the Early Universe that we would expect from the visible matter alone. This is just what we need to explain the timing of the formation of different sorts of objects at high redshifts (Destinations 19–21).

As far as the gas here at the Dark Ages is concerned, this picture suggests that much of it—perhaps all—will end up in stars and galaxies in the future Universe, drawn in by the gravitational force of these objects along with the DM they are associated with. But what about isolated H clouds located in initially underdense subregions of the Early Universe? This gas is likely to suffer two fates. As the first stars and AGN form in the Universe in a few hundred *Myr*—probably at around a redshift of $z = 10$–20—the intense ultraviolet radiation from these objects might ionize the gas in a process known as *Reionization.*[11]

11 As we shall see at Destination 23, the gas here in the Dark Ages had all been ionized during the initial evolution of the Universe and up through the time that the CMB radiation we see today was produced. Hence the name *Re*ionization.

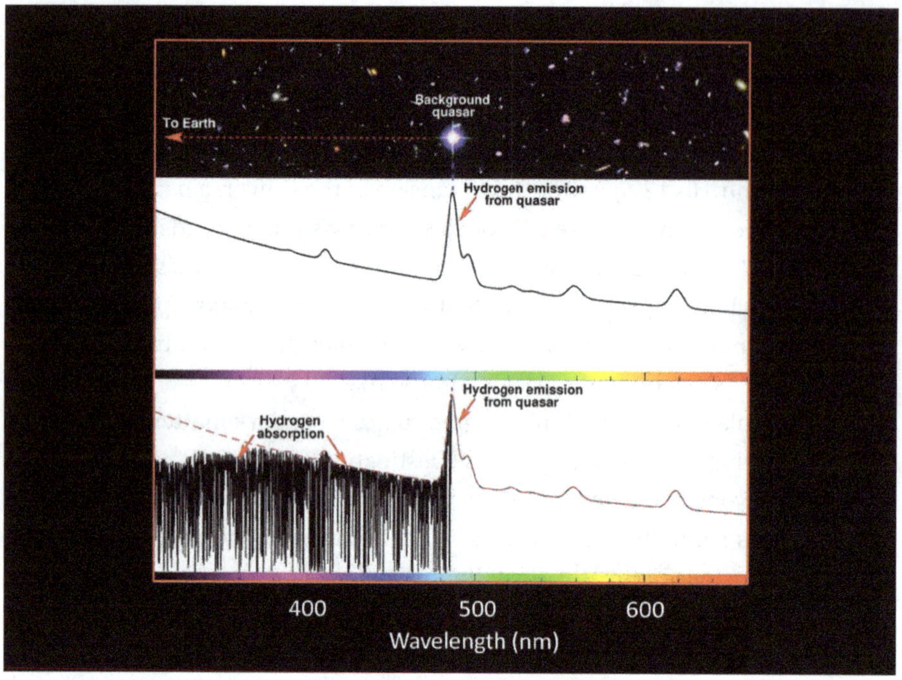

Fig. 22.2

(*Upper*) A schematic view of how we might see the light from a distant quasar (at center) pass by many intergalactic clouds of neutral H along our line of sight (red line). (*Middle*) This image shows how the quasar's spectrum would appear if there were no intervening absorption. (*Lower*) This image shows the absorption by individual neutral H clouds between us and the quasar. Each narrow absorption line represents a single cloud at a redshift smaller than that of the quasar (that is, a cloud between us and the quasar). Notice how there is no absorption to the right of the quasar's Lyman-*a* emission line. That would require clouds with higher redshifts than the quasar. But those would be further away and hence could not cause absorption. These absorption lines are often referred to collectively as the *Lyman-α Forest*. These intergalactic clouds may be the direct descendants of the neutral H clouds we see here in the Dark Ages.

Image by M. Murphy/ Swinburn University of Technology and Hubble Ultra-Deep Field image from STScI/NASA.

But some of the gas, particularly the stuff far from where the bulk of stars and galaxies are forming, may survive in the form of neutral clouds that are either chemically pristine—that is, contain the same mixture of elements we see here at the Dark Ages—or, possibly, polluted by gas spewed into intergalactic space by the winds of the first massive stars, supernovae, and AGN that form in the Early Universe.[12] Such clouds can been seen in the Present-Day Universe—not from their 21-*cm* emission—but from their absorption of radiation from distant sources, such as quasars (Destination 21), that passes through the clouds (see Figure 22.2 for an explanation of this process).

If this basic picture of how the neutral H gas clouds we see here during the Dark Ages evolve into the galaxies of the Present-Day Universe is correct, then DM is more than just some oddity of Nature that produces flat rotation curves (Destinations 9 and 14). Rather, it is the fundamental reason we exist now in the Present-Day Universe and not billions of years later. If at all. Without the accelerating effects of DM on the rate of the early contraction of subregions in the Universe, it would have taken far longer to transform the smoothness of visible matter we see here at $z = 100$ to the more extreme clumpiness that characterizes the Present-Day Universe.

But, whatever the correct interpretation or its details, the real wonder of the Dark Ages is that the Universe—somehow—pulled off the trick of transforming

12 In the latter case, the gas probably was originally ionized and has since cooled down to become neutral again. The same happens in our Galaxy. Once the hottest stars in the Orion Nebula fade away (Destination 6), the ionized gas of the Hot ISM will recombine with free electrons to produce neutral gas similar to what comprises the Warm ISM (Destination 5).

itself from utter smoothness to large-scale clumpiness. And it did this in a manner that ensured structures formed far sooner and more vigorously than we would have predicted from the distribution of visible matter alone. And it's a good thing, too. If they hadn't, we'd not be around today to appreciate the view.

22.3 THE PROBLEM WITH BEING SMOOTH, PART 2: WHY IS THE CMB SO UNIFORM?

Though we seem to have identified all of the matter that's present in the Dark Ages—the visible matter from the 21-*cm* radiation it emits and the DM from its role in forming structures—there's something else we'd expect to detect here. Something we had first noticed from the depths of the Boötes Void (Destination 18): The Cosmic Microwave Background (CMB). With that in mind, we tune our microwave detectors to the frequencies in the microwave portion of the EM spectrum (see Figure 5.8) where we 'know' the CMB must emit radiation. Back at the Boötes Void (Destination 18) we surmised that this radiation must be coming from some very early, hot period in the history of the Universe, a time that presumably occurs even before our current location at $z = 100$ here in the Early Universe. We should still see the radiation from the CMB shining on us from all directions in the sky. And as we confidently look around to detect the radiation from the CMB, we are startled to see ... nothing. If our interpretation of the origin of the CMB was correct, this result is very hard to understand. So we check our detectors and try again. Nothing. There is definitely no microwave background from the sky, or from any individual sources, here at $z = 100$.

We also know that there is no excess radiation at longer wavelengths—we have already explored the radio portion of the EM Spectrum as we studied the 21-*cm* radiation from the neutral H gas here—and so, somewhat desperately, we decide to explore shorter wavelengths corresponding to more energetic EM radiation. We've already discovered there is nothing to see optically, or in the ultraviolet or more energetic portions of the EM spectrum. Instead, we tune our receivers to the infrared portion of the EM spectrum, corresponding to radiation with wavelengths of about 10 *microns* (see Destination 5).[13] To our surprise, we do find a source of radiation at these wavelengths that emits remarkably uniformly over the entire sky. And, like the CMB, it has a nearly perfect thermal spectrum. But is this really the CMB? A fit to the spectrum reveals that it comes from a thermal source with a temperature of about 273 K, not the 2.7K that we observe for the CMB in the Present-Day Universe. No longer near absolute zero, the CMB radiates as a thermal sources at nearly room temperature! If this really is the CMB, why is it so much hotter?

The answer again lies in the expansion of the Universe and the nature of the Hubble Law. As we look at ever more distant objects, the Hubble Law tells us that they must exhibit corresponding larger redshifts. We've emphasized that these redshifts do not imply motions in the usual sense, but rather reflect the change in the scale of the Universe, R, as the radiation travels through space to reach us in the Present-Day Universe from remote and ancient sources (Destination 16). This is true of the light from distant galaxies or stars. And it is true of the CMB radiation. In the Present-Day Universe, the CMB has traveled nearly the full lifetime of the Observable Universe. And it has here too at the Dark Ages. But the Universe here at z = 100 is much younger than the Present-Day Universe, and so the era when the CMB was created must be much closer in time, much closer in 'distance'. Whatever the CMB is, we are clearly approaching it. And that means that the CMB radiation has not been as strongly redshifted here at z = 100 compared to the Present-Day Universe. This causes the CMB to peak at a bluer wavelength—the near infrared—here at z = 100, rather than in the microwave portion of the EM spectrum as we observe in the Present-Day Universe.[14] Since infrared wavelengths are considerably shorter than microwave radiation, Wien's Law (Destination 6) dictates that the temperature of the CMB will appear higher here at z = 100. Since the scale of the Universe—and hence the wavelength of radiation that travels through the Universe—varies as $R = 1/(1 + z)$, the observed peak wavelength of the CMB radiation is $(1 + z)$ = 101 times shorter than what we observe in the Present-Day Universe. Precisely the ratio of the wavelengths of microwave and infrared radiation (see Figure 5.8).

And yet, despite this important difference in the apparent temperature of the CMB, it's what has *not* changed that is more surprising. In particular, the broad features of the CMB that we see here at z = 100 are similar to those that we noted at Destination 18. The CMB still appears uniform over the entire sky.

---

13 Recall that one *micron* is one millionth of a *meter*, or, equivalently, one thousandth of a *millimeter*.

14 Strictly speaking, this means that the CMB is no longer microwave radiation and hence the name is no longer accurate. This will become even more apparent as we push on to the era when the radiation from the CMB was formed (Destination 23). Nonetheless, for the sake of simplicity, we'll retain the name Cosmic Microwave Background—CMB—throughout this discussion, even as we visit destinations when the radiation from the CMB is no longer microwave radiation.

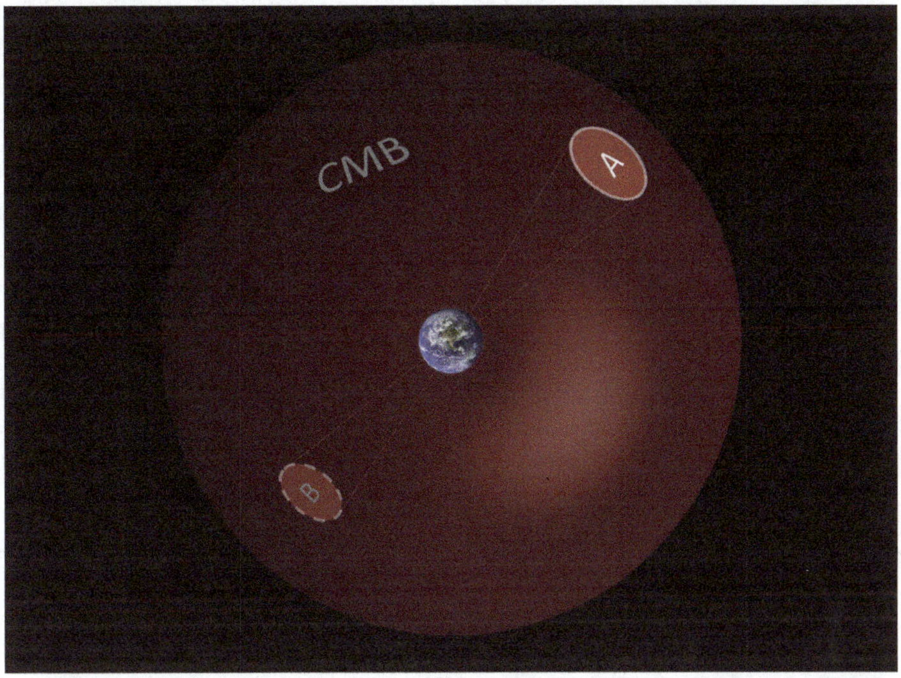

Fig. 22.3

A schematic plot of the Horizon Problem. If the radiation from the portion of the sky labeled 'A' requires the age of the Universe to reach us, then it requires *two times* the age of the Universe to reach the opposite side of the sky, labeled 'B.' That's fine, but why are the two opposite sides nearly identical in temperature? It is as if they have in the past communicated so that they could become synchronized, yet now find themselves so far apart that it is hard to understand how that synchronization could ever have occurred.

The spectrum of any one section of the sky appears indistinguishable to the spectrum of any other region of the sky. Hence, the underlying temperature of the CMB is the same everywhere we look. That is why the CMB is so remarkably smooth in structure over the entire sky. In one respect, this smoothness does not bother us as much as it once did. After all, we've seen that that the Universe manages to evolve from smooth conditions into clumpier conditions later on. And, with the help of dark matter, about as rapidly as observed. But as we ponder the uniformity of the CMB more thoughtfully, we realize there is something about it that has nagged at us from the first time we observed the CMB back at the Boötes Void (Destination 18).

To get a sense of the nature of this problem, imagine that we observe the CMB radiation at some time—exactly when doesn't really matter—in the history of the Universe. If we assume the CMB originates from a source in the very Early Universe—earlier even than our current vantage point at $z = 100$—then it should be clear that the time it takes light to reach us from the source of the CMB is essentially equal to the age of the Universe. But that means the source of the CMB is effectively at a 'distance' away from us that is equal to the Age of the Universe expressed in light-years (*ly*). For the Present-Day Universe, the source of the CMB radiation is about 13.5 billion *ly* away because the Age of the Present-Day Universe is 13.5 *Gyr*.

Since the CMB exhibits precisely the same temperature in all directions—at least to better than about 1 part in 100000 (we'll explore this more at Destination 23)—this means that points in the CMB located in opposite directions on the sky have the same spectrum, and hence the same temperature.

From our earlier argument, this means that these antipodal points are separated in time by about *twice* the age of the Universe. And that's the problem. There can't have been enough time—even over the entire Age of the Universe—for one side of the CMB to have synchronized its temperature with a region on the opposite side. But this is strange. If there has been no contact, either through direct physical contact or through the exchange of radiation[15], then the identical temperatures the CMB exhibits even for points on opposite sides of the sky must be a coincidence. A fantastically impossible coincidence. This classic dilemma is known as the *Horizon Problem* (Figure 22.3) since it is unclear how one side of the Universe can 'know' what another side—which is located beyond the 'horizon' of the first location's observable Universe—is doing. It is as if the entire CMB was once in sufficiently good contact to synchronize its temperature. Yet it is impossible to understand how this synchronization could have ever occurred. At least not if the Universal Expansion has proceeded in a manner that is similar to the way we observe in the Present-Day Universe. A possible solution of the Horizon Problem—one that has strange but profound implications—awaits us at our last destination. Even further back in time. Almost at the dawn of time itself.

15 Both approaches are standard ways that people use to make things equal in temperature. Think of refrigerators (cold air contacts the items one wants to cool) or heat radiators that use infrared radiation to heat nearby objects.

DESTINATION 22 QUESTIONS

22.1. Explain why molecular Hydrogen is not abundant at z = 100. **HINT:** How does the radiation temperature at z = 100 compare to regions with molecular gas? You may need to review the properties of the interstellar medium in earlier destinations.

22.2. How much denser is the Universe here at z = 100 compared with the Present-Day Universe? Consider the discussions in Destinations 20 and 21 to get a sense of how to calculate this. From this information and the value of the critical density (Destination 19), what is the density of matter at z = 100 if $\Omega = \Omega_{Matter} = 1.0$? How does this compare to the mean density of the gas in the warm ISM of the Milky Way (Destination 5)?

22.3. Imagine tracking our Galaxy back to a redshift of z = 100. On average, about how far away is the material that will ultimately become the Virgo Cluster? As a reminder, the Virgo Cluster is about 18 *Mpc* from Earth in the Present-Day Universe (Destination 14).

22.4. Following the logic described in this destination, calculate the temperature we would measure for the CMB at z = 2 (Destination 20) and z = 4 (Destination 21).

22.5. Back at Destination 6 we introduced an analysis developed by Sir James Jeans that describes how gas clouds can collapse under the influence of their own gravity. Following Jeans' lead, it is possible to calculate the 'Jeans Mass' corresponding to the mass of a gas cloud with a mean number density of n (in particles per cm^3), and temperature, T (in K) must have to be in perfect equilibrium between collapse (due to gravity) and expansion (due to its internal pressure). A cloud is unstable to collapse if its mass exceeds the Jeans Mass. In this question we will explore what the Jeans Mass is here at our current destination, the Universe at z = 100.

a) What is the mean number density of the gas at z = 100? You may assume the gas is entirely composed of H, which is actually not a bad guess.

b) What is the mean temperature of this gas at z = 100? You can assume the gas has the same temperature as the CMB as seen at that epoch.

c) The Jeans Mass is proportional to $T^{3/2}/n^{1/2}$. If the Jeans Mass is approximately $1M_\odot$ for $T = 5K$ and $n = 10^6\ cm^{-3}$ (conditions appropriate for star-forming regions in our Galaxy; see Destination 6), what is the Jeans Mass for the gas at z = 100?

d) What sorts of structures seen in the Present-Day Universe have masses similar to this and are sufficiently old to have plausibly started to form at z = 100?

22.6. The redshift of the Dark Ages is usually taken to be defined by the range from about z = 1000 to z = 30 or so.

a) Use the relation for the lookback time for a Critical Universe (Destination 19) to determine the time interval corresponding to this redshift range. That is, how long did the Dark Ages of the Universe last?

b) Between what two wavelengths would the 21-*cm* radiation from the atomic H gas that fills space the Dark Ages be detected on Earth? What kind of telescopes would be needed to detect this radiation? (**HINT:** See Figure 5.8.)

22.7. For this problem, refer to Destination 14 to get some basic data for M 31.

a) Assuming no other effects are at work, how far away would M 31 have been from our Galaxy at $z = 100$? You may also assume both galaxies were recognizable at that epoch, even though they definitely weren't! And you will need to adopt a present-day distance to M 31 (Destination 14).

b) The Hubble 'Constant' was about 38000 *km/s/Mpc* at $z = 100$ (see footnote 4). What recession velocity would we have measured for M 31 at that time? How does that compare to the actual relative speed of the galaxies observed in the Present-Day Universe? (Look this up on the web, or see Destination 14.)

22.8. Given the fraction, by mass, of H and He in the gas here at the Dark Ages at $z = 100$, what are the fractions of each element by number in the gas? How do these compare to the number fractions of H and He in the Present-Day Universe (Destination 5)? Comment on what may have happened to these elements since the Dark Ages to have caused these values to change over the intervening history of the Universe.

DESTINATION 23

THE ERA OF RECOMBINATION

23.1 INTO THE FOG

From utter darkness to this! Everywhere we look, it is dazzlingly bright. But not from a single source such as a star or galaxy, or even from a large number of discrete sources. Every spot of the sky emits nearly 10% as much radiation as the surface of the Sun. We have traveled back in time to the era in our Universe that—from Earth—has a redshift of $z = 1200$. An era when the Universe is literally aglow.

Fig. 23.1

In foggy conditions we can't always see everything we know exists around us.

But it's not merely the amount of light that fills this sky—or even the striking uniformity of the light—that is so remarkable about this place. It's how 'foggy' it is here. We can appreciate this by dropping beacons—ones that emit radio waves, X-rays, or any form of EM radiation ... it doesn't matter. As we move around in this era, we discover that the signal from every beacon gradually fades in brightness, eventually becoming too dim to be detected after we've traveled only a few thousand parsecs.[1] That may seem like a large distance—and it

1 There certainly was a time early in our journeys where 'a few thousand parsecs' would have seemed a daunting distance to travel (and, make no mistake, it *is* daunting). But since we set out to the Moon in Destination 1, we've learned how to travel over much longer distances than a few measly *kpc*. We've certainly taken advantage of this capability as we have explored many subsequent destinations, including this one.

certainly is—but it is only about 2% of the extent of the Observable Universe[2] here at z = 1200. This behavior is fundamentally different from anything we've encountered before. Up to now, it has *always* been possible to see a large fraction of the Observable Universe—and, usually, all of it—in one form or another from any of the destinations we've visited. For example, in our Galaxy, even in places where dust and gas can strongly obscure our optical view within the Galactic Plane, we can always penetrate that veil to see the remote reaches of the Universe by observing at radio or infrared waves (Destinations 9 and 10). Or we can simply direct our view away from the Galactic Disk (Destinations 11 and 13) to guarantee a fairly unobscured view of external galaxies—many located billions of *pc* from our Galaxy (Destinations 19–21)—that reside near the 'edge' of the Observable Universe.

The 'edge' of the Universe? Is that where we are now? It depends. If we define the 'edge' of the Universe as the time when the Universe commenced its expansion, then we certainly are not *there* ... yet (Destination 25).[3] We know this because as we look around us we still see the Hubble Law at work. The beacons we left behind earlier, for example, were all exhibiting significant redshifts at distances of only a few hundred *pc*, implying that the Hubble 'Constant' (denoted now as just '*H*,' without the subscript '0') is well over 1 *million km/s/Mpc* here! Since the age of the Universe at any time in its history is proportional to $1/H$, the mere fact that H is not infinite tells us that the Universe has a finite age at this location. Here, at a redshift of z = 1200, the Universe is only about 320000 *yr* old. Although tiny compared to the age of the Present-Day Age of the Universe—about 13.5 *Gyr* (Destination 19)—it's still not the earliest time we can visit. In that respect, we have not yet reached the ultimate 'edge' of the Universe.

But in another sense, we *have* gotten to some sort of edge within the Universe. If we can see only a few *kpc* around us, that means there is no way that anyone in the future Universe—say observers of the CMB on the Present-Day Earth—could possibly see us either. To reach them, any radiation produced in the Universe here at z = 1200 would have to travel millions of times further than it can travel through this murky place. It is as if we have reached a wall of fog that blocks radiation from redshifts higher than about z = 1200 from ever reaching the Present-Day Universe. A *luminous* wall of fog that shines all

2 Recall from Destination 18 that the radius of the 'Observable Universe' is defined as the distance light can travel over the age of the Universe at any given time. As we'll discuss further, here at a redshift of z = 1200, the age of the Universe is about 320000 *yr*. Its 'observable' extent is 320000 *ly*, or nearly 100 *kpc*. That's smaller than the size of the Local Group (Destination 14) in the Present-Day Universe! But it is a lot larger than the few *kpc* that we can see around us here at z = 1200.

3 Notice how we are starting to interchange time for space as we explore regions of higher and higher redshift. It is a trend we have noted before: 'When' we are is becoming more important than 'where' we as we explore the Early Universe. This is something we have to get used to.

around us. But this fogbank does not exist at a specific *place*. It exists at a fairly specific *time* in the early history of the Universe.

And then it hits us! We must have reached a time in the Early Universe *before* the Cosmic Microwave Background (CMB) was produced. The same CMB that we can see from Present-Day Universe (Destination 18) and that we noted from our visit to the Dark Ages (Destination 22). The luminous sky around us reveals an era in the Early Universe when conditions were drastically different from what we are used to in the Present-Day Universe. Here, at a redshift $z = 1200$, we find ourselves in some sense *inside* the CMB! Not at *the* ultimate limit of the Universe perhaps, but beyond a threshold we can never penetrate from any vantage point in the Present-Day Universe.[4]

The bizarre nature of the Alien Sky here at $z = 1200$ raises two obvious questions. First, why is the Universe so opaque—that is, foggy—at this time? Second, how did the murky Universe that surrounds us here transform into the comparatively transparent Universe we know exists at later times? The answers to these will reveal fundamental changes that occurred in the state of the Universe around this time. But more surprisingly, we will discover that the properties of the CMB also let us infer significant events that must have occurred at the earliest moments in the history of the Universe. Almost to the instant when the Universe began its expansion.

Let's start with the first question: Why can't we see very far around us here at $z = 1200$? In some respects, the gas that fills the Universe here appears similar to what we found at our last stop in the Dark Ages at $z = 100$ (Destination 22). The Universe here is still overwhelmingly composed of Hydrogen (H) and Helium (He), along with trace amounts of Lithium (Li) and Beryllium (Be). This makes sense. No stars will have formed between now—when the Universe is a few hundred thousand years old—and the Dark Ages a few million years into the future (Destination 22). But it's the state of the gas here at $z = 1200$ that makes this era fundamentally different from the Dark Ages. Though H still comprises 76% of all the mass of normal matter—that is, all the stuff that is not dark matter—we discover that it is now almost completely ionized. H atoms here are basically bare protons that share space with free electrons (and some He, Li, and Be nuclei). A gas like this that is composed largely of unbound charged particles is known as a *Plasma* (Figure 23.2), and here at $z = 1200$, an ionized, H-rich plasma fills all of space.

---

4 At least not with EM radiation. As we've seen and will discuss in more detail later in this destination, the photons that carry EM radiation can travel only a comparatively short distance here at $z = 1200$ before being scattered. However, some known particles such as neutrinos—which are nearly massless and produced during reactions involving protons and neutrons (see Destination 24)—are unaffected by the processes that scatter photons. In principle, it could be possible to penetrate the fog of the CMB by observing these sorts of particles directly. Unfortunately, we are a *long* way from being able to do this. Maybe someday.

The only way to understand this is to conclude that something in the Universe must be hot enough to ionize H and keep it ionized. The problem is that there doesn't seem to be anything in the Universe yet that can do this. No stars. No galaxies. No AGN (see Destination 22). Nothing that could conceivably produce energetic photons to ionize the gas here in the same way that hot stars ionize the gas of the Orion Nebula (Destination 6). Nothing at all. Well, nothing ... except the Universe itself!

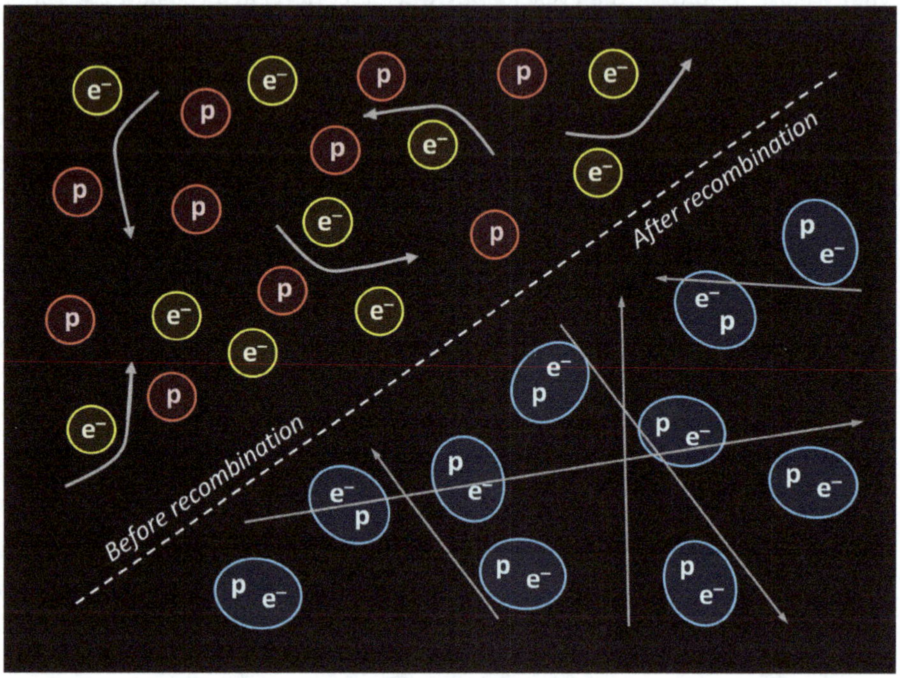

Fig. 23.2

At a time before recombination (*upper left*), the normal matter in the Universe was mostly in the form of free protons (red) and electrons (yellow), too hot to combine into neutral atoms. EM radiation that passes through this ionized gas—or plasma—is efficiently scattered by the lighter electrons (curved arrows) in a process known as Electron Scattering (or also Thomson Scattering). The net effect is that radiation can traverse only a small fraction of the Observable Universe before being scattered. That is, the Universe was 'foggy' before recombination. After recombination (*lower right*), the neutral H atoms (blue) no longer strongly interact with the radiation, apart from at specific wavelengths corresponding to valid energy transitions in the atoms (Destination 6). Consequently, virtually all forms of EM radiation can now freely travel through space (straight arrows) from their point of origin to all locations in the future—more transparent—Universe.

But hold on. How can the Universe ionize itself? When we measure it from Earth (or, equivalently, from the Boötes Void as at Destination 18), we found that the CMB emits radiation just as if it were a thermal source with a well-defined (but low!) temperature of about 2.73 *K*. That's far too cool to ionize *any* atom, or even to dissociate fragile molecules. But we also discovered that as we approached the CMB—that is, as we traveled to higher redshift (Destinations 19–22)—its spectrum shifted steadily to the blue. Toward higher temperatures. The reason is that the radiation from the CMB has passed through less expanding space the closer we get to its source. From our vantage point at the Dark Ages at $z = 100$ (Destination 22), we measured the CMB temperature to be 276 *K*, about 100 times higher than we observe in the Present-Day Universe. Although that's hot enough to destroy many molecules (see Destination 22), it is still far too low to ionize Hydrogen atoms either through radiation or collisions between the atoms (Destination 6).

But things are different here. By traveling to the time when the CMB was being created, we see its spectrum at the time it is produced. That is, with no redshift.[5] Consequently, the thermal spectrum we observe here at $z = 1200$ really *is* the emitted spectrum of the CMB (see Figure 18.8 from Destination 18), not some redshifted version of its spectrum. And what we now appreciate is that the *emitted* spectrum of the CMB does not peak in the microwaves, or far-infrared parts of the EM Spectrum, but as near-infrared radiation, just redward of what our eyes can detect.[6] From Wien's Law (Destination 6), we can infer that the temperature of the gas surrounding us must be about 3300 K, a factor of $(1 + z) = 1201$ higher than the temperature we measure for the CMB from Earth in the Present-Day Universe. It turns out that for gas containing practically no heavy elements, this temperature can ionize virtually all the H atoms in the Universe here at $z = 1200$. Under these conditions, H ionization can occur either when the atoms absorb sufficiently high-energy radiation—which is rare at this temperature—or, much more commonly, through collisions with other particles in the gas (Destination 6).

In this respect, the entire Universe is now somewhat like a gigantic H II region composed of ionized gas, superficially resembling the Orion Nebula (Destination 6). But there are some important differences. Since the gas here at $z = 1200$ contains essentially no heavy elements, its temperature balance is very different than a typical H II region in the Present-Day Universe.[7] The gas density is also comparatively much higher than in a typical H II region, with about 200 atoms—and electrons—per cubic *cm* compared to 1 atom per cubic *cm* in the hot gas of the Orion Nebula (Destination 6). Also, unlike an H II region such as Orion, the CMB is embedded in the fabric of the vigorously expanding space of the Early Universe. This alone ensures that conditions in the gas producing the CMB must change as the Universe itself expands and evolves. But perhaps the most important difference of all is the energy source. While H II regions are ultimately powered from ionizing radiation emitted from hot, massive stars, the Universe has no 'external' sources of energy here at $z = 1200$. Rather, the gas here is ionized from the thermal energy of the Universe itself. In this respect the CMB directly provides the clearest evidence that the Universe evolved from a hot, dense, compact initial state—the state we see

5 We can in fact see enough of the Universe from here at $z = 1200$ to appreciate that the light from distant regions that we can discern through the fog is redshifted. But, because we can see so little of the volume of the Observable Universe, the effect is quite mild. Nothing like the enormous redshift that the CMB exhibits as seen from Present-Day Earth.

6 Of course, since the thermal spectrum is broad (see Destination 6), that means that a lot of its radiation is emitted as optical radiation. The sky here would have a noticeably reddish tinge, but it would still be extremely bright.

7 Heavy elements in H II regions can act somewhat like thermocouples to regulate the gas temperature, mostly through the energy they can radiate from numerous emission lines. When there are no heavy elements at all in a gas, the temperature balance goes haywire. That's a big part of the reason the ionized gas here at $z = 1200$ is comparatively much cooler than the 10000 K gas found in Present-Day—and heavy-element-rich—H II regions like the Orion Nebula (Destination 6).

around us now—to the much cooler, more rarefied conditions characteristic of the Present-Day Universe. The CMB tells us that without a doubt, *the Universe had a hot and dense beginning.*[8]

The fact that the Early Universe is ionized is also important because it's the many free electrons at $z = 1200$ that make our view here appear so murky. The reason has to do with the distinct ways electrons interact with light, depending on whether they reside in or out of atoms. As we discovered at Destination 6, electrons bound to atoms or ions can exist only in orbits of fixed energy. The critical consequence of this is that bound electrons can absorb or emit radiation only at *specific* wavelengths that correspond to energy differences between the allowed orbits. That is, they can produce only absorption or emission lines (see Destinations 5, 6, and 8). But when electrons are unbound to atoms—as in the hot plasma here at $z = 1200$—they can effectively scatter EM Radiation of *any* wavelength. Radio waves. Infrared radiation. X-rays. *Any* EM Radiation. This process—known as *Electron Scattering* or sometimes *Thomson Scattering*—is what limits our view here (Figure 23.2). After photons have scattered just a few times off the free electrons that fill space here at $z = 1200$, it becomes impossible to tell where they originally came from. And so the sky looks uniformly bright in all directions. And very murky.

23.2 LIFTING THE VEIL: THE ERA OF 'RE'-COMBINATION

Of course, this situation *must* change. After all, by the time the Dark Ages at $z = 100$ roll around in just a few million years, we know that most of the normal matter of the Universe—that is, excluding the enigmatic dark matter—will almost entirely be in the form of *neutral* Hydrogen (Destination 22). We also know that the Universe itself will become essentially transparent to EM radiation. The fog of electron scattering that pervades the Universe here at $z = 1200$ must eventually clear.

So the question becomes, *why* did the Universe change from essentially all ionized to essentially all neutral? Part of the answer is tied to the evolution of the Universe itself. As it expands, the Universe cools[9], and as it cools, the free

---

8 You may recall we came to this same conclusion at Destination 16 by analyzing the implications of the Hubble Law. It is always reassuring when multiple lines of reasoning reveal the same conclusion, and worrisome when they don't.

9 Technically, this is a direct consequence of the *Perfect Gas Law* (see Footnote 19 of Destination 6). As a gas expands along with the Universe, it must cool down as long as

Fig. 23.3

The transition of the Universe out of the Era of Recombination is similar to travel from within a fog bank to a clear region in sunlight. Within the fog, it is technically still possible to see, just not nearly as far as in clear conditions. Notice that the transition is not instantaneous. Even in a sharp fog bank like the one depicted here, there's a region that is partly foggy, partly clear. Although this is a good analogy for the Universe as it recombined, what's fundamentally different is that the Universe also gets cooler as it evolves from its hotter, foggy state before recombination, to its cooler, more transparent state after recombination.

electrons and protons in the plasma can start to combine to produce neutral H atoms. The *first* neutral atoms ever produced in the Universe! Of course, as the free electrons disappear, electron scattering becomes steadily less effective and space becomes increasingly more transparent. And it will stay this way for some time. There won't be any hot, young stars forming in the Universe between here at $z = 1200$ and the Dark Ages (Destination 22) to re-ionize the gas. The CMB itself will have become far too cool—due to Universal expansion—to ionize the gas at later times. This process in which the Universe steps out of the fog of its youth into the transparency of its maturity is remarkable in itself (Figure 23.3). But as we'll now see, the transition of the Universe from its early ionized state to its later neutral state imprints the CMB with a pattern that reveals the conditions in the Universe at a time well under one second after the expansion of the Universe commenced (Destination 25).

This discussion suggests that our current vantage point at $z = 1200$ must be close in time to the era when neutral H atoms first formed in the Universe. And indeed it is. We are currently at the threshold of emerging from a fog bank when the first ghostly outlines of the world beyond starts to emerge (see Figure 23.3). We can be a bit more precise about this by formally defining the radius of the Observable Universe as

$$R_{obs} = cT,$$

where c is the speed of light, and T is the Age of the Universe at any given time (note that R_{obs} is the same as the Light-Travel Distance we introduced at Destination 19). We can also define R_{vis} as the distance we actually *can* see within the Universe after taking into account all sources of scattering

there is no input of new energy.

and absorption. We can call this the size of the 'Visible Universe.' Notice the difference: R_{obs} is how far we could see if the Universe were transparent to radiation, while R_{vis} is how far we can actually see due to electron scattering. What we can say with certainty is that here at $z = 1200$, the Observable Universe is far larger than the distance we can actually see around us; far larger than the Visible Universe. Mathematically, we can state this as $R_{vis} << R_{obs}$. It turns out that by a redshift of $z = 1000$, the opposite is true. At that point in time in the Universe, enough of the free electrons have combined with H nuclei so that we could potentially see much further than the size of the Observable Universe. That is, $R_{vis} >> R_{obs}$. The entire Observable Universe has become 'visible.' Obviously, at some time between these two redshifts it must be true that $R_{vis} = R_{obs}$. In terms of the state of the gas in the Universe, this transition from fog to clarity occurs when about half of the electrons and free protons in the hot gas of the Early Universe have combined to form neutral H atoms and half remain unbound.

Somewhat misleadingly, astronomers refer to this special period as the *Era of Recombination*. We'll follow convention and use this term—it would be too confusing with respect to most of what's written about the CMB to do otherwise—but it is important to appreciate that this really is the *Era of 'First'-combination*. As we've emphasized, this is the first time *ever* when stable, neutral H atoms can exist in the history of the Universe.[10] From the detailed properties of the CMB, we estimate that recombination occurred at a redshift of $z_{Rec} = 1090$, at a time when the Universe had an average temperature of $T_{Rec} = 2970\ K$. But the process was not instantaneous. Recombination can be said to have begun when neutral H atoms began to form in significant numbers. This occurred at a redshift of about $z = 1200$, our current destination. By a redshift of $z = 1000$—a mere 50000 *yr* in the future—just about all the free electrons combined to form neutral H atoms. By that time, the Era of Recombination effectively came to an end.

This picture helps explain two key features of the CMB that we've observed from earlier locations (Destinations 18 and 22). First, thermal radiation produced as the hot, young Universe becomes transparent (from $z = 1200$ to $z = 1000$) can

---

10 In general, the term 'recombination' refers to situations where electrons become bound to ions in regions where the gas has been ionized. Hydrogen atoms in the gas of the Orion Nebula (Destination 6), for example, will typically be ionized by radiation from hot stars in the region or by energetic collisions with other particles. After some time, those ions may then truly recombine with free electrons (though probably only to be ionized again soon afterwards). The point is that the term 'recombination' makes sense most times when electrons merge with ions to produce neutral atoms or less-ionized ions. This process is at work in the CMB. It's just that in that case, the protons and electrons had never been able to form stable neutral atoms in the earlier history of the Universe.

suddenly[11] travel freely from its point of origin to any location in the Universe, including ones that exist far in the future. Since the journey of this radiation is entirely through the expanding space of the Universe (Destination 19), it will be redshifted as it travels away from its origin, causing the spectrum to appear as if it had been produced by a lower-temperature thermal source.[12] By the time of the Dark Ages roll around—about 17 Myr after recombination—the CMB will appear about 10 times cooler than its emitted temperature (Destination 22). By the time it reaches the Present-Day Universe, the radiation will have shifted so far to the red—that is, to longer wavelengths—that it appears to have originated from a thermal source with a temperature of just 2.7 *K*.

Second, since the radiation from the CMB originates at a specific *time* in the history of the Universe, it will also appear to arrive from all directions in the sky no matter where it is observed from.[13] That too is consistent with what we've seen at previous destinations: The CMB always fills the entire sky. The sky of the Dark Ages (Destination 22). The sky from the Boötes Void (Destination 18). The sky from Earth (Destination 1). All because the CMB originates at a specific *time*, not a specific *place*. Due to the nature of the Hubble Law (Destination 16), we are guaranteed to encounter that time regardless of the direction we look. And, remarkably, the spectrum of the CMB—measured anywhere in the sky and from any location or time—is always consistent with that of a thermal source with a single, well-defined temperature.

We already noted at Destination 22 how strange this is. For one thing, it is hard to understand how the mean temperature of the CMB became so uniform over the entire sky. This is the Horizon Problem we discussed at Destination 22, but the answer to that particular paradox does not lie here at $z = 1200$. We must travel back almost to the time when the expansion of the Universe commenced (Destination 25) before we can address that issue. But there is a second 'problem' with the CMB: Why is it so smooth? As we poke around the Universe here at $z = 1200$, we appreciate that conditions here really are as remarkably uniform as the smoothness of the CMB radiation suggests. The temperature of the gas seems to be the same anywhere we go to better than 1 part in 10000. That's equivalent to the entire surface of the Earth having the same temperature to within 0.03 *C*, or about 1/20 of a degree Fahrenheit. And yet, despite the numbing monotony here at $z = 1200$, we have to keep reminding ourselves that this is the *same* Universe that will eventually be full of structure, ranging from superclusters, cosmic walls, galaxy clusters, galaxies,

11 'Suddenly' is a relative term here. It took the Universe about 50000 *yr* to change from 'mostly foggy' to 'mostly clear.' But, compared to the age of the Present-Day Universe—13.5 *Gyr*—that's a very short time.

12 It is not hard to prove that if the original source of radiation was a thermal body with a Planck spectrum (see Destination 6), the radiation will retain its thermal spectrum even as it passes through the expanding Universe.

13 And, of course, so long as it is at a time after the Era of Recombination.

stars, planets (Destination 18).[14] It's the same Universe that *we* will inhabit some 13.5 *Gyr* in the future. It seems inconceivable that the CMB is *perfectly* uniform. If it were, the Present-Day Universe would not be clumpy. And we would not exist to ponder the problem.

Ironically, our current vantage point at $z = 1200$ is not ideal to search for structure in the Early Universe. After all, conditions here are so foggy that we can see only about 1 millionth of the Observable Universe from any location. A better approach—though one with its own limitations—is to view the CMB from some future time. That way, we can get a comprehensive look at the CMB from the viewpoint of a comparatively transparent location. Since the CMB is visible everywhere and at every time after recombination, we could choose any location/era in the future Universe to study the CMB. But, as it happens, our best instruments to map the CMB are conveniently located back in the Present-Day Universe at Earth. What do we see?

At first, much the same as we see from our vantage point at $z = 1200$: Extreme smoothness. As we noted at the Boötes Void (Destination 18) and from the Dark Ages (Destination 22), a map of the CMB reveals that its temperature over the entire sky is uniform at the level $\delta T/T \approx 10^{-4}$, where δT represents the variations of the temperature of the CMB compared to its average temperature T (see Figure 18.8). But as we look harder, we discover that the CMB turns out not to be precisely the same temperature everywhere. With considerable effort, it *is* possible to measure minuscule temperature variations in the CMB at the level of *micro*-K. That is, to a few millionths of a

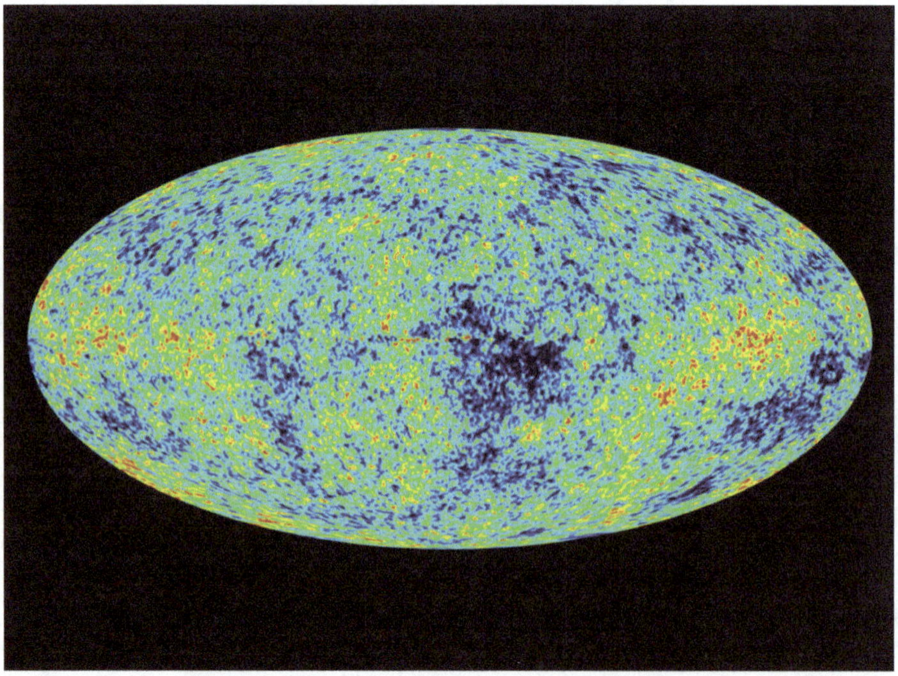

Fig. 23.4

A map of the entire sky as seen from Earth, showing the CMB temperature fluctuations measured by the Wilkenson Microwave Anisotropy Probe (WMAP) in 2005. The colors denote temperature differences with the red/yellow regions hottest, and the blue/dark blue regions coolest. This map reveals the distribution of density fluctuations at the Era of Recombination when the Universe was about 385000 years old, or about 0.003% of its current age. A careful analysis of this map reveals that the fluctuations are correlated on various angular scales (see Figure 23.5). These fluctuation scales, in turn, reveal conditions in the Universe prior to recombination. The smallest-scale structures visible in this map correspond to regions that span over 200 *Mpc* in the Present-Day Universe. Keep in mind that the temperature fluctuations in this map are typically only a few *micro-K*. If variations in the surface of the Earth varied by a comparable amount, the tallest hills would be only about 200 *feet* high. The map shown here plots the sky in the same way as Figures 9.3 to 9.6. The band of the Milky Way runs horizontally across the middle of this map, but its radiation has been removed to highlight the CMB fluctuations.

CMB image from WMAP/NASA.

14 Even bizarre things like manatees and Venus flytraps!

degree *K*! At this level of precision, the CMB is no longer the smooth, single-temperature structure we have become used to, but instead reveals itself to be a riot of temperature variations that cover the entire sky (Figure 23.4). These so-called *CMB Temperature Fluctuations* reveal for the first time that there is indeed structure in the CMB. Structure that had eluded us up to now because it is so subtle.[15]

What do these temperature fluctuations signify? The simplest interpretation is that they represent variations in the matter density at the epoch when the CMB was produced. If we denote these density fluctuations as $\delta\rho/\rho$, it is not hard to prove that

$$\frac{\delta\rho}{\rho} = 3\frac{\delta T}{T}.$$

This equation simply states that as the temperature varies from place to place in the Early Universe, the inferred density of the matter emitting the CMB radiation changes by a factor that's three times larger. Since each part of the Universe at this early time radiates as a thermal source and the gas acts as a perfect gas where temperature and density are proportional, this makes sense. We know from Planck's Law (Destination 6) that hotter, denser regions should be somewhat brighter than cooler, more rarified regions simply because the emitted flux from a thermal source varies as $F = \sigma T^4$. As a result, typical overdense regions here at $z = 1200$—the regions that appear hotter in the CMB—are denser by about 1 part per 100000 than the average density of the Universe at the time of recombination. That is, $\delta\rho/\rho \sim 10^{-5}$. Cooler regions are underdense by a similar factor.

But the temperature fluctuations in the CMB tell us more than this. They also reveal *patterns* in the fluctuations. To understand this, we have to define the distribution the temperature fluctuations on the sky in a more precise manner. One way to do this involves modeling the fluctuations in terms of periodic mathematical functions such as sines and cosines. This approach lets us produce the so-called CMB *Power Spectrum* of the fluctuations (Figure 23.5), a rather complicated plot that reveals the relative importance of the different sine and cosine curves that must be used to best describe how the fluctuations are distributed over the sky (see Figure 23.6 for a description of

15 To put the size of the CMB temperature fluctuations into perspective, a cue ball has surface variations equivalent to about $dR/R = 0.01$, about 1000 times 'lumpier' than the CMB where we have equated the CMB temperature fluctuations with the cue ball's surface roughness. Even the surface of the Earth is a lot rougher than the CMB. If scaled up to the size of the Earth, on average the CMB temperature fluctuations are equivalent to hills (or valleys) that are about 60 *m* high (or deep). That's about 100 times smoother than the actual surface of the Earth.

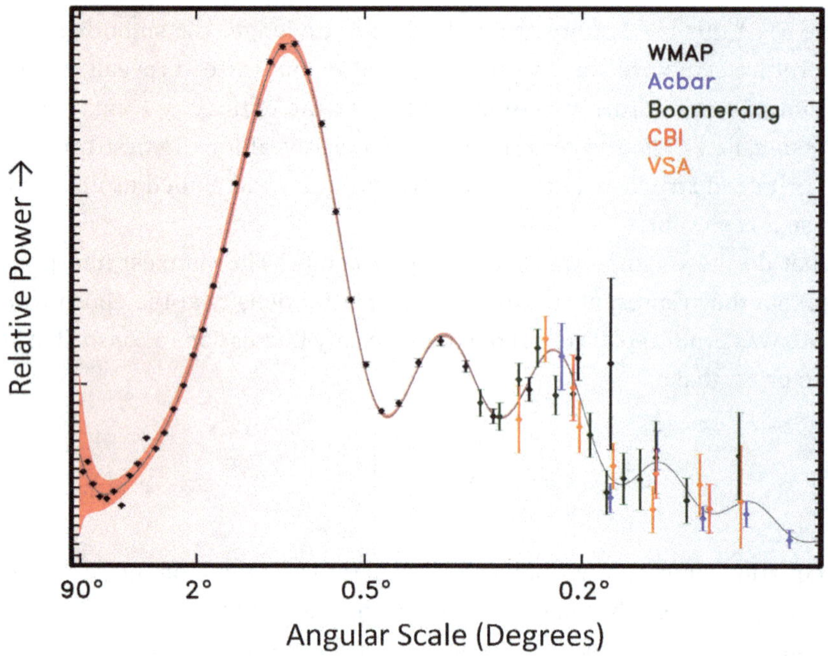

Fig. 23.5

This plot shows the CMB Power Spectrum measured from the map of the CMB temperature fluctuations measured by WMAP (Figure 23.4) and four other CMB experiments as listed in the upper right. The individual data points are shown representing the amount of 'power' at different angular scales in the sky. For example, the tall peak at about 1 *deg* scale (the *x*-axis) reveals that there are many structures in the CMB temperature fluctuation map that are about that size in the sky. The curved band/line shows the best fit of the data to a model for the Universe that incorporates parameters that describe the evolution of the Universe before and after recombination. The red background tint emphasizes that the features in this power spectrum that are at the left of the plot correspond to big structures (and characterized by long wavelengths in the Power Spectrum), while the right background tint (blue) corresponds to smaller features (characterized by shorter wavelengths). Features that correspond to the 1-degree scale of the leftmost peak have grown to sizes of around 200 *Mpc* in the Present-Day Universe (see Figure 23.4).

Image from WMAP/NASA.

how the power spectrum can be interpreted in terms of what is seen in the map of the CMB).

What's astonishing about this analysis is the discovery that temperature fluctuations at certain angular sizes are very clearly preferentially represented in the CMB. The most prominent size scales in the CMB are revealed by the peaks in the power spectrum seen at 0.91, 0.35, and 0.22 *deg* (Figure 23.5). Since we know how distant the CMB is from Earth from its redshift ($z \sim 1200$, we can convert these angles into the physical sizes of comparable structures within the Present-Day Universe.[16] When we do this, we discover that the three tallest peaks in the CMB power spectrum correspond to structures that are about 220, 80, and 50 *Mpc* in size in the Present-Day Universe, reminiscent of the sizes of structures ranging from Cosmic Walls to Superclusters (see Table 18.1; Destination 18).

But this does not entirely resolve our dilemma: The fact remains that the magnitude of the density fluctuations in the CMB—about one part in 100000—is still tiny compared to the range of the density fluctuations of the Present-Day Universe

16 This is equivalent to using the Small-Angle Formula from Destination 3, except in this case we have to account for the expansion of the Universe. This greatly complicates the calculation, but it is straightforward enough as long as we know the expansion properties of the Universe (Destination 19). However, it is bizarre. For any model of an expanding Universe (Destination 19), an object of fixed physical size would, at first, appear to get smaller with increasing distance, just as common sense and experience dictate. But, at large redshift, this trend reverses, and the object would actually start to appear *larger* (in terms of angular size) with increasing distance! This highly non-linear (and unintuitive) behavior occurs due the way light must propagate in an expanding space. The observation of this effect is one of the strongest pieces of evidence that space is indeed expanding on large scales.

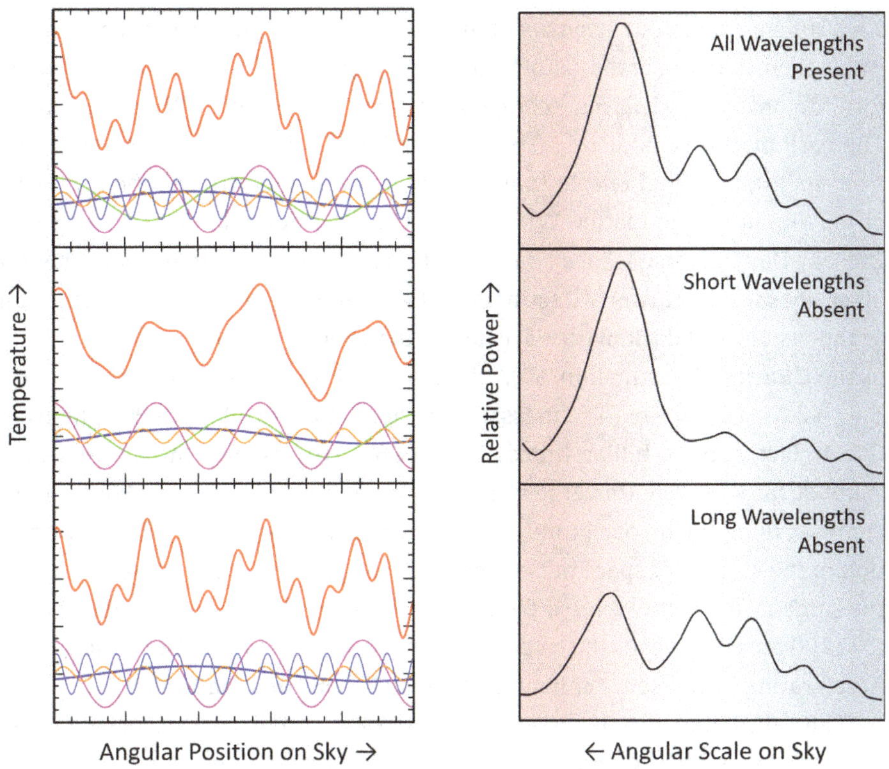

Fig. 23.6

(*Left*) The red curves in the three panels are the sums of the individual sine curves shown below each (the summed curves have been offset so they can be easily distinguished in this plot). (*Right*) The corresponding power spectra in each case. The top panels show the fluctuations and their power spectrum, assuming all the sine curves are present. If a short wavelength component is removed (middle panels), then power on the right side of the power corresponding to short wavelengths (blue) is diminished (middle panels). If a long wavelength component is removed (lower panels), there is less power on the left side of the spectrum corresponding to long wavelengths. A similar—but computationally much more complex—analysis can be done of the two-dimensional temperature fluctuation map (Figure 23.4) to produce its power spectrum (Figure 23.5).

where the density contrast is around 10000 to 1. Even taking into account the accelerating role of Dark Matter in the process (see Destination 22), it remains unclear in detail how such tiny density fluctuations can evolve—even over the 13.5 *Gyr* lifetime of the Universe—into the complex and extremely clumpy structures that characterize the distribution of matter in the Universe today. More fundamentally, nothing we've seen so far answers the most basic question regarding the nearly-but-not-quite-smooth structure of the CMB. Where, when and from what did the CMB density fluctuations arise in the first place?

23.3 THE SEEDS OF CREATION

To get a sense of how the smooth Early Universe could have transformed into the clumpy Present-Day Universe, let's return to our destination at redshift

$z = 1200$ and carry out a census of the content of the Universe at that time. As we've seen, the form of the visible matter here is pretty simple: It is mostly in the form of ionized gas (plasma) composed almost exclusively of H and He atoms. The dark matter is made up of, well ... who knows? But we can still determine from its gravitational effects that it is far more abundant—by about a factor of ten—than visible matter. When we last carried out a similar census in the Present-Day Universe—based on the motions of stars in individual galaxies and of galaxies within clusters (Destinations 14, 16 and 19)—we found that the sum of the visible and dark matter accounted for about 27% of the critical density in the Universe (Destination 19), which in the Present-Day Universe is about $\rho_{crit,0} = 2.6 \times 10^{-30}\ g/cm^3$. The mass density here at $z = 1200$ is *much* higher than this. In fact, about 1 billion times higher ($4.6 \times 10^{-21}\ g/cm^3$)! In one sense, this is not surprising. The Universe is now approximately $(1 + z)^3 \approx 10^9$ times more compact than during our home era. Since *density = mass/volume,* the contraction of the volume of space means that the mean density will increase by about the same factor. But is there a paradox here? It seems the density of matter at $z = 1200$ greatly exceeds the critical density, implying a Closed Universe, not the Accelerating Universe we concluded that we live in (Destination 19).

The solution is that the critical density (Destination 19) has also changed. The reason—which we noted earlier—is that the Hubble 'Constant' here at $z = 1200$ is astoundingly large, well over *one million km/s/Mpc*! It is clear that whatever density could halt the expansion in the Present-Day Universe when $H_0 = 72\ km/s/Mpc$ (Destination 16) will be completely inadequate here at $z = 1200$ where the expansion is far more vigorous. Consequently, the critical density must *also* be much higher here.[17]

And indeed it is, though perhaps not quite as high as we might have expected. Here at $z = 1200$ the mass-ratio parameter for matter, $\Omega_{Matter} = \rho_{Matter} / \rho_{Crit}$, is very close to the critical value of 1.0. But this is not surprising. It turns out that as we go back to high redshifts in the Universe—and as long as we do not live in a Wide-Open Universe with Ω_{Matter} equal to zero (Destination 19)—the value of Ω_{Matter} will get closer and closer to 1.0 regardless of what value it might have at later times in the history of the Universe.[18] We can see this directly if we write

17 The expression for the critical density (Destination 19) is $\rho_{Crit} = (3H^2)/(8\pi G)$, where H is the value of the Hubble 'Constant' at the redshift of interest. Clearly, as H increases (with increasing redshift), the critical density increases even more rapidly.

18 This analysis assumes that $\Omega \approx \Omega_{Matter}$ and that the density parameter includes the contributions by normal and dark matter since both contribute mass. This is reasonably correct for the era of the CMB here at $z = 1200$ because it turns out that Ω_Λ—the parameter that causes the acceleration of Universal expansion in the Present-Day Universe—tends toward zero the earlier we travel in the Universe. Moreover, Ω_{Rad} is still comparatively small at that time, which is somewhat ironic given how blindingly bright it is here. As we push to even earlier times in the Universe, however, the temperature will steadily increase and Ω_{Rad} becomes relatively larger, eventually dominating over Ω_{Matter}. At such times—in what is known as the *Radiation Era* of the Universe (Destination 25)—the

the (somewhat intimidating) expression for how Ω_{Matter} changes with redshift, z, given the value of $\Omega_{Matter,\,0}$ that we measure in the Present-Day Universe:

$$\frac{1}{\Omega_{Matter}} - 1 = \left(\frac{1}{\Omega_{Matter,0}} - 1 \right) \left(1+z\right)^{-1}.$$

Notice that as $(1 + z)$ gets large—that is, as we go further back in time toward when the Universe began its expansion—the right-hand side of this equation approaches zero regardless of the (non-zero) value of $\Omega_{Matter,\,0}$ (which, you may recall is around 0.27; see Destination 19). The only way the left-hand side of this equation can also approach zero is for Ω_{Matter} to get closer and closer to 1.0. This behavior is known as the *Flatness Problem* by astronomers[19] because it suggests that the Universe must have started out to be almost, but not exactly, flat at very early times (see Destination 19). That is, the initial value of Ω_{Matter} must have been precisely equal to 1.000... to at least 30 decimal places! If that were not the case, the geometry of the Present-Day Universe would not appear to be even remotely as flat as it is measured to be. Another way to frame the dilemma is by asking why $\Omega_{Matter,\,0}$ is so close to 1.0 in the Present-Day Universe, yet not exactly equal to one? As we shall see at Destination 25, the Universe has doubled in size far fewer times since recombination than over the time that preceded that era. But if that's so, why has Ω_{Matter} only relatively recently begun to stray away from 1.0? It could have done so much earlier in the expansion history of the Universe. We'll see that one possible solution of the Flatness Problem also helps resolve the Horizon Problem—the uniformity of the CMB temperature over the entire sky—that we encountered in the Dark Ages (Destination 22). But the possible solution does not appear until we reach a time corresponding almost to the instant the Universe commenced its expansion (Destination 25).

For our purposes, there's an important implication to the fact that Ω_{Matter} has a value very close to 1.0 here at $z = 1200$. In particular, it means that any regions that are even very slightly denser than average may locally have $\Omega_{Matter} > 1$. The importance of this is that such regions can begin the process of gravitational

---

equation describing how Ω changes remains valid, but one must realize that Ω_{Matter} has to be replaced by $\Omega_{Matter} + \Omega_{Rad}$.

19 The name comes from the fact that the geometry of the Universe for $\Omega = 1.0$ —which is the value of Ω for both the Critical Universe and for a Flat, Accelerating Universe—is that of 'flat' or Euclidean geometry (see also Destination 19). In other cases where $\Omega \neq 1.0$—that is, either an Open or Closed Universes (Destination 19)—the geometry of space would not obey Euclidean rules. In those Universes, weird things would happen. For example, parallel lines would cross. And the interior angles of triangles would not add up to 180 *deg*.

collapse even as they and the rest of the Universe expand vigorously.[20] Indeed, it appears that these slightly overdense regions here at $z = 1200$ are precisely the regions where structures such as galaxies, clusters, and superclusters will form in the future Universe. Recall that in the case of star formation (Destination 6), matter that eventually becomes a star comes from the interstellar medium (ISM). But since much of the ISM is hot, the internal gas pressure does not allow a typical region of the ISM to collapse under the force of its own gravity and the only way to overcome the outward force of the gas pressure is if the cloud is very dense. Though it is conceivable that this could happen in the hot ISM if some violent event compressed it (maybe the matter ejected by a supernova?), what's more common is for cold clouds—ones with low temperatures, low internal pressures, and high central densities—to collapse first.

The same basic process is likely acting on matter in the Early Universe but with some significant differences. Regions that can collapse must not only be overdense—they require that locally $\Omega_{Matter} > 1$—but they must also have the right temperature for their internal gravity to start the process. Since the temperature here at $z = 1200$ is high, the internal pressure of the gas is also substantial; it's clear we need a lot of mass. To determine which of these regions can really begin to collapse requires that we take into account the opposing effects of the hot gas—including the radiation—to disperse the matter. A classical analysis of this problem shows that only overdense regions with masses greater than around 10^6 to $10^7 M_{\odot}$ can start to collapse at $z = 1200$ *before* the pressure of the hot gas disrupts them. Lower-mass regions do not have enough matter to collapse in the first place, at least not at the high temperatures here. At the other extreme, large, overdense regions certainly have enough mass to

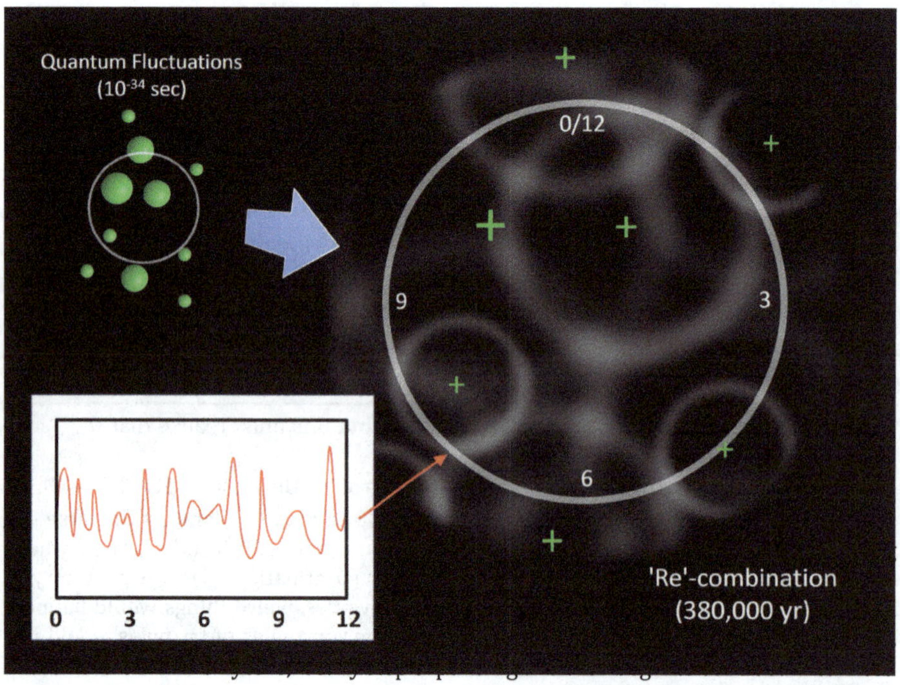

Fig. 23.7

The current interpretation of the CMB temperature fluctuations is that they arise from density enhancements caused by sound waves produced in the Early Universe. The idea is that quantum fluctuations from the very earliest periods in the history of the Universe (*upper left* and Destination 25), spawned density (sound) waves that expanded like the disturbance from a stone dropped in a pond (*right*). Those waves could propagate in the Universe only until recombination (denoted by the gray circle at right) at which point they are 'frozen' into the distribution of matter. A plot along the circle starting at 0 (at top) and running clockwise back 12 (again, at top) would reveal temperature (and hence, density) fluctuations as shown in the plot (*lower left*). In the case of the CMB, the white circle is replaced by the surface of a sphere and the plot is replaced by the temperature fluctuation map of the sky. But the idea is the same. By analyzing the power spectrum of temperature fluctuations and having a model that describes how the waves evolved in the Early Universe, it is possible to use the CMB to probe early times in the Universe and constrain how the expansion of the Universe evolved.

collapse, but they also fail to do so at this redshift. In these cases, the matter tends to get 'smoothed out' by large-scale motions of the hot gas—essentially turbulence (Destination 10)—and by radiation pressure (Destination 17) before the collapse can progress very far. As a result of these complicated effects, there's a 'sweet spot' corresponding to a few million solar masses that can begin to contract under its own gravity in the early Universe, and so long as the region where this mass is located is sufficiently overdense to promote collapse in the first place. It is remarkable that the favored mass range is similar to the masses of the smallest galaxies (for example, the Sculptor dwarf galaxy; see Destination 13) and the oldest globular clusters (such as M 92; see Destination 11) that we see in the Present-Day Universe.

Though this process implies that small objects formed first in the early Universe, big structures clearly clearly had to form, too. After all, we see galaxies much more massive than a few solar masses in the Present-Day Universe. In the picture we've just described, these larger objects likely started to form only after the original, smaller systems began to collide and merge. Over time, a progression of such mergers would eventually form bigger systems like Maffei 1, M 31 or our own Galaxy (Destinations 9, 14, and 15). Because some of the overdense regions at the era of the CMB have size scales ranging from tens to a few hundred *Mpc*, gravity causes individual galaxies to congregate into structures that will eventually evolve into clusters, superclusters, and cosmic walls in the Present-Day Universe (see Destination 18). The overall picture we derive from the properties of the CMB regarding how galaxies and larger structures form in the Early Universe is broadly consistent with the idea of Hierarchical Galaxy Formation we inferred from the properties of young galaxies at $z = 2$ (Destination 20).

But this description still fails to account for the *origin* of the temperature fluctuations. To understand that requires that we consider key events that must have occurred before the Era of Recombination here at z = 1200. As we'll see at Destination 25, one of these events was the period in the Early Universe—during what's known as the *Radiation Era*—when the mass density of radiation greatly exceeded that of matter itself. Another important event was *Inflation*, a process by which the Universe grew extremely rapidly in size for a short time just after the expansion commenced. We'll leave the details of all this for Destination 25, but the point to make here is that Inflation imprinted a random pattern of density fluctuations throughout space at a very early time in the history of the Universe, well under one second after Universal expansion commenced. The idea is that these initial density fluctuations were destined to become the temperature variations we observe in the CMB itself. But how?

While the Universe was expanding during the Radiation Era (see Destination 25), normal matter was highly ionized and hence very strongly coupled to the radiation field. What this means is that any process that tried to change to the distribution of normal matter would be strongly suppressed—perhaps even reversed—by the uniformity of the radiation field. But Dark Matter (DM), which by definition hardly interacts with radiation, was not coupled to the radiation or to normal matter. Consequently, DM got a 'head start' in the process of gravitational collapse at the overdense regions left behind by Inflation. As a result, such regions got denser faster, and slightly underdense regions became more rarified sooner, in effect amplifying the initial density variations (similar to DM's role described at Destination 22). As this occurred, the enhanced gravitational attraction of the densest regions started to pull in normal matter. However, the strong coupling of normal matter and the intense radiation field countered any tendency toward gravitational collapse. As a result of these opposing forces—inward toward the more compressed clumps of dark matter, but outward due to the strong coupling with the radiation field—the normal matter began to oscillate in density, and in the process produced ultra-low frequency sound waves. These waves then traveled outward from regions corresponding to the original density fluctuations produced in the Inflation Era very, very early in the history of the Universe. Over time—and in particular, by the time Recombination commenced at $z = 1200$—these steadily expanding disturbances became a tangle of crisscrossing waves much like the turbulent surface of a lake pelted by hail or debris falling into it (see Figure 23.7).

Since these waves relied on the close coupling between radiation and normal matter (but not dark matter) in a hot plasma, they could travel through the expanding Universe only as long as it was ionized. Once the Universe recombined—as it is starting to do here at $z = 1200$—the density fluctuations from these sound waves were effectively frozen into the distribution of normal matter, resulting in the characteristic features seen in the CMB power spectrum (Figures 23.5 and 23.6). These random density fluctuations that arise from events that occurred nearly at the birth of the Universe eventually evolve into the kinds of structures we see in the Present-Day Universe (see Destination 18).

It is a strange, complicated story. But it explains the distribution of the temperature fluctuations in the CMB to incredible precision (Figure 23.5). And it implies that the myriad structures we see in the Universe today—and, ultimately, even ourselves—result from the crests of waves produced by the interactions of dark and normal matter generated by random density fluctuations on microscopic scales in the infant Universe that occurred long before the CMB was formed at $z = 1200$. But this is *still* not the complete story. We must continue to press further back in time—long before the Era of Recombination—to fully appreciate how conditions in the Early Universe shaped what we see in the Present-Day Universe.

DESTINATION 23 QUESTIONS

23.1. Dwarf galaxies and globular clusters appear to have masses similar to the lower limit we inferred could form in the early Universe. What properties of these stellar systems are consistent with the notion that they formed soon after recombination? Be as specific as you can.

23.2. Calculate the theoretical size of the Observable Universe at $z = 1200$. If the typical photon travels only about 10 *kpc* before being scattered, what fraction of the volume of the Observable Universe can we actually see? **HINT:** You'll need to look up the age of the Universe at recombination given in this destination in order to estimate the volume of the Observable Universe.

23.3. Let's say you are a photon produced just as the Hydrogen in the Universe becomes neutral at a redshift of about 1200. Describe a plausible history of your life from that point into the future. That is, as a photon, where will you go? How far might you go? What might destroy or alter you as you travel through space?

23.4. Until you think about it carefully, it is hard to appreciate just how young the Universe was at the Era of Recombination compared to the age of the Present-Day Universe. Calculate your age after birth when the ratio of your age then to your current age was the same as the age of the Universe at recombination to the age of the Present-Day Universe.

23.5. As noted in the text, the mass density of the Universe at $z = 1200$ was about $4.6 \times 10^{-21}\, g/cm^3$.

a) Translate that density into the number of H atoms per cm^3.

b) Use the expression for the critical density from Destination 19 (section 19.3) to calculate the actual value of the Hubble 'Constant' at $z = 1200$. You may assume that the critical density of the Universe at this era is essentially equal to its mass density given in this problem.

23.6. From information you can find on the web, write a short description of one of the space-based or ground-based experiments that have measured the temperature fluctuations of the CMB from Earth. Be sure to summarize the key results/findings of the experiment and some details about how the experiment was carried out.

Destination 24

THE FIRST ELEMENTS IN THE UNIVERSE

24.1 THE EVOLUTION OF THE CHEMICAL ELEMENTS: GETTING STARTED

Underlying the spectacular Alien Skies we have explored so far is the persistent theme of *evolution*. Stars evolve through birth (Destination 6), middle age on the Main Sequence (Destination 8), and, ultimately, death (Destinations 11 and 12). Galaxies evolve too. The process of galaxy formation involves the steady accumulation of small objects building up into larger systems (Destinations 20–21). Even the Universe as a whole evolves. As its scale increases with time—that is as the Universe expands—it grows cooler and less dense, all the while providing the stage on which all the myriad evolution that occurs within it is played out.

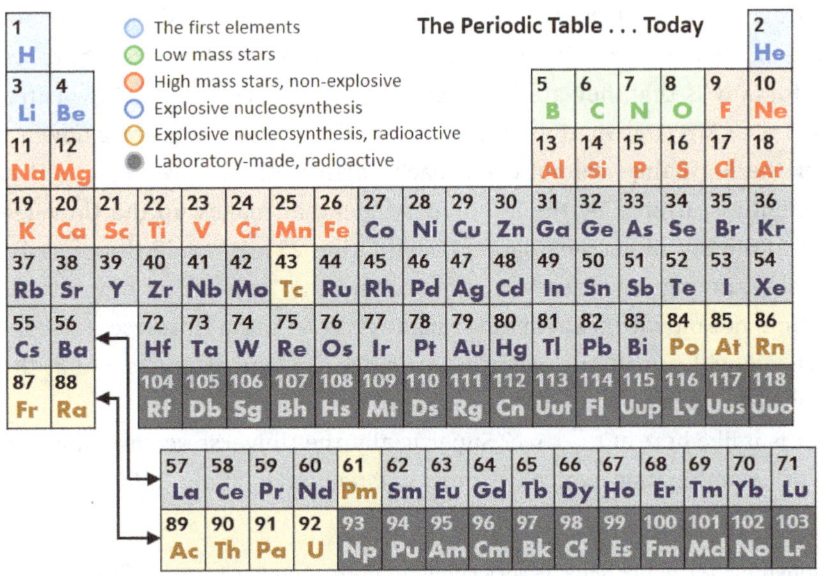

Fig. 24.1

A representation of the Periodic Table of the Elements, circa the early 21st Century on Earth. Each cell represents a distinct element which are all identified by their atomic number—equal to the number of protons in an element's nucleus—and their standard chemical abbreviation. Nearly 120 elements have been detected in nature or produced in laboratories. Everything that comprises 'normal'—that is, not 'dark'—matter is made of these basic building blocks. The legend in the figure summarizes some of the principal sources of the elements. In this destination we explore the formation of the very first elements—H, He, Li and Be—in the first few minutes after Universal Expansion had commenced.

And, remarkably, we have seen that the chemical elements also evolve. Stars like the Sun—in fact, the vast majority of all stars in galaxies such as ours—contain mostly Hydrogen (H) and Helium (He), but also varying quantities of the other elements known in the periodic table (Figure 24.1). Older stars contain progressively less mass in the form of heavier elements—that is, elements other than H and He. As we have traveled further back in time, we have seen this trend seems to apply to the Universe as a whole. There appear to have been fewer heavy elements, on average, in the Early Universe than we observe in the Present-Day Universe. The last time we checked—during the Dark Ages at a redshift of $z = 100$ (Destination 22)—the chemical make-up of the Universe had become particularly simple: 76% of the mass in the form of normal matter in the Universe—that is, excluding dark matter—was composed of H, while virtually all of the remaining 24% of the mass of normal matter was in the form of He. Apart from a trace fraction in the form of Lithium (Li) and Beryllium (Be)—both elements amounting to less than one part per billion by mass—that was it (Figure 24.2). Only the first four elements of the periodic table (Figure 24.1). The four elements with the simplest, lightest atomic nuclei. Since we encountered these elemental abundances *before* any stars had formed in the Universe, this begs an obvious question: Where and how did these initial elements form in the first place?

To answer this, we begin by exploring the Universe at a very early time in its evolution: Just one second after the expansion started in an event often referred to as the *Big Bang.*[1] Notice that we've changed how we represent 'when' we are from a time in the past, to a time since the beginning. This approach in terms of elapsed time since the expansion began starts to make much more sense than does using redshifts as we travel further back in time. Not only is the redshift here one second into the expansion of the Universe ridiculously high[2], but it is also changing very rapidly with time. During the time over which the Universe evolves at our current destination—about ten minutes!—the redshift will have changed by over a factor of 100. But the primary reason redshifts are less useful now is that they are less logical. As we know from having visited the era of the CMB (Destination 23), light from the era we are visiting now—1 *sec* after the Big Bang—cannot even reach Earth in the future because it is too efficiently scattered by the hot, ionized gas that makes up the Universe right up to the Era of Recombination (Destination 23). It would therefore be impossible—among other reasons—to measure the 'redshift' of our current destination, not only from Earth but from anywhere in the Universe. For convenience, we will specify the elapsed time since the expansion started at the Big Bang with the symbol t.

What is it like here at $t = 1$ *sec*? Superficially, the Universe seems similar to the era of recombination, which we know will happen in about 320000 *yr* into

1 Although the term 'Big Bang' is very common in the popular press and even in scientific papers, it leads to the rather unfortunate imagery of an explosion as the way the Universe began its expansion. In fact, all space and all matter in the Universe was part of the Big Bang. It was not an explosion in the usual sense involving something expanding into surrounding space. There was no space outside the Universe it was expanding into. And certainly no vantage point from which to observe it.

2 If you must know, the redshift of the Universe at $t = 1$ sec is well over 1 billion.

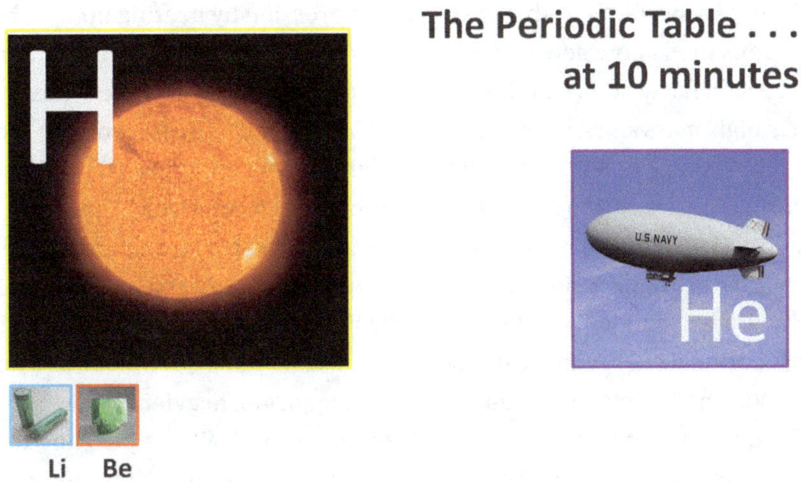

Fig. 24.2

The periodic table of the Early Universe. Only four elements—the four lightest—were present at the time, with H and He dominating. Chemistry would have been very simple back then! The relative abundances, by mass, of H and He are shown in proportion to the area of their boxes. The areas of the Li and Be boxes are shown about 10 million and 1 billion times larger than they should be if they too were to be displayed in proportion to their abundance relative to H in this diagram. The fundamental properties of the items shown in this periodic table derive from H (the Sun's nuclear fusion), He (makes a blimp lighter than air), Li (modern batteries), and Be (the key element in emeralds, a form of the mineral beryl).

Images from Earth Observatory/NASA (H/Sun), J. F. Williams/USN (He/blimp), and R. Lavinsky (Be/emerald). Concept from T. Gray and M. Whitby/periodictable.com.

the future (Destination 23). Normal matter is fully ionized, the electrons freely moving about, all unbound to any nuclei. This certainly makes sense because we can determine from the typical energies of the particles in the gas (see Destination 5) that the temperature of the Universe at $t = 1$ *sec* is a very balmy 10^{10} *K*! No atom can retain its electrons at such extreme temperatures. But there is one fundamental difference here compared to all our previous destinations. That is, the ones in the future Universe.[3] Apart from bare protons—which are essentially just H nuclei—there are no elements here at all. No Beryllium. No Lithium. Not even Helium. We have reached a time before *any* of the chemical elements—apart from the simplest one, Hydrogen—have formed. We have reached a time that predates the start of the process of chemical evolution in the Universe.

Certainly, the building blocks of atomic nuclei—protons and neutrons—are present here at $t = 1$ *sec*. And although the Universe at this era is characterized by an extraordinarily high temperature—10 *billion K* as noted earlier—conditions here are actually comparable to the temperatures we encountered during supernova explosions (Destination 12). In those environments, the sudden release of gravitational energy from the collapsing core of a massive star produced a fertile site for the formation of heavy elements. But here, one second after the expansion of the Universe started, there is nothing. Only the isolated protons and neutrons that make up the gas around us. And lots and lots of thermal energy.

To understand why this is strange, let's recall what happens as a star, particularly a high-mass star, evolves (Destination 12). As the massive stellar core of such a star runs out of the fuel required for a given fusion reaction, it begins to collapse gravitationally under its own weight. The collapse usually causes the

3 There's that confusing time-travel grammar again.

star's core to heat up and become denser since the matter in the core is still in the form of a 'perfect' gas that reacts to compression by heating up. The hotter conditions then allow new fusion sources to ignite, which can in turn maintain the star in a fairly stable configuration. At least until the new fuel runs out. In typical high-mass stars, the progression usually involves He fusion after the end of core H fusion when the star is on the Main Sequence. Oxygen (O) and Carbon (C) form after He fusion has run its course, while Neon (Ne), Silicon (Si), and even Iron (Fe) follow at later times. The cycle is broken by Iron due to the endothermic nature of its fusion reactions (Destination 12). At that point, a massive star's core will suddenly collapse, rapidly heating up from the sudden release of gravitational energy. In most cases, this causes the star to explode as a supernova. Within the first few seconds after the explosion, heavier elements can be produced in the hot, expanding gas of the supernova. What's puzzling here at t = 1 *sec* is that in many important ways, the conditions at this time in the history of the Universe seem reasonably similar to those in the core of a massive star. Both are extremely hot. And both are sufficiently dense that particle collisions—a prerequisite for fusion reactions—are very common. Stars can produce heavy elements copiously under these conditions. The Universe, apparently, does not. Why?

The answer, which seems circular, is that the Universe cannot form heavy elements because it does not yet *have* heavy elements. As the sequence of events just outlined illustrates, each new fusion reaction in a star uses as its fuel the heavy elements formed during prior stages. That is, fusion reactions—apart from most types of H fusion—require heavier elements to act as 'seeds.' As we shall see (Destination 25), no nuclear seeds are formed prior to t = 1 *sec* in the Universe. Only protons and neutrons. Thus, it's not possible to make Ne because there is no O. And it is not possible to make O because there is no C. And it is not possible to make C because there is no He. But there is H. Plenty, in fact, in the form of the protons that pervade the gas here. So it's clear that to initiate the production of the chemical elements, the Universe must use these protons to first make He nuclei. But *that's* the problem. To understand why, we have to delve a bit into the details of H fusion reactions.

The most common isotope of He contains nuclei composed of two protons (which is what defines the nucleus as He) and two neutrons. Since this isotope contains four nuclear particles it is denoted ^{4}He and typically pronounced 'Helium-four,' though also sometimes referred to as an 'α-particle' (that is the Greek letter 'alpha'). There is also an isotope of He with three particles—two protons and one neutron—known as ^{3}He ('Helium-three'). But there is no stable isotope of He that consists of just two protons. And so the simplest approach to producing He—simply mating two protons into one nucleus—does not work because that configuration instantly breaks apart. Instead, the process of producing He requires some intermediate steps, the first of which is to produce an isotope of Hydrogen known as *Deuterium*. Once that happens, two Deuterium nuclei can fuse together to produce ^{3}He, soon followed by ^{4}He and all the other elements (see Figure 24.3 for the details of the reaction chain). The problem is that Deuterium—which consists of one proton and one neutron and is denoted as ^{2}H or sometimes simply 'D'—is not stable at temperatures higher than a few hundred million *K*. That is, here at $t = 1$ *sec*, it is simply too hot for Deuterium nuclei to exist. Even if a proton and neutron do happen to collide to form a ^{2}H nucleus, it will very likely be destroyed instantly by absorption of an energetic photon of EM radiation or from the energy of a collision with another particle. But without Deuterium, the first vital fusion reaction cannot occur to produce He (see Figure 24.3). And without He, there can be no C, no O, no Si. In fact, no heavy elements at all.[4]

The only solution in this instance is to wait. After all, we know that as the Universe expands, its temperature decreases. Eventually, it must cool down sufficiently so that Deuterium nuclei, once produced, will be able to survive long enough to participate in subsequent reactions to produce He and other elements (Figure 24.3). One minute passes. By this time, the Universe has cooled by about a factor of seven, but it is still too hot to support stable Deuterium fusion. And so we wait some more. Two minutes. The temperature is still too high. Then, nearly three minutes after the beginning of the expansion of the Universe when the temperature has 'cooled down' to well under 1 *billion K*, the first Deuterium nuclei can finally start to form—and survive—through the reaction

$$p + n \rightarrow D,$$

where p denotes a single proton, n denotes a single neutron, and D is a Deuterium nucleus consisting of one proton and one neutron.

This start of this simple reaction represents a milestone event in the history of the Universe. Now that Deuterium is present, the fusion floodgates can finally open (Figure 24.3). In short order, other nuclei are produced. One reaction produces *Tritium*, denoted ^{3}H. This isotope of Hydrogen contains one proton and two neutrons and can be formed via a reaction involving Deuterium (denoted here as ^2H to emphasize the fact its nucleus is composed of two particles):

$$^2H + {}^2H \rightarrow {}^3H + p\,.$$

4 One could question this. For example, couldn't four particles collide to form a Helium nucleus? Yes, it could happen, but even in the early Universe, the probability of this is very low. More fundamentally, even if this reaction were to occur at a significant rate, the highly energetic radiation and particles present at t = 1 *sec* would destroy most of the resulting He nuclei. By the time these nuclei could survive, the Universe would be too cool to support this particular reaction in the first place. And don't even think about more extreme collisions—say, five, six, or even more particles—as a way to try to avoid He altogether and produce really heavy elements right off the bat. Such reactions are so unlikely, they won't occur at significant rates even in the extreme conditions we find here at t = 1 *sec*.

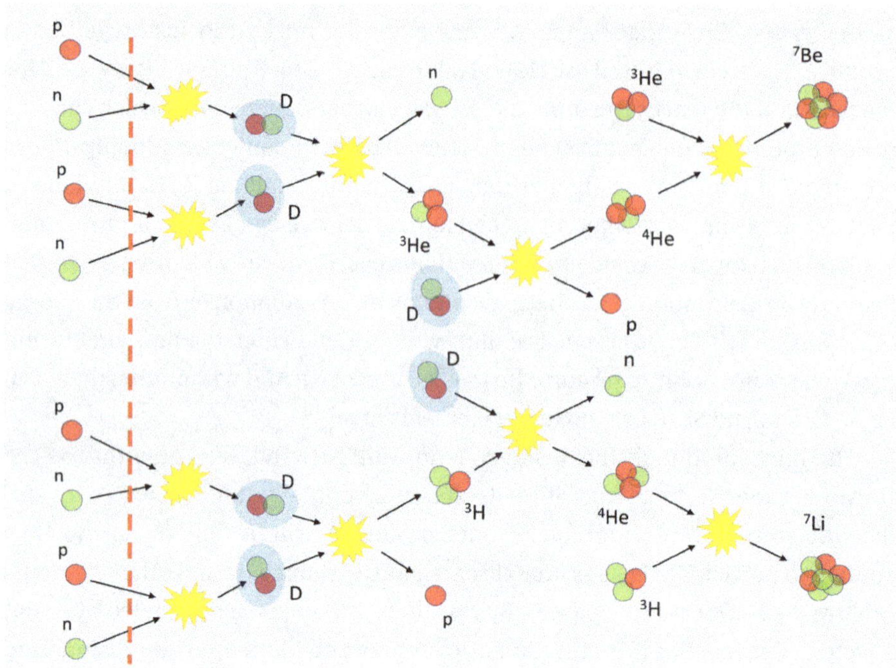

Fig. 24.3

A schematic view of some of the nuclear reactions that can occur in the Early Universe. The key to all these reactions is the production of Deuterium, shown in this network with blue ellipses. None of the reactions to the right of the red dashed line can occur until about three minutes after the start of the expansion of the Universe—the Big Bang—because at earlier times the Universe is too hot for Deuterium to survive. No elements more massive than ^7Li and ^7Be can be formed in this network of reactions because the Universe cools off too quickly after fusion commences to build up nuclei more massive than these.

A similar reaction can now produce ^3He:

$$^2H + {}^2H \rightarrow {}^3He + n.$$

Once these nuclei—each consisting of three particles—are produced, they can participate in further reactions to produce even larger, more complex nuclei. The most important of these subsequent steps involve producing 'normal' Helium (that is, ^4He) through reactions such as

$$^2H + {}^3He \rightarrow {}^4He + p,$$

and

$$^2H + {}^3H \rightarrow {}^4He + n.$$

Additional reactions involving these newly produced nuclei can even produce small amounts of Li and Be (see Figure 24.3)[5].

5 Most of the reactions described here also produce particles known as 'neutrinos' (Destination 25). The key properties of neutrinos are that they carry angular momentum and energy, so they help to 'balance' the reactions at a detailed level. They also have the amazing property that they travel from their point of origin here in the Early Universe all the way to the Present-Day Universe billiions of years into the future. The reason is that neutrinos are largely unimpeded by matter such as electrons or protons that are so effective at scattering radiation (Destination 23). If we ever develop suitable neutrino detectors, it may be feasible to 'see' the events described here directly. But even if that is possible, it is likely a long way off in the future. Current detectors are the size of small stadiums but can spot at most a handful of neutrinos per day!

This is more like it! Our understanding of how stars produce heavy elements (Destinations 8, 11, and 12) suggests we should now see a host of new elements forming as reactions such as these proliferate to form more and more massive nuclei.

24.2 FUSION IN AN EXPANDING UNIVERSE

But, of course, we don't. And the reason, though simple, is profound. As we've noted in our capsule summary, a star's core can contract and heat up in a series of processes that set the stages for subsequent fusion reactions. But the Universe does not behave this way. Indeed, it is doing just the opposite—expanding and cooling—as it evolves from the initial impetus of the Big Bang. Consequently, just at the time that fusion reactions are starting to take hold in the Early Universe, the conditions in the Universe quickly evolve to quench further reactions. Rather than forming copious amounts of heavier elements, the Universe is in a race to maintain sufficient temperatures and densities high enough to support any fusion reactions at all. As it turns out, by the time the temperature of the Universe becomes too low to support further nuclear reactions, only about 24% of the mass originally in the form of protons and neutrons will have been converted to He, with a trace amount of Li and Be. The rest remains in the form of Hydrogen atoms, and mostly as single protons (Figure 24.2) with just a little Deuterium (Figure 24.4). What's remarkable is that these proportions agree remarkably well with the amount of H, ^{2}H, ^{3}He, ^{4}He, and Li that we observe in stars and galaxies from the very Early Universe (Figure 24.4; Destinations 11, 20, and 21).

Why these specific percentages? In particular, why didn't all the protons at t = 1 *sec* manage to fuse with other particles to make heavier nuclei leaving no H behind? One reason is that the number of protons and neutrons at the start of this process was not equal. And the reason for this has to do with the small mass difference between the two particles that causes the protons to be relatively more abundant after t = 0.1 *sec* (we'll explore at Destination 25 why this is the case). Another reason has to do with a peculiar feature of the neutron itself. Left on its own—that is, outside of an atomic nucleus—a neutron

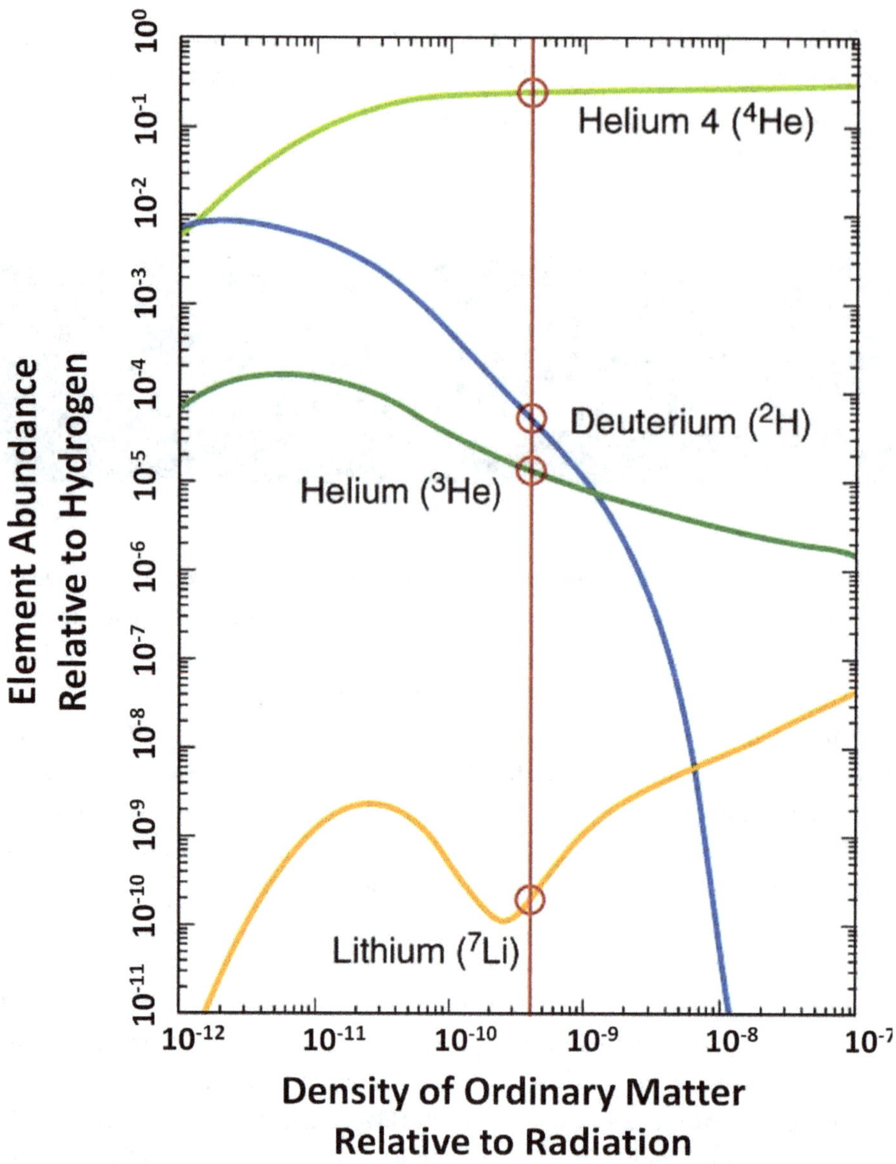

Fig. 24.4

The colored lines in this plot show estimated abundances (the *y*-axis) of four of the nuclei that can form in the very Early Universe, ^{4}He (light green), Deuterium (^{2}H, blue), ^{3}He (dark green), and ^{7}Li (yellow). The *x*-axis shows the relative fraction of ordinary matter to radiation at the time the elements were formed. As we'll see at Destination 25, the Universe was Radiation Dominated at this time. Notice that the plot allows a wide range of combinations of these nuclei. The only requirement is that they all occur at the same ratio of the density of ordinary matter to the radiation density (the values along the *x-axis* in this plot). As it turns out, these elements can also be measured in objects believed to reflect the abundances of these nuclei in the Early Universe. The circles correspond to these measurements and show that all four nuclei are consistent with a single matter-to-radiation ratio in the Early Universe. This not only strongly suggests these calculations may be correct, but also places strong limits on the overall quantity of ordinary matter in the Present-Day Universe.

will spontaneously decay into a proton and an electron.[6] Every 10 *min* (617 *sec* to be precise), half of any sample of free neutrons will decay in this way. By the time Deuterium fusion can commence (at about $t = 3$ *min*), many of the original neutrons are gone, having decayed into new protons. As a result, by the time Deuterium fusion can proceed, protons outnumber neutrons by a ratio of about 6 to 1. Taken together, these effects all limit the amount of He that can be produced since at least one neutron is required to make the nucleus of that element. As it happens, just about all the free neutrons do combine with protons, ultimately producing heavier nuclei (mostly ^{4}He). But because the neutrons are

6 And a particle known as an 'anti-neutrino.' This is the anti-particle twin of the normal neutrinos we discussed in the previous footnote. in this destination.

comparatively scarce compared to the number of protons, they can combine to produce only enough He atoms to account for 24% of the total mass that was originally present in the form of protons and neutrons.

There's one other important consequence of the way fusion proceeded in the Early Universe. In order to get the right *proportions* of Hydrogen and Helium, we must also constrain the total quantity of these two elements. That is, the total mass of Hydrogen and Helium produced in this process in the Early Universe amounts to between 1% and 5% of the total mass needed to achieve a Critical Universe with $\Omega_{Matter} = 1$. This confirms what we had already observed. The visible mass—the stuff that ultimately forms from the H and He formed here at the earliest times in the expansion of the Universe—does not by itself come close to producing a Critical Universe with $\Omega_{Matter} = 1$. But it is worse than this. If the value $\Omega_{Matter,\,0}$ (which denotes the value in the present-day Universe) exceeds 0.05, then we require that *most of the mass of the Universe is in the form of Dark Matter.* Since the amount of matter in the Universe does appear to exceed this total (recall from Destination 19 that $\Omega_{Matter,\,0} \sim 0.27$), this means that the DM *cannot* be some form of normal matter that gives off very little light. That is, it isn't composed of asteroids, some type of hard-to-detect gas, low-luminosity stars, or even ping pong balls. All those things are made up of normal matter. As a result, we still cannot say what the Dark Matter *is*, but the details of how the first elements were produced in the Universe do tell us what Dark Matter *isn't*. This conclusion is perfectly consistent with everything else we have seen in our earlier journeys: Dark Matter dominates the mass of individual galaxies. What we have learned here is that the proportion of the lightest elements in the Universe tells us that Dark Matter must dominate on large scales too. Scales as large as the Universe itself. Given how much of it there appears to be in the Universe, it would be nice to really understand what DM actually is. And yet, frustratingly, we still do not know.

Within 10 short *minutes* after the Big Bang—the moment the expansion of the Universe commenced—all of the pioneering fusion reactions that produced the first few elements have ceased. The process of *Big Bang Nucleosynthesis* is over. There will be no more fusion until the first stars form, many millions of *years* into the future. But we still got to see what we came for at this destination. We watched as the first elements formed. And we know that from these humble beginnings, all the other chemical elements will eventually follow. In a few billion years, on at least one small planet in one of the myriad galaxies of the Cosmos, these elements will produce life that will eventually develop the ability to study how the first elements formed in the Universe (Table 24.1).

TABLE 24.1 Major Events in the History of the Universe Since $t = 3$ min

EVENT (SEE NOTE 1)	REDSHIFT	TEMPERATURE (*K*)	TIME AFTER EXPANSION BEGAN	COMMENTS [DESTINATION(S) FOR MORE DETAILS]
Start of First Nuclear Reactions	$> 10^8$	10^9	3 *min*	The Universe was ready to do this at 1 *sec* after the Big Bang, but Deuterium is too fragile to have survived at the high temperatures at the time. [D24]
End of First Nuclear Reactions	5×10^7	6×10^8	10 *min*	Still hot enough for fusion, but the density in the expanding Universe is now too low for reactions to proceed. [D24]
End of Radiation Era/ Start of Matter Era	3200	8900	55000 *yr*	After this time, the density of matter exceeds that of radiation, causing the details of its expansion to change. We still live in the Matter-Dominated Era. [D25]
'Re'-combination	1200	3000	320000 *yr*	Atomic nuclei and electrons combined for the first time. The Universe became transparent to radiation. This era is seen today as the CMB. [D23]
Dark Ages	1000–50	2500–50	0.5–50 *Myr*	Initial linear growth of structure. Universe composed almost entirely of neutral gas clouds. Ended with re-ionization by first stars. [D22]
First Stars and First Galaxies	30–10??	50–30	100–500 *Myr*	Formation of first stars, likely massive, and likely in regions of high mean density. May have formed seeds for first supermassive BHs. [D21, D22]
Age of Quasars	10–0.5	30–4	0.5–9 *Gyr*	Likely concurrent with formation of First Stars and Galaxies, but may have preceded first stable galactic systems. [D17, D21]
Era of Galaxy Formation	8–0	24–2.7	0.6–13.5 *Gyr*	Broad period of star formation in galaxy environments. Still ongoing, but the rate of star formation has been steadily winding down since $z = 1$. [D14, D20, D21]
Growth of Structure, Galaxy Clusters	2–0	8–2.7	3.3–13.5 *Gyr*	This process is ongoing from the earliest times in the Universe, but the era of strong 'non-linear' growth is relatively recent. [D16, D18, D22]
The Accelerating Universe	< 0.7	< 5	7.5–13.5 *Gyr*	The accelerating nature of the Universe has become evident in its expansion evolution only during the latter half of its history. [D19]
Present-Day Universe	0	2.73	13.5 *Gyr*	Where we started our journeys [D1–D17]. The Present-Day Universe also contains the eras of Galaxy Formation, Growth of Structure, and the Accelerating Universe. And us!

DESTINATION 24 QUESTIONS

24.1. As noted at this destination, free neutrons are actually unstable (fortunately, they are stable when inside atomic nuclei). On average, about half of all neutrons in a sample will decay to a proton, an electron, and a neutrino about every ten minutes. We can express this behavior mathematically by saying that the number of neutrons, N_n, is given by the relation $N_n = N_{n,0}2^{-t/t_{1/2}}$. In this expression, $N_{n,0}$ is the number of neutrons originally present, t is the elapsed time, and $t_{1/2}$ is the 'half-life.' For the decay of an isolated neutron, $t_{1/2} = 617\ s$. If at $t = 1\ s$ in the history of the Universe there are equal numbers of protons and neutrons (this was *not* the case; protons outnumbered neutrons at this time by a small factor, but you can ignore that for this problem), what is the ratio of the number of neutrons and protons at $t = 229\ s$ when the fusion reactions in the Early Universe first kicked in? Keep in mind that the original protons do not decay, so their number remains constant. But the *total* number of protons does increase since each neutron transforms to a proton as it decays.

24.2. How does the abundance of He in the Universe today compare to when it was produced in the very earliest times of the history of the Universe? Is there more or less Helium now than at $t = 10\ min$? Explain your answer by noting where He may have been formed or destroyed over the history of the Universe.

24.3. If the age of the Universe is proportional to the square of the temperature, T^2, during the era when the first elements were formed, then what is the temperature of the Universe when the first fusion reactions commenced at $t = 229\ s$? How does the temperature at the onset of fusion compare to the core temperature of the Sun? (Destination 8)? **HINT:** Find the temperature at $t = 1\ sec$ in the text.

24.4. The oldest types of objects known in the Present-Day Universe are globular clusters. While these systems do have chemical abundances similar to what was produced very early in the Early Universe, the clusters do contain some heavy elements. That is, elements with atomic masses greater than H, He, Li, and Be. Speculate why this is the case. That is, if the Universe was composed of only the four lightest elements at the time of Big Bang Nucleosynthesis, how did globular clusters gain heavier elements?

24.5. Assume that a $10\ M_{\odot}$ star forms out of 'primordial gas' that has a composition equal to what is expected at the end of Big Bang Nucleosynthesis. Now, imagine adding the equivalent of 1 Earth mass of Iron (Fe) to that star.

a) What is the mass fraction of Fe in the star after the addition of this matter?

b) Recall that the Sun contains about 2% of heavy elements by mass, where 'heavy elements' is defined as all elements with atomic numbers greater than that of H (1) and He (2). By what factor is the mass fraction of heavy elements in the 10 $M_{\odot}$ lower than in the Sun?

c) The star with the lowest heavy-element content known has only 10^{-7} of its mass in heavy elements. Could such a low abundance be explained by the addition of an Earth mass of heavy elements as envisioned in part (a)?

d) Where might the heavy elements described in part (a) of this question have come from?

24.6. We saw at Destination 23 that the value of $\Omega_{Matter} + \Omega_{Rad}$ gets very close to 1.0 the closer we get to the time when the Universe began its expansion.

a) Based on the expression $\rho_{rad} = 4\sigma T^4 / c^3$ (Destination 19), what is the density of radiation at $t = 1$ *sec* assuming that the temperature of the Universe at that epoch was $T = 10^{10}\,K$?

b) From Figure 24.4, what was the approximate *matter* density at that time? **HINT:** Read the *x*-axis.

c) How does that matter density compare to the typical density at the core of a star with a mass of about ten times that of the Sun? **HINT:** You can estimate the central density of such a star using your basic knowledge of stellar properties (Destination 8) or from researching a suitable model of such a star that you find online.

d) Do your results support the assumption that conditions (specifically, temperature and mass density) in the early Universe were indeed similar to those in the core of a star? To the extent that they were different, what effect might such differences have had on the process of nucleosynthesis in the early Universe?

Destination 25

TO A TIME BEFORE TIME

25.1 THE GENESIS OF MATTER

Since we first embarked on our odyssey through the Universe, we have grown accustomed to things happening on gargantuan scales ranging from individual stars and planets all the way up to the size of Universe itself. But now, as we approach the earliest moments of the existence of the Universe, we begin to appreciate that key developments occur exclusively on microscopic scales. This was already apparent at our last destination where we saw that the process of forming the first chemical elements played out entirely on the scale of individual subatomic particles interacting in the hot, dense conditions of the Early Universe. Our current destination—the last of our long voyage—will draw us even deeper into the microscopic fabric of the Universe. But more astonishingly, as we approach the time when the Universe commenced its expansion, we shall find that the observable Universe itself becomes a microscopic entity. So small, in fact, that the laws of physics that we normally apply to atoms and their nuclei must be applied to the Universe as a whole.

Our goal at this destination is to travel back in time, approaching as close to $t = 0$ as we possibly can. On the way, we shall explore a complex sequence of events that occurred within *one second* after the expansion of the Universe began. One second after the Big Bang. But as we noted at our last destination, the connotations that the name 'Big Bang' engenders can be somewhat misleading.[1] As we approach $t = 0$, we shall not encounter an 'explosion' of any sort, at least none like any we are familiar with. It's true that explosions imply expansion and we've seen ample evidence the Universe is indeed expanding (Destinations 16–24). But a standard explosion propagates from a single point into a surrounding medium. That is, stuff moves away from other stuff and into some sort of volume encompassing surrounding stuff. But that's not how the Universe works. We've emphasized (most explicitly in Destinations 16 and 19) that Universal Expansion—from the moment it began to the Present-Day Universe—reflects the change in the scale of space itself. And of *everything* tied to space. Radiation. Matter. Energy. The fundamental forces of Nature. There is nothing 'outside' of the Universe into which it is expanding, at least nothing that we can directly observe or that would be recognizable to our inherent three-dimensional senses.

---

1 And even somewhat maligning. The name 'Big Bang' was originally coined by British astronomer Fred Hoyle in the 1960's as a derogatory description of the concept. At the time, he held that the Universe evolved more sedately than Big Bang models implied. He was certainly wrong about that, a point he eventually admitted. The evolutionary nature of the Universe is now completely beyond question given our present observations of the Cosmos. The concept of Universal evolution—as opposed to long-term equilibrium—serves as the underlying principle in our interpretation of nearly everything we've seen on our many journeys through the Universe.

For us, then, the 'Big Bang' is shorthand for the process involving the initial evolution of energy, matter, and forces in the remarkably eventful first fractions of a second in the history of our Universe (see Figure 25.1, which we'll refer to frequently). It is *not* a description of the ultimate 'prime mover' that initiated the expansion of the Universe in the first place.[2]

Let's begin our journey toward $t = 0$ by traveling to the time when the Universe was only $t = 10^{-4}$ *sec* old. That is, one *ten-thousandth* of a second.[3] The observable Universe is just 10^{-4} *light-sec* in radius, spanning a mere 60 *km* from end to end. What would we 'see' at this time in the Early Universe? In some respects, this question is problematic. The scale of the Universe, *R* (Destination 19), is now so small and the Hubble 'Constant' is so large that the expansion of space would quickly tear our bodies apart.[4] Ouch. But, assuming we could survive the expansion and observe our surroundings, it's probably the temperature here—0.7×10^{12} *K*, or nearly one *trillion* degrees!—that would really attract our attention. At this time in the Universe it is so hot that no particles such as molecules or complex atomic nuclei could possibly survive the bombardment of high-energy photons or the frequent extremely energetic collisions with other particles. The majority of the matter that exists here is exclusively in the form of basic particles such as neutrons, protons, and electrons. We know that in about three minutes some of these particles will begin to fuse together to form the first complex nuclei ever to exist in the Universe (Destination 24). But not now. Not yet. Here at $t = 10^{-4}$ *sec*, the Universe is still

2 This statement can be taken to be as profound or mundane as one wishes it to be. The important point is that as we push toward the start of the expansion of the Universe—to the start of the events that comprise the Big Bang—we'll discover that modern physical concepts will limit our ability to go arbitrarily far back in time. The reason is that the conditions in the early Universe are so extreme that they force us to stretch our everyday conceptions of 'time,' 'space,' and 'causality' up to—even perhaps beyond—their logical meanings. But it's not that the laws of physics 'break down' as we approach $t = 0$. Rather, the laws that govern subatomic scales have inescapable uncertainties built into them, a point we'll expand upon later on in this destination. These uncertainties place strong constraints on how far back in time we can travel. If one wishes to conclude it is impossible to understand the Universe at times smaller than where (that is, 'when') we end up at this destination, modern science cannot presently dispute that. But it is also the history of exploration that what is not known today—sometimes, what can't even be conceived of today—can become understood, even commonplace, tomorrow. At this destination we'll go as far back in time as our current understanding allows. But, as explorers and scientists, we'll be ready to press on further when our understanding of the Early Universe inevitably improves and lets us pass beyond our current boundaries.

3 Parameters such as redshift, *z*, or scale, *R*, are becoming less and less useful to help us specify where—that is 'when'—we are in the Universe. The redshift at $t = 10^{-4}$ *sec* is around 300 billion, implying that $R \sim 1/(3 \times 10^{11}) = 3 \times 10^{-12}$. The Universe is nearly 1 trillion times more compact in linear scale than in the Present-Day Universe, about 10^{35} times more compact in volume.

4 The Hubble 'Constant' here at $t = 10^{-4}$ *sec* is around 5.3×10^{18} *km/s/Mpc* (Destination 16). To put this in perspective, a Hubble 'Constant' of this magnitude implies that a person of average height would be growing in size at a rate of about 1 *foot per second* due to the expansion of the Universe!

far too hot for seed nuclei such as Deuterium to survive. If we really could travel back in time to see this Alien Sky, the view would be blindingly bright. Intense thermal radiation (Destination 6) from the incredibly hot, young Universe fills all of space at this very early time in its evolution.

For many of the particles here—for example, the electrons—existence is unusual. We usually think of subatomic particles as stable, long-lived entities (though recall from Destination 24 that we discovered that the neutron, when outside of atomic nuclei, is unstable). But as we'll see, such particles did not always exist in some stable form we'd be familiar with. Instead, each had to be 'born' at some distinct time during the evolution of the Early Universe before they could exist as long-lived particles. That does not mean particles could not be present in the Universe *before* birth as we define it here, but their existence would be quite different than what we usually associate with stable particles. When—and how—different particles came into stable existence depends sensitively on the nature and properties of the particles involved, as well as on the detailed evolution of the temperature and density in the Early Universe.

To understand why, we need to return to our earlier discussion from Destination 19 regarding how we determine the average mass density of the Universe. For any component that comprises the Universe—normal matter, radiation, planets, or beach balls—we can compare their equivalent mass densities with the critical density of the Universe and any given time to calculate the density ratio $\Omega_i = \rho_i / \rho_{crit}$. If we restrict ourselves to the major components of the Universe that we know about, the subscript 'i' can refer to matter (which we denote as 'Matter' and which, in this instance, comprises *both* visible and Dark Matter), radiation (denoted 'Rad'), or *Dark Energy* (denoted 'Λ,' the capital Greek letter lambda; see Destination 19).[5] As we saw earlier (Destination 23), the sum of the density ratios of these components, $\Omega = \Omega_{Matter} + \Omega_{Rad} + \Omega_{\Lambda}$, is very close to 1.0 here at $t = 10^{-4}$ *sec*, regardless of its value in the Present-Day Universe (though it seems to be quite close to 1.0 there as well; Destination 19).

We can use our estimate of the current temperature of the Universe—$T = 0.7 \times 10^{12}$ K—to determine that the radiation density is now about $\rho_{rad} \approx 4 \times 10^{12}$ g/cm^3 (Destination 19). When we compare this with the local density of both visible and Dark Matter here at $t = 10^{-4}$ *sec*, we discover that the radiation density, ρ_{rad}, is more than 100 million times higher than the density of matter at this time.[6] That means that here at $t = 10^{-4}$ *sec* the mass density of radiation is so high that $\Omega_{Rad} = 1.0$ to high precision, while the value of the matter-density ratio here at is about $\Omega_{Matter} \approx 10^{-8}$. When we last considered the relative mass densities of matter and radiation—back at $t = 2 \times 10^{17}$ *sec*, or $z = 0.8$ (Destination

5 We do not likely need to define the Ω parameter for beach balls. It's unlikely that they comprise a *major* component of the Universe. But if they did, we could simply add $\Omega_{beach\ ball}$ to our inventory.

6 This ratio is actually rather well determined by Big-Bang Nucleosynthesis models of the sorts described at Destination 24 (see Figure 24.4).

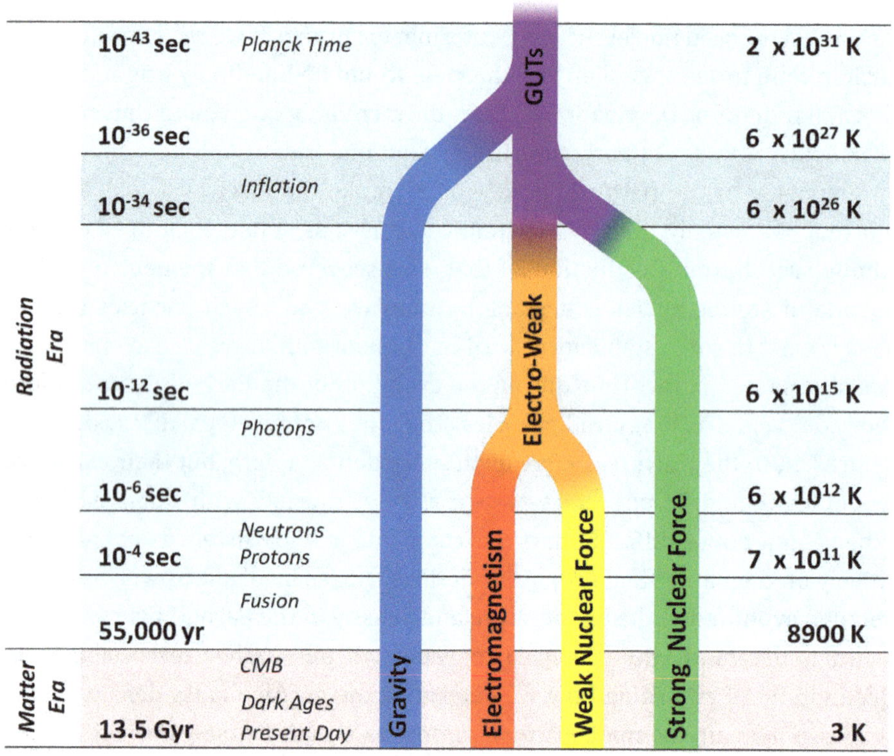

Fig. 25.1

A timeline of the Big Bang, starting at the Planck Time (at top) and extending to the Present-Day (at bottom). The age of the Universe (along the left) and its temperature (along the right) are given for some notable events that we describe in this destination. Some of those events—and some others from later eras in the Universe—are labeled in italics along the center-left side of the timeline. Most of the events plotted here occurred during the first 10^{-4} *sec* after the expansion of the Universe commenced (the Big Bang) and include the formation of the first particles in the Universe, the genesis of the forces in the Universe (colored, vertical bars) and the Inflation Era (shown as a narrow region between $t = 10^{-36}$ *sec* and 10^{-34}*sec* highlighted in blue). The light blue region covering the first 55000 *yr* of the Universe (the top of the timeline) represents the Radiation Era when the mass density of energy exceeded that of matter. The red region (at bottom) represents the Matter Era and covers nearly the entire 13.5 *Gyr* of the history of the Universe. Note that the timeline grows indistinct for times earlier than the Planck Time at $t = 10^{-43}$ *sec*, where modern physics is unable to penetrate further.

19)—we found that the radiation density was minuscule compared to the mass density of the Universe. But that's no longer true. Here at $t = 10^{-4}$ *sec* the tables have turned and we find ourselves deep inside what is known as the *Radiation Era* of the Universe (see Figure 25.1). During this era of the Early Universe, the density of radiation exceeded that of matter, which is equivalent to saying that $\Omega_{Rad} > \Omega_{Matter}$.

Why the change? Since radiation density is proportional to T^4 where T is the temperature of the Universe, it is easy to show that it decays more rapidly as the Universe expands than does the matter density.[7] A detailed calculation of this reveals that the Radiation Era was relatively short-lived, ending about 55000 *yr* after the expansion of the Universe commenced. At that time, the Universe entered the so-called *Matter Era* where the mass density of matter does indeed exceed that of radiation and so $\Omega_{Matter} > \Omega_{Rad}$ (see Figure 25.1).[8] The Earth, the

7 Expressed in terms of redshift, it is easy to show that the radiation density of the Universe is proportional to $(1 + z)^4$, while the matter density varies as $(1 + z)^3$. The exponent for the radiation density ultimately comes from the T^4 term in the Planck Equation (Destination 6), while the exponent for the matter density reflects that mass density is proportional to $1/R^3$ and that $R = 1 / (1 + z)$. That means the ratio of the radiation density to the matter density varies as $(1 + z)$. Thus, at high redshift—that is, when the Universe was very young—there had to be a time when the density of radiation dominated over the density of matter.

8 You may have noticed that in this discussion we are ignoring Ω_Λ, the density ratio for Dark Energy. The reason is that while the mass density of matter and radiation both increase—though at different rates—as we press toward earlier times in the Universe,

Sun, our Galaxy—even the entire Local Universe—all reside deep in the Matter Era of the Universe.[9]

The distinction between the Matter and Radiation Eras—while fundamentally important—can often seem fairly esoteric. But there is one way in which this distinction is crucial to understanding the basic properties of the young, evolving Universe. Because radiation has become so dominant, another way to characterize the Universe at these early times is by the characteristic energy, E_{char}, of its radiation field

$$E_{char} = kT,$$

where T represents the temperature of the Universe, and k is Boltzmann's constant (Destination 5). At a temperature of nearly $T = 10^{12}$ K, we discover that a typical photon of EM radiation here at $t = 10^{-4}$ sec carries quite a wallop, equivalent to about 100 times the energy contained in the mass of an electron.[10]

Under such extreme conditions and in the presence of such an intense radiation density, electrons can *spontaneously form from the energy field of the Universe.* But the conditions also imply that the opposite reaction is equally possible, and so when electrons are formed here, they are nearly instantly destroyed.[11] This behavior represents what's known as an *equilibrium reaction* where energy is continuously being converted into electrons and the electrons back into energy (see Figure 25.2). This equilibrium reaction underscores the equivalence of mass and energy stated by Einstein's Equation, $E = mc^2$ (Destination 17). Here at $t = 10^{-4}$ sec, the very high temperature of the Universe makes it 'easier' to store the mass of an electron in the form of energy than in the form of a particle.

---

the equivalent mass density of Dark Energy remains constant. Thus, since the critical density is also increasing as we go back in time (see Destination 19), Ω_Λ gets smaller and smaller as we approach $t = 0$. Here at our present location at $t = 10^{-4}$ sec, we can indeed safely ignore the effects of Dark Energy as we explore the workings of the Early Universe. Its effect is completely negligible at this time and will remain unimportant for many *Gyr* into the future.

9 Notice that the era of 're'-combination that we visited at Destination 23 was *not* in the Radiation Era despite our impression—for example the intense brightness of the sky!—that it was dominated by radiation. It turns out that the mass density of matter (visible and dark combined) exceeded the mass density of radiation at that time by about a factor of three. However, we were already well into the Radiation Era when the first fusion reactions commenced in the Universe (Destination 24; also see Figure 25.1).

10 Recall that the energy from Einstein's Equation, $E = m_e c^2$, where m_e is the mass of the electron. Get used to this. At this stage in the history of the Universe it is often more convenient to talk about the equivalent energy of a particle's mass than about the mass itself. The characteristic energy of the radiation is comparable to kT, where k is the Boltzmann Constant, and T is the temperature of the radiation field. So, to be precise, the characteristic energy of the radiation here at $t = 10^{-4}$ sec is about 100 times kT, or about 100 times $m_e c^2$.

11 The process is analogous to how ionized H can recombine with electrons in a H II region such as the Orion Nebula (Destination 6), only to be ionized almost instantly as the system absorbs an energetic photon from a nearby hot star.

For the electrons, this is a strange sort of existence. Collectively, there are many electrons present here in the very Early Universe, but no *individual* electron can survive for long at the temperature characteristic of the Universe here at $t = 10^{-4}$ *sec.* Not until the Universe cools off so that $kT < m_e c^2$ will it become more favorable to store the energy of electrons in the form of particles (Figure 25.2). At that point, individual electrons are much more likely to survive indefinitely and much less likely to be destroyed by the energy in the radiation field or from collisions with other particles. When does this transition from equilibrium reactions to long-term survival occur for electrons? Not until the Universe has reached the ripe old age of around $t = 3.7$ *sec.* From our vantage point here at $t = 10^{-4}$ *sec*, that seems ages away!

But we are not traveling forward in time, but rather back in time through the processes that constitute the Big Bang and toward the beginning of Universal Expansion at $t = 0$. And so we press on to $t = 10^{-6}$ *sec*, one millionth of a second—one *micro-sec*—after the Universe commenced its expansion. The temperature of the Universe is now 6×10^{12} *K*, and its characteristic energy, kT, is comparable to the masses of neutrons and protons, the particles that comprise atomic nuclei. We might imagine that as we push further back in time these particles would behave in a manner similar to the way we discovered for electrons. But they don't. Instead, as the characteristic energy of the Universe exceeds the masses of the neutrons and protons (that is, when $kT > m_{n,p}c^2$, where $m_{n,p}$ refers to the mass of either a neutron, n, or a proton, p), we find that neither type of particle exists at earlier times in the Universe! Not even temporarily in equilibrium reactions as electrons do (Figure 25.2). Why the big difference?

The answer lies in the nature of subatomic particles, the building blocks of matter. In general terms, these particles can be classified into two broad categories: *Fundamental Particles* (Figures 25.2 and 25.3) or *Composite Particles.* As their name implies, composite particles can be subdivided into smaller particles that may themselves be either composite or fundamental. But fundamental particles—including electrons—cannot.[12] Although there is a large 'zoo' of composite particles known, current models of matter have identified only sixteen Fundamental Particles (see Figure 25.3).[13] These particles can be combined to

12 Strictly speaking, this is not quite true. As we'll see, when a particle and its anti-particle meet, they annihilate to form radiation—that is, photons. Likewise, some fundamental particles are unstable—for example, muons—and they can decay on their own into energy (photons) and other fundamental particles. The distinctions arise from the fact that there are 'generations' of fundamental particles. Higher-generation particles—such as muons—can decay into lower-generation particles—such as electrons, but nothing else (see Figure 25.3).

13 This too is not strictly true. Referring to Figure 25.3, each particle in the first three columns has its own *anti-particle* (we'll be discussing *Anti-Matter* shortly, but also see Figure 25.2). Four of the particles in the final column of Figure 25.3 (the photon, the gluon, the Z^0, and the W^0) are their own anti-particles, while the anti-particle of the W^+

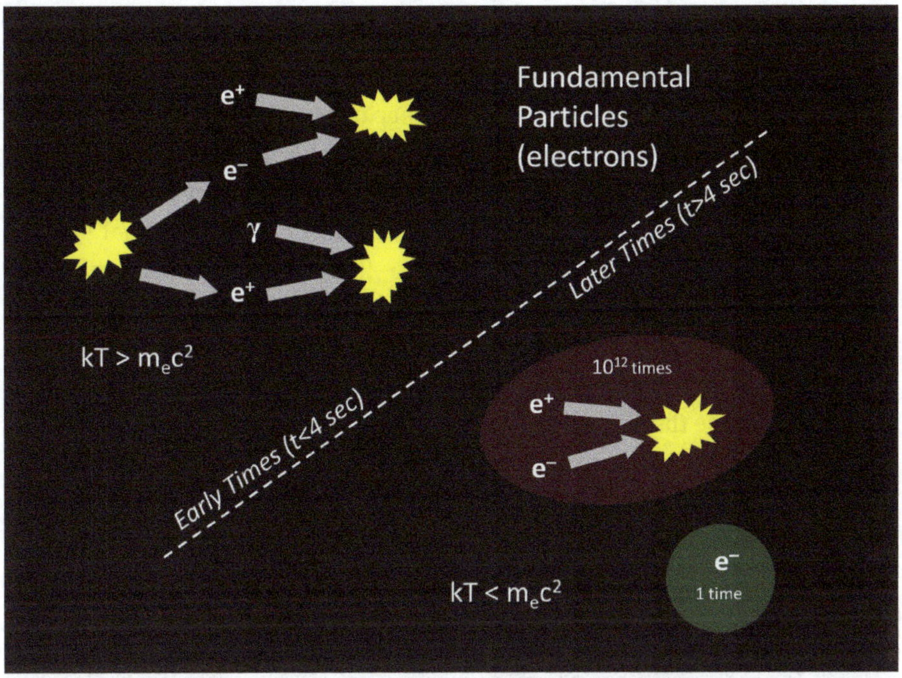

Fig. 25.2

At early times when the energy density of the Universe exceeds the mass of a fundamental particle (that is, $kT > m_{Fund}c^2$, where m_{Fund} is the particle mass), these particles exist only momentarily within equilibrium reactions such as those shown schematically at the upper left. In the case shown here, the fundamental particle is the electron (other fundamental particles are listed in Figure 25.3). Under these conditions, electrons can spontaneously appear from the energy field (denoted by the yellow explosion symbols), only to annihilate with a positron (the anti-particle of the electron, and also just formed from the energy field) or be destroyed by an energetic photon (denoted with the Greek letter gamma, γ). In either case, no single electron survives for long, even though there are always lots of electrons present. Once the Universe cools sufficiently so that $kT < m_{Fund}c^2$ (lower right in the figure), all the free electrons and positrons that can stably survive rapidly meet and annihilate, returning their mass to the radiation field. But since the Universe is cooler, there is no return reaction. If the number of electrons and positrons were precisely equal at this time, there would be none left in the Universe after that time. Fortunately, due to some as-yet-unknown asymmetry in how normal matter and anti-matter form, it turns out that there is one excess electron

build all of the known composite particles. For example, a neutron is actually a cluster of two 'down' and one 'up' quarks, while a proton consists of one 'down' and two 'up' quarks (Figures 25.3 and 25.4). Neutrons and protons can then be combined to build up atomic nuclei. And these can be combined with electrons—also fundamental particles—to produce the atoms of all the chemical elements (see Figure 24.1). These atoms, in turn, can serve as building blocks to produce all known molecules. In short, all forms of normal matter—that is, excluding Dark Matter which we know is not 'normal' matter, even if it is abundant—can be built up from the Fundamental Particles shown in Figure 25.3.

What this means is that when the Universe was too hot and its radiation field too energetic, neutrons and protons did not exist. Such conditions existed immediately before our current 'location' at $t = 10^{-6}$ *sec*. In the younger, hotter Universe, neutrons and protons were disassociated into to their constituent quarks, similar to how high gas temperatures or energetic radiation can destroy molecules (Destinations 6 and 22). That makes our current location at $t = 10^{-6}$ *sec* the *actual* epoch when the chemical elements first formed. It was not until now, about one micro-second after the Universe began its expansion, that the first protons—the first H atoms—made their appearance in the Universe.

A similar pattern of events occurs for all types of particles. As the Universe expands, it cools. As it cools, the characteristic energy of the Universe first

is the W^-. As a result, the array in Figure 25.3 actually represents thirty particles in all. Finally, there is one particle missing from Figure 25.3—the *Higgs Boson*—that has only recently been observed to exist. That particle is also its own anti-particle. Once these issues are considered, there are actually a total of thirty-one Fundamental Particles known to exist.

exceeds, then equals, then dips below the mass of each type of particle. During the period when the Universe was too hot for a given particle to exist stably—that is, when $kT > m_{particle}c^2$—it either did not exist at all (as for composite particles; see Figure 25.4) or only existed intermittently in equilibrium reactions similar to what we saw for electrons (as for fundamental particles; see Figure 25.3). Once the Universe cooled so that $kT < m_{particle}c^2$, that specific particle could finally exist stably, or at least—if it is a fundamentally unstable particle like a free neutron—as stably as its nature allows. One by one, all of the subatomic particles, starting with the most massive, seem to appear out of nowhere in the Universe. But it's not magic. These particles come into existence because the characteristic energy of the young, hot Universe eventually becomes comparable to the mass of the particles themselves. And as the Universe cools, it becomes preferable to store the energy in the form of mass rather than radiation.[14]

for every 1 trillion (10^{12}) electron–positron pairs. This tiny imbalance—which applies to all types of particles that exist in the Present-Day Universe—causes normal matter to survive this era, thus allowing all the structures we see in the Present-Day Universe—ourselves included—to have formed. For electrons, the era when they 'freeze out' of the energy field as long-lived stable particles (lower right) is at about $t = 3.7$ *sec* after Universal Expansion started. More massive fundamental particles freeze out at progressively earlier/hotter times.

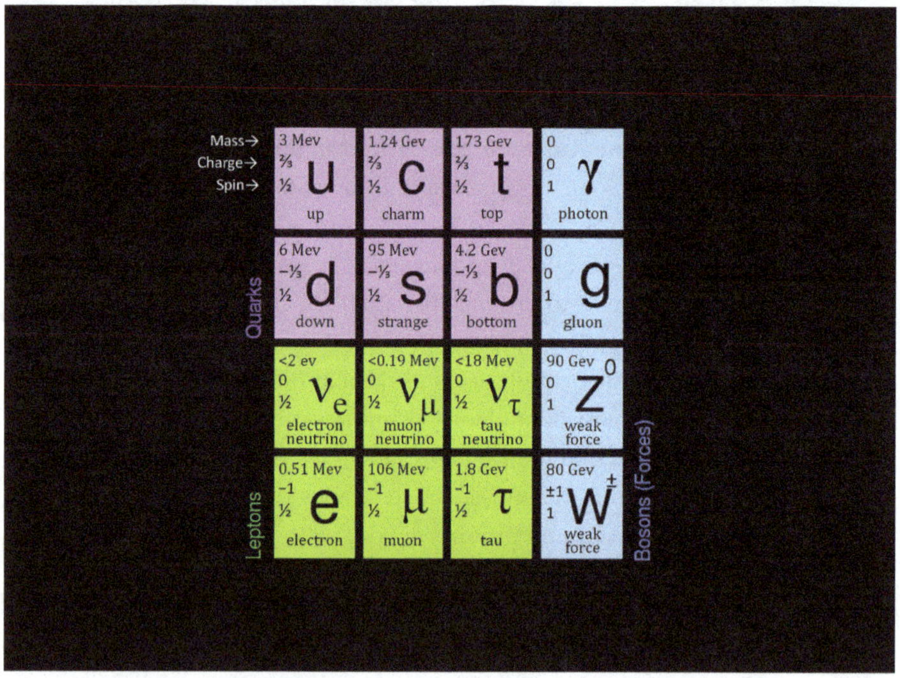

Fig. 25.3

A table of almost all of the currently known Fundamental Particles. The *Leptons* (in green) include the electron, the muon, and the tau, along with their associated neutrinos. The *Quarks* (in purple) include three pairs that can combine to make composite particles such as neutrons and protons (see Figure 25.4). These first three columns represent the three known *generations* of matter. The fourth column represents the *Bosons* or force carriers. Two of these are massless: The photon, which transmits the Electromagnetic Force and the gluon, which transmits the *Strong Nuclear Force* that holds atomic nuclei together. The two other bosons have mass (the *W* and *Z* bosons)

Another very important event occurred around $t = 10^{-6}$ *sec*. As it happens (see footnote 13), virtually all particles in the sub-atomic world have matching *Anti-Particles.*[15] In general, anti-particles have the same mass and spin as their

14 Why that's true is somewhat of a mystery. However, one idea is that there is a Universal field that promotes particles to have mass and hence to actually appear as particles. The 'enabler' of this field is known as the *Higgs Boson* (see previous footnote), the existence of which was confirmed in 2013.

15 Although frequently used by science fiction writers, anti-matter really does exist and various anti-particles have been produced and observed on Earth. Even anti-atoms—positronium composed of an electron and positron (the anti-particle of the electron),

and transmit the *Weak Nuclear Force*. During the earliest evolution of the Universe, these particles—and the forces they transmit—systematically came into existence (see Figure 25.1). For each particle in this Figure, its mass (in energy units that can be calculated using Einstein's Equation, $E = mc^2$), electric charge, and spin (angular momentum) is provided. The one particle missing from this image is the so-called *Higgs Boson*, which was only recently confirmed, in 2013, to exist. It appears to be the fundamental carrier of the field that imposes mass on (some of) the other fundamental particles listed in this table.

normal-matter counterparts, but opposite electric charges. Thus, the anti-particle of the negatively charged electron is the positively charged *positron*. The *Anti-Proton* is the negatively charged anti-particle of the normal, positively charged proton. And so on. But there is a problem with this. When particles of normal matter and their corresponding anti-particles collide, they spontaneously *annihilate* and convert their entire mass into an equivalent amount of energy as given by Einstein's Equation, $E = mc^2$. The problem is that there seems to be no clear reason that one type of particle—comprising what we call normal matter—would be preferred over the other—what we call anti-matter. And yet, when we look around us, there is no evidence of large reservoirs of anti-matter in the Universe today. No merging galaxies with immense X-ray or γ-ray luminosities arising from the annihilation of vast quantities of matter and anti-matter. No interacting gas clouds in our Galaxy or in other galaxies that show evidence of matter/anti-matter interactions. These observations seem to indicate that virtually all visible matter in the Universe is composed of normal matter. The tiny bit of anti-matter that does exist naturally in the Present-Day Universe seems to form only under very unusual circumstances such as near the event horizon of a black hole (see Destinations 10 and 17) or due to collisions of ultra-high energy atoms produced in Supernovae (Destination 12).

The solution to the riddle of the predominance of normal matter is not understood. But the root of the answer likely lies in a fundamental *Asymmetry* that

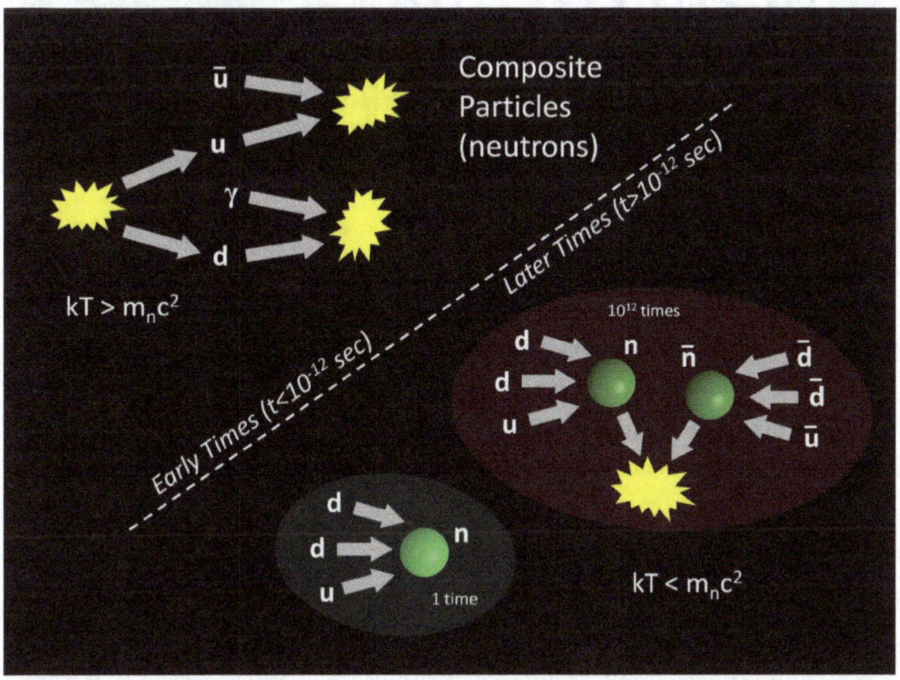

Fig. 25.4

This figure is analogous to Figure 25.2, but now for Composite Particles. Since these particles are composed of Fundamental Particles, their evolution in the Early Universe is different. In particular, at earliest times (when $kT > m_{Comp}c^2$), composite particles do not exist in any form. However, their constituent fundamental particles do, and these participate in the same sorts of equilibrium reactions described in Figure 25.2 (the diagram at upper left of this Figure shows such reactions for Down (*d*) and Up (*u*) quarks). Once the Universe cools sufficiently so that $kT < m_{Comp}c^2$, the fundamental particles combine to produce long-lived composite particles. One example—the formation of neutrons—is shown in the

anti-Hydrogen composed of a positron and anti-proton, and even anti-Helium—have been produced in the lab. But anti-matter can be pricey: The estimated cost of anti-H is about $10 billion per milligram! Why anti-matter does not exist naturally in greater quantities remains a mystery. But, as we'll see, a fortuitous mystery.

exists in how certain forces operate. The net result is that some of the processes that can produce normal matter—again, excluding Dark Matter—are very slightly favored in the Universe compared to the (nearly) identical processes that can produce anti-matter. The idea is that the quantity of normal stuff that formed in the Early Universe—that is, the matter we and all that we see around us is made of—very slightly exceeded the quantity of the stuff we call anti-matter. But the required excess was tiny: For every trillion anti-particles produced in the Early Universe, one trillion and *one* normal particles had to be produced. As individual particles and their anti-particles began to form in the cooling Universe, all trillion of the matter/anti-matter pairs would have annihilated, leaving just one particle of normal matter (see Figures 25.2 and 25.4). Remarkably, this tiny asymmetry is sufficient to account for the quantity of normal matter in the Present-Day Universe, and the apparent complete lack of anti-matter.[16] Good thing, too. Without an asymmetry that favors normal matter over anti-matter, there'd be no particles in the Universe. No matter at all. And certainly no 'us' to study the Universe. Just a pervasive sea of energy in the form of radiation.

lower right of the Figure. This composite particle consists of one Up and two Down quarks (you can verify that the total charge of this combination is zero using the data in Figure 25.3). As with Fundamental Particles, there is a tiny excess of normal neutrons when they first form, sufficient to account for the amount of normal matter in the Present-Day Universe (see Figure 25.2).

25.2 THE GENESIS OF PHYSICAL FORCES AND THE INFLATION ERA

There is one very interesting implication of the sequence of events we have seen as we approach $t = 0$. According to modern physical theory, the four forces of nature—Electromagnetism, the Weak Nuclear Force, the Strong Nuclear Force, and Gravity (see Figure 25.1)—rely on 'transmitter' particles of some sort to relay the respective force between the 'parent' particles. For example, photons—which have no mass—transmit the electromagnetic force.

16 One candidate asymmetry is the so-called *Charge-Parity* (or *CP*) *Violation*, which is observed to occur for some subatomic particles. This asymmetry implies that there is a tiny difference in the physics that applies to anti-matter compared to regular matter. It can be seen to occur as some types of particles spontaneously change into their own anti-particles or when some energetic particles decay into other particles. Why this asymmetry occurs—and what basic physical process drives it—remains unknown. Yet it may represent the key reason we exist! Or, perhaps not. At present, experiments that have tried to probe the frequency of CP Violations suggest that it occurs too rarely to explain the observed low abundance of anti-matter today. And so the search continues.

Similarly, particles with mass—known as *bosons*—transmit the weak nuclear force.[17] The key point is that these forces rely on the existence and exchange of *particles*, analogous to how two people tossing a ball stay close to one another as they exchange a particle (the ball). As we've seen, particles form out of the energy field as the young, hot Universe evolves. But this must also be true of the force (transmitter) particles. Consequently, this means that before the force particles came into existence, *the forces they transmit did not exist* (Figure 25.1)!

We can see this by traveling back to a time of $t = 10^{-12}$ *sec* when a species of bosons that had existed at earlier times evolves into a state where it no longer has mass. These particles become the first photons, and with their arrival, the Electromagnetic Force as we know it is 'born.'[18] Prior to this time, the Weak Nuclear Force and the Electromagnetic Force—which have distinct properties in the Present-Day Universe—were merged into a single force dubbed the *Electro-Weak Force*, characterized by a single set of bosons that no longer exists. Do not be (too) confused by all this. The lack of photons in the Universe before around $t = 10^{-12}$ *sec* doesn't mean the Universe suddenly lacks energy at earlier times. The energy in the Universe is most definitely still present; it is just no longer in the form of EM Radiation. Instead, it has been relegated to particles—mostly in the form of highly relativistic quarks and electrons—and into the fabric of space itself.

Strange as all this is, the really weird stuff begins at our next stop, just 10^{-36} *sec* after the expansion of the Universe commenced. This was the time when the ancestors of the particles that transmit the Strong Nuclear Force[19]—gluons (Figure 25.3)—first begin to form in the Universe. Though this sounds similar to what we've seen before as other particles and forces come into existence in the Early Universe, what's different now is that 10^{-36} *sec* is a very, very short time even for subatomic particles. Specifically, the particles that are now coming into existence in this era actually require considerably more time than this—about 10^{-34} *sec*—to form.

What could possibly happen over the seemingly infinitesimal time interval from $t = 10^{-36}$ *sec* to $t = 10^{-34}$ *sec*? Under normal circumstances, the Universe would expand by about a factor of about 30 and it would also cool down by

17 These particles are known as the 'force carriers' and their properties are summarized in Figure 25.3. Bosons differ in that they have mass, a feature that photons (Electromagnetism), gluons (the Strong Nuclear Force), and gravitons (Gravity; this particle has not been detected or well understood theoretically, so it is not shown in Figure 25.3) all lack. The two known fundamental Bosons—the *W* and *Z* particles (see Figure 25.3)—are related to the Weak Nuclear Force.

18 This is also the time when the Weak Nuclear Force—the force that regulates the radioactivity of atomic nuclei—first appears. The Weak Force is transmitted by the *W* and *Z* bosons (Figure 25.3), both of which, unlike the photon, have mass.

19 This is the force that holds atomic nuclei together against the very strong electric forces of the protons that try to disrupt the nuclei.

a factor of about 10. As we've seen, temperature (in particular, kT) is the key parameter that dictates the timing of the formation of particles in the Early Universe, and it is this large drop in temperature during this tiny time interval that causes a problem. If the temperature was just right to form particles at 10^{-36} *sec*, then by 10^{-34} *sec*—when the particles actually do come into existence—the Universe would be *much* cooler than when the process began. This leads to an unstable situation where the energy field of the Universe should have created the particles required for the Strong Nuclear Force but hasn't yet been able to do so. This situation is known technically as a *Supercooled* state. Though the specific case we are considering here in the Early Universe is unusual, supercooling itself is not some esoteric phenomenon. It can happen in everyday life. For example, a supercooled liquid—including water—can, under the right circumstances, instantly turn solid when it's disturbed.[20]

As the Universe reacts to its unstable, supercooled state at $t = 10^{-36}$ *sec*, it is not the particles that react, but space itself. The result—which can be predicted using the same sort of analysis that describes how the Universe expands under more normal conditions—is that the expansion of space must accelerate rapidly to a phenomenally high rate. Thus, instead of expanding by a 'mere' factor of 30 over this time interval, the Universe responds by expanding by over a factor of 10^{40} (that's ten thousand trillion trillion trillion for anyone keeping score; see Figure 25.5). This era of extreme growth is known as *Inflation*, and it ends when the supercooled state is lifted. In the present case, this happens when the particles that started to form at $t = 10^{-36}$ *sec* finally come into existence at $t = 10^{-34}$ *sec*. As they do, they release energy that heats the previously supercooled Universe, allowing it to continue to expand in much the same way it had before Inflation (see Figure 25.5).

Inflation stamps some very important properties on the Universe, including some we can observe at later times. For instance, imagine two points within the expanding Universe just before Inflation began (Figure 25.6). If they are closer together than the distance light can travel between them in the lifetime of the Universe—which, here at $t = 10^{-36}$ *sec*, is about 3×10^{-25} *cm*, a size smaller than the radius of an electron!—the characteristic temperature of the Universe at the two locations could conceivably have had enough time to become identical.[21] But now, if the Universe expands super-rapidly through a process like

20 If a supercooled liquid happens to be a drink in a can that was just opened, this phenomenon can be inconvenient—though fascinating—when it happens. One cannot drink the drink until the entire thing melts! Another example of a supercooled liquid is freezing rain. Falling from the sky as supercooled water, it instantly turns to ice on contact with a solid surface.

21 The reason for this arises from fundamental thermodynamics, which states that if two bodies of unequal temperature come into contact, they will evolve to a common temperature.

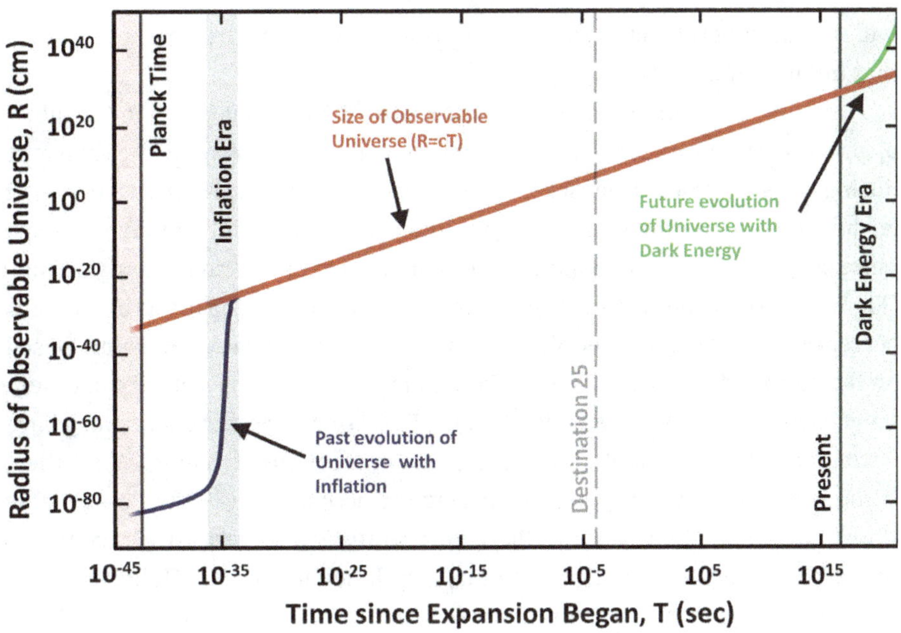

Fig. 25.5

This plot illustrates the change of the scale size of the Universe over time. The *x*-axis measures the logarithm of the time, *T*, from the Planck Time (around 10^{-43} *sec* at left) to beyond the Present Day (at around 4×10^{17} *sec*, at right). The *y*-axis measures the logarithm of the size scale of the Universe, *R*. The red line shows the radius of the Observable Universe defined as $R_{obs} = cT$, where c is the speed of light and *T* is the age of the Universe. The dark blue line shows the evolution of the size of a volume of space that corresponds to the size of the Present-Day Universe. Notice that during the time of Inflation (highlighted as the vertical light-gray bar), this volume shrinks by an enormous factor—over 10^{40}!—as we travel back in time through the time interval spanning $T = 10^{-34}$ to 10^{-36} *sec*. What this means is that what we call the Observable Universe today was a microscopic entity in a far-larger Observable Universe before Inflation. For this reason, the properties in the Present-Day Universe can be much more homogeneous than we'd have any reason to expect based on the arguments surrounding the Horizon Problem that we first encountered at Destination 22. We now apparently live at the start of a second era of Inflation caused by Dark Energy (green shaded region to the right). This implies that another acceleration of the Universe has begun,

Inflation, these same two points will be, at the end of 10^{-34} *sec*, separated by a distance that is around 10^{41} times greater than their original separation. That is, about 10^{16} *cm* (Figure 25.6), equivalent to a region much larger than the diameter of the Solar System! What used to be the Observable Universe at $t = 10^{-36}$ *sec* is now—thanks to Inflation—much, much larger than the size of the Observable Universe at $t = 10^{-34}$ *sec* (see Figure 25.6). This instantly helps explain the Horizon Problem (Destination 18), or why the CMB has such a uniform temperature everywhere in the sky (see Figure 18.8). In a nutshell, the Observable Present-Day Universe corresponds to just a minuscule fraction of the Observable Universe at $t = 10^{-36}$ *sec*. A fraction so small it was able to easily come into temperature equilibrium *before* Inflation commenced.

Inflation also affects the value of the Universal density ratio, Ω. By carefully considering what happens to space during Inflation, it is possible to show that if Ω started out with a value greater than zero—even barely greater than zero—it would very nearly equal 1.0000 ... after Inflation. This behavior persuades many people believe that if Ω was that close to 1.0 just as the Universe completed the Inflation Era, it is likely the ratio was precisely equal to 1.0 and so still is. Since $\Omega_{Matter, 0} = 0.27$ (a total that includes both visible and Dark Matter; Destination 19), then there must still be something else contributing to the mass ratio in the Present-Day Universe. We noted earlier that radiation by itself cannot account for the difference (Destination 19). Only a third component—the so-called *Dark Energy* represented by the Ω_Λ term (Destination 19)—can make up the difference. In this picture, about 97% of the Universe must be in the form of either Dark Matter or Dark Energy—neither of which has been detected or understood theoretically—in order to bring the value of Ω to the value of 1.0 that Inflation demands.

This concept of Inflation implies that there are potential forces in the fabric of space that can effectively accelerate its expansion. Early in the Universe during its supercooled phase, this force rocketed the scale of the Universe far beyond what it would have been without Inflation. Oddly enough, a much more subtle and weaker version of the same effect appears to be going on now (Figure 25.5). Recall that in Destination 19 we concluded that the Universe is currently expanding at an accelerating rate relative to what we'd expect based on the observed quantity of matter. Inflation implies that space itself can potentially have energy associated with it. In the Early Universe, this energy may have become available to accelerate the expansion from supercooling. What then, is the source of the energy that is driving the acceleration of the Present-Day Universe? No one knows. But if this interpretation is correct, the Present-Day Universe is experiencing its *second* major inflationary period. Unlike the first one, we are around to witness the show.

though this time far more sedately. The current anomalous acceleration of the Universe will result in shrinking the fraction of the Universe visible from the location of the Sun in the far future (green line). Before the Planck Time (shaded red region to the left), the evolution of the Universe cannot be described with modern physics. All of the events we will discuss here at Destination 25 occur to the left of the vertical dashed line in this figure.

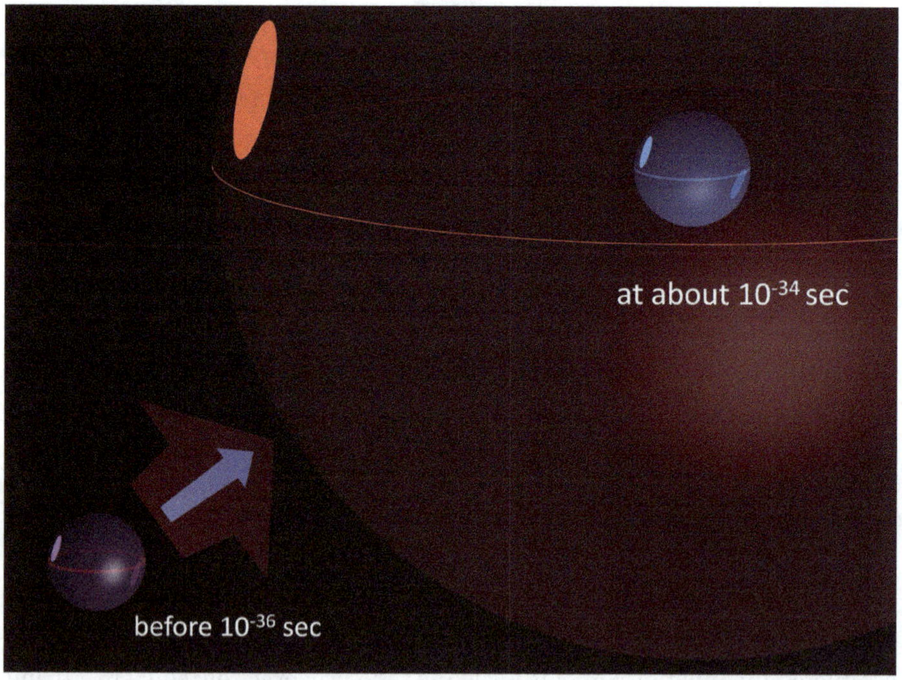

Fig. 25.6

This schematic figure illustrates Inflation from a different—but equivalent—perspective as shown in Figure 25.5. At the lower left, the observable Universe is defined by two coincident spheres—one red and one blue—that exist before the Inflation Era began. The two spots at opposite ends of this volume are not in 'causal contact.' That is, they are separated by more than the distance light can travel over the age of the Universe. After Inflation, the boundary of the Universe grows by a gargantuan factor, denoted here in the upper right as a much larger red sphere. But the size of the Observable Universe—the blue sphere—is only slightly larger than before. Notice that points on opposite sides of the blue sphere are now buried deep within the red sphere and hence were in close contact before Inflation. The points that are

There's one other loose end that Inflation helps to address. It's related to a problem we began to notice when we first saw how oddly smooth the CMB appears in comparison with the very lumpy distribution of matter in the Present-Day Universe (Destinations 18 and 22). The solution to this paradox is based on the fact that the Observable Universe was subatomic in size just as Inflation was starting to kick in at $t = 10^{-36}$ *sec*. As such, it would have been subject to random variations in temperature and density—known as *Quantum*

not in causal contact—one of which is shown as a large red dot on the larger sphere—are now far beyond the boundaries of the Observable Universe in blue. Notice too that the red spot has grown considerably in size, meaning that all structures in the red sphere expand relative to observers in the blue sphere. In this way, microscopic quantum fluctuations can grow to huge size during Inflation, eventually to be 'locked in' as the temperature and density fluctuations seen in the CMB (Destination 23), and the seeds of the large-scale structure that pervades the Present-Day Universe (Destinations 18 and 22).

Fluctuations—that are characteristic of all microscopic systems (more on this below). Before Inflation, these fluctuations were tiny. After Inflation, they would also have grown vastly larger (Figure 25.7). As the post-inflationary Universe resumed its comparatively sedate expansion after Inflation ended (Figure 25.5), these expanded Quantum Fluctuations would have become the seeds for the temperature and density variations that we found in the CMB (Destination 23). In turn, these fluctuations—via linear, then non-linear gravitational growth processes (Destination 22)—eventually led to the formation of the first stars, galaxies, quasars, clusters (Destination 20 and 23). In short, all the structures found in the Present-Day Universe arose from random quantum fluctuations in the ultra-young Universe before Inflation. In this sense, we are here just by chance.

Fig. 25.7

This Australian Indigenous painting depicts the formation of the world according to lore (*lower left*). A blow-up of the central region (*upper right*) shows clearly the individual dots that comprise the painting. In a similar way, Inflation caused a microscopic portion of the Early Universe to grow so large than the quantum fluctuations became macroscopic in size. These fluctuations then propagated through the fabric of the Universe and began to evolve with the expansion of the Universe until we could see them as temperature fluctuations in the CMB (background). In this respect, all the structure of the Present-Day Universe—ourselves included—arose from the random fluctuations in the fabric of a microscopic portion of the ultra-young Universe.

25.3 AS FAR AS WE CAN GO: THE PLANCK TIME

We are not quite done. We can still journey a bit deeper into the past. But as we do so, we enter a strange, very poorly charted territory. The four forces of Nature we know in the Present-Day Universe have merged into two: Gravity, still separate from the others, while the Electro-Weak Force and the Strong Nuclear Force have merged into one. The details of how this merger might happen—or indeed, whether it really does happen—remain unclear. As we go further back to the younger, hotter, denser Universe, even these two forces may merge into one overarching force (Figure 25.1). Ideas that attempt to describe how this might occur are called *Grand Unified Theories*, or *GUTs*, and the period we have now entered is known as the *GUTs Era* of the Universe, corresponding to the interval before $t = 10^{-36}$ *sec* after Universal Expansion began. At this time just about all the energy in the Universe is held in the fabric of space, fundamental particles such as electrons, neutrinos, and quarks (Figure 25.3) have not had time to form, the expansion rate of the Universe is phenomenally high, and the four distinct forces we know today have merged into one. The Universe is far simpler at this stage than at any later time. And yet, compared to the many other destinations we have visited on our previous voyages through the Universe, it is fundamentally more alien (Figure 25.8). Nothing we are used to exists.

But we do finally reach an end: At $t = 10^{-43}$ sec we discover we can go no further back in time. This very early moment in the evolution of the Universe is known as the *Planck Time*, the closest we can hope to get to the moment that the expansion of the Universe commenced. The reason for this has to do with a fundamental feature of the *subatomic* world known as the *Uncertainty Principle*. This principle states that in order to specify the energy of a particle precisely, you must accept that you lose all information about the duration of the events affecting that particle.[22] Here at the very earliest times, the observable Universe

22 Sound absurd? Well, although we did not discuss it at the time, the concept of degenerate matter—the stuff that white dwarfs and neutron stars are made of—intimately relies on the validity of the Uncertainty Principle. And we know that white dwarfs and neutron stars exist. Thus, the strange behavior that the Uncertainty Principle implies really does occur. Indeed, it represents a fundamental characteristic of nature, not some 'flaw' in our understanding. Another more common way of describing the Uncertainty Principle is as a relation between position and velocity: The more precisely you try to determine the location of a subatomic particle, the less certain its momen-tum becomes. Likewise, if you determine the motion of a particle precisely (its momentum), you lose all information regarding where the particle is located. The subatomic world is an inherently 'fuzzy' place.

is about the size of a proton (!), so we can apply the Uncertainty Principle to the Universe itself. Mathematically, we can state all this as

$$\Delta E \ \Delta t \geq \frac{h}{4\pi},$$

where ΔE is the energy fluctuations in the Universe, Δt is the age of the Universe, and h is Planck's constant (which we first encountered in Destination 5). The Uncertainty Principle is the source of the Quantum Fluctuations we found earlier were imprinted on the Early Universe by Inflation. The same fluctuations that served as the seeds for the formation of structure later on in the evolution of the Universe.

But now, the Uncertainty Principle bars us from peering further back in time in the history of the Universe. As we have noted already many times at this destination, at any given time the characteristic energy of the Universe can be calculated as kT, where k is the same Boltzmann constant that we used to estimate the motions of particles in a hot gas (Destination 5). If we set Δt to be 10^{-43} *sec* (the Planck Time) and, noting that $\Delta E = k\Delta T$, we discover that the maximum value of ΔT is about 10^{31} *K*. This happens to also be the temperature of the Universe at the Planck Time.

Pretty wild. But what does this all really mean? Because of the Uncertainty Principle, the random variations of the energy content of the Universe. That is, here at the Planck Time, the Universe could legitimately have an infinite total energy ... or none at all. Under such conditions, the physical laws that have guided us to the very edge of Creation cannot—by their own admission—make logical predictions for earlier epochs in the history of the Universe. Before the Planck Time, the meanings of space, distance, mass, even time itself become undefined (Figure 25.8). With the tools we have now—the physical laws that have brought us this far—we simply can't follow the evolution of the Universe any closer to its origin at t = 0. We've reached the end of our travels. For now.

As we arrive at the threshold of the beginning of time—an apt conclusion to our long journey through the Universe—it's worth pondering where we started. The serenity and beauty of the night sky from Earth could hardly have prepared us for what we have encountered. We have been exposed to distances, forces, and time-scales that stretch far beyond the range of human experience and comprehension. But we have also seen that with the proper tools—the scientific method, some mathematics, physics, and a lot of curiosity—we can still begin to understand some of the amazing things that go on in our Universe. Things we can barely imagine in our mind's eye. In the end, the sights we've seen in these Alien Skies should not humble us ... too much. After all, we humans—tiny and flawed as we are—have proven capable of contemplating the Universe in detail. Perhaps the only part of the Universe that can. *That* alone makes us special.

Fig. 25.8

An amazing, comprehensible Universe. Stranger than any fiction.

I sincerely hope that as we traveled the Universe you could *really* start to appreciate what could be seen at the amazing destinations we have visited. Maybe someday, some of you—or perhaps your kids or grandchildren—may get the chance to undertake some of these fantastical trips. But beware. The real Alien Skies our descendants will one day see in person are sure to be far, far stranger than any we've encountered on our voyages.

DESTINATION 25 QUESTIONS

25.1. What are the characteristic 'temperatures' of a proton and a neutron? Based on your answer, which particle formed first as the Universe cooled down after its initial expansion started? **HINT:** Use Einstein's Equation to calculate the energy of these particles given their masses. Then use the relation $E_{char} = kT$ to estimate the temperature.

25.2. Since it is not possible to see the earliest times in the Universe directly (due to the scattering of radiation by free electrons at the era of the 're'-combination), explain how we might be able to test whether or not the events that occurred in the first few seconds of the Universe really happened. **HINT:** If you were a detective, in general terms, how might you go about trying to discover who committed a crime?

25.3. We discussed how to make neutrons from combining a set of three Up and Down quarks. Protons are also composed of only Up and Down quarks. Based on the data in Figure 25.3, how could you combine three of these types of quarks to produce a particle—the proton—with a net charge of +1?

25.4. Use the Uncertainty Principle to determine over what timescale you would legitimately be able to disappear. For this problem, calculate your equivalent energy using Einstein's Equation, $E = mc^2$.

25.5. Calculate the change in the expansion velocity over the following length scales, assuming the Present-Day value of the Hubble Constant. Supply reasonable estimates for the relevant lengths in each case.

a) The height of the house or apartment you live in (best guess).
b) The diameter of the Earth.
c) The 'diameter' of the Galactic Disk (see Destination 9).
d) The diameter of the Local Universe (see Destination 18).
e) Which of these (if any) are likely to clearly exhibit the Hubble (Universal) Expansion? Explain your reasoning.

25.6. The expansion factor that is believed to have occurred during Inflation is pretty hard to grasp. Look up the classical radius of an electron. How big would an electron be post-Inflation for the expansion factor noted in the text?

25.7. It is easy to think of the Universe when it was 10^{-4} *sec* old as 'young.' However in one important sense, it was already quite mature.

a) From Figure 25.5, estimate the factor by which the Universe expanded from the Planck Time to $t = 10^{-4}$ *sec*. You may consider the evolutionary track with or without inflation when you make this estimate.

b) Likewise, estimate the factor by which the Universe has expanded since $t = 10^{-4}$ *sec* to the present time.

c) How do these results imply the Universe is already 'old' by $t = 10^{-4}$ *sec*?

d) From Figure 25.5, estimate the time, t_{mid}, when the Universe had expanded by the same factor before and after t_{mid}. Calculate this for both the inflationary and non-inflationary evolutionary tracks.

APPENDICES

A: SUMMARY OF DESTINATIONS

B: UNITS, SYMBOLS, AND CONVERSION FACTORS

C: CONSTANTS

D: MATHEMATICAL FORMULAE

E: BASIC DATA FOR SELECTED PLANETS AND DWARF PLANETS OF THE SOLAR SYSTEM

F: THE CONSTELLATIONS OF EARTH'S SKY

G: THE NATURAL CHEMICAL ELEMENTS

APPENDIX A: SUMMARY OF DESTINATIONS

DESTINATION 1: THE MOON: OUR FIRST STEP OFF EARTH

Distance: 384000 *km* on average.

Vicinity: Orbiting Earth.

Location: Anywhere on near side of Moon. Fra Mauro Highlands/Apollo 17 landing site would be a suitable—and scenic—choice.

Highlights of Visit: The appearance of the sky from Earth and the motions of some of the more prominent celestial bodies. An introduction to scale of the Universe. The Moon and how its sky differs—and is the same as—the sky from Earth. The view of Earth from the Moon.

DESTINATION 2: THE VIEW FROM MARS

Distance: Ranges from 56 million *km* to 250 million *km.*

Vicinity: Orbiting the Sun on Mars.

Location: Anywhere on the surface of Mars. Locations near the equator of Mars are probably most convenient as they allow easier observations of its moons.

Highlights of Visit: Basic properties of Mars. The view back to Earth from another planet and the puny scale of the solar system relative to the distances to stars. The unusual motions of the moons of Mars and retrograde motion of the planets. The underlying order of those strange motions embodied in Kepler's Third Law. The nature of scientific hypotheses.

DESTINATION 3: THE SKY FROM EARTH: A LAST LOOK FROM HOME

Distance: Zero.

Vicinity: Orbiting the Sun back on Earth.

Location: Anywhere where celestial sources can be observed year-round and, preferably in a good climate and away from interfering city lights.

Highlights of Visit: Navigation in the sky using stars, constellations, and celestial coordinates. Angular measurements in the sky. How to use angular positions of stars to determine their motions. The method of trigonometric parallax and defining the scale of interstellar distances. The parsec, a fundamental unit of distance. Identifying the closest stars.

DESTINATION 4: THE SKY FROM α CENTAURI

Distance: 1.26 *pc* = 4.11 *ly*.
Vicinity: Orbiting the nearby star α Cen A within the α Cen triple-star system.
Location: On a hypothetical planet located 1 AU from α Cen A.
Highlights of Visit: The Sun as a mere nighttime star due to the effects of the inverse-square law of brightness. The concept of center of mass of a system and how that relates to orbital motions. A new version of Kepler's Third Law that incorporates the masses of orbiting bodies. A first chance to systematically compare the basic properties of different stars.

DESTINATION 5: THE 'EMPTY' SKY OF DEEP SPACE: AT 'ANTI-CENTAURI'

Distance: 1.26 *pc* = 4.11 *ly*.
Vicinity: An empty region of space located opposite α Cen as seen from Earth toward the constellation Cassiopeia.
Location: Nowhere in particular; just floating through interstellar space.
Highlights of Visit: The Interstellar Medium (ISM), matter that pervades 'empty' space between stars. The distribution of the chemical elements in interstellar space and how that compares to the Sun and Earth. The physical properties of the 'Warm' ISM which is comprised mostly of neutral hydrogen atoms. Electromagnetic (EM) radiation. How neutral hydrogen in the ISM produces characteristic radio radiation that can be used to detect the warm ISM from Earth.

DESTINATION 6: THE ORION NEBULA

Distance: Approximately 500 *pc*.
Vicinity: Within the inner regions of the Orion Nebula.
Location: On the surface of a hypothetical free-floating planetesimal body in the nebula.
Highlights of Visit: The Orion Nebula and its optically bright sky. How atoms can emit optical line radiation. The Doppler effect and its use to measure motions of stars and gas. The properties of the 'Hot' or ionized ISM that pervades the Orion Nebula. The dark clouds of the Orion Nebula, examples of the 'Cold' or molecular ISM. The role of dark clouds in star and planet formation, a process regulated by gravitational collapse and the conservation of angular momentum. Basic properties of the brightest, hottest stars in the Orion Nebula and of hot bodies in general. Wien's Law and Planck's Law, fundamental relations regarding how hot bodies emit EM radiation. The Orion Nebula as a nursery for star formation and the intricate interplay between the stars, and the Hot and Cold ISM of the nebula.

DESTINATION 7: EXTRASOLAR PLANETARY SYSTEMS: 51 PEG

Distance: 14.7 *pc*.
Vicinity: In the vicinity of a non-descript, solar-like star in the constellation Pegasus as seen from Earth.

Location: On a hypothetical planet orbiting about 1 AU away from the star 51 Peg.

Highlights of Visit: A deeper look at the concept of Luminosity, or the total rate of energy output—or power—from a star. How luminosity affects the brightnesses of stars in the sky and alters the inverse-square law of brightness. The first extrasolar planet known, 51 Peg b, an example of a so-called Hot Jupiter unlike any planets in the Solar System. How 51 Peg b was discovered through use of the Doppler effect. The properties of other extrasolar planetary systems. A summary of various methods used to discover such systems. Some ideas of how planetary systems form, including unusual cases such as the 51 Peg system. The Drake Equation and some speculation about the frequency of life—intelligent or otherwise—within our Galaxy.

DESTINATION 8: THE PLEIADES

Distance: 140 *pc*.

Vicinity: Somewhere inside the central core of the Pleiades star cluster in the direction of the constellation Taurus, the Bull, in Earth's sky.

Location: On a hypothetical planet orbiting some star near the cluster center.

Highlights of Visit: Star clusters and how they can be used to measure basic stellar properties. The HR Diagram and the stellar Main Sequence. How the HR diagram is used to map a range of basic stellar properties including luminosity, temperature, radius and—from binary stars—mass. The spectra of stars and how these are produced in stellar atmospheres. Spectral types and how these vary primarily as a function of stellar surface temperature. Stellar structure and evolution. How to get energy from thermonuclear reactions that fuse small atomic nuclei into larger ones. How main sequence stars generate their luminosity through thermonuclear fusion reactions that convert H into He in their cores. How these concepts can be used to determine the ages of stars. The mass-luminosity and mass-age relations for main sequence stars.

DESTINATION 9: EXPLORING THE MILKY WAY

Distance: Ranging from very near the Sun to over 10000 $pc = 10\ kpc$ away.

Vicinity: Throughout our Galactic Disk as mapped out by the Milky Way. The specific journeys are toward the constellations Sculptor and Auriga in Earth's sky.

Location: Various.

Highlights of Visit: The basic properties of the Milky Way as seen from Earth, including its path in the sky as a Great Circle and its visibility in all forms of EM radiation. The Galactic equator and poles. A journey to the South Galactic Pole (SGP) and how that reveals that the Milky Way represents a distribution of stars and gas known as the Galactic Disk. The location of the center of our Galaxy toward the constellation Sagittarius. Density laws in general, and the exponential density law in particular as a model for our Galaxy. A journey toward the Galactic 'Anti-Center' (away from Sagittarius) and how that reveals the Galactic Disk to be an extended thin structure that is also exponential in form outward from the Galactic

Center. The stellar and gas content of the outer Galactic Disk and how that compares to the region near the Sun. The rotation of the Galactic Disk and how that is used to determine the mass of the Galaxy. A first hint of anomalous motions in the Galaxy and the inference of the presence of 'dark matter' though a careful analysis of Galactic rotation.

DESTINATION 10: INWARD BOUND: THE GALACTIC CENTER

Distance: From very near the Sun to 8.5 *kpc*.
Vicinity: Toward the center of the Galaxy in the direction of the constellation Sagittarius.
Location: Various, but ultimately some good viewing location near the Galactic Center.
Highlights of Visit: The properties of the stars and gas of the Galactic Disk as observed during a journey inward toward the Galactic Center in the direction of the constellation Sagittarius in Earth's sky. The Galactic Bulge, a new component of the Galaxy that is concentrated within the inner 1-2 *kpc* of the Galaxy and is composed mostly of older stars and little interstellar gas. The bizarre properties in the immediate vicinity of the Galactic Center, including the presence of a rich, varied ISM, active star formation, uniquely young, massive star clusters, unusual energetic phenomena, and, at the very center of the Galaxy, a supermassive black hole.

DESTINATION 11: A BLAZING SKY OF STARS: THE GLOBULAR CLUSTER M92

Distance: About 8.5 *kpc*.
Vicinity: In the direction of the constellation Hercules, and well outside the Galactic Disk.
Location: Some planetesimal body located about 10 *pc* from the center of the cluster M 92.
Highlights of Visit: Globular clusters. How the distribution of globular clusters can tell us where and how far away the Galactic Center is located. A journey to one such cluster, M 92. The majestic view of our Galaxy from well outside the Galactic Disk. The awesome view of the nearly 1 million stars of the cluster itself. The HR diagram of M 92 and how it differs from the HR diagram of the Pleiades. What happens when main sequence stars exhaust H in their cores that had been supporting thermonuclear reactions. Red giant stars and the process of initiating He fusion reactions in the cores of stars. The properties of degenerate matter and the relation to the Helium Flash. How stars evolve in M 92 through the process of 'low mass' stellar evolution and how stars like those in M 92—and like the Sun—eventually die. White dwarfs, the corpses of low-mass stars. M 92 as an example of a self-gravitating stellar system that—unlike the Galactic Disk—does not need to rotate to be stable.

DESTINATION 12: THE MAGELLANIC CLOUDS: THE NEAREST GALAXIES

Distance: 50 *kpc*.
Vicinity: Within the Large Magellanic Cloud, a satellite galactic system of our Galaxy.

Location: Various locations including star clusters NGC 1818, NGC 1850, Supernova 1987A and the x-ray source LMC X-3.

Highlights of Visit: The properties of the Magellanic Clouds, nearby dwarf galaxies. The remarkable star clusters of the Large Magellanic Cloud (LMC) and journeys to two of these systems, NGC 1818 and NGC 1850. The mass that separates 'low-mass' stellar evolution observed in M 92 and 'high-mass' evolution which first encounter in NGC 1850. The HR diagrams of the two clusters. A detailed description of high-mass stellar evolution, highlighting how this differs fundamentally from the evolution of low-mass stars. The sequence of thermonuclear reactions that power high mass stars during their evolution. Why such reactions can only proceed to the fusion of Iron nuclei and what happens as heavier elements begin to form. Stellar core collapse and the formation of exploding stars (supernovae) and high-mass stellar remnants such as neutron stars and black holes. Journeys to specific locations within the LMC that display some of these key stages of high-mass stellar evolution. Visits to a supernovae, a neutron stars/pulsar, and a bizarre binary star system consisting of a massive young star and a stellar-mass sized black hole, all predicted outcomes of models describing high-mass stellar evolution.

DESTINATION 13: THROUGH THE HEART OF DARKNESS: THE GALACTIC HALO

Distance: Out to 70-100 *kpc* from the Galactic Center.

Vicinity: In the outer reaches of the Galactic Halo.

Location: Near the Sculptor dwarf galaxy, an inconspicuous satellite of our home Galaxy.

Highlights of Visit: The nature of the Galactic Halo, the outermost component of our Galaxy. The properties of stars and dwarf galaxies in the Halo and the presence of the 'Very Hot' or coronal gas in the ISM throughout the Halo. The concept of velocity dispersion and its use in the Virial Theorem as a way to determine masses of non-rotating, stable self-gravitating systems. How to use the motions of dwarf galaxies in the Halo to determine that the vast majority of the mass of the Galaxy is virtually invisible. Mass-to-light ratios as a way to characterize how 'dark' a system is. The apparent dominance of dark matter in the outer Galaxy. Dwarf galaxies are also dominated by dark matter; the definition of galaxies? The tidal destruction of dwarf systems in the Galactic Halo as a process that actively contributes stars, gas and dark matter to the outer Galaxy.

DESTINATION 14: THE NEARBY SPIRAL GALAXIES M31 AND M33

Distance: About 780-860 *kpc* from the Sun.

Vicinity: Throughout the Local Group of galaxies that contains our Galaxy and M 31 and M 33.

Location: Within 5-10 *kpc* from the nearby spiral galaxies M 31 and M 33.

Highlights of Visit: Cepheid variables and their Period-Luminosity Relation. How this is used to determine extragalactic distances. The vast distances to the nearest spiral/disk galaxies, M 31 and M 33. The stellar and gas contents of these two galaxies and how these differ from our Galaxy, another example of a spiral/disk system. The rotation of M 31 and M 33 and how this is used to explore the dark matter content in these galaxies. General properties of disk galaxies as derived from our Galaxy, the LMC (also a disk system), and M 31 and M 32. The origin of spiral arms in disk galaxies and how that relates to Galactic rotation. A description of the Local Group of galaxies, consisting of our Galaxy, the LMC, M 31, M 33 and dozens of dwarf systems associated with these galaxies or the group as a whole. The Tully-Fisher relation for disk galaxies that relates their rotation properties and luminosities.

DESTINATION 15: THE GIANT ELLIPTICAL GALAXY MAFFEI 1

Distance: About 2400 *kpc* = 2.4 *Mpc* from the Sun.

Vicinity: Beyond the Local Group in the direction of the constellation Cassiopeia to the nearby Maffei Group of galaxies.

Location: As close as 10 *kpc* from the center of Maffei 1.

Highlights of Visit: The properties of galaxy groups beyond the Local Group. The existence of one nearby group, the Maffei Group, which contains a fundamentally new type of galaxy: Maffei 1, an example of a 'giant' elliptical galaxy. A journey to Maffei 1 to explore its stellar and gas content and how that compares to the Galaxy. The similarities of Maffei 1 and the Galactic Bulge. The motions of stars and gas in Maffei 1 and the use of the Virial Theorem to determine its mass. The unusual nuclear region of Maffei 1. A first comparison of properties of elliptical/spheroidal galaxies, including Maffei 1, the Galactic Bulge, dwarf spheroidal galaxies and M 32, an elliptical dwarf satellite of M 31. The introduction of the Hubble galaxy classification system. How elliptical/spheroid galaxies and spiral/disk galaxies are classified in this scheme. Barred spiral galaxies, S0 galaxies and irregular galaxies. How global properties of galaxies relate to their Hubble types. The Faber-Jackson relation between the luminosities and internal motions of elliptical/spheroidal systems.

DESTINATION 16: THE VIRGO CLUSTER: A SKY OF GALAXIES

Distance: 17 *Mpc*.

Vicinity: Toward the constellation Virgo, far beyond the confines of the Local Group.

Location: Very close to the center of mass of the Virgo Cluster.

Highlights of Visit: An introduction to galaxy clusters and a survey of the properties of galaxies within the nearest such cluster in the constellation Virgo. How cluster galaxies differ from systems in groups. The propensity of elliptical/spheroid galaxies in the core of the Virgo Cluster and the unusually large size of some of these system. The morphology-density relation in Virgo causing a strong segregation of galaxy types within the cluster. The common presence of peculiar galaxies in clusters such as Virgo as evidence for galaxy interactions in regions of high galaxy density. The transformation of galaxy types due to interactions. Other examples of interacting galaxies in the nearby Universe. The discovery of the Hubble Law relating the distances to galaxies and their apparent Doppler recession velocity. The simple form of the Hubble Law and this has profound implications regarding the evolution of the Universe, including the realization that the Universe has a finite, well-defined age.

DESTINATION 17: THERE'S SOMETHING WRONG HERE: THE SEYFERT GALAXY M 77

Distance: 16 *Mpc*.

Vicinity: Toward the constellation Cetus, but not within any particular cluster or prominent group.

Location: En route, then progressively closer to the very center of M 77.

Highlights of Visit: M 77 as an example of an unusual type of system known as a Seyfert galaxy. The 'normal' nature of Seyfert galaxies apart from the properties of their cores which are examples of Active Galactic Nuclei (AGN). The unusual properties of the spectrum of the AGN in M 77 and its time variability. How this implies extremely high luminosity and high mass in the AGN within a region similar in size to the Solar System. How the properties of the spectra of AGN vary depending on the viewing angle of the region. A model for AGNs related to the presence of supermassive black hole that are accreting mass at high rates at the centers of galaxies. How this model explains the broad properties of AGN. The role of General Relativity to explain how this accretion occurs in the vicinity of the central black hole. An exploration by a volunteer of a 'naked' supermassive black hole to observe some of the unusual implications of General Relativity on time and how light travels in the vicinity of strong gravitational fields. The appearance of an accreting supermassive black hole from close proximity.

DESTINATION 18: THE BOÖTES VOID

Distance: 210 *Mpc*.

Vicinity: Toward the constellation Boötes, a region in Earth's sky that is particularly devoid of relatively bright galaxies.

Location: The center of the void.

Highlights of Visit: Using the Hubble Law as a way to get galaxy distances. Redshift surveys and mapping the distribution of galaxies. The existence of 'structure' in the distribution of galaxies, including clusters, super clusters, cosmic walls and voids. Quantifying the 'lumpiness' of the Present-day Universe inferred from the existence of this structure. The view from the closest void. Understanding the subtle but profound implication that the nighttime sky is dark. The Cosmic Microwave Background (CMB), a source of long-wavelength radiation from all over the sky. The CMB is from a very distant source, but what is that source?

DESTINATION 19: BACK IN TIME TO A DISTANT GALAXY CLUSTER

Distance: 2100 *Mpc* = 2.1 *Gpc*, corresponding to a redshift of $z = 0.8$.

Vicinity: Toward the constellation Sextans but at a time when the Universe was about 50% its current age.

Location: The center of the cluster as it was in the past.

Highlights of Visit: The need to travel in time to see extremely distant objects as they were when their light was emitted. How to navigate in time within the Universe. The concept of redshift—which can be measured directly from galaxy spectra—and how that relates to the expansion scale of the Universe. Different types of Universe models depending on how much matter is present. The simplest models: Wide-Open, Open, Critical and Closed Universes. How to characterize different types of Universes depending on the mean density of matter. A census of visible and dark matter: ours is an Open Universe? Problems with this conclusion revealed from distance measurements to remote galaxies at high redshift. An alternative model: The Accelerating Universe and why this appears likely to be the Universe we live in. Exploring a very distant galaxy cluster, MS1054-032. This cluster is much younger than the Virgo Cluster and does not exhibit a strong morphology-density relation while containing an even larger number of clearly interacting galaxies. Evidence that galaxies and clusters evolve significantly over cosmic time.

DESTINATION 20: HIGH REDSHIFT GALAXIES: THE YOUNG AND RESTLESS UNIVERSE

Distance: Redshift of $z = 2.0$.

Vicinity: A time when the Universe was about 25% its current age. The precise location matters very little.

Location: No place in particular. 'When' we are located matters more than 'where' we are located.

Highlights of Visit: Universe is more compressed since the scale varies as $1/(1 + z)$, where z is the redshift. All else equal, galaxies, gas, stars, everything more compactly distributed in this younger, smaller Universe. The unusual nature of galaxies in the Early Universe. Galaxies are younger, more numerous, smaller and more metal poor at this redshift than compared to the present-day Universe. These properties and the proximity of galaxies suggest galaxies here are forming independently but are likely to soon merge together to form larger systems as part of a process known as Hierarchical Galaxy Formation. The rate of galaxy formation in the early Universe and how much that has declined from that time to the present day.

DESTINATION 21: QUASARS AT THE DAWN OF GALAXY FORMATION

Distance: Redshift of $z = 4.0$.

Vicinity: When the Unvierse was about 15–20% its current age.

Location: As for Destination 20, 'when' matters more than 'where'.

Highlights of Visit: The appearance of star-like objects in the sky known as Quasars. From their spectra and distances, we find these to be exceptionally luminous AGN. Much more common at this redshift in the Early Universe than in the Present-day Universe. A summary of the many different types of AGN and how they mostly differ in their mass accretion rates. Why quasars cannot shine for a significant fraction of the age of the Universe and the evolution of AGN. Finding the supermassive black holes that used to power quasars at the centers of nearby galaxies in the Present-day Universe. The strong correlation of supermassive black hole masses and the luminosities of the galaxies they inhabit. How the unusual properties of star formation in the Early Universe may have led to the rapid formation of supermassive black holes to power the quasars from that era. How AGN can affect the environments of the host galaxies in which they reside.

DESTINATION 22: THE DARK AGES

Distance: Redshift of 100.

Vicinity: When Universe was 17 *Myr* old.

Location: No place in particular, but we have chosen to be near the matter that will ultimately coalesce into our Galaxy and the Local Group.

Highlights of Visit: A dark sky from a time before galaxies or even any stars had formed. The only elements present are hydrogen, helium, and some lithium and beryllium, the lightest elements of the periodic table. Virtually no molecules either. Just about all matter is in the form of neutral gas, dominated by H and He. Universe here is full of radio radiation from neutral H just like the warm ISM of the galaxy and is very compact Detecting this gas in the early Universe. The only other radiation is from the CMB which here radiates primarily in the infrared. We are closer to the source of the CMB and so the redshift is smaller. Implies a hot source earlier in the history of the Universe. The Universe is very smooth, implying that structure formation cannot happen by the time we know it does in the Present-day Universe without the help of dark matter. The Horizon Problem: Why are opposite sides of the CMB at the same temperature to high precision? Does not appear to be enough time in the Universe to achieve this.

DESTINATION 23: THE ERA OF RECOMBINATION

Distance: Redshift ranging from about 1200–1050

Vicinity: Starting at about 320000 *yr* after the expansion of the Universe commenced.

Location: No place in particular, but at a particular time.

Highlights of Visit: The gas in the Universe is fully ionized. The many free electrons efficiently scatter all forms of Electromagnetic Radiation causing the Universe to be opaque. Since by $z = 100$ the gas is mostly neutral, this must be close to the time of 're'combination (really, 'first' combination) that transforms the ionized gas to neutral. The properties of CMB. Temperature fluctuations and how those relate to density variations in the Early Universe. Possible origins of these fluctuations. How we measure fluctuations and how they relate to modern structure. The role of DM in generating these fluctuations early enough to ensure the formation of structure in the Universe today.

DESTINATION 24: THE FIRST ELEMENTS IN THE UNIVERSE

Distance: Anywhere but quite early in the history of Universal expansion compared to the age of the Present-day Universe.

Vicinity: Between about 1 second after Universal expansion began to about 10 minutes later.

Location: No place in particular; all regions of the observable Universe would be the same.

Highlights of Visit: The conditions in the very early Universe. The lack of any elements apart from protons which are equivalent to hydrogen nuclei. Conditions appear suitable for element fusion but the high temperature restricts the formation of Deuterium nuclei, crucially limiting the process of element formation to proceed. What happens before the Universe cools enough for Deuterium formation, and the resulting thermonuclear fusion. The formation of just H, He and some Li and Be in the first few minutes. What we can learn about the Early Universe from the initial quantities of these crucial elements.

DESTINATION 25: TO A TIME BEFORE TIME

Distance: Anywhere.

Vicinity: To the very start of the expansion of the Universe.

Location: Undefined.

Highlights of Visit: The Radiation Era of the Universe and how that compares to the later Matter Era of the Present-Day Universe. The production and destruction of particles from the energy field. Fundamental and Composite particles and how they form in the Early Universe. The creation of fundamental forces as the particles that transmit the forces are formed. The formation of the Strong Nuclear Force and the onset of the Inflation Era. A possible solution to the Horizon Problem and the production of the initial density fluctuations in the Universe, both due to Inflation. The Planck Time and the role of the Uncertainty Principle in the evolution of the very Early Universe at time when its extent is microscopic in size. The end of our journeys using modern physics.

APPENDIX B: UNITS, SYMBOLS, AND CONVERSION FACTORS

cgs units *use centimeters for length, grams for mass, and seconds for time.*
mks or SI units *use meters for length, kilograms for mass, and seconds for time.*

B.1 UNIT ABBREVIATIONS

Distance

nanometer	→	nm
micron	→	μm
millimeter	→	mm
centimeter	→	cm
meter	→	m
kilometer	→	km
Astronomical Unit	→	AU
light-year	→	ly
parsec	→	pc
kiloparsec	→	kpc
Megaparsec	→	Mpc
Gigaparsec	→	Gpc
Earth Radius	→	$R_\oplus$
Jupiter Radius	→	R_{Jup}
Solar Radius	→	$R_\odot$

Time

hour	→	hr
solar day	→	d
solar year	→	yr
Megayear	→	Myr
Gigayear	→	Gyr
Hertz (frequency)	→	$Hz = 1/s$
MegaHertz	→	MHz
GigaHertz	→	GHz

Mass

gram	→	g
kilogram	→	kg

Earth Mass	→	$M_\oplus$
Jupiter Mass	→	M_{Jup}
Solar Mass	→	$M_\odot$

Angular Measurement

degree	→	*deg*
minute of arc	→	*arcmin*
second of arc	→	*arcsec*
radian	→	*rad*

Energy

Electron volt	→	eV
erg	→	$erg = g\ cm^2/s^2$
Joule	→	$J = kg\ m^2/s^2$
Watt	→	$W = J/sec$
Solar Luminosity	→	$L_\odot$
Horsepower	→	*hp*

Temperature

Degree Kelvin (absolute)	→	K
Degree Celsius	→	C
Degree Farenheit	→	F

Density

Mass density	→	g/cm^3
Number density	→	cm^{-3}

B.2 PARAMETER SYMBOLS AND UNITS

Parameter		**Symbol**		**Typical Units**
Wavelength	→	λ (Greek lower-case 'lambda')	→	*nm, mm, cm*
Frequency	→	ν (Greek lower-case 'nu')	→	*Hz, MHz, GHz*
Energy	→	E	→	*erg, J*
Kinetic Energy	→	KE	→	*erg, J*
Potential Energy	→	U	→	*erg, J*
Temperature	→	T (Note context!)	→	*K, C*
Distance	→	D, d	→	*cm, m, AU, pc, kpc, Mpc*
Separation	→	R, r	→	*cm, m, AU, pc, kpc, Mpc*
Galactic Plane Distance	→	z (Note context!)	→	*pc, kpc*
Luminosity/Power	→	L	→	*erg/s, W*
Flux/Brightness	→	F	→	$erg/s/cm^2$
Semi-Major Axis	→	a	→	*cm, AU*
Period	→	P	→	*s, d, yr, Myr*
Elapsed time	→	t	→	*s, yr, Myr*
Velocity	→	V	→	*km/s*
Velocity Dispersion	→	σ_V (Greek lower-case 'sigma')	→	*km/s*
Mass	→	M, m	→	$g, M_\odot$
Momentum	→	p (Note context!) or mV	→	*g cm/s*
Mass density	→	ρ (Greek lower-case 'rho')	→	g/cm^3
Number density	→	n (Note context!)	→	number/cm^3 or just cm^{-3}
Mass-to-Light Ratio	→	M/L	→	$M_\odot/L_\odot$
General Angle	→	θ (Greek lower-case 'theta')	→	*rad, deg,*
Parallax Angle	→	p (Note context!)	→	*arcsec*
Efficiency Factor	→	η (Greek lower-case 'eta')	→	(Dimensionless)
Redshift	→	z (Note context!)	→	(Dimensionless)
Age/Lookback Time	→	τ (Greek lower-case 'tau')	→	*s, yr, Gyr*
Age of Universe	→	T (Note context!)	→	*s, Gyr*
Dark Energy Symbol	→	Λ (Greek upper-case 'Lambda')	→	(Dimensionless)
Density Parameter	→	Ω (Greek upper-case 'Omega')	→	$\Omega_i = \rho_i/\rho_{c,i}$
Principal Quantum Number	→	n_i (Note context!)	→	(Dimensionless)
Difference Operator	→	Δ (Greek upper-case 'Delta')	→	$\Delta x = (x_1 - x_2)$
Summing Operator	→	Σ (Greek upper-case 'Sigma')	→	$\Sigma x = x_1 + x_2 + \cdots$
Average Operator	→	$\bar{x}$ or $\langle x \rangle$	→	$\bar{x} = \langle x \rangle =$ avg of x values
Proportionality	→	$\propto$ ('proportional to')	→	if $a \propto b$, then $a = Cb$ for some constant C

B.3 SELECTED UNIT CONVERSION FACTORS

Distance

$1\ nm = 1 \times 10^{-9}\ m = 1 \times 10^{-7}\ cm$

$1\ cm = 10\ mm = 0.01\ m$

$1\ m = 100\ cm = 1000\ mm$

$1\ km = 1000\ m$

$1\ inch = 25.4\ mm$

$1\ mile = 1.61\ km = 1609\ m$

$1\ AU = 1.50 \times 10^{8}\ km = 9.30 \times 10^{7}\ miles$

$1\ pc = 3.09 \times 10^{18}\ cm = 3.09 \times 10^{13}\ km = 1.92 \times 10^{13}\ miles = 2.06 \times 10^{5}\ AU = 3.26\ ly$

$1\ ly$ = speed of light × duration of one year $= 9.46 \times 10^{12}\ km = 0.307\ pc$

$1\ kpc = 1000\ pc = 3.09 \times 10^{21}\ cm = 3.09 \times 10^{16}\ km$

$1\ Mpc = 10^{6}\ pc = 1000\ kpc = 3.09 \times 10^{24}\ cm = 3.09 \times 10^{19}\ km$

$1\ Gpc = 1 \times 10^{9}\ pc = 10^{6}\ kpc = 1000\ Mpc$

$1\ R_{\oplus} = 6.38 \times 10^{8}\ cm = 6.38 \times 10^{3}\ km = 9.17 \times 10^{-3}\ R_{\odot}$

$1\ R_{Jup} = 6.99 \times 10^{9}\ cm = 6.99 \times 10^{4}\ km = 1.00 \times 10^{-1}\ R_{\odot}$

$1\ R_{\odot} = 6.96 \times 10^{10}\ cm = 6.96 \times 10^{5}\ km = 1.09 \times 10^{2}\ R_{\oplus}$

Time

$1\ solar\ day = 1\ d = 24\ hr = 86400\ s$

$1\ solar\ year = 365.25\ d = 3.16 \times 10^{7}\ sec$

$1\ Myr = 10^{6}\ yr$

$1\ Gyr = 10^{9}\ yr$

Mass

$1\ kg = 1000\ g$

$1\ M_{\oplus} = 5.97 \times 10^{27}\ g = 5.97 \times 10^{24}\ kg = 3.00 \times 10^{-6}\ M_{\odot}$

$1\ M_{Jup} = 1.90 \times 10^{30}\ g = 1.90 \times 10^{27}\ kg = 9.55 \times 10^{-5}\ M_{\odot}$

$1\ M_{\odot} = 1.99 \times 10^{33}\ g = 1.99 \times 10^{30}\ kg = 3.3 \times 10^{5}\ M_{\oplus} = 1.0 \times 10^{3}\ M_{Jup}$

Angular Measure

$1\ deg = 60\ arcmin = 3600\ arcsec$

$1\ arcmin = 60\ arcsec$

$1\ rad = 360/2\pi = 57.30\ deg = 206265\ arcsec$

Full circle $= 360\ deg = 1.296 \times 10^{6}\ arcsec = 2\pi\ rad = 24$ *hrs of Right Ascension*

Energy

$1\ eV = 1.6 \times 10^{-12}\ erg = 1.6 \times 10^{-19}\ J$

$1\ J = 10^{7}\ erg$

$1\ W = J/sec = 10^{7}\ erg/sec$

$1\ hp = 746\ W = 7.46 \times 10^{9}\ erg/s$

$1\ L_{\odot} = 3.85 \times 10^{33}\ erg/s = 3.85 \times 10^{26}\ W$

Temperature

$C = (5/9) \times (C - 32)$

$F = [(9/5) \times C] + 32$

$K = C + 274.2$

Frequency

$MHz = 10^{6}\ Hz$

$GHz = 10^{9}\ Hz$

APPENDIX C: CONSTANTS

C. 1 PHYSICAL CONSTANTS (CGS UNITS)

Boltzmann constant (k):	$1.381 \times 10^{-16}\ erg/K$
Speed of light (c):	$2.995 \times 10^{10}\ cm/s$
Planck constant (h):	$6.626 \times 10^{-27}\ erg \times sec$
Stefan-Boltzmann constant (σ):	$5.671 \times 10^{-5}\ erg/cm^2/s/K^4$
Gravitational Constant (G):	$6.673 \times 10^{-8}\ cm^3/g/s^2$
Mass of Proton (m_p):	$1.673 \times 10^{-24}\ g$
Mass of Neutron (m_N):	$1.675 \times 10^{-24}\ g$
Mass of Electron (m_e):	$9.109 \times 10^{-28}\ g$

C. 2 ASTRONOMICAL CONSTANTS

Earth-Moon Parameters

Solar Day of Earth (d):	$24\ hr$
Sidereal Day of Earth (d_{sid})	$23.934\ hr = 23\ hr\ 56\ min\ 4\ sec$
Radius of Earth ($R_\oplus$):	$6371\ km$
Semi-major axis of Earth's orbit:	$1.496 \times 10^8\ km = 1.00\ AU$
Sidereal orbital period of Earth:	$365.3\ d = 1.00\ yr$
Average orbital velocity of Earth ($V_\oplus$):	$29.8\ km/s$
Earth Mass ($M_\oplus$):	$5.97 \times 10^{27}\ g = 3.00 \times 10^{-6}\ M_\odot$
Lunar Month:	$29.53\ d$
Radius of Moon:	$1740\ km$
Mean distance Earth to Moon:	$239000\ miles = 384000\ km$

Selected Solar System Parameters (see also Appendix E and Table 2.1)

Radius of Mars:	$3390\ km$
Solar Martian Day:	$24.62\ hrs = 24\ hr\ 37\ min\ 23\ sec$
Semi-major axis of Mars' orbit:	$2.28 \times 10^8\ km = 1.524\ AU$
Sidereal orbital period of Mars:	$687\ d = 1.881\ yr$
Semi-major axis of Jupiter's orbit:	$7.786 \times 10^8\ km = 5.204\ AU$
Sidereal orbital period of Jupiter:	$4333\ d = 11.86\ yr$
Jupiter Mass (M_{Jup}):	$1.898 \times 10^{30}\ g = 9.55 \times 10^{-4}\ M_\odot = 318\ M_\oplus$

Selected Solar Properties (see also Appendix A and Table 4.1)

Solar Mass: ($M_\odot$):	$1.989 \times 10^{33}\ g = 1.989 \times 10^{30}\ kg$
Solar Luminosity ($L_\odot$):	$3.827 \times 10^{33}\ erg/s = 3.827 \times 10^{26}\ W$
Age of the Sun:	$4.7 \pm 0.3\ Gyr$
Estimated Lifetime of the Sun ($\tau_\odot$):	$10\ Gyr$

Selected Parameters for Nearby Stars and the Galaxy (see also Table 4.1)

Distance to α Cen system:	$1.26\ pc = 4.11\ ly = 3.91 \times 10^{19}\ cm$
Orbital Period of α Cen A and B:	$79.9\ yr$
Semi-major axis of α Cen system:	$23.2\ AU$
Solar Distance to Galactic Center (R_0):	$8.5\ kpc$ (adopted)
Solar Velocity about Galactic Center (V_0):	$220\ km/s$ (adopted)

Selected Cosmological Parameters

Hubble Constant (H_0):	$72 \pm 4\ km/s/Mpc$
Present-Day CMB Temperature (T_{CMB}):	$2.725\ K$
Present-Day Critical Density ($\rho_{c,0}$):	$9.7 \times 10^{-30}\ g/cm^3$
Parameters of Accelerating Universe:	$\Omega_{\Lambda,0} = 0.73$; $\Omega_{Matter,0} = 0.27$; $H_0 = 72\ km/s/Mpc$ (see also Table 19.1)
Redshift at recombination (Z_{Rec}):	1090
Temperature at recombination (T_{Rec}):	$2970\ K$

APPENDIX D: MATHEMATICAL FORMULAE

D. 1 GENERAL MATHEMATICAL RELATIONS AND TOOLS

Pi (π):	$\pi = 3.14159 \ldots$
Circumference (C) of circle of radius R:	$C = 2\pi R$
Area (A) of circle of radius R:	$A = \pi R^2$
Volume (V) of sphere of radius R:	$V = (4/3)\pi R^3$
Surface area (SA) of sphere of radius R:	$SA = 4\pi R^2$
Velocity (V), time (t), distance (d) relation:	$V = d/t$
Logarithm (x) of a number (a):	$\log(a) = x$ means $a = 10^x$
Difference (Δ) of two values (x_1, x_2):	$\Delta x = x_2 - x_1$
Summing (Σ) of a sequence of n values ($x_1, x_2 \ldots x_n$):	$\Sigma x = x_1 + x_2 + \cdots + x_n$
Average of a set of n values ($x_1, x_2 \ldots x_n$):	$\bar{x} = \langle x \rangle = (1/n)\Sigma x$
Proportionality (∝) of two values, x, y:	$x \propto y$ means $x = Cy$ for some constant C.
Power law relation between x, y:	$x^a \propto y^b$ for exponents a, b
Exponential relation between x, y:	$y = e^x$ where $e = 2.7183 \ldots$

D. 2 MATHEMATICAL FORMULAE IN TEXT BY DESTINATION

Units for these relations must be consistent. Constant values, where noted, are in the cgs system. See Appendix C for alternative values using different units.

DESTINATION 2: THE VIEW FROM MARS

Kepler's Third Law (General Form): $a^3 = CP^2$

a = Semi-major axis of orbit
P = Sidereal orbital period
C = Constant that depends on the body about which the planets or moons orbit and the units adopted for a and P.

Kepler's Third Law (Solar-System Form): $a^3 = P^2$

a = Semi-major axis of orbit in *AU*
P = Sidereal orbital period in *yr*

DESTINATION 3: THE SKY FROM EARTH: A LAST LOOK FROM HOME

Parallax Equation (Astronomer's Form): $D = 1/p$

D = Distance to source in *pc*

p = Parallax angle in *arcsec*

Parallax Equation (General Units): $D = d/p$

d = Distance from Earth to Sun in convenient units (*cm, km, fathoms, AU*, etc.)

D = Distance to source in same units as d

p = Parallax angle in *radians*

DESTINATION 4: THE SKY FROM α CENTAURI

Kepler's Third Law (Newton's Version, Astronomy Units): $a^3 = (M_1 + M_2)P^2$

a = Semi-major axis of system in *AU*

P = Orbital period of system in *yr*

M_1, M_2 = Masses of bodies in system in $M_\odot$

Center of Mass Relation: $a_1 M_1 = a_2 M_2$

a_1, a_2 = Separation of bodies from the location of the system's center of mass (same distance units)

M_1, M_2 = Masses of bodies in a binary system (same mass units)

Small-Angle Formula: $d = D\theta$

d = Size of object in convenient units (*cm, km, fathoms, AU*, etc.)

D = Distance to object in same units as d

θ = Angular extent of object in *radians*

DESTINATION 5: THE 'EMPTY' SKY OF DEEP SPACE: AT 'ANTI-CENTAURI'

Boltzmann Relation: $KE = (1/2)\, mV^2 = (3/2)\, kT$

KE = Average kinetic energy of gas particles

m = Average mass of the particles

V = Average speed of the particles

T = Temperature of the gas

k = Boltzmann constant = 1.381×10^{-16} *erg/K* (cgs)

Frequency-Wavelength Relation for EM Radiation: $c = \lambda v$

λ = Wavelength of the radiation

v = Frequency of the radiation

c = Speed of light = 2.998×10^{10} *cm/s* (cgs)

Energy of Photons of EM Radiation: $E = h\nu = hc/\lambda$

E = Energy of a photon of a given wavelength or, equivalently, frequency
ν = Frequency of the radiation
λ = Wavelength of the radiation
h = Planck constant = 6.626×10^{-27} $erg \times s$ (cgs)
c = Speed of light = 2.998×10^{10} cm/s (cgs)

DESTINATION 6: THE ORION NEBULA

de Broglie Equation: $\lambda = h/p$

λ = Wavelength of the particle (can be a matter particle or a photon)
p = Particle momentum (which is equal to mass × velocity for matter particles)
h = Planck constant = 6.626×10^{-27} $erg \times s$ (cgs)

Bohr Atom Condition: $n\lambda_e = 2\pi R_n$

n = Principal quantum number; this is just an integer from 1 to ∞
λ_e = de Broglie wavelength of an electron
R_n = Orbital radius for electron corresponding to principal quantum number n

Energy Change in the Bohr Atom: $\Delta E = E_{n_2} - E_{n_1} = -C\left(\frac{1}{n_2^2} - \frac{1}{n_1^2}\right)$

n_1, n_2 = Principal quantum numbers of initial orbit (n_1) and the final orbit (n_2)
E_{n_1}, E_{n_2} = Energies corresponding to the initial and final orbits
C = Constant that has energy units that depends on basic parameters of the H atom

Doppler Equation (Solving for Observed Wavelength): $\lambda_{obs} = \lambda_0\,(1 + V/c)$

λ_{obs} = Observed wavelength
λ_0 = Emitted wavelength
V = Source speed toward/away from observer
c = Speed of light = 2.998×10^{10} cm/s (cgs)

Doppler Equation (Solving for Source Speed): $V = c(\lambda_{obs} - \lambda_0)/\lambda_0$

λ_{obs} = Observed wavelength
λ_0 = Emitted wavelength
V = Source speed toward/away from observer
c = Speed of light = 2.998×10^{10} cm/s (cgs)

Definition of Angular Momentum: $J = mV_{rot}R$

J = Angular momentum
m = Mass of rotating object
V_{rot} = Characteristic rotation speed of object
R = Characteristic size of rotating object

Wien's Law: $\lambda_{max} T = 0.290\ cm\ K$

λ_{max} = Wavelength at which a body emits the most thermal radiation in *cm*

T = Temperature of the body in K

Planck's Law: $F = \sigma T^4$

F = Surface flux of a thermal source

T = Temperature of the body

σ = Stefan-Boltzmann constant = $5.671 \times 10^{-5}\ erg/cm^2/s/K^4$ (cgs)

DESTINATION 7: EXTRASOLAR PLANETARY SYSTEMS: 51 PEG

Flux-Luminosity-Distance Relation: $F = L/4\pi D^2$

L = Luminosity of a source

F = Flux of a source

D = Distance to the source

Drake Equation: $N = R_* f_p n_e f_l f_i f_c L$

N = Number of planets with life

R_* = Star formation rate in yr^{-1} in our Galaxy

f_p = Fraction of stars with planets

n_e = Typical number of Earth-like planets formed in extrasolar planetary systems

f_l = Fraction of planets on which life arises

f_i = Fraction of planets with intelligent life (if applicable)

f_c = Fraction of planets with civilizations able and willing to search for extra-planetary life (if applicable)

L = Timescale over which life is present on a planet in *yr*

DESTINATION 8: THE PLEIADES

Luminosity and Radius of a Star:

$$L = 4\pi R^2 \times \sigma T^4$$

$$R = \sqrt{L / 4\pi\sigma T^4}$$

L = Luminosity of a star assumed to be spherical in shape and emitting radiation approximately like a thermal source

R = Radius of the star

T = Surface temperature of the star

σ = Stefan-Boltzmann constant = $5.671 \times 10^{-5}\ erg/cm^2/s/K^4$ (cgs)

Main-Sequence Mass-Luminosity Relation: $L = \begin{cases} M^{3.6}, for\ M > 1M_{\odot} \\ M^{2.5}, for\ M \leq 1M_{\odot} \end{cases}$

L = Luminosity of star in *solar units*, $L_{\odot}$
M = Mass of star in *solar units*, $M_{\odot}$

Gravitational Energy Content of a Star: $E_{grav} = GM^2/R$

E_{grav} = Gravitational energy content of a star
M = Mass of star
R = Radius of star
G = Gravitational constant = $6.673 \times 10^{-8}\ cm^3/g/s^2$ (cgs)

Dynamical Timescale of a Star: $\tau_{gravity} = GM^2/RL$

M = Mass of star
R = Radius of star
L = Luminosity of star
$\tau_{gravity}$ = Time it takes for a star to collapse under the influence of gravity alone, or, equivalently, how long a star can shine at luminosity L by contracting under its own gravity.
G = Gravitational constant = $6.673 \times 10^{-8}\ cm^3/g/s^2$ (cgs)

Einstein's Equation: $E = mc^2$

E = Energy
m = Mass
c = Speed of light = $2.998 \times 10^{10}\ cm/s$ (cgs)

Nuclear Timescale: $\tau_{nuclear} = E_{nuclear}/L$

$E_{nuclear}$ = Total energy available from fusion in a star
L = Luminosity of star
$\tau_{nuclear}$ = Time star takes to run out of fuel for fusion reactions

Main-Sequence Lifetime: $\tau_* = \begin{cases} \frac{M}{M^{3.6}} = M^{-2.6},\ for\ \ M > 1M_{\odot} \\ \frac{M}{M^{2.5}} = M^{-1.5},\ for\ \ M \leq 1M_{\odot} \end{cases}$

M = Mass of star in *solar units*, $M_{\odot}$
τ_* = Main-sequence lifetime of star in units of solar main-sequence lifetime (~10 *Gyr*)

DESTINATION 9: EXPLORING THE MILKY WAY

Exponential Density Law (z Direction): $\rho_i(z) = \rho_{i,0} e^{(-|Z|/h_i)}$

z = Distance from the Galactic Plane

$\rho_i(z)$ = Space or mass density of stars of a given spectral type or a component of the ISM, denoted by the index '*i*', at a distance z from the Galactic Plane

$\rho_{i,0}$ = Space or mass density of stars or ISM of type '*i*' at $z = 0$ near the Sun

h_i = Scale height of stars or ISM of type '*i*' in same units as z.

Note: The units of the lengths z and h_i must be the same.

Exponential Density Law (R Direction): $\rho_i(R) = \rho_{i,0} e^{(-(R-R_0)/H)}$

R = Distance from the Galactic Center

R_0 = Distance of the Sun from the Galactic Center

$\rho_i(R)$ = Space or mass density of stars of a given spectral type or a component of the ISM, denoted by the index '*i*', at a distance R from the Galactic Center

$\rho_{i,0}$ = Space or mass density of stars or ISM of type '*i*' at $z = 0$ near the Sun

H = Scale length of stars or ISM in same units as R.

Newton's 'Universal' Law of Gravitation: $F_{grav} = \dfrac{GM_1M_2}{R^2}$

F_{grav} = Gravitational force between two bodies, assumed here to be point masses

M_1 = Mass of one of the bodies

M_2 = Mass of the other body

R = Distance between the two bodies

G = Gravitational constant = $6.673 \times 10^{-8}\ cm^3/g/s^2$ (cgs)

Mass of Body Determined from Orbiting Tracer: $M = V^2_{circ} R/G$

$$V_{circ} = \sqrt{GM/R}$$

M = Mass of body

V_{circ} = Orbital velocity of tracer, assumed to be in a circular orbit

R = Radius of orbit of tracer

G = Gravitational constant = $6.673 \times 10^{-8}\ cm^3/g/s^2$ (cgs)

DESTINATION 10: INWARD BOUND: THE GALACTIC CENTER

Power Law Density Distribution for Galactic Bulge: $$\rho_B(R) = \rho_0\left(1 + \left(\frac{R}{R_c}\right)\right)^{-4}$$

R = Distance from the Galactic Center

R_c = 'Core radius' of the density law in same units as R

$\rho_B(R)$ = Space or mass density of stars in the R direction that are components of the Galactic Bulge

ρ_0 = Space or mass density of stars at the Galactic Center that are associated with the Galactic Bulge

Schwarzschild Radius of a Black Hole: $$R_s = 2GM / c^2$$

R_s = Schwarzschild radius of a non-rotating black hole

M = Mass of the black hole

c = Speed of light = 2.998×10^{10} cm/s (cgs)

G = Gravitational constant = 6.673×10^{-8} $cm^3/g/s^2$ (cgs)

DESTINATION 12: THE MAGELLANIC CLOUDS: THE NEAREST GALAXIES

Kepler's Law for Binary Velocities: $$M_1 + M_2 = PV^3$$

M_1, M_2 = Masses of body 1 and body 2, respectively, in *solar units*, $M_\odot$

P = Orbital period of the system in *yr*

V = Orbital velocity of one of the bodies relative to the other body in units of the Earth's average orbital velocity, $V_\oplus$ = 29.8 km/s

DESTINATION 13: THROUGH THE HEART OF DARKNESS: THE GALACTIC HALO

Virial Theorem: $$2K = -U$$

K = Total kinetic energy of a self-gravitating system
U = Gravitational potential energy of system

Note: The Virial Theorem explicitly demands that the system be in equilibrium but it can still be used for systems that are close to an equilibrium state.

Mass Estimation with the Virial Theorem: $$M = \bar{R}\,\sigma_V^2 / G$$

M = Mass of a self-gravitating system
$\bar{R}$ = Characteristic radius of the system
σ_V = Velocity dispersion of the particles within the system
G = Gravitational constant = 6.673×10^{-8} $cm^3/g/s^2$ (cgs)

Power Law Density Distribution for Galactic Halo: $$\rho_H(R) = \rho_0\left(1+\left(\frac{R}{R_c}\right)\right)^{-3}$$

R = Distance from the Galactic Center
R_c = 'Core radius' of the density law in same units as R
$\rho_H(R)$ = Space or mass density of stars in the R direction that are components of the Galactic Halo
ρ_0 = Space or mass density of stars at the Galactic Center that are associated with the Galactic Halo

DESTINATION 14: THE NEARBY SPIRAL GALAXIES M31 AND M33

Cepheid Period-Luminosity Relation: $$\log_{10}\left(L_{Ceph}\right) = 1.1\log_{10}\left(P_{Ceph}\right) + 2.5$$

L_{Ceph} = Mean luminosity of a Cepheid variable star in *solar units*, $L_\odot$
P_{Ceph} = Pulsation period of the star in *days*

Tully-Fisher Relation: $$L_{Disk} = 6.2\times10^9 L_\odot\left(\frac{V_{rot}}{220\ km/s}\right)^4$$

L_{Disk} = Luminosity of a disk (spiral) galaxy in *solar units*, $L_\odot$
V_{rot} = Disk rotation velocity in *km/s*

DESTINATION 15: THE GIANT ELLIPTICAL GALAXY MAFFEI 1

Faber-Jackson Relation: $$L_E = 5.9 \times 10^9 L_\odot \left(\frac{\sigma_V}{220\ km/s} \right)^4$$

L_{Disk} = Luminosity of an ellipsoidal system in *solar units*, $L_\odot$
σ_V = Velocity dispersion of the system in *km/s*

DESTINATION 16: THE VIRGO CLUSTER: A SKY OF GALAXIES

Definition of Redshift: $$z = \left(\frac{\Delta\lambda}{\lambda_0} \right)$$

$\Delta\lambda = \lambda_{obs} - \lambda_0$ where λ_{obs} is the observed wavelength of a feature in a galaxy's spectrum, and λ_0 is the
true wavelength of that feature
z = Redshift of the galaxy

The Hubble Law (Astronomer Units): $$V = H_0 D$$

V = Recessional velocity of a galaxy in *km/s*
D = Distance to the galaxy in *Mpc*
H_0 = Hubble Constant of the Present-Day Universe in units of *km/s/Mpc*

Note: The units in this form of the Hubble Law are 'astronomer units', convenient for actual observations of galaxies.

DESTINATION 17: THERE'S SOMETHING WRONG HERE: THE SEYFERT GALAXY M77

Accretion Luminosity Relation: $$L = \eta c^2 \times \left(\frac{M}{\Delta t} \right)$$

L = Luminosity of an accreting source
η = Efficiency factor (between 0.0 and 1.0)
$\Delta M/\Delta t$ = Mass accretion rate
c = Speed of light = 2.998×10^{10} *cm/s* (cgs)

Constraint on Source Size from Luminosity Variability: $$R \leq c\,\Delta t$$

R = Approximate maximum size of the source
Δt = Timescale over which the source exhibits coherent variability
c = Speed of light = 2.998×10^{10} *cm/s* (cgs)

DESTINATION 19: BACK IN TIME TO A DISTANT GALAXY CLUSTER

Lookback Time (General Definition): $\tau = D_L / c$

τ = Lookback time to a source
D_L = Light-travel distance to the source
c = Speed of light = $2.998 \times 10^{10}\ cm/s$ (cgs)

Scale of the Universe: $R = \lambda_{em} / \lambda_{obs} = \frac{1}{1+z}$

R = Scale of the Universe (value between 0.0 and 1.0 for all past times up to the present)
λ_{em} = Wavelength of radiation when emitted from a source
λ_{obs} = Wavelength of radiation observed from the source
$z = (\lambda_{obs} - \lambda_{em})/\lambda_{em}$ = the redshift of the source

Lookback Time (Wide-Open Universe): $\tau = \frac{1}{H_0} \times \frac{z}{(1+z)}$

τ = Lookback time to a source
z = Redshift of the source
H_0 = Present day value of the Hubble Constant

Lookback Time (Critical Universe): $\tau = \left(\frac{2}{3H_0}\right) \times \left[1-(1+z)^{-\frac{3}{2}}\right]$

τ = Lookback time to a source
z = Redshift of the source
H_0 = Present day value of the Hubble Constant

Note: For the previous two relations, be sure to use the proper physical units, particularly for the Hubble Constant, H_0.

Critical Density of the Universe: $\rho_{c,z} = 3H_z^2 / 8\pi G$

$\rho_{c,z}$ = Critical density at redshift z
H_z = Hubble Constant at redshift z
G = Gravitational constant = $6.673 \times 10^{-8}\ cm^3/g/s^2$ (cgs)

Note: Be sure to use consistent units, especially for the Hubble Constant. The value of the critical density at the Present Day ($z = 0$) is denoted as $\rho_{c,0}$ and is calculated using the Present-Day Hubble Constant, H_0.

Universal Mass Density Parameter: $\Omega_z = \frac{\rho_z}{\rho_{c,z}}$

Ω_z = Density parameter for Universe at redshift z
ρ_z = Mean density of Universe at redshift z
$\rho_{c,z}$ = Critical density of Universe at redshift z

Note: Present-day values for the actual and critical density (for $z = 0$) can be used to calculate Ω_0. For models of the Universe that only include gravity(mass), an Open Universe corresponds to $\Omega < 1.0$, a Critical Universe corresponds to $\Omega = 1.0$, and a Closed Universe corresponds to $\Omega > 1.0$.

Radiation Density of the CMB: $$\rho_{rad,z} = \frac{4\sigma T^4_{CMB,z}}{c^3}$$

$\rho_{rad,z}$ = Radiation density from the Cosmic Microwave Background (CMB) at redshift z
$T_{CMB,z}$ = Observed temperature of the CMB at redshift z
c = Speed of light = 3.00×10^{10} cm/s (cgs)
σ = Stefan-Boltzmann constant = 5.671×10^{-5} $erg/cm^2/s/K^4$ (cgs)

Note: The present-day radiation density, $\rho_{rad,0}$, can be calculated using the present-day CMB temperature, $T_{CMB,0} = 2.725$ K.

DESTINATION 21: QUASARS AT THE DAWN OF GALAXY FORMATION

Accretion Timescale for an AGN: $$\tau_{acc} = \left(\frac{1}{1-\eta}\right)\frac{M}{(\Delta M / \Delta t)}$$

τ_{acc} = Timescale for accretion to form a central black hole of mass M
M = Mass of the central black hole in the AGN
η = Energy conversion efficiency factor (between 0.0 and 1.0)
$(\Delta M/\Delta t)$ = Mass accretion rate

Growth Rate of a Black Hole in an AGN: $$M_{BH}(T) = T\left(\frac{\Delta M}{\Delta t}\right)(1-\eta) + M_{BH,0}$$

$M_{BH}(T)$ = Mass of an AGN's central black hole as a function of time, T
T = Time since the AGN began accreting onto its central black hole
η = Energy conversion efficiency factor (between 0.0 and 1.0)
$M_{BH,0}$ = Mass of the black hole when the AGN accretion started (at $T = 0$)
$(\Delta M/\Delta t)$ = Mass accretion rate

DESTINATION 23: THE ERA OF RECOMBINATION

Extent of Observable Universe: $$R_{obs} = cT$$

R_{obs} = Extent of the observable Universe at a given age
T = Age of the Universe at any given point in its evolution
c = Speed of light = 2.998×10^{10} cm/s (cgs)

DESTINATION 25: TO A TIME BEFORE TIME

Uncertainty Principle (Energy-Time): $$\Delta E \, \Delta t \geq \frac{h}{4\pi}$$

ΔE = Precision to which the energy of a system is determined
Δt = Precision to which the time is specified for a system
h = Planck constant = 6.626×10^{-10} $erg \times s$ (cgs)

APPENDIX E: BASIC DATA FOR SELECTED PLANETS AND DWARF PLANETS OF THE SOLAR SYSTEM

PLANET NAME	SEMI-MAJOR AXIS (a) IN *AU*	ORBITAL PERIOD (*P*) IN *yr*	PLANET MASS IN UNITS OF EARTH MASS ($M_\oplus$)	NUMBER OF KNOWN MOONS
Major Planets				
Mercury	0.387	0.241	0.0553	0
Venus	0.723	0.615	0.815	0
Earth	1.00	1.00	1.00	1
Mars	1.52	1.88	0.107	2
Jupiter	5.20	11.9	318	63
Saturn	9.54	29.4	95.2	61
Uranus	19.2	84.0	14.5	27
Neptune	30.1	165	17.2	14
Selected Minor Planets				
Ceres	2.77	4.60	0.00016	0
Pluto	39.5	248.0	0.0022	5
Haumea	43.3	285.1	0.0007	2
Makemake	45.8	309.9	0.0007	0
Eris	67.7	557.2	0.0028	1

APPENDIX F: THE CONSTELLATIONS OF EARTH'S SKY

CONSTELLATION NAME	OFFICIAL ABBREVIATION	POSSESSIVE FORM OF CONSTELLATION NAME	WHAT THE CONSTELLATION REPRESENTS	BRIGHTEST STAR IN CONSTELLATION
Andromeda	And	Andromedae	Andromeda (The Chained Princess)	Alpheratz
Antlia	Ant	Antliae	The Air Pump	α Antliae
Apus	Aps	Apodis	Bird of Paradise	α Apodis
Aquarius	Aqr	Aquarii	The Water Bearer (Zodiac)	Sadalsuud
Aquila	Aql	Aquilae	The Eagle	Altair
Ara	Ara	Arae	The Altar	β Arae
Aries	Ari	Arietis	The Ram (Zodiac)	Hamal
Auriga	Aur	Aurigae	The Charioteer	Capella
Boötes	Boo	Boötis	The Herdsman	Arcturus
Caelum	Cae	Caeli	The Chisel	α Caeli
Camelopardalis	Cam	Camelopardalis	The Giraffe	β Camelopardalis
Cancer	Cnc	Cancri	The Crab (Zodiac)	Tarf

(*CONTINUED*)

(*CONTINUED*)

Canes Venatici	CVn	Canum Venaticorum	The Hunting Dogs	Cor Caroli
Canis Major	CMa	Canis Majoris	The Greater Dog	Sirius
Canis Minor	CMi	Canis Minoris	The Lesser Dog	Procyon
Capricornus	Cap	Capricorni	The Sea Goat (Zodiac)	Deneb Algiedi
Carina	Car	Carinae	The Keel of Argo Navis	Canopus
Cassiopeia	Cas	Cassiopeiae	Queen Cassiopeia	Shedir
Centaurus	Cen	Centauri	The Centaur	α Centauri
Cepheus	Cep	Cephei	Prince Cepheus	Alderamin
Cetus	Cet	Ceti	The Whale	Deneb Kaitos
Chamaeleon	Cha	Chamaeleontis	The Chameleon	α Chamaeleontis
Circinus	Cir	Circini	The Drafting Compass	α Circini
Columba	Col	Columbae	The Dove	Phact
Coma Berenices	Com	Comae Berenices	Berenice's Hair	β Comae Berenices
Corona Australis	CrA	Coronae Australis	The Southern Crown	Alphekka Meridiana
Corona Borealis	CrB	Coronae Borealis	The Northern Crown	Alphecca
Corvus	Crv	Corvi	The Crow	Gienah
Crater	Crt	Crateris	The Cup	Labrum

(*CONTINUED*)

Crux	Cru	Crucis	The Southern Cross	Acrux
Cygnus	Cyg	Cygni	The Swan	Deneb
Delphinus	Del	Delphini	The Dolphin	Rotanev
Dorado	Dor	Doradus	The Goldfish	α Doradus
Draco	Dra	Draconis	The Dragon	Etamin
Equuleus	Equ	Equulei	The Small Horse	Kitalpha
Eridanus	Eri	Eridani	The Celestial River	Achernar
Fornax	For	Fornacis	The Chemical Furnace	Fornacis
Gemini	Gem	Geminorum	The Twins (Zodiac)	Pollux
Grus	Gru	Gruis	The Crane	Alnair
Hercules	Her	Herculis	Hercules	Kornephoros
Horologium	Hor	Horologii	The Pendumum Clock	α Horologii
Hydra	Hya	Hydrae	Hydra (Mythological Water Monster)	Alphard
Hydrus	Hyi	Hydri	The Lesser Water Snake	β Hydri
Indus	Ind	Indi	An American Indian	The Persian
Lacerta	Lac	Lacertae	The Lizard	α Lacertae
Leo	Leo	Leonis	The Lion (Zodiac)	Regulus

(*CONTINUED*)

Leo Minor	LMi	Leonis Minoris	The Lesser Lion	Praecipua
Lepus	Lep	Leporis	The Hare	Arneb
Libra	Lib	Librae	The Balancing Scale (Zodiac)	Zubeneshamali
Lupus	Lup	Lupi	The Wolf	Men
Lynx	Lyn	Lyncis	The Lynx	Elvashak
Lyra	Lyr	Lyrae	The Lyre or Harp	Vega
Mensa	Men	Mensae	Table Mountain	α Mensae
Microscopium	Mic	Microscopii	The Microscope	γ Microscopii
Monoceros	Mon	Monocerotis	The Unicorn	β Monocerotis
Musca	Mus	Muscae	The Fly	α Muscae
Norma	Nor	Normae	The Carpenter's Square	γ^2 Normae
Octans	Oct	Octantis	The Octant	υ Oct
Ophiuchus	Oph	Ophiuchi	The Serpent Bearer	Ras Alhague
Orion	Ori	Orionis	Orion the Hunter	Rigel
Pavo	Pav	Pavonis	The Peacock	Peacock
Pegasus	Peg	Pegasi	Pegasus the Flying Horse	Enif
Perseus	Per	Persei	King Perseus	Mirfak

(*CONTINUED*)

Phoenix	Phe	Phoenicis	The Phoenix	Ankaa
Pictor	Pic	Pictoris	The Artist's Easel	α Pictoris
Pisces	Psc	Piscium	The Fishes (Zodiac)	Alpherg
Piscis Austrinus	PsA	Piscis Austrini	The Southern Fish	Fomalhaut
Puppis	Pup	Puppis	The Poop Deck of Argo Navis	Naos
Pyxis	Pyx	Pyxidis	The Mariner's Compass	α Pyxidis
Reticulum	Ret	Reticuli	The Eyepiece	α Reticuli
Sagitta	Sge	Sagittae	The Arrow	γ Sagittae
Sagittarius	Sgr	Sagittarii	The Centaur Archer (Zodiac)	Kaus Australis
Scorpius	Sco	Scorpii	The Scorpion (Zodiac)	Antares
Sculptor	Scl	Sculptoris	The Sculptor	α Sculptoris
Scutum	Sct	Scuti	The Warrior's Shield	α Scuti
Serpens[9]	Ser	Serpentis	The Snake	Unukalhai
Sextans	Sex	Sextantis	The Sextant	α Sextantis
Taurus	Tau	Tauri	The Bull (Zodiac)	Aldebaran
Telescopium	Tel	Telescopii	The Telescope	α Telescopii
Triangulum	Tri	Trianguli	The Triangle	β Trianguli

(CONTINUED)

Triangulum Australe	TrA	Trianguli Australis	The Southern Triangle	Atria
Tucana	Tuc	Tucanae	The Toucan	α Tucanae
Ursa Major	UMa	Ursae Majoris	The Great Bear	Alioth
Ursa Minor	UMi	Ursae Minoris	The Lesser Bear	Polaris
Vela	Vel	Velorum	The Sails of Argo Navis	Regor
Virgo	Vir	Virginis	The Virgin (Zodiac)	Spica
Volans	Vol	Volantis	The Flying Fish	β Volantis
Vulpecula	Vul	Vulpeculae	The Fox	Anser

APPENDIX G: THE NATURAL CHEMICAL ELEMENTS

ATOMIC NUMBER [SEE NOTE 1]	ELEMENT NAME	CHEMICAL ABBREVIATION	DENSITY @ STP (g/cm^3) [SEE NOTE 2]	YEAR OF DISCOVERY
1	Hydrogen	H	0.09×10^{-3}	1776
2	Helium	He	0.18×10^{-3}	1895
3	Lithium	Li	0.53	1817
4	Beryllium	Be	1.85	1797
5	Boron	B	2.34	1808
6	Carbon	C	2.26	ancient
7	Nitrogen	N	1.25×10^{-3}	1772
8	Oxygen	O	1.43×10^{-3}	1774
9	Fluorine	F	1.7×10^{-3}	1886
10	Neon	Ne	0.9×10^{-3}	1898
11	Sodium	Na	0.97	1807
12	Magnesium	Mg	1.74	1755
13	Aluminum	Al	2.70	1825
14	Silicon	Si	2.33	1824
15	Phosphorus	P	1.82	1669
16	Sulfur	S	2.07	ancient
17	Chlorine	Cl	3.21×10^{-3}	1774
18	Argon	Ar	1.78×10^{-3}	1894
19	Potassium	K	0.86	1807
20	Calcium	Ca	1.55	1808
21	Scandium	Sc	2.99	1879
22	Titanium	Ti	4.54	1791
23	Vanadium	V	6.11	1830
24	Chromium	Cr	7.19	1797

(*CONTINUED*)

(*CONTINUED*)

25	Manganese	Mn	7.43	1774
26	Iron	Fe	7.87	ancient
27	Cobalt	Co	8.90	1735
28	Nickel	Ni	8.90	1751
29	Copper	Cu	8.96	ancient
30	Zinc	Zn	7.13	ancient
31	Gallium	Ga	5.91	1875
32	Germanium	Ge	5.32	1886
33	Arsenic	As	5.72	ancient
34	Selenium	Se	4.79	1817
35	Bromine	Br	3.12	1826
36	Krypton	Kr	3.75×10^{-3}	1898
37	Rubidium	Rb	1.63	1861
38	Strontium	Sr	2.54	1790
39	Yttrium	Y	4.47	1794
40	Zirconium	Zr	6.51	1789
41	Niobium	Nb	8.57	1801
42	Molybdenum	Mo	10.22	1781
43	Technetium	Tc	11.50	1937
44	Ruthenium	Ru	12.37	1844
45	Rhodium	Rh	12.41	1803
46	Palladium	Pd	12.02	1803
47	Silver	Ag	10.50	ancient
48	Cadmium	Cd	8.65	1817
49	Indium	In	7.31	1863
50	Tin	Sn	7.31	ancient
51	Antimony	Sb	6.68	ancient
52	Tellurium	Te	6.24	1783
53	Iodine	I	4.93	1811
54	Xenon	Xe	5.9×10^{-3}	1898
55	Cesium	Cs	1.87	1860
56	Barium	Ba	3.59	1808

(*CONTINUED*)

57	Lanthanum	La	6.15	1839
58	Cerium	Ce	6.77	1803
59	Praseodymium	Pr	6.77	1885
60	Neodymium	Nd	7.01	1885
61	Promethium	Pm	7.30	1945
62	Samarium	Sm	7.52	1879
63	Europium	Eu	5.24	1901
64	Gadolinium	Gd	7.90	1880
65	Terbium	Tb	8.23	1843
66	Dysprosium	Dy	8.55	1886
67	Holmium	Ho	8.80	1867
68	Erbium	Er	9.07	1842
69	Thulium	Tm	9.32	1879
70	Ytterbium	Yb	6.90	1878
71	Lutetium	Lu	9.84	1907
72	Hafnium	Hf	13.31	1923
73	Tantalum	Ta	16.65	1802
74	Tungsten	W	19.35	1783
75	Rhenium	Re	21.04	1925
76	Osmium	Os	22.60	1803
77	Iridium	Ir	22.40	1803
78	Platinum	Pt	21.45	1735
79	Gold	Au	19.32	ancient
80	Mercury	Hg	13.55	ancient
81	Thallium	Tl	11.85	1861
82	Lead	Pb	11.35	ancient
83	Bismuth	Bi	9.75	ancient
84	Polonium	Po	9.30	1898
85	Astatine	At	(6.35)	1940
86	Radon	Rn	9.73×10^{-3}	1900
87	Francium	Fr	(1.87)	1939
88	Radium	Ra	5.50	1898
89	Actinium	Ac	10.07	1899

(*CONTINUED*)

90	Thorium	Th	11.72	1829
91	Protactinium	Pa	15.40	1913
92	Uranium	U	18.95	1789
93	Neptunium	Np	20.20	1940
94	Plutonium	Pu	19.84	1940
95	Americium	Am	13.67	1944
96	Curium	Cm	13.50	1944
97	Berkelium	Bk	14.78	1949
98	Californium	Cf	15.10	1950

Notes to Table:

1. The atomic number equals the number of protons in the nucleus of an atom of a given element.
2. 'STP' stands for 'Standard Temperature and Pressure' and corresponds to a temperature of 0 C and a pressure of 1 bar, or about 1 atmosphere.

FIGURE CREDITS

Fig 1.1 Abraham Cresques, http://commons.wikimedia.org/wiki/File:Caravane_Marco_Polo.jpg. Copyright in the Public Domain.

Fig 1.3 Copyright © Gregory H. Revera (CC BY-SA 3.0) at http://home.arcor-online.de/axel.mellinger/.

Fig 1.4 Copyright © Naskies (CC BY-SA 3.0) at http://commons.wikimedia.org/wiki/File:Comet_Lovejoy_2011_Milky_Way_Wide_Field.jpg.

Fig 1.5 Copyright © Joe Orman. Reprinted with permission.

Fig 1.6 Adapted from: Fastfission, http://commons.wikimedia.org/wiki/File:Earth_to_moon_vertical.png. Copyright in the Public Domain.

Fig 1.7 NASA, http://www.hq.nasa.gov/alsj/a15/a15pan1231715EvM.jpg. Copyright in the Public Domain.

Fig 1.8a Copyright © H. Raab (CC BY-SA 3.0) at http://commons.wikimedia.org/wiki/File:Clavius_Crater.jpg.

Fig 1.8b NASA, http://earth.jsc.nasa.gov/scripts/sseop/photo.pl?frame=86&mission=STS61A&roll=35. Copyright in the Public Domain.

Fig 1.9 NASA, http://commons.wikimedia.org/wiki/File:Apollo17.jpg. Copyright in the Public Domain.

Fig 2.1a Micha L. Rieser/NASA, http://commons.wikimedia.org/wiki/File:Mars_Valles_Marineris-Edit-MichaLR.jpg. Copyright in the Public Domain.

Fig 2.1b NASA, http://nssdc.gsfc.nasa.gov/image/planetary/mars/vikinglander2-2.jpg. Copyright in the Public Domain.

Fig 2.1c NASA/JPL/University of Arizona, https://commons.wikimedia.org/wiki/File:Icy_Crater_on_Mars_ESP_016954_2245_subimage_2.jpg. Copyright in the Public Domain.

Fig 2.1d NASA/JPL/Cornell, http://www.nasa.gov/images/content/138225main_20051121_Larry-s_Lookout.jpg. Copyright in the Public Domain.

Fig 2.2 NASA/JPL/Texas A&M/Cornell, http://www.nasa.gov/images/content/154221main_20060807_PIA07997_060802183716.jpg. Copyright in the Public Domain.

Fig 2.3 Copyright © Ron Wayman.

Fig 2.4a NASA/JPL-Caltech/University of Arizona, http://commons.wikimedia.org/wiki/File:Deimos-MRO.jpg. Copyright in the Public Domain.

Fig 2.4b NASA/JPL-Caltech/University of Arizona, http://commons.wikimedia.org/wiki/File:Phobos_edit_2.jpg. Copyright in the Public Domain.

Fig 2.5a NASA/JPL/Texas A&M/Cornell, http://marsrovers.jpl.nasa.gov/gallery/press/spirit/20050909a.html. Copyright in the Public Domain.

Fig 2.5b NASA/JPL/Texas A&M/Cornell, http://mars.nasa.gov/allaboutmars/extreme/moons/. Copyright in the Public Domain.

Fig 2.6 NASA/JPL/Texas A&M/Cornell, http://commons.wikimedia.org/wiki/File:PIA05547-Spirit_Rover-Earth_seen_from_Mars.png. Copyright in the Public Domain.

Fig 2.7 Copyright © StarryNight Education. Reprinted with permission.

Fig 2.8 Copyright © StarryNight Education. Reprinted with permission.

Fig 2.9 http://commons.wikimedia.org/wiki/File:Johannes_Kepler_1610.jpg. Copyright in the Public Domain.

Fig 2.10 NASA, http://commons.wikimedia.org/wiki/File:Olympus_Mons.jpg. Copyright in the Public Domain.

Fig 3.1 Willibald Pirckheimer, Geographia Universalis, 1545 (http://commons.wikimedia.org/wiki/File:Claudius_Ptolemaeus_-_Geographia_universalis.jpg). Copyright in the Public Domain.

Fig 3.2 Copyright © StarryNight Education. Reprinted with permission.

Fig 3.3a Akira Fujii, http://www.hubblesite.org/newscenter/archive/releases/2001/12/image/b. Copyright © Akira Fujii.

Fig 3.3b Copyright © Torsten Bronger (CC BY-SA 3.0) at http://commons.wikimedia.org/wiki/File:Orion_constellation_map.png.

Fig 3.4 U.S. Department of Defense, https://commons.wikimedia.org/wiki/File:Globe_Atlantic.svg. Copyright in the Public Domain.

Fig 3.5a Australian Astronomical Observatory, http://ftp.aao.gov.au/images/captions/misc015.html. Copyright © Australian Astronomical Observatory.

Fig 3.5b Australian Astronomical Observatory, http://ftp.aao.gov.au/images/captions/misc015.html. Copyright © Australian Astronomical Observatory.

Fig 3.6 Copyright © Azcolvin429/NASA (CC BY-SA 3.0) at https://commons.wikimedia.org/wiki/File:1_Earth_%28blank_2%29.png.

Fig 4.1 NASA/JPL/Space Science Institute, http://photojournal.jpl.nasa.gov/jpeg/PIA10406.jpg. Copyright in the Public Domain.

Fig 4.2a Copyright © StarryNight Education. Reprinted with permission.

Fig 4.2b Copyright © StarryNight Education. Reprinted with permission.

Fig 4.4 Sid Meier's Alpha Centauri, Firaxis Games. Copyright © Firaxis Games.

Fig 4.5 Godfrey Kneller, http://commons.wikimedia.org/wiki/File:Sir_Isaac_Newton_(1643-1727).jpg. Copyright in the Public Domain.

Fig 4.7 David Malin, http://apod.nasa.gov/apod/ap051204.html. Copyright © David Malin. Reprinted with permission.

Fig 4.8 Copyright © Steve Quirk. Reprinted with permission.

Fig 5.1 Copyright © Michael Padilla. Reprinted with permission.

Fig 5.2 Adapted from: Skatebiker (CC BY-SA 3.0) at http://en.wikipedia.org/wiki/File:Alpha,_Beta_and_Proxima_Centauri.jpg.

Fig 5.3 Copyright © Carl Buell and John Gatesy. Reprinted with permission.

Fig 5.6a Copyright © babyknight (CC BY 2.0) at http://commons.wikimedia.org/wiki/File:New_Meadowlands_Stadium_Mezz_Corner.jpg.

Fig 5.6b Copyright © Randi Hausken (CC BY-SA 2.0) at http://commons.wikimedia.org/wiki/File:By_the_road_in_%C3%85motsdal_(3845346438).jpg.

Fig 5.6c Copyright © Tage Olsin (CC BY-SA 2.0) at http://commons.wikimedia.org/wiki/File:Baseball_(crop).jpg.

Fig 5.6d Copyright © Barfooz (CC BY-SA 3.0) at http://commons.wikimedia.org/wiki/File:Paramecium.jpg.

Fig 5.6e Janice Haney Carr/Centers for Disease Control and Prevention, http://commons.wikimedia.org/wiki/File:Pseudomonas.jpg. Copyright in the Public Domain.

Fig 5.6f Centers for Disease Control and Prevention, http://commons.wikimedia.org/wiki/File:H1N1_influenza_virus.jpg. Copyright in the Public Domain.

Fig 5.6g Copyright © Rogerdodd (CC BY-SA 3.0) at http://commons.wikimedia.org/wiki/File:Ubiquitin_spheres.png.

Fig 5.6h Solkoll, http://commons.wikimedia.org/wiki/File:H2O_(water_molecule)_white.png. Copyright in the Public Domain.

Fig 5.6i Copyright © Marekich (CC BY-SA 3.0) at https://commons.wikimedia.org/wiki/File:Nucleus_drawing.svg.

Fig 5.6j Copyright © David Neale (CC BY-SA 2.0) at http://commons.wikimedia.org/wiki/File:Comb_Hill_-_geograph.org.uk_-_37593.jpg.

Fig 5.6k Rfc1394, http://commons.wikimedia.org/wiki/File:Old_television_set.png. Copyright in the Public Domain.

Fig 5.6l Copyright © Apoltix (CC BY-SA 3.0) at http://commons.wikimedia.org/wiki/File:Microwave_oven.jpg.

Fig 5.6m Copyright © Shawn (CC BY-SA 2.0) at http://commons.wikimedia.org/wiki/File:737NG_Navigation_Display_with_weather_radar_showing._(5734080397).jpg.

Fig 5.6n Copyright © KMJ (CC BY-SA 3.0) at http://commons.wikimedia.org/wiki/File:Gluehlampe_01_KMJ.jpg.

Fig 5.6o Copyright © epimetheus (CC BY-SA 2.0) at http://commons.wikimedia.org/wiki/File:Lumi%C3%A8res_noires.jpg.

Fig 5.6p Copyright © Nevit Dilmen (CC BY-SA 2.0) at http://commons.wikimedia.org/wiki/File:Medical_X-Ray_imaging_NZW06_nevit.jpg.

Fig 5.6q Yann, http://commons.wikimedia.org/wiki/File:Radioactivity_symbol.png. Copyright in the Public Domain.

Fig 6.1a Copyright © StarryNight Education. Reprinted with permission.

Fig 6.1b Copyright © Jens Hackmann. Reprinted with permission.

Fig 6.2a Copyright © ESO/J. Emerson/VISTA (CC BY 3.0) at http://www.eso.org/public/images/eso1006a/.

Fig 6.2b NASA/C.R. O'Dell and S.K. Wong, http://commons.wikimedia.org/wiki/File:Orion.nebula.arp.750pix.jpg. Copyright in the Public Domain.

Fig 6.3a Jon Christensen, http://christensenastroimages.com/nebula/m42_2005.htm. Copyright © Jon Christensen.

Fig 6.3b Copyright © NASA/ESA/J. Bally (CC BY 3.0) at http://www.spacetelescope.org/images/opo0113b/.

Fig 6.3c C. R. O'Dell, "An HST Survey of the Inner Regions of the Orion Nebula," Science with the Hubble Space Telescope, II, ed. P. Benvenuti, F. D. Macchetto, and E. J. Schreier. Space Telescope Science Institute, 1996. Copyright in the Public Domain.

Fig 6.3d C. R. O'Dell, "An HST Survey of the Inner Regions of the Orion Nebula," Science with the Hubble Space Telescope, II, ed. P. Benvenuti, F. D. Macchetto, and E. J. Schreier. Space Telescope Science Institute, 1996. Copyright in the Public Domain.

Fig 6.4a http://www.shokabo.co.jp/sp_e/optical/milky/nebula/m42.htm. Copyright © Okayama Astrophysical Observatory. Reprinted with permission.

Fig 6.4b Adapted from: http://www.shokabo.co.jp/sp_e/optical/milky/nebula/m42.htm. Copyright © Gunma Astronomical Observatory.

Fig 6.5 Nobel Foundation, https://commons.wikimedia.org/wiki/File:Niels_Bohr.jpg. Copyright in the Public Domain.

Fig 6.7 C. R. O'Dell, "An HST Survey of the Inner Regions of the Orion Nebula," Science with the Hubble Space Telescope, II, ed. P. Benvenuti, F. D. Macchetto, and E. J. Schreier. Space Telescope Science Institute, 1996. Copyright in the Public Domain.

Fig 6.8 NASA/ESA/C.R. O'Dell, https://commons.wikimedia.org/wiki/File:M42proplyds.jpg. Copyright in the Public Domain.

Fig 6.9a Benjah-bmm27, https://commons.wikimedia.org/wiki/File:Acetic-acid-3D-vdW.png. Copyright in the Public Domain.

Fig 6.9b Benjah-bmm27, https://commons.wikimedia.org/wiki/File:Carbon-monoxide-3D-vdW.png. Copyright in the Public Domain.

Fig 6.9c Dbc334/Jynto, https://commons.wikimedia.org/wiki/File:Methane-3D-space-filling.svg. Copyright in the Public Domain.

Fig 6.9d Benjah-bmm27, https://commons.wikimedia.org/wiki/File:Trihydrogen-cation-3D-vdW.png. Copyright in the Public Domain.

Fig 6.9e Benjah-bmm27, https://commons.wikimedia.org/wiki/File:Formaldehyde-3D-vdW.png. Copyright in the Public Domain.

Fig 6.9f Dbc334, https://upload.wikimedia.org/wikipedia/commons/archive/1/1c/20100521193723!Water_molecule_3D.svg. Copyright in the Public Domain.

Fig 6.12 John Bally, http://casa.colorado.edu/~bally/. Copyright © John Bally.

Fig 6.15 Nobel Foundation, https://nov.wikipedia.org/wiki/File:Max_Planck_%28Nobel_1918%29.jpg. Copyright in the Public Domain.

Fig 6.16 Copyright © ESO/M.McCaughrean et al. (AIP) (CC BY 3.0) at http://www.eso.org/public/images/eso0104a/.

Fig 7.1 Copyright © StarryNight Education. Reprinted with permission.

Fig 7.2 Copyright © Debivort (CC BY 3.0) at http://commons.wikimedia.org/wiki/File:51_Pegasi_b_v2.jpg.

Fig 7.4 Data from: Paul Butler/Geoffry Marcy, exoplanets.org.

Fig 7.5 Data from: exoplanets.org.

Fig 7.6 NASA/ESA/ G. Bacon, http://www.nasa.gov/mission_pages/hubble/science/exoplanet_transit.html. Copyright in the Public Domain.

Fig 7.7 Copyright © Azcolvin429/NASA (CC BY-SA 3.0) at https://commons.wikimedia.org/wiki/File:1_Earth_%28blank_2%29.png.

Fig 7.8 Gregory Laughlin, http://www.space.com/1204-strange-world.html. Copyright © Gregory Laughlin. Reprinted with permission.

Fig 7.9 Copyright © ESO/A.-M. Lagrange et al. (CC BY 4.0) at https://commons.wikimedia.org/wiki/File:ESO_-_Beta_Pictoris_planet_finally_imaged_%28by%29.jpg.

Fig 7.10 Copyright © ESO/L. Calçada (CC BY 3.0) at http://commons.wikimedia.org/wiki/File:Artist%E2%80%99s_Impression_of_a_Baby_Star_Still_Surrounded_by_a_Protoplanetary_Disc.jpg.

Fig 7.11 NASA, http://commons.wikimedia.org/wiki/File:Extrasolar_planet_NASA2.jpg. Copyright in the Public Domain.

Fig 7.12 Copyright © Christian Thrower.

Fig 8.1a NASA/ESA/AURA/Caltech/Palomar Observatory, http://commons.wikimedia.org/wiki/File:Pleiades_large.jpg. Copyright in the Public Domain.

Fig 8.1b Elihu Vedder, https://commons.wikimedia.org/wiki/File:Pleiades_Elihu_Vedder.jpg. Copyright in the Public Domain.

Fig 8.2 Sylvain Weiller, http://sweiller.free.fr/Conjunctions.html. Copyright © Sylvain Weiller. Reprinted with permission.

Fig 8.3 NASA/The Hubble Heritage Team (STScl/AURA), http://heritage.stsci.edu/2000/36/index.html. Copyright in the Public Domain.

Fig 8.4a N. A. Sharp, http://apod.nasa.gov/apod/ap021129.html. Copyright © N. A. Sharp (NOAO)/AURA/NSF.

Fig 8.4b Copyright © Jean-Charles Cuillandre (CFHT) & Giovanni Anselmi (Coelum). Reprinted with permission.

Fig 8.4c NASA/ESA/Hubble Heritage (STScI/AURA), http://en.wikipedia.org/wiki/File:NGC_3603_HST_ACS.jpg. Copyright in the Public Domain.

Fig 8.5a Copyright © European Southern Observatory (CC BY 3.0) at http://www.phys-astro.sonoma.edu/bruceMedalists/.

Fig 8.5b The World's Work, 1921 (http://www.phys-astro.sonoma.edu/bruceMedalists/). Copyright in the Public Domain.

Fig 8.7 Michael Lemke (Universitaet Sternwarte Bamberg) and Simon Jeffery (Armagh Observatory), http://www.arm.ac.uk/~csj/pus/spectra/sim.html. Copyright © Michael Lemke and Simon Jeffery. Reprinted with permission.

Fig 8.8a CSIRO, http://outreach.atnf.csiro.au/education/senior/astrophysics/spectral_class.html. Copyright © CSIRO.

Fig 8.8b Adapted from: http://www.jb.man.ac.uk/distance/life/sample/stars/index.html. Copyright © Tim O'Brien (The University of Manchester). Reprinted with permission.

Fig 8.8c Smithsonian Institution, https://en.wikipedia.org/wiki/File:Cecilia_Helena_Payne_Gaposchkin_%281900-1979%29_%283%29.jpg.

Fig 8.9 N. A.Sharp, http://www.noao.edu/image_gallery/html/im0600.html. Copyright © N. A.Sharp/NOAO/NSO/Kitt Peak FTS/AURA/NSF.

Fig 8.10a National Nuclear Security Administration, http://en.wikipedia.org/wiki/File:Operation_Upshot-Knothole_-_Badger_001.jpg. Copyright in the Public Domain.

Fig 8.10b Scott Bauer/U.S. Department of Agriculture, https://commons.wikimedia.org/wiki/File:Leaf-cutter_ant.png. Copyright in the Public Domain

Fig 8.10c Ferdinand Schmutzer, https://commons.wikimedia.org/wiki/File:Albert_Einstein_1921_by_F_Schmutzer.jpg. Copyright in the Public Domain.

Fig 8.11 NASA/JHU/Caltech, http://www.solarviews.com/cap/ds/gliese229.htm. Copyright in the Public Domain.

Fig 8.12 Copyright © DiBgd (CC BY-SA 3.0) at http://en.wikipedia.org/wiki/File:Ceratosaurus_nasicornis_DB.jpg.

Fig 9.1 Dan Duriscoe/U.S. National Park Service, http://commons.wikimedia.org/wiki/File:Deathvalleysky_nps_big.jpg. Copyright in the Public Domain.

Fig 9.2 NASA, http://commons.wikimedia.org/wiki/File:Earth_night.jpg. Copyright in the Public Domain.

Fig 9.3 Knut Lundmark, http://apod.nasa.gov/apod/ap980523.html. Copyright © Lund Observatory. Reprinted with permission.

Fig 9.4 NASA, http://mwmw.gsfc.nasa.gov/mmw_product.html#viewgraph. Copyright in the Public Domain.

Fig 9.5 Copyright © Axel Mellinger. Reprinted with permission.

Fig 9.6 2MASS/J. Carpenter, T. H. Jarrett, & R. Hurt, http://www.ipac.caltech.edu/2mass/gallery/showcase/allsky_stars/index.html. Copyright in the Public Domain.

Fig 9.7 NASA, http://commons.wikimedia.org/wiki/File:Top_of_Atmosphere.jpg. Copyright in the Public Domain.

Fig 9.9 Serge Brunier, http://antwrp.gsfc.nasa.gov/apod/ap051004.html. Copyright © Serge Brunier. Reprinted with permission.

Fig 9.10a Copyright © Ken Crawford (CC BY-SA 3.0) at http://en.wikipedia.org/wiki/File:Needle_Galaxy_4565.jpeg.

Fig 9.10b NASA/JPL-Caltech/ESO/R. Hurt, http://commons.wikimedia.org/wiki/File:Artist%E2%80%99s_impression_of_the_Milky_Way.jpg. Copyright in the Public Domain.

Fig 9.11 Copyright © Yuri Beletsky. Reprinted with permission.

Fig 9.12 Copyright © psmithy (CC BY 2.0) at http://www.flickr.com/photos/22798311@N00/2538425267.

Fig 9.13 Copyright © Axel Mellinger. Reprinted with permission.

Fig 9.14 Copyright © Joop van Bilsen (CC BY-SA 3.0) at https://en.wikipedia.org/wiki/File:JanOort.jpg.

Fig 9.15 Copyright © Michael Padilla. Reprinted with permission.

Fig 10.1a Silas Laycock, http://news.harvard.edu/gazette/story/2005/01/see-through-galaxy/. Copyright © Silas Laycock. Reprinted with permission.

Fig 10.1b Copyright © Stephane Guisard / Los Cielos de America. Reprinted with permission.

Fig 10.1c Serge Brunier, http://antwrp.gsfc.nasa.gov/apod/ap051004.html. Copyright © Serge Brunier. Reprinted with permission.

Fig 10.2 Copyright © Axel Mellinger. Reprinted with permission.

Fig 10.3 NASA/JPL-Caltech/ESO/R. Hurt, http://www.eso.org/public/images/eso1339g/. Copyright in the Public Domain.

Fig 10.4a Copyright © Axel Mellinger. Reprinted with permission.

Fig 10.4b 2MASS/J. Carpenter, T. H. Jarrett, & R. Hurt, http://www.ipac.caltech.edu/2mass/gallery/showcase/allsky_stars/index.html. Copyright in the Public Domain.

Fig 10.5 2MASS/J. Carpenter, T. H. Jarrett, & R. Hurt, http://www.ipac.caltech.edu/2mass/gallery/showcase/allsky_stars/index.html. Copyright in the Public Domain.

Fig 10.6 Copyright © Michael Padilla. Reprinted with permission.

Fig 10.7 NASA/ESA/Q.D. Wang (University of Massachusetts, Amherst) & NASA/JPL/S. Stolovy (Spitzer Science Center/Caltech), http://hubblesite.org/gallery/album/pr2009002a. Copyright in the Public Domain.

Fig 10.8a NASA/Don Figer/STScI, http://hubblesite.org/newscenter/archive/releases/1999/30/image/a/. Copyright in the Public Domain.

Fig 10.8b NASA/Don Figer/STScI, http://hubblesite.org/newscenter/archive/releases/1999/30/image/a/. Copyright in the Public Domain.

Fig 10.9a Copyright © NRL/SBC Galactic Center Radio Group (CC BY-SA 2.5) at http://commons.wikimedia.org/wiki/File:GCRT_J1745-3009_2.jpg.

Fig 10.9b Copyright © Farhad Zadeh et al. (Northwestern)/VLA/NRAO (CC BY 3.0) at http://apod.nasa.gov/apod/ap080427.html.

Fig 10.9c NASA/CXC/MIT/F.K.Baganoff et al., http://chandra.harvard.edu/photo/2003/0203long/index.html. Copyright in the Public Domain.

Fig 10.10a George Reike and Marcia Rieke, http://ircamera.as.arizona.edu/NatSci102/NatSci102/lectures/galcenter.htm. Copyright © George Reike and Marcia Rieke. Reprinted with permission.

Fig 10.10b George Reike and Marcia Rieke, http://ircamera.as.arizona.edu/NatSci102/NatSci102/lectures/galcenter.htm. Copyright © George Reike and Marcia Rieke. Reprinted with permission.

Fig 10.11a Copyright © Andrea Ghez (UCLA). Reprinted with permission. This image was created by Prof. Andrea Ghez and her research team at UCLA and are from data sets obtained with the W. M. Keck Telescopes.

Fig 10.11b Copyright © ESO (CC BY 3.0) at http://www.eso.org/public/images/eso0226c/.

Fig 11.1a Ansiaume, https://commons.wikimedia.org/wiki/File:Charles_Messier.jpg. Copyright in the Public Domain.

Fig 11.1b Copyright © Emiel Kempen. Reprinted with permission.

Fig 11.2a Copyright © Bob Birket (SkyPi Observatory). Reprinted with permission.

Fig 11.2b Adapted from: Copyright © Bob Birket (SkyPi Observatory). Reprinted with permission.

Fig 11.3a Copyright © Axel Mellinger. Reprinted with permission.

Fig 11.3b Smithsonian Institution, https://commons.wikimedia.org/wiki/File:Officials_of_the_AAS_in_1947.jpg. Copyright in the Public Domain.

Fig 11.4a Copyright © Adam Block/Mount Lemmon SkyCenter/University of Arizona (CC BY-SA 3.0 US) at http://commons.wikimedia.org/wiki/File:M92s.jpg.

Fig 11.4b NASA/ESA, http://commons.wikimedia.org/wiki/File:M92_Hubble_WikiSky.jpg. Copyright in the Public Domain.

Fig 11.5 Janusz Kaluzny and Barbara Mochejska, http://apod.nasa.gov/apod/ap010223.html. Copyright © Janusz Kaluzny and Barbara Mochejska. Reprinted with permission.

Fig 11.6 Copyright © Oona Räisänen/Mrsanitazier (CC BY-SA 3.0) at https://commons.wikimedia.org/wiki/File:Sun_red_giant.svg.

Fig 11.7a George Jacoby (WIYN Obs.) et al., http://apod.nasa.gov/apod/ap010123.html. Copyright © George Jacoby (WIYN Obs.) et al./WIYN/AURA/NOAO/NSF.

Fig 11.7b The Hubble Heritage Team (AURA/STScI/NASA), http://apod.nasa.gov/apod/ap990107.html. Copyright in the Public Domain.

Fig 11.7c Bruce Balick (University of Washington)/Jason Alexander (University of Washington)/Arsen Hajian (U.S. Naval Observatory)/Yervant Terzian (Cornell University)/Mario Perinotto (University of Florence, Italy)/Patrizio Patriarchi (Arcetri Observatory, Italy)/NASA, http://apod.nasa.gov/apod/ap971219.html. Copyright in the Public Domain.

Fig 11.7d NASA/ESA/HEIC/The Hubble Heritage Team (STScl/AURA), http://heritage.stsci.edu/2004/27/index.html. Copyright in the Public Domain.

Fig 11.7e B. Balick (University of Washington)/NASA, http://apod.nasa.gov/apod/ap020106.html. Copyright in the Public Domain.

Fig 11.7f NASA/R. Sahai and J. Trauger (JPL)/The WFPC2 Science Team, http://www.spacetelescope.org/images/opo9607a/. Copyright in the Public Domain.

Fig 11.8 NASA/H. Richer (University of British Columbia)/NOAO/AURA/NSF, http://www.jpl.nasa.gov/spaceimages/details.php?id=PIA04231. Copyright in the Public Domain.

Fig 11.9 Adapted from: Harvey B. Richer, et al., " Deep Advanced Camera for Surveys Imaging in the Globular Cluster NGC 6397: The Cluster Color-Magnitude Diagram and Luminosity Function," The Astronomical Journal, vol. 135, issue 6, The American Astronomical Society, 2008.

Fig 11.10a NASA/ESA, H. Bond (STScI)/M. Barstow (University of Leicester), http://www.nasa.gov/multimedia/imagegallery/image_feature_468_prt.htm. Copyright in the Public Domain.

Fig 11.10b ESA/NASA, http://www.spacetelescope.org/images/heic0516c/. Copyright in the Public Domain.

Fig 11.11a NASA/ESA, http://commons.wikimedia.org/wiki/File:Messier_107_Hubble_WikiSky.jpg. Copyright in the Public Domain.

Fig 11.11b Copyright © ESO/M.-R. Cioni/VISTA Magellanic Cloud Survey (CC BY 3.0) at http://www.eso.org/public/usa/images/eso1302a/.

Fig 11.11c ESA/Hubble/NASA, https://commons.wikimedia.org/wiki/File:Messier_5_Hubble_WikiSky.jpg. Copyright in the Public Domain.

Fig 11.11d Copyright © Bob Birket. Reprinted with permission.

Fig 12.1 Copyright © Axel Mellinger. Reprinted with permission.

Fig 12.2a Mike Bessell, http://www.mso.anu.edu.au/~bessell/. Copyright © Mike Bessell. Reprinted with permission.

Fig 12.2b Source: https://en.wikipedia.org/wiki/File:Ferdinand_Magellan.jpg. Copyright in the Public Domain.

Fig 12.3 Wei-Hao Wang, http://apod.nasa.gov/apod/ap060510.html. Copyright © Wei-Hao Wang. Reprinted with permission.

Fig 12.4a C. Smith, S. Points, the MCELS Team and NOAO/AURA/NSF, http://www.noao.edu/image_gallery/html/im0994.html. Copyright © C. Smith, S. Points, the MCELS Team and NOAO/AURA/NSF.

Fig 12.4b Copyright © Astrodon Imaging. Reprinted with permission.

Fig 12.5 NASA/ESA, https://commons.wikimedia.org/wiki/File:Ngc1818_hst_big.jpg. Copyright in the Public Domain.

Fig 12.6 ESA/NASA/Martino Romaniello (ESO), https://commons.wikimedia.org/wiki/File:NGC1850.jpg. Copyright in the Public Domain.

Fig 12.9 David Malin, http://ftp.aao.gov.au/images/captions/aat050.html. Copyright © Australian Astronomical Observatory/David Malin Images. Reprinted with permission.

Fig 12.10a The Hubble Heritage Team (AURA/STScI/NASA), http://hubblesite.org/newscenter/archive/releases/1999/04/image/e/. Copyright in the Public Domain.

Fig 12.10b Christopher Burrows/ESA/STScI/NASA, http://hubblesite.org/newscenter/archive/releases/1995/49/image/a/. Copyright in the Public Domain.

Fig 12.11 M. Aschwanden et al. (LMSAL)/TRACE/NASA, http://apod.nasa.gov/apod/ap000928.html. Copyright in the Public Domain.

Fig 12.12 N.A. Sharp/NOAO/AURA/NSF, http://www.noao.edu/image_gallery/html/im0565.html. Copyright © N.A. Sharp/NOAO/AURA/NSF.

Fig 12.13 NASA/CXC/M.Weiss, https://commons.wikimedia.org/wiki/File:Black_hole_Cygnus_X-1.jpg. Copyright in the Public Domain.

Fig 13.1a Copyright © ESO (CC BY 3.0) at https://commons.wikimedia.org/wiki/File:Irregular_Galaxy_NGC_55_%28ESO_0914a%29.jpg.

Fig 13.1b Wei-Hao Wang, http://apod.nasa.gov/apod/ap060510.html. Copyright © Wei-Hao Wang. Reprinted with permission.

Fig 13.2 NASA/WikiSky, https://en.wikipedia.org/wiki/File:Sculptor_Dwarf_Galaxy.jpg. Copyright in the Public Domain.

Fig 13.2 Copyright © Adam Block/Mount Lemmon SkyCenter/University of Arizona (CC BY-SA 3.0 US) at http://commons.wikimedia.org/wiki/File:M92s.jpg.

Fig 13.3a David Malin, http://apod.nasa.gov/apod/ap991003.html. Copyright © Australian Astronomical Observatory/David Malin Images. Reprinted with permission.

Fig 13.3b Hubble Heritage Team (AURA/STScl/NASA), http://heritage.stsci.edu/1999/25/images/9925a.jpg. Copyright in the Public Domain.

Fig 13.4 Copyright © Axel Mellinger. Reprinted with permission.

Fig 13.7 Copyright © European Space Agency /NASA (CC BY 3.0) at http://commons.wikimedia.org/wiki/File:M101_hires_STScI-PRC2006-10a.jpg.

Fig 13.8 Copyright © European Space Agency /NASA (CC BY 3.0) at http://commons.wikimedia.org/wiki/File:M101_hires_STScI-PRC2006-10a.jpg.

Fig 13.9 Vasily Belokurov, http://www.ast.cam.ac.uk/~vasily/sdss/field_of_streams/dr6/fos_dr6_marked.jpg. Copyright © Vasily Belokurov. Reprinted with permission.

Fig 13.10 David Martinez-Delgado (MPIA) & Gabriel Perez (IAC), http://apod.nasa.gov/apod/ap030930.html. Copyright © David Martinez-Delgado (MPIA) & Gabriel Perez (IAC). Reprinted with permission.

Fig 14.1a Emiel Kempen, http://apod.nasa.gov/apod/ap080102.html. Copyright © Emiel Kempen. Reprinted with permission.

Fig 14.1b Copyright © Rolando Ligustri.

Fig 14.2a Barbara Mochejska Hedberg (Warsaw University)/The DIRECT Project/FLWO/MDM, http://apod.nasa.gov/apod/ap990402.html. Copyright © Barbara Mochejska Hedberg. Reprinted with permission.

Fig 14.2b Michael Rich, Kenneth Mighell, and James D. Neill (Columbia University)/Wendy Freedman (Carnegie Observatories)/NASA/ESA, http://www.spacetelescope.org/images/opo9734t/. Copyright in the Public Domain.

Fig 14.2c H. Yang (UIUC)/J. Hester (ASU)/NASA, http://apod.nasa.gov/apod/ap960816.html. Copyright in the Public Domain.

Fig 14.3a Bill Schoening, Vanessa Harvey/REU program/NOAO/AURA/NSF, http://www.noao.edu/image_gallery/html/im0424.html. Copyright © Bill Schoening, Vanessa Harvey/REU program/NOAO/AURA/NSF.

Fig 14.3b Davide De Martin, http://www.skyfactory.org/m33dss/index.php. Copyright © Davide De Martin. Reprinted with permission.

Fig 14.4 Wendy L. Freedman/Observatories of the Carnegie Institution of Washington/NASA, http://hubblesite.org/newscenter/archive/releases/1994/49/image/b/. Copyright in the Public Domain.

Fig 14.5 Source: https://commons.wikimedia.org/wiki/File:Leavitt_aavso.jpg. Copyright in the Public Domain.

Fig 14.6a NASA/JPL-Caltech, http://commons.wikimedia.org/wiki/File:Andromeda_Galaxy_in_Ultraviolet.jpg. Copyright in the Public Domain.

Fig 14.6b Bill Schoening, Vanessa Harvey/REU program/NOAO/AURA/NSF, http://www.noao.edu/image_gallery/html/im0424.html. Copyright © Bill Schoening, Vanessa Harvey/REU program/NOAO/AURA/NSF.

Fig 14.6c 2MASS, www.ipac.caltech.edu/2mass/gallery/m31atlas.jpg. Copyright in the Public Domain.

Fig 14.6d R. Braun, E. Corbelli, R.A.M. Walterbos and D. Thilker, "Gas and Dark Matter in the Outskirts of M31" ASTRON Newsletter 17. Copyright © ASTRON.

Fig 14.7a NASA/JPL-Caltech, https://commons.wikimedia.org/wiki/File:Galaxy_M33_in_Ultraviolet.jpg. Copyright in the Public Domain.

Fig 14.7b 2MASS, http://www.ipac.caltech.edu/2mass/gallery/m33atlas.jpg. Copyright in the Public Domain.

Fig 14.8a Phil Hart, http://www.philhart.com/gallery/Astrophotography/Phenomena/BioLuminescenceThu_323.jpg.html. Copyright © Phil Hart. Reprinted with permission.

Fig 14.8b NASA/The Hubble Heritage team (STScl/AURA), http://heritage.stsci.edu/2001/10/big.html. Copyright in the Public Domain.

Fig 14.9a Bill Schoening, Vanessa Harvey/REU program/NOAO/AURA/NSF, http://www.noao.edu/image_gallery/html/im0424.html. Copyright © Bill Schoening, Vanessa Harvey/REU program/NOAO/AURA/NSF.

Fig 14.9b Copyright © NRAO (CC BY 3.0) at http://www.nrao.edu/pr/2001/m33gas/.

Fig 14.10a Phil Massey, http://www2.lowell.edu/users/massey/lgsurvey/. Copyright © Phil Massey. Reprinted with permission.

Fig 14.10b Jean-Charles Cuillandre (CFHT) & Giovanni Anselmi (Coelum Astronomia), http://apod.nasa.gov/apod/ap080909.html. Copyright © Jean-Charles Cuillandre (CFHT) & Giovanni Anselmi (Coelum Astronomia)/Hawaiian Starlight. Reprinted with permission.

Fig 14.10c Local Group Galaxies Survey Team/NOAO/AURA/NSF, http://www.noao.edu/image_gallery/html/im0698.html. Copyright © Local Group Galaxies Survey Team/NOAO/AURA/NSF.

Fig 15.1a NASA/JPL/California Institute of Technology, http://www.jpl.nasa.gov/spaceimages/details.php?id=PIA04924. Copyright in the Public Domain.

Fig 15.1b Copyright © ESO (CC BY 3.0) at http://www.eso.org/public/images/eso1037a/.

Fig 15.1c Copyright © ESO/INAF-VST (CC BY 4.0) at http://www.eso.org/public/images/eso1152a/.

Fig 15.1d NASA/2MASS/Jarrett et al., http://irsa.ipac.caltech.edu/data/LGA/images/ngc253/ngc253_mosaic_JHK.jpg. Copyright in the Public Domain.

Fig 15.2a NASA/JPL-Caltech/S. Willner (Harvard-Smithsonian CfA), http://www.nasa.gov/centers/ames/images/content/126032main_spitzer.jpg. Copyright in the Public Domain.

Fig 15.2b Stefan Seip/Adam Block/NOAO/AURA/NSF, http://www.noao.edu/kpvc/observers/m81.html. Copyright © Stefan Seip/Adam Block/NOAO/AURA/NSF.

Fig 15.2c Copyright © NRAO/AUI/Tom Oosterloo (CC BY 3.0) at http://images.nrao.edu/object/index.php?id=510.

Fig 15.2d Copyright © Adam Block/Mount Lemmon SkyCenter/University of Arizona (CC BY-SA 3.0 US) at https://commons.wikimedia.org/wiki/File:N2403s.jpg.

Fig 15.3a Bill Schoening, Vanessa Harvey/REU program/NOAO/AURA/NSF, http://www.noao.edu/image_gallery/html/im0424.html. Copyright © Bill Schoening, Vanessa Harvey/REU program/NOAO/AURA/NSF.

Fig 15.3b Copyright © Wolfgang Promper. Reprinted with permission.

Fig 15.4 Sean Walker & Sheldon Faworski, http://masil-astro-imaging.netfirms.com/8%20f3.68/Maffei%201-2.jpg. Copyright © Sean Walker & Sheldon Faworski. Reprinted with permission.

Fig 15.5a 2MASS/T. Jarrett, http://www.ipac.caltech.edu/2mass/gallery/showcase/maffei/. Copyright in the Public Domain.

Fig 15.5b 2MASS/T. Jarrett, http://www.ipac.caltech.edu/2mass/gallery/showcase/maffei/. Copyright in the Public Domain.

Fig 15.6a Wolfgang Promper, http://www.astro-pics.com/m32.htm. Copyright © Wolfgang Promper. Reprinted with permission.

Fig 15.6b NASA/T. Brown, http://apod.nasa.gov/apod/ap991103.html. Copyright in the Public Domain.

Fig 15.7a Daniel Verschatse, http://www.astrosurf.com/antilhue/ngc4697.htm. Copyright © Daniel Verschatse. Reprinted with permission.

Fig 15.7b NASA/CXC/UVa/C.Sarazin et al., http://chandra.harvard.edu/photo/2002/1140/. Copyright in the Public Domain.

Fig 15.7c NASA/CXC/UVa/E.Blanton et al., http://chandra.harvard.edu/photo/2002/1058/. Copyright in the Public Domain.

Fig 15.7d DSS/STScl & Courtney Seligman, http://cseligman.com/text/atlas/ngc15a.htm#1553. Copyright © Courtney Seligman. Reprinted with permission.

Fig 15.7e NASA/ESA/STScI/M. West (Optical) and NASA/CXC/Penn State/G. Garmire (X-ray), http://www.nasa.gov/multimedia/imagegallery/image_feature_1012_prt.htm. Copyright in the Public Domain.

Fig 15.8a DSS/STScl & Courtney Seligman, http://cseligman.com/text/atlas/ngc15a.htm#1553. Copyright © Courtney Seligman. Reprinted with permission.

Fig 15.8b Daniel Verschatse, http://www.astrosurf.com/antilhue/ngc4697.htm. Copyright © Daniel Verschatse. Reprinted with permission.

Fig 15.9a Copyright © Adam Block/Mount Lemmon SkyCenter/University of Arizona (CC BY-SA 3.0 US) at http://skycenter.arizona.edu/gallery/Galaxies/NGC2841.

Fig 15.9b Copyright © ESO/IDA/Danish 1.5 m/R. Gendler, J-E. Ovaldsen, C. Thöne, and C. Feron (CC BY 3.0) at http://commons.wikimedia.org/wiki/File:The_Great_Barred_Spiral_Galaxy.jpg.

Fig 15.9c Copyright © ESO/IDA/Danish 1.5 m/R.Gendler and A. Hornstrup (CC BY 3.0) at http://www.eso.org/public/images/ngc1232b/.

Fig 15.9d Phil Massey, http://www2.lowell.edu/rsch/LMI/gallery.html. Copyright © Phil Massey. Reprinted with permission.

Fig 15.10a NASA/ESA/W. Harris, https://commons.wikimedia.org/wiki/File:NGC_7049aa.jpg. Copyright in the Public Domain.

Fig 15.10b NASA/ESA/The Hubble Heritage Team (STScI/AURA), https://commons.wikimedia.org/wiki/File:File-Ngc5866_hst_big.png. Copyright in the Public Domain.

Fig 15.11 NASA, ESA, and A. Feild (STScI), http://heritage.stsci.edu/2006/24/supplemental.html. Copyright in the Public Domain.

Fig 15.12 Jim Misti, http://www.mistisoftware.com/astronomy/Galaxies_m105.htm. Copyright © Jim Misti. Reprinted with permission.

Fig 16.1a Copyright © Richard Powell (CC BY-SA 2.5) at https://commons.wikimedia.org/wiki/File:Foam_-_big.jpg. http://www.atlasoftheuniverse.com/galgrps/wvir.gif.

Fig 16.1b Michael Theusner, http://www.theusner.eu/astro/images/deepsky/Virgo01.jpg. Copyright © Michael Theusner. Reprinted with permission.

Fig 16.2a Greg Morgan, http://apod.nasa.gov/apod/ap080425.html. Copyright © Greg Morgan. Reprinted with permission.

Fig 16.2b Adam Block, http://apod.nasa.gov/apod/ap100520.html. Copyright © Adam Block. Reprinted with permission.

Fig 16.3a NOAO/AURA/NSF, http://www.noao.edu/image_gallery/html/im0706.html. Copyright © NOAO/AURA/NSF.

Fig 16.3b NOAO/AURA/NSF, http://www.noao.edu/image_gallery/html/im0419.html. Copyright © NOAO/AURA/NSF.

Fig 16.3c NOAO/AURA/NSF, http://www.noao.edu/image_gallery/html/im0704.html. Copyright © NOAO/AURA/NSF.

Fig 16.4a NOAO/AURA/NSF, http://www.noao.edu/image_gallery/html/im0654.html. Copyright © NOAO/AURA/NSF.

Fig 16.4b ESA/NASA/E. Peng, https://commons.wikimedia.org/wiki/File:NGC_4660HST.jpg. Copyright in the Public Domain.

Fig 16.4c Ken Crawford, http://apod.nasa.gov/apod/ap070608.html. Copyright © Ken Crawford. Reprinted with permission.

Fig 16.4d Tomer Tal and Jeffrey Kenney/Yale University and NOAO/AURA/NSF, http://www.noao.edu/outreach/press/pr08/pr0807.html. Copyright © Tomer Tal and Jeffrey Kenney/Yale University and NOAO/AURA/NSF. Reprinted with permission.

Fig 16.5a Chris Mihos (CWRU) et al./KPNO/NOAO/ NSF, http://apod.nasa.gov/apod/ap050926.html. Copyright © Chris Mihos (CWRU) et al./KPNO/NOAO/ NSF. Reprinted with permission.

Fig 16.5b NOAO/AURA/NSF/B. Twardy, B. Twardy, and A. Block (NOAO), http://www.esa.int/spaceinimages/Images/2006/10/Wide_ground_view_of_the_Antennae_galaxies. Copyright © NOAO/AURA/NSF/B. Twardy, B. Twardy, and A. Block (NOAO).

Fig 16.5c NASA/H. Ford (JHU), G. Illingworth (UCSC/LO), M. Clampin (STScl), G. Hartig (STScl) and the ACS Science Team, http://antwrp.gsfc.nasa.gov/apod/image/0205/ngc4676_hst_big.jpg. Copyright in the Public Domain.

Fig 16.6 NASA/ESA/The Hubble Heritage (STScI/AURA)-ESA/Hubble Collaboration/A. Evans (University of Virginia, Charlottesville/NRAO/Stony Brook University), http://hubblesite.org/newscenter/archive/releases/2008/16/. Copyright in the Public Domain.

Fig 16.7a Fred Calvert/Adam Block/NOAO/AURA/NSF, http://www.caelumobservatory.com/obs/n2403.html/. Copyright © Fred Calvert/Adam Block/NOAO/AURA/NSF.

Fig 16.7b The Hubble Heritage Team (AURA/STScI/NASA), http://hubblesite.org/newscenter/archive/releases/1999/25/image/a/. Copyright in the Public Domain.

Fig 16.7c NASA/ESA/Hubble Heritage Team (STScI/AURA)-ESA/Hubble Collaboration, http://hubblesite.org/gallery/album/entire/pr2006046a/. Copyright in the Public Domain.

Fig 16.7d Daniel Verschatse, http://www.astrosurf.com/antilhue/ngc4697.htm. Copyright © Daniel Verschatse. Reprinted with permission.

Fig 16.7e Copyright © Adam Block/Mount Lemmon SkyCenter/University of Arizona (CC BY-SA 3.0 US) at http://skycenter.arizona.edu/gallery/Galaxies/NGC2841.

Fig 16.8 Copyright © John Dubinski (University of Toronto). Reprinted with permission.

Fig 16.9a Copyright © ESA (CC BY 3.0) at http://www.spacetelescope.org/about/history/the_man_behind_the_name/.

Fig 16.9b Mark SubbaRao, http://cas.sdss.org/dr6/en/sdss/data/data.asp. Copyright © Mark SubbaRao. Reprinted with permission.

Fig 16.10a The Hubble Heritage Team (AURA/STScI/NASA), http://hubblesite.org/newscenter/archive/releases/1999/25/image/a/. Copyright in the Public Domain.

Fig 16.10b Adam Block, http://apod.nasa.gov/apod/ap100520.html. Copyright © Adam Block. Reprinted with permission.

Fig 16.10c NOAO/AURA/NSF, https://www.noao.edu/image_gallery/html/im0654.html. Copyright © NOAO/AURA/NSF.

Fig 16.10d ESA/NASA/E. Peng, https://commons.wikimedia.org/wiki/File:NGC_4660HST.jpg. Copyright in the Public Domain.

Fig 16.11a MPE (Max Planck Institute), http://www.mpe.mpg.de/xray/wave/rosat/publications/highlights/virgo_1.gif. Copyright © MPE (Max Planck Institute). Reprinted with permission.

Fig 16.11b NASA/CXC/Columbia University/C.Scharf et al., http://www.nasa.gov/centers/marshall/multimedia/photos/2004/photos04-231.html. Copyright in the Public Domain.

Fig 16.11c NASA/CXC/SAO, https://commons.wikimedia.org/wiki/File:Hydra_A_gas_cloud.jpg. Copyright in the Public Domain.

Fig 16.11d NASA/CXC/Ohio University/B.McNamara et al., http://chandra.harvard.edu/photo/2002/1269/. Copyright in the Public Domain.

Fig 17.1a Scott Rosen, http://www.astronomersdoitinthedark.com/index.php?c=121&p=471. Copyright © Scott Rosen. Reprinted with permission.

Fig 17.1b Francois and Shelley Pelletier/Adam Block, http://www.noao.edu/kpvc/observers/m77.html. Copyright © Francois and Shelley Pelletier/Adam Block/NOAO/AURA/NSF.

Fig 17.2a Montage by William Keel. Reprinted with permission.

Fig 17.2b STScI/NASA, http://www.astr.ua.edu/gifimages/ngc5548.html. Copyright in the Public Domain.

Fig 17.2c STScI/NASA, http://www.astr.ua.edu/gifimages/ngc5548.html. Copyright in the Public Domain.

Fig 17.3a Copyright © NVSS/NRAO (CC BY 3.0) at http://coolcosmos.ipac.caltech.edu/cosmic_classroom/multiwavelength_astronomy/multiwavelength_museum/m77.html.

Fig 17.3b IRAS/ NASA/NIVR/SERC, http://coolcosmos.ipac.caltech.edu/cosmic_classroom/multiwavelength_astronomy/multiwavelength_museum/m77.html. Copyright in the Public Domain.

Fig 17.3c 2MASS, http://coolcosmos.ipac.caltech.edu/cosmic_classroom/multiwavelength_astronomy/multiwavelength_museum/m77.html. Copyright in the Public Domain.

Fig 17.3d Francois and Shelley Pelletier/Adam Block, http://www.noao.edu/kpvc/observers/m77.html. Copyright © Francois and Shelley Pelletier/Adam Block/NOAO/AURA/NSF.

Fig 17.3e ASTRO-1/NASA, http://coolcosmos.ipac.caltech.edu/cosmic_classroom/multiwavelength_astronomy/multiwavelength_museum/m77.html. Copyright in the Public Domain.

Fig 17.3f ROSAT, http://coolcosmos.ipac.caltech.edu/cosmic_classroom/multiwavelength_astronomy/multiwavelength_museum/m77.html. Copyright in the Public Domain.

Fig 17.4 Adapted from: http://www.astr.ua.edu/keel/agn/spectra.html. Copyright © William C. Keel.

Fig 17.5 Adapted from: http://www.astr.ua.edu/keel/agn/spectra.html. Copyright © William C. Keel.

Fig 17.6a Adapted from: Jack F. Gallimore, Stefi A. Baum, and Christopher P. O'Dea, "A Direct Image of the Obscuring Disk Surrounding an Active Galactic Nucleus," Nature, 1997. Copyright © Macmillan Publishers Limited.

Fig 17.6b Adapted from: A. L. Roy, A. S. Wilson, J. S. Ulvestad, and E. J. M. Colbert, "Slow Jets in Seyfert Galaxies: NGC 1068," pp. 3, 2000. Copyright © A. L. Roy, A. S. Wilson, J. S. Ulvestad, and E. J. M. Colbert.

Fig 17.6c D. Macchetto (ESA, StScI), W. Sparks, A. Capetti (StScI) et al./FOC/HST/NASA, http://apod.nasa.gov/apod/ap960911.html. Copyright in the Public Domain.

Fig 17.6d Francois and Shelley Pelletier/Adam Block, http://www.noao.edu/kpvc/observers/m77.html. Copyright © Francois and Shelley Pelletier/Adam Block/NOAO/AURA/NSF.

Fig 17.7 C. Megan Urry and Paolo Padovani, "Unified Schemes for Radio-Loud Active Galactic Nuclei," Publications of the Astronomical Society of the Pacific, vol. 107, no. 715, 1995. Copyright © The University of Chicago Press.

Fig 17.9 NASA, http://commons.wikimedia.org/wiki/File:GPS_Satellite_NASA_art-iif.jpg. Copyright in the Public Domain.

Fig 17.10a Copyright © Andrew Hamilton and Axel Mellinger. Reprinted with permission.

Fig 17.10b NASA/ESA/A. Bolton (Harvard-Smithsonian CfA)/SLACS Team, http://einsteinstelescope.com/images/C1thumb.jpg. Copyright in the Public Domain.

Fig 18.1a 2MASS/T. H. Jarrett, J. Carpenter, and R. Hurt, http://www.ipac.caltech.edu/2mass/gallery/showcase/allsky_gal/index.html. Copyright in the Public Domain.

Fig 18.1b S. Maddox (Nottingham U.) et al. APM Survey, Astrophys. Dept. Oxford University, http://apod.nasa.gov/apod/ap071007.html. Copyright © S. Maddox (Nottingham U.) et al. APM Survey, Astrophys. Dept. Oxford University.

Fig 18.2 Copyright © Matthew Colless (The 2dF Galaxy Redshift Survey team). Reprinted with permission.

Fig 18.4 Adapted from: 2MASS/T. H. Jarrett, J. Carpenter, and R. Hurt, http://www.ipac.caltech.edu/2mass/gallery/showcase/allsky_gal/index.html. Copyright in the Public Domain.

Fig 18.5a Copyright © André Karwath (CC BY-SA 2.5) at https://commons.wikimedia.org/wiki/File:Foam_-_big.jpg.

Fig 18.5b Copyright © Matthew Colless (The 2dF Galaxy Redshift Survey team). Reprinted with permission.

Fig 18.6 Copyright © ESO (CC BY 4.0) at http://commons.wikimedia.org/wiki/File:Wide-field_view_of_the_region_around_Sun-like_star_HIP_102152.jpg.

Fig 18.7a Adapted from: Copyright © Richard Powell (CC BY-SA 2.5) at http://www.atlasoftheuniverse.com/universe.html.

Fig 18.7b Adapted from: Copyright © SDSS-III (Sloan Digital Sky Survey)/Astrophysical Research Consortium.

Fig 19.1 Copyright © ESO/Davide De Martin (CC BY 3.0) at http://apod.nasa.gov/apod/ap080927.html.

Fig 19.2 SockPuppetForTomruen, http://commons.wikimedia.org/wiki/File:Lunar_eclipse_from_moon-1991Dec21.png. Copyright in the Public Domain.

Fig 19.4a Copyright © K. Swartz/Great Circle Mapper & Bue Marble/NASA.

Fig 19.4b Copyright © K. Swartz/Great Circle Mapper & Bue Marble/NASA.

Fig 19.5a NOAO/AURA/NSF, http://www.noao.edu/image_gallery/html/im0654.html. Copyright © NOAO/AURA/NSF.

Fig 19.5b ESA/NASA/E. Peng (Peking University, Beijing), http://www.nasa.gov/mission_pages/hubble/science/virgo_cluster_prt.htm. Copyright in the Public Domain.

Fig 19.5c The Hubble Heritage Team (AURA/STScI/NASA), http://hubblesite.org/newscenter/archive/releases/1999/25/image/a/. Copyright in the Public Domain.

Fig 19.5d NOAO/AURA/NSF, http://www.noao.edu/image_gallery/html/im0419.html. Copyright in the Public Domain.

Fig 19.6 Adapted from: http://cas.sdss.org/dr6/en/sdss/data/data.asp. Copyright © Sloan Digital Sky Survey.

Fig 19.9 Copyright © Azcolvin429/NASA (CC BY-SA 3.0) at https://commons.wikimedia.org/wiki/File:1_Earth_%28blank_2%29.png.

Fig 19.10a Pieter van Dokkum, Marijn Franx (University of Groningen/Leiden)/ESA/NASA, http://hubblesite.org/newscenter/archive/releases/1999/28. Copyright in the Public Domain.

Fig 19.10b Isabella Gioia (Univ. of Hawaii)/NASA/ESA, http://www.spacetelescope.org/images/opo9826d/. Copyright in the Public Domain.

Fig 19.11a NASA/ESA/Hubble Heritage (STScl/AURA), http://heritage.stsci.edu/2008/28/images/p0828aa.jpg. Copyright in the Public Domain.

Fig 19.11b NASA/ESA/Hubble Heritage (STScl/AURA), http://hubblesite.org/gallery/album/pr2005011a. Copyright in the Public Domain.

Fig 20.1 NASA/ESA, http://en.wikipedia.org/wiki/File:Hubble_ultra_deep_field_high_rez_edit1.jpg. Copyright in the Public Domain.

Fig 20.2 NASA/ESA, http://en.wikipedia.org/wiki/File:Hubble_ultra_deep_field_high_rez_edit1.jpg. Copyright in the Public Domain.

Fig 20.3 Adapted from: NASA/ESA, http://en.wikipedia.org/wiki/File:Hubble_ultra_deep_field_high_rez_edit1.jpg. Copyright in the Public Domain.

Fig 20.4 NASA/ESA, http://en.wikipedia.org/wiki/File:Hubble_ultra_deep_field_high_rez_edit1.jpg. Copyright in the Public Domain.

Fig 21.1a NASA/ESA/S. Beckwith (STScI)/HUDF Team, http://en.wikipedia.org/wiki/File:Hubble_ultra_deep_field_high_rez_edit1.jpg. Copyright in the Public Domain.

Fig 21.1b Hubble Heritage Team (AURA/STScl/NASA), http://heritage.stsci.edu/1999/25/index.html. Copyright in the Public Domain.

Fig 21.1c NASA/Swift/Stefan Immler (GSFC) and Erin Grand (UMCP), Bill Schoening, Vanessa Harvey/REU program/NOAO/AURA/NSF, http://www.noao.edu/image_gallery/images/d4/androx.jpg. Copyright in the Public Domain.

Fig 21.2 Charles Steidel (California Institute of Technology)/NASA, http://hubblesite.org/newscenter/archive/releases/1996/25/text/. Copyright in the Public Domain.

Fig 21.3 Copyright © Astrophysics Research Institute of Liverpool JMU. Reprinted with permission.

Fig 21.4 Copyright © Astrophysics Research Institute of Liverpool JMU. Reprinted with permission.

Fig 21.5a NASA/J. Bahcall (IAS), http://hubblesite.org/newscenter/archive/releases/2003/03/image/b/. Copyright in the Public Domain.

Fig 21.5b NASA/A. Martel (JHU), H. Ford (JHU), M. Clampin (STScI), G. Hartig (STScI), G. Illingworth (UCO/Lick Observatory)/ACS Science Team/ESA, http://hubblesite.org/newscenter/archive/releases/2003/03/image/b/. Copyright in the Public Domain.

Fig 21.5c John Hutchings/Dominion Astrophysical Observatory/NASA, http://hubblesite.org/newscenter/archive/releases/1994/16/image/a/. Copyright in the Public Domain.

Fig 21.5d NASA/ESA/G. Canalizo (University of California, Riverside), http://hubblesite.org/newscenter/archive/releases/2007/39/image/b/. Copyright in the Public Domain.

Fig 21.5e John Bahcall/Institute for Advanced Study/NASA, http://hubblesite.org/newscenter/archive/releases/1995/04/image/a/. Copyright in the Public Domain.

Fig 21.6 Holland Ford, Space Telescope Science Institute, Johns Hopkins University/Richard Harms, Applied Research Corp./Zlatan Tsvetanov, Arthur Davidsen, and Gerard Kriss at Johns Hopkins/Ralph Bohlin and George Hartig at Space Telescope Science Institute/Linda Dressel and Ajay K. Kochhar at Applied Research Corp./Bruce Margon, University of Washington in Seattle/NASA, http://hubblesite.org/gallery/album/galaxy/pr1994023d/npp/all/titles/true/hires/true/. Copyright in the Public Domain.

Fig 21.7a NASA/CXC/SAO/H.Marshall et al., http://chandra.harvard.edu/photo/2000/0131/. Copyright in the Public Domain.

Fig 21.7b NASA/STScI, http://chandra.harvard.edu/photo/2000/0131/more.html. Copyright in the Public Domain.

Fig 21.7c MERLIN/VLBI National Facility/Jodrell Bank Observatory/The University of Manchester, http://chandra.harvard.edu/photo/2000/0131/more.html. Copyright © MERLIN/VLBI National Facility/Jodrell Bank Observatory/The University of Manchester.

Fig 21.7d Copyright © NRAO/AUI (CC BY 3.0) at http://images.nrao.edu/AGN/Radio_Galaxies/328.

Fig 21.7e Copyright © NRAO/AUI (CC BY 3.0) at http://images.nrao.edu/AGN/Quasars/132.

Fig 21.7f Copyright © NRAO/AUI (CC BY 3.0) at http://www.cv.nrao.edu/~abridle/images/3c219_lb_sum_large.jpg.

Fig 21.7g Copyright © NRAO/AUI (CC BY 3.0) at http://images.nrao.edu/AGN/Radio_Galaxies/110.

Fig 21.8a Josef Pöpsel and Stefan Binnewies, http://www.capella-observatory.com/ImageHTMLs/Galaxies/NGC5128.htm. Copyright © Capella University. Reprinted with permission.

Fig 21.8b NASA/CXC/CfA/R.Kraft et al., http://chandra.harvard.edu/photo/2008/cena/. Copyright in the Public Domain.

Fig 21.8c NASA/JPL-Caltech, http://www.galex.caltech.edu/media/glx2005-02r_img13.html. Copyright in the Public Domain.

Fig 22.1 Copyright © Anatoly Klypin and Andrey Kravtsov. Reprinted with permission.

Fig 22.2 Michael Murphy, http://www.eso.org/~jliske/qsoal/. Copyright © Michael Murphy. Reprinted with permission.

Fig 22.3a Copyright © Azcolvin429/NASA (CC BY-SA 3.0) at https://commons.wikimedia.org/wiki/File:1_Earth_%28blank_2%29.png.

Fig 23.3b Copyright © Grombo (CC BY-SA 3.0) at https://commons.wikimedia.org/wiki/File:Morning_Fog_at_GGB.JPG.

Fig 23.4 NASA/Goddard/WMAP Science Team, https://commons.wikimedia.org/wiki/File:WMAP.jpg. Copyright in the Public Domain.

Fig 23.5 NASA/WMAP Science Team, https://commons.wikimedia.org/wiki/File:PowerSpectrumExt.svg. Copyright in the Public Domain.

Fig 24.3a NASA, https://commons.wikimedia.org/wiki/File:Solar_Prominence_STEREO_Behind.jpg. Copyright in the Public Domain.

Fig 24.3b Adapted from: John F. Williams/U.S. Department of Defense, http://commons.wikimedia.org/wiki/File:The_U.S._Navy_MZ-3A_manned_airship_conducts_an_orientation_flight_over_Naval_Air_Station_Patuxent_River,_Md.,_Nov._6,_2013_131106-N-PO203-262.jpg. Copyright in the Public Domain.

Fig 24.3c Copyright © Lead holder (CC BY-SA 3.0) at http://commons.wikimedia.org/wiki/File:LiIon-18650-match.jpg.

Fig 24.3d Copyright © Rob Lavinsky/iRocks.com (CC BY-SA 3.0) at https://commons.wikimedia.org/wiki/File:Beryl-122885.jpg.

Fig 24.4 Adapted from: NASA, http://map.gsfc.nasa.gov/universe/bb_tests_ele.html. Copyright in the Public Domain.

Fig 25.7 NASA/Goddard/WMAP Science Team, https://commons.wikimedia.org/wiki/File:WMAP.jpg. Copyright in the Public Domain.

Fig 25.8 Camille Flammarion, L'Atmosphere: Météorologie Populaire, pp. 163, 1888 (http://en.wikipedia.org/wiki/File:Flammarion.jpg). Copyright in the Public Domain.

INDEX

Legend: *Fnn.mm* refers to footnote *mm* in Destination *nn.*
Tnn.mm refers to table *mm* in Destination *nn.*

— A —

— B —

— C —

— D —

— E —

— F —

— G —

— K —

— L —

— M —

— N —

— O —

— P —

— Q —

— R —

— S —

— T —

— U —

— V —

— W —

CPSIA information can be obtained
at www.ICGtesting.com
Printed in the USA
LVHW06s2147160918
590348LV00007B/186/P

9 781516 5